AF574446

HEAT TRANSFER IN COLD CLIMATES

HEAT TRANSFER IN COLD CLIMATES

Virgil J. Lunardini

U.S. Army Cold Regions Research and
Engineering Laboratory
Hanover, N.H.

VNR VAN NOSTRAND REINHOLD COMPANY
NEW YORK CINCINNATI ATLANTA DALLAS SAN FRANCISCO
LONDON TORONTO MELBOURNE

Van Nostrand Reinhold Company Regional Offices:
New York Cincinnati Atlanta Dallas San Francisco
Van Nostrand Reinhold Company International Offices:
London Toronto Melbourne

Library of Congress Catalog Card Number: 80–19257
ISBN: 0–442–26250–7

Manufactured in the United States of America
Published by Van Nostrand Reinhold Company
135 West 50th Street, New York, N.Y. 10020
Published simultaneously in Canada by Van Nostrand Reinhold Ltd.
15 14 13 12 11 10 9 8 7 6 5 4 3 2 1

Library of Congress Cataloging in Publication Data
Lunardini, Virgil J
Heat transfer in cold climates.
Includes index.
1. Heat—Transmission. 2. Building—Cold weather conditions. 3. Permafrost I. Title.
TJ260.L78 621.402'2'0911 80–19257
ISBN 0–442–26250–7

To My Mother and Father

Preface

The cold climate regions of North America have recently received increased interest, due mainly to the discovery of fossil fuels. The utilization of these resources has led to the design and construction of roads, airfields, pipelines, and numerous types of structures. In many cases, the thermal interaction of these engineering systems with the environment has resulted in unexpected failures of the system. In most cases, these failures are associated with attempts to utilize the technology of temperate climates without regard for the special conditions and problems of the cold regions. The temperate zone engineer must be aware of these problems so as to avoid repeating them. The correct application of thermal principles requires an interdisciplinary approach among mechanical and civil engineers and physical scientists. It is unfortunate, although understandable, that past trends in engineering education have divorced civil engineering students from thermal concepts, and mechanical engineering students have had little contact with soil systems.

This book is an introduction to the heat transfer problems that often arise in cold regions. The literature in this field is extensive and diffused among reports, journals, books, and notes from numerous, disparate disciplines. No claim is made that the totality of this literature has been covered. Rather, an attempt has been made to introduce the reader, in a coordinated fashion, to the somewhat bewildering array of primary sources. The goal has been to allow the beginning student to follow the theory without formal supervision and to aid the experienced practitioner in adapting standard solutions for special applications. To this end, illustrations and worked examples are presented, of typical solution methods, based on fundamental principles. Every effort has been made to present complete derivations of important relations, even if some repetition of the basic equations has been required. This book is intended primarily for mechanical and civil engineers faced with the thermal design of complex systems, suitable for low-temperature environments. It is expected that other engineers, as well as earth scientists, will find relations useful for their specialities.

Both SI and British Engineering units have been used, as required, since the literature data bases use both systems, and the reader will be forced to use both systems of units for many years.

Chapters 1 and 2 define the cold regions, introduce some peculiarities of these regions, and examine the data needed for typical problems. Chapter

5 is a brief look at the natural form and shape of the cold regions landscape. A number of specialized definitions and concepts, which occur frequently in the literature, are found here. It is assumed that most readers will be somewhat familiar with heat transfer concepts, but Chapter 3 examines the basic laws of conduction, convection, and radiation, along with some particularly useful calculation methods. Permafrost is a critical component of the cold climate regions, and Chapter 4 shows how it can be treated as an engineering material. The thermal properties of the frozen soils are presented along with some elementary permafrost heat transfer relations. The energy balance at the earth-atmosphere interface presents the engineer with some new concepts, discussed in Chapter 6. The flux of sensible atmospheric energy is examined in relation to buoyancy and turbulence. The convective heat transfer can be very different from that of the usual engineering systems. Because freezing and thawing introduce major difficulties, analytic methods and solutions for conduction with phase change are detailed in Chapter 8. The material has been organized in terms of soil systems, but the fundamental relations and methods can be applied to solar storage systems, metallurgical applications, food preservation, and ablation design. Because complete, analytic solutions are rare, considerable effort is devoted to approximate methods, such as the quasi-stationary and the heat balance integral concepts. Chapter 9 completes the introductory study of conduction phase change with a presentation of commonly used, finite difference methods. Chapters 10 and 11 apply phase change theory to typical problems involving geotechnical design and the thermal aspects of structures and pipelines. Finally, Chapter 12 concludes with a brief glimpse at the way man reacts in cold climates. As has always been the case, man attempts to engineer the environment for his benefit or even survival. With new knowledge, there appears to be no reason why this manipulation cannot be to the benefit of man and his natural surroundings.

The author would like to acknowledge the often stimulating and beneficial comments from former students who have followed lecture notes that eventually evolved into this text. The advice of colleagues has been valuable, and the author is grateful to the many practicing engineers who have made possible the immense literature in this field.

Thanks are due also to Denise Champion-Demers, who typed several drafts, and to my wife, Rosemary, who typed the final manuscript and provided general, editorial assistance.

V. J. L.

Hanover, New Hampshire

Contents

HEAT TRANSFER IN COLD CLIMATES

1
Cold Climates

1.0 INTRODUCTION

One of the intriguing aspects of cold climate engineering is the question of what is meant by a cold climate. This is not an easy question to answer, and, in fact, there is no one absolute answer, since the concept of cold is itself subjective. Generally, we think of cold in terms of temperature (cold is a state of low temperature), and this is certainly the most important single concept associated with cold.

Cold climates can be defined in relation to

(a) temperature
(b) polar circles
(c) tree limits
(d) frost penetration
(e) sea ice limits

It should be noted that most of these definitions are static in that they do not take into account the seasonal, and daily, transient meteorological phenomena that can inflict the adverse characteristics of the cold regions on areas that are far outside the accepted cold regions. From an engineering viewpoint, the effects of cold are significant, and we shall define cold regions in terms of operational aspects, as will be discussed later.

1.1 TEMPERATURE DEFINITIONS OF COLD CLIMATES

1.1.1 Koppen

One of the most widely used definitions of the cold regions is that of Koppen (1936):[1]

(i) The mean temperature for the coldest month is less than −3°C (26.6°F). (This is the highest temperature at which snow can exist for an appreciable length of time.)

[1] References are listed at the end of each chapter.

(ii) Not more than 4 months have a mean temperature greater than 10°C (50°F).

The temperature definition of cold regions is often more restrictive with regard to the warm months as compared with the cold months. For example, based on the Koppen definition, Ottawa, Ontario, would not be classed in a cold climate region since Ottawa has 5 months (May–September) with mean temperatures above 10°C. Ottawa, however, has 4 months with a mean temperature less than −3°C. This illustrates the shortcomings of a static temperature definition, since most of its inhabitants would describe Ottawa, which has the lowest mean annual air temperature of the major capital cities, as cold indeed.

1.1.2 Polar/Snow-Forest

Wilson (1967) and Haurwitz and Austin (1944) describe the cold regions based upon temperature and precipitation.

A description of the climatic limits of the cold regions is shown in Fig. 1.1.

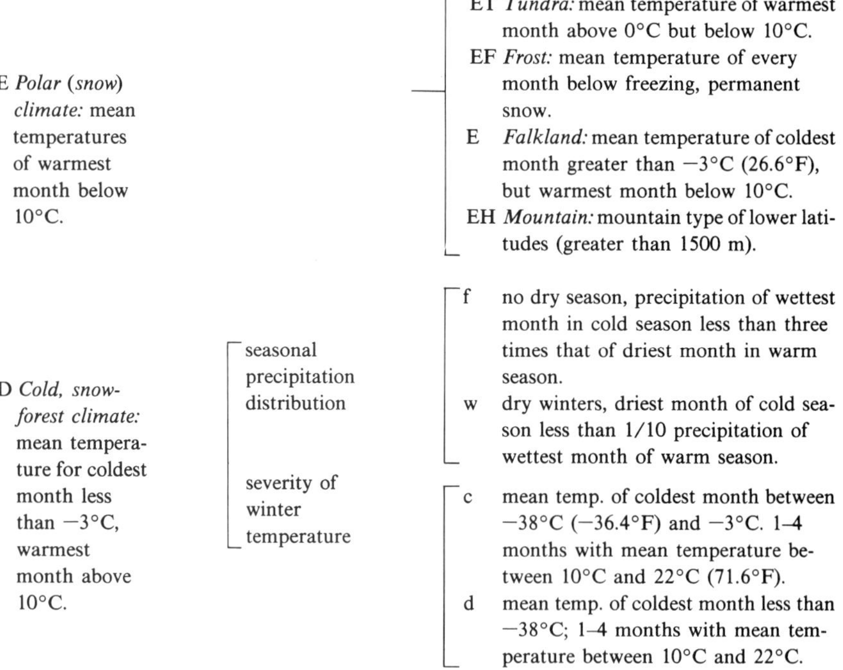

Fig. 1.1. Cold Regions (after Wilson 1967).

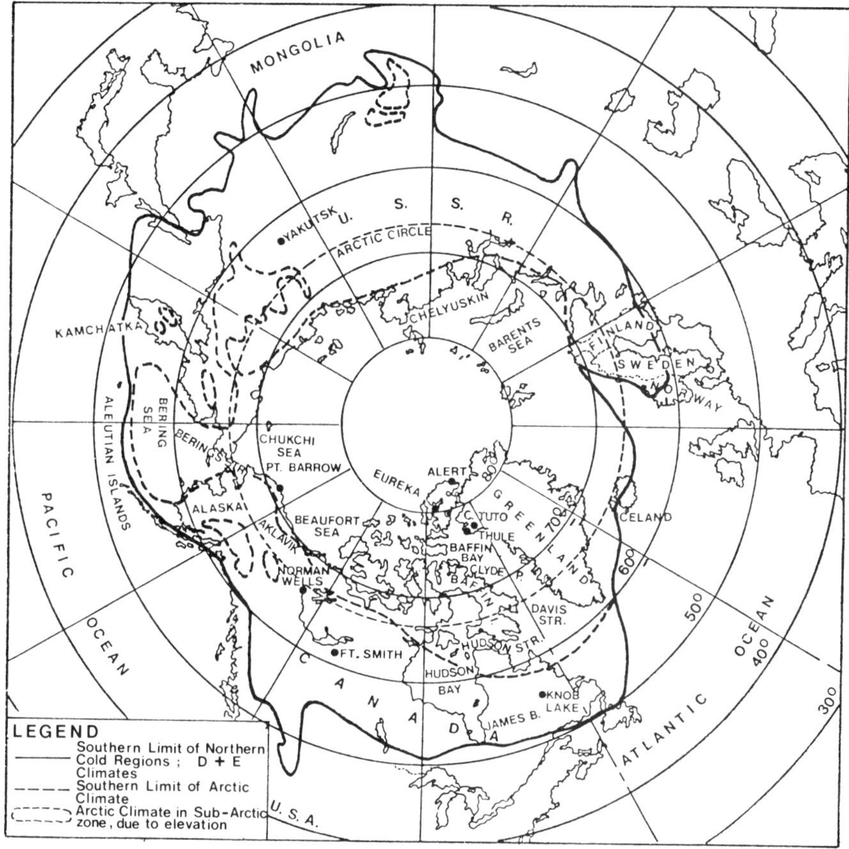

Fig. 1.2. Northern Cold Regions (from Wilson 1967).

This description allows the cold regions to be described very conveniently as EF, Dwd, etc.

Figure 1.2, taken from Wilson (1967), shows the cold regions of the northern hemisphere based upon the above definition. The southern limit of the cold snow-forest climate agrees fairly well with Koppen's definition.

1.1.3 Arctic/Subarctic

Taylor (1957) and Stearns (1965) define the familiar terms "arctic" and "subarctic" as follows:

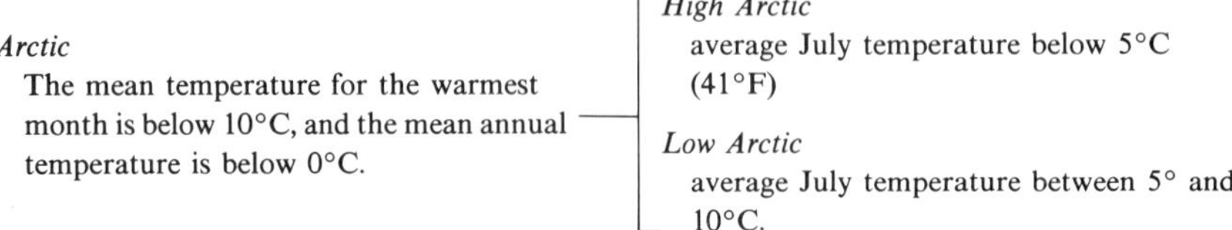

Subarctic
The mean temperature for the coldest month is below 0°C; the mean temperature of the warmest month is above 10°C, but not more than 4 months have a mean temperature above 10°C.

These definitions clearly are quite similar to the previous definitions, as may be seen on Fig. 1.3.

The southern limit of the subarctic region extends somewhat farther south than that of the polar/snow-forest definition.

The word "arctic" derives from the Greek *arktos,* the Great Bear (also the Big Dipper), and denotes those regions where this constellation is always visible. The arctic regions, although not always so, have remained cold since the last interglacial warm period, which ended about 70,000 years ago. Evidence for this comes from radiocarbon dating and animal remains found preserved in the near surface layers of frozen ground.

The Arctic, in North America, largely delineates the original territory occupied by Inuit (Eskimos). These nomadic peoples had one of the most sophisticated technological societies of any of the hunter-gatherer peoples. It is ironic that their culture has been largely displaced by modern technology and not surprising that they have adapted readily to occupations requiring technical skills.

1.1.4 Freezing Isotherm

Bates and Bilello (1966) define the southern limit of the cold regions as the 0°C (32°F) isotherm for the coldest month of the year. They divide the cold regions as follows:

Cold winter—mean temperature during the coldest month is between 0°C and −17.8°C (0°F).

Very cold winter—mean temperature during the coldest month is between −17.8°C and −32°C (−25°F).

Extremely cold winter—mean temperature during the coldest month is between −32°C (−25°F) and −62.2°C (−80°F).

These regions can be seen in Fig. 1.4 and represent a good engineering definition of the cold regions.

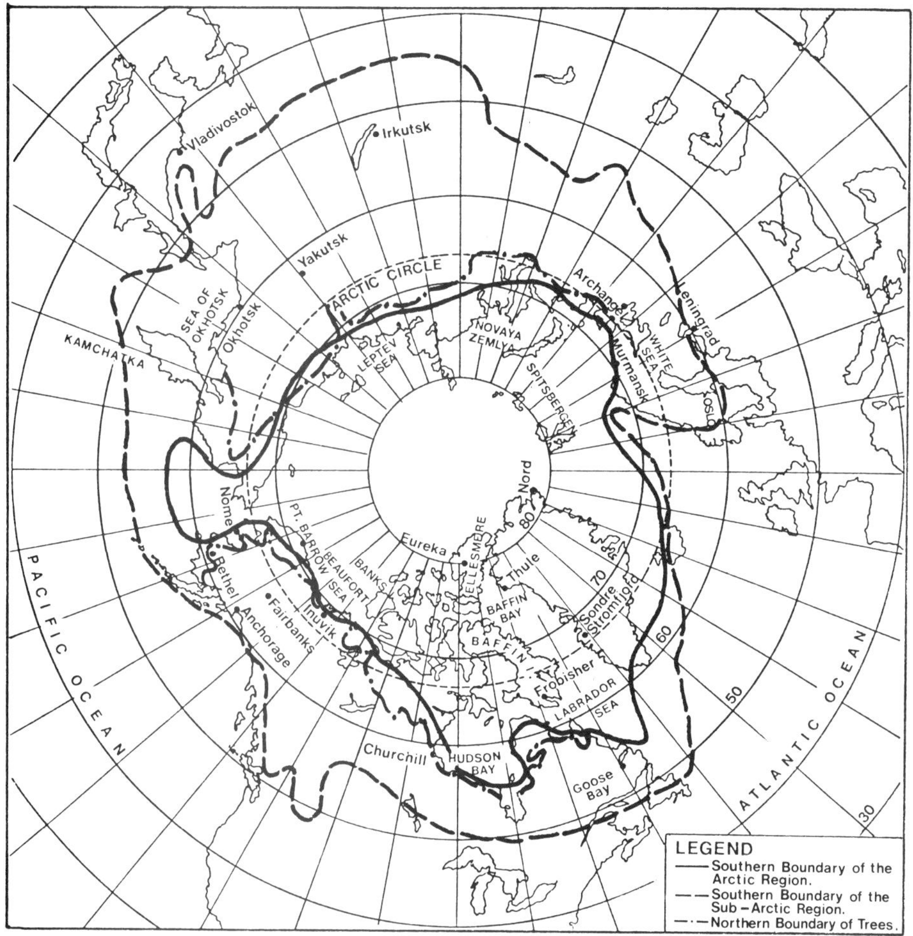

Fig. 1.3. North Cold Regions (from Stearns 1965).

1.1.5 Temperature Frequency

Yet another definition of the cold regions is the 1% occurrence of −32°C in January. This is equivalent to a January temperature of −32°C being recorded more than once every 3 years (Wilson 1967).

1.2 FROST PENETRATION

As has been mentioned, the various definitions of the cold regions often fail to include many areas that experience grave engineering difficulties due

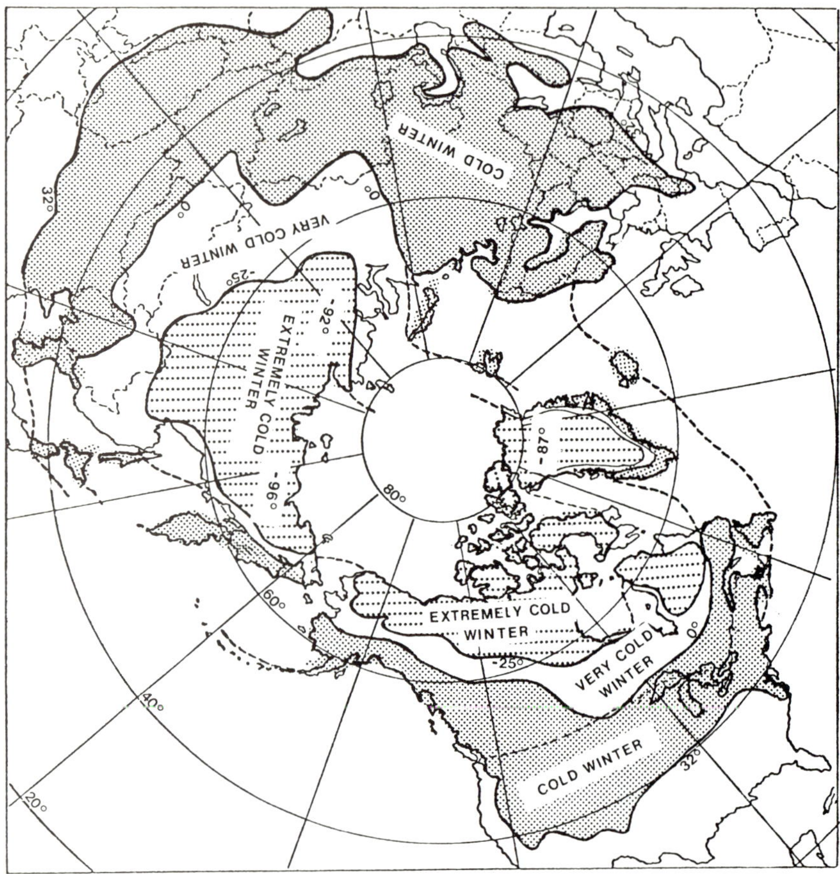

Fig. 1.4. Climatic Zones of the Cold Regions of the Northern Hemisphere (from Gerdel 1969).

to the effects of snow, ice, freezing soil, etc. For this reason, engineers have attempted to broaden the definition to include these regions. The 40th parallel of longitude delineates quite well those northern regions in North America that experience the problems of a cold environment (Gerdel 1969). Engineers have defined the southern limit of the cold regions in terms of the 6 to 12-in. depths of frost penetration. Figure 1.5 demonstrates the close agreement of this limit and the 40°N latitude. The 1-ft frost depth is roughly related to about 100 degree-days of temperatures below freezing (32°F).

1.2.1 Permafrost Limits

Permafrost, or perennially frozen ground, is a distinct characteristic of parts of the cold regions. Permafrost that exists in an unbroken fashion over an

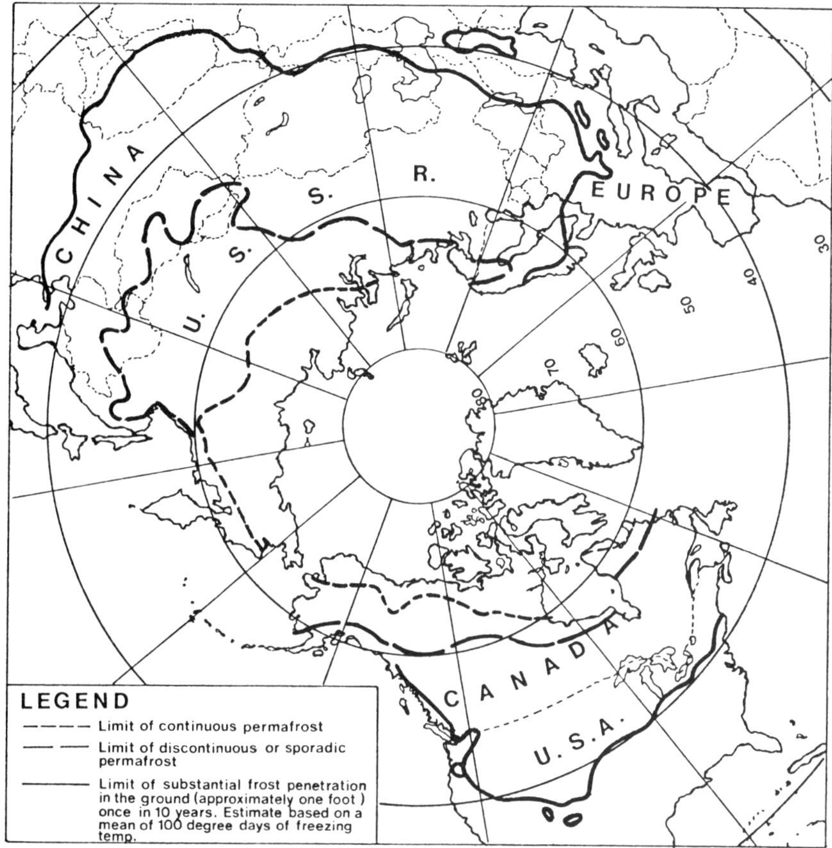

Fig. 1.5. The Distribution of Seasonally Frozen Ground and Permafrost in the Northern Hemisphere (adapted from Bates 1966).

entire region is called "continuous permafrost," and permafrost that is patchy or in isolated islands is called "discontinuous permafrost." The lines of continuous and discontinuous permafrost correspond roughly to the arctic and subarctic regions, as can be seen on Fig. 1.6. Clearly, there is a vast region outside the permafrost zones that must be characterized as cold. Permafrost can also exist far south of these zones in small patches dependent upon the local climatic, vegetative, and soil conditions, often related to high elevations.

1.3 TREE LINE

There is an imprecisely defined region, beyond which trees cannot normally exist, which may vary from a few miles to dozens of miles in width. Generally,

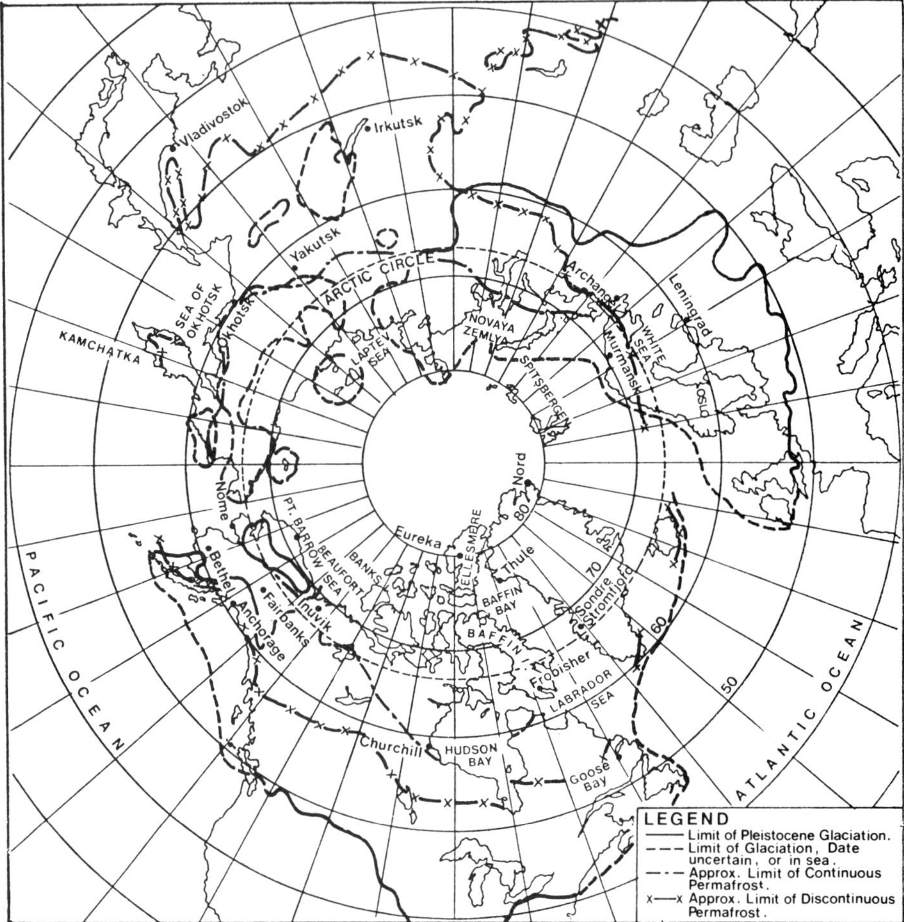

Fig. 1.6. North Cold Regions: Glaciation and Limits of Permafrost (adapted from Stearns 1965).

the trees of a given species, i.e., black spruce, will become progressively smaller and sparser over this region until they fail to exist at all, except in pockets with a favorable microclimate. This belt is called the limit of tree growth, or the "tree line." It also exists as one goes up a mountain. The lack of trees is related to the general climate, and this is reflected in its rough correspondence to the 10°C (50°F) mean July temperature, as can be seen in Fig. 1.7. January and July are used for temperature references because these are generally, but not always, the coldest and warmest months at a given location. The tree line is related to soil type and moisture, bodies

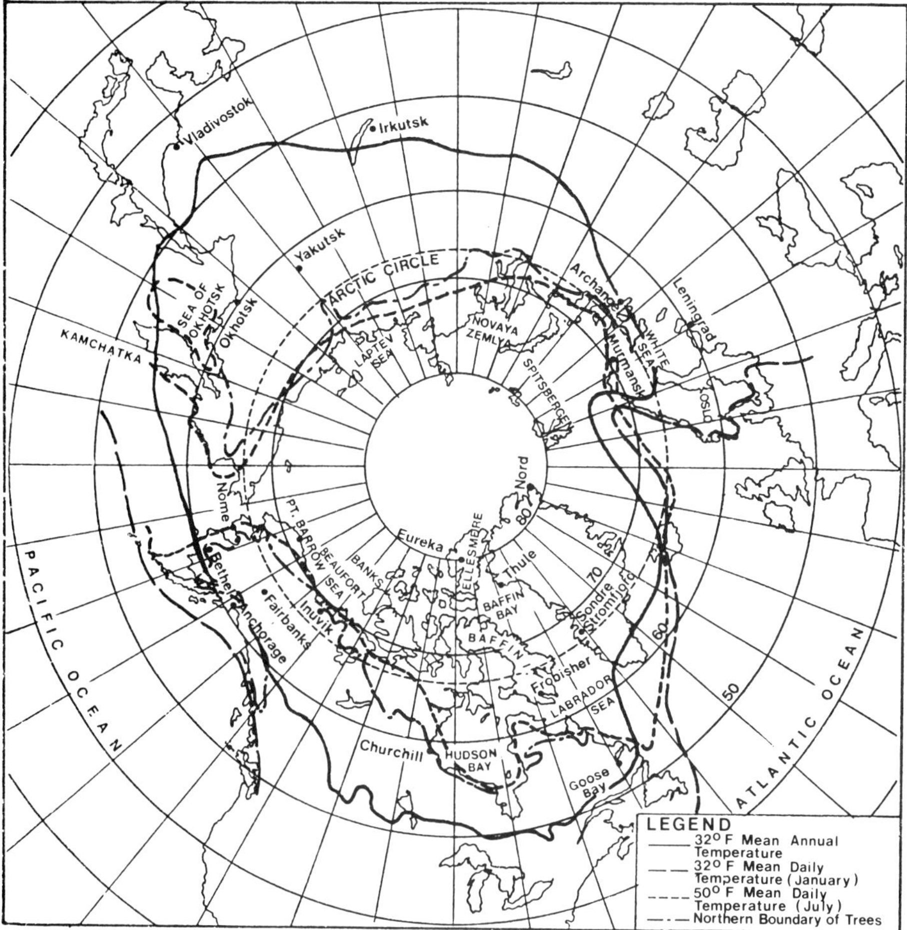

Fig. 1.7. North Cold Regions: Tree Line and Temperatures (adapted from Stearns 1965).

of water, wind, elevation, and snow cover, but is apparently not related to the isotherms of the coldest month. It does define a reasonable northern limit of the subarctic region, but there is no clear southern subarctic limit based on tree growths, although distinct associations of trees tend to grow in distinct climatic habitats.

1.4 POLAR CIRCLES

Our discussions are confined to the northern cold regions, and we will not discuss the Antarctic Circle, which has the same significance as the Arctic

Circle. The Arctic Circle is the 66°33' north circle of latitude. Although it is often used as a limit of the arctic regions, it actually defines only those regions that have 24 hours of light or darkness during certain periods of the year. Regions south of the Arctic Circle have some daylight throughout the entire year. As one goes north, the period of total darkness increases until there is 6 months of total darkness at the north pole. Figure 1.2 shows that regions with an arctic climate lie both north and south of the Arctic Circle, and this division of the northern regions is not particularly useful.

1.5 LIMITS OF SEA ICE

The boundaries of the Arctic and Sub-Arctic can be described in terms of water temperature and the ice cover variations of the surrounding oceans. The ice cover and its thickness and age have a great engineering influence on transportation and also modify the surrounding climate. The maximum and minimum ice covers are shown in Fig. 1.8. The Sub-Arctic is defined as the area between the northern limit of nonarctic water and the southern

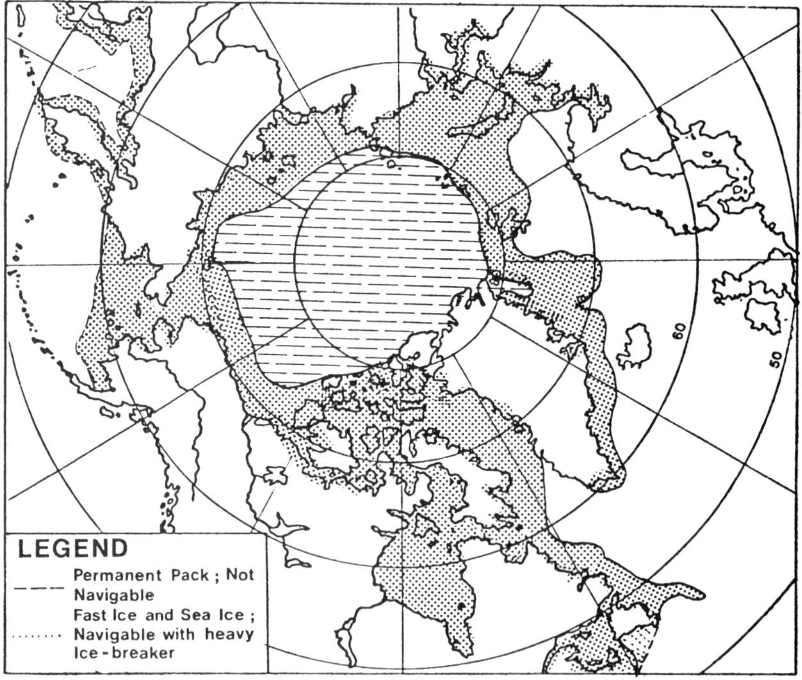

Fig. 1.8. Maximum Extent of Ice Cover, February–March (from Stearns 1965).

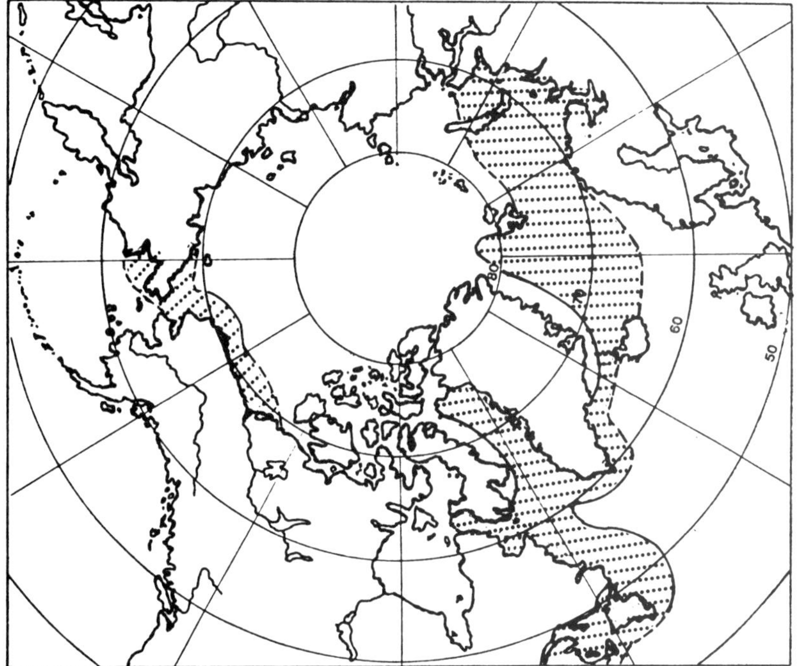

Fig. 1.9. Zones of the Marine Environment; the Sub-Arctic Zone is Stippled (adapted from Stearns 1965).

limit of arctic water, as shown in Fig. 1.9. Arctic water has quite distinctive temperature, density, salinity, and biological productivity. The density, temperature, and salinity are lower, and the productivity is higher than nonarctic water.

1.6 ENGINEERING CONCEPT OF COLD CLIMATES

From an engineering viewpoint, the most reasonable delineation of cold climates will include those regions north of the 40°N latitude in North America. This coincides, more or less, with the freezing isotherm definition and the frost penetration concept.

In general, a cold climate environment exists wherever frost affects engineering systems; ice on rivers, lakes, and harbors interferes with transportation or damages structures; snow loads must be considered in the design of structures; low temperatures affect the efficiency of man or machines; the manifestations of low temperature affect the economics of engineering design. The cost of snow and ice control in the United States is well over $100 million per year in direct costs.

1.7 CHARACTERISTICS AND STRESSES OF COLD CLIMATES

1.7.1 Temperature

Certainly, temperature is one of the most evident stresses associated with cold regions. One might tend to relate all the problems to temperature, but in fact, the temperature itself is a manifestation of the energy exchanges that occur naturally.

In North America, the temperature tends to drop as one moves north, but this must be qualified. In general, there is an east-west asymmetry in the isotherms, such that the isotherms of mean annual air temperature dip from west to east from the Mackenzie delta to the Hudson's Bay region and then jump northward again through Quebec; this is a good example of the maritime effect on air temperature. In other words, the isotherms don't follow the lines of latitude. One can say, however, that the average temperatures decrease as one goes farther north.

The mean temperatures tend to obscure the temperature extremes, and it is somewhat surprising to learn that the minimum temperatures recorded did not occur at the most northerly locations. It is well known that the oceans and other large bodies of water moderate the air temperature and that large, continental land masses have no such buffering effect. For this reason, the extremes of temperatures in Canada tend to occur in central Canada. Table 1.1 shows that there is no great difference in the minimum low temperatures recorded between Winnipeg and Baker Lake or Coppermine, although Winnipeg is more than 1300 miles south of these towns. The minimum temperature ever recorded in North America was −81°F at Snag, Yukon Territory, and the minimum in the northern hemisphere was −96°F at Verkhovansk, Siberia.

Thus, the major characteristic of the temperature, as one goes from south to north, is not the extremes of temperature, but the duration of the low temperatures. In other words, the low temperatures of southern Canada and the northern United States have a much shorter duration than those of northern Canada, and, of course, the "winters" are correspondingly shorter and the heating of structures in the north will be more expensive.

A temperature of −25°F is close to the critical limit below which man and equipment have a limited operational capability. If the temperature is below −25°F for several days, most capabilities are reduced 50%, and several additional days at these temperatures may result in complete immobilization. It can be expected that in the extremely cold winter region, most of man's

Table 1.1. Temperature Extremes in Canada (from data in Kendall and Currie 1955).

Location	Lat(N)	Long(W)	Alt., ft.	Yearly mean, °F	Max. temp., °F	Min. temp., °F	Precip., in.	$\Delta T =$ Mean max. − mean min.
Fort Good Hope (NWT)	66°15	128°38	214	17	95	−79	9.9	83
Fort Resolution (NWT)	61°09	113°37	520	23	90	−63	11.1	77
Fort Simpson (NWT)	61°52	121°15	415	24	95	−69	13.0	80
Fort Smith (NWT)	60°	111°52	680	26	103	−71	12.1	75
Aklavik (NWT)	68°14	134°50	25	16	93	−62	9.2	74
Baker Lake (NWT)	64°18	96°05	30	9	82	−58	3.8	79
Chesterfield (NWT)	63°20	90°43	24	11	86	−60	12.1	73
Churchill (Man)	58°47	94°11	43	18	96	−57	15.3	73
Coppermine (NWT)	67°49	115°10	13	12	87	−58	11.0	68
Beaverlodge (Alta)	55°12	119°19	2500	36	98	−54	16.8	54
Bentton River (BC)	57°23	121°25	2755	30	88	−54	16.5	56
Edmonton (Alta)	53°33	113°30	2158	37	99	−57	18.0	56
Fort Nelson (BC)	58°50	122°35	1230	31	98	−61	16.7	67
Fort St. John (BC)	56°14	120°44	2275	34	92	−53	14.9	56
North Battleford (Sask)	52°46	108°15	1796	34	98	−51	10.1	62
Prince Albert (Sask)	53°10	105°45	1432	33	103	−70	16.1	67
Saskatoon (Sask)	52°08	106°38	1690	34	104	−55	14.5	66
Calgary (Alta)	51°01	114°08	3675	39	97	−49	16.4	49
Medicine Hat (Alta)	50°01	110°45	2365	42	108	−51	13.0	55
Regina (Sask)	50°26	104°30	1884	35	111	−56	14.7	66
Swift Current (Sask)	50°18	107°42	2440	39	107	−54	14.4	58
Brandon (Man)	49°57	99°57	1200	34	110	−52	15.7	68
Norway House (Man)	53°59	97°50	720	29	94	−63	15.7	74
Rivers (Man)	50°01	100°19	1553	35	98	−47	12.1	66
The Pas (Man)	53°49	101°15	890	31	100	−54	15.4	74
Winnipeg (Man)	49°53	97°17	786	35	108	−54	20.5	70

time may be spent merely in survival, with no useful work done. Notice that the duration of the low temperature is as important as the low temperature itself. Figures 1.10–1.12 show the frequencies of low temperature in the northern hemisphere and indicate where difficulties may be expected. These remarks do not pertain to Eskimos under natural conditions—as opposed to Eskimos in new, permanent villages—who have adapted clothing that will still outperform modern clothing where mobility is necessary.

1.7.1.1 Equivalent Temperature

The minimum temperature alone is not the critical aspect of the heat loss associated with cold climates. The wind velocity, which occurs simultaneously,

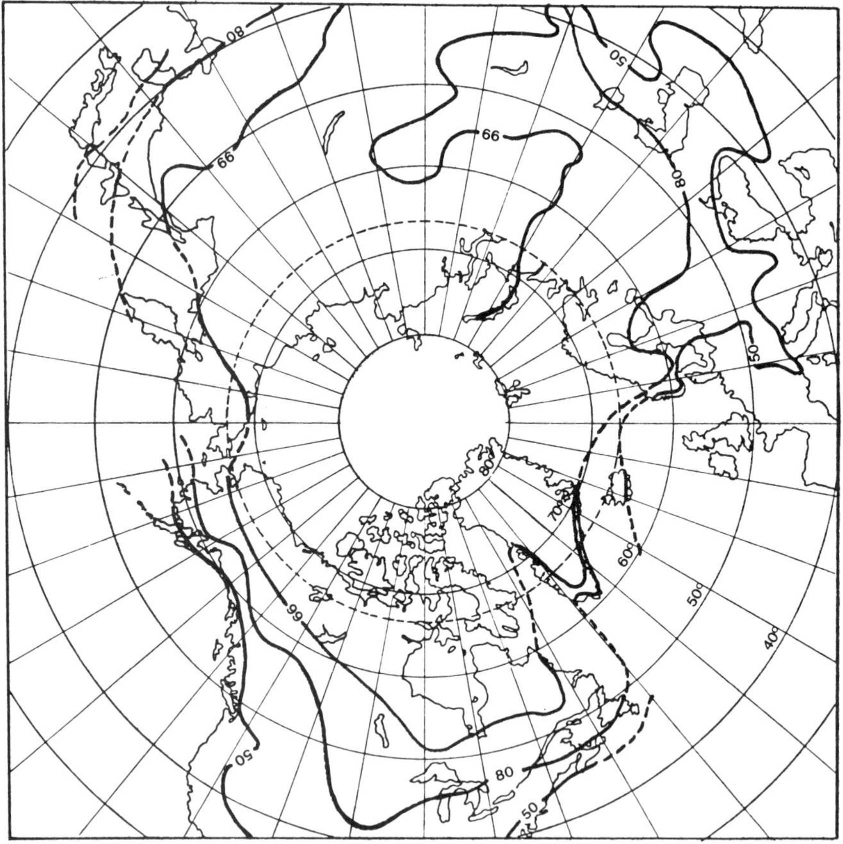

Fig. 1.10. Percentage Frequency of Temperature below 32°F during January in the Cold Regions of the Northern Hemisphere (after Gerdel 1969).

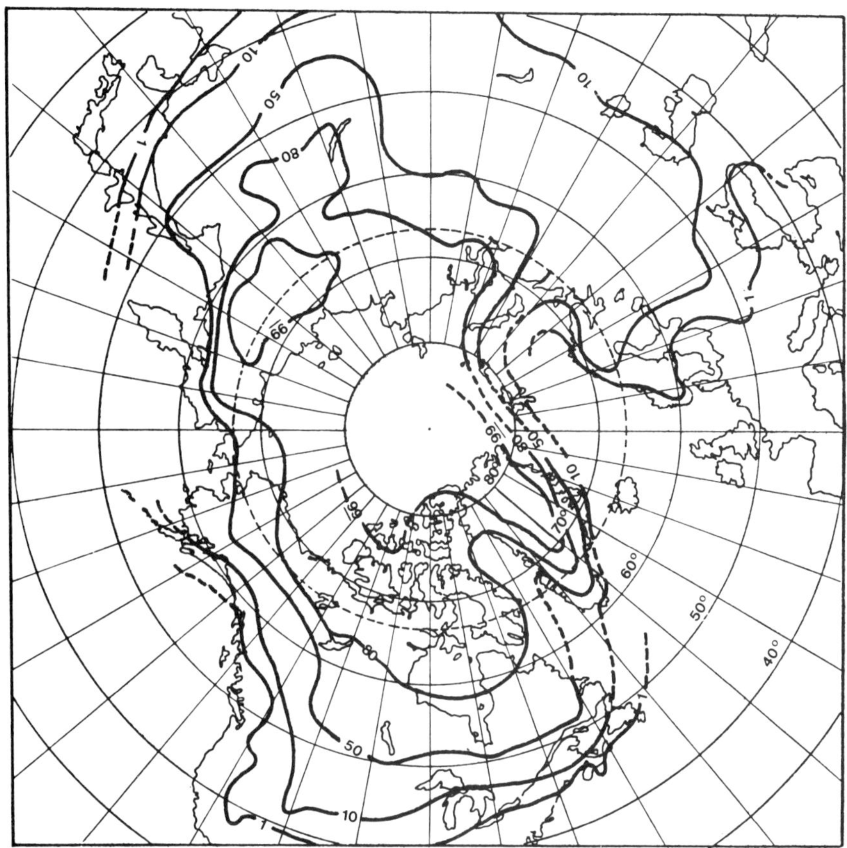

Fig. 1.11. Percentage Frequency of Temperature below 0°F during January in the Cold Regions of the Northern Hemisphere (after Gerdel 1969).

is also significant. The equivalent temperature is defined as that temperature, with zero wind speed, that will produce the same heat loss, from exposed flesh, as the actual combination of temperature and wind speed. Some extreme values may be noted in Table 1.2, and it is interesting that the lowest values are rarely associated with the minimum temperatures recorded. This is because minimum temperatures normally occur at very low wind speeds. More information on equivalent temperature is found in Chap. 12.

1.7.1.2 Temperature Inversion

Normally, the temperature of the air near the ground decreases as the elevation increases. Such a temperature gradient promotes mixing of the air and disper-

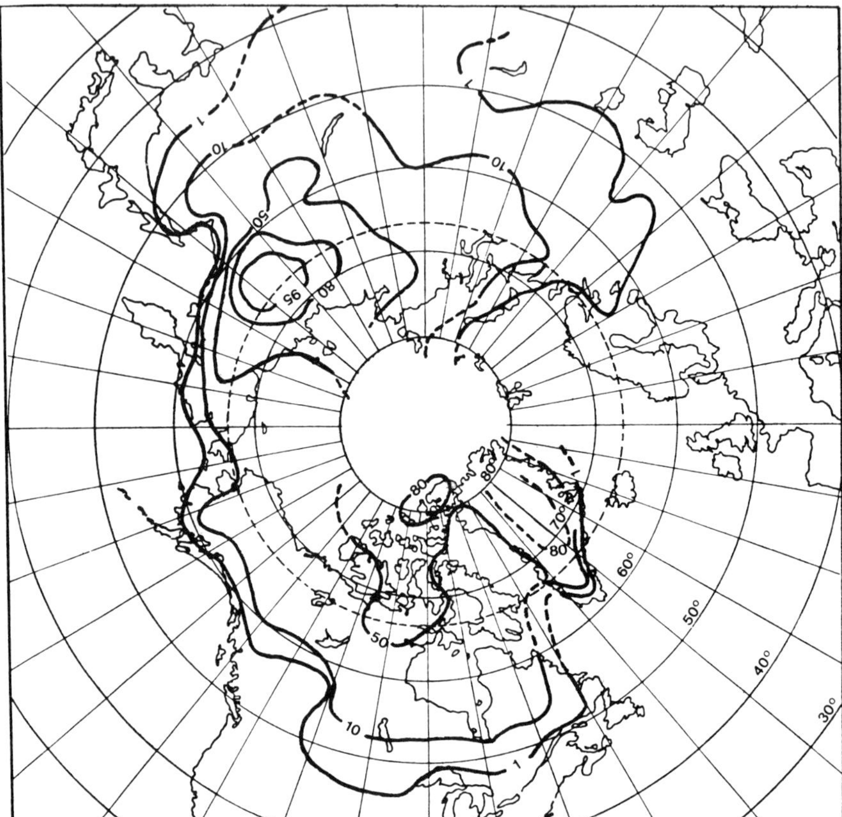

Fig. 1.12. Percentage Frequency of Temperature below −25°F during January in the Cold Regions of the Northern Hemisphere (after Gerdel 1969).

sion of gases, particles, etc. An inflow of cold air or a rapid cooling of the ground surface may cause the air temperature at the ground to be less than it is at elevation. This is called a "temperature inversion" and it results in a stable layer of air near the ground that traps water, ice particles, pollutants, etc. Although temperature inversions can occur anywhere, they are most frequent in cold climates.

1.7.2 Precipitation

A second characteristic of the north, which is not well known, is its aridity. Many parts of the north are classed as semiarid regions, with total precipita-

Table 1.2. Central Canadian Extreme Equivalent Temperatures (calculated from data of Kendrew and Currie 1955)

Location	Actual temp., °F	Wind speed, mph	Equiv. temp., °F
Fort Nelson	−36	26	−98.7
Churchill	−31	50	−101.6
Chesterfield	−44	26	−112.8
Repulse Bay	−39	60	−116
Fort McPherson	−50	11	−86.2
Fort Good Hope	−67	10	−100
	−25	40	−92.5
Hay River	−33	30	−98.5
Fort Chipewyan	−28	30	−91.0
Battleford	−31	23	−86.0
Saskatoon	−20	33	−80.8
Minnedosa	−19	30	−77.4

tion of less than 10 in. compared with 30–50 in. in southern Canada and the United States. The reason for this is that there is no large moisture supply available. The Arctic Ocean is ice-covered for much of the year and cannot evaporate water into the air to be precipitated as rain or snow. The snow-covered landscape of the north is misleading, since this is really a yearly accumulation as there is virtually no melting until late spring or early summer. Also, the snow is crystalline and tends to blow easily until compacted, and it may give the appearance of a snowfall when it is merely air-borne movement. During the summer, the appearance of water everywhere is again illusory, since this represents melted water that cannot drain due to the underlying permafrost. Were it not for this effect, the terrain would be dry and perhaps desertlike. The low snowfall can be melted even during the short arctic summers, and there is no net accumulation of snow and thus no glaciation. If the mean temperature increases, it may lead to glaciation rather than a greening of the arctic. This is especially true if the ocean should lose its ice cover for a year or more due to the increased temperature.

1.7.3 Snow and Ice

Snow is a familiar substance in all parts of the cold regions. It presents one of the most severe hazards of a cold climate in terms of transportation and daily operations in populated areas.

Snow has a vast variety of physical characteristics, however. The Eskimo had well over a hundred different words for snow types and dozens more for ice forms. These words describe the world of snow and ice to a people who depended upon these forms of water for their daily survival. Arctic snow is not like the snow of the northern United States. Compacted by the wind, it may form a surface as solid as concrete for transportation or it may form an immovable barrier against doors opening outward. Driven by the wind, fine snow may infiltrate buildings like dust.

1.7.3.1 Annual Snowfall and Accumulation

The total precipitation that occurs as snow in an average year is the annual snowfall. Much of this snow may melt during the cold season. The accumulation of snow is the amount of snow recorded on the ground at various times during the year. Measurements are made at the end of each month, and the maximum values are listed for design purposes.

Snow accumulation is important for transportation. Snow up to 18 in. is no problem for heavy, tracked vehicles if it overlays solid surfaces, such as frozen ground, old dense snow, ice, etc. Light snow over 2 ft deep prevents cross-country travel for all but very lightly loaded vehicles or men on skis or snowshoes. If the snow has a density less than 0.2 g/cm^3, even light vehicles and ski-equipped planes may be immobilized, and movement on foot may be of a strictly survival type.

The snow load is the weight of the accumulated snow on structures, etc. A value of 1.5 lb/ft^2 per inch of snow is reasonable for the northern regions, but may not account for late spring falls of heavy, wet snow.

1.7.3.2 Ice

There are many meteorological and climatological phenomena associated with water or water vapor, freezing.

Glaze is rain freezing to a surface that is below freezing temperature. It is transparent and dense, 0.9 g/cm^3, and usually forms around 25°F.

Rime is the water vapor of clouds freezing on cold surfaces; it is opaque and less dense than glaze, 0.1–0.6 g/cm^3. It generally builds up on the windward side of surfaces.

Hoarfrost, the low temperature equivalent of dew, forms on clear, calm days. Generally the density is less than 0.2 g/cm^3, and it presents few structural problems. It may be initiated on windshields of moving vehicles if the air is saturated.

Rain that freezes before hitting the ground, essentially small particles of ice, is the familiar sleet.

A. Sea Ice. Freezing of the ocean (at 28°F) leads to ice with a brine content less than seawater, which weakens the ice. At 28°F, the sea ice loses strength more quickly than freshwater ice and is rapidly broken up. There is much pure ice in the ocean ice pack due to precipitation. Leads are open-water areas among the pack ice of oceans or large lakes. They often cause transportation problems.

B. Frazil Ice. If water is agitated and in contact with cold air, a supercooled layer of water may exist, which then freezes by nucleation and forms a dense mixture of water and ice particles. Such ice tends to agglomerate and form cohesive masses of ice on surfaces, called "anchor ice." The dendritic, or frazil, ice can lead to flow blockage in pipes, water intakes, etc.

1.7.4 Visibility

Poor visibility is a common hazard in northern regions and is associated with a number of factors.

1.7.4.1 Atmospheric Boil (Scintillation, Shimmer)

A refraction phenomenon leading to image distortion and apparent image motion, it is due to atmospheric density changes along the line of sight. Such changes may be related to incomplete turbulent mixing of thermally stratified layers of air near the ground. Resolution becomes poorer as the wind increases up to about 5 mph and then improves at higher wind speeds (Gerdel 1969).

1.7.4.2 Blowing Snow

Visibility may be reduced to zero due to snow either falling or being blown by the wind. Snow will begin to blow at about 8 mph, and as much as 95% of the snow may be within 1 ft of the ground (Gold 1964).

1.7.4.3 Ice Fog

Due to large radiation losses from the ground in the north, the ground temperature frequently drops rapidly, causing temperature inversions. If the dew point temperature of the air is greater than 0°C (Dp_2 in Fig. 1.13), then a normal water fog, or cloud, occurs; if the dew point is below 0°C (Dp_1), however, an ice fog can result. These ice or water fogs can be supplied with moisture from combustion of fuel for heating and transportation near cities, camps, etc. The result can be a serious impairment of visibility for extended periods of time. The most severe fogs are due to man-caused pollution of the air with water and occur at low temperatures (−40°F).

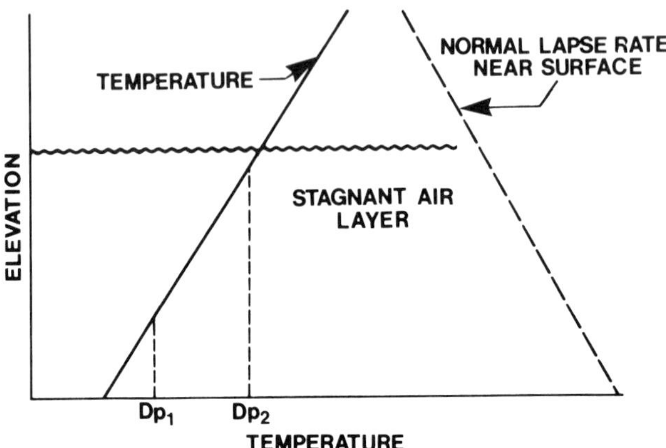

Fig. 1.13. Temperature Inversion and Fog.

1.7.4.4 Arctic Whiteout

A temperature inversion may be altered by relatively weak radiation warming the air near the ground, so that the stagnant layer lifts a short distance, but the convection is too weak to dissipate the layer entirely (Fig. 1.14). The radiation may then be scattered and re-reflected from snow surface to fog layer, resulting in a loss of shadows, horizon, and visual orientation. Judgments of distance and size are very difficult, although the horizontal

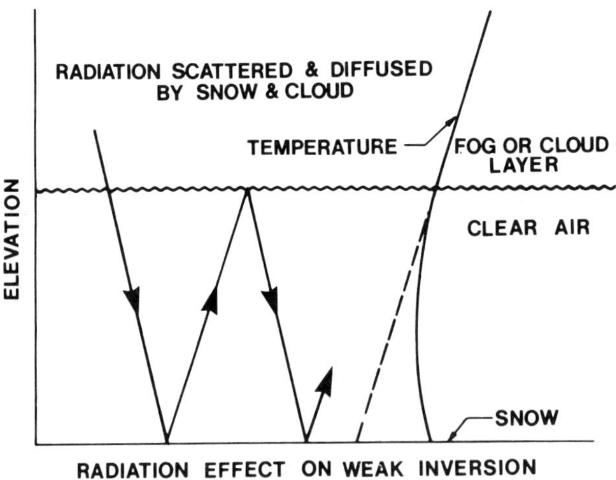

Fig. 1.14. Radiation Effect on Weak Inversion.

visibility of a dark object is not changed appreciably. Movement under such conditions is hazardous. Whiteouts are well known in the arctic and can be aggravated by blowing snow.

1.7.4.5 Snow Blindness

Mention should be made here of the severe effects of solar radiation, reflected off snow and ice covers, which can seriously damage the eye. Severe cases require days of complete eye rest with attendant immobility. Eye protection is a necessity under strong solar conditions.

1.7.5 Cycles of Light and Dark

Another characteristic of the cold regions, especially in high latitudes, is the daily cycles of light and darkness. As one moves north, the periods of light and dark become progressively longer—not simultaneously—until at the Arctic Circle, there is 24 hours of light and darkness. For example, at 70°N latitude, there is 24 hours of daylight from late May until early August and 24 hours of darkness from late November to late January. The daylight periods seem to present no difficulties, but the long hours of darkness definitely lead to psychological problems. These problems can severely affect man's performance and seem to be suffered by a large percentage of outsiders to the northern regions and even some of the native people. Any large-scale settlement north of the 50°N latitude in North America will have to take cognizance of this phenomenon.

At high latitudes, the solar elevation is relatively low, and the sun always seems fairly low on the horizon. Even during the season of the midnight sun (24 hours of daylight), the sun never seems to be directly overhead. This low elevation, coupled with the albedo characteristics of snow and ice, leads to very short seasons of spring and autumn. The albedo—the fraction of the solar radiation reflected from a surface—of snow and ice is quite high, about 80%, unlike that of land, vegetation, etc., which is about 20%. Thus, while there is snow on the ground, most of the solar energy is reflected off its surface, and little is available for melting snow or thawing ground. As soon as the snow melts and leaves patches of bare ground, warming of the ground and further snow melting is quite rapid. Thus, the spring season in the far north may be just a few weeks from snow melt to flower blooming, but it will be rather retarded—June or early July.

REFERENCES

Bates, R. E., and Bilello, M. A. 1966. *Defining the cold regions of the northern hemisphere.* Hanover, New Hampshire: CRREL, TR 178.

Boyd, D. W. 1961. Maximum snow depths and snow loads on roofs in Canada. *Proceedings, western snow conference.* Spokane, Washington.

Gerdel, R. W. 1969. *Characteristics of the Cold Regions.* CRREL, 1-A.

Gold, L. W. 1964. Characteristics of snow and ice relevant to snow removal and ice control. *Snow removal and ice control conference.* Ottawa: NRC 8146.

Haurwitz, B., and Austin, J. M. 1944. *Climatology.* New York: McGraw-Hill.

Kendrew, W. G., and Currie, B. W. 1955. *The climate of central Canada.* Ottawa: Queen's Printer.

Koppen, D. W. 1936. The geographical system of climate. *Handbuch der Klimatologie.* VI. Part C, pp. 5–44.

Stearns, S. R. 1965. *Selected aspects of geology and physiography of the cold regions.* CRREL, 1-A-1.

Taylor, T. G. 1957. *Geography in the twentieth century.* New York: Philosophical Library.

Wilson, C. 1967. *Cold regions science and engineering, introduction northern hemisphere I.* CRREL, 1-A3.

PROBLEMS

1. Discuss the relation of Ottawa and Norman Wells to cold climates or regions defined in terms of:
   (a) Koppen's limit
   (b) Wilson and Haurwitz
   (c) Bates and Bilello
   (d) polar circles
   (e) tree limits
   (f) permafrost limits
   (g) sea ice limits
2. What is the lowest mean temperature listed for Canada? Where does it occur?
3. What are the extreme temperatures ever recorded in Canada?
4. Estimate the percentage frequency of below −25°F January temperatures for Churchill, Manitoba.
5. Discuss why the operational concept of a cold climate is superior to a temperature definition.
6. How does the climatology of the north of Canada differ from that of the south, in terms of engineering problems?
7. Define what is meant by a "cold climate." How does this encompass terms such as arctic and subarctic? In engineering aspects, does Canada have a cold climate?

2 Design Data for Cold Climates

2.0 INTRODUCTION

The design and analysis of engineering systems that will function in cold climates require accurate and extensive meteorological data. Typically, data are collected and averaged over 30-year periods, where possible, to avoid short-term fluctuations, which may exceed ±20% of the average values. Many areas of northern North America are still not densely surveyed in terms of meteorological stations, and some of the stations have been functioning for only 10–20 years, especially in Canada. The situation has improved greatly, however, and no doubt the data will become still more reliable. Varotta (1972) has compiled and condensed the data from 118 weather stations north of the 50°N latitude across Canada. Most of the data have been taken from the references at the end of this chapter. The required design information is presented as basic data and quantities of interest calculated from the data. Not all the required information has been compiled; for example, soil surveys and ground temperature profiles are so variable as to defy large-scale compilation. Each of the important datum will be briefly discussed.

2.1 TEMPERATURE

As was discussed in Chap. 1, temperature is probably the single most important property of northern regions. The effects and ramifications of low temperature are manifested in virtually all phases of engineering design and in the proper function of engineering systems. Nearly all classes of weather stations in Canada and the United States record some temperatures.

2.1.1 Basic Data

(a) Mean Daily Temperature. Each day, the maximum and minimum temperatures are recorded. The mean daily maximum temperature for a month is the mean of all the daily maximum temperatures recorded for that month during the period of observation (30 years, if available). The mean daily minimum temperature is similarly derived from all daily minimum temperatures.

(b) Mean Monthly Temperature. The mean monthly temperature is the mean of the mean maximum and mean minimum temperatures for the month.

(c) Mean Annual Temperature. The mean annual temperature is the arithmetic mean of the mean monthly temperatures.

(d) Temperature Extremes. These are the absolute maximum and minimum temperatures ever recorded at a given location. The absolute minimum temperature recorded for North America, −81°F, was obtained at Snag Airport, Yukon Territories, February 1947, but temperatures not much higher have been recorded in Alaska.

2.1.2 Calculated Quantities

There are a number of important quantities used for engineering purposes that are calculated from the basic temperature data.

(a) Winter Design Temperatures. The 1% design temperature is the value that the air temperature does not remain below for more than 1% of the month of January.[1] The 2½% design temperature is the temperature that the ambient temperature remains above for 97½% of the time. The 1% temperature will be lower than the 2½% value.

(b) Summer Design Temperatures. The 1% design temperature is that temperature not exceeded for more than 1% of the month of July;[1] 2½ and 5% values can also be calculated. These relations apply to both the dry-bulb and wet-bulb temperatures.

(c) Degree-Days. The degree-day concept has found widespread use in calculating heating energy requirements, depths of thaw and freeze, ice breakup, etc. There are several degree-day definitions, of which the most significant will be discussed.

Heating Degree-Days. The most common use of the heating degree-day is in estimating the fuel costs for a given heating season. For heating, it is known that if the outside dry-bulb temperature is above 65°F (actually 66°F is closer, but 65°F has long been used) then no indoor heating is required. A degree-day is defined as the difference between the mean daily temperature and 65°F.

$$D \equiv (65 - \bar{t}_i) \tag{2.1}$$

A temperature difference of $(65 - \bar{t})$ is used, since any outside temperature above 65°F will not require heating. This is due to solar radiation and internal heat loads of equipment and people. Thus, if for January 17 the

[1] The winter period for the United States is December–February and the summer period is June–August.

mean temperature is 5°F, then there will be 60 degree-days for January 17. The summation of all the degree-days is the seasonal degree-days or degree-days/season. Table 2.1 gives examples of typical heating degree-days for Canada. Extensive lists of heating degree-days for the United States are given in ASHRAE (1977). One can estimate the average energy consumption for any period if one knows the degree-days for the period and the heating load of the structure.

The heat load is calculated using the design data

$$Q = UA(t_i - t_o) \tag{2.2}$$

where

Q = heat load of structure, Btu/hr
U = overall heat transfer coefficient Btu/(hr–ft²–°F); this may include an infiltration effect
A = area, ft²
t_i, t_o = inside and outside design temperatures.

Now the actual heat transfer for the heating season is

$$Q_a = UAN(t_i - \bar{t}) \tag{2.3}$$

where

$\bar{t}$ = average air temperature during the entire heating season
N = length of heating season, days, used to calculate $\bar{t}$.

Substituting for (UA) from Eq. (2.2)

$$Q_a = \frac{Q}{(t_i - t_o)} N(t_i - \bar{t}) \tag{2.4}$$

But from Eq. (2.1), the seasonal degree-days D is defined below as

$$D = N(65 - \bar{t})$$

Thus, the actual heat transfer or energy consumption for the season may be written

$$Q_a = \frac{Q}{(t_i - t_o)} \frac{D(t_i - \bar{t})}{(65 - \bar{t})} \tag{2.5}$$

Table 2.1. Heating Degree-Days, Canada.

Prov.	City	Jan.	Feb.	Mar.	Apr.	May	June	July	Aug.	Sep.	Oct.	Nov.	Dec.	Year
Alta	Calgary	1609	1355	1215	750	480	264	108	167	432	722	1122	1426	9650
	Edmonton	1810	1504	1299	777	428	222	70	167	441	750	1224	1593	10,285
BC	Vancouver	893	736	682	498	326	162	43	65	234	459	657	818	5573
	Victoria	815	689	651	504	366	237	155	158	264	446	462	738	5485
	Prince Rupert	933	804	809	645	521	357	282	229	342	546	702	893	7063
Man	Churchill	2604	2288	2204	1530	1097	672	350	391	696	1187	1773	2356	17,148
	Winnipeg	2111	1775	1531	822	397	78	27	34	330	744	1302	1829	10,980
NB	Moncton	1528	1392	1190	798	474	180	8	56	282	595	936	1373	8812
	Saint John	1417	1266	1132	792	505	261	124	112	270	558	870	1271	8578
NS	Halifax	1280	1168	1065	768	493	210	12	12	189	493	783	1141	7614
Ont	Fort William	1807	1582	1386	888	567	234	62	155	354	722	1143	1596	10,496
	Hamilton	1305	1187	1063	651	322	6	3	9	108	471	816	1178	7119
	London	1336	1240	1073	642	307	105	34	11	126	508	843	1200	7425
	Ottawa	1646	1459	1256	726	310	91	29	22	204	595	984	1494	8816
	Toronto	1304	1206	1072	669	341	87	16	35	168	504	817	1155	7374
	Windsor	1283	1148	995	582	251	73	19	5	54	425	795	1172	6802
PEI	Charlottetown	1463	1338	1203	858	536	213	5	57	222	539	858	1246	8538
PQ	Montreal	1587	1392	1209	702	298	88	7	20	180	561	948	1407	8399
	Québec	1696	1481	1311	849	428	102	17	43	276	651	1050	1534	9438
Sask	Saskatoon	2027	1635	1445	725	397	138	16	68	396	781	1305	1767	10,700
YT	Dawson	2666	2159	1879	1092	580	246	167	322	687	1209	1908	2440	15,355

Since t_i is often close to 65°F, Eq. (2.5) is often simplified to

$$Q_a = \frac{QD}{(t_i - t_o)} \tag{2.6}$$

Example 1: The heat loss from a house is calculated as 70,000 Btu/hr with $t_i = 72°F$, $t_o = -20°F$, $D = 8816$ °F-day/season. The cost of heating oil is 74.6¢/gal (the cost of fossil fuels is changing so rapidly that no example can be current.) The furnace efficiency is 75%, and the heating value of the oil is 140,000 Btu/gal. Calculate the estimated cost of heating this house for one season.

From Eq. (2.6)

$$Q_a = \frac{24(70{,}000)(8816)}{92} = 1.615 \times 10^8 \text{ Btu/season}$$

$$\text{Seasonal cost} = \frac{1.615 \times 10^8}{1.40 \times 10^5} \times \frac{.746 \text{ dollar}}{.75 \text{ gal}} = \$1147$$

The gallons of fuel required (1538 gallons) can also be calculated.

Freezing Degree-Days or Freezing Index (FI). There is a good correlation between the depth of freeze and the freezing degree-day. The FI is defined as the difference between the mean daily temperature and 32°F when $\bar{t}_d < 32°F$. Thus

$$\text{FI} = (32 - \bar{t}_d) \text{ per day} \tag{2.7}$$

The freezing index also may be used to predict ice thickness and growth.

Thawing Degree-Days or Thawing Index (TI). This is the same as FI, except that $\bar{t}_d > 32°F$. This can be used to predict thawing depth of a frozen terrain or ice breakup dates. Figures 2.2 and 2.3 and Tables 2.2 and 2.3 give values of the freezing and thawing indices for northern Canada.

Length of the Thawing and Freezing Seasons. These time spans are very important in engineering calculations dealing with freezing and thawing. The thawing season is defined as the time between the date of the end of the freezing season and the beginning of the freezing season. The end of the freezing season is the date on which the seasonal accumulation of freezing degree-days is maximum, and the beginning of the freezing season is the mean date of the maximum seasonal accumulation of thawing degree-days. A fair estimate can be obtained by adding the months without

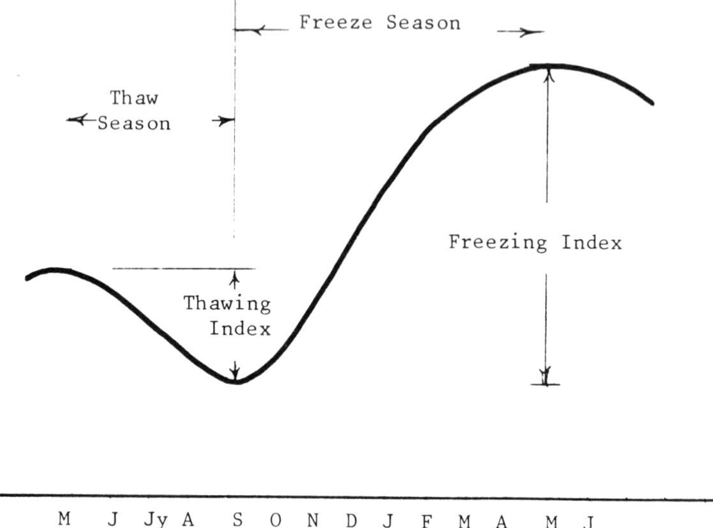

Fig. 2.1. Freezing and Thawing Degree-Days.

freezing degree-days to the proportional amounts of those months that have both freezing and thawing degree-days. The length of the freezing season will be 365 minus the length of the thawing seasons.

Figures 2.4 and 2.5 show the mean values of the beginning and end of the freezing season for Canada. It is also possible to estimate the lengths of the seasons, using Fig. 2.6, which is simply a correlation of data for North America.

The values of the freezing and thawing indices and the season length can be calculated by plotting the cumulative values calculated with the mean daily temperature at a location. This is shown in Fig. 2.1, for an idealized set of temperatures. The absolute value of the cumulative total is not significant.

2.2 PRECIPITATION

The total precipitation, as well as the amounts of snow, rain, etc., are important characteristics for any region. One of the peculiarities of far northern regions is the scanty annual precipitation, a fact that is often overlooked due to the large amount of surface water present. This phenomenon has been referred to in Chap. 1.

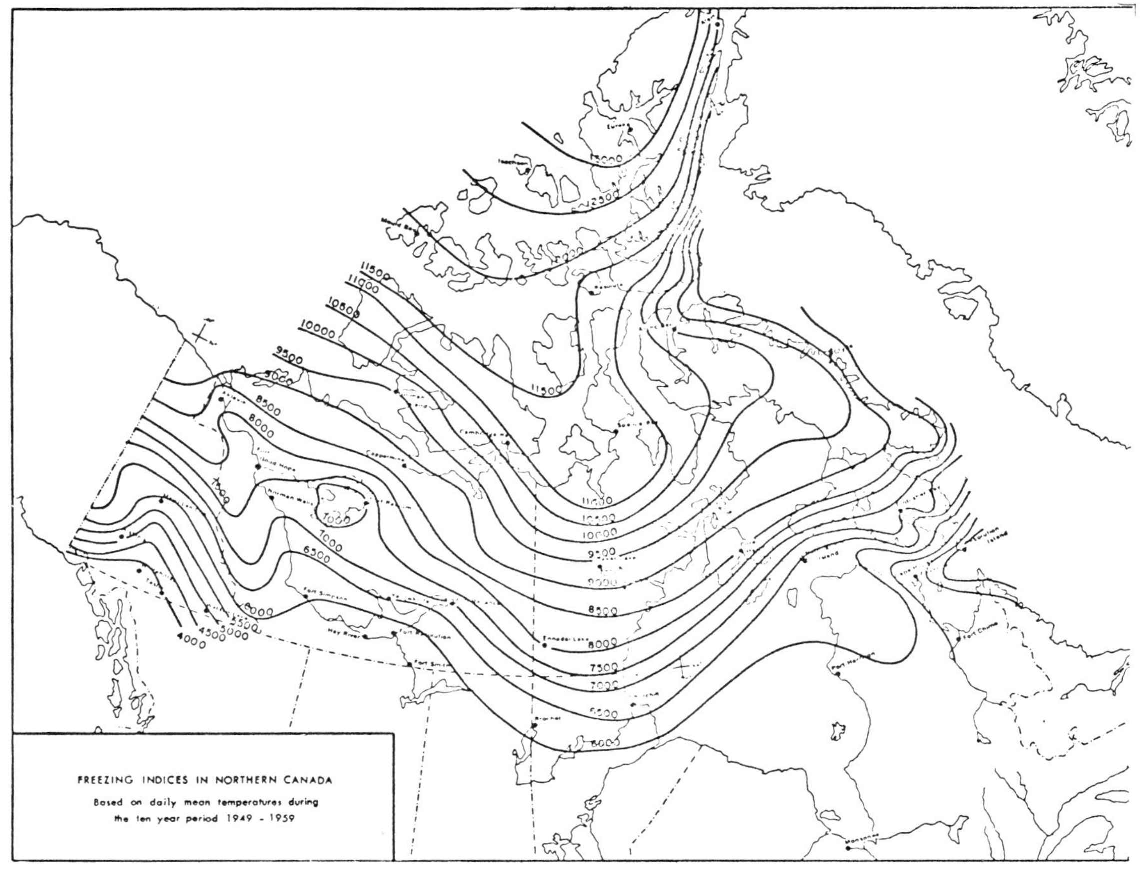

Fig. 2.2. Freezing Indices in Northern Canada; Based on Daily Mean Temperatures during the 10-Year Period 1949–1959 (adapted from Thompson 1963). Reproduced from *Permafrost International Conference*, 1963, with the permission of the National Academy of Sciences, Washington, D.C.

Table 2.2. Freezing and Thawing Indices for Canadian Locations (adapted from Thompson 1963). Reproduced from *Permafrost International Conference,* 1963, with the permission of the National Academy of Sciences, Washington, D.C.

Station	10-year average		High		Low	
	FI	TI	FI	TI	FI	TI
Aishihik A	5302	2412	6291	2715	4114	2035
Aklavik	8037	2261	8594	2859	6811	1761
Alert	12,093	387	12,544	612	11,397	141
Arctic Bay	9923	808	10,674	990	8849	699
Baker Lake	9422	1515	10,072	1947	8891	1193
Brochet	6216	2935	7040	3412	5417	2525
Cambridge Bay	10,860	1016	11,311	1406	10,302	767
Cape Hopes Advance	5443	974	6309	1198	4562	822
Chesterfield	8750	1321	9449	1673	8151	1098
Churchill A	6718	2056	7755	2380	6253	1764
Clyde	8671	655	9365	795	7943	550
Coppermine	8882	1362	9439	1620	7869	954
Coral Harbour A	8539	1175	9510	1487	7666	1021
Ennadai Lake	8130	1979	8840	2483	7660	1620
Eureka	13,322	701	14,243	847	12,519	417
Fort Chimo A	5381	2206	6176	2501	4518	1787
Fort Good Hope	7860	2908	8638	3214	6877	2385
Fort Reliance	7172	2465	7726	2924	6337	1768
Fort Resolution	5776	3176	6354	3571	4618	2633
Fort Simpson	6040	3476	6895	3777	4916	2995
Fort Smith A	5613	3365	6235	3685	4567	2806
Frobisher Bay A	7052	1262	8096	1541	6226	1073
Hay River	5548	3171	6332	3472	4287	2573
Holman Island	9015	1100	9677	1660	7931	820
Isachsen	12,753	402	13,486	649	11,983	136
Mayo Landing	5881	3168	7020	3475	4480	2816
Moosonee	3790	3611	4626	4365	2810	3053
Mould Bay	11,912	422	12,529	691	10,987	129
Norman Wells A	7220	2996	7973	3354	6137	2531
Nottingham Island	6401	928	7227	1131	5547	780
Port Harrison	5804	1677	6527	2293	5165	1288
Port Radium	6982	2220	7776	2799	5713	1582
Resolute A	11,204	536	11,591	888	10,292	322
Resolution Island	4434	552	5483	686	3791	429
Spence Bay	11,093	973	11,578	1150	10,524	833
Teslin A	4048	2991	5224	3354	2931	2568
Watson Lake A	5402	3388	6510	3683	4515	2965
Whitehorse A	3861	3271	5105	3607	2852	2865
Yellowknife A	6623	3079	7159	3354	5318	2483

Table 2.3. Accumulation of Freezing and Thawing Degree-Days during First 30 Days of Freezing Season (adapted from Thompson 1963). Reproduced from *Permafrost International Conference,* 1963, with the permission of the National Academy of Sciences, Washington, D.C.

	Average		Lowest		Highest	
Location	FI	TI	FI	TI	FI	TI
Aishihik A	289	232	31	168	615	307
Aklavik	314	236	159	72	604	383
Alert	330	162	173	22	597	315
Arctic Bay	255	243	122	157	376	384
Baker Lake	287	272	96	190	617	362
Brochet	419	201	142	83	782	312
Cambridge Bay	338	258	178	139	549	520
Cape Hopes Advance	160	169	98	33	242	270
Chesterfield	296	227	83	167	620	316
Churchill A	356	254	206	108	536	518
Clyde	199	153	72	15	273	263
Coppermine	310	243	73	69	580	529
Coral Harbour A	262	201	32	97	501	309
Ennadai Lake	249	289	25	86	672	515
Eureka	407	247	210	120	564	350
Fort Chimo A	254	242	75	85	442	450
Fort Good Hope	356	325	144	207	749	537
Fort Reliance	357	254	34	89	643	448
Fort Resolution	339	305	141	169	594	430
Fort Simpson	326	334	39	81	570	574
Fort Smith A	346	309	102	90	615	480
Frobisher Bay A	321	138	146	51	713	334
Hay River	362	279	66	162	680	455
Holman Island	290	239	71	60	507	563
Isachsen	334	160	144	57	611	346
Mayo Landing	395	267	138	83	745	444
Moosonee	252	211	36	53	519	426
Mould Bay	211	164	115	42	352	281
Norman Wells A	339	307	92	133	645	500
Nottingham Island	143	143	51	50	210	238
Port Harrison	260	183	104	3	395	345
Port Radium	301	247	64	103	509	543
Resolute A	311	191	129	120	471	270
Resolution Island	119	67	57	15	242	124
Spence Bay	279	297	135	225	493	364
Teslin A	325	210	74	96	451	364
Watson Lake A	444	267	66	165	691	449
Whitehorse A	345	186	148	98	584	373
Yellowknife A	357	348	79	107	508	546

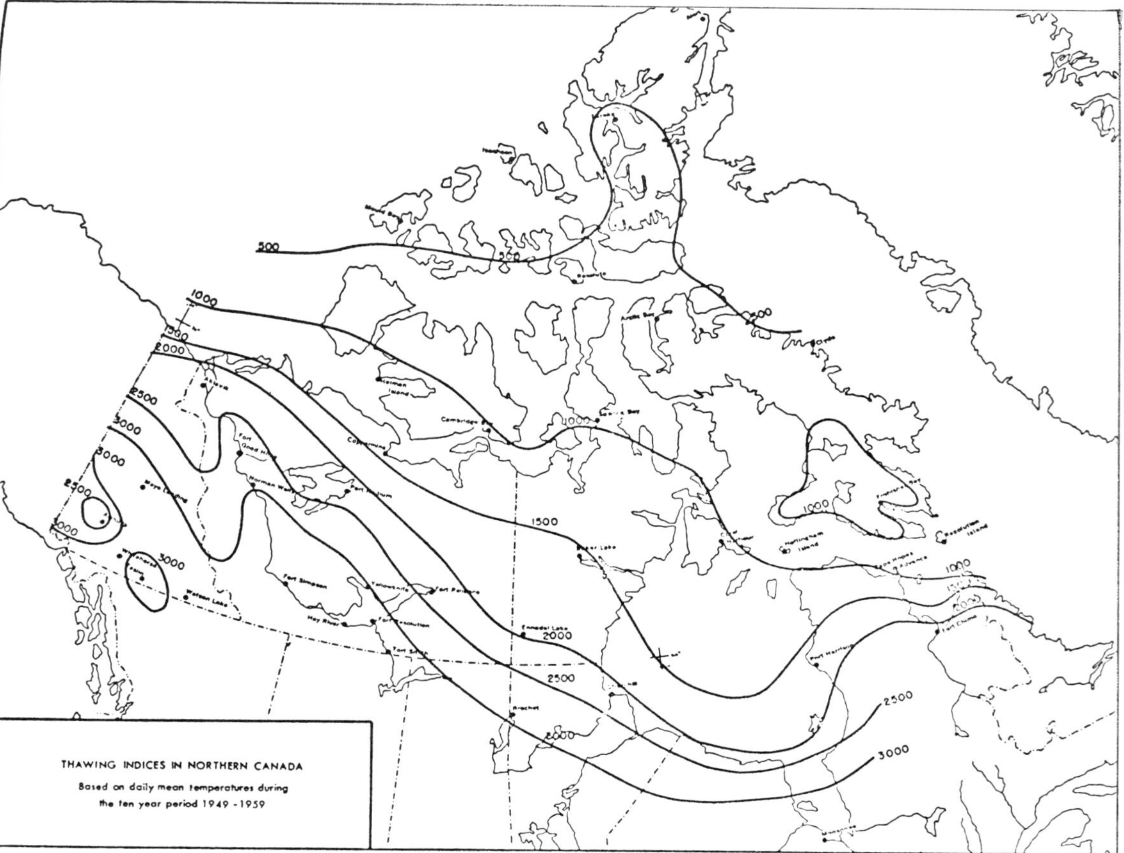

Fig. 2.3. Thawing Indices in Northern Canada; Based on Daily Mean Temperatures during the 10-Year Period 1949–1959 (adapted from Thompson 1963). Reproduced from *Permafrost International Conference*, 1963, with the permission of the National Academy of Sciences, Washington, D.C.

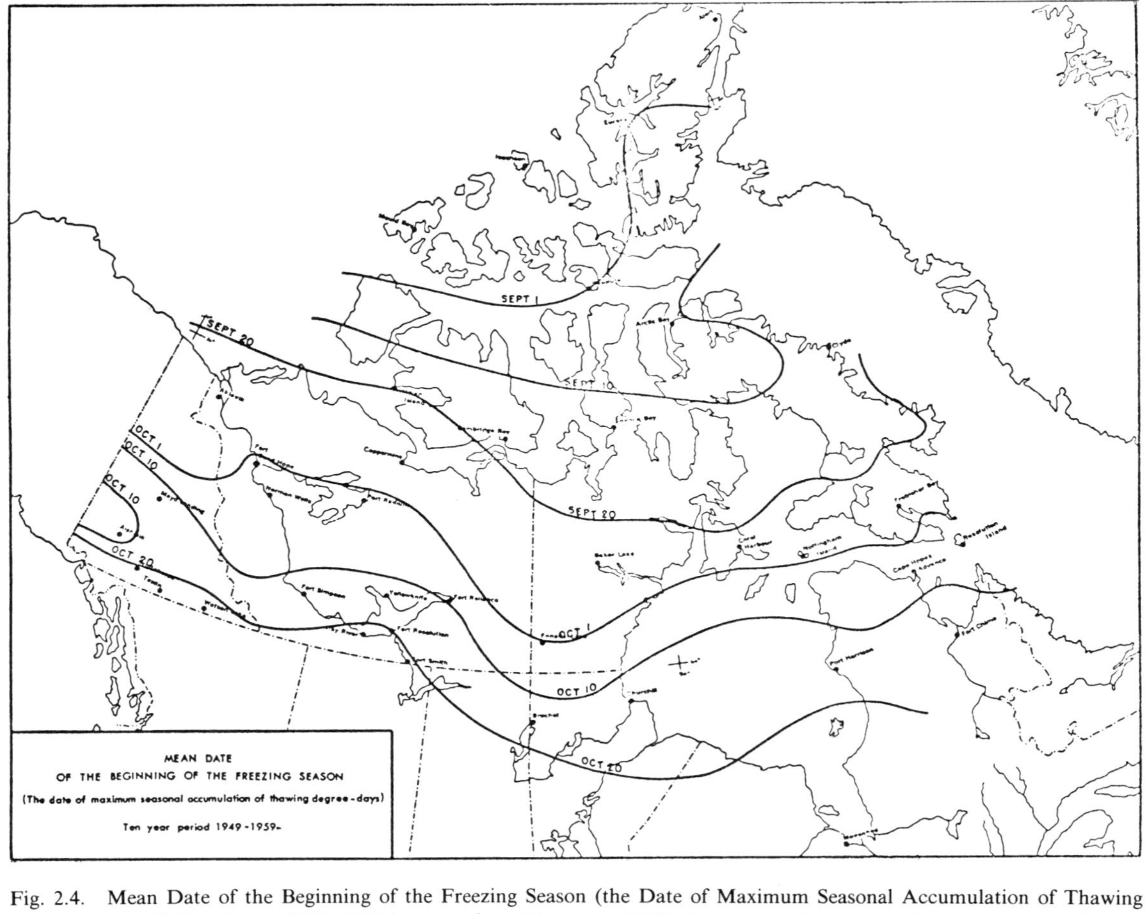

Fig. 2.4. Mean Date of the Beginning of the Freezing Season (the Date of Maximum Seasonal Accumulation of Thawing Degree-Days), 10-Year Period 1949–1959 (adapted from Thompson 1963). Reproduced from *Permafrost International Conference*, 1963, with the permission of the National Academy of Sciences, Washington, D.C.

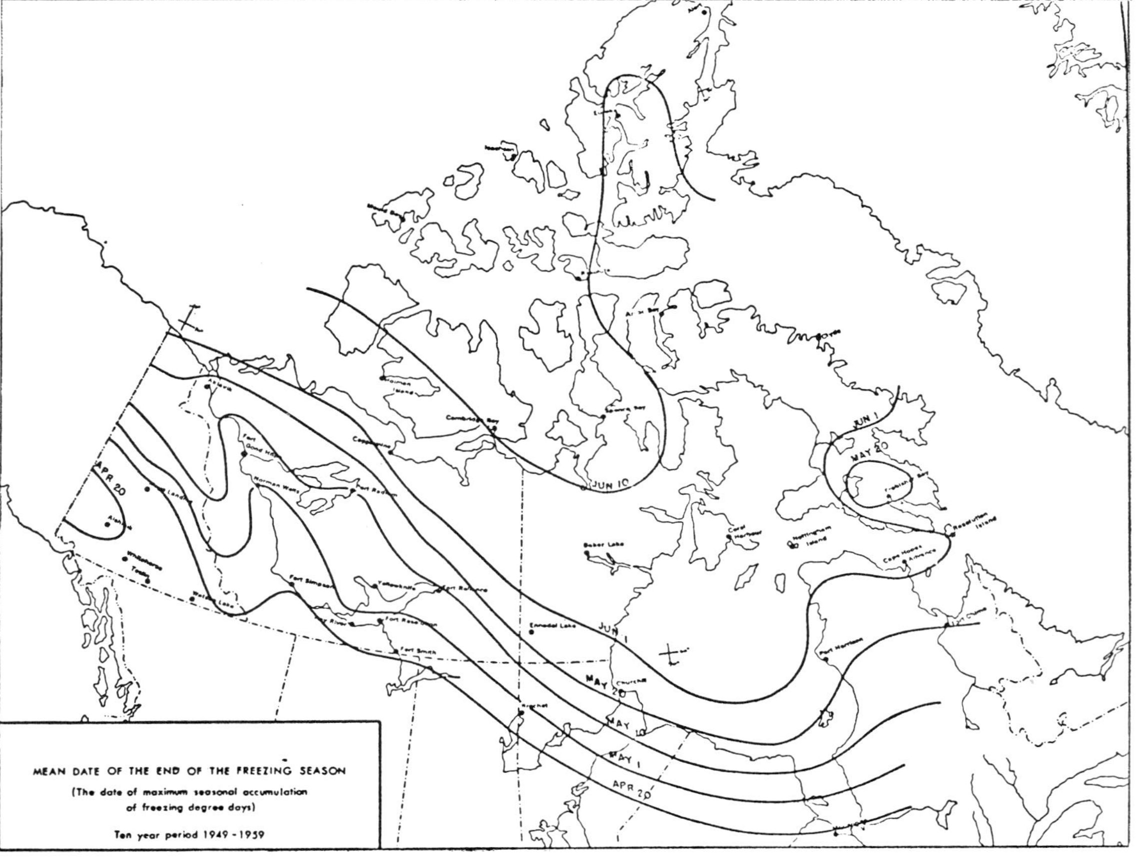

Fig. 2.5. Mean Date of the End of the Freezing Season (the Date of Maximum Seasonal Accumulation of Freezing Degree-Days), 10-Year Period 1949–1959 (adapted from Thompson 1963). Reproduced from *Permafrost International Conference*, 1963, with the permission of the National Academy of Sciences, Washington, D.C.

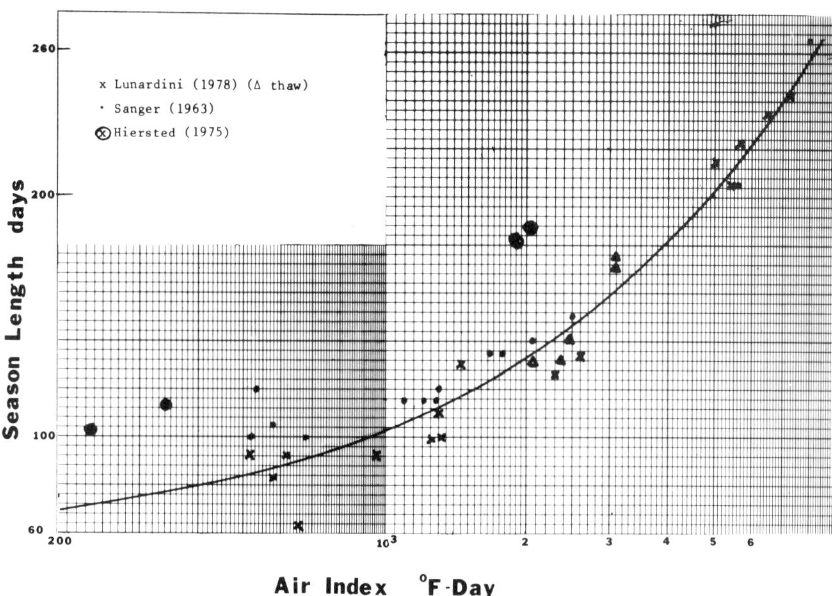

Fig. 2.6. Season Length and Air Index, Freezing or Thawing.

2.2.1 Basic Data

(a) Mean Monthly and Annual Rainfall. Rain is measured to the nearest .01 in. and averaged each month over the time period. The annual values are the summation of the mean monthly values.

(b) Mean Monthly and Annual Snowfall. The amount of water resulting from melting of freshly fallen snow is taken as 1/10 the depth of the snow. The total precipitation is equal to the rain plus 1/10 the snowfall.

(c) Days of Measurable Precipitation. A day with measurable rain is one on which more than .01 in. has fallen, and a day with snow is one with at least 1/10 in. of snowfall. The number of days each month with precipitation are then totaled and averaged over the time period.

(d) Maximum Precipitation. Each location records the maximum total precipitation recorded for a 15-minute interval and a 24-hour interval.

2.2.2 Calculated Quantities

(a) Ground Snow Load. The depth of snow on the ground that will not be exceeded once in 30 years is used with the assumption that 1 in. of snow cover corresponds to 1 lb/ft^2 (this is equal to a specific gravity of 0.192).

To this is added the load due to the maximum 1-day rainfall for the location in late winter or early spring.

2.3 RADIATION

The thermal regime and equilibrium of permafrost, which often directly supports engineering structures, are dependent upon the net flows of radiant energy at the interface between the surface and the atmosphere. Average and, sometimes, instantaneous values of long- and shortwave radiation, both direct and indirect, are needed. Not all these data have been or are likely to be available on a large-scale basis.

2.3.1 Basic Data

(a) Hours of Sunshine. This information is recorded continuously and reduced to hours of sunshine per month or year.

(b) Mean Cloud Amount and Frequency. The data are reported as a frequency and can be thought of as the fraction of time the sky is more or less covered by clouds.

(c) Global Solar Radiation (formerly referred to as "insolation"). The total shortwave radiation at a location is based on both data and calculations. The calculated values are computed so as to agree with the data and are used to draw continuous curves over areas for which no data exist. The global radiation does not differentiate between direct and indirect shortwave radiation. The information is presented as monthly and yearly averages.

2.4 WIND

The movement of air near the surface of the earth has important engineering implications in regard to energy balance, visibility, hypothermia, etc. Wind data have been collected extensively, but many areas of northern Canada lack data.

2.4.1 Basic Data

(a) Wind Speed and Probability, Monthly and Yearly. Generally, the values of wind speed, direction, and duration, are recorded at a height 33 ft above the ground surface. Wind direction refers to the true direction from which the wind is coming. The absence of any measurable wind is denoted as calm.

2.4.2 Calculated Quantities

(a) Wind Pressure. The force exerted upon structures by the wind is calculated from the gusts at a location that, in turn, depend upon the wind speed and frequency. The value of the gust at a given location is given by the National Building Code of Canada (1968)

$$G = 5.8 + 1.29\,V \tag{2.8}$$

where

G = gust speed, mph and
V = hourly speed, mph.

The velocity pressure is then given by

$$P = CG^2 \tag{2.9}$$

where

P = wind pressure, lb/ft^2
C = .0027 (depends on temperature and pressure) and
G = ft/sec.

2.5 SEASONS

The use of length of significant periods of the year, such as the heating season, growing season, etc., is quite common in many engineering analyses. Some of the quantities are known or calculated only imprecisely. The most important are the length of freeze or thaw, which have been discussed.

REFERENCES

American Society of Heating, Refrigeration, and Air Conditioning Engineers. 1977. *ASHRAE handbook.* New York.

Department of Transport, Meteorological Branch. 1967. *Temperature and precipitation tables for the north—Y. T. and N. W. T.* Toronto.

Department of Transport, Meteorological Branch. 1968. *Wind, climatic normals,* vol. 5. Toronto.

Department of Transport, Meteorological Branch. 1970. *Climatological station data catalogue—Y. T. and N. W. T.* Toronto, CLI 3–70.

Frazer, W. C. 1964. *A study of winds and blowing snow in the Canadian Arctic.* Toronto: Department of Transport, Meteorological Branch.

Heiersted, S. R. S. 1975. Thermal climate regime on road and ground surface. *Frost I Jord* NR 10.

Kendall, G. R., and Anderson, S. R. 1966. Standard deviations of monthly and annual mean temperatures. *Climatological Studies, no. 4.* Toronto: Department of Transport, Meteorological Branch.

Lunardini, V. J. 1978. Theory of N-factors and correlation of data. Edmonton, Alberta: *Proceedings, Third International Conference on Permafrost,* vol. 1, pp. 40–46.

National Building Code of Canada. 1968. *Supplement to the building code for the north.* NRC 10368.

Potter, J. G. 1966. *Snow cover.* Ottawa: Queen's Printer.

Sanger, F. J. 1963. Degree days and heat conduction in soils. *Proceedings Permafrost International Conference,* Lafayette, Indiana. NRC Pub. 1287, pp. 253–262.

Thomas, M. K. 1960. *Canadian arctic temperatures.* Toronto: Department of Transport, Meteorological Branch, CIR-3334, CLI-24.

______. 1964. *Snowfall in Canada.* Toronto: Department of Transport, Meteorological Branch, CIR 3977, TEC 503.

Titus, R. L., and E. J. Truhler. 1969. *A new estimate of average global solar radiation in Canada.* Toronto: Department of Transport, Meteorological Branch, CLI 7–69.

Thompson, H. A. 1963. Air temperatures in northern Canada with emphasis on freezing and thawing index. *Proceedings Permafrost International Conference.* NRC 1287, Washington, D.C., pp. 272–280.

Thompson, H. A. 1967. *The climate of the Canadian arctic.* Dominion Bureau of Statistics, Ottawa.

Varotta, R. F. 1972. *Design data of the Canadian north,* vols. 1 and 2. Fourth year thesis, Univ. of Ottawa, Mechanical Engineering.

PROBLEMS

1. A house with a heat loss of 120,000 Btu/hr is located at Alert. Estimate the gallons of No. 2 oil needed per year and the cost at a furnace efficiency of 70%.
2. A house in Churchill and one in Ottawa have identical design heat losses. What is the relation between the fuel costs?
3. Using the data available for Whitehorse, Y. T., estimate the length of the thawing and freezing seasons.
4. The town of Inuvik (Aklavik data) has 600 dwellings with an average heat loss of 85,000 Btu/hr. Number 2 fuel oil is to be used for heating purposes. Estimate the size of the oil reservoirs needed for a 2-year period.
5. Repeat Problem 4 if natural gas is used.
6. Derive Eq. 2.8 from fluid mechanic principles.
7. Discuss the climatological data needed for engineering design in northern Canada.
8. A location has the following mean monthly temperatures. Calculate the freezing index. Would you expect this to be accurate?

Month	Jan.	Feb.	Mar.	Apr.	May	June	July	Aug.	Sept.	Oct.	Nov.	Dec.
Temp. °F	5	4	18	41	54	60	61	58	58	46	20	8

9. The following table notes the freezing and thawing indices for a location. Estimate the length of the thaw season in days.

	Jan.	Feb.	Mar.	Apr.	May	June	July	Aug.	Sept.	Oct.	Nov.	Dec.
FI	1556	1456	1217	883	145	0	0	0	0	496	1329	1457
TI	0	0	0	0	67	632	759	689	290	50	0	0

10. A house in Alert, NWT, has a heat loss of 100,000 Btu/hr. Estimate the savings in No. 2 oil if the house insulation is augmented such that the overall heat transfer coefficient is reduced by 48%

$$\text{Alert: design temperature} - 45°\text{F, degree-days} = 25{,}000\,\frac{°\text{F-days}}{\text{season}}$$

11. A location has the following mean monthly temperatures. Calculate the thawing index.

Month	Jan.	Feb.	Mar.	Apr.	May	June	July	Aug.	Sept.	Oct.	Nov.	Dec.
Temp. °F	12	13	25	41	54	64	69	66	58	46	33	17

12. Define a degree-day and list some of the more common degree-days. What use can be made of such quantities?

3
Heat Transfer

3.0 INTRODUCTION

The emphasis of this chapter will be on the special methods available for the solution of permafrost heat transfer problems of interest to the engineer. A brief discussion of heat transfer is presented here for the reader unfamiliar with the concept; further details can be found in the standard texts listed at the end of the chapter. In the space available, it is not possible to tabulate the thermal properties of all engineering materials. Standard material properties are readily available in handbooks and the aforementioned books. For unusual materials or systems, an attempt has been made to provide adequate data on the thermal properties in this book.

In the general case, the transfer of energy may take place by one or more of the three modes of heat transfer. These are conduction, radiation, and convection. The first two methods are not associated with mass transfer, whereas the third is dependent on mass transfer. In addition to these modes of energy transfer, we may encounter energy transfer due to mechanical work, and energy liberation by electrical, chemical, or other means. An energy balance for any selected thermodynamic system may be established for the case of a medium in motion. From this energy balance, an equation of the temperature as a function of the space coordinates, time, and the velocity components, along with some properties of the medium, may be written.

From such a general equation, the equations for special cases may be developed by choosing the proper values for the variables. For example, if the system is a stationary solid, the velocities are zero, and terms in which the velocities appear may be set equal to zero.

3.1 CONDUCTION

If a body or a series of bodies are joined, and if different portions of the body are maintained at different temperatures, then there will be a flux of energy from one point within the body to another. This flux is called "conductive heat transfer." The process depends upon the movement of atomic particles within the material and lattice waves, which transmit energy by atomic vibrations. The mutual interaction of molecules and electrons will result in a transfer of kinetic energy from one location to another. The different phases of matter conduct energy differently, as may be noted

Gases: mainly by molecular collision
Nonmetallic liquids: lattice vibrations and molecular collision
Solids: lattice vibrations and molecular collision
Metallic solids: electron flow plus lattice waves

The basic law of conduction is an empirical relation known as "Fourier's law" or "Biot's law" and is (Fig. 3.1)

$$q = -kA\frac{\Delta T}{\Delta L} \tag{3.1}$$

In words, the flow of energy q in the direction of the temperature change is proportional to the cross-sectional area A, and the temperature gradient $\Delta T/\Delta L$. The proportionality coefficient k is the thermal conductivity of the material, and it is a transport property depending mainly on the temperature for any given substance. The negative sign is a convention that makes a flow of heat positive in the direction of decreasing temperature. If the thickness ΔL of the material becomes very small, then the conduction law may be written such that it is valid at a point

$$q = -kA\frac{dT}{dx} \tag{3.2}$$

if the temperature varies only in the x-direction. Also

$$q_x = -kA\frac{\partial T}{\partial x} \tag{3.3}$$

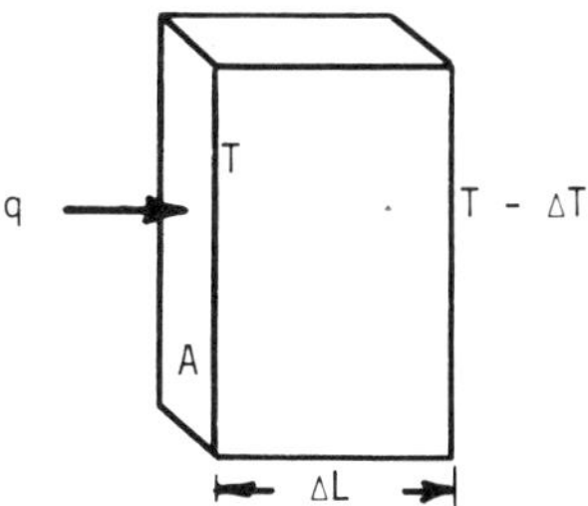

Fig. 3.1. Simple Conduction Model.

Here the temperature may vary in any direction, and q_x is the heat flow in the x-direction at the point x.

The thermal conductivity is actually a tensor quantity for an anisotropic material. The general relation is

$$\bar{\bar{k}} = \begin{bmatrix} k_{11} & k_{12} & k_{13} \\ k_{21} & k_{22} & k_{23} \\ k_{31} & k_{32} & k_{33} \end{bmatrix} \tag{3.4}$$

Each of the components k_{ij} can be a function of the thermodynamic state of the material.

For an orthotropic material, the principal axes of the conductivity tensor will coincide with a set of coordinate axes such that

$$\bar{\bar{k}} = \begin{bmatrix} k_x & 0 & 0 \\ 0 & k_y & 0 \\ 0 & 0 & k_z \end{bmatrix} \tag{3.5}$$

Finally, for an isotropic material (the usual assumption)

$$\bar{\bar{k}} = k \tag{3.6}$$

The conductivity is now simply a scalar function, as implied by Eq. (3.1). The general Fourier law is

$$\bar{q} = \bar{\bar{k}} \text{ grad } T \tag{3.7}$$

The magnitude of the thermal conductivity varies over a wide range for different materials. Typical values in units of Btu/hr-ft-°F are

Metals and alloys	10–100
Dielectric solids, crystalline	.1–10
Liquids	.01–.1
Gases, insulators	.001–.01

Equation 3.1 may be rewritten as

$$q = \frac{\Delta T}{R} \tag{3.8}$$

R is called the thermal resistance and is given, for conduction only, by

$$R = \frac{\Delta L}{kA} \tag{3.9}$$

A more general relation is

$$R = \frac{1}{k_m} \int_{x_1}^{x_2} \frac{dx}{A(x)} \tag{3.10}$$

In the latter case, the mean thermal conductivity between temperatures T_1 and T_2 is given by

$$k_m = \frac{1}{(T_2 - T_1)} \int_{T_1}^{T_2} k(T)\, dT \tag{3.11}$$

A conductive system can usually be broken into parts such that it is called a "series system," if the total heat flow traverses each part, or a "parallel system," if a part of the total heat flow crosses each point. The resistance of a series structure is

$$R = \sum_{i=1}^{n} R_i \tag{3.12}$$

where R_i is the thermal resistance of each of the n parts. For a parallel system

$$R = \frac{1}{\sum_{i=1}^{n} \frac{1}{R_i}} \tag{3.13}$$

The resistance concept can be used to good advantage in many heat transfer applications.

3.2 CONVECTION

According to thermodynamics, the term "heat" has a precise definition: heat is the transfer of energy across the boundaries of a system, without a transfer of mass, and due to a temperature potential. Conduction and radiation are not associated with macroscopic mass transfer and, thus, are in accord with the thermodynamic definition of heat. Convection, however, involves one or more fluids that interact with each other or with a solid boundary in such a way that energy is transferred into or out of the fluid system. For convection to be present, it is necessary that the energy be carried by discrete, macroscopic, aggregates of mass. In other words, the transport of energy by convection is necessarily linked to the transport of mass within a fluid system. Thus, convection is not really a heat transfer process in the thermodynamic sense, but rather an energy transport phenomenon. For this reason, there is no basic law of convection, as there is for conduction and radiation. Convection is essentially an interrelationship between conduction and radiation energy from a solid body into a fluid and the diffusion of this energy into the bulk of the fluid by mass transport.

3.2.1 Newton's Law of Cooling

For convenience, the heat flow between a solid surface and the fluid surrounding it is often written as

$$q = hA\Delta T \tag{3.14}$$

where

q = rate of heat transfer
h = surface coefficient of convection
A = surface area and
ΔT = temperature difference between solid and fluid.

This is not a basic law, but merely a definition of h. The surface coefficient is a complex function of the fluid flow conditions and the geometric and thermal conditions of the solid and fluid. In general, h may vary from point to point on the surface and also may vary in time. Thus, it is more correct to denote Eq. (3.14) as a differential

$$\frac{dq}{dA} = h\Delta T \tag{3.15}$$

The integration may be carried out if the variation of h and ΔT are known as functions of A.

It should be emphasized that h is not a fluid property, as is k. It is merely a convenient parameter defined by Eq. (3.14), and it varies very greatly even for one fluid. Some typical values of h are given below.

Convective mode	h (Btu/hr-ft^2-$^\circ$F)
Free convection	1–5

Forced convection	10–1000
Boiling	500–10,000
Condensing	1000–20,000

3.2.2 Convection and Fluid Flow

From the definition of convection, it is obvious that a knowledge of the flow system is vital to the calculation of convective transport problems.

We may imagine that energy is conducted from a solid to a fluid through a stationary layer at the solid-liquid interface. There will be temperature variation in this layer, which will depend upon the flow pattern, among other considerations. Now at the surface of a solid in a fluid (Fig. 3.2)

$$\frac{dq}{dA} = k\left(\frac{\partial T}{\partial n}\right)_{n=0} \tag{3.16}$$

where n is the direction normal to the surface. If this is equated to Eq. (3.15), then

$$h = \frac{k\left(\dfrac{\partial T}{\partial n}\right)_{n=0}}{\Delta T} \tag{3.17}$$

Thus, in order to evaluate h as a function of position and time, it is necessary to evaluate the temperature profile in the fluid, which, in turn, depends upon the flow pattern. Any analytic treatment of convection must be a combination of the heat transfer mechanism and the fluid flow situation. For this reason, one must have an adequate fluid mechanics background.

In the general case, the equations governing convection are coupled and

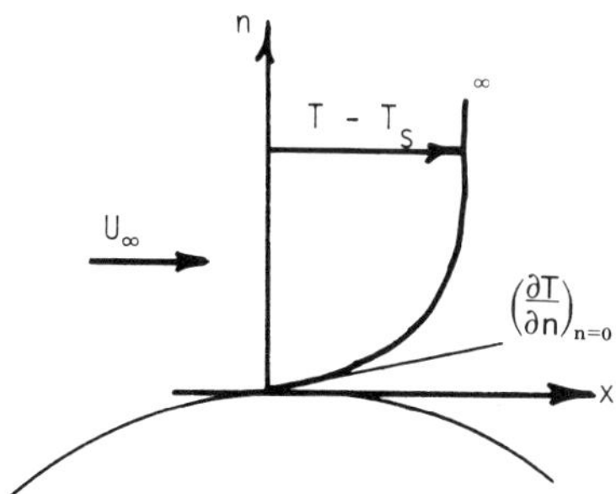

Fig. 3.2. Convection at a Surface.

cannot be solved independently. By appropriately classifying the field of convection into forced and free convection, however, it is possible to simplify the situation. In forced convection, the flow conditions are brought about by forces external to the situation, such as a blower, fan, pump, etc. Thus, the body forces involved in the momentum equation can be neglected in comparison with other forces. This allows the fluid field to be established independently of the temperature field as long as the fluid properties are constant. With the velocity field known, the energy equation can then be solved for the temperature, and finally the heat transfer may be calculated.

In free convection, the motion of the fluid is due directly to the density variations in the fluid, which are dependent upon the temperature variations. Thus, the velocity and temperature fields must be solved simultaneously, with great mathematical complications. Fortunately, free convection is of lesser import than forced convection, since most engineering systems attempt to maximize the heat transfer.

In virtually all convective systems, two conditions are assumed.

The fluid velocity at a solid boundary is zero. This is a direct consequence of the fluid viscosity. From this arises the concept of the boundary layer, as hypothesized by L. Prandtl in 1904. Within the boundary layer, the fluid velocity changes rapidly from the free stream value to zero at the wall. Flow outside the boundary layer is not influenced by the solid boundary and is termed "potential flow." Classical hydrodynamics, which considers inviscid fluids, is the study of potential flow. Applications of classical results to real flows must be carried out with great care. A flow situation in which the fluid velocity is not zero at a solid boundary arises for very low density (molecular) cases. This is called "slip flow," and although it has great importance for satellites, etc., we shall not consider it.

The fluid temperature at the solid boundary is exactly the same as the surface temperature. Due to the fluid continuity and the mechanism of heat conduction, this condition is quite general. In certain cases, perhaps low density again, a thermal jump may occur, giving rise to a temperature discontinuity. We shall not consider this possibility.

3.2.3 Boundary Layers

A velocity and a temperature boundary layer will build up for all convective flow situations. The following relations may be noted with the aid of Fig. 3.3.

1. Velocity is proportional to distance (y) from the surface in the boundary layer.
2. A laminar sublayer exists in turbulent flow.

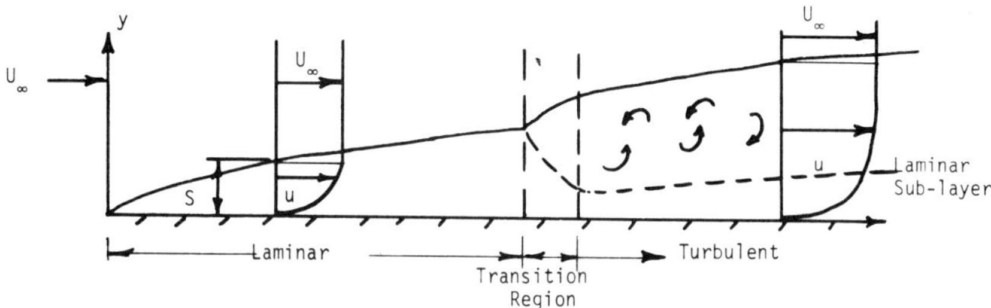

Fig. 3.3. Boundary Layer Relations.

3. Most of the velocity change occurs in the laminar sublayer.
4. The transition region is not sharply defined.

In 1883, O. Reynolds first demonstrated the existence of two distinct types of flow: laminar and turbulent. Laminar flow is characterized by:

(a) No transverse velocity components superimposed upon the mean directed velocity.
(b) Individual laminae slide over each other at different velocities, with no mixing of fluid elements.
(c) The pattern of streamlines at a fixed point is stationary with time.
(d) A fluid particle maintains a constant distance from a solid boundary.

Turbulent flow is noted by:

(a) Strong mixing of fluid particles.
(b) Transverse velocity components exist.
(c) Pattern of streamlines undergoes continuous fluctuation.
(d) Velocity distribution is more uniform than laminar.

Turbulent flow is often characterized as in Fig. 3.4. The velocity at any time is

$$u = \bar{u} + u'$$

The flow is steady if $\overline{u'} \equiv 0$. The concepts of laminar and turbulent flow refer only to a macroscopic view of the fluid.

Energy Transport Mechanisms. In the convective process, energy will be transported by conduction through stationary layers (molecular diffusivity)

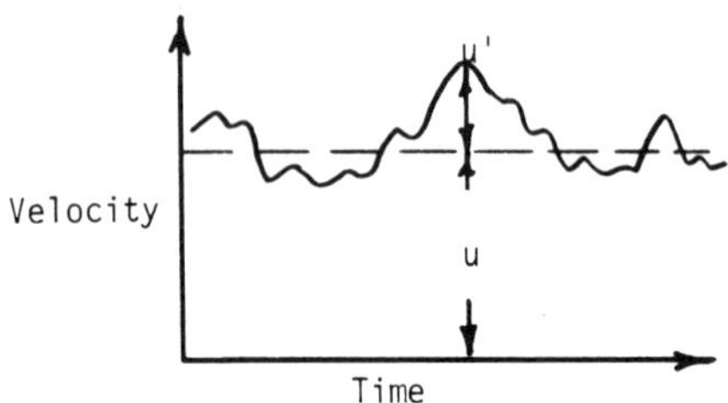

Fig. 3.4. Velocity Components in Turbulent Flow.

and by enthalpy transport with mass transport (eddy diffusivity). A general flow system can be divided into three regions:

(a) Laminar sublayer—thin layer in which laminar flow takes place. Molecular diffusivity predominates.
(b) Buffer zone—layer between sublayer and the turbulent core. Both molecular and eddy diffusivity may be important.
(c) Turbulent core—central region in which flow is entirely turbulent. Eddy diffusivity is the most important transport mechanism.

3.2.4 Solution of Convection Problems

Most engineering problems may be attacked by a variety of methods. Some of the more important methods for convection will be discussed briefly.

1. Analytical—very difficult for other than laminar flow and linear equations.
2. Analogy
3. Boundary layer
4. Similarity

(items 2–4) usually depend upon a combination of experiments and analysis.

5. Experiment—often expensive, but applies to any system.

Analytic Method. This method depends upon an exact knowledge of the mechanism involved for the system. A mathematical model must be derived from the real situation to facilitate analysis. The final results will apply to the real case only insofar as the mathematical model applies. For convective problems, the following system of equations must be solved.

Equations of flow—Newton's second law
Energy equation—First law of thermodynamics
Continuity equation—Conservation of mass
Equation of state—empirical

In many practical problems, it is acceptable to find the velocity field from the flow equations without considering the effect of temperature. This velocity field is then substituted into the energy equation to obtain the temperature distribution. The heat transfer by conduction may then be evaluated. This method can be carried out, unfortunately, only on a small class of problems, due to mathematical difficulties.

Analogy. A comparison of the equations of energy transport, momentum transport, and mass transfer reveals that a direct analogy can be made among them. This method is valuable when friction data are available.

Boundary Layer Technique. Approximate results can be obtained by writing the basic equations for the boundary layer. These equations are simpler than the general equations and can be combined with empirical data to yield approximate heat transfer results. The method is most useful for one- or two-dimensional flow systems.

Similarity. A dimensional analysis can be carried out on the system if all the important physical parameters are known. This can be done only if the mechanisms of transport are known. Functional relations can then be found for a complete set of dimensionless numbers. For example, many convective problems are described by

$$N_u = f(N_{re}, N_{pr}) \tag{3.18}$$

where

$N_u = hd/k$ Nusselt number
$N_{re} = \rho ud/\mu$ Reynolds number and
$N_{pr} = \nu/\alpha$ Prandtl number.

The functional relation must be found by experimental techniques.

3.2.5 Temperature Differences

In the equation defining the surface coefficient h, the particular temperature difference ΔT chosen will determine the value of h. There are many possible choices for ΔT. Some of the more important choices will be mentioned below. In all cases, the temperatures are considered point values and may vary from one position to another.

External flow

$$\Delta T \equiv (T_w - T_\infty) \tag{3.19}$$

where

T_w = solid surface temperature and
T_∞ = temperature of the stream undisturbed by the solid.

For atmospheric flows, the proper choice of T_∞ (and the velocity at this location) does not follow the usual methods. Chapter 6 examines the appropriate relations for this important case.

Internal flow

$$\Delta T = (T_w - T_c) \tag{3.20}$$

where

T_c = center line temperature of fluid. Also used is

$$\Delta T = T_w - T_{mm} \tag{3.21}$$

where

T_{mm} = mixed mean temperature. This is the temperature a sample would attain if thoroughly mixed and allowed to come to equilibrium in an adiabatic fashion.

3.3 RADIATION

All bodies that have a temperature greater than absolute zero emit energy from their surfaces. The rate of this energy transfer is given by

$$q = \sigma A T^4 \tag{3.22}$$

where

q = energy emitted, Btu/hr
A = surface area, ft^2
T = absolute temperature, °R and
$\sigma = 0.1714 \times 10^{-8}$ Btu/hr-ft^2-°R^4

This is known as the Stefan-Boltzmann law and was discovered empirically by Stefan and later derived from first principles by Boltzmann. A body that emits energy according to Eq. (3.22) is termed a black body. All real bodies emit energy at a lesser rate given by

$$q = \sigma\epsilon\, AT^4 \tag{3.23}$$

where ϵ is the emissivity of the surface and is less than or equal to one. The radiant energy emitted by a body is in the form of electromagnetic waves and travels with the speed of light. It can be transferred across a vacuum, as well as through gases, liquids, and solids. For solids and liquids, most of the radiant energy emitted is absorbed by the medium within a very short distance. For this reason radiation is usually treated as a surface phenomenon and is not significant in the interior of a body.

The radiant energy that is emitted by one body and falls upon another body is a complicated function of geometry, the rules of which are given in standard heat transfer texts.

3.4 GENERAL ENERGY EQUATION

Using the fundamental laws previously described, it is possible to write a general equation for the temperature of a body at any point and time. A thermodynamic analysis requires that we select the boundaries of the system we wish to investigate and account for all the energy crossing the boundaries or generated within the system. The net flow of this energy into such a system in a given time interval must be exactly equal to the change in the stored energy in the system.

Let us consider an element of volume dV having side lengths dx, dy, and dz, as shown in Fig. 3.5. The faces forming the boundaries of this volume are fixed in space; however, these are not physical boundaries and both energy and mass may cross them.

The characteristics of the thermodynamic system we have chosen are as follows.

1. Mass may be transferred across its boundaries, and energy transfer may be associated with this mass transfer. Energy may also be transferred across the boundaries without mass transfer.
2. Mechanical work may be done by frictional processes resulting from shearing stresses only, since the boundaries of our system are fixed.
3. Energy resulting from electrical, magnetic, chemical, or biochemical means may be developed within the boundaries of the system.
4. The stored energy within the boundaries of the system may change.

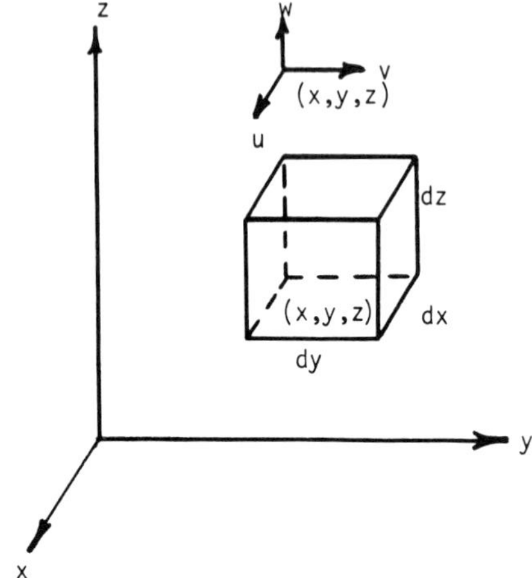

Fig. 3.5. Thermodynamic System for General Energy Equation.

The first law of thermodynamics applied to the system may be written in the form

$$Q - W + (E_i - E_o) + Q_g = \Delta E_s \tag{3.24}$$

where

Q is the net heat energy transferred across the boundaries
W is the net mechanical work across the boundaries
E_i, E_o are the energies, in all forms, carried into and out of the system, with the mass flow
Q_g is the energy developed within the system by electrical, chemical, magnetic, nuclear, etc., means and
ΔE_s is the change in stored energy.

The coordinate system and the fixed space lattice are as shown. The side lengths of the element are dx, dy, and dz. The time interval over which our observations will be made is dt. The face perpendicular to the x axis and closest to the origin passes through the point x, y, z, and the velocity components at this point are u, v, and w, respectively.

3.4.1 Energy Transfer without Mass Transfer

Energy transfer across the boundaries in the form of heat may take place by conduction and by radiation. The energy transfer by conduction may be computed by applying Fourier's equation to each of the three directions. The net energy into the system will be the sum of the energies into the system in each of the three directions.

Applying Fourier's equation to the face perpendicular to the x-axis and passing through the point x, y, z, the energy transfer by conduction for an orthotropic material, in the time dt is

$$dQ_{cx} = -k_x\, dy\, dz \frac{\partial T}{\partial x}\, dt$$

The energy transfer by conduction in the x-direction at $x + dx$ is

$$dQ_{c(x+dx)} = \left[-k_x \frac{\partial T}{\partial x} + \frac{\partial}{\partial x}\left(-k_x \frac{\partial T}{\partial x}\right) dx \right] dy\, dz\, dt$$

The net energy into the system in the x-direction is the difference between these two values

$$dQ_{xc} = \frac{\partial}{\partial x}\left(k_x \frac{\partial T}{\partial x}\right) dV\, dt$$

Similar equations may be written for dQ_{yc} and dQ_{zc} for the net energy transfer by conduction in the y and z direction. Then the net energy into the system by conduction is

$$dQ_c = \left[\frac{\partial}{\partial x}\left(k_x \frac{\partial T}{\partial x}\right) + \frac{\partial}{\partial y}\left(k_y \frac{\partial T}{\partial y}\right) + \frac{\partial}{\partial z}\left(k_z \frac{\partial T}{\partial z}\right)\right] dV\, dt \qquad (3.25)$$

The energy transfer by the radiation process will depend on the "radiation geometry," the emissivities and temperatures of the surroundings and the material within the boundaries. Since this will be a complex function, it will be left in a general form. The net energy transferred by radiation will be

$$dQ_r = q_r\, dV\, dt \qquad (3.26)$$

where q_r is the net radiant energy into the system per unit volume, per unit time.

The term Q in Eq. (3.20) may now be written

$$Q = dQ_c + dQ_r$$

In addition to energy transfer in the form of heat, mechanical work not associated with mass transfer may also occur.

This work is due to friction caused by the shearing or tangential stresses only, since the normal stresses cannot move the faces of the system chosen and, therefore, can do no work. The frictional work will be a function of the shearing stresses, the velocities, and the viscosity. Relationships involving these variables are necessary to write explicit expressions for the frictional work. We express this term in the general form

$$W = q_f \, dV \, dt \tag{3.27}$$

where q_f is the frictional energy per unit volume and per unit time. The derivation of q_f can be found in standard texts, i.e., Kreith (1973), and q_f is

$$q_f = -\left\{\frac{\rho}{2}\frac{Dq^2}{Dt} + u\frac{\partial P}{\partial x} + v\frac{\partial P}{\partial y} + w\frac{\partial P}{\partial z}\right\} - \mu\phi \tag{3.28}$$

where $D/Dt = (\partial/\partial t) + \bar{q}\cdot\nabla$ is the total or substantive derivative. The dissipation function ϕ is

$$\phi = 2\left[\left(\frac{\partial u}{\partial x}\right)^2 + \left(\frac{\partial v}{\partial y}\right)^2 + \left(\frac{\partial w}{\partial z}\right)^2\right] - \frac{2}{3}\Delta^2 + \left(\frac{\partial v}{\partial x} + \frac{\partial u}{\partial y}\right)^2 + \left(\frac{\partial w}{\partial x} + \frac{\partial u}{\partial z}\right)^2 + \left(\frac{\partial w}{\partial y} + \frac{\partial v}{\partial z}\right)^2 \tag{3.29}$$

where

$\Delta = \text{div}\ \bar{q} = (\partial u/\partial x) + (\partial v/\partial y) + (\partial w/\partial z)$ and
$\bar{q}$ is the velocity.

3.4.2 Energy Transfer Associated with Mass Flow

The energy entering the system, associated with the mass flow, may be considered to be made up of three parts. These are the microscopic energies called the "internal energy," the macroscopic energies resulting from flow velocities and gravitational potentials, and the "flow energy," or work done in transporting the mass into the system.

The internal energy is due to the microscopic kinetic energy resulting from the various motions of the atoms and their parts and the microscopic potential energy due to the force fields existing between the atoms and their parts. The symbol U is used to indicate the internal energy per unit of mass.

The macroscopic energy consists of a kinetic energy term and a potential energy term. The kinetic energy is that due to the gross velocity of the aggregate of material involved in the flow and is designated, for unit mass, by the symbols $C^2/2J$, where C is the flow velocity and J is a factor for converting mechanical work units to heat units (in SI units, J may be dropped).

The flow energy is that due to the work done in moving the mass and is expressed as the product of the pressure p and the specific volume v_s or pv_s. The macroscopic potential energy is that due to external force fields, such as gravity. In our problem, we will assume that the changes in this form of energy will be negligibly small. The sum of these macroscopic energy terms is the total enthalpy designated by the symbol H. For our system then

$$H = U + \frac{C^2}{2J} + \frac{pv_s}{J} + \frac{gZ}{g_o J} \tag{3.30}$$

The value of E_i may be computed by the following equation

$$E_{ix} = \rho u \, dy \, dz \, H_x \, dt$$

Similarly the value of E_o is

$$E_{ox} = \left[\rho u \, dy \, dz \, H_x + \frac{\partial}{\partial x} (\rho u H_x) \, dx \, dy \, dz \right] dt$$

The net inflow of energy due to mass flow is

$$E_i - E_o = (E_{ix} + E_{iy} + E_{iz}) - (E_{ox} + E_{oy} + E_{oz})$$

or

$$E_i - E_o = -\left[\frac{\partial}{\partial x} (\rho u H_x) + \frac{\partial}{\partial y} (\rho v H_y) + \frac{\partial}{\partial z} (\rho w H_z) \right] dV \, dt \tag{3.31}$$

The value of H depends on the type of material and the flow velocities. For example, if the material is a perfect gas, and assuming the macroscopic potential energy changes may be neglected, the total enthalpy is

$$H_x = c_p T + \frac{u^2}{2J}$$

where c_p is the specific heat at constant pressure.

For solids at rest with no relative motion, the macroscopic energy terms of Eq. (3.24) will be zero since

$$u = v = w = 0$$

3.4.3 Energy Generated

In any substance, it is possible that energy may be released or absorbed by electrical, magnetic, or chemical means. For example, an electric conductor subjected to a voltage potential, as well as temperature potentials, will generate energy that will be proportional to the square of the current flow and to the electrical resistance. Similarly, if a chemical reaction is involved, the energy released or absorbed will depend upon the type of reaction and the reaction rate. For our purpose, we will express this energy term in a general form. Thus, the generation term becomes

$$Q_g = q_g \, dV \, dt \tag{3.32}$$

where q_g is the energy released per unit volume and per unit time.

3.4.4 Stored Energy

The energy stored within the volume is made up of internal energy and macroscopic kinetic and potential energy terms. The macroscopic kinetic energy is that due to the velocity $\bar{q}$ and is equal to $q^2/2J$ for unit mass. The changes in macroscopic potential energy are assumed to be so small that they may be neglected. Thus, the change in stored energy ΔE_s may be written

$$\Delta E_s = \frac{\partial}{\partial t}\left[\rho\left(U + \frac{q^2}{2J}\right)\right] dV \, dt \tag{3.33}$$

where $q^2 = u^2 + v^2 + w^2$

If the material is a perfect gas, the value of U may be taken as

$$U = c_v T + \text{constant}$$

where c_v is the specific heat at constant volume.

3.4.5 The Energy Equation

The general energy equation may now be written

$$\frac{\partial}{\partial x}\left(k_x\frac{\partial T}{\partial x}\right)+\frac{\partial}{\partial y}\left(k_y\frac{\partial T}{\partial y}\right)+\frac{\partial}{\partial z}\left(k_z\frac{\partial T}{\partial z}\right)+q_r-q_f+q_g$$
$$=\frac{\partial}{\partial x}(\rho u H_x)+\frac{\partial}{\partial y}(\rho v H_y)+\frac{\partial}{\partial z}(\rho w H_z)+\frac{\partial}{\partial t}\left(\rho U+\frac{\rho q^2}{2J}\right) \quad (3.34)$$

After some manipulation, Eq. (3.34) can be rewritten

$$\text{div}\,(\bar{\bar{k}}\,\text{grad}\,T)+q_r+q_g=\rho c_p\frac{DT}{Dt}-T\beta\frac{Dp}{Dt}-\mu\phi \quad (3.35)$$

where

$\beta=-(1/\rho)(\partial\rho/\partial T)_p$ *is the coefficient of thermal expansion.*

For a perfect gas, $\beta T=1$, and

$$\text{div}\,(\bar{\bar{k}}\,\text{grad}\,T)+q_r+q_g=\rho c_p\frac{DT}{Dt}-\frac{Dp}{Dt}-\mu\phi \quad (3.36)$$

The incompressible fluid has $\beta=0$, and

$$\text{div}\,(\bar{\bar{k}}\,\text{grad}\,T)+q_r+q_g=\rho c_p\frac{DT}{Dt}-\mu\phi \quad (3.37)$$

For flow through permeable materials, such as soils, the viscous dissipation is negligible. If the material is also isotropic, then Eq. (3.37) can be written

$$k\nabla^2T+q_r+q_g=\rho c_p\frac{DT}{Dt} \quad (3.38)$$

where ∇^2 is the usual nabla operator. Vector operators for the cartesian and cylindrical systems are most useful
Cartesian coordinates

$$\nabla=\frac{\partial}{\partial x}e_x+\frac{\partial}{\partial y}e_y+\frac{\partial}{\partial z}e_z \quad (3.39)$$

$$\nabla^2 = \frac{\partial^2}{\partial x^2} + \frac{\partial^2}{\partial y^2} + \frac{\partial^2}{\partial z^2} \tag{3.40}$$

Cylindrical coordinates

$$\nabla = \frac{\partial}{\partial r} e_r + \frac{1}{r}\frac{\partial}{\partial \theta} e_\theta + \frac{\partial}{\partial z} e_z \tag{3.41}$$

$$\nabla^2 = \frac{\partial^2}{\partial r^2} + \frac{1}{r}\frac{\partial}{\partial r} + \frac{1}{r^2}\frac{\partial^2}{\partial \theta^2} + \frac{\partial^2}{\partial z^2} \tag{3.42}$$

In addition to Eq. (3.35), it is necessary to use the equation of momentum, or the Navier Stokes equation, to find the velocity field. This equation will not be derived, except as needed for special applications.

3.4.6 Conduction Equation

Equation (3.35) is the most general form of the energy equation. It must be simplified if solutions are to be found. For a solid where the material is not in motion relative to itself and all the velocities are zero, the dissipation term and the pressure term may be dropped. Also

$$\frac{DT}{Dt} = \frac{\partial T}{\partial t} + \bar{q}\nabla \cdot T = \frac{\partial T}{\partial t} \tag{3.43}$$

Equation (3.35) is then

$$\text{div}\,(\bar{\bar{k}}\,\text{grad}\,T) + q_r + q_g = \rho c_p \frac{\partial T}{\partial t} \tag{3.44}$$

Assume the material is orthotropic and the radiation term will be handled as a boundary condition. Then Eq. (3.44) becomes the general conduction equation

$$\frac{\partial}{\partial x}\left(k_x \frac{\partial T}{\partial x}\right) + \frac{\partial}{\partial y}\left(k_y \frac{\partial T}{\partial y}\right) + \frac{\partial}{\partial z}\left(k_z \frac{\partial T}{\partial z}\right) + q_g = \rho c \frac{\partial T}{\partial t} \tag{3.45}$$

This can be further simplified if all the properties are assumed constant. Then

$$\nabla^2 T + \frac{q_g}{k} = \frac{1}{\alpha}\frac{\partial T}{\partial t} \tag{3.46}$$

where $\alpha = k/\rho c$ is the thermal diffusivity. Equation (3.46) is the equation one often begins with in solving conduction problems. No subscript is used on the specific heat, since $c_p \simeq c_v$ for solids or liquids.

3.5 UNSTEADY PROBLEMS IN CONDUCTION

The unsteady flow of heat can be classified as follows

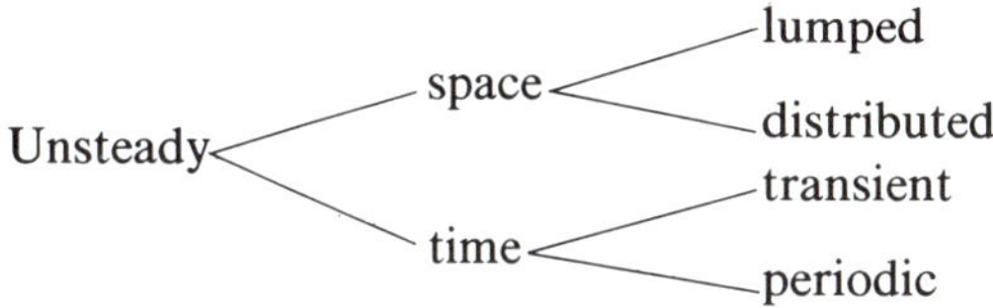

Lumping a space coordinate eliminates it from the analysis and thus leads to an initial value problem (conditions specified at any location), whereas the distributed parameter approach leads to an initial value-boundary value problem. Here we are more concerned with the time aspect, which may be portrayed as shown in Fig. 3.6.

For the steady periodic heat transfer, the initial conditions have no effect upon the final results. This fact is taken advantage of in the method of separation of variables and the complex temperature method. Many periodic problems lend themselves to the Laplace transform and Duhamel's integral; in these cases the initial transient may also be obtained, as well as the steady periodic portion. These methods will be applied to problems of a geothermal nature in the following chapters.

3.5.1 Steady Periodic Heat Transfer

One of the methods available for solving this type of problem is the complex temperature method. The general energy equation may be written following Eq. (3.38) as

$$\frac{\partial \theta}{\partial t} + \bar{q} \cdot \nabla \theta = \alpha \nabla^2 \theta + \frac{q_g}{\rho c} \tag{3.47}$$

where θ is a temperature with respect to a reference level. In general the initial condition is

$$\theta(r,0) = f(r)$$

This is unimportant, however, because the initial condition will have no effect upon the steady periodic solution.

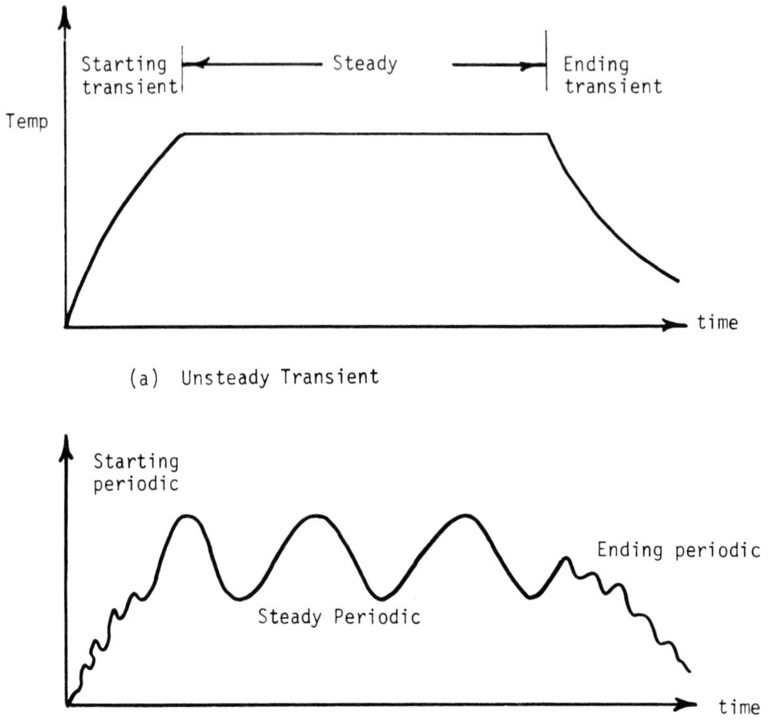

Fig. 3.6 Transient Relations.

Part of the boundary S_o is subject to one of the steady homogenous conditions

$$\theta(S_o, t) = 0, \quad \frac{\partial\theta(S_o,t)}{\partial n} = 0, \quad \pm k\frac{\partial\theta(S_o,t)}{\partial n} = h\theta(S_o,t) \tag{3.48a}$$

and the remainder of the boundary S_1 is subject to one of the periodic conditions

$$\theta(S_1,t) = \theta_o \cos m\omega t, \quad \pm\frac{\partial\theta(S_1,t)}{\partial n} = \frac{q_o''}{k} \cos m\omega t,$$
$$\pm k\frac{\partial\theta(S_1,t)}{\partial n} = h\left[\theta(S_1, t) - \theta_\infty \cos m\omega t\right] \tag{3.48b}$$

More generally, any surface disturbance can be reduced to a sum of harmonic terms, similar to the above.

Let us pose a supplementary problem for the system. Let $\theta^*(r,t)$ be the temperature associated with the same problem, except that the disburbance has a phase angle of $\pi/2$ relative to the actual problem. Then

$$\frac{\partial \theta^*}{\partial t} + \bar{q} \cdot \nabla \theta^* = \alpha \nabla^2 \theta^* + \frac{q_g}{\rho c} \tag{3.49}$$

$$\theta^*(r,0) = f(r)$$

$$\theta^*(S_o,t) = 0, \quad \frac{\partial \theta^*(S_o,t)}{\partial n} = 0 \ \text{or} \pm k \frac{\partial \theta^*(S_o,t)}{\partial n} = h\theta^*(S_o,t)$$

Now

$$\theta^*(S_1,t) = \theta_o \cos\left(m\omega t - \frac{\pi}{2}\right) = \theta_o \sin m\omega t$$

or

$$\pm \frac{\partial \theta^*(S_1,t)}{\partial n} = \frac{q_o''}{k} \sin m\omega t$$

or

$$\pm k \frac{\partial \theta^*(S_1,t)}{\partial n} = h\left[\theta^*(S_1,t) - \theta_\infty \sin m\omega t\right]$$

Now we define the complex temperature as

$$\psi = \theta + i\theta^* \tag{3.50}$$

The problem may now be posed in terms of the complex temperature

$$\frac{\partial \psi}{\partial t} + \bar{q} \cdot \nabla \psi = \alpha \nabla^2 \psi + \frac{q_g}{\rho c}(1+i)$$

$$\psi(r,0) = f(r) + i f(r) = (1+i) f(r)$$

$$\psi(S_o,t) = 0 \ \text{or} \ \frac{\partial \psi(S_o,t)}{\partial n} = 0 \ \text{or} \ \pm k \frac{\partial \psi(S_o,t)}{\partial n} = h\psi(S_o,t)$$

$$\psi(S_1,t) = \theta(S_1,t) + i\theta^*(S_1,t) = \theta_o \cos m\omega t + i\theta_o \sin m\omega t = \theta_o e^{im\omega t}$$

or

$$\pm k \frac{\partial \psi(S_1,t)}{\partial n} = q''_o e^{im\omega t}$$

or

$$\pm k \frac{\partial \psi(S_1,t)}{\partial n} = h\{\psi(S_1,t) - \theta_\infty e^{im\omega t}\}$$

We solve for ψ, which is the complex temperature. The real part of ψ is θ, the temperature actually sought.

$$\theta = \text{real}\ (\psi) = R\psi$$

Now use separation of variables as usual

$$\psi(r,t) = \phi(r)\ \tau(t)$$

But since the initial condition is ignored, the period of the response will follow that of the disturbance. Thus let $\psi = \phi(r)\ e^{im\omega t}$.

Then

$$\alpha \nabla^2 \phi = i\ m\omega\phi + \bar{q} \cdot \nabla\phi$$

The periodic boundary condition is

$$\theta_o e^{i\ m\omega t} = \phi(S_1)\ e^{im\omega t}$$

Therefore $\phi(S_1) = \theta_o$, and the problem reduces to a steady problem in ϕ.

Example 1: Consider the case of sinusoidal surface temperature for a semi-infinite medium. The temperature equation is

$$\frac{\partial \theta}{\partial t} = \alpha \frac{\partial^2 \theta}{\partial x^2} \tag{3.51}$$

$$\theta(x,0) =$$

$$\theta(0,t) = \theta_o \cos \omega t$$

$$\theta(\infty,t) = 0$$

The supplementary problem is

$$\frac{\partial \theta^*}{\partial t} = \alpha \frac{\partial^2 \theta^*}{\partial x^2} \tag{3.52}$$

$$\theta^*(x,t) = 0$$

$$\theta^*(0,t) = \theta_0 \sin \omega t$$

$$\theta^*(\infty,t) = 0$$

The complex temperature problem is

$$\frac{\partial \psi}{\partial t} = \alpha \frac{\partial^2 \psi}{\partial x^2} \tag{3.53}$$

$$\psi(x,0) = 0$$

$$\psi(\infty,t) = 0$$

$$\psi(0,t) = \theta_0 e^{i\omega t}$$

Then $\psi = \phi\ (x)\ e^{i\omega t}$

$$\alpha \frac{\partial^2 \phi}{\partial x^2} - i\omega\phi = 0$$

$$\phi\ (0) = \theta_0$$

$$\phi\ (\infty) = 0$$

Now the differential equation for ϕ is

$$\frac{d^2\phi}{dx^2} - \frac{i\omega}{\alpha}\phi = 0$$

The solution to this problem is well known

$$\phi = ae^{x\sqrt{\frac{i\omega}{\alpha}}} + b\, e^{-x\sqrt{\frac{i\omega}{\alpha}}}$$

Since $\phi(\infty) = 0$, then $a = 0$. Also $b = \theta_o$, and

$$\phi = \theta_o e^{-x\sqrt{\frac{i\omega}{\alpha}}}$$

The complex temperature is

$$\psi = \theta_o e^{-x\sqrt{\frac{i\omega}{\alpha}}} e^{i\omega t}$$

This may be rewritten in terms of real and imaginary parts as

$$\psi = \theta_o e^{-x\sqrt{\frac{\omega}{2\alpha}}}\left[\cos\left(\omega t - x\sqrt{\frac{\omega}{2\alpha}}\right) + i\sin\left(\omega t - x\sqrt{\frac{\omega}{2\alpha}}\right)\right]$$

Finally, the real part of this equation is the temperature actually sought

$$\theta(x,t) = \theta_o e^{-x\sqrt{\frac{\omega}{2\alpha}}} \cos\left(\omega t - x\sqrt{\frac{\omega}{2\alpha}}\right)$$

This equation has been used in Chap. 4 in the section dealing with permafrost temperatures. The imaginary part has no physical meaning. This method has allowed an easy solution to a fairly complicated problem.

3.5.2 Duhamel Integral

Many significant problems in engineering deal with systems for which the surface temperature is varying with time. In general, simple methods, such as separation of variables, cannot be applied to these problems. There is a powerful technique, Duhamel's integral, which uses simpler solutions, available fairly easily, as the starting point for the time-dependent problem. For example, the solution for the case where the surface temperature changes as a step function is known and can be used to generate a formal solution for time-dependent cases.

If the surface disturbance is time-dependent, rather than a step function, then the actual problem is (Fig. 3.7)

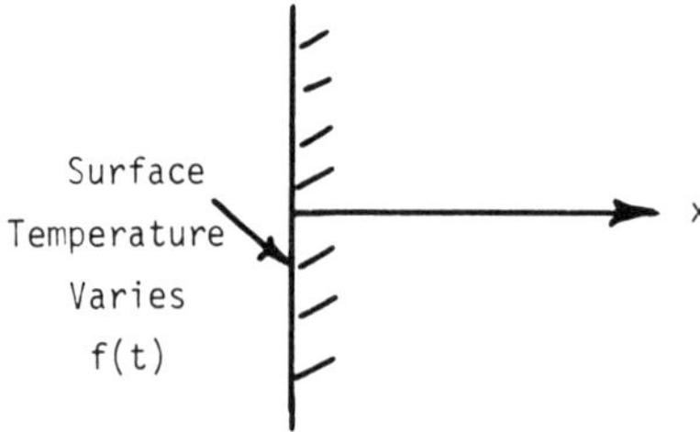

Fig. 3.7. Variable Surface Temperature.

$$\frac{\partial \theta}{\partial t} = \alpha \frac{\partial^2 \theta}{\partial x^2} \tag{3.54}$$

$$\theta(0,t) = f(t)$$

$$\theta(x,0) = 0$$

Now consider the simpler problem

$$\frac{\partial \psi}{\partial t} = \alpha \frac{\partial^2 \psi}{\partial x^2} \tag{3.55}$$

$$\psi(0,t) = 1$$
$$\psi(x,0) = 0$$

Here $\psi(r,t)$ is the solution of the unit step-change problem, available from separation of variable technique.

Now let us define a function ϕ such that

$$\phi(r,t) = \begin{cases} 0 & t < s \\ \psi(r,t-s) & t > s \end{cases}$$

where s is an arbitrary time > 0. The disturbance is zero until $t = s$, and then at $t = s$ it jumps to one. The solution is then $\phi(r,t)$, which represents the case of a unit step change of temperature at $t = s$.

Now consider an arbitrary variation of the surface temperature for example $f(t)$, as in the problem of Eq. (3.54). We can break this into a series of step functions, each part of which will satisfy the following equations, see Fig. 3.8,

$$\frac{\partial \psi}{\partial t} = \alpha \frac{\partial^2 \psi}{\partial x^2} \tag{3.55}$$

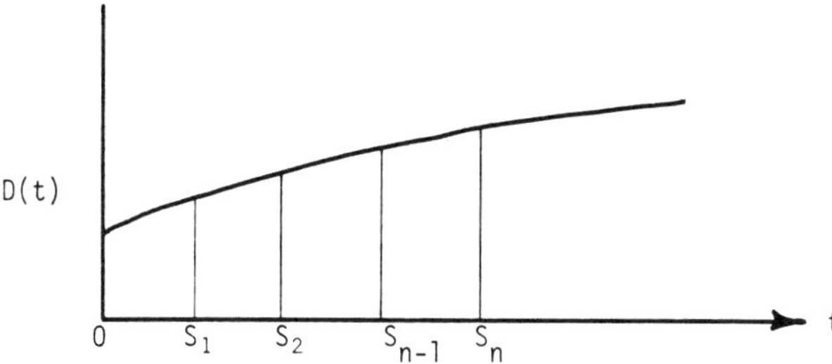

Fig. 3.8. Superposition of Solutions.

$$\psi(0,t-s)=D$$

$$\psi(x,0)=0$$

where D is a constant. The sum of all these solutions (the system is linear and solutions can be superimposed) will be the desired solution. First let the disturbance jump to $D(0)$ at $t=0$ and hold there until $t=s_1$, then jump to $D(s_1)$ and so on.

Now from the superposition concept, it follows that

$$\phi(r,t)=D(0)\psi(r,t)+[D(s_1)-D(0)]\,\psi(r,t-s_1)+[D(s_2)-D(s_1)]\,\psi(r,t-s_2) + \ldots [D(s_n)-D(s_{n-1})]\,\psi(r,t-s_n) \qquad (3.56)$$

Let $\Delta D_m = D(s_m)-D(s_{m-1})$ $\Delta S_m = S_m - S_{m-1}$

$$\phi(r,t)=D(0)\;\psi\;(r,t)+\sum_{m=1}^{n}\left(\frac{\Delta D_m}{\Delta S_m}\right)\Delta S_m\;\psi(r,t-s_m) \qquad (3.57)$$

From the definition of an integral

$$\lim_{\substack{n\to\infty\\ \Delta x\to 0}}\sum_{m=1}^{n}\psi(r,t-s_m)\left(\frac{\Delta D_m}{\Delta S_m}\right)\Delta Sm=\int_0^t\psi(r,t-s)\,\frac{dD(s)}{ds}\,ds$$

and it follows that Eq. (3.57) becomes

$$\phi(r,t)=D(0)\;\psi(r,t)+\int_0^t\psi(r,t-s)\,\frac{dD(s)}{ds}\,ds \qquad (3.58)$$

where $\psi(r,t)$ is known or can be found simply from the separation of variables methods, etc.

Equation (3.58) is a special case of the Duhamel integral valid only when the surface varies as a function of time.

3.5.2.1 General Duhamel Integral

Let $V = F(x,y,z,\lambda,t)$ be the temperature at any time at any point (x,y,z) in a solid for which the initial temperature is zero and the surface temperature is $\phi(x,y,z,\lambda)$.

Then the solution to the problem for which the initial temperature is zero and the surface temperature is $\phi(x,y,z,t)$, is

$$V = \int_0^t \frac{\partial}{\partial t} F(x,y,z,t-\lambda)\ d\lambda \tag{3.59}$$

If the surface temperature is a function of time only $\phi(t)$, then

$$F(x,y,z,\lambda,t) = F(x,y,z,t)\ \phi(\lambda)$$

Thus

$$V = \int_0^t \frac{\partial}{\partial t} [\phi(\lambda) F(x,y,z,t-\lambda)]\ d\lambda = \int_0^t \phi(\lambda)\ \frac{\partial F(x,y,z,t-\lambda)}{\partial t}\ d\lambda \tag{3.60}$$

One can easily show that Eq. (3.60) is equivalent to Eq. (3.58) by using integration by parts on Eq. (3.60). The result is

$$V = \phi(0) F(x,y,z,t) + \int_0^t F(x,y,z,t-\lambda)\ \frac{d\phi(\lambda)}{d\lambda} d\lambda$$

which is the same as Eq. (3.58) with a change of notation.

3.6 LAPLACE TRANSFORMS

Many transient conduction equations can be handled fairly easily by this method. A general transform may be written as

$$T\{F(t)\} = \int_a^b K(p,t)\ F(t)\ dt$$

The Laplace transform is a special case of the above general transform

$$L\{F(t)\} = \int_0^\infty e^{-pt}\ F(t)\ dt \equiv f(p) \tag{3.61}$$

The function f is called the Laplace transform of F.

3.6.1 Properties of the Laplace Transform

The following operational properties of the Laplace transform are useful. They are presented without proof; see, for example, Churchill (1958).

1. $L\{C_1F(t) + C_2G(t)\} = C_1f(p) + C_2g(p)$ (linear transform)
2. $L\left\{\frac{dF(t)}{dt}\right\} = p\,f(p) - f(0)$

3. $L\left\{\dfrac{\partial^n F(x_i,t)}{\partial x_i^n}\right\} = \dfrac{\partial^n}{\partial x_i^n} f(x_i, p)$ (x_i does not contain t)

4. $L\left\{\int_o^t F(\tau)\, d\tau\right\} = \dfrac{1}{p} f(p)$

5. $L\{F(\alpha t)\} = \dfrac{1}{\alpha} f\left(\dfrac{p}{\alpha}\right)$

6. $L\{e^{-\beta t}F(t)\} = F(p+\beta)$

7. $L\{F(t-\alpha)\} = e^{-\alpha p} f(p)$

   $F(t) = 0 \quad t < 0$

8. (Convolution integral)

$$L\left\{\int_o^t F(t-\tau)\, G(\tau)\right\} = f(p)\, g(p)$$

Duhamel's integral may be obtained from relation 8 as follows. Let

$$\psi(r,t) = \int_o^t G(\tau)\, F(r,\, t-\tau)\, d\tau$$

$$\phi(r,t) = \frac{\partial\psi}{\partial t} = \int_o^t G(\tau)\, \frac{\partial F(r,\, t-\tau)}{\partial t}\, d\tau + F(r,o)$$

Now, if $F(r,o) = 0$

$$\phi(r,t) = \int_0^t G(\tau)\, \frac{\partial F(r,\, t-r)}{\partial t}\, d\tau = \int_0^t G(s)\, \frac{\partial F(r,\, t-s)}{\partial t}\, ds$$

which is Duhamel's integral, as discussed earlier.

Example 2: Consider the simultaneous solution of two coupled equations.

$$\frac{d\psi}{dt} = -b_1(\psi - \phi)$$

$$\frac{d\phi}{dt} = b_2(\psi - \phi) - b_3\phi$$

$$\psi(0) = \phi(0) = T_0 - T_\infty = \psi_0$$

Let $\overline{\psi}$, $\overline{\phi}$ be the Laplace transforms of ψ and ϕ

$$p\,\overline{\psi}(p) - \psi_o = -b_1\overline{\psi}(p) + b_1\overline{\phi}(p)$$
$$p\,\overline{\phi}(p) - \psi_o = b_2\overline{\psi}(p) - b_3\overline{\phi}(p)$$

Solve these algebraic equations for $\overline{\psi}$ and $\overline{\phi}$

$$\frac{\overline{\psi}(p)}{\psi_o} = \frac{p + b_1 + b_2 + b_3}{p_2 + (b_1 + b_2 + b_3)\,p + b_1 b_3};\quad \frac{\overline{\phi}(p)}{\psi_o} = \frac{p + (b_1 + b_2)}{p_2 + (b_1 + b_2 + b_3)\,p + b_1 b_3}$$

The inverses of these functions can now be found in tables of Laplace transforms.

3.6.2 Fourier Integral

A function may be represented over a semi-infinite interval by

$$f(x) = \int_0^\infty b(\lambda)\sin\lambda x\,dx \qquad 0 < x < \infty \tag{3.62}$$

The functional relation for $b(\lambda)$ can be found. Let

$$F \equiv \int_0^L f(x)\sin\lambda_1 x\,dx = \int_0^\infty \sin\lambda_1 x\left(\int_0^L b(\lambda)\sin\lambda x\,d\lambda\right)dx$$

Now let

$$\beta \equiv (\lambda - \lambda_1)L \qquad \gamma = (\lambda + \lambda_1)L$$

If

$$\lambda = 0 \qquad \beta = -\lambda_1 L \qquad \gamma = \lambda_1 L$$

Now replacing the dummy variable λ by β or γ, one has

$$F = \tfrac{1}{2}\int_{-\lambda_1 L}^\infty b\left(\lambda_1 + \frac{\beta}{L}\right)\frac{\sin\beta}{\beta}\,d\beta - \tfrac{1}{2}\int_{\lambda_1 L}^\infty b\left(\frac{\gamma}{L} - \lambda_1\right)\frac{\sin\gamma}{\gamma}\,d\gamma$$

As L approaches infinity, the second term above approaches zero. Thus, in the limit

$$\int_{-L}^L f(x)\sin\lambda_1 x\,dx = \tfrac{1}{2}\int_{-\infty}^\infty b(\lambda_1)\frac{\sin\beta}{\beta}\,d\beta = \frac{b(\lambda_1)\,\pi}{2}$$

Therefore

$$b(\lambda)=\frac{2}{\pi}\int_0^\infty f(\xi)\sin\lambda_i\xi\,d\xi$$

Note that the procedure here is identical to the usual Fourier series details

$$f(x)=\frac{2}{\pi}\int_0^\infty \sin\lambda x\int_0^\infty f(\xi)\sin\lambda\xi\,d\lambda \qquad x>0$$

Complete Fourier Integral. The result may be extended over the entire x plane as

$$f(x)=\frac{1}{\pi}\int_0^\infty\left[\int_{-\infty}^\infty f(\xi)\cos\lambda(x-\xi)\,d\xi\right]d\lambda \quad -\infty<x<\infty \tag{3.63}$$

or

$$f(x)=\frac{1}{2\pi}\int_{-\infty}^\infty\int_{-\infty}^\infty f(\xi)\cos\lambda\,(x-\xi)\,d\xi\,d\lambda$$

Example 3: Consider a semi-infinite medium with a step change in surface temperature. Use the Fourier integral to solve for the temperature

$$\frac{\partial\theta}{\partial t}=\alpha\frac{\partial^2\theta}{\partial x^2} \tag{3.64}$$

$$\theta(x,0)=\theta_o \tag{3.64a}$$

$$\theta(0,t)=0 \tag{3.64b}$$

$$\theta(\infty,t)=\theta_o \tag{3.64c}$$

When one has infinite space coordinates, it is logical to employ the Fourier integral rather than Fourier series.

Now $\theta_\lambda(x,t) = b(\lambda)\,e^{-\alpha\lambda^2 t}\sin\lambda x$ satisfies Eqs. (3.64) and 3.64*b*), and

$$\theta(x,t)=\int_0^\infty b(\lambda)\,e^{-\alpha\lambda^2 t}\sin\lambda x\,d\lambda$$

is also a solution, since the system is linear and any sum of solutions is a solution. The initial condition, Eq. (3.64*a*) may be written as

$$\theta_o = \int_0^\infty b(\lambda) \sin \lambda x \, d\lambda$$

This means that the constant θ_o must be expressed as a Fourier integral. One can immediately write down the equation for $b(\lambda)$ as

$$b(\lambda) = \frac{2}{\pi} \int_0^\infty \theta_o \sin \lambda \xi \, d\xi = \frac{2\theta_o}{\pi} \int_0^\infty \sin \lambda \xi \, d\xi$$

Thus, a complete solution is

$$\frac{\theta(x,t)}{\theta_o} = \frac{2}{\pi} \int_0^\infty \int_0^\infty e^{-\alpha\lambda^2 t} \sin \lambda \xi \sin \lambda x \, d\xi \, d\lambda$$

This solution may be transformed into an important form of solution for conduction problems

$$\frac{\theta(x,t)}{\theta_o} = \frac{2}{\pi} \int_0^\infty \int_0^\infty e^{-\alpha\lambda^2 t} \tfrac{1}{2} [\cos(\lambda x - \lambda \xi) - \cos(\lambda x + \lambda \xi)] \, d\lambda \, d\xi$$

Integrate with respect to λ

$$\frac{\theta(x,t)}{\theta_o} = \frac{1}{2\sqrt{\pi \alpha t}} \int_0^\infty [e^{-(x-\xi)^2/4\alpha t} - e^{-(x+\xi)^2/4\alpha t}] \, d\xi$$

Introduce a new variable $\eta = (x - \epsilon)/2\sqrt{\alpha t}$

$$\frac{\theta(x,t)}{\theta_o} = \frac{1}{\sqrt{\pi}} \int_{-x/2\sqrt{\alpha t}}^\infty e^{-n^2} \, d\eta - \int_{x/2\sqrt{\alpha t}}^\infty e^{-\eta^2} d\eta$$

Finally

$$\frac{\theta(x,t)}{\theta_o} = \frac{2}{\sqrt{\pi}} \int_0^{\frac{x}{2\sqrt{\alpha t}}} e^{-\eta^2} \, d\eta = \operatorname{erf}\left(\frac{x}{2\sqrt{\alpha t}}\right)$$

Here erf is the error function, which is tabulated and used in phase change problems. The same result could be obtained directly by use of Laplace transforms.

3.6.3 Inversion Theorem

Although many inverse functions are tabulated and can be looked up, there are conduction problems that have inverses that are not in tables. In these cases, use can be made of a general equation to find the inverse of any Laplace transform.

If $\bar{f}$ is the Laplace transform of f, then the function f can be found from

$$f(t) = \frac{1}{2\pi i} \lim_{L\to\infty} \int_{\gamma - iL}^{\gamma + iL} e^{pt} \bar{f}(p)\, dp \tag{3.65}$$

In this equation, p is complex and is often written as z. Finding the inverse is then a matter of evaluating a complex integral over an infinite domain.

Two cases arise frequently and can be discussed. $\overline{T}(r,p)$ is the Laplace transform of $T(r,t)$.

Case I: $\overline{T}(r,p)$ has N poles to the left of some vertical line; γ may be any finite value. From the inversion theorem, one has

$$T(r,t) = \frac{1}{2\pi i} \lim_{L\to\infty} \int_{\gamma - iL}^{\gamma + iL} e^{+pt}\, \overline{T}(r,p)\, dp$$

Now referring to Fig. 3.9, let

$$f(z) = e^{+zt}\, \overline{T}(r,z)$$

$$z = p + iq$$

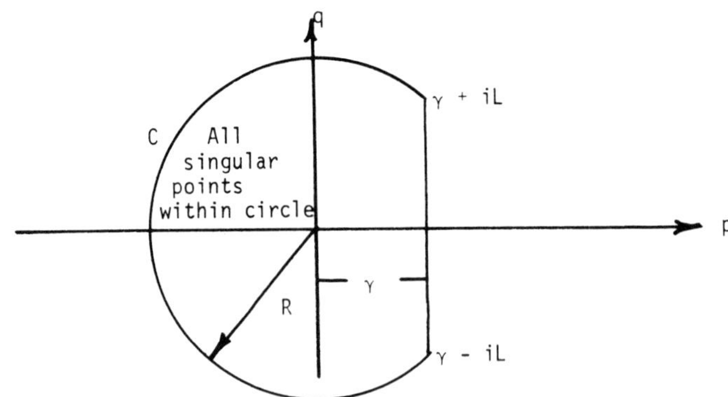

Fig. 3.9. Geometry for Inversion Theorem.

Note $\int_{\gamma-iL}^{\gamma+iL} e^{pt}\,\overline{T}(r,p)\,dp$ is the line integral of

$$\int_{\gamma-il}^{\gamma+il} e^{zt}\,\overline{T}(r,z)\,dz$$

along the line $p = \gamma$. It is clear that

$$\oint f(z)\,dz = \int_{\gamma-iL}^{\gamma+iL} f(p)dp + \int_{c} f(z)dz = 2\pi i \sum_{n=1}^{N} Res(a_n)$$

This follows since it was assumed that $\overline{T}$ has N poles to the left of the line $p = \gamma$. Now let R approach infinity. Then

$$\lim_{L\to\infty} \int_{\gamma-iL}^{\gamma+il} f(p)dp + \lim_{R\to\infty} \int_{c} f(z)dz = 2\pi i \lim_{R\to\infty} \sum_{n=1}^{N} Res(a_n)$$

Now it can be shown that for many cases, but not all

$$\lim_{R\to\infty} \int_{c} f(z)dz = 0$$

In any given case, it is necessary to verify this assumption. Thus, for this case

$$T(r,t) = \lim_{R\to\infty} \sum_{n=1}^{N} Res(a_n) \tag{3.66}$$

Case II: Branch point at $z = 0$

Since at a branch point the function is multivalued, consider the contours in Fig. 3.10. If we eliminate the branch point, then Cauchy's theorem will be valid. Thus,

$$\oint f(z)\,dz = \int_{\gamma-iL}^{\gamma+iL} f(p)\,dp + \int_{c_1+c_2} f(z)\,dz - \int_{L_1} f(z)\,dz = 2\pi i \sum_{n=1}^{N} Res(a_n)$$

Now

$$\lim_{L\to\infty} \int_{\gamma-iL}^{\gamma+iL} f(p)\,dp + \lim_{R\to\infty} \int_{c_1+c_2} f(z)\,dz - \lim_{\substack{R\to\infty \\ \rho\to 0}} \int_{L_1} f(z)\,dz$$

$$= 2\pi i \lim_{R\to\infty} \sum_{n=1}^{N} Res(a_n)$$

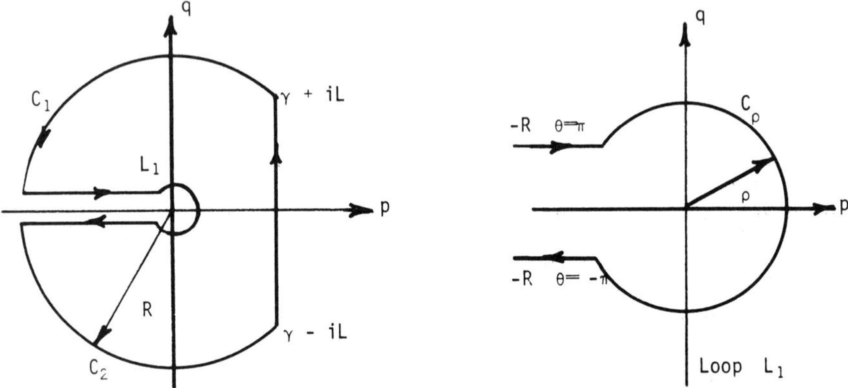

Fig. 3.10. Branch Point Geometry.

As usual, it follows that

$$T(r,t)=\frac{1}{2\pi i}\lim_{L\to\infty}\int_{\gamma-iL}^{\gamma+iL} f(p)\,dp=\lim_{R\to\infty}\sum_{n=1}^{N} Res(a_n)+\frac{1}{2\pi i}\lim_{\substack{R\to\infty\\ \rho\to 0}}\int_{L_1} f(z)\,dz$$

where it is assumed that

$$\lim_{R\to\infty}\int_{c_1+c_2} f(z)\,dz=0$$

$$f(z)=e^{zt}\,\overline{T}(r,z)$$

On the upper part of the loop

$$z_u=Re^{i\theta},\qquad \lim_{\rho\to o} e^{i\theta}=e^{i\pi}=-1$$

Then the integral is

$$-\int_{o}^{\infty} e^{-Rt}\,\overline{T}(r,Re^{i\pi})\,dR=\lim_{\substack{R\to\infty\\ \rho\to 0}}\int_{Re^{i\theta}}^{\rho} e^{Re^{i\theta}t}\,T(r,Re^{i\theta})\,e^{i\theta}\,dR$$

On the lower part of the loop

$$z_e=Re^{i\theta}\lim_{\rho\to o} e^{i\theta}=e^{-i\pi}=-1$$

As above, one has

$$\int_0^\infty e^{-Rt}\,\overline{T}(r,Re^{-i\pi})\,dR$$

The remaining part of L_1 is the circle of radius ρ

$$z=\rho e^{i\theta} \qquad dz=i\rho e^{i\theta}\,d\theta \qquad \pi<\theta<-\pi$$

The integral to be evaluated is

$$\lim_{\rho\to 0}\int_{-\pi}^{\pi} e^{t\rho e^{i\theta}}\,\overline{T}(r,\rho e^{i\theta})\,i\rho\, e^{i\theta}\,d\theta$$

Replacing these integrals into the inversion equation gives the general result

$$T(r,t)=\sum_{n=1}^{\infty} Res(a_n)+\frac{1}{2\pi i}\int_0^\infty e^{-Rt}[\overline{T}(r,\,Re^{-i\pi})-\overline{T}(r,\,Re^{i\pi})]\,dR$$
$$+\frac{1}{2\pi i}\lim_{\rho\to 0}\int_{-\pi}^{\pi} \rho e^{t\rho e^{i\theta}}\,\overline{T}(r,\rho e^{i\theta})\,d\theta \tag{3.67}$$

In this equation, each of the integrals must be evaluated by the usual methods. In some cases they will reduce to zero.

3.7 CONFORMAL MAPPING

Two-dimensional, steady-state, conduction problems involve the solution of the Laplace equation. These problems are ideally suited for the use of complex variables and, in particular, conformal mapping. The idea behind conformal mapping is simply to find a transformation function that will change a complex temperature problem into a readily solvable one. This transformation function is called a "mapping function." Consider the function

$$w=e^z \tag{3.68}$$

This function can be thought of as changing a given region in the z plane into a different region of the w plane. See Fig. 3.11. The rectangular region $ABCD$ maps into the annular segment $A'\,B'\,C'\,D'$ in the w plane. Rewrite Eq. (3.68) as

$$\rho e^{i\phi}=e^{x+iy} \tag{3.69}$$

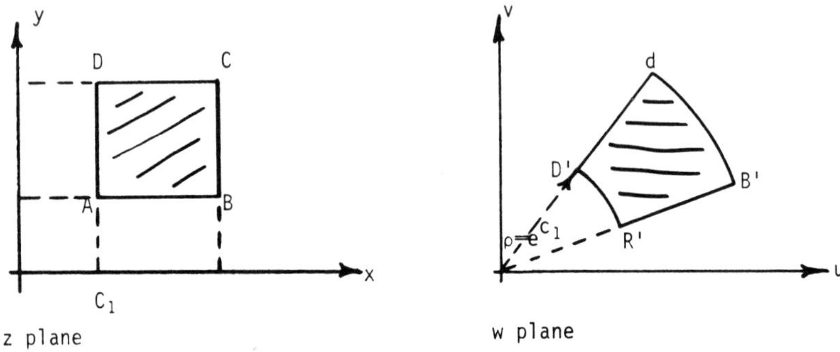

Fig. 3.11. Physical and Transformation Planes.

Then the line $x = C_1$ in the z plane becomes $\rho = e^{C_1}$ in the w plane: a constant radius of value e^{C_1}. The other parts of the boundary can be similarly mapped. There are numerous tables of functions available that will map regions of interest into simpler regions.

A conformal mapping is one for which the angles formed in a plane will not change under the mapping. Thus orthogonal lines in one plane (isothermal and heat flow lines) will remain orthogonal in the new plane. All elementary functions (actually all analytic functions) will be conformal. This means that all solutions of the Laplace equation, called "harmonic functions," are analytic or vice versa. A solution of the Laplace equation in one region will remain a solution of the Laplace equation in the transformed region. Also, the following boundary conditions remain unchanged for a conformal mapping

$$H = \text{constant or } \frac{dH}{dn} = 0 \tag{3.70}$$

That is, the value of a function or its derivative at a boundary will remain unchanged.

Example 4: Show that a mapping is conformal. Consider a function in the z plane

$$H = e^{-x} \cos y$$

This is a harmonic function, as can readily be verified. For this function

$$\frac{\partial H}{\partial y} = 0 \text{ when } y = 0$$

Now transform the function according to $z = w^2$. Then the function is

$$H = e^{-(u^2 - v^2)} \cos 2\,uv$$

The derivative of this function gives

$$\frac{\partial H}{\partial u} = 0 \qquad u = 0$$

Also

$$\frac{\partial H}{\partial v} = 0 \qquad v = 0$$

Thus, the derivative of the function remains zero, i.e., the mapping according to $z = w^2$ is conformal.

Example 5: Find the temperature in the wall shown in Fig. 3.12.
The problem is

$$\nabla^2 T = 0$$

$$T\left(-\frac{\pi}{2}, y\right) = 0 = T\left(\frac{\pi}{2}, y\right)$$

$$T(x,0) = 1$$

A harmonic function that will satisfy the boundary conditions is not obvious in the z plane. Consider the mapping function

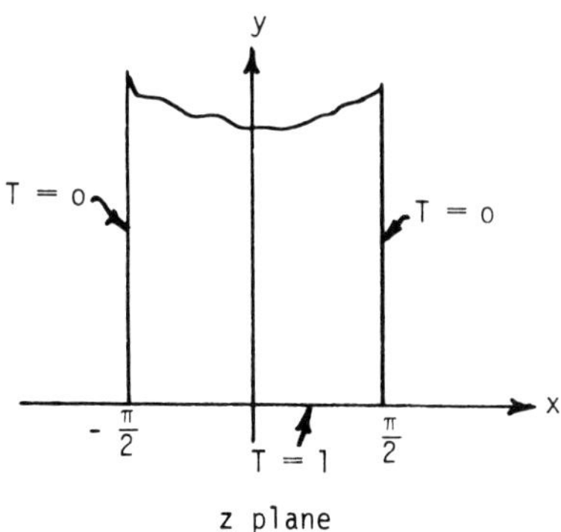

Fig. 3.12. Semi-Infinite Wall.

$$w = \ln \frac{\sin z - 1}{\sin z + 1} \tag{3.71}$$

This is a complicated function and is chosen from tables because it reduces the region to the w plane, as shown in Fig. 3.13. The values on the boundaries can be verified by examining the mapping function and the boundary conditions. In the w plane, a harmonic function that satisfies the boundary conditions is obvious. Take

$$T(u,v) = \frac{v}{\pi}$$

This satisfies the Laplace equation and the boundary conditions. It is now necessary to use Eq. (3.71) to find the relation between v and x, y in order to obtain a solution in the desired z plane. After algebraic manipulation it can be shown that

$$v = 2 \arctan \frac{\cos x}{\sinh y}$$

and our solution is

$$T(x,y) = \frac{2}{\pi} \arctan \frac{\cos x}{\sinh y}$$

It can be verified that this does, in fact, satisfy the original problem. Notice that the solution is very simple, once an appropriate mapping function has been found.

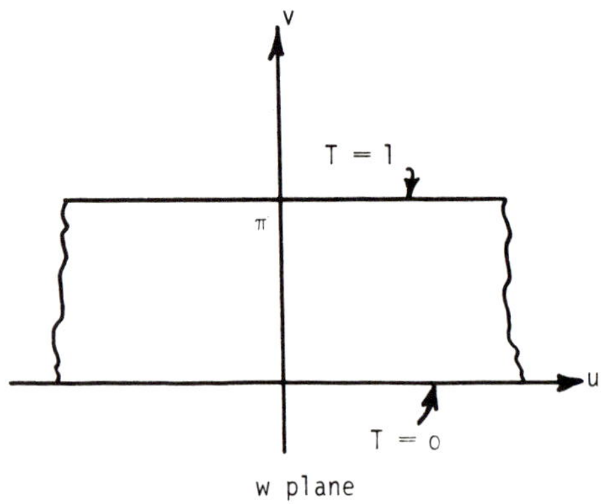

Fig. 3.13. Transformed Semi-Infinite Wall.

REFERENCES

Arpaci, V. S. 1966. *Conduction heat transfer.* Reading, Massachusetts: Addison-Wesley.

Carslaw, H. S., and Jaeger, J. C. 1959. *Conduction of heat in solids.* 2nd ed. Oxford: Clarendon Press.

Churchill, R. V. 1960. *Complex variables and applications.* 2nd ed. New York: McGraw-Hill.

Churchill, R. V. 1958. *Operational mathematics.* 2nd ed. New York: McGraw-Hill.

Eckert, E. R. G., and Drake, R. M. 1959. *Heat and mass transfer.* 2nd ed. New York: McGraw-Hill.

Ingersoll, L. R., Zobel, O. J., and Ingersoll, A. C. 1954. *Heat conduction.* Madison: Univ. of Wisconsin Press.

Kreith, F. 1973. *Principles of heat transfer.* 3rd ed. New York: Intext Educational Publishers.

4
Permafrost and Soils

4.0 INTRODUCTION

Permafrost is defined in terms of the thermal state of ordinary soil systems. Thus it will be useful to examine some details of soils because the engineering characteristics of permafrost can be understood only in regard to soils. No attempt will be made here to present a complete discussion of soil systems.

A soil system is a mechanical combination of solid materials, liquids (usually water), and gases (usually air and water vapor). The solid materials can be arranged in various geometries and form the skeleton of the soil, with voids between the solid grains or particles (Fig. 4.1). These voids may contain liquids, gases, or both. The solid material of soils is usually made up of minerals, but it can also include organic material, such as peat or humus.

The soil minerals arise mainly from the disintegration of rocks by two distinct phenomena: physical or mechanical processes and chemical processes. Rocks can be fractured due to inertial, thermal, or contact stresses, that is, various physical forces acting on the rock material. These forces are usually proportional to the mass of the rock system, except for the grinding action of glacier movements. As the size of the rock fragments decreases, a limit is reached where the mechanically induced stresses are less than the fracture strength of the mineral, and further decrease in the grain size ceases. This size is typically on the order of .05 mm. Grain sizes greater than this are called "coarse particles." Quartz, ferromagnesiums, feldspars and allied aluminosilicates, and other primary minerals typically make up the silt and sand size fractions. These minerals are rarely found in the clay fractions. Below the coarse particle size, chemical reactions, which proceed proportional to surface area rather than mass, become significant in the further reduction of grain sizes. New minerals formed by the chemical action of groundwater, rain, etc., on rock material are called "clay minerals," and their grain size is such that they are fine-grained material. These secondary minerals, the clays, are all platy aluminosilicates.

The clays are classified into four main groups: kaolin, hydrous-mica, mectite (montmorillonite), and polygorskite (fibrous-clay-mineral). The clay fraction makes up the surface-active part of a soil due to the large surface area per unit mass. Mineral solids-water-ice interfaces, which are functions of

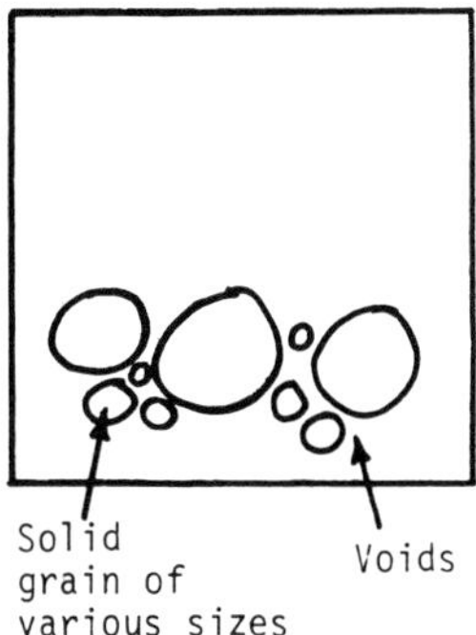

Fig. 4.1. Granular Soil.

the surface area, have a profound effect upon the enigneering properties of a soil system. The specific surface area is a function of the size and geometry of the solid particles. It cannot be calculated exactly for an actual soil, as the geometry is rarely known. For a single, spherical, soil particle, however, the specific surface area is

$$S = \frac{3}{\rho r} \tag{4.1}$$

where

r = particle radius and
ρ = density of the solids.

Assuming a solid density of 2700 kg/m^3, the specific surface area can be estimated according to Table 4.1. Those figures are simply estimates that give an idea of the relative surface area to be expected. A typical kaolin clay has $S = 10$ m^2/g, and a montmorillonite has $S = 800$ m^2/g.

The microstructure of clays gives rise to important physical properties.

Table 4.1. Specific Surface Area of Spherical Particles.

Type of soil	r (radius), mm	S (specific surface area), m^2/g
Gravel	>2	$<5 \times 10^{-4}$
Sand	.06–2	$5 \times 10^{-4} - .019$
Silt	.002–.06	.019–.56
Clay	$<.002$	$>.56$

Typically, clay is made up of layers of silicon and oxygen, in a tetrahedral array, alternating with layers of aluminum and oxygen in an octohedral formation. These platelets make up units called "tactoids." The substitution of elements with a less positive charge, such as iron or magnesium, for the aluminum or aluminum for the silicon, in the octahedral and tetrahedral layers, respectively, leads to an overall negative charge for the tactoids. This charge is assumed to be distributed over the particle surface. The negative charge is counterbalanced by adsorbed, exchangeable cations (positively charged ions), which migrate to the particle surface leading to a diffuse electric double layer. The cations are usually Na, Ca, K, etc. These ions have a significant effect upon the surface activity of the clays. These electrical relations help explain the motion of solid particles in electric fields and the enormous surface area and adsorption of the colloidal state (Fig. 4.2).

4.1 CLASSIFICATION OF SOILS

Mineral soils are classified on the basis of their grain size and the frequency distribution of the mass of material in given grain size ranges. For example, the frequency distribution of the grain sizes in a soil can be measured and plotted as shown in Fig. 4.3. Such a soil can then be described in terms of the proportion of fine and coarse-grained material it contains. The Massachusetts Institute of Technology grain size classification is readily used and assigns familiar names to the various grain sizes (Table 4.2).

If a soil is made up predominantly of one size, for example, particles of 0.6 mm, it is described as "poorly graded," but a soil with a smooth range of sizes is designated as "well-graded." The soil is usually given the name associated with the grain sizes in the majority. For example a soil with more than 50% mass in the range .06–2 mm would be called a "sandy" soil. The grain size range making up the second most common percentage of the soil is used as an adjective, for example, silty sand.

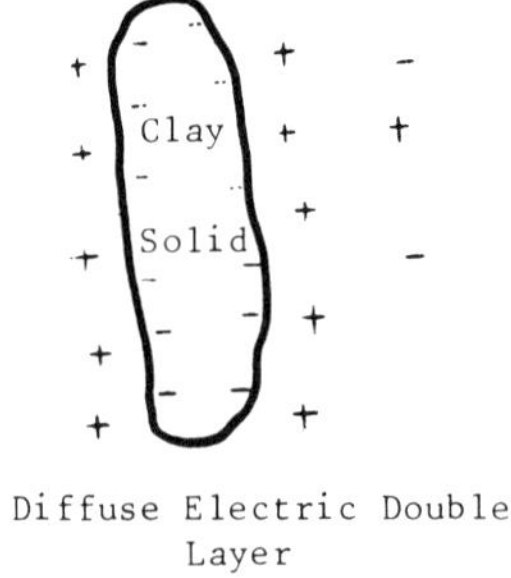

Fig. 4.2. Electric Field near Clay Solid.

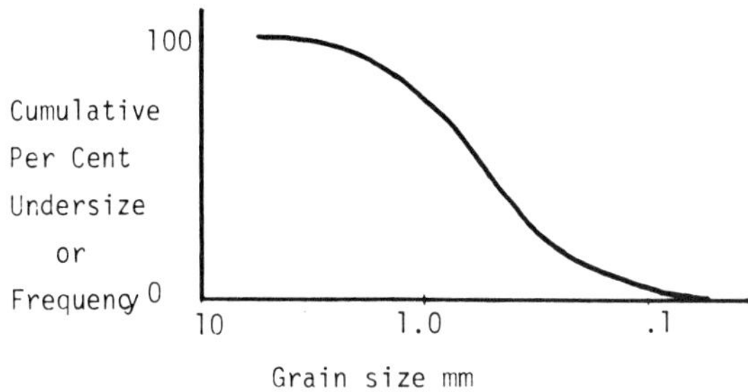

Fig. 4.3. Typical Grain Size Distribution.

Table 4.2. MIT Grain Size Classification.

<table>
<tr><td></td><td colspan="6">Grain size, mm
2 .6 .2 .06 .02 .006 .002</td><td></td></tr>
<tr><td rowspan="2">Gravel</td><td>Coarse</td><td>Medium</td><td>Fine</td><td>Coarse</td><td>Medium</td><td>Fine</td><td rowspan="2">Clay</td></tr>
<tr><td colspan="3">Sand</td><td colspan="3">Silt</td></tr>
</table>

The silts and clays cannot be graded with sieves because of their fine sizes, and they are often graded by the hydrometer method. The soil is mixed with water and the rate of settling out of the particles can be correlated with hydrometer measurements of the water-soil mixture to give the size ranges of the particles.

The classification used by the U.S. Corps of Engineers was adopted in 1942 and will be described here as it is a widely accepted system. The definitions are essentially the same as the MIT system, as can be seen in Table 4.3.

Table 4.4 describes the Unified Soil Classification System with the symbols commonly used in practice. The three main soil types are thus described as coarse-grained, fine-grained, and organic. While the mineral soils (fine and coarse-grained) are well described by Table 4.4, the organic soils require

Table 4.3. Unified Soil Classification Grain Size.

Name	Passing (sieve size in parentheses)	Retained on
Gravel	3 in.	2.0 mm (#10)
Sand	2.0 mm (#10)	.074 mm (#200)
Fines (silt or clay)	.074 mm (#200)	–

Table 4.4*a*. The Unified Soil Classification System (after *Soil Mechanics* by A. R. Jumikis 1962).

	Major divisions	Symbol letter		Name	Value as subgrade when no frost action	Value as subbase when no frost action	Value as base when not subject to frost	Protection needed from frost action	Drainage
Coarse-grained Soils	Gravel and Gravelly Soils	GW		Well-graded gravels or gravel-sand mixtures, little or no fines	Excellent	E	Good	None to very slight	E
		GP		Poorly graded gravels or g-s mixtures, little or no fines	G-E	G	Fair-G	N-VS	E
		GM	D	Silty gravels, gravel-sand-silt	G-E	G	F-G	Slight to medium	F-Poor
			U	Mixtures	G	F	Poor to not suitable	S-M	P to practically impervious
		GC		Clayey gravels, gravel-sand-clay mixtures	G	F	P-NS	S-M	P-PI
	Sand and Sandy Soils	SW		Well-graded sand or gravelly sands, little or no fines	G	F-G	P	N-VS	E
		SP		Poorly graded sands or gravelly sands, little or no fines	F-G	F	P-NS	N-VS	E
		SM	D	Silty sands, sand-silt	F-G	F-G	P	S-High	F-P
			U	Mixtures	F	P-F	NS	S-H	P-PI
		SC		Clayey sands, sand-clay mixtures	P-F	P	NS	S-H	P-PI

Table 4.4*b*. The Unified Soil Classification System (after *Soil Mechanics* by A. R. Jumikis 1962).

	Major divisions	Symbol letter	Name	Value as subgrade when no frost action	Value as subbase when no frost action	Value as base when not subject to frost	Protection needed from frost action	Drainage
Fine-grained Soils	Silts and Clays $W_{LL} < 50$	ML	Inorganic silts and very fine sand, rock flour, silty or clayey silts with slight plasticity	P-F	NS	NS	M-VH	F-P
		CL	Inorganic clays of low to medium plast., gravelly clays, sandy clays, silty clays, and lean clays	P-F	NS	NS	M-H	PI
		OL	Organic silts and organic silt clays of low plasticity	P	NS	NS	M-H	P
	Silts and Clays $W_{LL} > 50$	MH	Inorganic silts, micaceous or diatomaceous fine sandy or silty soils, elastic silts	P	NS	NS	M-VH	F-P
		CH	Inorganic clays of high-plasticity, fat clays	P-F	NS	NS	M	PI
		OH	Organic clays of medium to high plasticity, organic silts	P-VP	NS	NS	M	PI
	Highly Organic Soils	Pt	Peat and other highly organic soils	NS	NS	NS	S	F-P

further discussion. An organic soil, in general, is a mineral soil with the inclusion of a large amount of organic material. The organic material, or solids, may be peat or other fossilized detritus. The proportions of the mineral and organic materials will influence the soil properties. Peat consists of more or less fragmented remains of vegetable matter, sequentially deposited and fossilized, and is itself often considered as a soil type. If the peat contains

Table 4.5. Peat Classification (after Radforth 1969). Reprinted from *Muskeg Engineering Handbook,* Ivan C. MacFarlane, ed., by permission of University of Toronto Press, 1969.

Predominant characteristic	Category	Name
Amorphous-granular	1	Amorphous-granular peat
	2	Nonwoody, fine-fibrous peat
	3	Amorphous-granular peat containing nonwoody fine fibers
	4	Amorphous-granular peat containing woody fine fibers
	5	Peat, predominantly amorphous-granular, containing nonwoody, fine fibers, held in a woody, fine-fibrous framework
	6	Peat, predominantly amorphous-granular containing woody fine fibers, held in a woody, coarse-fibrous framework
	7	Alternate layering of nonwoody, fine-fibrous peat and amorphous-granular peat containing nonwoody fine fibers
Fine-fibrous	8	Nonwoody, fine-fibrous peat containing a mound of coarse fibers
	9	Woody, fine-fibrous peat held in a woody, coarse-fibrous framework
	10	Woody particles held in nonwoody, fine-fibrous peat
	11	Woody and nonwoody particles held in fine-fibrous peat
Coarse-fibrous	12	Woody, coarse-fibrous peat
	13	Coarse fibers criss-crossing fine-fibrous peat
	14	Nonwoody and woody fine-fibrous peat held in a coarse-fibrous framework
	15	Woody mesh of fibers and particles enclosing amorphous-granular peat containing fine fibers
	16	Wood, coarse-fibrous peat containing scattered woody chunks
	17	Mesh of closely applied logs and roots enclosing woody coarse-fibrous peat with woody chunks

mineral soils, it can be considered an organic soil. A highly or completely organic soil should have characteristics similar to peat. Radforth (1969) has classified peat into 17 categories, which are described in Table 4.5.

In order to classify the peat, or organic soils, quantitatively in accordance with Table 4.5, it is necessary to use a series of photographs given by Radforth (1969). Thus, the organic soils may be classified and correlated as to their engineering properties.

The water content of fine-grained soils is important in relation to their engineering properties. It has been observed that a clay will exhibit characteristics ranging from a viscous liquid at high water content to a brittle solid at very low moisture. Tests for the liquid limit, above which a clay-water mixture acts like liquid, and the plastic limit, above which it can be molded into and maintain a new shape without fracturing of the soil, have been standardized, but still depend upon the technique of individual investigators to a certain extent. Descriptions of these tests are given by Scott and Schoustra (1968), Capper and Cassie (1969), and Jumikis (1962). The liquid limit is the minimum water content at which the soil will exhibit essentially liquid behavior. The plastic limit is the minimum moisture content at which the soil will exhibit plastic behavior. The plasticity index of the soil is the difference, in percentage, between the liquid limit and the plastic limit. If the volume of a clay is measured while the water content is being decreased, it is found that the volume will decrease to a minimum level, at which the water content is nonzero. Further decrease in the moisture level will not affect the soil volume. This is called the "shrinkage limit" of the soil. The shrinkage limit is the maximum water content at which the soil volume will remain constant. The relations between these limits may be seen in Fig. 4.4. Table 4.6 gives the plasticity ratings of soils.

Figure 4.5 is a plasticity chart for fine-grained soils. The *A* line represents the boundary between typical inorganic clays, which are generally above the line, and the plastic soils containing organic colloids.

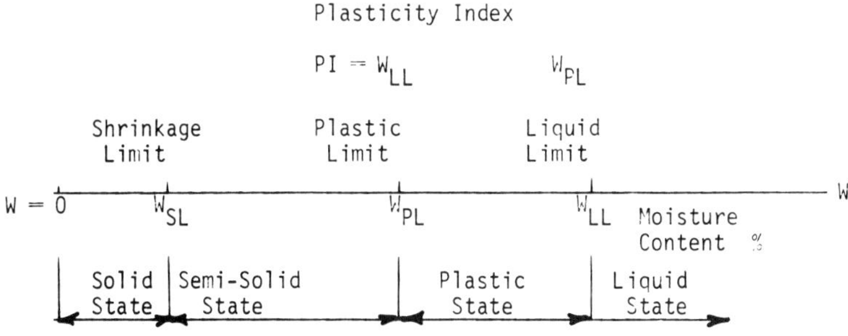

Fig. 4.4 Moisture Relations for Fine-Grained Soils.

Table 4.6. Plasticity Characteristics of Soils, PI = $W_{LL} - W_{PL}$.

Plasticity index	Soil characteristics	Soil type	Cohesiveness
0	Nonplastic	Sand	Noncohesive
<7	Low Plastic	Silt	Partly cohesive
7–17	Medium Plastic	Silty-Clay	Cohesive
>17	High Plastic	Clay	Cohesive

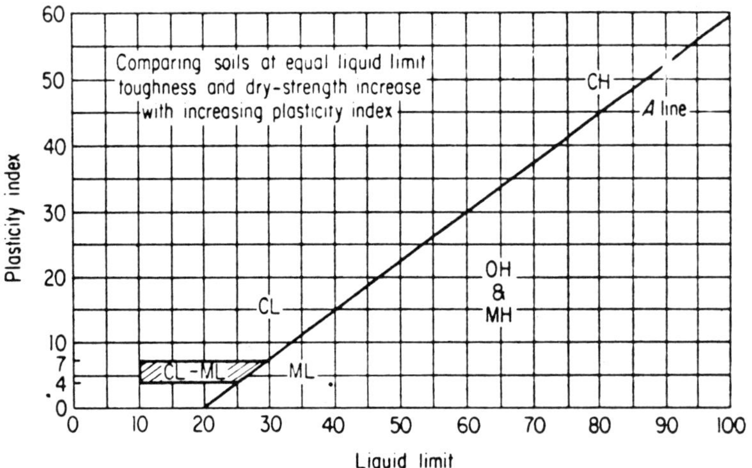

Fig. 4.5. Plasticity Chart for Fine-Grained Soils. From Scott and Schoustra, *Soil Mechanics and Engineering*, © 1968, McGraw-Hill. Used with permission of McGraw-Hill Book Co.

4.2 PHYSICAL PROPERTIES OF SOILS

A soil system can be analyzed as a mechanical combination of solids (s), liquids (l), and gases (g). Let the total mass of volume V of a soil be m, and let the constitutents have masses and volumes as noted in Fig. 4.6.

m_ℓ, V_ℓ, ρ_ℓ
m_g, V_g, ρ_g
m_s, V_s, ρ_s

Fig. 4.6. Soil Properties.

Now

$$m = m_s + m_l + m_g \tag{4.2}$$

$$V = V_s + V_l + V_g \tag{4.3}$$

On a volumetric basis, one has

$$x_s = \frac{V_s}{V} \tag{4.4}$$

$$x_l = \frac{V_l}{V} \tag{4.5}$$

$$x_g = \frac{V_g}{V} \tag{4.6}$$

and from Eq. (4.3)

$$x_s + x_l + x_g = 1 \tag{4.7}$$

The x_i terms denote the volumetric fraction of each soil component. It is possible that the liquid portion, usually water, may be present as liquid water and ice. In this case it will be necessary to use

$$x_i = \frac{V_i}{V} \tag{4.8a}$$

$$x_w = \frac{V_w}{V} \tag{4.8b}$$

If one starts with a given volume of water and part of it freezes, then $x_i + x_w \neq x_l$, since the density of ice is less than that of water.

Porosity. The porosity is defined as the ratio of the volume of the voids to the total volume

$$n = \frac{V_v}{V} \tag{4.9}$$

Now

$$n = \frac{V - V_s}{V} = 1 - x_s \tag{4.10}$$

Void Ratio. The void ratio is the ratio of the volume of voids to the volume of solids

$$e = \frac{V_v}{V_s} \tag{4.11}$$

also

$$e = \frac{1}{x_s} - 1 = \frac{n}{x_s} \tag{4.12}$$

Degree of Saturation. If the total void volume is occupied by a liquid, the soil is said to be saturated. The percentage of saturation can be calculated as

$$S = \frac{V_l}{V_v} \times 100 \tag{4.13}$$

Again

$$S = \frac{x_l}{x_l + x_g} 100 \tag{4.14}$$

and

$$S = \frac{x_l}{n} 100 \tag{4.15}$$

The previous relations all deal with the volumetric parameters of the soil system, but one can also use mass relationships.

Dry Unit Weight (Dry Density). In terms of engineering properties, the masses of the constituents are often important. The unit weights are, in effect, density relations, and the dry unit weight is defined as

$$\gamma_d = \frac{m_s}{V} \tag{4.16}$$

Solid Density (Compact Density). The density of the soil solids is defined as

$$\gamma_s = \frac{m_s}{V_s} \tag{4.17}$$

Total Unit Weight (Bulk Density). The density of the total soil system is

$$\gamma_t = \frac{m}{V} = \frac{m_s + m_l}{V} \tag{4.18}$$

where the mass of the gases is neglected.

Buoyant Unit Weight. If the entire system is immersed in water, the buoyancy results in

$$\gamma_b = \gamma_t - \gamma_w \tag{4.19}$$

where γ_w is the specific weight of water.

Liquid Percentage. It is convenient to describe the mass of liquid as a percentage of the solid mass

$$W = \frac{m_l}{m_s}\, 100 \tag{4.20}$$

From Eqs. (4.16, 4.18, and 4.20)

$$\gamma_d = \gamma_b/(1 + W/100)$$

Typical values of the properties of soils and minerals are listed in Tables 4.7 and 4.8.

4.2.1 Relation between Volumetric and Mass Parameters

Soil systems are often characterized by the two parameters γ_d and W for purposes of property tabulation.

The specific gravity of the soil solids is defined as

$$G_s = \frac{\text{specific weight of solids}}{\text{specific weight of water}} = \frac{\rho_s}{\rho_w} = \frac{\gamma_s}{\gamma_w} \tag{4.21}$$

The specific gravity of many minerals is about 2.6. If the system is composed of water, ice, air, and solids, then the following relations between mass and volume parameters are valid.

Table 4.7. Typical Soil Property Values.

Soil	Bulk density, water soluted γ_t, kg/m³	Void ratio, e	Porosity, %	Saturated soil, W%
Peat	1000	>5	>85	>500
Mud	1000–1300	3–6	75–85	150–300
Clay, Silt	1400–2000	–	–	–
Clay Soft	–	1–3	50–75	40–100
Clay Stiff	–	.3–.8	25–45	10–40
Silt	–	.3–1.4	25–60	10–50
Sand, Gravel	1700–2300	–	–	10–35
Sand, poorly graded	–	.5–.9	35–45	–
Sand, well graded	–	.15–.4	15–30	–
Till	2000–2400	.1–.3	10–25	5–10

Table 4.8. Density and Surface Area of Primary and Secondary Minerals.

Mineral	Solid density γ_s, kg/m³	Specific surface area, m²/g	Internal surface area, %
Calcite	2700	–	–
Quartz	2600–2660	–	–
Feldspar	2500–2900	–	–
Pyrite	4500–5100	–	–
Illite	2600–2700	100–200	0
Kaolinite	2600–2700	25–50	0
Montmorillonite	2400–3000	200–800	10–90
Organic material	1400–1700	–	–
Vermiculite	–	700–800	80–90

$$\gamma_d = G_s \rho_w x_s \tag{4.22}$$

$$W_i = \frac{100}{G_s} \frac{x_i}{x_s} \frac{\rho_i}{\rho_w} \tag{4.23}$$

$$W_w = \frac{x_w}{G_s x_s} 100 \tag{4.24}$$

where ρ_i is the density of ice.
Also

$$x_s = \frac{\gamma_d}{\rho_w G_s} \tag{4.25}$$

$$x_i = \frac{W_i \gamma_d}{100\ \rho_i} \tag{4.26}$$

$$x_w = \frac{W_w \gamma_d}{100} \tag{4.27}$$

If the system is made up of only water, air, and solids, then

$$\gamma_d = G_s \rho_w x_s \tag{4.28}$$

$$W = \frac{x_w}{G_s x_s}\ 100 \tag{4.29}$$

It is of interest to note the volume change of the system due to the phase change of the water. Let δ be the ratio of the final to the initial mass of ice, and assume that ice is always initially present. Then the volume change is

$$\frac{\Delta V}{V} = x_i\ (1 - \delta) \left(\frac{\rho_i}{\rho_w} - 1\right)$$

One may examine the phase change effect for a few cases. Let $\rho_i/\rho_w \simeq 0.9$. Then

Case	δ	$\Delta V/(Vx_i)$
(1) Complete melt	0	−.1
(2) No melt	1	0
(3) Additional freeze	2	+.1

Of course, if the densities of ice and water were equal there would be no volume change effects. The volume change effects due to water-ice density differences are relatively minor and cannot explain the adverse effects of frost heaving as discussed in Chap. 5.1.

4.3 EQUILIBRIUM STATES FOR WATER

The moisture contained within a soil system plays a major role in the soil properties and also in the phenomena associated with phase change. It is worthwhile to review the thermodynamic relations for pure water, in bulk, before examining the more complicated relations for soil water.

A pure substance, such as water, can be described by an equation of state that is a functional relation between pressure, temperature, and volume. The equilibrium diagram for water is sketched in Fig. 4.7.

Cross-sections of the PVT diagram are shown in Fig. 4.8. The intersections of the phase regions are equilibrium surfaces for which the presence of two or more phases is allowed. For example, Fig. 4.7 shows the solid-liquid surface ($S + L$). A change of phase is a series of states traversing this equilibrium surface as the water changes from a saturated liquid to a saturated solid (ice). In the pressure-temperature plane, Fig. 4.8*a*, these surfaces project as lines. Note that the $S - L$ surface slopes upward to the left. This is an uncommon thermodynamic characteristic, along with the increase in volume as water changes from a liquid to a solid. Despite its abundance, water is

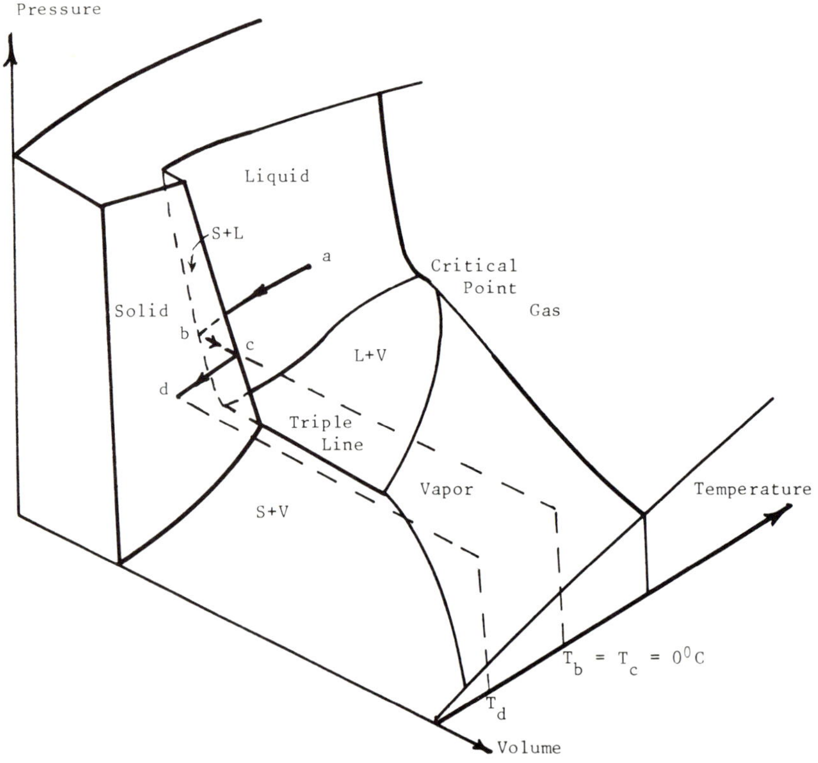

Fig. 4.7. Equilibrium States for Water.

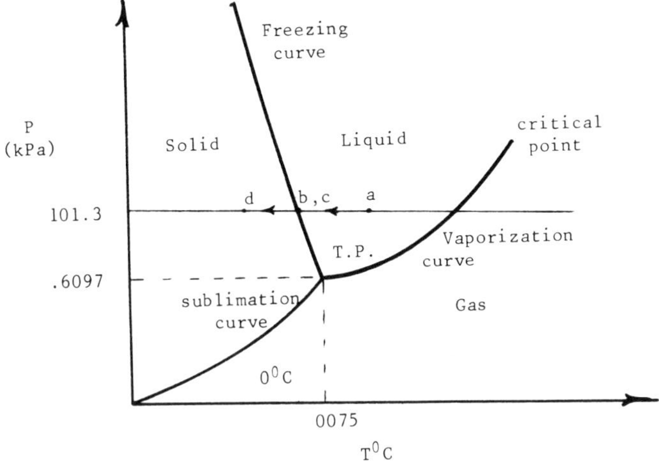

Fig. 4.8*a*. Pressure-Temperature Diagram for Water.

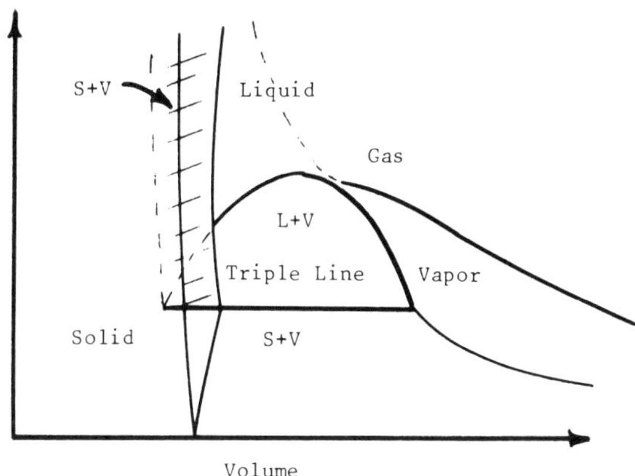

Fig. 4.8*b*. Pressure-Volume Diagram for Water.

classified as an uncommon substance thermodynamically, due to perculiarities such as these.

Water can exist simultaneously as a liquid, solid, and vapor in a series of equilibrium states defined by the intersection of the solid, liquid, and vapor states. This is denoted in Fig. 4.7 as the triple line. On the $P - T$ surface, this is a single point called the triple point, at a pressure of .6097 kPa and temperature of .0075°C, with a variable volume. At equilibrium, water cannot exist as a liquid at pressures less than that of the triple point.

The solid form of water, ice, can exist in a number of different lattice

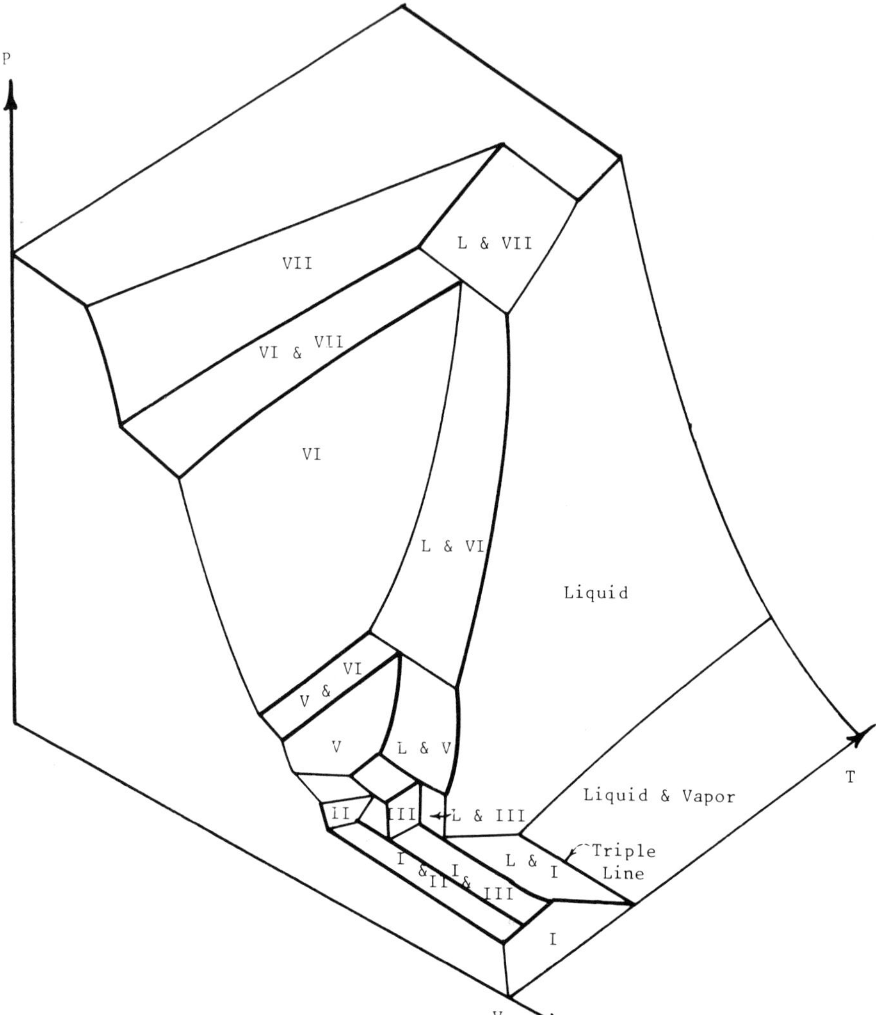

Fig. 4.9. Solid Forms of Water.

structures, each with its own properties, sketched in Figs. 4.9 and 4.10. Phase transformations (allotropic transformations) occur in going from one solid phase to another, with latent heat involved. Water has several triple points, as can be seen in Fig. 4.10. The common form of solid water is Ice I, and there is no evidence that any other form of ice is involved in soil systems. Ice I is a crystalline solid with a loosely bound hexagonal structure and

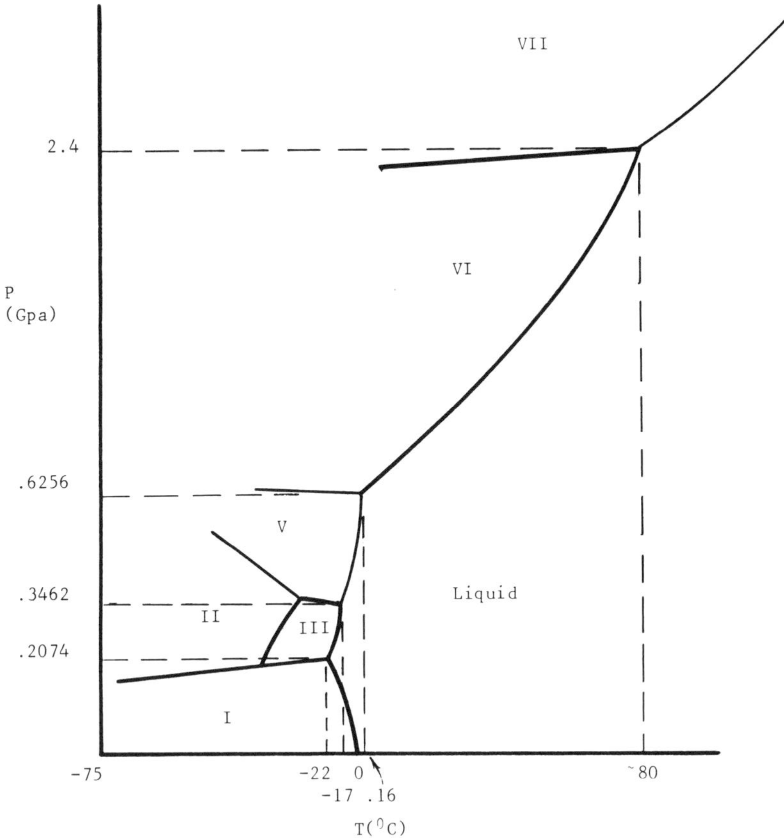

Fig. 4.10. Triple Points of Ice, Allotropic Forms.

anisotropic properties. It is unusual in that its density is less than that of liquid water in equilibrium with it.

Referring to Fig. 4.7, let us examine the state changes of water starting at point *a*. Assume the pressure is atmospheric (101.3 kPa) and constant and that the temperature is initially above 0°C. Thus, state *a* is a compressed liquid. As heat is extracted from the liquid, the temperature decreases until the fusion temperature of water is reached. This value is defined as 0°C for ice in equilibrium with air-saturated liquid water. The saturated liquid, at *b*, now changes to Ice I at constant temperature until *c* is reached. At this state all the water has changed phase to the solid, saturated Ice I. Further extraction of energy will now result in a decrease of temperature below 0°C. As noted earlier, the volume increases from state *b* to *c*.

Liquid water, in an equilibrium state, cannot exist at pressures less than the triple point value. In metastable equilibrium, which cannot be shown on a PVT diagram, however, liquid water (or ice) can exist in states not allowed by the stable equilibrium surfaces. The metastable state will, however, rapidly change to a stable state if disturbed, as by a slight energy addition due to vibration, etc., or if nucleation centers exist to induce phase transformation.

Liquid water at $p = 1$ atm, $T < 0°C$, can and often does occur, as shown in Fig. 4.11. This is called "supercooled water" and denotes a metastable state not characterized by the usual PVT surface, but which is not excluded by such surfaces.

4.4 PORE WATER

The water in a soil system exists in a number of phases and thermodynamic states. The liquid water can be divided into two classes: bound water and free water.

Bound water has been altered in some way that affects its equilibrium thermodynamic characteristics. This tends to alter the ability of the water to change phase or to move under various force fields. The water can be bound due to chemical bonding or surface force phenomena. Chemical bonding is exemplified by such minerals as the feldspars, where a water molecule is bonded to other elements to form the mineral. The bound water can also be absorbed within the body of a hygroscopic solid. Surface phenomena are associated with adsorbed water, which can be bound either strongly or weakly. The solid particle surface, in Fig. 4.12, has a negative charge as noted earlier. The energy of the chemical and electrical bonds between the mineral surface and the surrounding medium binds a layer of liquid water to the soil particles. The excess activation energy of the mineral surfaces hinders the ability of water to form the hexagonal lattice of Ice I, even at temperatures as low as −70°C. At greater distances from the solid interfaces, the water is less tightly bound and forms a layer (region) that is important for diffusion phe-

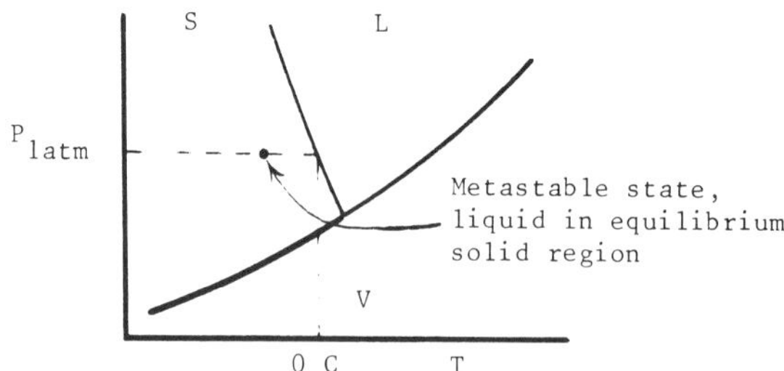

Fig. 4.11. Metastable Equilibrium of Water.

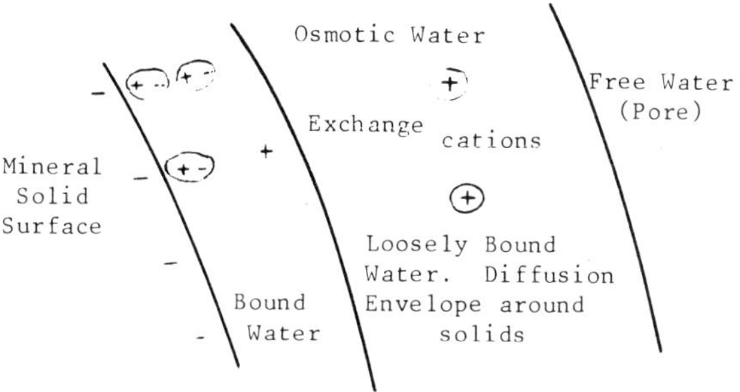

Fig. 4.12. Water in Soil Pores.

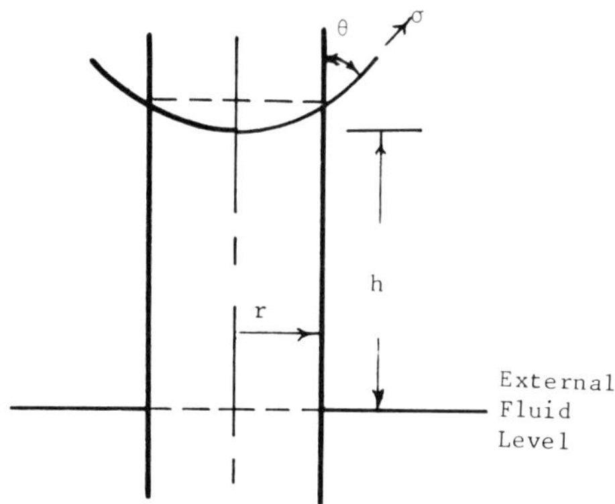

Fig. 4.13. Capillary Rise in Circular Cylinder.

nomena. The water in this osmotic layer is of variable phase, probably with a deficiency of activation energy, and is capable of releasing heat of fusion and freezing at temperatures less than 0°C. The transformation temperature will decrease as the thickness of the bound water layer decreases, due to the stronger mineral surface effects. There will be no distinct temperature of fusion for the bound water, but rather a temperature zone of solidification.

A familiar form of surface effect is that associated with the capillary rise of a liquid in a tube (Fig. 4.13). The fluid is pulled up in the tube by the surface tension force σ, defined as the force per unit length of the control surface perimeter. At equilibrium, the vertical surface tension force will just balance the weight of the liquid

$$2\, r\pi\sigma \cos\theta = \pi\, r^2 h\rho g$$

Thus

$$h = \frac{2\sigma \cos \theta}{r\rho g} \tag{4.30}$$

The contact angle θ is a complicated function of the force of adhesion exerted on the liquid molecules near the contact surface by the solid surface and the force of cohesion exerted by the bulk liquid molecules (Fig. 4.14). The net force is perpendicular to the slope of the free surface at any point.

Some typical capillary rise values are given below:

Soil	h (meters)
Sand	.03–3.5
Silts	1.5–12
Clays	>8

The water in the capillary is in a state of tension (negative pressure, or suction; Fig. 4.15) given by

$$P - P_a = -\rho g h \tag{4.31}$$

Then the tension can be estimated as:

h (ft)	Tension (atm)
1	.0302
10	.302
100	3.02

Tensions on liquid water, without rupture, have been measured at up to 50 atm (Dorsey 1940). Thus, the capillary rise is not out of line with the

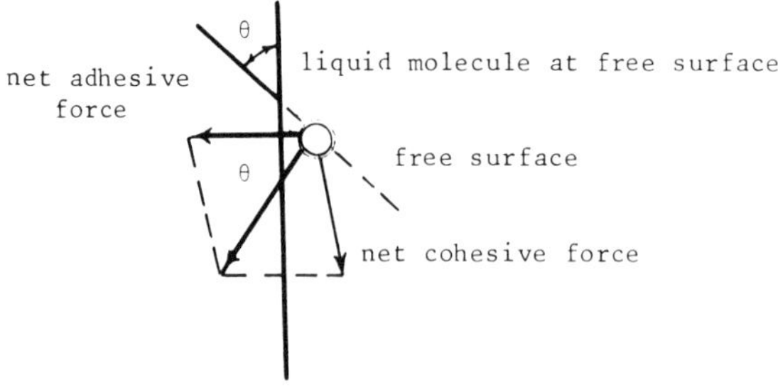

Fig. 4.14. Forces at Liquid-Solid Interface.

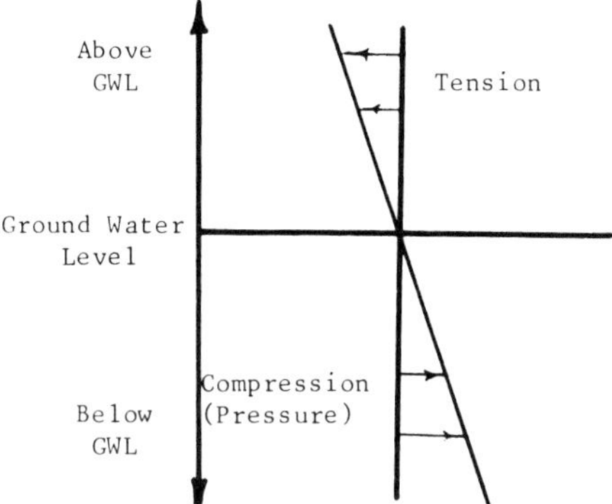

Fig. 4.15. Stresses in Capillary Liquid.

ability of water to resist rupture or cavitation. It also can lead to significant pressure differentials, which can activate water movement in soils.

The surface electromolecular forces of ice are stronger than the molecular bonds of bulk water, and thus free water can be adsorbed or absorbed by ice surfaces. These ice-water interfaces may be important in keeping some water in the liquid state.

The water in the soil that is not bound is free water. It is free to move, change phase, etc., and in general acts as normal liquid water subject to the usual effects of the solutes associated with the solid particles. If the free water is pure it will solidify very near the 0°C value.

If the soil pores are not filled with water, water vapor will exist. It will be displaced from locations of high vapor pressure (temperature-dependent) to regions of low vapor pressure. Vapor diffusion may be the main case of moisture redistribution, due to temperature gradients, in unsaturated soils.

4.5 SOIL WATER MOVEMENT

The potential energy of soil water is defined as the work needed to move a unit mass of water from a standard reference state (pure water at 1 atm, specified elevation and temperature) to the location and state of interest.

The total potential of the soil water can be expressed symbolically as

$$\psi = \psi_m + \psi_g + \psi_p + \psi_o + \psi_B \tag{4.32}$$

where

ψ = total soil water potential
ψ_m = matric potential. This is due to capillary and interfacial interactions

associated with the soil solid interfaces. Above the water table (phreatic surface) it is negative (suction or tension), and below the water table it is zero.

ψ_g = gravity potential, associated with the usual gravitational force

ψ_p = pressure potential. This is the pressure due to the hydrostatic weight of the water and any gas pressures active on the soil-water menisci.

ψ_o = osmotic potential, due to the different concentrations of solutes in the pore water and to the exchangeable cation complex. This can be significant for membranelike situations at interfaces.

ψ_B = overburden potential. The force that can be exerted on the soil water by the total soil system above a given point. This is difficult to estimate due to the geometry effect of the soil matrix.

In addition to the above phenomena, soil water can move due to imposed or induced electric fields (electroosmosis) or due to thermal fields (thermoosmosis). Unfortunately, there is no generally accepted method to describe the actual potential difference between two states of a soil system in a quantitative formula of the type of Eq. (4.32). The potential between the soil water in a thawed state and the soil water near or at a freezing zone will induce a flow of water to the freezing zone. This flux of water has a significant effect upon the freezing rate and is mainly responsible for frost-heaving effects. A comprehensive theory of the interaction of the diffusion of mass and energy in soil systems with phase change, which leads to quantitative results, is not yet available.

4.6 SOIL SYSTEM TEMPERATURE CHANGES

An important relation for calculating phase change is the relative amounts of liquid, solid, and vapor phases of water that exist in a soil at any given state. It is now widely accepted that a layer of liquid water separates the ice from the solid grains in frozen soil. Evidence for this layer comes from experiments that show that electric fields can induce a movement of solutes and water in frozen soil. With high ice content, the solid soil particles may themselves move slowly through the ice matrix. Thus, electroosmosis occurs in frozen, as well as unfrozen, soils. Ionic diffusion also takes place in frozen soils, and the path is presumably the unfrozen interface. The layer of bound water (liquid water), or bound water in general, can be thought of as having physical properties different from those of the bulk pore water, which tends to freeze at 0°C.

It is instructive to examine the cooling curve of a soil-water system as sketched in Fig. 4.16. When the temperature of the system drops to T_f^* (for free water at 1 atm, this would be 0°C), the pore water does not start

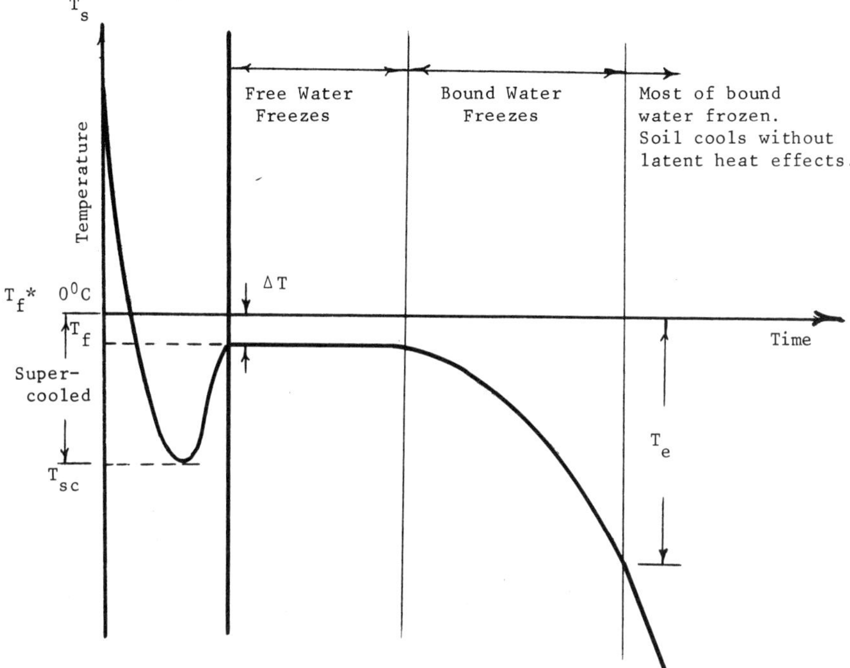

Fig. 4.16. Typical Freezing Curve of Soil Water.

to freeze immediately. The water will supercool down to T_{SC}. This is a metastable equilibrium state related, in an unclear way, to the thermodynamic equilibrium states for soil-water-ice interfaces. Nucleation centers can trigger an abrupt transformation of the free water to ice. These nuclei can be soil particles or even aggregations of water molecules. The value of T_{SC} is related to the nature of the nuclei and the interfaces. Pure water, carefully cooled, can be supercooled 10–30°C, but normally the supercooling is only a few degrees due to abundant nuclei. The formation of the ice releases latent heat, and the system temperature will rise to T_f. This value is the initial freezing temperature of the particular soil system. For soils with small specific surface areas, T_f may be close to T_f*, but for silts and clays the temperature depression can be as much as 5°C.

The free water in the soil will now freeze at an essentially constant temperature. After the free water solidifies, the temperature of the system will start to decrease, but the release of latent heat, as bound water freezes, will cause the rate of cooling to increase until the temperature T_e is reached. At this point, the bound water is almost all frozen, except for a minimum associated with hygroscopic effects. The temperature, T_e, at which there is little or no unfrozen water can be as low as −70°C, and thus a significant amount of liquid water can exist at higher temperatures. Again, this is particularly true for soils with high specific surface areas, which can bind large amounts of water.

The freezing temperature depression is also related to the salts in the pore water. As the water freezes, it forces the solutes into a smaller and smaller volume of solution, and thus the salt concentration increases. Banin and Anderson (1974) used the Clausius-Clapeyron equation to relate the concentration to the freezing point depression

$$T_o - T = \frac{-RT\,T_o}{L_f} \ln X_1 \tag{4.33}$$

where

X_1 = mole fraction of solvent (water) in the solution
R = universal gas constant
T = freezing point of solution
T_o = freezing point of pure solvent and
L_f = molar latent heat of solvent.

Assuming that all the salt is in solution and no precipitation occurs, the freezing point depression is given by

$$T_o - T = \frac{-RT\,T_o}{L_f} \ln \left(\frac{1000}{1000 + \dfrac{M_w n_m \rho_{se}}{z \rho_s W_u} \displaystyle\sum_{i=1}^{N} \frac{g_i}{M_i}} \right) \tag{4.34}$$

where

M_w = molecular weight of solvent
n_m = number of ions to which solute molecule dissociates in solution
z = valence of ions
ρ_{se} = density of solution in pores
ρ_s = density of solvent in solution
W_u = unfrozen water
g_i = total content of ion i in the solid and
M_i = equivalent weight of ion i.

Example:
Suffield silty clay

.072 g 1.5 N NaCl solution/g soil added to soil solution

$L_f = 1435.22$ cal/mole

$M_w = 18.015$

$R = 1.9867$ cal/g-mole-°K

$\rho_{se} = \rho_s$

$n_m = 2$

$z = 1$

Then

$$\sum_{i=1}^{N} \frac{g_i}{M_i} = (.072)(1.5) = 0.108 \text{ meq NaCl/g}_s$$

$$X_1 = 0.9809$$

$$T_o - T = 1.976°\,C \qquad \textit{Measured value} = 2.09°\,C$$

4.7 UNFROZEN SOIL WATER

It is convenient to specify a thermodynamic reference state in terms of either

(a) the dry solid at 1 atm and a standard temperature or
(b) a water-Ice I mixture (or supercooled liquid) at 1 atm and a standard temperature.

The water content of the frozen soil, at any given state, can be written as

$$W = W_u + W_i \tag{4.35}$$

The mass of the vapor phase can be safely ignored. The heterogeneity and variation in chemical composition are usually accounted for by specifying a specific unfrozen water content for each soil. In clay systems, for example, the unfrozen water depends strongly upon the saturation of the clay with different exchange cations.

Tsytovich (1975) postulated that the unfrozen water is practically independent of the total soil moisture content for a given soil. Despite the wide use of this relation, it has been shown that the unfrozen water content tends to increase somewhat, at a given temperature, with an increase in water content (Tice et al. 1978). The Tsytovich assumption is acceptable as a first approximation, and the unfrozen water content is then principally a function of specific surface area and temperature. Lesser effects are exerted by pressure, chemical and mineralogical composition, such physicochemical characteristics as exchange ions, solute content, and composition, and total water content.

The specific surface area represents the effects of the soil texture and matrix geometry and seems to be the dominant variable. A typical curve of unfrozen water versus temperature is sketched in Fig. 4.17. The region of significant phase change (1) occurs when W_u changes by more than 1% per degree centigrade. All the free water freezes, as does some loosely bound water. The transition region has W_u changing by .1–1.0%/°C, with all the loosely bound water frozen. Region (3) is an essentially frozen zone. Only strongly bound water exists, and the unfrozen water approaches the maximum hygroscopicity of the soil. Eventually most of the water will freeze, but at very low temperatures.

The unfrozen water content has been measured by dilatometry, adiabatic and isothermal calorimetry, x-ray diffraction, heat capacity measurements, nuclear magnetic resonance, and differential thermal analysis. Anderson and Tice (1973) have described these methods and discussed the merits of each. Most of the methods tend to give comparable results when properly carried out. The unfrozen water has been correlated by

$$W_u = a\,\theta^{\,b} \tag{4.36}$$

where

W_u = percentage of unfrozen water based on dry weight of solid
θ = temperature below 0°C and
a, b = parameters for each soil.

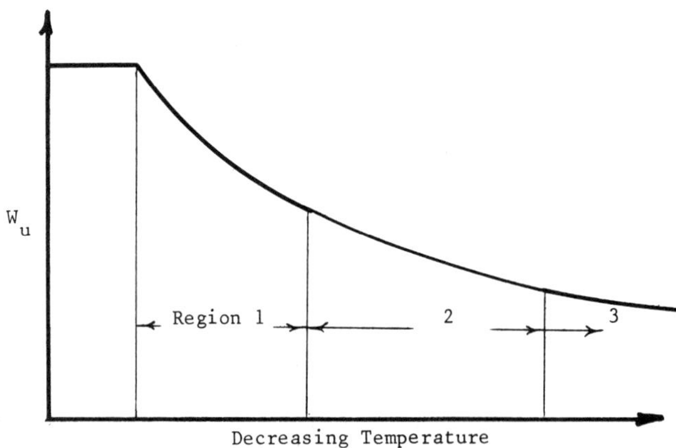

Fig. 4.17 Unfrozen Water versus Temperature.

Table 4.9. Correlation Coefficients for Unfrozen Water (from Anderson and Tice 1972).

Soil	Surface area, m^2/g	a	b
Basalt	6	3.45	−1.13
Rust	10	11.05	−0.80
West Lebanon gravel	18	3.82	−0.64
Limonite	26	8.82	−0.83
Fairbanks silt	40	4.81	−0.33
Dow Field Silty clay	50	10.35	−0.61
Kaolinite	84	23.80	−0.36
Suffield silty clay	140	13.92	−0.31
Hawaiian clay	382	32.42	−0.24
Wyoming bentonite	800	55.99	−0.29
Umiat bentonite	800	67.55	−0.34

Values of a and b are given in Table 4.9. Using the data of Table 4.9, Anderson and Tice (1972) were able to correlate a and b with the specific surface area S. These correlations then yielded a simple empirical equation for the unfrozen water content.

$$\ln W_u = .2618 + .5519 \ln S - 1.4495\, S^{-.2640} \ln \theta \tag{4.37}$$

where

W_u = percent
S = m^2/g and
θ = °C.

Equation (4.37) will give acceptable estimates of the phase composition of a broad range of soils. The major uncertainty is in the value of S for a given soil. This can be measured or estimated as noted by Dillon and Andersland (1966)

$$S = \frac{\sum_{n=1}^{N} S_n \dfrac{P_n}{d_n}}{\sum_{n=1}^{N} \dfrac{P_n}{d_n}} \tag{4.38}$$

where

P_n = size fraction of particles in a given size range (% by weight)
d_n = average particle size diameter for the size fraction and
S_n = specific surface area for particles in the size range.

4.8 PERMAFROST

Permafrost is a widespread phenomenon that has been and still is greatly misunderstood. The term "permafrost" is generally attributed to S. W. Muller (1945), who apparently coined the name in place of the more awkward terms: permanently frozen ground or permanent frost. Bryan (1946) suggested the term "pergelisol," but this has not been adopted except in the French literature. In order to understand the concept, let us look at some definitions in general use. Brown (1970) states

> It is a term used to describe the thermal condition of earth materials such as soil and rock, when their temperature remains below 32°F continuously for a number of years. . . . Permafrost includes ground that freezes in one winter, and remains frozen through the following summer and into the next winter.

Pihlainen and Johnston (1963) also describe permafrost

> Permafrost is defined as the thermal condition under which earth materials exist at a temperature below 32°F continuously for a number of years.

Swinzow's (1969) definition is

> Permafrost may be defined as a deposit of earth materials such as sand, gravel, silt or glacial till, etc., cemented by ice (frozen) and remaining in that state for a significantly long time, but not necessarily for a whole geological period.

Finally, Brown and Johnston (1964) state that permafrost is

> the thermal condition of earth materials such as rock and soil when their temperature remains below 0°C continuously for a number of years, which may be as few as two or as many as tens of thousands.

Clearly permafrost is not so much a material as it is the thermal state of ordinary soil systems. The following definition is a slight modification of the above definitions: Permafrost describes the thermal condition of earth materials (sand, glacial till, organic matter, etc.) when their temperature remains at or below 32°F (0°C) continuously for a significantly long time, but not necessarily for an entire geological period. It does not include earth materials that drop below 32°F during one winter and remain below 32°F through the following summer and into the next winter, although for practical engineering purposes such materials may be included.

This definition includes all the significant concepts of permafrost. It includes soil systems that contain no moisture (or ice), such as solid rock (often termed "dry permafrost"), as well as systems that have moisture and are below 32°F, but may contain no ice, due to the chemical nature of the water and the surface forces involved, and thus no cementing effect of ice. It does not include systems that are at or below 32°F, but contain no earth materials. Such systems exist commonly as ice caps, glaciers, and icebergs. There is no agreement on the minimum time during which the material must remain below 32°F in order to qualify as permafrost. Soils that freeze during an exceptionally severe winter and survive for 1 or 2 more years are called "pereletoks," (Soviet 1960), and often are not classified as permafrost (Swinzow 1969). The problem here is in determining how long the frozen material has existed or will exist; although the concepts can be clearly defined, it may be impossible in practice to clearly delineate pereletoks and permafrost, especially on a short-time basis. Radforth (1954) introduced the term "climafrost" to denote that part of the permafrost that is affected by seasonal climatic factors, but this term has not gained acceptance.

4.8.1 Continuous and Discontinuous Permafrost

Field studies have shown that permafrost may underly a given area on a virtually continuous basis, that is, there will be no regions where the ground is completely thawed at the time of the maximum thaw penetration—usually in late summer or early autumn. This refers to an areal continuity of the permafrost and is not related to its thickness. Under such conditions, the permafrost is called "continuous." Thus, in the zone of continuous permafrost, one would expect to find permafrost everywhere, with certain exceptions. As one investigates the extent of permafrost, it becomes evident that the thickness of the permafrost becomes smaller, and eventually there may be gaps of thawed material as shown in Fig. 4.18. Outside the continuous zone, the permafrost is patchy, and its thickness decreases as the distance from the continuous zone increases. This region is the discontinuous permafrost zone. Within this zone, the permafrost will exist as islands of frozen material surrounded by thawed and frozen material as the season dictates. In Russia (Bondarev 1959) and the United States (Ferrians 1964), the discontinuous zone is defined in terms of the ground temperature at the depth of zero annual amplitude change. If the temperature at this depth is greater than 23°F (−5°C), then the permafrost is discontinuous. Such a delineation requires extensive temperature data. Eventually the permafrost will disappear entirely except for very small local patches dependent upon special microme-

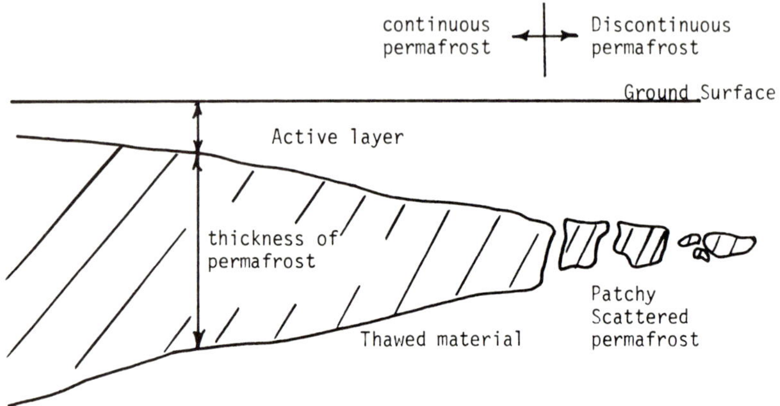

Fig. 4.18. Continuous and Discontinuous Zones of Permafrost.

teorological, vegetative, and geomorphological conditions. Finally, there may be permafrost associated with high elevations far outside the normal permafrost zones (alpine permafrost).

4.8.2 Active Layer

Even in the most severe climatic regions of continuous permafrost, there is usually a layer of the ground that undergoes seasonal thaw and freeze effects. This layer is called the "active layer," and it varies from one location to another. The active layer may be defined for a given location as the maximum depth, determined over a number of years, to which seasonal frost effects will penetrate. For any one year the maximum depth of thaw may not penetrate down to the upper surface of the permafrost, called the "permafrost table." This condition is shown in Fig. 4.19*a*, with a pereletok or temporarily frozen zone. It is also possible to have the frost penetration for a given year fail to reach the permafrost table as shown in Fig. 4.19*b*. This will leave a layer of thawed soil that may persist for one or more years. Such a zone is called a "talik," from the Russian literature (Soviet 1960). The conditions in 4.19*a* and *b* are sometimes referred to as "nonblending permafrost." Another possibility is shown in Fig. 4.19*c*, where the frost penetrates down to the permafrost table, a condition sometimes called "blending permafrost." This condition is typical of the continuous permafrost zone; that in 4.19*b* is most likely in the discontinuous zone; and the condition in 4.19*a* may be expected in either zone. It is important to keep in mind that these are all dynamic conditions which are variable, and they all depend on the thickness of the active layer or the depth to the permafrost table. These latter two

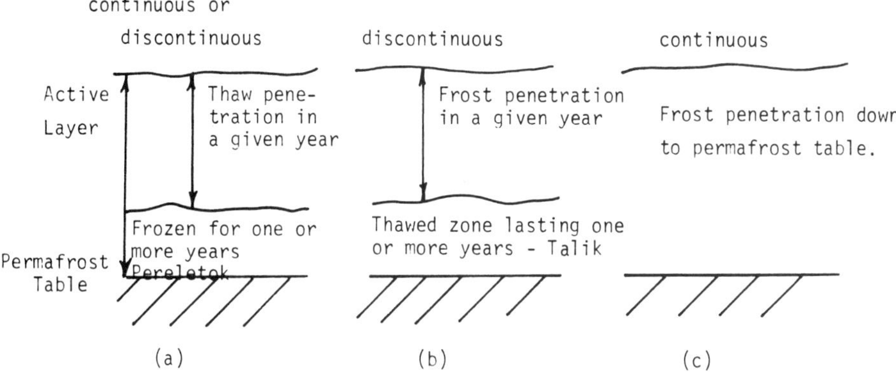

Fig. 4.19. Relation of Active Layer and Permafrost Table.

concepts are related, in that one can define the depth to the permafrost table as the thickness of the maximum observed thaw depth for a given location, which is constant with time, and the active layer thickness would then denote simply the thickness of the seasonal frost action layer, which will vary from year to year. Which concept is being referred to is often not clear in the literature.

4.9 DISTRIBUTION OF PERMAFROST

4.9.1 Northern Hemisphere

Permafrost underlies about 20% of the land surface of the world, and it is thus a very common phenomenon (Leggett 1954). Canada and Russia have most of the land area underlain by permafrost, each having about 50% of their total land underlain by permafrost. The areal extent of permafrost in Russia is about 2½ times that of Canada (Brown 1967*a*). Figure 4.20 shows the areal distribution of permafrost, at the present time, in the northern hemisphere apart from mountainous regions (see also Fig. 1.7). In general, there is a talik beneath bodies of water that do not freeze to the bottom in the winter.

The existence of permafrost is a result of the history and the present state of the energy balance at the earth's surface and the deep earth heat flow. If permafrost exists and the net gain of energy by the surface layers is equal to the net loss of energy, then the permafrost will remain stationary, although an excess heat gain over heat loss will result in a net loss of frozen material. Given the same energy balance, however, i.e., net gain of energy over the year, one region may have permafrost (albeit degrading) while another

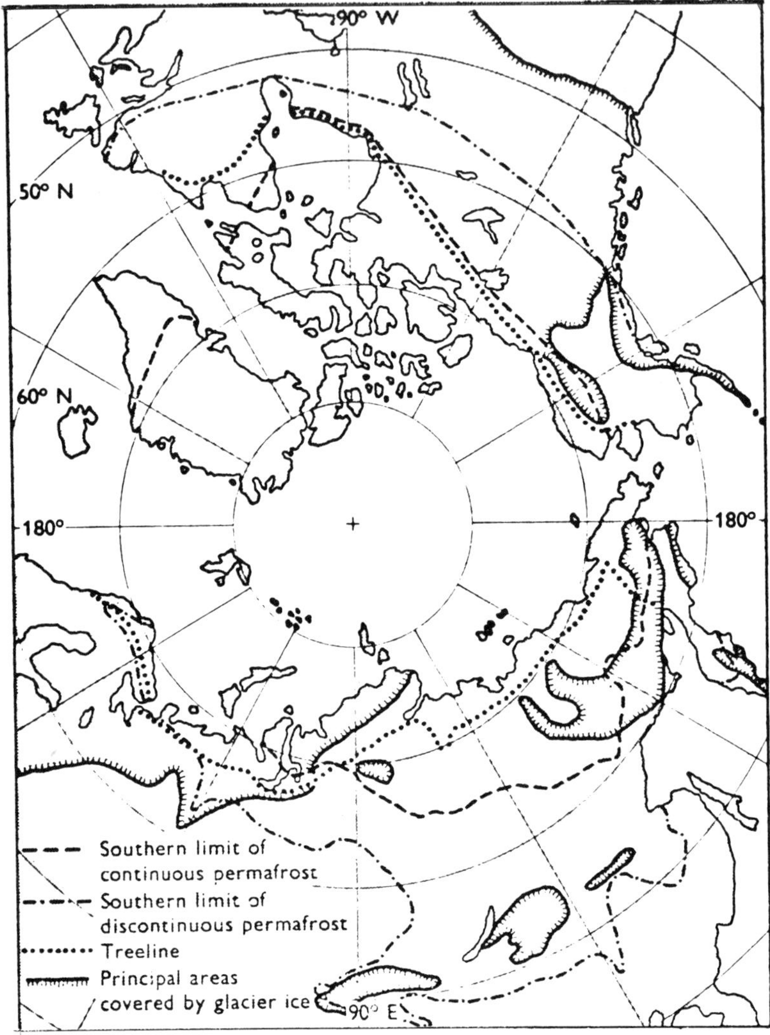

Fig. 4.20. Permafrost in the Northern Hemisphere, (after Brown 1967*a*). From Ecology of Subartic Regions (Proceedings of the Helsinki symposium) (c) Unesco 1970. Reproduced by permission of Unesco.

will not. This is due to the thermal history of the frozen ground in the two areas. Though both are losing permafrost one region may have started with a larger volume of permafrost than the other. Thus, the present energy flow conditions may be such that permafrost cannot exist in one region, whereas it will subsist, although in a receding form often referred to as "relic permafrost," in another area. In this sense previous glaciation has very likely played an important role in the present existence of permafrost in marginal areas. It is safe to conclude that little, if any, permafrost exists beneath glaciers, but once they withdraw, a rapid formation and growth of permafrost may take place. Thus, previously glaciated regions will show a lesser volume of frozen ground than unglaciated regions with similar climatic histories. In this regard it is significant that Canada was heavily glaciated while Russia (Siberia) had little permanent glaciation (Fig. 4.21). Thus, the permafrost thickness in Siberia is much greater than in Canada, although the climates are similar. Since the energy balance involves meteorological conditions, surface vegetation, topography, and soil conditions, it can be anticipated that no simple correlation will be found for the existence of permafrost in the marginal or discontinuous regions. Permafrost does exist far south of the usually accepted limits in scattered patches almost always associated with high altitudes and, thus, microclimates similar to the usual permafrost regions.

Countries in the northern hemisphere with significant amounts of permafrost are noted in Fig. 4.20 as Russia, Canada, United States (Alaska), Mongolia (Nakaya and Sugaya 1973), China, Finland, Iceland, Denmark (Greenland), Norway, and Sweden (small amounts).

4.9.2 Permafrost in Canada

The distribution of permafrost has been presented in the form of maps by Brown (1968, 1970). Figure 4.22 is taken from the permafrost map of Canada and illustrates the close relation between the distribution of permafrost and the isotherms of air temperature. It will be shown that the mean annual ground surface temperature must be less than 0°C for permafrost to exist in a widespread condition. Since the average ground surface temperature is greater than or equal to the mean annual air temperature, it follows that air isotherms less than 0°C will characterize the permafrost zones. Naturally permafrost can and does exist outside the widespread zones, as is clear from Fig. 4.22. The precise limits of the permafrost zones, in North America, have not been investigated completely by ground drilling, etc., and many regions of the map simply follow the air isotherms for the reasons noted above. The southern limit of discontinuous permafrost follows quite closely the 30°F air isotherm, and the southern limit of the continuous zone follows closely the 17°F isotherm except along the southern edge of Hudson Bay, where it exists at an air temperature close to 20°F. Figure 4.21 shows the same features, with the areas of glaciation also noted. Clearly glaciation does not correlate with the present existence of permafrost. Note that the Pleistocene glaciation extended far south of the present limits of permafrost in

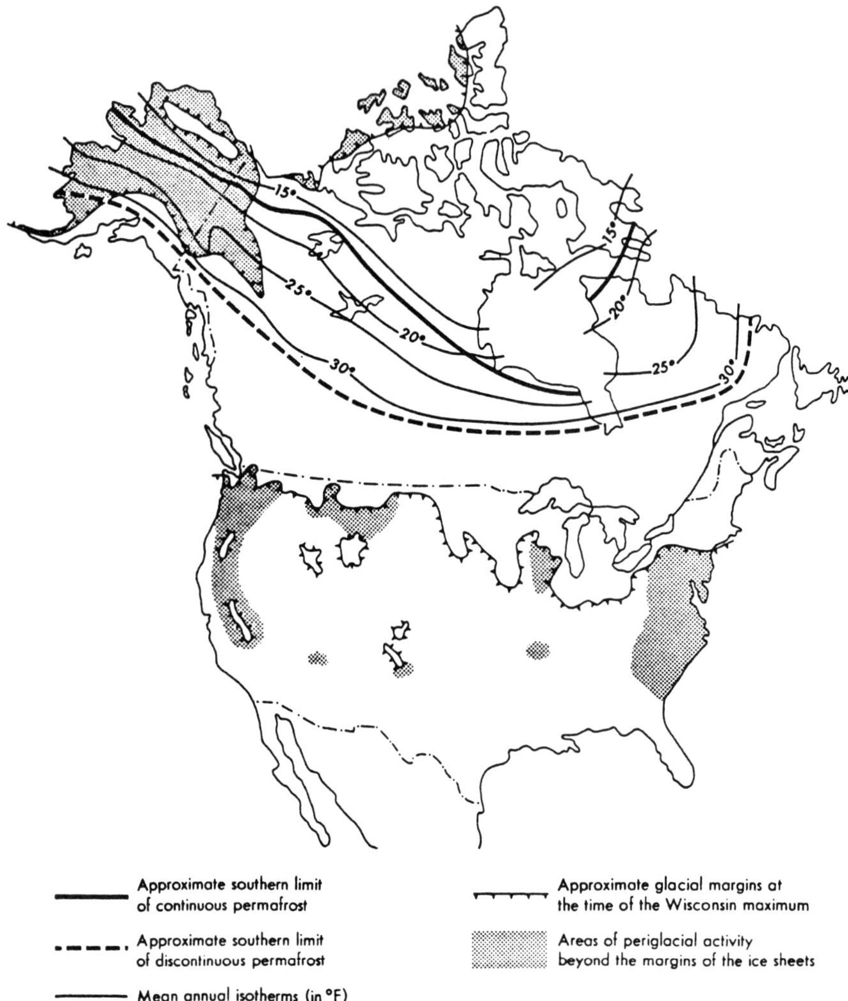

Fig. 4.21. Permafrost and its Relation to Pleistocene Glaciation and Periglacial Phenomena in North America.

North America and was absent over a large area of Alaska and the Yukon territory. The glaciers themselves were certainly not responsible for permafrost and, in fact, very likely prevented its growth over large areas of Canada. In general, the thickness of the permafrost is greatest in those areas where glaciation was absent.

It should not be inferred that the thickness of the permafrost in these zones is solely related to the ground surface temperature. Vegetation alone

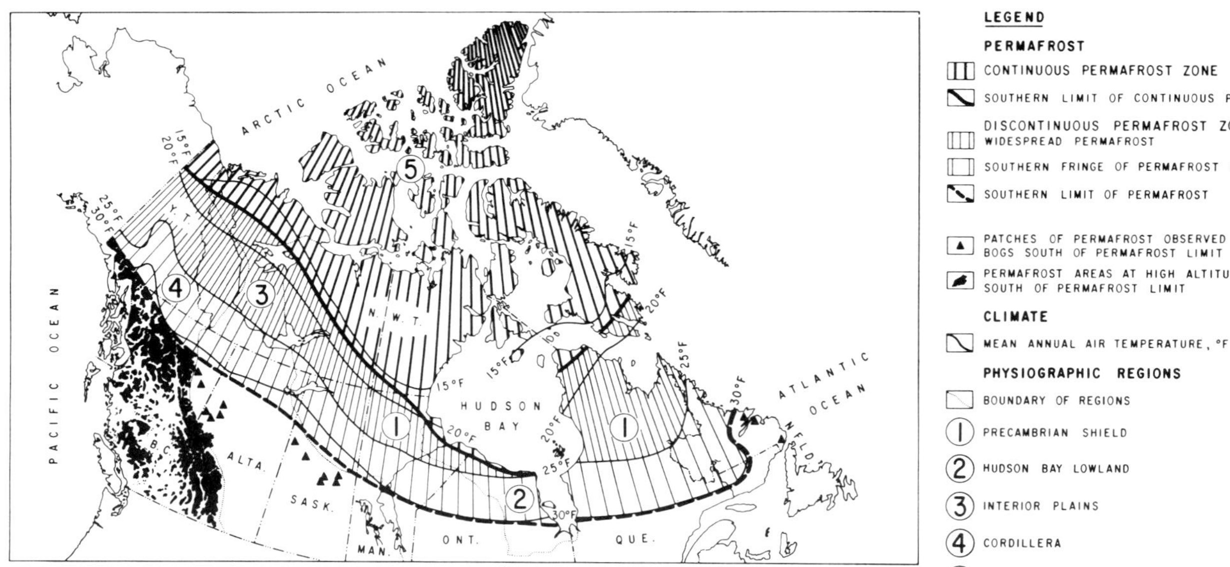

Fig. 4.22. Permafrost in Canada (from Brown 1970) by permission. Reprinted from *Permafrost in Canada: Its Influence on Northern Development*, R. J. E. Brown, © University of Toronto Press, 1969.

is also not indicative of the presence of permafrost, although it is important in the existence of scattered patches of permafrost at the southern limit of permafrost. Regions south of the discontinuous permafrost zone have essentially the same vegetative types as those within the permafrost zone.

The general conclusions on permafrost indicators in this area may be summarized as follows.

Climate: this is clearly of first importance in the sense that climate is a manifestation of the net energy exchange in an area. If the energy exchange is such that the mean annual ground temperature is below 0°C, then permafrost may occur. The combination of vegetative types, surface topography, water regimes, soil type, snow depth, and time of occurrence, will lead to ground temperatures several degrees warmer than the air temperature. Thus, when the air temperature is less than 25°F, permafrost is widespread, and for air temperatures between 25–30°F it is patchy and depends critically on the combination of climate, vegetation, etc. It seems that factors other than climate will be significant only in the patchy, southern limit of the discontinuous permafrost zones.

Tree types: not reliable indicators by themselves.

Sphagnum and lichens: most of the sites investigated with permafrost have sphagnum, lichens, or both, but many sites without permafrost also have these vegetative types. They are not reliable, but they increase the possibilities of finding permafrost.

Drainage: quite significant in these regions, since it is intimately associated with the formation and growth of peat plateaus and palsas. No permafrost was found in the discontinuous zone where the water table was at or near the surface.

Snow cover: important in regard to the formation of palsas along with the ground water regime. In the southern areas permafrost is very characteristic of peat bogs, and Fig. 4.23 shows a typical profile found in such areas.

4.9.3 Permafrost in the United States

Virtually all the significant permafrost in the United States is located in Alaska, the distribution of which is shown in Fig. 4.24. It can be seen that almost the entire state of Alaska is underlain by permafrost, and this must, of course, be considered in terms of resource development. Figure 4.25 shows the relation of Alaskan permafrost with the recent glaciation of North America. Again it can be noted that the existence of permafrost is independent of the previous state of glaciation.

Outside Alaska, the permafrost in the United States is limited to high altitudes. This permafrost is called "alpine," and, since it usually exists in small patches at inaccessible locations, it is of almost no engineering signifi-

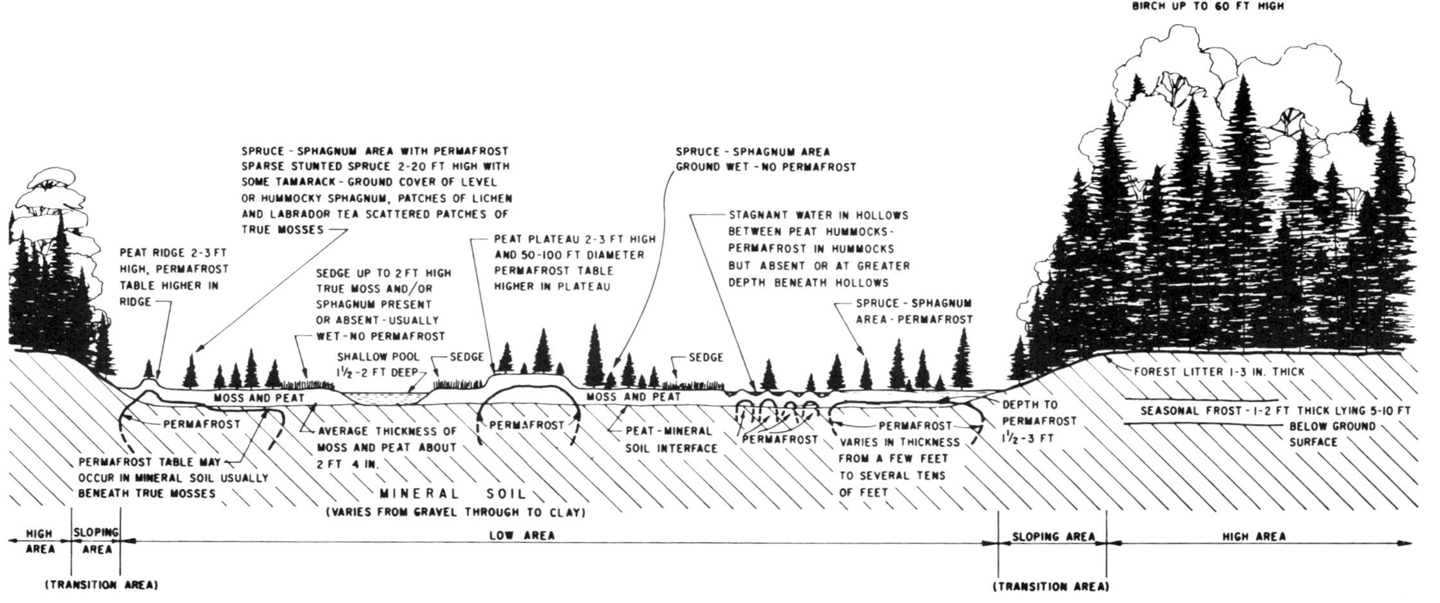

Fig. 4.23. Profile through a Typical Peat Bog in Southern Fringe of Discontinuous Zone (from Brown 1965). Used with permission of Division of Building Research, National Research Council of Canada.

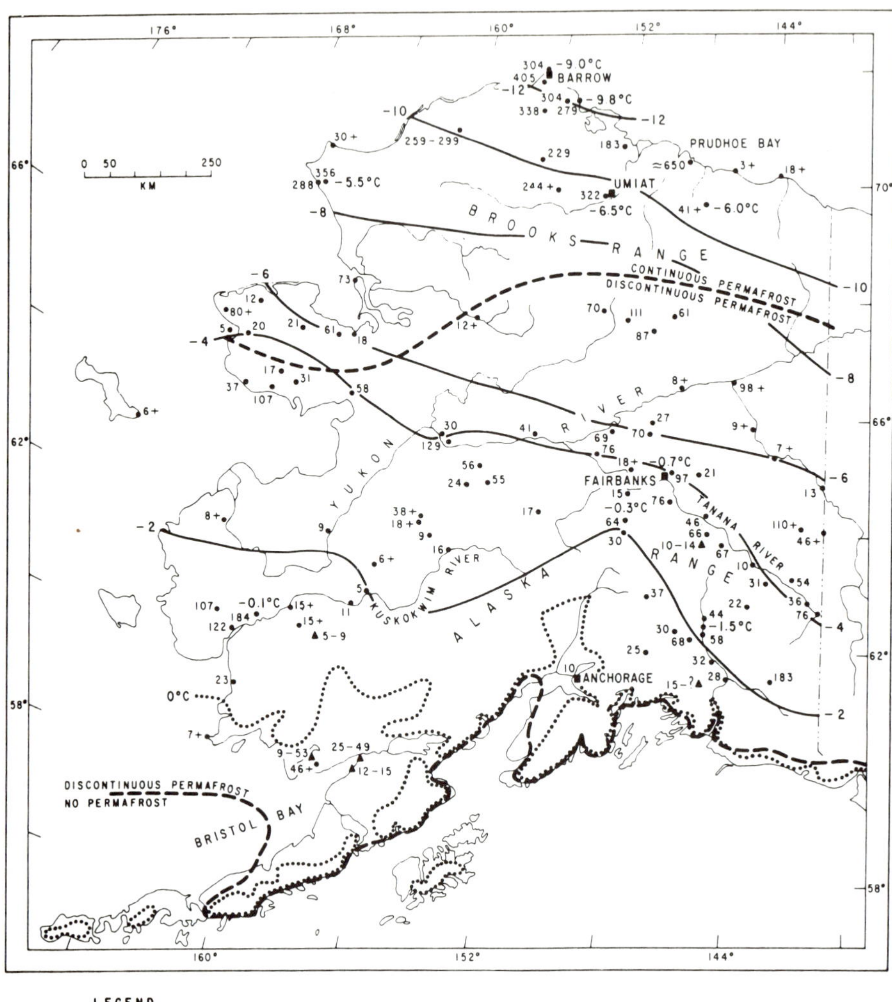

LEGEND

PERMAFROST, DATA IN METERS

122 • THICKNESS, DEPTH TO BASE

18+ • MINIMUM THICKNESS, BASE UNKNOWN

12–15 ▲ RELIC PERMAFROST, GROUND FROZEN BETWEEN DEPTHS SHOWN

-6.5°C • TEMPERATURE OF PERMAFROST AT 15 TO 25 METERS DEPTH

— — — PERMAFROST ZONE BOUNDARY

CLIMATE

········ APPROXIMATE POSITION OF MEAN ANNUAL AIR TEMPERATURE ISOTHERM, 0°C

——— MEAN ANNUAL AIR TEMPERATURE °C

Fig. 4.24. Permafrost in Alaska (adapted from Brown and Pewe 1973). Reproduced from *Permafrost: The North American Contribution to the Second International Conference,* 1973, with the permission of the National Academy of Sciences, Washington, D.C.

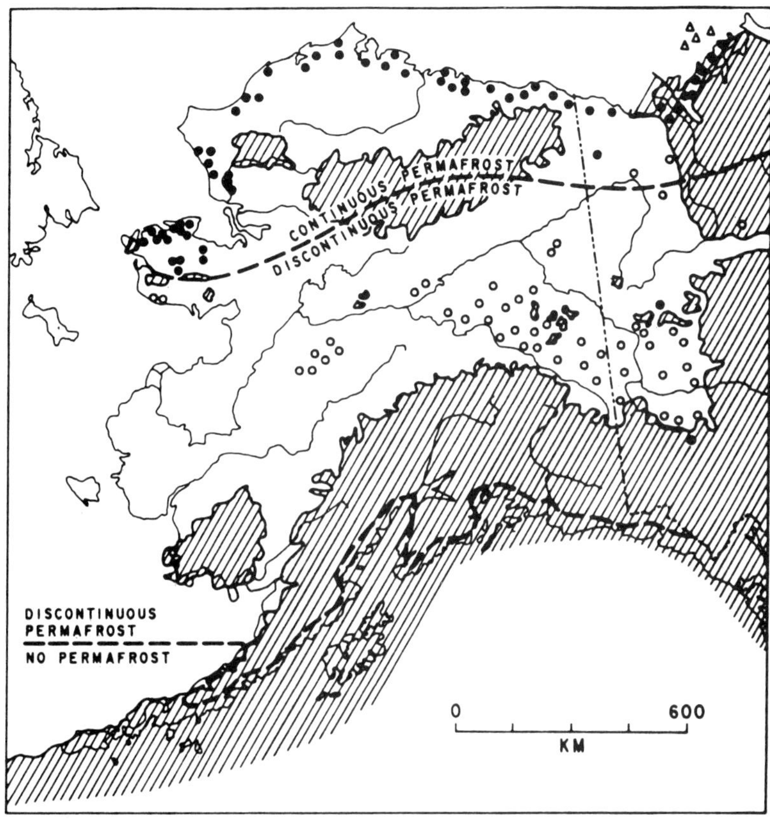

Fig. 4.25. Distribution of Open and Closed System Pingos in Relation to Permafrost Zones and Areas Covered by Late Wisconsinian Glacial Ice in North America (adapted from Brown and Pewe 1973). Reproduced from *Permafrost: The North American Contribution to the Second International Conference,* 1973, with the permission of the National Academy of Sciences, Washington, D.C.

cance. The most common locations are naturally the Rocky Mountains (Ives and Fahey 1971, Retzer 1965), but it also occurs on Mt. Washington, New Hampshire (Antevs 1932).

4.9.4 Subsea Permafrost

Since the presence of an unfrozen body of water will maintain a thawed region beneath the water, it was originally speculated that permafrost would not occur beneath the northern oceans. It is now well known, however, that permafrost does exist beneath the ocean, at least for relatively small distances from shore (Are 1978; Lewellen 1973; Chamberlain et al. 1978). Brown and Johnston (1964) suggest that permafrost exists beneath the ocean at Resolute, NWT, Canada, based on theoretical considerations.

Offshore permafrost can be due to relic permafrost left after the ocean has risen and submerged land areas, or it may actually be formed in situ.

The mechanism for the formation of permafrost under ocean water seems to have been first described by Dzens-Litovskii (Shvetsov 1969). Initially, the bottom sediments are saturated with saltwater at a fusion temperature of $T_1 < 0°C$ (Fig. 4.26). As ice forms, it forces the salt out of the ice into the layer just beneath the ice. This layer is dense and has a fusion temperature of $T_2 < T_1$. This cold, heavy water forms a layer of water with a temperature less than the fusion temperature of the bottom sediments. Conduction then causes freezing of the sediments, which can remain frozen for many years.

Relic permafrost can be explained with the help of simple heat transfer relations. The theory of transient temperature change beneath a large area, such as the ocean, is discussed in Chap. 7. The transient behavior of the 0°C isotherm, without considering phase change, is described after a sudden

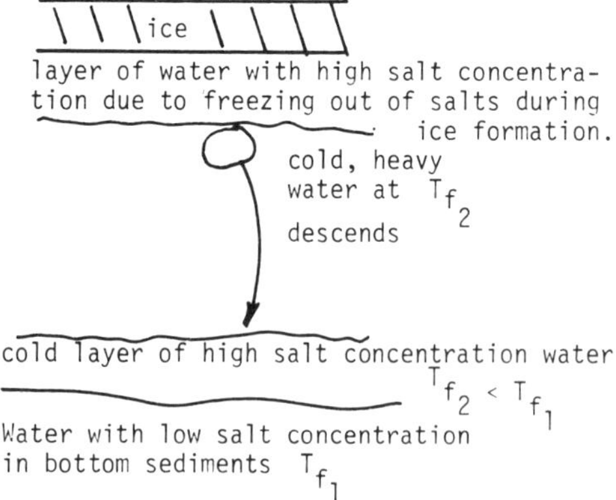

Fig. 4.26. Ocean Bottom Freezing.

submergence of a shoreline. For Pt. Barrow, Alaska, the following data have been used to examine the change in the permafrost:

Mean annual ground surface temperature (air temp.)	—9°C
Mean annual ocean bottom temperature	—0.5°C
Undisturbed depth of permafrost	1300 ft
Assumed thermal diffusivity of ground	.01 cm^2/sec

Figure 4.27 shows the shift of the permafrost (the 0°C isotherm) as calculated by the simple theory. After an infinite amount of time (new steady-state condition) the original 1300-ft depth of permafrost has decreased to about 70 ft, far out from shore. The ocean temperature is —0.5°C, and thus a layer of permafrost will be maintained indefinitely. Even 10,000 years after the assumed submergence, however, the 0°C isotherm is still far from equilibrium, and it is possible that even with an ocean temperature above 0°C, there will be slowly degrading relic permafrost, or steady-state permafrost, as can be seen in Fig. 4.28. This illustrates the slow changes that can occur in permafrost regions and the difficulty in interpreting field data. The values shown in Figs. 4.27 and 4.28 are not to be considered exact, but they should be representative of actual conditions.

4.10 ORIGIN AND EXISTENCE OF PERMAFROST

In discussing the present distribution of permafrost, questions often arise relating to the origin and age of the permafrost. These two concepts should

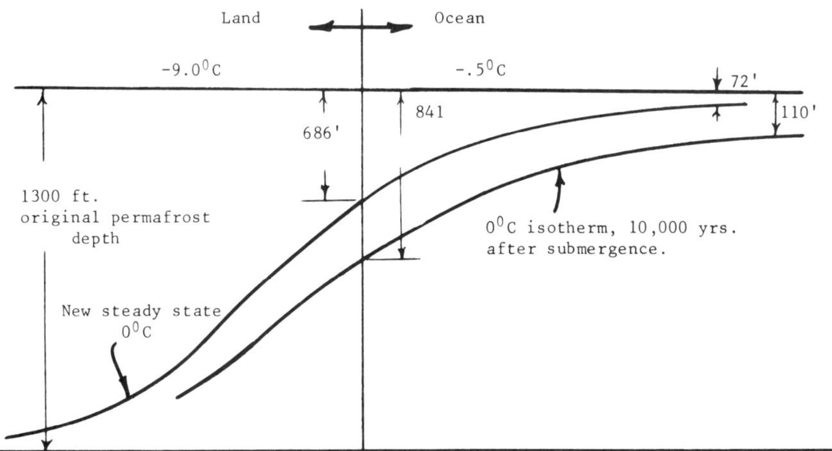

Fig. 4.27. Permafrost Disturbance due to Land Submergence, $T_s = -0.5°C$.

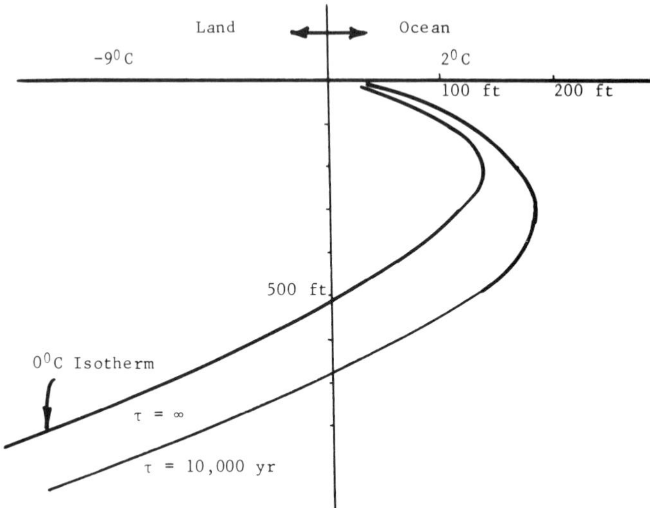

Fig. 4.28. Permafrost Disturbance under Ocean, $T_s = 2°C$.

be clearly differentiated since they deal with two separate phenomena. The age of a particular deposit of permafrost is the time that has elapsed since the freezing of the soil system. Actually it may be very difficult or impossible to determine this age because cycles of thawing and freezing may occur at long intervals and different frequencies in different regions of the earth. Thus, the age of two "similar" deposits of permafrost may be quite different. In this regard, the presence of preserved animal remains may be a reliable clue to the age of a deposit of frozen ground. It would not appear that the age of permafrost is a question of great significance to engineers, although it may be useful to paleontologists, paleobotanists, etc. The present thermal state of the permafrost: temperature, degradation, aggradation, etc., is of interest to engineers.

The origin of permafrost involves the question of the conditions under which it can form and grow. These same conditions will explain the present existence of permafrost at a given location. As the conditions for the origin of permafrost are dynamic, it is certainly possible that areas now lacking permafrost once had these underlying frozen strata and that the present area of permafrost could once have been thawed. In other words, the present existence of permafrost depends upon two things: the proper energy exchange conditions and the thermodynamic state of the permafrost mass itself. The first of these conditions has little if anything to do with past climatic conditions, and the second is a function of the complete thermal history of the permafrost and is thus related to past climate. In this sense, it is incorrect to describe permafrost simply as a legacy of the last great ice age as is stated in Canadian (1957). It is possible that some, perhaps most, permafrost had its origin at the beginning of the Pleistocene era (Brown 1964), but this should not imply that the intervening thermal conditions were without significance. Shvetsov (1959) quotes Tumel as follows "the problem of permafrost cannot be solved only from a consideration of paleoclimatic investigations." He also notes the recent formation of permafrost in newly deposited islands, and other investigations have recorded the recent appearance or disappearance of permafrost in Canada (Thie 1974). Permafrost deposits can be classed into two types: syngenetic, where the material is laid down continuously and freezes during gradual surface deposition, and heterogenetic, where the material is deposited before any freezing takes place.

From the above discussion, it should be clear that the formation and existence of permafrost are related to the present and historical conditions of energy exchange between the soil and the atmosphere. Nevertheless, it is not possible to state these conditions in a simple, precise manner that will allow one to define unique permafrost indicators. We shall attempt to look at the broad aspects of the energy exchange and leave the details for a later study.

Over a sufficiently long time span, the energy exchange will be periodic, and, averaged over a number of periods, the net energy flow for a given soil volume will determine its thermodynamic state. Dynamic equilibrium of the energy flows may exist such that the soil is perennially frozen or the thermodynamic state of the soil element may be varying due to an imbalance of the cyclic energy flows. The original formation of permafrost depends upon a net periodic (yearly) loss of energy from the soil volume that must persist for many, many years for the permafrost to attain great thicknesses. The maintenance of the present thermodynamic state of permafrost, requires only that the net energy flow averaged over a number of years be zero.

One of the difficulties in discussing the energy flows is the definition of the system to be analyzed. The system must be chosen so that it illuminates the physical processes occurring and does not combine them into an indistinguishable overall energy balance that is simple, but not instructive. Figure 4.29 shows a typical permafrost region, divided into a number of subsystems or layers. This particular model is chosen on the basis of thermal processes occurring periodically within the layers. Each layer will, in general, be nonhomogeneous with respect to properties. Table 4.10 describes each layer and, combined with Fig. 4.29 should explain the choice of subsystems. For example, layers 3 and 4 make up the active layer, and layers 5, 6, 7, and 8 form the permafrost volume. It should be noted that the boundaries of some subsystems, even if the permafrost is stationary in time, may be varying periodically over the time period of significance, typically 1 year. An energy balance

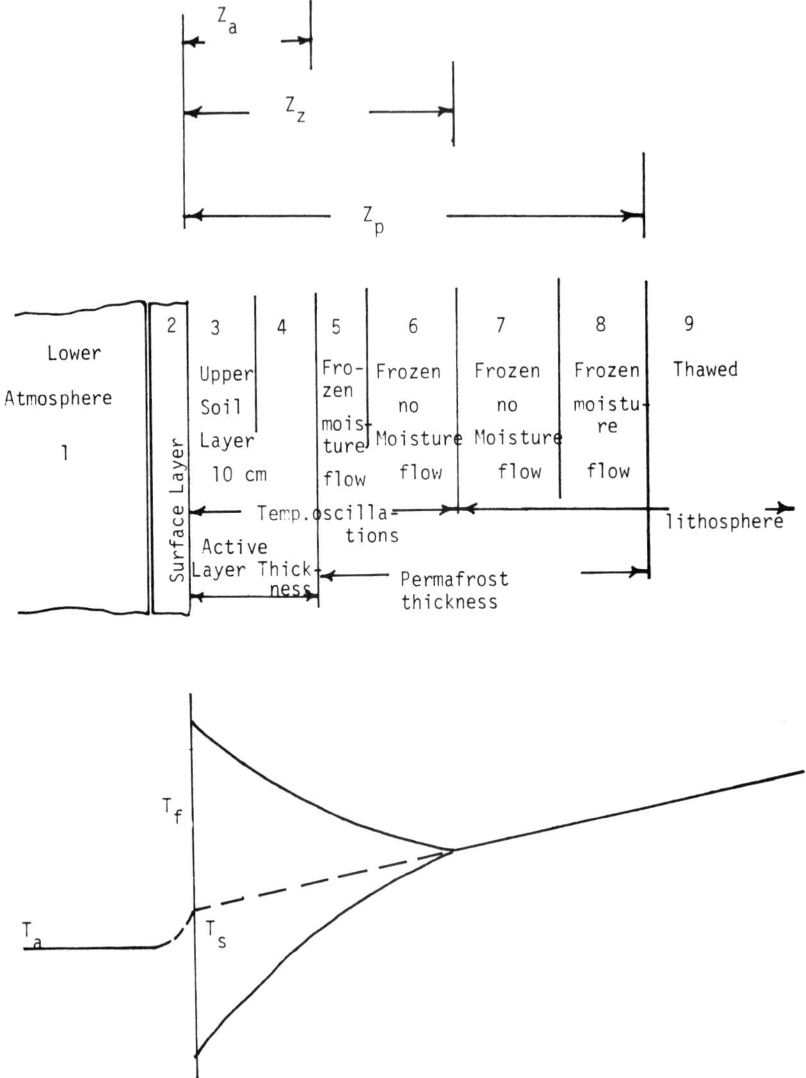

Fig. 4.29. Possible Energy Balance Layers in Permafrost Zone.

may be applied to each subsystem and will yield a set of nine coupled, nonlinear equations for the time- and space-varying temperature of each layer. Provided one knew all the required boundary and initial conditions, it would be theoretically possible to predict the complete thermal history of each layer and, hence, the existence of permafrost in a given locality. In fact, the bound-

Table 4.10. Description of Layers in Model.

Layer	Description of layer
1	Portion of lower atmosphere adjacent to surface layer
2	Contact layer between soil and atmosphere, may include vegetation, snow, etc. This layer is difficult to specify exactly for all possibilities.
3	Thin layer of soil adjacent to contact layer, arbitrarily 10 cm
4	When added to layer 3, this constitutes the active layer
5	Region of frozen soil in which water may be liquid
6	Frozen zone with no liquid water; 3, 4, 5, and 6 constitute the soil layer in which temperature oscillations occur
7	Frozen zone with no liquid water
8	Frozen zone with liquid water, same criteria as layer 5. Layers 5, 6, 7, and 8 constitute the thickness of permafrost
9	Thawed layer below permafrost—the lithosphere.

ary and initial conditions cannot be specified uniquely, and even if they could, the equations could not be solved completely. Nevertheless, it is worth examining some thermal details of the model.

The lithosphere is the region below the depth of zero, annual amplitude, comprising layers 7–9 in Table 4.10. In the lithosphere, the geothermal heat flux is dominant.

4.10.1 Geophysical Processes Involved in Energy Transfer

The physical processes significant in the energy transfer can be listed as follows:

1. Surface radiation—the net sum of all the radiant heat flows occurring at the surface, including solar shortwave radiation and the longwave radiation of the surface and the atmosphere. These terms make up the most significant energy flows in the permafrost regions.
2. Atmospheric convection—turbulent heat transfer between the atmosphere and the surface layer.
3. Moisture flow—this includes the energy flows associated with the humidity of the atmosphere, i.e., evaporation and condensation, as well as the movement of moisture in the various layers.
4. Phase change—this refers to the physical change of state of the moisture in any of the layers except the atmosphere, which already involves the phase change of evaporation-condensation.
5. Geothermal heat transfer—flow of energy by conduction and convection in layers below the zone of oscillating temperature, the lithosphere.

6. Conduction and convection—modes of energy transfer that may exist in any of the layers.

The likely energy modes for each subsystem are described in Table 4.11.

4.10.2 Geophysical Conditions Affecting Permafrost

In addition to the above-mentioned physical processes, other phenomena are significant in the composition, structure, spatial distribution, and thickness of the permafrost. In fact, all the processes and conditions interact to control the thermal state of the permafrost.

1. Physical location—latitude and longitude. This would be reflected in the circulation of air and water masses, area of the land, and distribution of land portions with respect to lakes, oceans, etc.
2. Height above sea level.
3. Relief of the land—will play a significant role in the movement of moisture.
4. Vegetation—a major characteristic of the surface or contact layer.
5. Snow cover—depth, density, time of appearance and disappearance. This will also be a character of the contact layer.
6. Cloud cover.
7. Surface water.
8. Soil properties—permeability, moisture content, etc.; will influence the thermal properties of the layers and the modes of energy transfer.
9. Groundwater.

4.10.3 Surface Energy Balance

The net radiation at the surface of the contact layer may be written as

$$Q_N = Q_{TC} - Q_R - R \tag{4.39}$$

Table 4.11. Possible Energy Modes in Layers.

Mode of energy transfer	Layer 1	2	3	4	5	6	7	8	9
Radiation	X	–	–	–	–	–	–	–	–
Conduction	–	X	X	X	X	X	X	X	X
Convection	X	X	X	X	X	–	–	X	X
Evaporation-condens.	X	X	–	–	–	–	–	–	–
Phase change	–	X	X	X	–	–	–	–	–
Moisture flow	–	X	X	X	X	–	–	X	X

where

Q_N = net radiative flux
Q_{TC} = direct and scattered incoming solar radiation
R = net longwave radiation between surface and atmosphere and
Q_R = reflected shortwave radiation.

If one neglects the thermal mass of the contact layer, then the energy balance (Fig. 4.30) can be written

$$Q_G = Q_N - Q_E - Q_H \tag{4.40}$$

where

Q_G = flux of energy to or from soil
Q_E = latent heat of evaporation and
Q_H = turbulent convection.

If the underlying permafrost is stationary, then over many years there will be a net heat flow out of the upper soil that will exactly balance the net heat input due to the geothermal heat flow. This quantity Q_G controls the rate of permafrost growth or decay. Unfortunately, the value of Q_G is small compared to the other terms in the equation, and this illustrates the difficulty of calculating the time required to form a given thickness of permafrost, since Q_G is imprecisely known and can be in considerable error. Methods for calculating the quantities in Eqs. (4.39) and (4.40) are given in Chap. 6.

4.10.3.1 Cooling and Heating Cycle of Upper Soil

Consider the heat flows that occur during the heating and cooling of the layer of soil with an oscillating temperature (layers 3–6, Fig. 4.29).
Cooling cycle

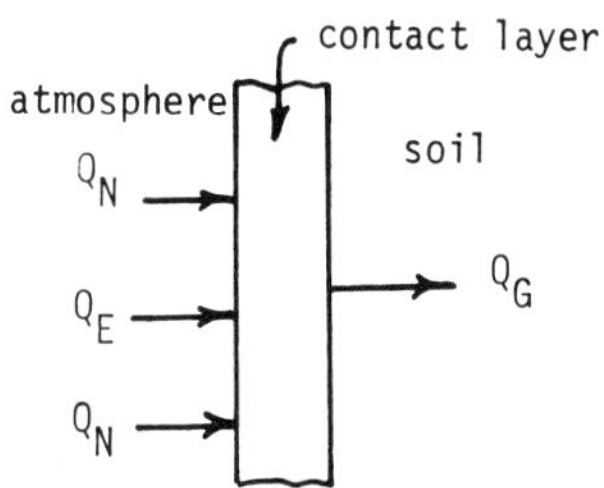

Fig. 4.30. Surface Energy Balance.

$$q_c = \int_0^h C_c \gamma_d (T_2 - T_1)\, dh - \int_0^h l\, W_c\, dh + q_g/2 \tag{4.41}$$

where

C = specific heat
γ_d = dry unit weight
T_1, T_2 = maximum and minimum yearly soil temperatures
h = thickness of soil layer
l = latent heat of fusion
W = water content of soil and
q_g = geothermal energy flow.

Heating cycle

$$q_h = \int_0^h C_h \gamma_d (T_1 - T_2)\, dh + \int_0^h l\, W_h\, dh + q_g/2 \tag{4.42}$$

Now the net flow of energy to the soil is

$$Q_G = q_h + q_c = \int_0^h \gamma_d (T_1 - T_2)(C_h - C_c)\, dh + \int_0^h l(W_h - W_c)\, dh + q_g \tag{4.43}$$

It is not necessarily true, but if $C_h = C_c$ and $W_h = W_c$, then $Q_G = q_g$. Thus, the net annual heat flow at the surface of the soil must equal the geothermal heat flow if the soil is to have a dynamic equilibrium condition.

The heat cycle is defined as

$$\Sigma_q = \int_0^h C_\gamma (T_1 - T_2)\, dh + \int_0^h l\, W\, dh + \int_0^\omega L\, e\, d\omega \tag{4.44}$$

In the high latitudes the value of Σ_q (negative) is greater than the inputs of flux (positive) due to insolation, convection, etc. Presumably, this means that the negative surface fluxes must balance these terms.

4.11 THERMAL REGIME OF PERMAFROST

From its very definition, permafrost is dependent upon the temperature or thermal regime of the ground. The thermal regime can be looked at from two viewpoints: the steady-state heat flow and the transient surface effects. Actually, these two phenomena are not independent, but they can be separated for discussion purposes.

4.11.1 Steady State

Figure 4.31 shows a mean temperature profile in a homogeneous system. Here G denotes the geothermal gradient, which is constant for a given locality over fairly long time periods. This gradient is associated with the steady flow of energy q_g from deep within the earth to the surface. The extrapolation of the geothermal gradient is taken as the mean surface temperature T_s. The temperature will vary periodically in the zone of nonzero temperature amplitude, but this will have little or no effect on q_g. Figure 4.31 denotes typical mean maximum and minimum monthly temperatures at various depths. The mean annual temperature at any depth would lie along the dotted line if the simple steady-state assumptions were valid. It can be seen from Fig. 4.31 that if T_s is greater than the freezing temperature of the ground T_f (usually taken as 0°C or 32°F), then no volume of ground will remain continually frozen, i.e., no permafrost will exist. There may be a zone of seasonal frost action, as is shown, or even this may not exist if the minimum temperature is always above T_f. Figure 4.32 denotes a typical temperature profile in a permafrost zone. Here the mean temperature is less than T_f, and, as can be seen, a layer of soil is at a temperature less than T_f continuously. Thus, the thermal condition for the existence of permafrost is satisfied. The depth to the permafrost layer is determined by the location at which the maximum temperature never exceeds T_f and likewise at the lower depth,

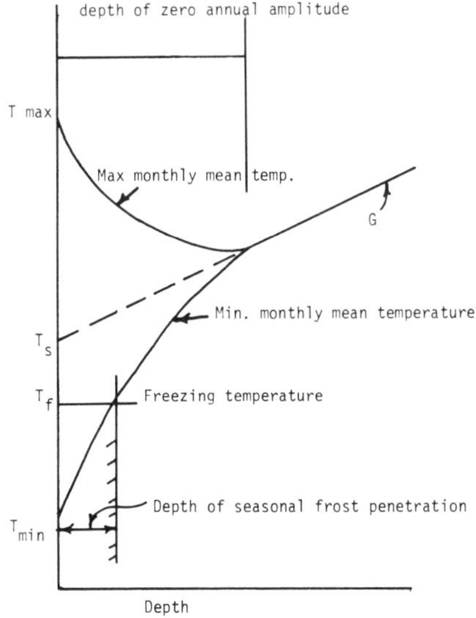

Fig. 4.31. Temperature Climate, No Permafrost.

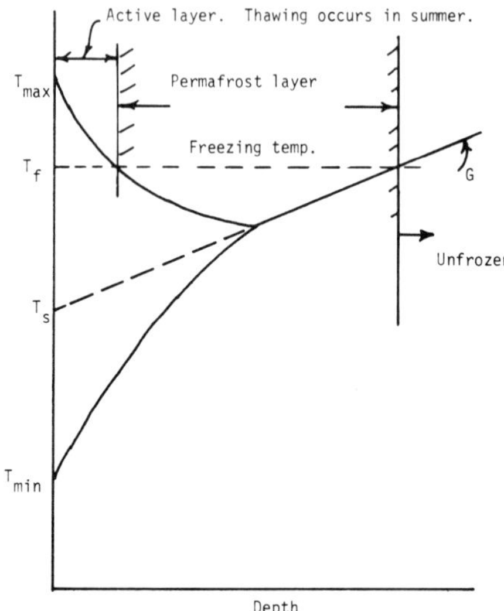

Fig. 4.32. Temperature Profile in Permafrost.

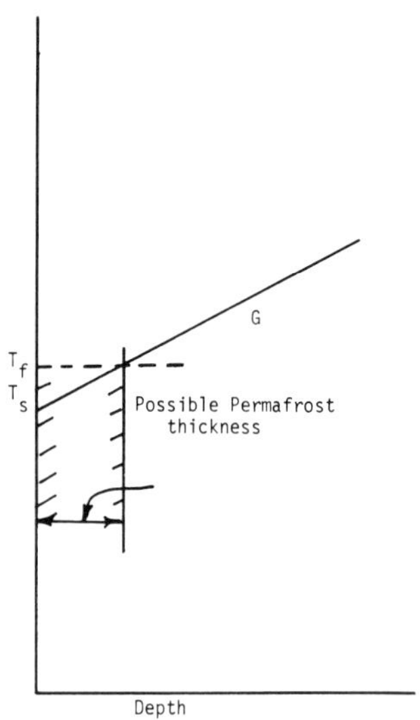

Fig. 4.33. Temperature Profile beneath Surface Ice Layer.

but here the geothermal gradient controls the location where the temperature equals T_f. Figure 4.33 shows a possible condition that might occur beneath an ice cap or glacier. The surface temperature T_s would likely be continuously less than T_f, and a relatively thin layer where the temperature is always less than T_f could exist. In this case, however, due to the ice pressure, T_f may be less than 0°C, and the water in the soil may exist as a liquid. This would still be defined as permafrost, but the removal of the ice overburden would leave little or no permafrost in the region. Another possible case could occur where the maximum surface temperature never exceeds T_f. This would result in permafrost with no active layer, but the conditions for this are rather remote, since radiation gains can increase the surface temperature above T_f even if the air temperature is always considerably below T_f. Figure 4.34 shows temperature variations with depth that might occur in permafrost regions.

4.11.1.1 Depth to the Bottom of Permafrost, Homogeneous System

The depth to the bottom of the permafrost layer is controlled by the steady-state (on a relative time scale) heat transfer from the deep layers to the surface.

Consider the simple system shown in Fig. 4.35, where the soil system is assumed to have constant thermal properties for $z \geq 0$. Neglect any effects

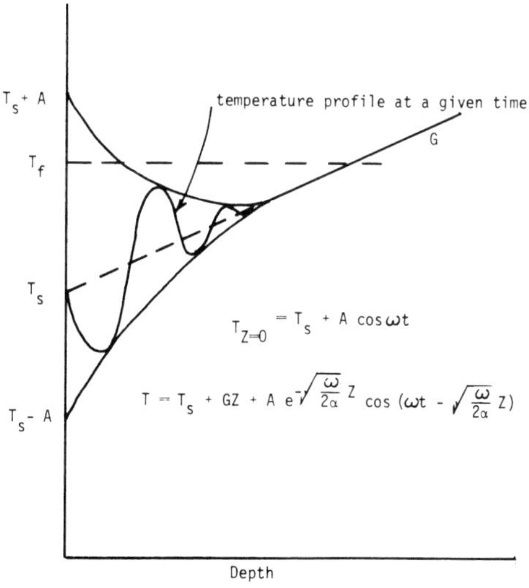

Fig. 4.34. Temperature Oscillation near Surface.

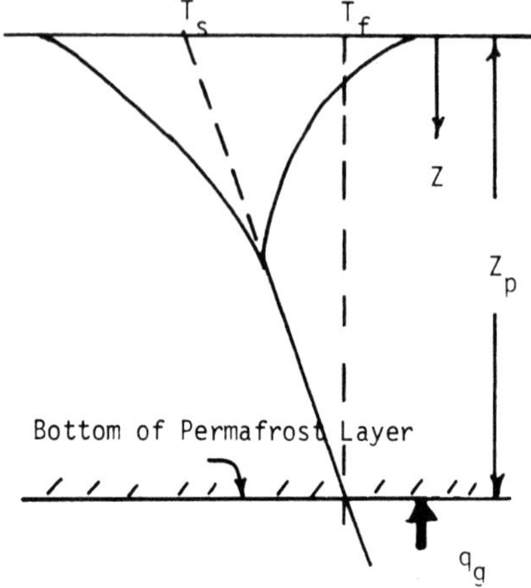

Fig. 4.35. Permafrost Thickness.

of phase change in the active layer. The deep earth heat flux, q_g, is a constant value, and the conduction equation is

$$q_g = -k\frac{dT}{dZ} \tag{4.45}$$

Integrating from the surface to an arbitrary depth Z gives

$$\int_{T_s}^{T} dT = -\frac{q_g}{k}\int_{o}^{z} dZ$$

Then

$$T - T_s = -\frac{q_g}{k} Z$$

and

$$T = -\frac{q_g}{k} Z + T_s \tag{4.46}$$

Now when the temperature equals the fusion temperature T_f, one has the distance Z_p to the bottom of the permafrost layer

$$T_f = -\frac{q_g}{k} Z_p + T_s$$

$$Z_p = (T_s - T_f)\frac{k}{q_g} \tag{4.47}$$

where q_g and T_s are negative.
If $T_f = 0°C$

$$Z_p = T_s \frac{k}{q_g} \tag{4.48}$$

The geothermal heat flux may also be written in terms of the geothermal gradient G as

$$q_g = -kG$$

Then the permafrost depth is

$$Z_p = \frac{T_f - T_s}{G}$$

One can calculate Z_p if the surface temperature and G are known.

Equation (4.48) shows that the depth to the bottom of the permafrost is controlled by the surface temperature, the deep earth heat flux, and the thermal conductivity of the permafrost. Thus, it is not simply related to the surface temperature unless k and q_g are identical for two locations. Any effect (volcanic, hot springs) that will increase the deep earth flow will decrease the permafrost thickness. Notice in Fig. 4.36 the striking effect on the permafrost thickness of a change in the mean surface temperature T_s. An increase in T_s will result in a degradation of the permafrost at both the upper and lower surfaces, although the larger effect is likely to occur at the bottom surface due to the sensitivity of Z_p to the surface temperature. From Eq. (4.48), at the same location, where q_g and k are constants

$$\frac{Z_p^1}{Z_p^0} = \frac{T_s^1}{T_s^0}$$

and, thus, a 10% increase in T_s will eventually result in a 10% decrease in the depth to the bottom of the permafrost.

The results are based on steady-state temperature profiles, and, although the surface temperature may change fairly quickly (on the order of 1°C in a century), it may take many centuries for the permafrost layer to adjust to the new surface conditions. Gold and Lachenbruch (1973) discuss some

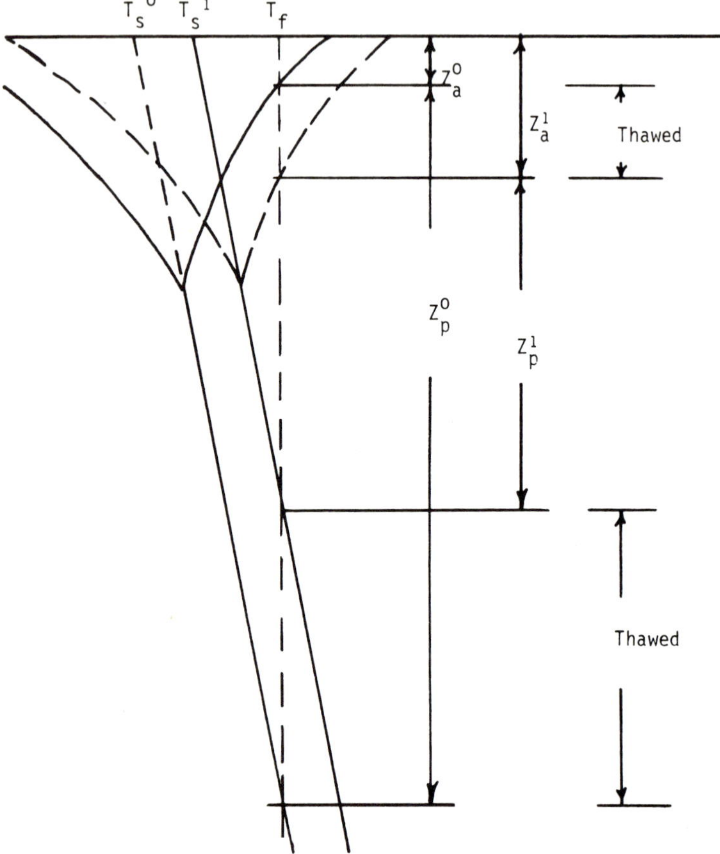

Fig. 4.36. Effect of Mean Surface Temperature on Permafrost Thickness.

of the calculation procedures in transient studies of this kind. It is possible that surface temperature changes are not significant as far as permafrost depth is concerned, if the change is of a periodic nature with a short period, say a century. Then the surface temperature fluctuations may never penetrate to the lower strata. Table 4.12 notes some temperature data for Canadian sites.

4.11.1.2 Nonhomogeneous System

It is likely that the ground will be stratified with zones of varying thermal properties. This case can be considered rather simply. For a simple series system (see Fig. 4.37), one may write

Table 4.12. Thickness and Temperature of Permafrost. Reprinted by permission from *Permafrost in Canada: Its Influence on Northern Development,* R. J. E. Brown, © University of Toronto Press, 1969.

Location	Thickness of permafrost, ft	Ground temperature °F at various depths, ft	Mean annual air temperature, °F
1. Aishihik, YT	50–100	28.3 (20)	24.5
2. Asbestos Hill, PQ	>900	19–20 (50–200)	17
3. Churchill, Man	100–200	27.5–28.9 (25–54)	19
4. Dawson, YT	200	—	23.6
5. Fort Simpson, NWT	40	35.4–33.2 (0–5)	25.0
6. Fort Smith, NWT	unknown	about 32 (15)	26.2
7. Fort Vermillion, Alta	nil	39.8–38.9 (0–5)	28.2
8. Inuvik, NWT	>300	26 (25–100)	15.6 (Aklavik)
9. Keg River, Alta	5	31–32 (5)	31
10. Kelsey, Man	50	30.5–31.5 (30)	25.5
11. Mackenzie Delta, NWT	300	23.8–26.5 (0–100)	15.6 (Aklavik)
12. Mary River, NWT-Baffin Island	unknown	10 (30)	6.3 (Pond Inlet)
13. Milne Inlet, NWT-Baffin Island	unknown	10 (50)	6.3 (Pond Inlet)
14. Norman Wells, NWT	150–200	26–28.5 (50–100)	20.8
15. Port Radium, NWT	350	—	19.2
16. Rankin Inlet, NWT	1000	15–17 (100)	11.2 (Chesterfield Inlet)
17. Resolute, NWT-Cornwallis Island	1300	10–8.5 (50–100)	2.8
18. Schefferville, PQ	>250	30–31.5 (25–190)	23.9
19. Thompson, Man	50	31–32 (25)	24.9
20. Tundra Mines Ltd., NWT	900	29 (325)	17
21. Uranium City, Sask	30	31–32 (30)	24

Table 4.12. Cont.

Location	Thickness of permafrost, ft	Ground temperature °F at various depths, ft	Mean annual air temperature, °F
22. United Keno Hill Mines Ltd., YT	450	28–29.3 (100–200)	24.2 (Elsa)
23. Winter Harbour, NWT-Viscount Melville Island	1500	—	—
24. Yellowknife, NWT	200–300	31.4 (40)	22.2

$$q_g = \frac{\Delta T_t}{R_t} \tag{4.49}$$

where

$$\Delta T_t = T_s - T_f$$

$$R_t = \sum_{i=1}^{n} \frac{d_i}{k_i} \tag{4.49a}$$

is the total thermal resistance.

The heat flux may be written in terms of a mean thermal conductivity as

$$q_g = \frac{\bar{k}\,\Delta T_t}{Z_p} \tag{4.50}$$

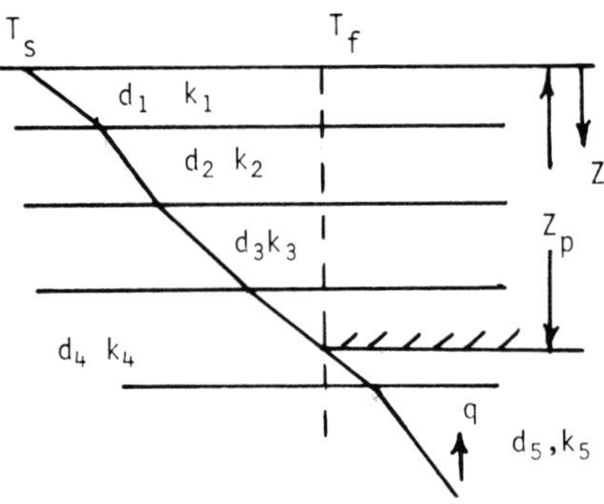

Fig. 4.37. Stratified Soil.

which, when equated to Eq. (4.49), yields the harmonic mean thermal conductivity

$$\bar{k} = \left\{ Z_p^{-1} \sum_{i=1}^{n} \frac{d_i}{k_i} \right\}^{-1} \quad (4.51)$$

The distance to the bottom of the permafrost is then

$$Z_p = \bar{k}\,\frac{(T_s - T_f)}{q_g} \quad (4.52)$$

Actually, this result is not very convenient to use for calculations, and one can solve Eq. (4.49) for R_t and then find

$$Z_p = \sum_{i=1}^{n} d_i \quad (4.53)$$

from Eqs. (4.49 and 4.50).

Consider a location (Prudhoe Bay) where $T_s = -11°\text{C}$

$$q_g = 1.4 \times 10^{-6}\,\text{cal/sec-cm}^2$$

$$d_1 = 640\text{ m}$$

$$k_1 = 3.2\,\frac{\text{kcal}}{\text{hr} - \text{m} - °\text{C}} \qquad Z \leq 640\,\text{m}$$

$$k_2 = 1.6 \qquad Z > 640\text{ m}$$

Note at this depth (640 m) the overburden pressure may be such that the ground temperature is less than 0°C, but the water is not frozen. Thus the value of k will drop to about one-half the value of the soil with ice. Then from Eq. (4.49)

$$R_t = \frac{11}{.0504} = 218.25\,\frac{\text{hr} - \text{m}^2 - °\text{C}}{\text{kcal}}$$

and from Eq. (4.49a)

$$\frac{d_1}{k_1} + \frac{d_2}{k_2} = R_t$$

Thus

$$d_2 = 29.2 \text{ m}$$

and

$$Z_p = d_1 + d_2 = 669.2 \text{ m}$$

It may be noted that, based on data from all the continents, the value of q_g does not vary very significantly, being about $1\text{–}2 \times 10^6$ cal/sec-cm².

4.11.2 Transient Relations

4.11.2.1 Equilibrium Time without Phase Change

The time for the temperature profile to reach a new equilibrium value, after a change in the surface temperature, can be estimated with no phase change (Fig. 4.38). Consider a semi-infinite system with an initial temperature given by the straight line

$$T^o = Gx + T_s^o \tag{4.54}$$

Initially, the surface temperature is T_s^o and the geothermal gradient is G. If the surface temperature suddenly changes to T_s^1, then the new steady-state temperature will be

$$T^1 = Gx + T_s^1 \tag{4.55}$$

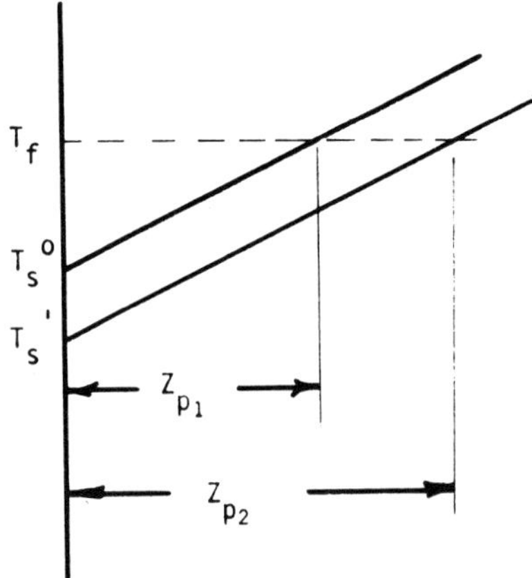

Fig. 4.38. Temperature Shifts.

How long will it take for the new temperature profile to be reached? This problem can be stated mathematically as follows. We must find the transient temperature such that

$$\frac{\partial^2 v}{\partial x^2} = \frac{1}{\alpha}\frac{\partial v}{\partial t} \tag{4.56}$$

$$v = Gx + V \qquad t = 0 \tag{4.56a}$$

$$v = 0 \qquad x = 0 \tag{4.56b}$$

$$\lim_{x\to\infty} \frac{\partial v}{\partial x} \to G \tag{4.56c}$$

where

$$v = T - T_s^1$$

$$V = T_s^o - T_s^1$$

The solution to this problem is given by Carslaw and Jaeger (1959) as

$$v = V \operatorname{erf}\left(\frac{x}{2\sqrt{\alpha t}}\right) + Gx$$

or

$$T = T_s^1 + (T_s^o - T_s^1) \operatorname{erf}\left(\frac{x}{2\sqrt{\alpha t}}\right) + Gx \tag{4.57}$$

Now the time required for the temperature at any depth x to reach within 1% of its steady-state value can be found from

$$\operatorname{erf}\frac{x}{2\sqrt{\alpha t}} = \frac{.01(Gx + T_s^1)}{(T_s^o - T_s^1)} \tag{4.58}$$

Let us consider some typical data for Inuvik, NWT.

$T_s^o = 25.5°F \qquad Z_{p1} = 344\ ft\ (104.9\ m) \qquad G = .0189°F/ft\ (.0344°C/m)$

The value of the thermal diffusivity of the soil will vary, but take a typical value of $\alpha = 13.14\ m^2/year$. Now if the surface temperature decreases suddenly by 1°F, we have

$$T_s^1 = 24.5 \qquad Z_{p2} = 397 \text{ ft } (121.0 \text{ m})$$

The following table notes the time required to reach equilibrium for different values of x, after a surface temperature change of 1°F.

Depth, m	Time, yr
10	38.4
50	795
100	2570
105	2773
121	3463

The effect of the thermal diffusivity may be noted in the following table at a depth equal to the original permafrost thickness of 104.9 m.

Thermal diffusivity, m^2/yr	Equilibrium time for different soils, yr
1.314	27,730
13.14	2773
131.4	277

4.11.2.2 Equilibrium Time with Phase Change

A simple estimate of the time to reach equilibrium, when phase change occurs, can be found as follows (Fig. 4.39). Initially the temperature is a straight line with gradient G and thickness of permafrost X^o. At $t = 0$, the surface temperature drops from T_s^0 to T_s^1, and after time Δt, the temperature is again a straight line with gradient G and permafrost thickness X^1. At any intermediate time the permafrost thickness is X, and if the growth of the permafrost is very slow, then the temperature of the frozen layer X will be given by a linear curve in x.

$$T = (T_f - T_s^1)\frac{x}{X} + T_s^1 \tag{4.59}$$

The difference between the conduction of energy to the surface and the geothermal heat flow must equal the latent heat given up in freezing a layer dX. Thus with the aid of Fig. 4.40

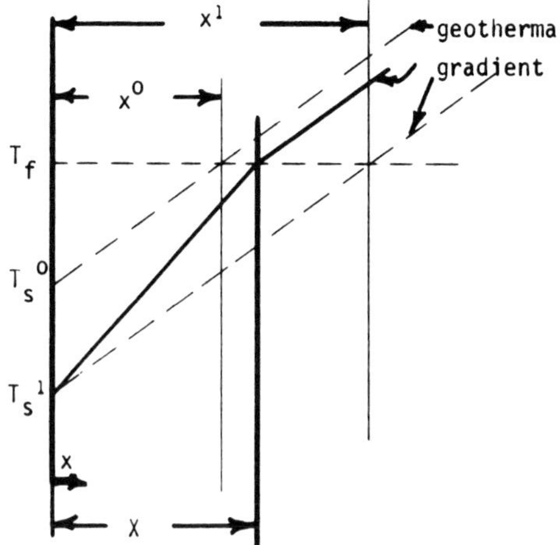

Fig. 4.39. Effect of Surface Temperature Change.

$$k_f \frac{\partial T}{\partial x} - k_u G = L \frac{dX}{dt} \tag{4.60}$$

is evaluated at $x = X$.

This neglects any sensible heat effects, and thus the actual time will be longer than this.

Assume that the thermal conductivities of the frozen and unfrozen layers have the same value k. Then, using Eq. (4.59) to solve for the temperature gradient of the frozen layer, one obtains a differential equation for X

$$\frac{L}{k} X \frac{dX}{dt} + GX = (T_f - T_s^1) \tag{4.61}$$

Solving this equation for the time for which the permafrost thickness is X/X^1 gives

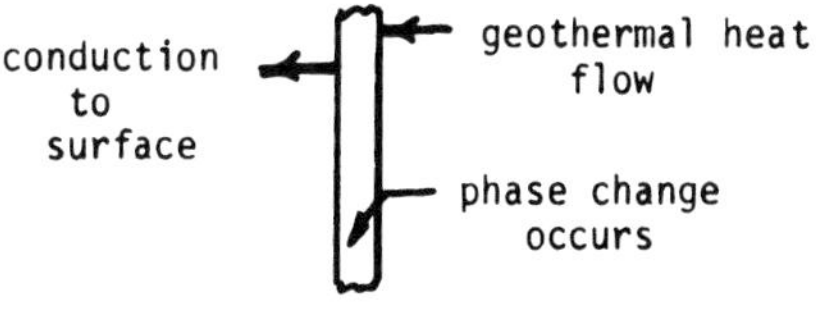

Fig. 4.40. Energy Balance at $x = X$.

$$\Delta t = \frac{LX^1}{Gk}\left\{\left(\frac{X^o}{X^1} - \frac{X}{X^1}\right) - \ln\left(\frac{1 - \dfrac{X}{X^1}}{1 - \dfrac{X^o}{X^1}}\right)\right\} \tag{4.62}$$

$$X^0 = \frac{T_f - T_s^1}{G} = 104.97 \text{ m}$$

$$X^1 = \frac{T_f - T_s^1}{G} = 121.12 \text{ m}$$

Again, considering the case of Inuvik, with

$$k = 0.3 \frac{\text{kcal}}{\text{hr} - \text{m} - {}^\circ\text{C}}$$

$$L = 5 \text{ cal/cm}^3$$

the following table gives the time for the permafrost to reach a given thickness.

$\frac{X}{X^1}$	Time, yr
.87	0
.9	1557
.95	5865
.99	16,379
1.0	∞

Of course, the time to reach equilibrium will not be infinite, this is simply the exponential nature of the approximate solution. The time, in this case, is greater than in the previous estimate. Clearly, the time is a function of the thermal properties, but these values are representative. One can surmise from these estimates that surface temperature fluctuations of short period, less than a century, will not affect the permafrost thickness significantly. It will require several centuries for the surface effects to penetrate deeply into the permafrost.

4.11.2.3 Transient Surface Effects

In the upper layers of the soil system, the thermal regime is closely controlled by the ground surface energy balance, with the deep earth flow playing a minor role. The energy fluxes at the surface are related to systematic diurnal and yearly periodicities that cause periodic temperature variations in the zone of temperature change. At any one location, the energy flux tends to display somewhat random variations from periodicity (annual) over a number of years.

A simple, homogeneous system with no phase change may be investigated. Let the surface temperature be a simple harmonic function of time with a period of 1 year. Then (see Chap. 7) the temperature at any time and depth is given by the well-known equation

$$T = T_s + Gz + T_a e^{-rz} \cos\left(\frac{2\pi t}{P} - rz\right) \tag{4.63}$$

where

T_a = amplitude of the surface temperature
t = time
z = depth
$r = \sqrt{\dfrac{\omega}{2\alpha}} = \sqrt{\dfrac{\pi}{\alpha P}}$
T_s = mean, annual ground surface temperature
P = period, 1 year
α = thermal diffusivity of soil and
G = geothermal gradient.

Figure 4.34 shows a sketch of this temperature. The upper and lower temperature envelopes denote the maximum and minimum yearly temperatures attained at any depth. The temperature oscillations tend to die out rather quickly with depth, on the order of 3 ft for diurnal variations and 50 ft for yearly fluctuations. Although the effect of the geothermal gradient is taken into account in Eq. (4.63), it is really not warranted for two reasons. In the fluctuating zone, the uncertainties in the approximations of the simple model are greater than the effect of the deep earth heat flow, and at lower depths we are not concerned with the oscillating temperatures. Thus, let us drop the geothermal gradient term and deal with

$$T = T_s + T_a e^{-rZ} \cos\left(\frac{2\pi t}{P} - rZ\right) \tag{4.64}$$

From Fig. 4.34, it can be seen that the curve of the maximum yearly temperature will delineate the depth to the top of the permafrost or the depth of the active layer. The maximum yearly temperature at any depth T_m will occur when the cosine term in Eq. (4.64) is equal to one; thus

$$T_m = T_s + T_a e^{-rZ} \tag{4.65}$$

When $T_m = T_f$, then the distance to the top of the permafrost, Z_a, is reached. Solving Eq. (4.65) for this depth gives

$$Z_a = \sqrt{\frac{\alpha P}{\pi}} \ln\left(\frac{T_a}{T_f - T_s}\right) \tag{4.66}$$

Consider an example for Inuvik, NWT, where $T_a/(T_f - T_s) \sim 2.6$ and $\alpha \sim 15$ cm²/hr, then $Z_a = 1.95$ m. This value is about double the measured active layer depths in Inuvik and would be modified considerably if phase change were important.

Figure 4.41 demonstrates that a change in the amplitude of the surface temperature has no effect on the depth to the bottom of the permafrost, but will change the active layer depth. This latter effect is also given by Eq. (4.66). The depth to the zero annual amplitude will also vary with a surface temperature amplitude change and can be calculated from

$$Z_z = \frac{1}{r} \ln\left(\frac{T_a}{D}\right) \tag{4.67}$$

where D is an arbitrarily small amplitude chosen to define Z_z. The theory does not allow $D = 0$ unless $Z \to \infty$, but this will not alter the results, since D may be chosen as .01 or .001, etc.

It should be kept in mind that the details of this section depend upon a simple, linear theory of heat transfer and will not be valid for other conditions such as phase change, nonconstant properties, and nonperiodic boundary conditions (periodic surface effects can be handled by Fourier series methods).

4.11.3 Ground Temperatures

It has already been noted that the ground temperature at the zero annual variation level will be greater than the mean annual air temperature. Brown (1963) has tabulated the difference between the mean annual air temperature and the ground temperature, and the results for Canadian stations are given in Table 4.13.

The temperature difference between the ground at Z_z and the air varies between 2 and 11°F, for a great range of climates, but this temperature

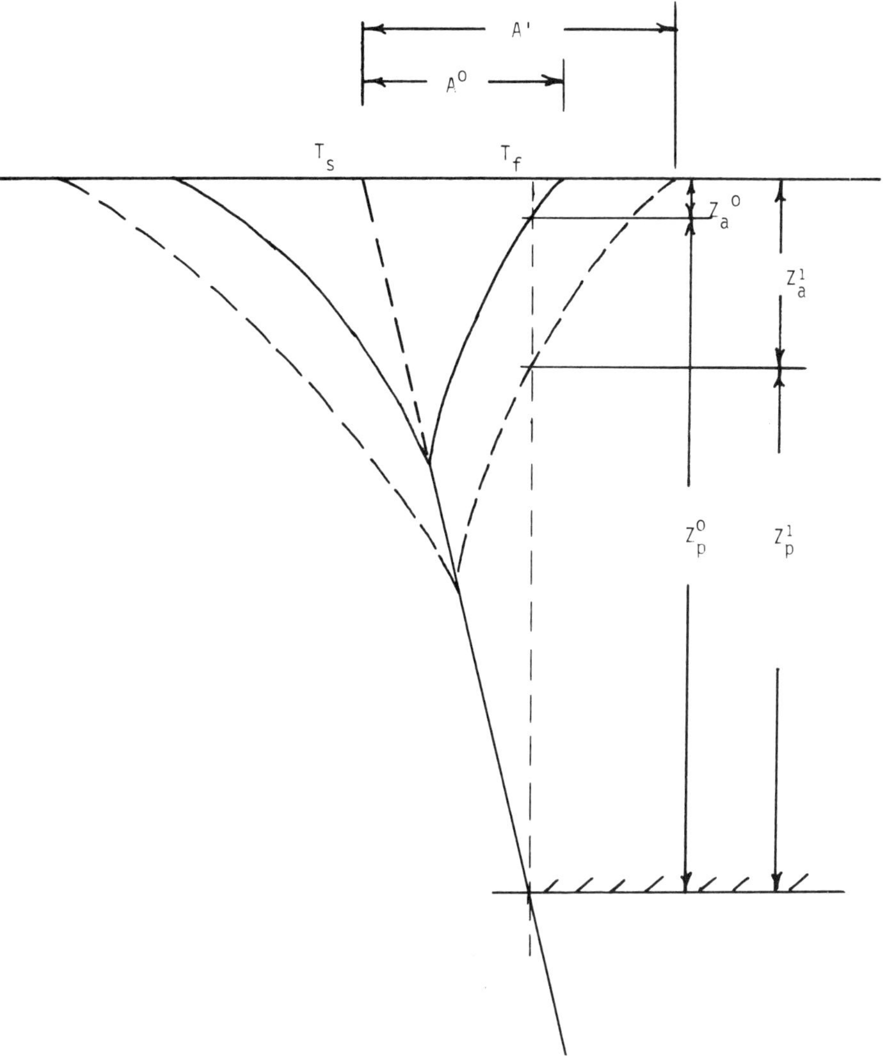

Fig. 4.41. Effect of Surface Temperature Amplitude on Permafrost Thickness.

difference does not correlate with the thickness of permafrost or mean annual air temperature. The simple theory shows, from Eq. (4.63), that the temperature difference is given by

$$\Delta T = GZ_z = G\frac{1}{r}\ln\left(\frac{T_a}{D}\right) = T_{Z_z} - T_s; \qquad T_s > T_a \tag{4.68}$$

Table 4.13. Ground Temperatures in Canada (from Brown 1963). Reproduced from *Permafrost International Conference,* 1963, with the permission of the National Academy of Sciences, Washington, D.C.

Location	Permafrost zone	Permafrost thickness, ft	Mean ann. air temp., °F	Ground temp., °F, and depth, ft	Air-ground difference, °F
Aishihik, YT	Discont.	50–100	24.5	28.3 (20)	4
Asbestos Hill, PQ	Cont.	>930	17	19–20(50–200)	2–3
Churchill, Man	C	100–200	19	27.5–28.9(25–54)	8–10
Fort Simpson, NWT	D(patchy)	40	25.0	35.4–33.2(0–4.9)	8–10
Fort Smith, NWT	None	–	26.2	32(15)	6
Fort Vermillion, Alta	None	–	28.2	39.8–38.9(0–4.9)	10–11
Inuvik	C	>300	15.6	26(25–100)	10
Kelsey, Man	D(patchy)	30	25.5	30.5–31.5(0–30)	5–6
Mackenzie Delta Lake	C	200–300	15.6	23.8–26.5(0–100)	8–11
Yellowknife, NWT	D	200–300	22.2	33–31.7(2.3–8.3)	9–11
Norman Wells, NWT	C or D	150–200	20.8	26–28.5(50–100)	5–8
Rankin Inlet, NWT	C	1000	11.2	15–17(100)	4–6
Resolute, NWT	C	1300	2.8	10–8.5(50–100)	6–7
Schefferville, PQ	D	>250	23.9	30–31.5(25–190)	6–8
Taurcanis, NWT	C	900	17	29(325)	12
Thompson, Man	D	50	24.9	31–32(0–25)	6–7
Uranium City, Sask	D	30	24	31–32(0–30)	7–8
Ottawa	None	–	41.6	49.6–48(0–4.9)	6–8

Now with the depth of $Z_z \sim 50$ ft, this could give $\Delta T \sim 1°F$. Thus, the ground surface mean temperature must differ from the mean air temperature in general. Some of the factors accounting for this are: net radiation, vegetation, snow cover, ground thermal properties, relief, slope, and subsurface drainage. These local effects can also account for the patchy conditions in the southern fringe of the permafrost zone.

Table 4.14. Selected Values of the Geothermal Gradient.

Location	dT/dx, °F/ft	Reference
Inuvik	.0189	W. G. Brown et al. (1964)
Resolute NWT (Can. Archipelego)	.0217	Misener (1955)
Point Barrow (oil well), Alaska	.0238	Brewer (1958*a*)
Point Barrow (beneath small lake)	.0189	Brewer (1958*b*)
Lena River Basin (Siberia)	.00309–.0139	Brown (1963)
Asbestos Hill, PQ	.031[a]	Samson and Tordion (1959)
Milne Inlet, Baffin Island, NWT	.088[a]	

[a] Data limited to 55 ft.

Table 4.14 lists some values for the geothermal gradient at a number of sites. These values depend strongly upon the thermal conductivity of the permafrost, as the geothermal heat flow is relatively constant. The high values at Milne Inlet may be due to marine effects or incomplete data.

4.12 THERMAL PROPERTIES OF PERMAFROST

As has been noted, permafrost is a thermal state of soil systems and as such will have the properties of soil systems at the appropriate temperatures.

Thermal properties are important in the study of heat transfer where temperature profiles, heat fluxes, energy balances, etc., are to be calculated or measured. For steady-state conduction problems, only the thermal conductivity, k, need be considered, though transient phenomena require the volumetric specific heat C as well; the thermal diffusivity, $\alpha = k/C$, arises naturally in transient problems. In those problems where phase change occurs, the latent heat of fusion, L, is also required. Thus, the most significant thermal properties to consider are k, C, α and L, only three being independent. The radiative properties, which are basically surface properties for soil systems, although often of great importance, will be considered later. It is important to emphasize the difference between thermal properties and thermal processes. A property is defined if the thermodynamic state of the system is fixed and the property may then be tabulated. The property, such as k, α, C_p, etc., will always be the same if the state of the system is invariant. Thus, given a conduction process with a given temperature gradient and mean temperature, one may calculate the heat transfer as a function of the thermal conductivity. If convection is significant, however, then the heat transfer is related to a nonproperty, usually called the "surface coefficient." One cannot tabulate properties involving a convection process, as the same thermodynamic state may involve a number of different values of the surface coefficient. This does not mean that convection is unimportant in soil systems, but that if there is significant fluid motion, then the pure conduction effects must be augmented by the convective heat transfer.

Quite often in geophysical heat transfer problems, one is interested only in approximate temperatures, and rather crude property values are acceptable; even in such cases, however, the best data available should be used. In other cases, considerable design uncertainty can arise due to inaccurate thermal data (W. G. Brown 1964). Frozen and unfrozen soil systems are examined here since even problems in permafrost regions, such as depth of thaw, require data on both conditions (Sanger 1969).

A soil system is a complex combination of solid earth materials, ice, water, and air that can be arranged in a vast number of ways. In addition, the solids may be of various sizes, shapes, and materials.

4.12.1 Thermal Conductivity

The conductivity (electrical or thermal) has been studied extensively, and a number of formulae have evolved. Let us look at some simple relations first.

4.12.1.1 Simple Geometries

Consider Fig. 4.42, where two phases are in series. Now for such a system it is well known that

$$q = k\,A\frac{\Delta T}{d} = \frac{A\,\Delta T}{\dfrac{d_s}{k_s} + \dfrac{d_a}{k_a}} \tag{4.69}$$

The mean conductivity follows directly as

$$k = \frac{k_s k_a}{x_a k_s + x_s k_a} \tag{4.70}$$

$$x_a + x_s = 1.0$$

It is easy to show that when the phases are such that only parallel heat flow occurs

$$k = (x_a k_a + x_s k_s) \tag{4.71}$$

Equations (4.70) and (4.71) represent the minimum and maximum values, respectively, of the conductivity of the mixture. Equation (4.70) is the weighted harmonic mean of the conductivities, and Eq. (4.71) is the weighted arithmetic mean.

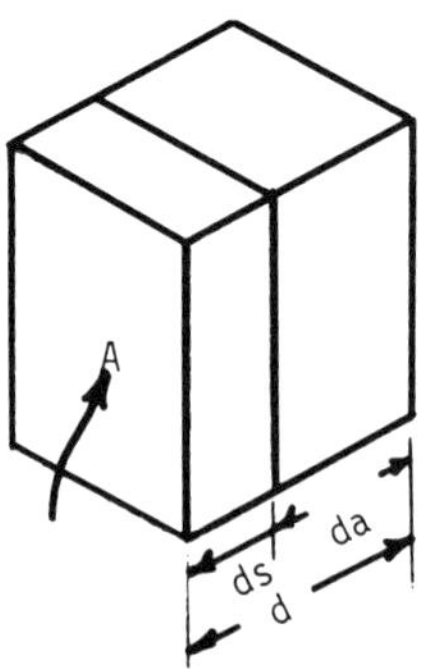

Fig. 4.42. Series Conduction.

An intermediate value of k is given by the weighted geometric mean written as

$$k = k_a^{x_a} k_s^{x_s} \tag{4.72}$$

which corresponds to the weighted arithmetic mean of the logarithms of the separate conductivities. This relation, though simple, can be used with good accuracy for complicated systems. The above results can be extended to any number of constituents.

Examine the case of a saturated soil. The void volume is completely filled with a liquid, and the thermal conductivity relations can be written in terms of the porosity and the ratio of the solid to liquid conductivities, σ

$$\frac{k}{k_w} = \frac{\sigma}{(1-n) + n\sigma} \qquad \textit{series} \tag{4.73}$$

$$\frac{k}{k_w} = n + (1-n)\sigma \qquad \textit{parallel} \tag{4.74}$$

$$\frac{k}{k_w} = \sigma^{1-n} \qquad \textit{geometric mean} \tag{4.75}$$

These equations are plotted on Fig. 4.43, which shows the significant effect of geometry upon the thermal conductivity.

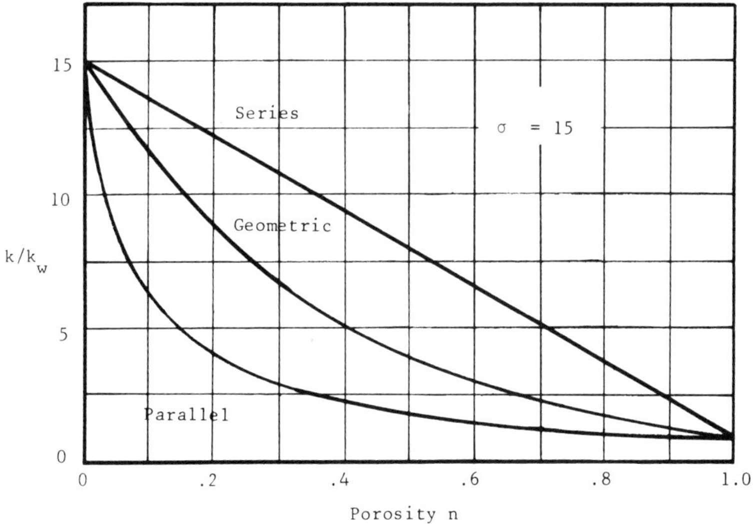

Fig. 4.43. Effect of Geometry on Thermal Conductivity of Saturated Soil.

4.12.1.2 Complex Configurations

Consider a system consisting of a mixture of two substances with thermal conductivities k_0 and k_1, where zero is the continuous medium. The heat flux as usual is

$$q = -k\left(\frac{dT}{dx}\right)$$

where k is intermediate between k_0 and k_1. It was shown theoretically by Burger (1915) that

$$k = \frac{x_o k_o + x_1 k_1 \lambda_1}{x_o + \lambda_1 x_1} \tag{4.76}$$

where λ depends upon the average temperature gradients in the continuous and discontinuous media and the ratio of their conductivities. Equation (4.76) is an extension of and reduces to the theoretical relation for the electrical conductivity of a random distribution of solid spheres in a continuous medium (Maxwell 1904), see Fig. 4.44.

DeVries (1963) applied Eq. (4.76) to soils, and if there are several types of granules with different shapes then

$$k = \frac{\sum_{i=0}^{N} \lambda_i \, x_i \, k_i}{\sum_{i=0}^{N} \lambda_i \, x_i} \tag{4.77}$$

$$\lambda_i = \frac{1}{3} \sum_{j=1}^{3} \left[1 + \left(\frac{k_i}{k_o} - 1\right) g_j\right]^{-1} \tag{4.78}$$

where N is the number of types of granules. Now for ellipsoids of axes a_i, a_j, a_k

Fig. 4.44. Complex Geometry.

$$g_i = \frac{1}{2} a_i a_j a_k \int_0^\infty \frac{du}{(a_i^2 + u)^{3/2} (a_j^2 + u)^{1/2} (a_k^2 + u)^{1/2}} \tag{4.79}$$

The quantities g_i depend only on the shape of the ellipsoid, not its size, with

$$g_a + g_b + g_c = 1 \tag{4.80}$$

Some values for practical shapes are given in Table 4.15, calculated with Eq. (4.79). Equation (4.78) can be written explicitly for special soil systems, such as a highly organic soil. Consider the case of a system made up of elongated cylinders in which air or water is the continuous medium.

Case I: Water is the continuous medium, $x_a < 0.5$. Note here that, referring to Fig. 4.45

$$g_a = g_b = \frac{1}{2}, \; g_c = 0$$

Table 4.15. Geometric Shape Factors for Soil Solids.

Shape	g_a	g_b	g_c
Sphere	$\frac{1}{3}$	$\frac{1}{3}$	$\frac{1}{3}$
Elongated cylinder, circular cross-section	$\frac{1}{2}$	$\frac{1}{2}$	0
Lamellae, with $\frac{b}{a} = \frac{c}{a} \to \infty$	1	0	0

Fig. 4.45 Organic Soil with Continuous Water.

Then from Eq. (4.77)

$$k = \frac{k_w x_w + \lambda_s k_s x_s + \lambda_a x_a k_a}{x_w + \lambda_x x_s + \lambda_a x_a} \tag{4.81}$$

and from Eq. (4.78)

$$\lambda_s = \frac{1}{3}\left[\frac{4}{\left(1 + \frac{k_s}{k_w}\right)} + 1\right] \tag{4.82}$$

$$\lambda_a = \frac{1}{3}\left[\frac{4}{\left(1 + \frac{k_a}{k_w}\right)} + 1\right] \tag{4.83}$$

Equation (4.81) can be rewritten using the soil parameters introduced earlier

$$k = \frac{\left(\frac{k_w}{\lambda_a} - k_a\right)\frac{\gamma_d W}{100} - \left(\frac{\lambda_s}{\lambda_a} k_s - k_a\right)\frac{\gamma_d}{\rho_s} - k_a}{\frac{\gamma_d W}{100}\left(\frac{1}{\lambda_a} - 1\right) + \left(\frac{\lambda_s}{\lambda_a} - 1\right)\frac{\gamma_d}{\rho_s} + 1} \tag{4.84}$$

Case II: Air is the continuous medium, $x_a > 0.5$

$$k = \frac{k_a x_a + \lambda_s k_s x_s + \lambda_w k_w x_w}{x_a + \lambda_s x_s + \lambda_w x_w} \tag{4.85}$$

$$\lambda_w = \frac{1}{3}\left[\frac{4}{\left(1 + \frac{k_w}{k_a}\right)} + 1\right] \tag{4.86}$$

Equation (4.85) can also be written as

$$k = \frac{\gamma_d\left(\frac{W}{100} + \frac{1}{\rho_s}\right)(\lambda_{sw} k_{sw} - k_a) + k_a}{\gamma_d\left(\frac{W}{100} + \frac{1}{\rho_s}\right)(\lambda_{sw} - 1) + 1} \tag{4.87}$$

where

$$k_{sw} = \frac{\frac{W}{100} k_w + \frac{k_s \lambda_s}{\rho_s}}{\frac{W}{100} + \frac{\lambda_s}{\rho_s}}$$

$$\lambda_{sw} = \frac{1}{3} \left(\frac{4}{1 + \frac{k_{sw}}{k_a}} + 1 \right)$$

In these equations the apparent thermal conductivity of the air in the voids, k_a, is calculated as described in Section 4.12.1.4.

4.12.1.3 Saturated Granular Soil

A combination of series and parallel paths can be envisaged as shown in Fig. 4.46. Then

$$k = \frac{a\ k_s k_w}{(1-d)\, k_s + a k_w} + b k_s + c k_w \tag{4.88}$$

Woodside and Messmer (1961) using data, estimated

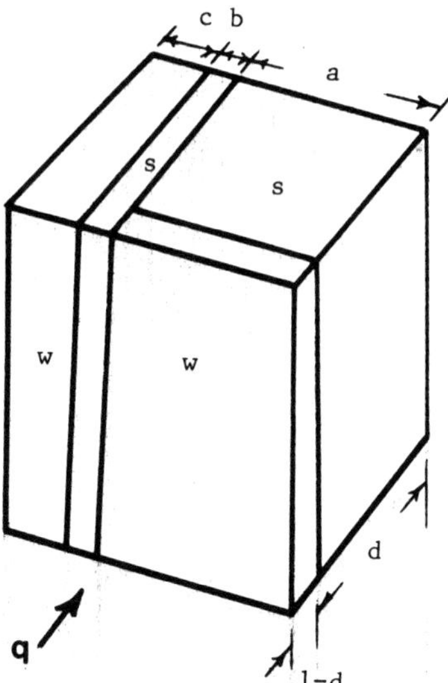

Fig. 4.46. Combination of Parallel and Series Heat Flow.

$$a = 1 - c$$
$$b = 0$$
$$c = n - 0.03$$
$$d = (1 - n)/a$$

Then

$$k = (n - 0.03)\,k_w + (x_s + 0.3)\left[\frac{x_s}{x_s + .03}\frac{1}{k_s} + \frac{.03}{x_s + .03}\frac{1}{k_w}\right]^{-1} \tag{4.89}$$

McGaw (1969) has shown that this semiempirical relation fits data better than most other proposed equations. He proposed a relation for the thermal conductivity of the system shown in Fig. 4.47

$$k = f(n)\,k_w + f(x_s)\,k_{gs} \tag{4.90}$$

where

$f(n), f(x_s)$ = functions of the series path through liquid alone and through the liquid and solid and

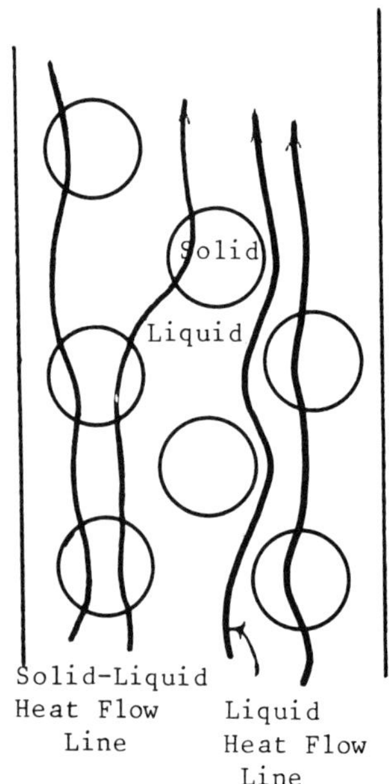

Fig. 4.47. Heat Flow Paths in Saturated Granular Soil.

k_{gs} = effective thermal conductivity of the material in the solid-liquid series path.

Assuming a contact resistance exists at each interface (Fig. 4.48), one has

$$k_{gs} = \epsilon k_s \left(1 + \frac{f}{s}\right)^2 \left(1 + \frac{V_f}{V_s}\right)^{-1} \left[1 + \sigma \frac{V_f}{V_s}\left(\frac{f}{s}\right)^2\right]^{-1} \tag{4.91}$$

where

$\epsilon \equiv 1 - (\Delta T_i / \Delta T)$
ΔT_i = temperature drop associated with the solid-liquid interfaces
f, s = length of series path across the fluid and solid respectively and
V_f, V_s = volume of the fluid and solids in the series path

McGaw (1969) made an arbitrary assumption about the volumes and lengths of the flow paths such that

$$\frac{f}{s} = \frac{V_f}{V_s} \equiv \frac{n_c}{x_s} \tag{4.92}$$

where n_c — volume fraction of fluid utilized as a series path. This will be a small part of n for saturated systems.

Then Eq. (4.91) is

$$k = (n - n_c)\, k_w + (x_s + n_c) \frac{\epsilon\, k_s (x_s + n_c)}{x_s + \sigma\, n_c} \tag{4.93}$$

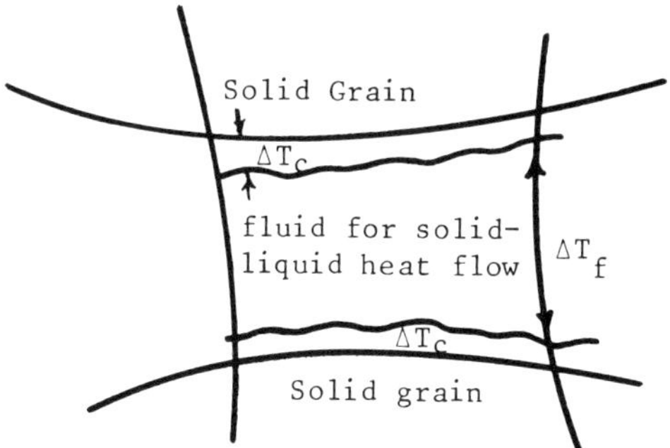

Fig. 4.48. Interface Contact Resistance.

The values of n_c and ϵ will vary with n, for each system but McGaw (1969) plotted data to show

$$n_c = \begin{cases} 0.2\ nx_s & \text{water} \\ 0.14\ nx_s & \text{oil} \end{cases}$$

The value of ϵ will normally be close to, but less than, one. The basic ideas for this model are physically sound, but the details have not been established.

4.12.1.4 Conductivity of Void Material

If the sample contains moisture, then the thermal conductivity of the gas-filled voids may be increased due to the heat flux associated with a moisture flux across the gas voids. Krischner and Rohnalter (1940) give the value of the apparent conductivity of the voids due to the transport of latent heat as

$$k_{ap} = k_a + k_v \tag{4.94}$$

where

k_{ap} = apparent conductivity of the void
k_a = conductivity of the gas in the void and
k_v = conductivity due to the vapor flux.

$$k_{vs} = \frac{LDP}{R\theta\,(P - P_{ws})} \frac{dP_{ws}}{d\,\theta} \tag{4.95}$$

k_{vs} = value of k_v for saturated vapor
L = latent heat of vaporization of water
R = gas constant of water
D = diffusion coefficient of water vapor in air
P = total pressure
P_{ws} = saturation vapor pressure and
θ = temperature.

Equations (4.94) and (4.95) do not consider any sensible heat effects. The diffusion coefficient is given by

$$D = \frac{17.6}{P} \left(\frac{\theta}{273}\right)^{2.3} \tag{4.96}$$

$D = \text{cm}^2/\text{sec}$
$P = \text{mm Hg}$ and
$\theta = {}^\circ\text{K}$.

If the air in the void is not saturated, Philip and DeVries (1957) suggest that

$$k_v = \phi \ k_{vs} \tag{4.97}$$

be used where ϕ is the relative humidity of air in void (as a fraction).

4.12.1.5 Effect of Radiation in Voids

Radiation across a void will increase the apparent thermal conductivity of the void. Consider a simple estimate as follows (Fig. 4.49). The total heat flux across the void, omitting convection, may be written as

$$k_a \ A \frac{T_2 - T_3}{d_g} + \sigma \ A(T_2^4 - T_3^4) = k_{ar} \frac{A(T_2 - T_3)}{d_g} \tag{4.98}$$

If the temperature difference across the void is not large

$$k_{ar} = k_a + 4 \, \sigma \ T^3 \, d_g \tag{4.99}$$

Then the ratio of the apparent thermal conductivity of the void is

$$\frac{k_{ar}}{k_a} = 1 + \frac{4 \, \sigma \ T^3 \, d_g}{k_a} \tag{4.100}$$

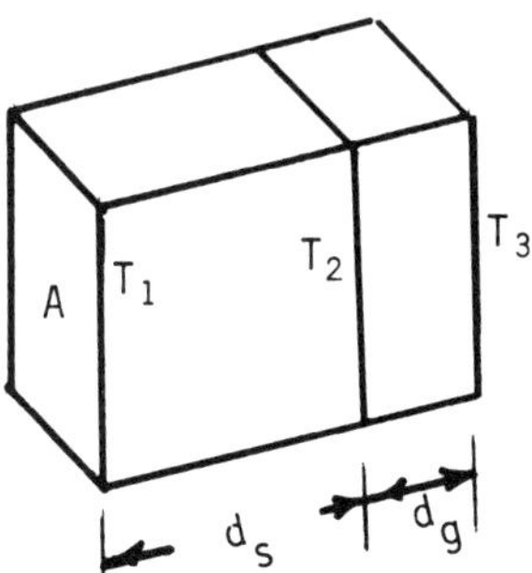

Fig. 4.49. Radiation across Pores.

Now if the average void gap is about 1 mm, then $(k_{ar}/k_a) = 1.26$ at 23°C and 1.12 at $-$ 40°C. Thus, the small effect of radiation can be neglected at low temperatures typical of permafrost problems.

4.12.2 Specific Heat

The specific heat of soils is rather difficult to measure accurately, especially if phase changes occur. If phase changes do occur, the measured quantity is called the "effective specific heat" and includes a latent heat effect. Unless one has detailed data on the effective specific heat, it is better to consider the phase change effect separately; this is usually a relatively large effect.

For moist soils the specific heat may be calculated from the weight fractions of the mixture constituents. Thus

$$mc = m_s c_s + m_w c_w + m_a c_a + m_i c_i \tag{4.101}$$

where

m, c = mass and specific heat of mixture and
s, w, a, i = refer to solids, water, air, and ice, respectively.

Replacing the mass with the densities and volumes

$$\rho c = \frac{1}{V}\,(m_s c_s + m_i c_i + m_w c_w + m_a c_a) \tag{4.102}$$

Now using the mass variables γ_d and W, one has

$$\rho c = C = \gamma_d\left(c_s + \frac{W_i}{100}\,c_i + \frac{W_w}{100}\,c_w\right) + \rho_a c_a\left[1 - \gamma_d\left(\frac{1}{\rho_w G_s} + \frac{W_w}{100\rho_w} + \frac{W_i}{100\rho_i}\right)\right] \tag{4.103}$$

where $\rho c = C$ is the volumetric specific heat of the soil system. The density of air is small, and the second term on the right of Eq. (4.103) may be dropped, and the general equation is

$$C = \gamma_d\left(c_s + \frac{W_i}{100}\,c_i + \frac{W_w}{100}\,c_w\right) \tag{4.104}$$

4.12.3 Thermal Diffusivity

Since the thermal diffusivity is simply a combination of the specific heat and thermal conductivity, it can be evaluated easily once these are known. The relation for the thermal diffusivity is

$$\alpha = \frac{k}{C} \tag{4.105}$$

4.12.4 Latent Heat of Fusion

Unlike the other properties, the latent heat of fusion is basically a function of the amount of water present and the latent heat of fusion of the water. Thus, the same relation will hold for all soil systems if one neglects any secondary effects of contamination. The equation is

$$L = \frac{\gamma_d \, W \, h_{if}}{100} \tag{4.106}$$

where

L = Latent heat of fusion of the soil system and
h_{if} = latent heat of fusion of water at 0°C. This value is taken as 79.71 cal/g.

Equation (4.106) assumes that all the liquid water freezes at 0°C, which is reasonable for coarse-grained and possibly organic soils, but certainly not for fine-grained soils. With this reservation, the latent heat of fusion is plotted as Fig. 4.60 for all soil systems. The ratio of unfrozen to frozen water content can be used for fine-grained soils if this seems warranted; see Section 4.7.

4.12.5 Specific Data for Soil Types

Fortunately, for purposes of thermal data, the permafrost can be classified into three broad types following the usual soil classification.

4.12.5.1 Organic Soils

This soil is a combination of minerals, partially decomposed vegetable matter, and humus. Most peaty soils are largely of an organic nature, with little mineral composition. These soils are of particular interest since North America contains at least 500,000 mi² of muskeg (MacFarlane 1969), much of which lies in the permafrost zones. A particular kind of organic soil called "Fairbanks Peat" was investigated by Kersten (1949) and described as "a fibrous, brown peat from the vicinity of Fairbanks, Alaska," with few further physical details. It probably falls into the fine-fibrous classification of Radforth (1969). This soil, though not necessarily typical of all organic soils, should possess thermal properties that will be a useful guide in heat transfer work.

The temperature of the frozen material is −3.89°C for all the results presented here, as this is the temperature used by Kersten (1949) in his measurements. It is not critical to use different temperatures because the temperature effect is of secondary importance for the properties, as long as the material remains either frozen or unfrozen. Care must be taken near the transition point from liquid to solid, where a knowledge of the liquid-solid phases is necessary to obtain accurate thermal data. The data of Kersten (1949) and Skaven-Haug (1963) are plotted as Fig. 4.50*a* for frozen organic material.

The unfrozen organic material was measured by Kersten at a temperature of 4.44°C. Again, Kersten's data were correlated, but they covered a very small range of the variables and could not be meaningfully extrapolated (Lunardini 1971). The thermal conductivity can, however, be calculated, using the details outlined earlier. When the volume fraction of air is less than or equal to 0.5, Eq. (4.84) was used to calculate the thermal conductivity, and when $x_a > 0.5$, Eq. (4.87) was applicable. The calculations used

$k_s = 0.216$ kcal/(hr-m-°C), thermal conductivity of solid material, DeVries (1963)
$k_w = 0.4838$, thermal conductivity of water
$\rho_s = 1.36$ g/cm³ assumed density of the organic solid material and
$k_a =$ thermal conductivity of voids.

The apparent thermal conductivity of the air voids is a combination of the actual thermal conductivity of air and the heat flux due to vapor movement in the voids. Equation (4.94), applied to this system, yields a value for k_a of .04473. When the water fraction is less than 0.1, the heat flux due to vapor movement is neglected, and k_a is 0.0210 (the value for air).

These equations are plotted as Fig. 4.50*b*. When the theoretical values were compared with the available data, it was found that the agreement was excellent, and since the theoretical values cover a much more significant range, they may be used with some confidence. The unfrozen peat will have a value of k of about 0.5 for the entire region above the $k = 0.5$ line (essentially the value for water).

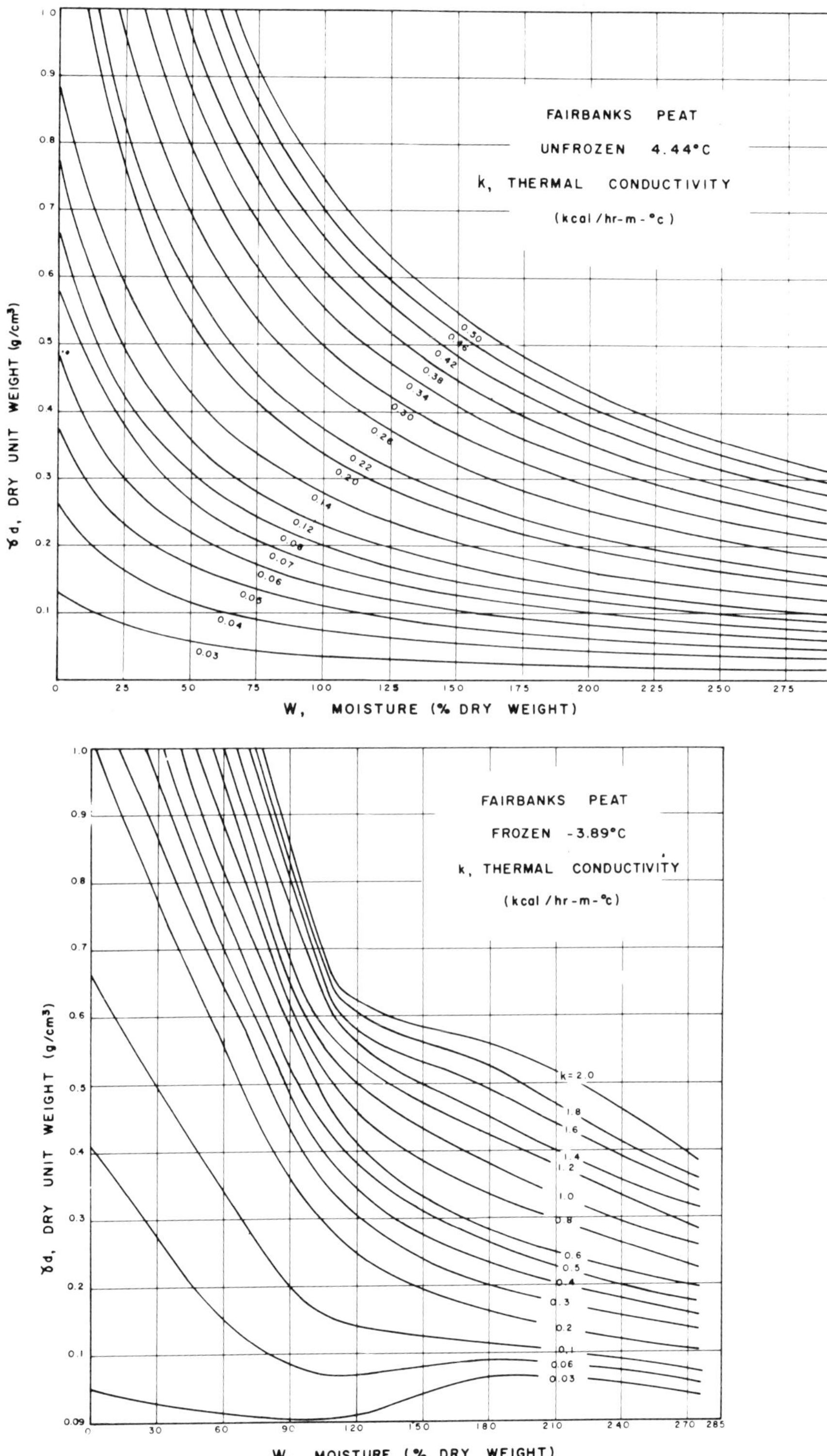

Fig. 4.50. Thermal Conductivity of Peat.

DeVries (1963) notes that the specific heat of organic soil matter is about 0.46 cal/(g—°C), and since c_i = .4963, at −3.89°C, then from Eq. (4.104)

$$C = \gamma_d(0.46 + .004963W) \tag{4.107}$$

is the specific heat for the frozen peat, and the specific heat of the unfrozen peat is

$$C = \gamma_d(0.46 + .0100403\ W) \tag{4.108}$$

Kay and Goit (1975) give a relation for the specific heat of a dry sphagnum peat as

$$c = 1.15 \times 10^{-3}\,T - .0245 \tag{4.109}$$

where

c = cal/(g—°C) and
T = °K

At −3.89°C, this gave a value of 0.290, though at 4.44°C, the value is 0.295. These values are much less than those quoted earlier, but at the high water content of peat soils the specific heats of Eqs. (4.107) and (4.108) will be changed by less than 20%.

Equation (4.107) is plotted as Fig. 4.51*a*, and Eq. (4.108) is plotted as Fig. 4.51*b* for the ranges of density and moisture of interest. With the equation of thermal conductivity and specific heat as noted, the values of α can be determined and are shown in Fig. 4.52. Note that the diffusivity of unfrozen peat shows remarkably little variation over a wide range of density and moisture and is much less than the values for frozen peat; thus there will generally be a slower rate of temperature change in the unfrozen peat.

4.12.5.2 Coarse-Grained Soils

The data of Kersten (1949) are used for the thermal conductivity of these soils. These data appear to be the best available for a large range of the variables γ_d and W, even though the soils are reconstituted and do not represent the in situ soil matrix. The equations used represent values for a soil composed of an average mineral content and can depart ±25% from the data for a particular mineral soil, although, in most cases, the accuracy is better than this.

Figure 4.53 is a plot of the thermal conductivity. These results agree

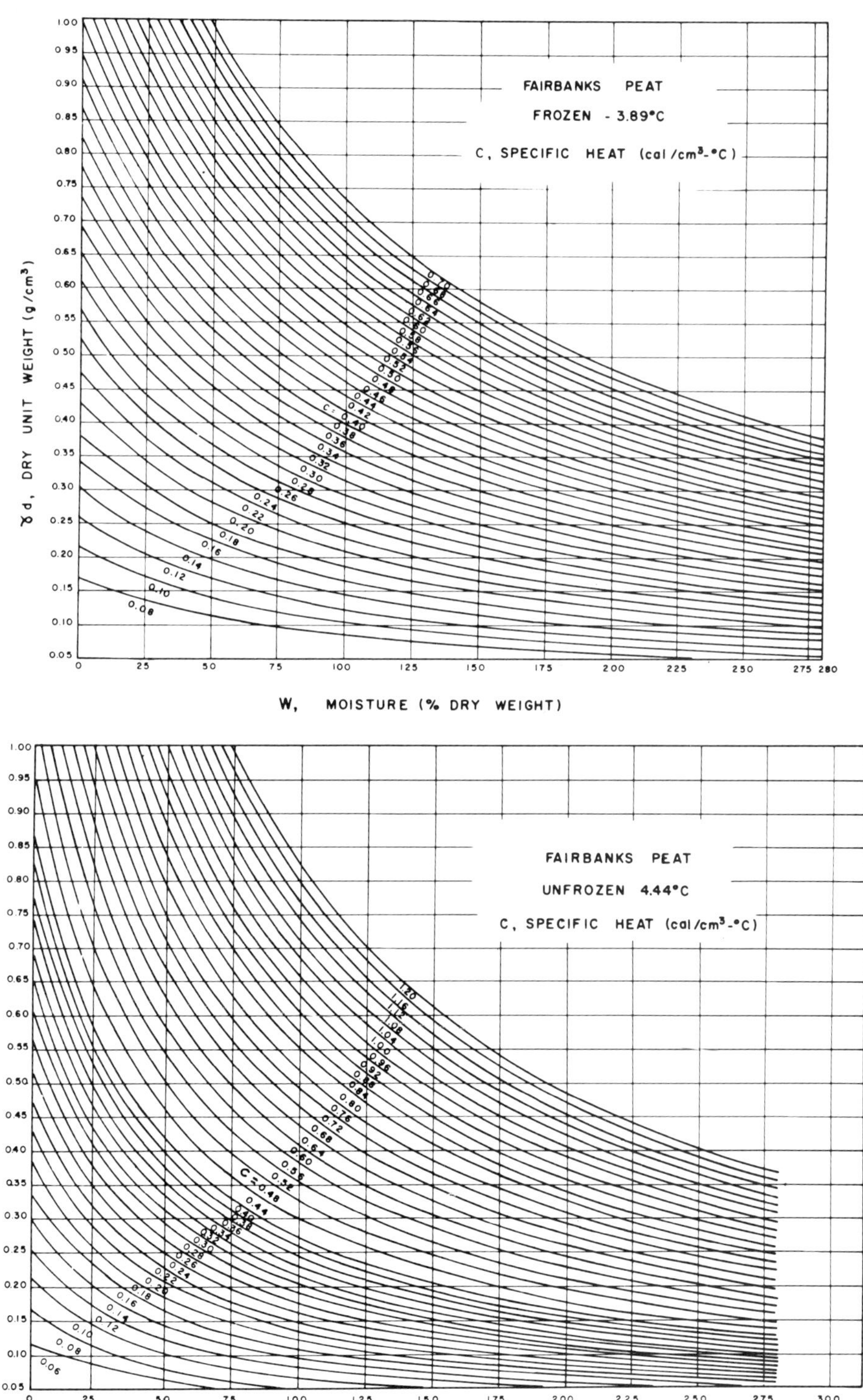

Fig. 4.51. Specific Heat of Peat.

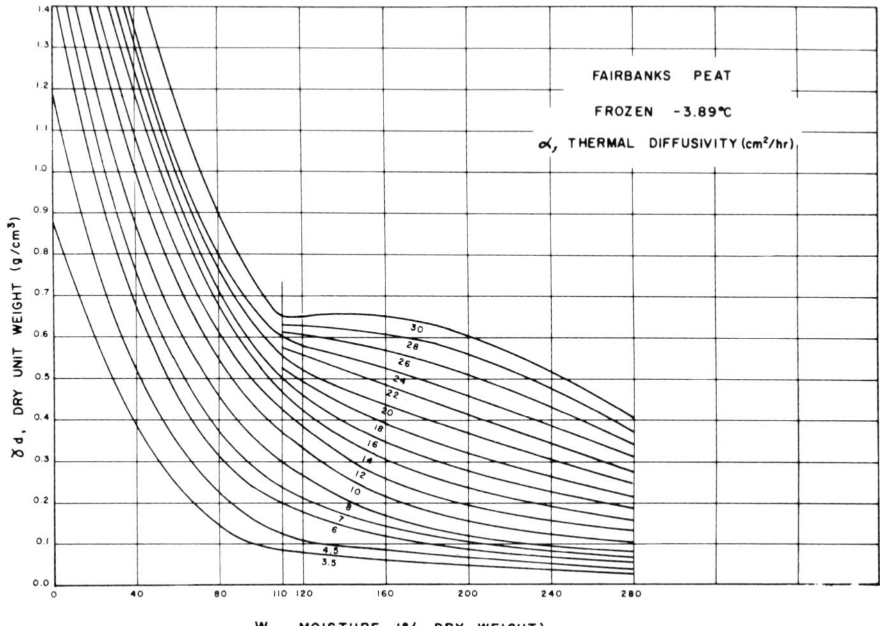

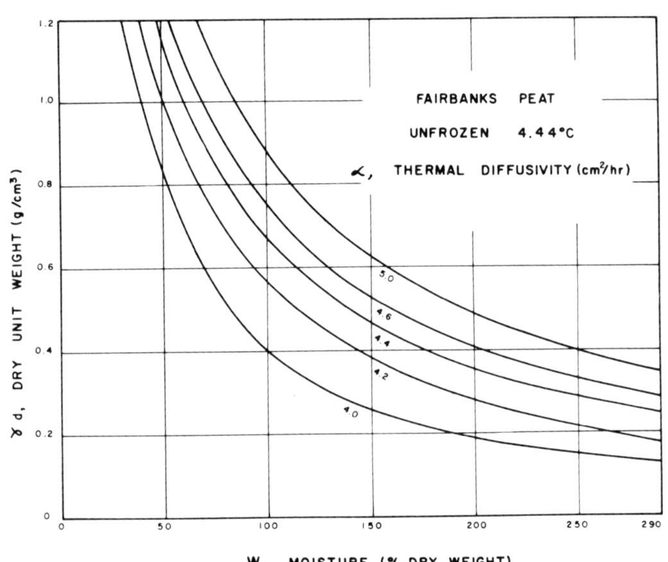

Fig. 4.52. Thermal Diffusivity of Peat.

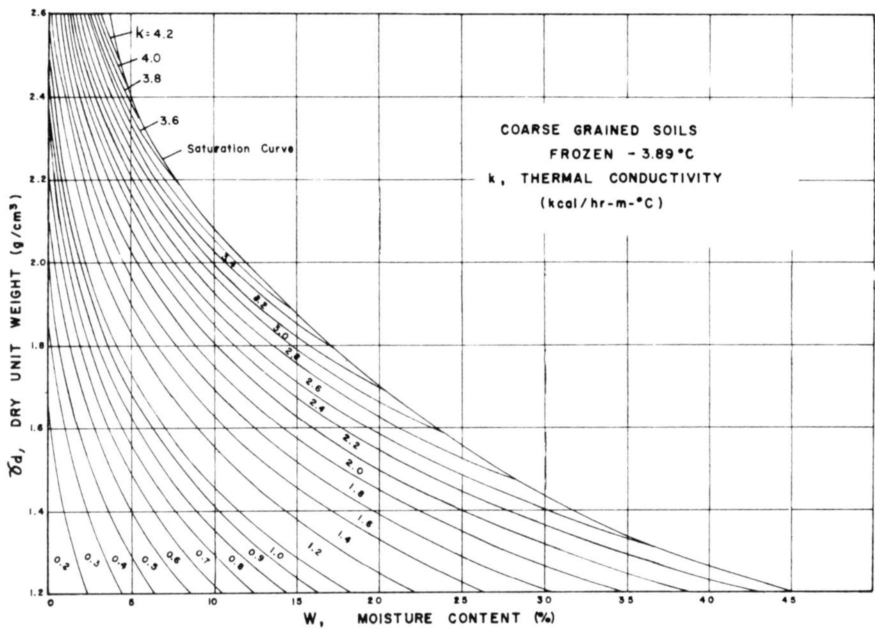

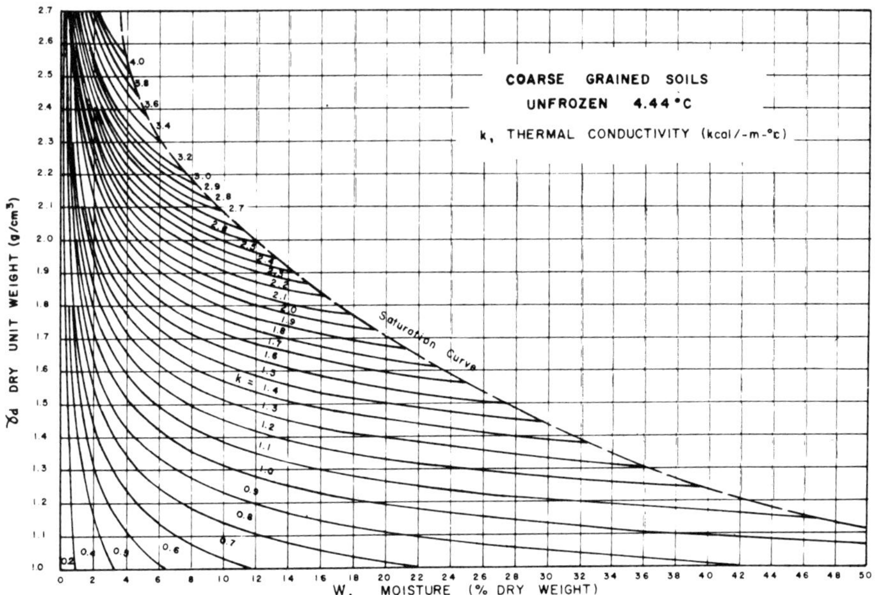

Fig. 4.53. Thermal Conductivity of Coarse-Grained Soils.

fairly well with Russian data, as reported by Temkin (Soviet 1960); Chudnovskii 1959), where the soils are classified as sandy and clayey. In the unfrozen case, the two sources of data are within 10%, except at low moistures where the difference can be as high as 20%, and the frozen state data are within 20% except at low moistures. These results are in line with the variation in the thermal conductivities of pure minerals.

The specific heat of the solid material has been obtained by Kersten, DeVries, and others. There is a small difference between the specific heats of the mineral solids for coarse- and fine-grained soils.

The following average equation may be used for the soil solids

$$c_s = 0.16 + 2.14 \times 10^{-4}\, T \quad 0 < T < 140°\text{F} \tag{4.110}$$

At $T = -3.87°\text{C}$, the specific heat of the solid is 0.1616 cal/(g−°C), and the specific heat of the frozen soil is

$$C = \gamma_d(0.1616 + .004963\, W) \tag{4.111}$$

At $T = 4.44°\text{C}$, $c_s = 0.1652$, and for the unfrozen soil

$$C = \gamma_d(.1652 + .0100403\, W) \tag{4.112}$$

Equations (4.111) and (4.112) are plotted in Fig. 4.54, and the results agree well with the Russian data mentioned earlier, since the calculation method is essentially the same.

The thermal diffusivity is plotted in Fig. 4.55. It is interesting to note the difference in the form of these curves. The unfrozen state has double values of α, and the frozen data are single-valued.

4.12.5.3 Fine-Grained Soils

Again, Kersten's data are used to give average values for this type of soil. The data are plotted in Fig. 4.56 for the thermal conductivity. Here the agreement with the Russian data mentioned earlier is within 15% for the unfrozen and frozen states.

In the unfrozen state the conductivity of the coarse soil can be double that of the fine soil, but in the frozen state the coarse soil has 25–50% higher values. As the ratio of the conductivity of quartz to clay is about 3, these results are reasonable.

The specific heat of the solid material is taken from Eq. (4.110),

$$C = \gamma_d(.1692 + .004963\, W) \tag{4.113}$$

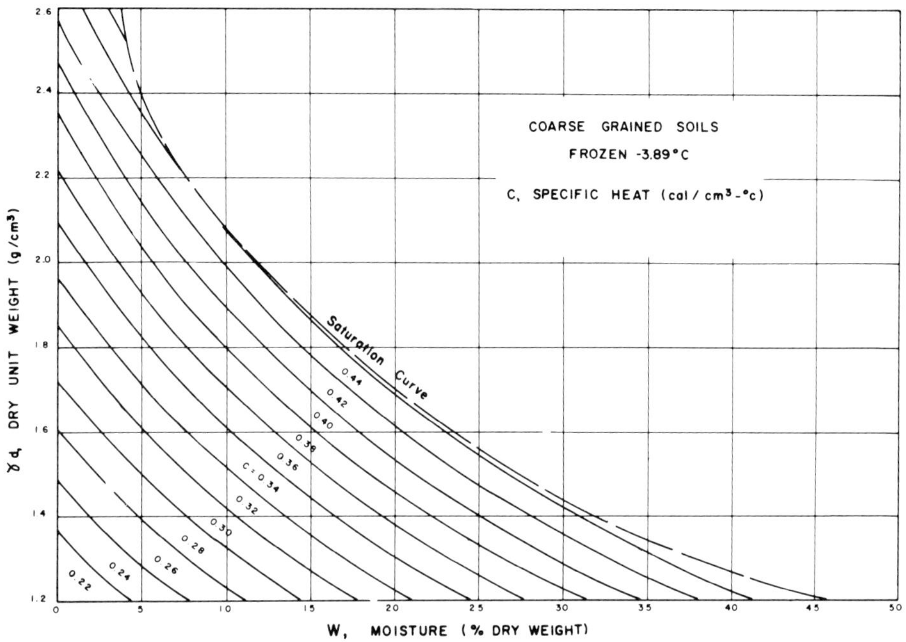

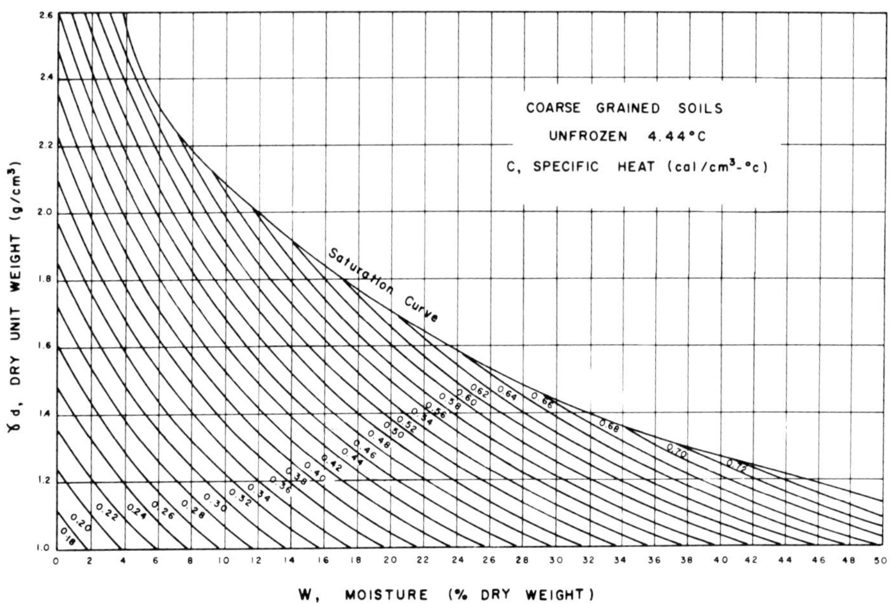

Fig. 4.54. Specific Heat of Coarse-Grained Soils.

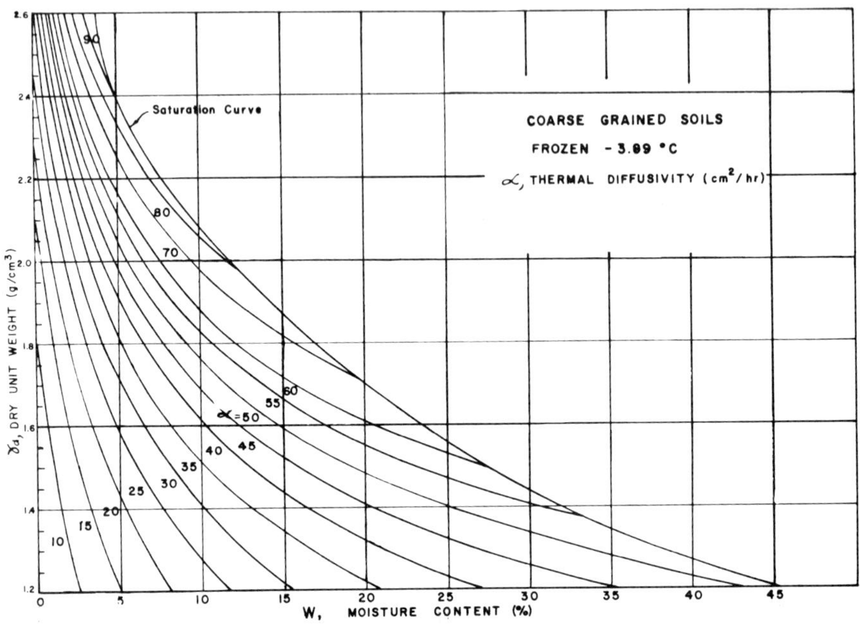

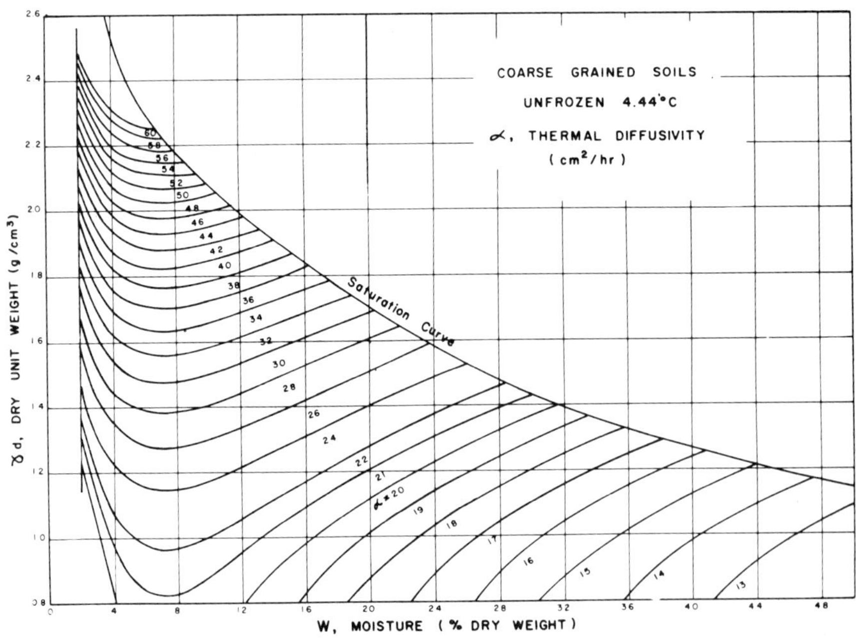

Fig. 4.55. Thermal Diffusivity of Coarse-Grained Soils.

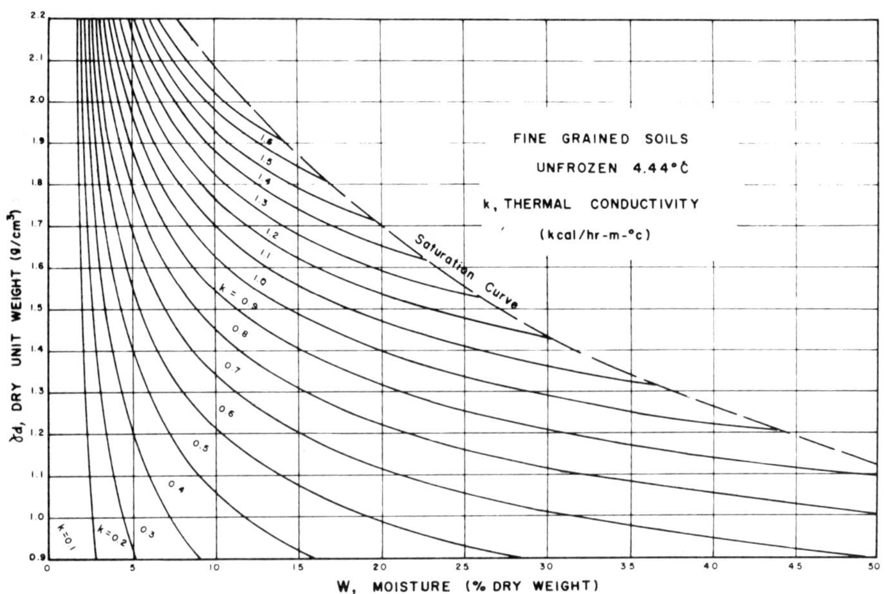

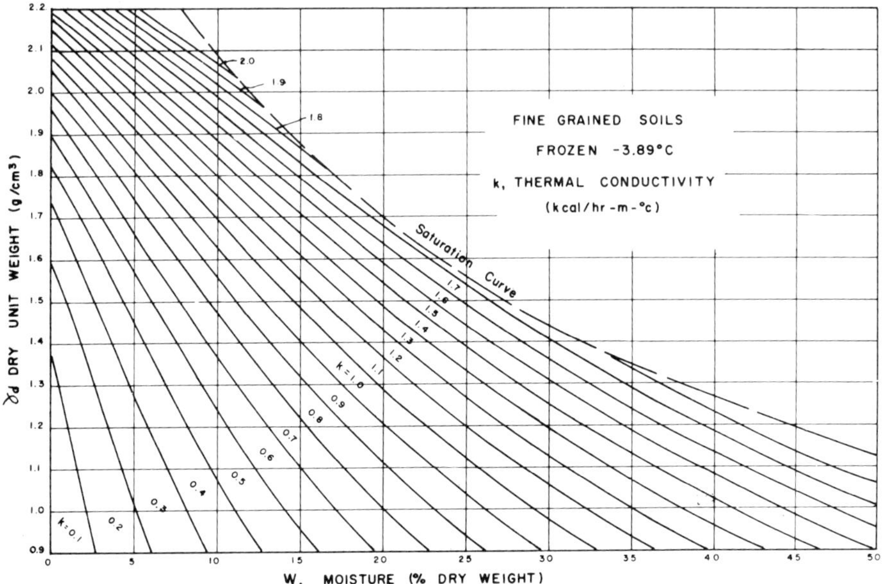

Fig. 4.56. Thermal Conductivity of Fine-Grained Soils.

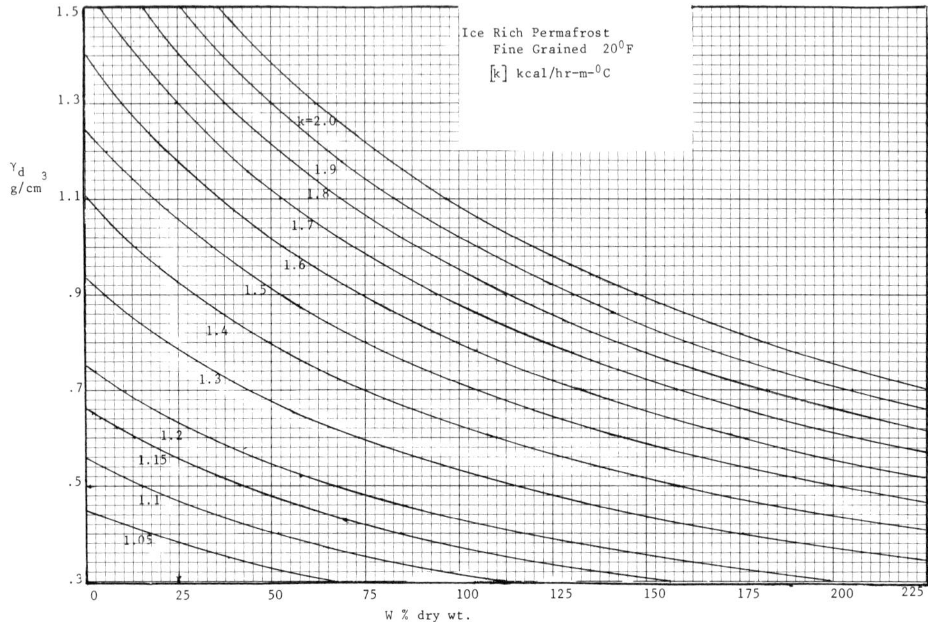

Fig. 4.57. Thermal Conductivity of Ice-Rich Permafrost.

is the equation for the frozen material, and

$$C = \gamma_d(.1724 + .010043\ W) \tag{4.114}$$

is valid for the unfrozen soil. These equations are plotted and shown in Fig. 4.58.

The thermal diffusivity follows, as discussed earlier, and is presented as Fig. 4.59.

4.12.6 Steady-State and Transient Measurements

Most of the comprehensive data in the literature were obtained by the steady-state method. This is true for the data presented here, in particular that of Kersten (1949). It is of interest to compare these data with results obtained by the thermal probe or transient method. There are few comparable data available, but Penner (1970) has carried out a careful study of the thermal conductivity of frozen fine-grained soils using a thermal probe, and his data can be compared at −4.0°C. If one assumes that the specific gravity of the solids is 2.6, then one has

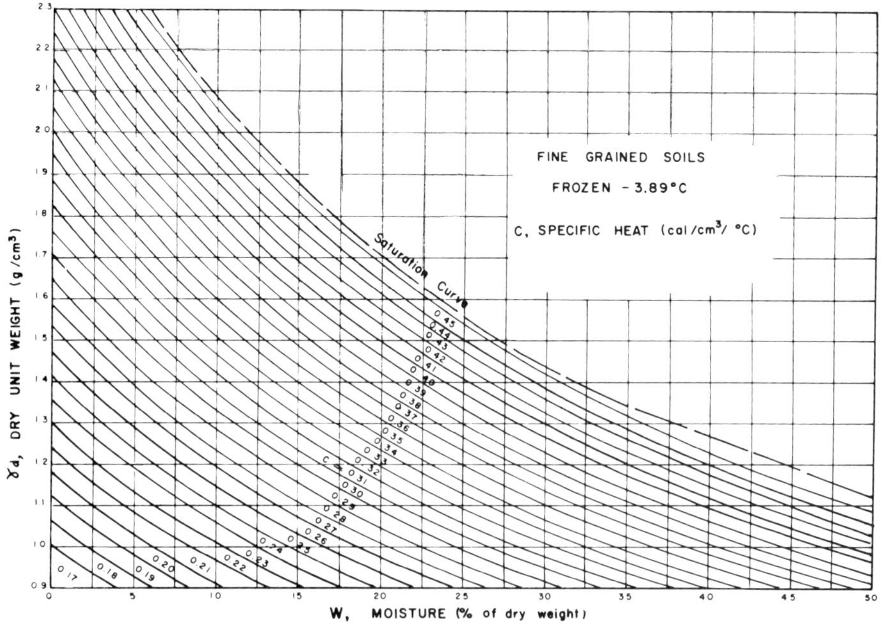

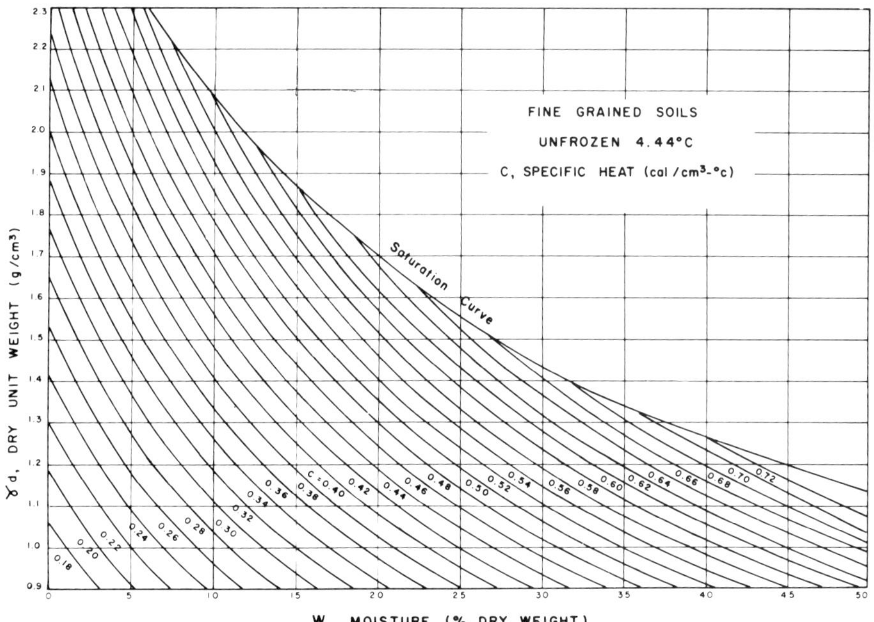

Fig. 4.58. Specific Heat of Fine-Grained Soils.

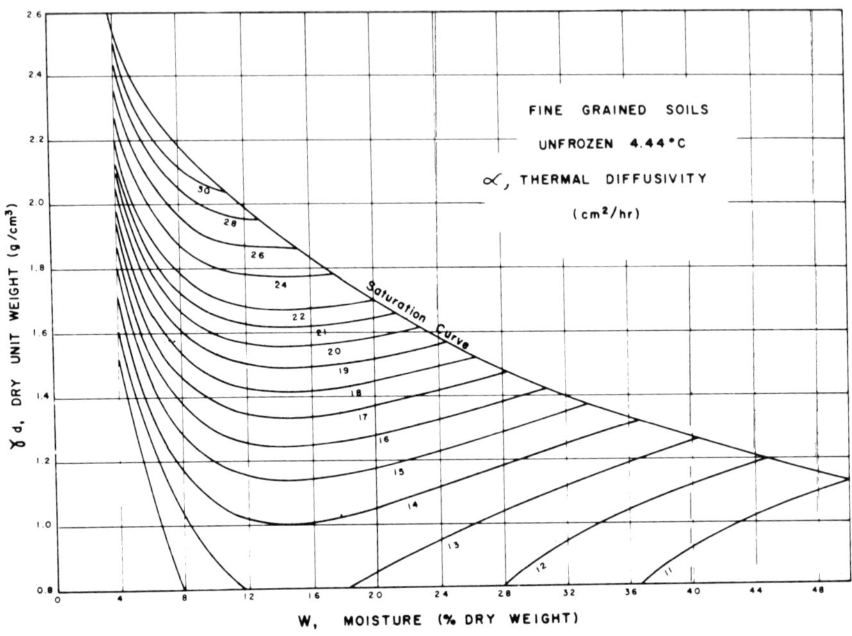

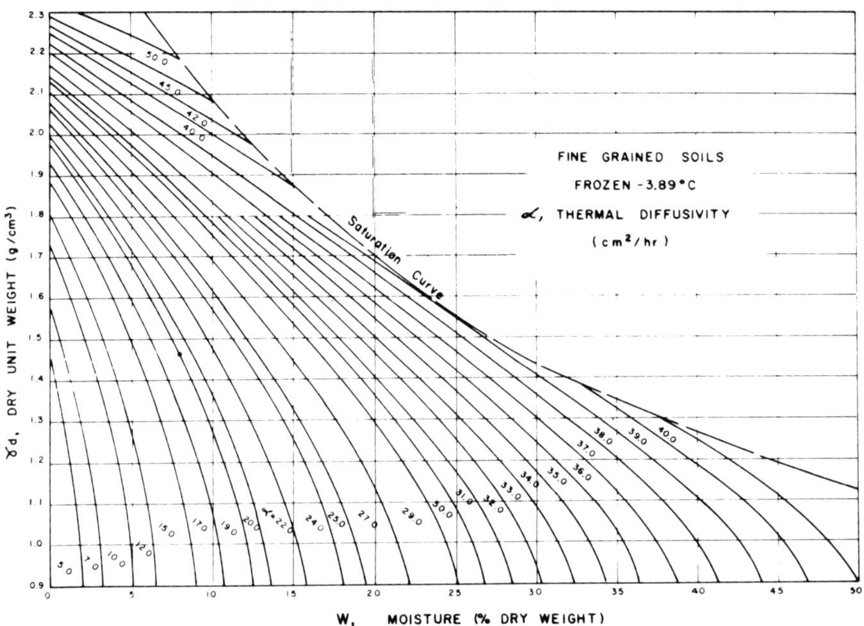

Fig. 4.59. Thermal Diffusivity of Fine-Grained Soils.

	γ_d	W	k (Fig. 4.56)	k (Penner, 1970)	% Difference
Soil					
Leda Clay	.78	90	1.7	1.5	13.4
Sudbury Silty Clay	1.43	31.4	1.8	1.8	0

When one considers that the density and moisture content of the Leda Clay were considerably outside the range measured by Kersten (1949), this is encouraging agreement between the two methods, although generalizations cannot be made from a few data points.

Reconstituting soils does not seem to seriously affect the thermal conductivity of the undisturbed soils. Watson et al. (1973) used the transient technique to measure the conductivity of an ice-rich, organic, clayey silt, permafrost soil, in an undisturbed condition. The measured values were compatible with the values taken from Fig. 4.50 for the appropriate density and water content. Penner et al. (1975) transiently measured the thermal conductivity of 10 natural soil types from the Mackenzie Valley, NWT, and found that the values were well within 25% of Kersten's data. This tends to instill further confidence in the use of Kersten's data for natural soil systems. Slusarchuk

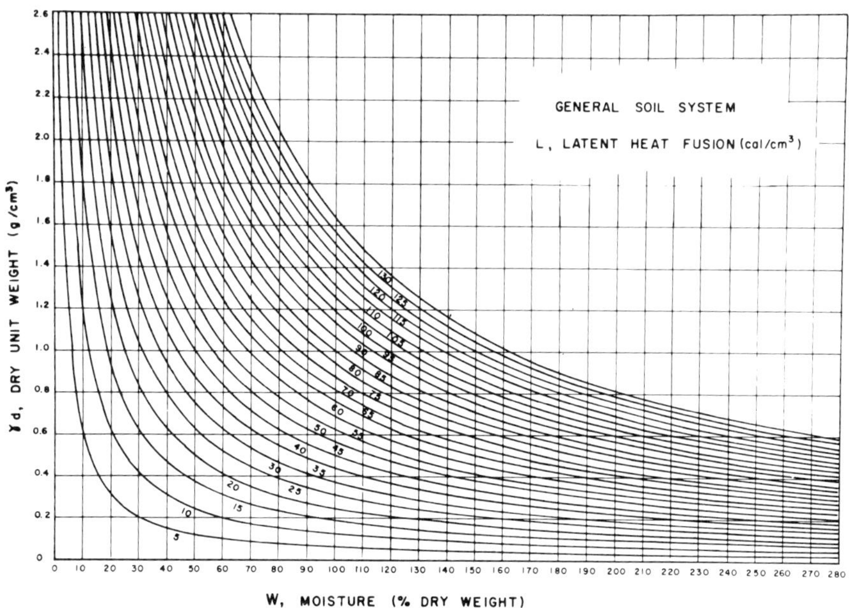

Fig. 4.60. Latent Heat of Fusion for a General Soil System.

and Watson (1975) measured the thermal conductivity of ice-rich permafrost soils and again agreed with Kersten's data for the ranges of γ_d and W where the measurements coincided. In addition, they presented data outside the range of previous data for frozen fine grained (MC-CL) soil, which can be correlated (present author's values), based on a geometric mean conductivity, as

$$k = [1.0071(1.0054)^{W/100}]^{\gamma_d} \tag{4.115}$$

This equation has been plotted for convenience as Fig. 4.57.

It is recommended that actual thermal property data be obtained whenever possible but the data and graphs presented here can be used with reasonable confidence if no data are available.

REFERENCES

Anderson, D. M., and Tice, A. R. 1973. The unfrozen interfacial phase in frozen soil water systems. In *Analysis and synthesis.* Ecological Studies, vol. 4, eds., A. Nodos et al., New York: Springer-Verlag, pp. 107–124.

______. 1972. Predicting unfrozen water content in frozen soils from surface area measurements. *Highw. Res. Rec.* 393:12–18.

Antevs, E. 1932. *Alpine zone of Mt. Washington, N. H.* Auburn, Maine: Merrill and Webber.

Are, F. E. 1978. Offshore permafrost in the Asiatic Arctic. *Third international conference on permafrost, proceedings.* Vol. 1. National Res. Council of Canada, Ottawa, pp. 335–340.

Banin, A., and Anderson, D. M. 1974. Effects of salt concentration changes during freezing on the unfrozen water content of porous materials. *Water Resour. Res.* 10:124–128.

Bondarev, P. D. 1959. A general engineering-geocryological survey on the permafrost regions of USSR and methods of construction in permafrost areas. *Problems of the North.* No. 3, pp. 23–47.

Brewer, M. C. 1958*a*. Some results of geothermal investigations of permafrost in northern Alaska. *Trans. Am. Geophys. Union* 39:19–26.

______. 1958*b*. The thermal regime of an Arctic lake. *Trans. Am. Geophys.* Union 39:278–84.

Brown, R. J. E. 1956. Permafrost investigations in the Mackenzie Delta. *Can. Geogr.* 7:21–26.

______. 1962. A review of permafrost investigations in Canada. *Can. Geogr.* 6:162–165.

______. 1963. Relation between mean annual air and ground temperature in the permafrost region of Canada. *Proceedings international conference on permafrost.* Washington, D.C.: NAC-NRC.

______. 1964. *Permafrost investigation on the Mackenzie Highway in Alberta and Mackenzie District.* NRC 7885, Ottawa, Canada.

______. 1964*a*. Distribution of permafrost in the discontinuous zone of western Canada. *Proceedings of the Canadian regional permafrost conference.* Tech. Memo No. 86, Ottawa, Canada.

______. 1965. *Permafrost investigations in Saskatchewan and Manitoba.* NRC 8375, Ottawa, Canada.

______. 1967. Comparison of permafrost conditions in Canada and U.S.S.R. *Polar Rec.* 13 (87):741–51.

______. 1967*a*. *Permafrost investigation in British Columbia and Yukon Territory.* NRC 9762, Ottawa, Canada.

———. 1968. *Permafrost map of Canada.* NRC 10322, Ottawa, Canada.

———. 1968*a*. *Permafrost investigation in northern Ontario and northeastern Manitoba.* NRC 10465, Ottawa, Canada.

———. 1970. *Permafrost in Canada.* Toronto: Univ. of Toronto Press.

———, and Johnston, G. N. 1964. Permafrost and related engineering problems. *Endeavour* 18, No. 89.

———, and Péwé, T. L. 1973. Distribution of permafrost in North America and its relationship to the environment: A review 1963–1973. *Permafrost, second international conference.* Washington, D.C.: National Academy of Science, pp. 71–100.

Brown, W. G. 1964. Difficulties associated with predicting depth of freeze or thaw. *Can. Geotech. J.* 1 (4):215–226.

———, Johnston, G. N., and Brown, R. J. E. 1964. Comparison of observed and calculated ground temperatures with permafrost distribution under a northern lake. *Can. Geotech. J.* 1 (3):147–154.

Bryan, K. 1946. Cryopedology—the study of frozen ground and intensive frost action with suggestions on nomenclature. *Am. J. Sci.* 244:622–642.

Burger, H. C. 1915. Dos Leitvermogen verdunnter mischkristallfreier Losungen. *Phys. Z.* 20:73–76.

Canadian. 1957. *Permafrost, a digest of current information.* Tech. Memo No. 49, Associate Committee on Soil and Snow Mechanics, Ottawa, Canada.

Capper, P. L., and Cassie, W. R. 1969. *The mechanics of engineering soils.* London: E. and F. N. Spon.

Carslaw, H. S., and Jaeger, J. C. 1959. *Conduction of heat in solids.* 2nd ed. Oxford: Clarendon Press.

Chamberlain, E. J., Sellmann, P. V., Blouin, S. E., Hopkins, D. M., and Lewellen, R. I. 1978. Engineering properties of subsea permafrost in the Prudhoe Bay region of the Beaufort Sea. *Third international conference on permafrost proceedings.* Vol. 1. Ottawa: NRC of Canada, p. 629–635.

Charles, J. L. 1959. Permafrost aspects of Hudson Bay Railroad. *Proc. Am. Soc. Civ. Eng.* 85 (6):125–135.

Chudnovskii, A. F. 1959. Heat exchange in dispersed media. *Gostekhizdat.* Reported in *Principles of geocryology.* Part 1, Chap. 6. G. A. Martynov, Moscow. NRC Translation TT-1065, Ottawa, 1963.

DeVries, D. A. 1963. Thermal properties of soils. Chap. 7 in *Physics of plant environment,* ed. W. R. Van Wijk. New York: Wiley.

———. 1958. Simultaneous transfer of heat and moisture in porous media. *Trans. Am. Geophys. Union* 39:909–916.

Dillon, H. B., and Andersland, O. B. 1966. Predicting unfrozen water contents in frozen soils. *Can. Geotech. J.* 3 (2):53–60.

Dorsey, N. E. 1940. *Properties of ordinary water substance.* New York: Rheinhold.

Eckert, E. R. G., and Drake, R. M. 1959. *Heat and mass transfer.* New York: McGraw-Hill.

Ferrians, O. J. 1964. Distribution and character of permafrost in the discontinuous zone of Alaska. *Proceedings Canadian regional permafrost conference.* TM No. 86, Ottawa, Canada.

Gold, L. W., and Lachenbruch, A. H. 1973. Thermal conditions in permafrost—A review of North American literature. *Second international conference on permafrost.* Yakutsk, Siberia, pp. 3–25.

Ives, J. D., and Fahey, B. D. 1971. Permafrost occurrence in the Front Range, Colorado Rocky Mountains, U.S.A. *J. Glaciol.* 10:105–111.

Johnston, G. H. 1965. *Permafrost studies at the Kelsey hydro-electric generating station—Research and instrumentation.* NRC 7943, Ottawa, Canada.

Johnston, G. N., and Brown, R. J. E. 1964. *Some observations on permafrost distribution at a lake in the Mackenzie Delta, NWT, Canada.* NRC 8252, Ottawa, Canada.

Jumikis, A. R. 1962. *Soil mechanics.* Princeton, New Jersey: Van Nostrand.

Kavianipour, A., and Beck, J. V. 1977. Thermal property estimation utilizing the Laplace transform with application to asphaltic pavement. *Int. J. Heat Mass Transfer.* 20:259–267.

Kay, B. D., and Goit, J. B. 1975. Temperature-dependent specific heats of dry soil materials. *Can. Geotech. J.* 12:209–212.

Kersten, M. S. 1949. *Thermal properties of soils.* Univ. of Minnesota Experiment Station, Bulletin No. 28.

Krischner, O., and Rohnalter, H. 1940. Warmeleitung und Dampfdiffusion in Feuchten Gutern. *VDI Forschungsh.* 402.

Lachenbruch, A. H. 1959. *Periodic heat flow in a stratified medium with application to permafrost problems.* U.S. Geological Survey Bulletin 1083-A.

Legget, R. F. 1954. Permafrost research. *Arctic* 7 (3 & 4):153–158.

______, Dickens, H. B., and Brown, R. J. E. 1961. Permafrost investigation in Canada. *Geology of the Arctic* (2):956–969.

Lewellen, R. I. 1973. The occurrence and characteristics of nearshore permafrost, Northern Alaska. *Permafrost, second international conference, North American Contribution.* National Academy of Sciences, Washington, D.C., pp. 131–135.

Lunardini, V. J. 1971. Presentation of some thermal properties of soil systems. CSME Paper 71-CSME-38, EIC General Meeting, Quebec City.

MacFarlane, I. C., ed. 1969. *Muskeg engineering handbook.* Toronto: Univ. of Toronto Press.

Maxwell, J. C. 1904. *A treatise on electricity and magnetism.* 3rd ed., vol. 1. Oxford: Clarendon Press.

McGaw, R. 1969. Heat conduction in saturated granular materials. Highway Research Board, Special Report 103, Washington, D.C., pp. 114–131.

Misener, A. D. 1955. Heat flow and depth of permafrost at Resolute Bay, Cornwallis Island, NWT, Canada. *Trans. Am. Geophys. Union* 36:1055–1060.

Muller, S. W. 1945. *Permafrost or permanently frozen ground and related engineering problems.* Spec. Rept. Strat. Engineering Study N062, U.S. Army Engineers, Milit. Intell. Div. Office.

Nakaya, U., and Sugaya, J. 1943. *A report on permafrost surveying (Manchuria, 1943).* NRC TT-382. (From *Teion Kagoku* 2:119–128, 1949).

Penner, E. 1970. Thermal conductivity of frozen soils. *Can. J. Earth Sci.* 7 (3):982–987.

______, Johnston, G. H., and Goodrich, L. E. 1975. Thermal conductivity laboratory studies of some Mackenzie Highway soils. *Can. Geotech. J.* 12:271–288.

Phillip, J. R., and DeVries, D. A. 1957. Moisture movement in porous materials under temperature gradients. *Trans. Am. Geophys. Union* 38:222–232.

Pihlainen, J. A., and Johnston, G. H. 1963. *Guide to a field description of permafrost for engineering purposes.* NRC 7596, Tech. Memo 79, Ottawa, Canada.

Radforth, N. W. 1954. Palaeobotanical method in the prediction of sub-surface summer ice conditions in northern organic terrain. *Trans. R. Soc. Can.* vol. 48, series 3.

______. 1962. The ice factor in muskeg. *Proceedings first Canadian conference on permafrost.* Tech. Memo No. 76, Ottawa, Canada, pp. 57–70.

______. 1969. Classification of muskeg. Chap. 2 in *Muskeg engineering handbook,* ed. I. C. MacFarlane. Toronto: Univ. Toronto Press.

Retzer, J. L. 1965. Present soil forming factors and processes in arctic and alpine regions. *J. Soil Sci.* 16:38–44.

Saltykov, N. I. 1959. *Principles of geocryology.* Part 2, chap. 1, Moscow: Academy of Science of the USSR.

Samson, L., and Tordion, F. 1959. Experience with engineering site investigations in northern

Quebec and northern Baffin Island. *Proceedings 3rd Canadian conference on permafrost,* pp. 21–38.

Sanger, F. J. 1963. Degree-days and heat conduction in soils. *Proceedings permafrost international conference.* NAS-NRC Pub. No. 1287, Washington, D.C., pp. 253–262.

———. 1969. *Foundations of structures in cold regions.* CRREL monograph III-C4.

Sass, J. H., Lachenbruch, A. H., and Munroe, R. J. 1971. Thermal conductivity of rocks from measurements on fragments and its application to heat flow determination. *J. Geophys. Res.* 76:3391–3401.

Scott, R. F., and Schoustra, J. J. 1968. *Soil mechanics and engineering.* New York: McGraw-Hill.

Shvetsov, P. F. 1959. *Principles of geocryology.* Part 1, Chap. 4, Moscow: Academy of Science of the USSR.

Skaven-Haug, Sv. 1963. Control of frost penetration in Norway. *Proceedings permafrost international conference.* NRC Pub. No. 1287, Washington, D.C., pp. 268–272.

Slusarchuk, W. A., and Watson, G. H. 1975. Thermal conductivity of some ice rich permafrost soils. *Can. Geotech. J.* 12:413–424.

Soviet, 1960. *Technical considerations in designing foundations in permafrost.* State Publishing House of Literature on Building Architecture and Building Materials, Moscow SN 91–60, NRC, TT-1033, Ottawa, Canada.

Swinzow, G. K. 1969. Certain aspects of engineering geology in permafrost. *J. Eng. Geology* 3:177–215.

Thie, J. 1974. Distribution and thawing of permafrost in the southern part of the discontinuous permafrost zone in Manitoba. *Arctic* 27 (3):189–200.

Tice, A. R. Burrows, C. M., and Anderson, D. M. 1978. Phase composition measurements on soils at very high water content by the pulsed nuclear magnetic resonance technique. *Moisture and frost-related soil properties.* Transportation Research Board, Nat. Acad. Sciences, pp. 11–14.

Tsytovich, N. A. 1975. *The mechanics of frozen ground,* ed. G. K. Swinzow. New York: McGraw-Hill.

Watson, G. H., Slusarchuk, W. A., and Rowley, R. K. 1973. Determination of some frozen and thawed properties of permafrost soils. *Can. Geotech. J.* 10:592–606.

Williams, P. J. 1963. Specific heats and unfrozen water content of frozen soils. *Proceedings first Canadian permafrost conference.* Tech. Memo No. 76, Ottawa, pp. 109–126.

———. Ice distribution in permafrost profiles. *Can. J. Earth Sci.* 5 (12):1381.

Woodside, W., and Messmer, J. N. 1961. Thermal conductivity of porous media I. *J. Appl. Phys.* 32 (9):1688–1699.

———. 1961*a*. Thermal conductivity of porous media: II. Consolidated rocks. *J. Appl. Phys.* 32:1699–1706.

Yong, R. 1968. Research on fundamental properties and characteristics of frozen soils. *Proceedings first Canadian permafrost conference.* Tech Memo No. 76, Ottawa, pp. 84–108.

PROBLEMS

1. Calculate the permafrost thickness at Inuvik. Neglect the active layer thickness. How does this compare with measured values?
2. At Inuvik calculate the effect on the permafrost thickness if the mean annual air temperature (mean ground surface temperature) increases by 1°F.
3. State reasons why the calculated and the measured permafrost thickness may differ at a given locality.

4. In a certain region a layer of ground froze during one winter and did not completely thaw during the following summer, but it thawed completely the second summer. Should this have been classified as permafrost? What is the engineering significance of permafrost?
5. A gravel soil with $\gamma_d = 1.22$ g/cc, W = 18% changes temperature from 40°F to 25°F. Discuss quantitatively the changes in thermal conductivity, specific heat, and latent heat of the soil.
6. An unfrozen peaty soil has a specific gravity of the solids of 1.4 and a volumetric analysis of the solids of 12% and water of 64.9% (the remaining volume is air). List the values of the thermal properties of this unfrozen soil.
7. A is the mean annual ground temperature. Curves are limits of maximum and minimum ground temperatures. Three different locations are indicated by the points a, b, c which denote the temperature curve at a site relative to the 32°F temperature. Which locations are permafrost regions? Sketch and label on a figure the active layer, and depth of permafrost (if any) for each of locations a, b, c.

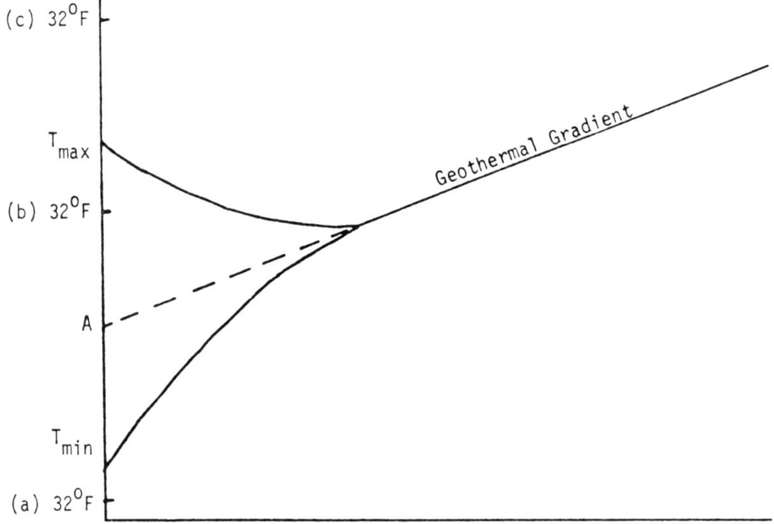

5 Geomorphology of Cold Climate Regions

5.0 INTRODUCTION

Geomorphology is the study of the form and shape of the land, rivers, lakes, etc., in a given region. Many regions of the world exhibit strange and often beautiful landforms, but perhaps nowhere, outside the cold regions, are the surface features so striking and peculiar as they are in arctic and subarctic areas. Many of the features occur outside the cold regions, but usually the effects are exaggerated in the cold regions. In many cases, the patterns are so geometrically regular as to appear artificial. Very often the surface patterns are significant from an engineering viewpoint, as they give clues to underlying features, such as ground ice, that may present grave engineering hazards. Most of the landforms can be explained on the basis of frost action, and it is useful to examine some of the mechanisms by which material can be shaped, moved, and sorted by size, volume, or mass.

In order to understand the origin and significance of the geomorphology of cold regions, it is necessary to look at the physical processes that allow material to be moved from one location to another and also to be segregated, or sorted, by size and volume. As can be expected, the outstanding and perhaps most unusual mechanisms in the cold regions are related to frost action.

5.1 FROST-HEAVE

One of the truly difficult phenomena associated with cold climates is that of frost-heave. Heave is the movement of soil masses during phase change. These movements can be so severe as to damage or even destroy engineering structures.

It is known that during the freezing process, water can migrate to the freezing interface from the surrounding soil volume (Fig. 5.1). The migration may be in the form of the liquid or vapor phases of water; ice movement can be neglected. Lenses of pure ice, 1–100 mm, commonly form at the freezing interface. The soil above the interface is then displaced to make room for the ice formed. This displacement is the frost-heave, and its magni-

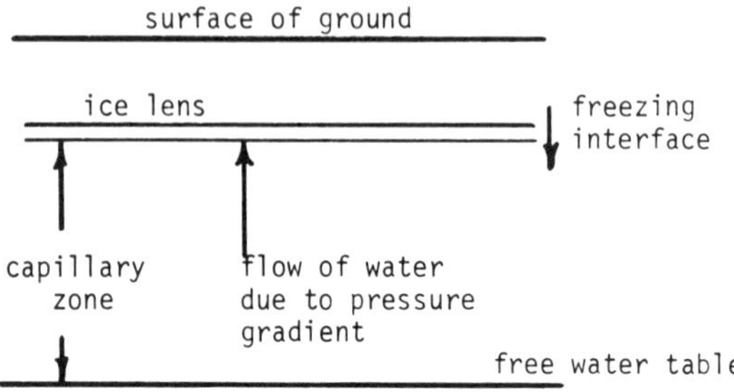

Fig. 5.1. Simple Frost-Heave.

tude depends upon the volume of ice formed and the deformation and pressures of the soil mass. Frost-heave can then be described in terms of four interrelated phenomena: the freezing process, moisture migration, ice segregation, and the deformation of the soil system.

5.1.1 Freezing Process

A freezing interface is required in order for frost-heave to develop. The characteristics of the freezing process are a function of the diffusion of energy and mass for the system. Ultimately, the heave is controlled by the energy transfer at the interface, as there must be sufficient cooling capacity to freeze the water.

5.1.2 Moisture Migration

The movement of water to the freezing interface is the least understood aspect of the heaving phenomenon. Why does water move toward a freezing interface? Assume that a mixture of ice, water, and soil particles exists in dynamic equilibrium. As long as the equilibrium persists, there will be no change in the state or position of the soil water. The freezing process disturbs the equilibrium and sets up an unbalanced force system that tends to drive water toward the interface where phase change is occurring. The exact nature of this motive force is not totally understood, but a number of theories have been propounded.

The flux of water can be described as

$$v = -D \operatorname{grad} F \tag{5.1}$$

where

D = permeability of the soil system through which the water flows
v = rate of water movement and
F = generalized motive force causing water movement.

It can be seen from Eq. (5.1) that, even with a strong gradient, the flux of water can be minor if the soil permeability is low. The ability of a soil to generate a motive force and the permeability of the soil seem to be inversely proportional. For this reason, silts, which have intermediate values of motive force and permeability, seem to be most susceptible to frost-heave.

Some of the postulated generalized forces or potentials can be examined in a qualitative fashion.

5.1.2.1 Absorption Potential of Soil Skeleton—Ice—Water Interfaces

The interaction of the free surface energies of the mineral particle surfaces, film water, and ice gives rise to a motive potential (or suction force). The properties of the bound water in films of various thicknesses, augmented by the mineral surface effects, set the water in motion from thick films with mobile water molecules to thin films, depleted by the formation of ice. This is essentially the adsorption-film theory of water movement, which has the most support at present. It should be noted that this theory, as for most of the theories, is difficult to cast into a quantitative form that can be used for predictions.

5.1.2.2 Pore Capillarity

The surface tension effects of the soil pores and water give rise to forces that can pull water toward a freezing interface. The presence of capillary menisci in the freezing zone is assumed, and the interfacial surface tensions give rise to the motive force. Combined in various ways with the motion of film water, this concept is called the "capillary theory" of water movement and is also very popular.

5.1.2.3 Water Vapor Pressure

Water vapor pressure is directly related to the surrounding temperature or the presence of different water phases. Thus, a temperature gradient can induce a vapor pressure gradient that will cause a flow of vapor from one part of the soil to another, in particular, to the freezing interface. This effect is most significant for soils with a low moisture content.

5.1.2.4 Osmotic Potential

The osmotic potential is similar, in concept, to the film water movement, but it relies on large amounts of solutes in the pore water to exert a significant force differential.

5.1.2.5 Ice Crystallization Potential

Nonequilibrium effects of ice formation induce motive forces. This concept is related to the adsorption-film idea. It aids in the formation of segregation ice by adding migrating water to the ice layer already formed.

5.1.2.6 Hydrostatic Pressure

External pressures and internal hydrostatic pressure can, of course, lead to a flux of water. Potentials of this type are most significant for water-saturated soils in closed systems.

These theories of motive force, and others not mentioned, are not mutually exclusive in most cases. It is likely that they all play a part in the movement of soil water. At this time, although the underlying physics is being clarified, there is no complete, satisfactory theory of soil water movement. Most analyses attempted, have relied on empirical factors, which tend to obscure the basic mechanisms.

5.1.3 Ice Segregation

The freezing of nonmotive water in the soil pores is called "in situ freezing." If there is no pore water expulsion or deformation of the soil skeleton, the in situ freezing will only contribute a 9% increase in the soil volume, which is not normally a severe problem. In situ heaving can be expressed as

$$h_i = \frac{\gamma_d(W - W_u)}{100\ \rho_w} X\alpha \tag{5.2}$$

where

X = depth of frozen soil
W, W_u = percentages of total and unfrozen water
ρ_w, ρ_i = densities of water and ice and
$\alpha = \rho_w/\rho_i - 1$

The more serious heaving problem is due to the formation or segregation of pure ice lenses in the soil. If the potential water supply convects energy (sensible and latent) to the phase change interface at a rate that equals or exceeds the heat removal rate, which is determined by the heat transfer characteristics of the system, a lens of ice will grow at a fixed location for an extended time. If the energy carried with the moisture flow is less than the rate of cooling, then the soil at the interface will freeze, and the interface will move into the unfrozen soil mass. The interface moves until a location is reached that again favors an adequate moisture supply, and a new lens starts to grow. The process is rhythmic and repetitive and leads to layers of ice and frozen soil. The thickness and spacing of the lenses is a function of the heat flow rate, moisture flow, and soil characteristics.

The heave, or increase in volume, associated with the formation of segregated ice is

$$h_s = (1 + \alpha) \int_0^t v \, dt \tag{5.3}$$

where v is the moisture velocity at the phase change interface. The total heave is then

$$h = \frac{\gamma_d (W - W_u)}{100 \rho_w} \alpha X + (1 + \alpha) \int_0^t v \, dt \tag{5.4}$$

The rate of heaving can be written as

$$\frac{dh}{dt} = \frac{\gamma_d (W - W_u)}{100 \rho_w} \alpha \frac{dX}{dt} + (1 + \alpha) v \tag{5.5}$$

The flow of water to the interface is a function of the heat transfer and the soil properties, and thus v is related to dX/dt. The mass and energy fluxes must be evaluated simultaneously. Reliable predictions are not yet available.

5.1.3.1 Ice Segregation Efficiency

An energy balance at the phase change interface gives

$$Q_c = l(\sigma + \rho_f \, dX/dt) \tag{5.6}$$

where

$Q_c = k_f \, \partial T_f/\partial x - k_t \, \partial T_t/\partial x$ is the net energy removal by conduction
l = latent heat and
$\sigma = \rho_w \, v$ is the rate of formation of segregated ice.

The ice segregation efficiency is defined as

$$E = \frac{\sigma l}{Q_c} \tag{5.7}$$

Thus

$$E = 1 - \frac{\rho_f l \dfrac{dX}{dt}}{Q_c} \tag{5.8}$$

For pure segregated ice formation, $E = 1$, and $dX/dt = 0$, but for in situ freezing only, $E = 0$. The efficiency will then, in general, be between 0 and 1.

5.1.4 Energy and Work

Work must be done to lift the overburden and counteract any surface surcharge to provide space for the growing ice. Work is also needed to induce and maintain the force gradient needed to sustain a water flow to the interface. This energy must come from the latent heat of the water in the phase change region. It is easy to show that the latent heat released is 20–50 times the work needed to lift an overburden (neglecting frictional effects). A surcharge can cause water to cease migrating to the interface and eventually cause the water to be expelled from the freezing front.

The accurate prediction of frost-heave must quantitatively consider the characteristics of the freeze front, the overburden and surcharge pressures, the rate of heat removal, the soil consolidation, and the flow of soil water. These effects combine to form a very difficult mathematical and physical system.

5.2 GRAVITY SORTING

In all parts of the world gravity is a potent force in moving and sorting bodies. The tendency of objects to move toward positions of minimum potential energy is well known and constantly acting. Thus, rocks tend to slide, roll, or tumble down slopes, with a natural tendency for the more massive sizes to travel farthest. The gravitational effects in cold regions are no different than in other regions, with the possible exception of solifluction.

5.2.1 Solifluction

Solifluction is the slow, gravitational, downslope motion of thawed soil materials behaving as a viscous fluid flowing over a frozen material. Rates of flow may be on the order of 2–10 cm/yr (Washburn 1973). It is most important during the spring, when meltwater is available to form the viscous material, and in areas where sufficient slope occurs to give rise to gravitational forces that will exceed the plastic limit of the saturated soil mass. Solifluction is assisted by an impermeable substratum and is thus to be expected in permafrost regions. The material involved is principally fines, but even boulders may be moved with the flow.

5.2.2 Artesian Pressure

Artesian pressure arises from differences in hydrostatic head and is slope-induced. Water-saturated soil may move within a soil volume, and the formation of earth mounds up to 60 cm high has been observed due to this process. Artesian pressures may be involved in the phenomenon of icing, which will be discussed later.

5.2.3 Nivation

Nivation is the process whereby snow affects the disintegration of some landforms and the formation of others. The melting of snow is a source of the lubricating fluid of solifluction. In particular, the gravitational action of snow may lead to a nivation protalus. This is the process whereby rocks may slide downslope over a layer of snow and build up a rampart at the base of the slope. The process leads to sorting because different masses of rock tend to move different distances under the forces of gravity and friction.

5.3 FROST ACTION (THERMAL) SORTING

There is no doubt that the principal forces acting in cold regions are those associated with the freeze-thaw cycles of water and the temperature-induced changes in other materials. These effects are manifested in a great number of expected and unexpected phenomena.

5.3.1 Frost Expansion

Repeated cycles of freezing and thawing can provide powerful forces for differentially moving solids of different sizes and materials. Frost expansion is a term that covers both frost-heaving and frost-thrusting. Frost-heaving is the vertical, or upward, movement due to pressures generated by freezing water, and frost-thrusting is the horizontal, or lateral, manifestation of the same phenomenon.

5.3.2 Frost-Heave

Frost-heave is due to the expansion of water as it turns to ice, as well as to the inflow of water to a freezing interface, as has been discussed. Corte (1963) demonstrated that fine-grained material moves ahead of a freezing interface, and, thus, a vertical sorting can take place during frost-heaving. The main sorting effect, however, is related to the movement of large particles (stones, etc.). The well-known phenomenon of a stone that is initially embedded in soil being pushed out of the soil, after repeated thaw-freeze cycles, can be explained on the basis of heaving. Two theories have been put forth: the frost-pull and the frost-push mechanisms.

5.3.3 Frost-Pull Mechanism

Fines, first freezing around the upper part of the stone, tend to pull up the stone as the fines expand due to freezing, as shown in Fig. 5.2. During the thaw part of the cycle, a thawing commences at the upper surface. The thawed fines tend to slump down while the stone is held in place at the elevated position by the frozen layers now at its base.

The original upthrust movement would tend to leave a cavity that presumably would partially fill with saturated fines and block the tendency of the stone to drop back when completely thawed. This mechanism is a bit vague during the thawing part of the cycle, but the effect is no doubt present.

5.3.4 Frost-Push Mechanism

The thermal conductivity of stones (quartz, etc.) is generally greater than that of fines (clay, silt), and as freezing occurs, the temperature of the stone

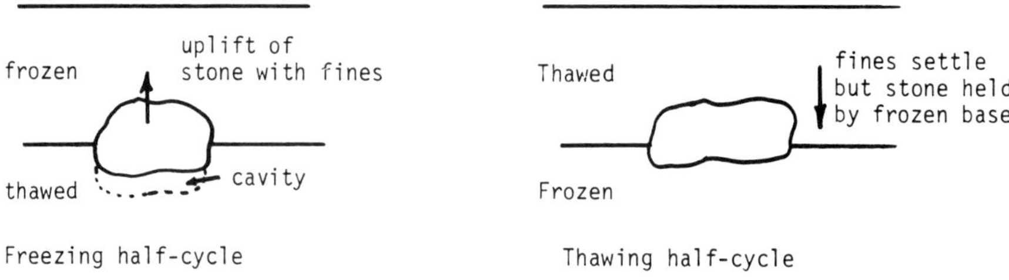

Fig. 5.2. Frost-Pull Mechanism.

and the adjacent fines will reach 0°C faster than the more distant, surrounding, fines. Thus, the freezing interface will tend to surround the stone and develop an ice lens under and around the stone due to moisture migration (Fig. 5.3). This ice may push the stone up even if the fines above the stone are frozen. The entire frozen zone over the stone may be pushed up, especially if the stone is near the surface, or the fines may be below freezing, but contain considerable unfrozen water and, hence, may allow the stone to be pushed into the fines by plastic deformation. During thawing, the latent heat of the ice would slow the thawing near the base of the stone, even though the temperature of the base of the stone should reach the thawing value more rapidly than the adjacent fines. Thus, the thawed fines can slump down around the stone, which is held in place by the pedestal of ice and frozen soil, with a little settling due to volume change of the ice as it melts and the possible squeezing out of some of the meltwater. The process would tend to be more efficient if the fines at the level of the stone were dry and, hence, melted rapidly as compared with the ice around the stone. As the ice melted completely, the fines would flow in due to the saturation effect of the meltwater and keep the stone from sinking back. It would seem that frost-pull may well occur during the initial freezing, with frost-push occurring later; the two effects are not incompatible.

In addition to the purely upward movement, the stones may tend to be rotated due to nonuniform ice action, and, of course, the shape of a stone can have a significant effect upon its movement through the soil.

5.3.5 Frost-Thrusting

The same general mechanism can cause lateral movement of the stones, even though the pressure generated is normally perpendicular to the freezing front.

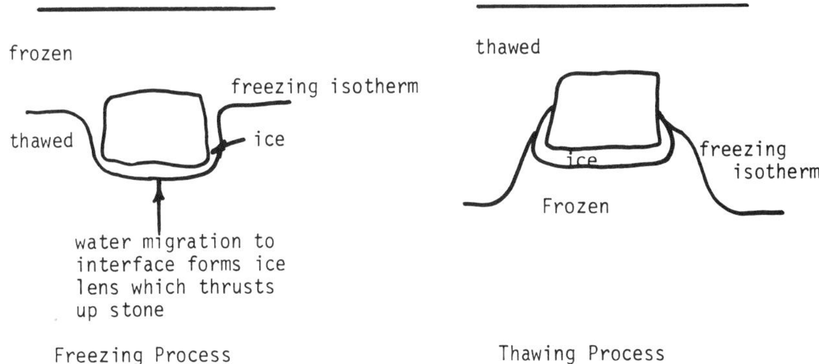

Fig. 5.3. Frost-Push Mechanism.

The variable thermal conductivity of the materials will vary the direction of the freezing interface, and different coefficients of expansion may initiate thermal stresses that are relieved by lateral as well as vertical movement. The expansion of ice as its temperature moves toward the freezing value during thawing may also lead to thrusting.

The above phenomena demonstrate that considerable vertical and lateral sorting of materials is possible due to freeze-thaw cycles.

5.3.6 Needle Ice

Slender, hairlike ice crystals may form at or just beneath the soil surface. The elongation is parallel to the temperature gradient, and the needle ice can lift fairly large surface stones; the subsequent thawing and collapse of needles may lead to significant lateral surface sorting of material. Lateral sorting is aided by the fact that rocks of different mass tend to roll different distances after being tumbled by the needle ice. Three conditions are necessary for the formation of needle ice.

(a) Surface temperature as low as −2°C for ice nucleation.

(b) Surface tension of soil water of 2 σ/r, where σ is the interfacial energy ice-water and r is the effective radius of soil pores.

(c) Heat flow out of the freezing zone balanced by latent heat release from the freezing soil water.

5.4 MASS DISPLACEMENT DUE TO FREEZING

The local motion of masses of unfrozen material within a soil volume due to frost effects is attributed to a number of related phenomena.

5.4.1 Cryostatic Pressure

The internal pressure generated by the increase in volume of freezing saturated soil trapped between a freezing interface and permafrost or bedrock is called "cryostatic pressure." Cryostatic pressure can produce heaving and thrusting, but it seems relatively unimportant.

5.4.2 Intergranular Pressure

The touching of solid grains in a soil gives rise to intergranular pressure, and the water in the pores gives rise to pore water pressure. Only the intergranular pressure can induce changes in the volume of soil mass or frictional forces. Freezing of the soil can induce high pore water pressure and at the same time, by expansion, lessen the solid contact or intergranular pressure. Upon melting, liquefaction of the material may occur, with a squeezing of excess water and fines and even stones to an area of lesser pore water pressure at the surface. Larger stones would tend to sink. Surface pressure, even that of walking upon the soil can induce an upwelling of fines to the surface in the form of a circle of mud or fines around the area of surface pressure.

5.4.3 Frost Cracking

At subfreezing temperatures, the contraction of the soil mass can lead to the fracture of the soil, if the induced tension stresses exceed the strength of the soil. Such cracks can form the boundaries of geometrical surface patterns and can cause sorting by filling with sand, rocks, or sediment. The details of the stress cracking will be discussed under ice wedges.

5.4.4 Frost-Fracturing

The well-known expansion of water upon freezing into ice, leads to high pressures if the volume of the system is held constant. When the water freezes in the voids or cracks of a rock, the pressures may be great enough to shatter or fracture the rock. This process is very efficient and occurs on a large scale in cold climate regions and can lead to large areas of more or less uniform material size and also to the formation of piles of rock debris (talus slopes) at the base of rock cliffs or slopes.

5.4.5 Frost-Churning

The repeated freezing and thawing of soils can lead to a churning, or mixing, of soil layers due to repeated expansion and contraction. These processes, also called "cryoturbation" or "congeliturbation," leave behind what might be called "fixed eddies" of material in the soil.

5.4.6 Ice Lenses

The process of forming ice lenses, which has been discussed, can itself lead to sorting of material due to differential migration at the freezing front. The ice lenses themselves are also examples of sorting of material; in this case the moisture of the soil system is separated out of the material.

5.4.7 Cryconites

Dust and other solid particles that deposit on a snow surface induce differential thawing of the snow due to differences in the emissivity of the dust and snow. These thaw holes allow the solid material to settle upon the ground surface in a differential manner, which can cause sorting of surface material.

5.5 MASS DISPLACEMENT DUE TO CRACKING

5.5.1 Dessication

The cracking of material due to drying is one of the most widespread of the cracking phenomena. It is not restricted to cold climates, and it occurs frequently in cement, concrete, paint, and any sort of drying process. Drying may be due to evaporation, drainage, or moisture flow to ice fronts. Dessication cracks are thought to give rise to many ground patterns less than 1 m in size.

5.5.2 Dilation

Dilation cracking is the rupture of ground due to stretching of the soil surface layers. In this sense it is similar to frost-cracking, since both lead to tensile stresses in the layer, but dilation is due to differential frost-heaving, which tends to push up one part of the surface layer in relation to surrounding areas. This mechanism probably does not lead to regular patterns of the soil surface.

5.5.3 Salt

Salt cracking describes the fracture of hard rock salt or salt crusts, and may be due to thermal contraction. The amount of expansion of a soil when it undergoes wetting and drying is strongly related to its salt content. Heaving of the soil may well be augmented by the effect of salt expansion.

5.6 MASS DISPLACEMENT OF FLOW PROCESSES

All the usual processes of flow are present in cold climates including water, wind, snow, and evaporation. Eluviation is the movement of dissolved or suspended soil materials from one place to another within the soil when rainfall exceeds evaporation. In cold climates, this mechanism would tend to be local and restricted in time, as rainfall is generally rather low. Rillwork is the process that starts small channels and is related to flow processes. Initially, small cracks may be widened by gravitational flow of water or water-soil mixtures.

The action of the mechanisms discussed above leads to the formation of various surface features that are characteristic of, but not confined to, the cold climate regions. A brief description of some of these phenomena will be presented here, but no attempt at an exhaustive discussion will be made.

5.7 GROUND ICE IN COLD CLIMATE

Ground ice takes many forms, some of which have already been discussed. A classification of ground ice, due to Mackay (1972), is presented in Fig. 5.4. This is based upon the origin of the water and the various processes whereby the ice was formed.

Aggradational ice is the newly formed ground ice due to the increase of the permafrost thickness. Buried ice describes ground ice that has formed at the surface and then has been buried. It includes glaciers, lake, river, and sea ice, and icings. It is part of the permafrost as is all permanent ground ice. Closed-cavity ice is ice that has formed in a completely closed space, or cave, in permafrost. Underground cavities along the western coast of arctic

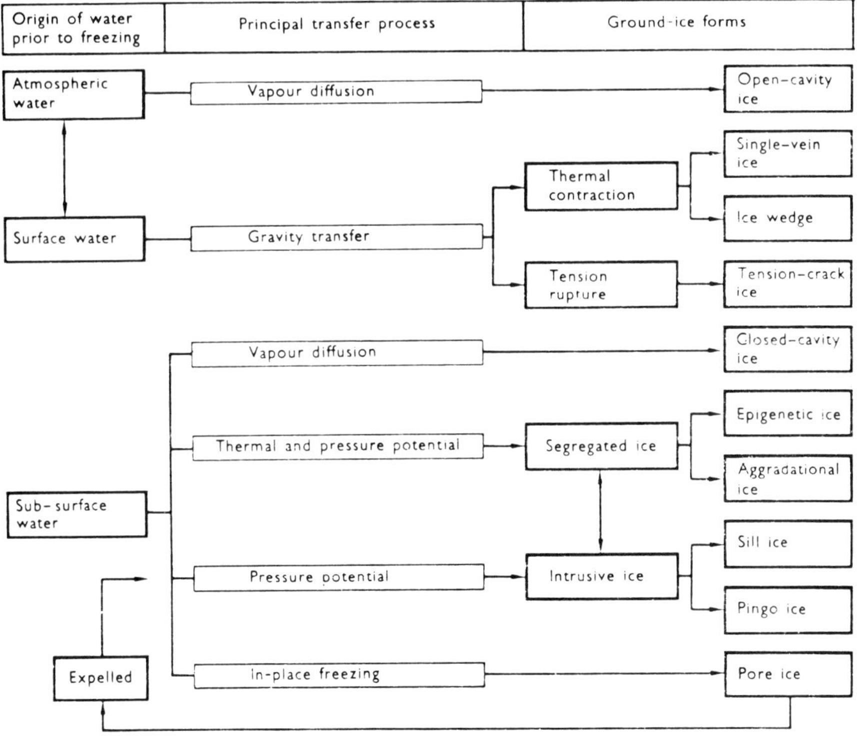

Fig. 5.4. Classification of Ground Ice. Reproduced by permission from Mackay 1972, *Annals of the Association of American Geographers,* 62:4, Fig. 2.

Canada have been found filled with ice crystals. Open-cavity ice is ice that forms in open cracks in the ground by sublimation of water vapor.

Excess ice refers to ground ice that exceeds the pore volume that the unfrozen ground would normally have. This excess ice gives rise to the problems of frost-heave and differential ground settlement. Injection ice describes the ice that has formed by the intrusion or injection of water under pressure into a porous soil system. The ice formation may lift the ground above it, giving rise to various topographic forms.

Massive ice is ice of any sort that is relatively large in size. The ice has dimensions of at least 10–100 cm and includes pingo ice, ice wedges, ice lenses, etc. Pingo ice is the core of ice found in pingos. It may be nearly pure or it may be mixed with various sediments. Pore ice is the ice found in the pores of a soil or rock. On melting it does not form excess pore water.

Reticulate ice is that in a network of horizontal and vertical ice veins forming a three-dimensional lattice normally found in glaciolacustrine sediments. The ice seems to form in shrinkage cracks, with the water coming from the clay sediments.

Segregated ice refers to the ice that forms in lenses at freezing fronts due to the migration of soil water to the freezing interface. It may be hairline or more than 10 m in thickness. Ice lenses are predominantly horizontal, with bubbles tending to be elongated normal to the ice lens. Syngenetic ice is ground ice that formed simultaneously with the deposition of the soil system in which it is found. Epigenetic ice forms after the deposition of the earth material.

Tension crack ice is the banded or layered ice that forms in the tension cracks of the ground, mainly due to the growth of segregated or injection ice.

Vein ice refers to a tabular or wedgelike body of ice, hair-thin to 5 cm thick, whose vertical dimension is predominant. It may be several meters in length. The thinness of its horizontal dimension distinguishes it from ice wedges, but the term is purely descriptive. New ice wedges would, in fact, be ice veins.

5.8 ICE WEDGE POLYGONS

The ice wedge polygon is one of the most striking surface features in arctic or subarctic regions, especially when viewed from aircraft. The massive ice wedges underlying these polygons are also the most common form of ground ice. These ice wedges present engineering problems for roads, pipelines, etc., as their thawing can induce considerable slumping and high stress in the structure. The sides of these polygons may be curved rather than the straight lines of a mathematical polygon. The polygons may be 10–300 ft in diameter. Large fissures commonly form the outline of the polygons, and they can be up to 2 ft wide, 1–2 ft deep, and several hundred feet in length. Beneath these fissures or troughs, massive ice wedges may occur that are from 4 in. to 10 ft wide at the top and 3–30 ft deep (Hamelin and Cook 1967). Some wedges may extend downward as deep as 25 m and have shapes other than wedges (Brown 1974).

The origin of these polygons is generally believed to be explained by the thermal contraction crack theory of Leffingwell (1915). This can be described by referring to Fig. 5.5. Vertical fractures on the order of 0.1 in. wide and several feet deep are known to occur in the tundra. Such a crack will occur during the winter if the thermal stress exceeds the strength of the ground

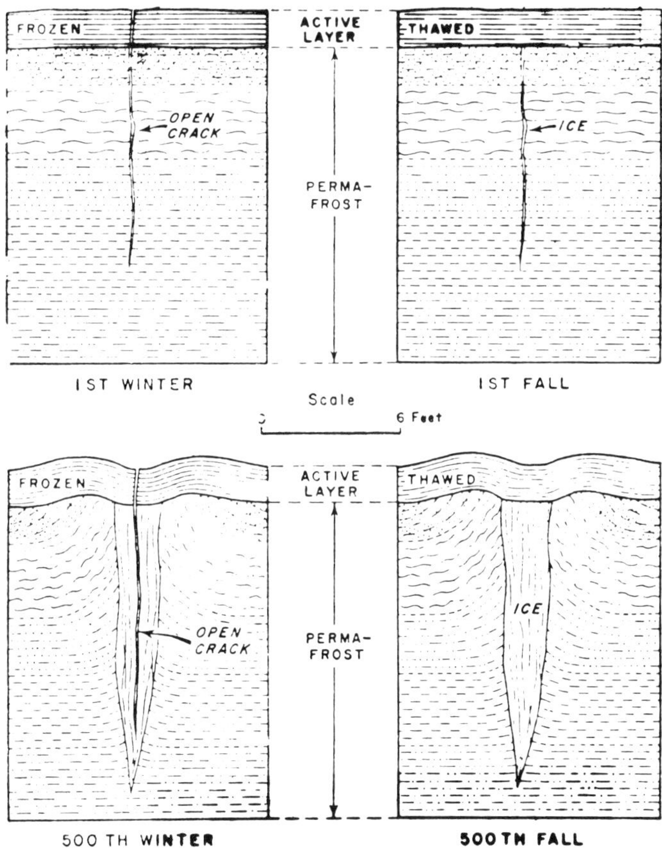

Fig. 5.5. Formation of Ice Wedges (adapted from Lachenbruch 1962).

material. During the early spring, meltwater will fill the crack and freeze into a thin sheet of ice. The crack will disappear in the active layer after the thawing season, but may reappear during the following winter when the crack will widen and more melted water will cause it to grow. Repeated cycles of this type will lead to a wedge of ice that can grow to the large sizes mentioned. The ice in these wedges appears layered, under a microscope, in support of the theory. The expansion of the permafrost will cause a pushing up of the ground surface by plastic deformation, which will cause ridges to occur. Differential thawing over the ice wedge, due to differences in snow level, may cause a slumping with the characteristic trough or fissure.

Lachenbruch (1962) has examined this theory from a quantitative view, and the application of physical laws explains convincingly the formation and existence of the polygons. He speculates that the crack reopens at the top of the permafrost table or the top of the ice wedge, rather than the surface of the ground. Calculations showed that cooling rates of .5–1°C/day for 2–3 days will cause sufficient stress to crack the material. The formation of regular patterns (polygons) is explained in Fig. 5.6. After a crack has opened, there is a zone of stress relief where the stress will be too low to cause fracture. The stress zone is on the order of two or three times the crack depth. These initial cracks will tend to be parallel as shown. Along the loci, the stress is still high enough for cracking, and as a crack will propagate perpendicular to the direction of maximum stress, a secondary crack opens that will meet the initial crack at a right angle. Thus, ideally, a rectangular area will be delineated or, more realistically, a distorted rectangle or polygon will form. Ice wedges will tend to grow each year because they

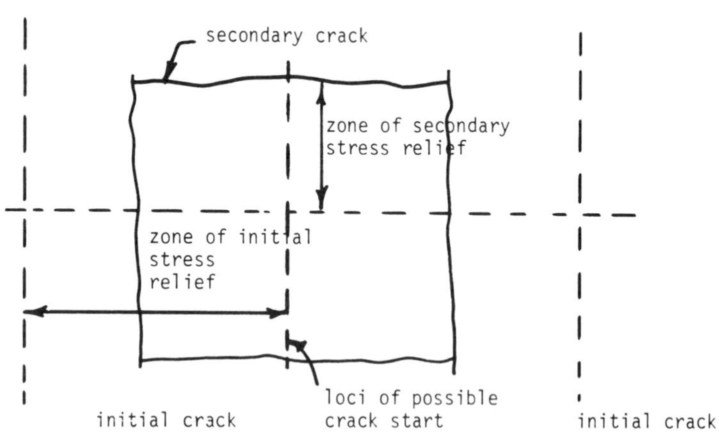

Fig. 5.6. Cracking of Permafrost in Regular Patterns.

are weak points in the system and will grow as long as they continue to crack each year.

The theory indicates that ice wedges will continue to grow as long as the stress induced each year is sufficiently large. Ice wedges may either be actively growing, inactive, or they may have vanished by melting. Hamelin (1967) points out that active ice wedges are predominant in the continuous permafrost zone where the minimum temperature at the permafrost table may be 5 to −22°F. Inactive ice wedges are characteristic of the northern part of the discontinuous zone where minimum permafrost temperatures are 26–30°F. The duration and rate of cooling of the permafrost must also be significant, however. Ice wedges that have melted and been replaced by soil, debris, or sediments are called "ice wedge casts" or "fossil ice wedges." They are common in North America and can be expected even in temperate zones. Sand wedges may also form due to the same cracking mechanism, except that sand or stones may fill in the crack rather than ice.

5.8.1 Low-Center Polygons

A growing ice wedge (active) tends to turn up the soil adjacent to it, and this soil layer also tends to expand during the summer. This causes the boundary to be higher than the center. Because active ice wedges are found in the continuous permafrost zone, it can be expected that low-center polygons will be predominant there also.

5.8.2 High-Center Polygons

When erosion, thawing, or depositions are more important than the upthrust of soil along a fissure, then the ridges will be absent, and there may be no polygons at the surface or the polygons may be higher at the center than at the boundaries. These polygons may be indicative of inactive ice wedges or slowly growing active wedges. As the ice wedges may be buried without any surface expression, these forms of ground ice present particular difficulties from an engineering viewpoint.

5.9 PINGO

"Pingo" is an Eskimo term for a symmetrical, conical hill with an oval base. They are ice-cored and may rise 150 ft above the surrounding terrain. The summit is often ruptured with a star-shaped fissure. They represent a striking feature of the continuous and discontinuous permafrost zones.

Two genetic types of pingo have been classified by Muller (1963): the open system, or East Greenland type, and the closed system, or Mackenzie type.

5.9.1 Open System

During the thaw season, melting of the active layer occurs. Below this the permafrost is impermeable to water, and meltwater or surface water rests on the permafrost or flows parallel to the permafrost table. Freezing commences at the surface, and liquid water is trapped between two horizontal, impermeable layers. The pressure buildup may force some of the water toward the surface at a weak point of the soil. This is called a "frost boil" if considerable sediment is carried with the water. If the trapped water lens is large enough, the pressure gradual enough, and the weakest portion of the soil can undergo plastic deformation, a blistering of the soil may occur. The water is only a few degrees warmer than the permafrost and will freeze within the blister, which may reach up to 100 ft in height with a massive ice core. The water is continually supplied from outside the pingo system. Their formation may take place over several seasons. These pingos are common in East Greenland, hence the name.

5.9.2 Mackenzie Type

Closed type pingos are common in the Mackenzie Delta, where over 1400 have been identified (Mackay, 1963). The name refers to the state of the water that forms the ice core of the pingo. The origin is related to the talik, or thawed bowl, commonly found beneath lakes. If the lake is drained, the thaw bowl will begin to freeze downward, as shown in Fig. 5.7*b*. This will trap a volume of water in the saturated soil originally beneath the lake. As the volume decreases, the pressure increases, and an uplift force is developed that will lift the soil layer lying above the water volume and cause it to blister. Water will fill this blister and freeze to form the ice core. The time scale for the formation of these pingos may be tens of years to several centuries,

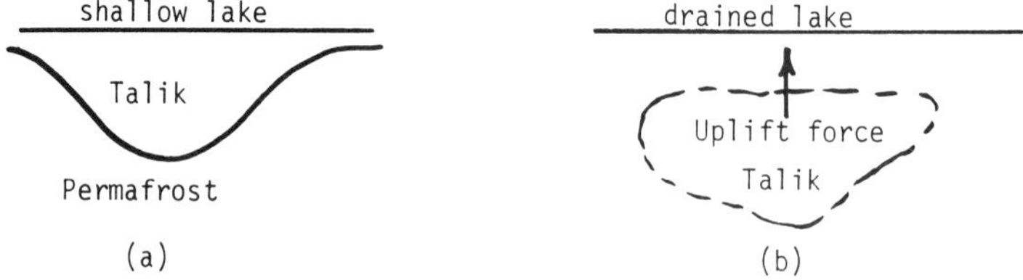

Fig. 5.7. Formation of Closed Type Pingo.

depending upon the temperature regime. Mackenzie pingos commonly are found at the center of drained lakes and may be destroyed due to subsequent melting of the ice core and collapse of the crown.

Pingos are found in the Mackenzie Delta, Yukon, East Greenland, commonly in Alaska, Baffin Island, and scattered widely throughout Siberia (Hamelin 1967).

The Eskimos of Tuktoyaktuk, NWT, Canada, and probably elsewhere, use pingos as natural cold-storage rooms, digging out spacious and picturesque chambers in the ice cores.

5.10 PALSA

This is a Finnoscandian term for a rounded hill or mound of 10 m maximum height, with a peat layer up to 7 m, overlying mineral soil. It has a permafrost core that reaches from the overlying peat down toward the mineral soil (Brown 1968). Palsas appear mainly in the discontinuous permafrost zone and are often the only evidence of permafrost in a given area. Peat mounds and peat plateaus generally refer to the same phenomenon, although the plateaus are more extensive areally.

The origin of palsa is uncertain, but Brown (1968) notes that originally low mounds of peat may protrude a few inches above the water level, in the middle of shallow ponds, due to intensive frost action and ice lensing (differential frost-heaving) when the pond freezes to the bottom in the winter. During the summer, the exposed peat dries and protects the underlying frozen soil from thawing, due to its low thermal conductivity, marking the beginning of a permafrost condition. For example, the thermal conductivity of dry, thawed peat is about 0.2 kcal/hr-m-°C, and frozen, saturated peat has a value of about 2.0. This change in thermal properties can lead to a significant net energy loss from the underlying soil. In addition, winter winds remove the insulating snow cover from the higher peat mounds, favoring deeper frost penetration and lower ground temperatures. Lindqvist (1965) states that the critical snowfall accumulation for the winter may not exceed 12 in. Peat accumulates yearly, and the mounds grow perhaps due to frost-heaving and water suction; water is important for the growth of palsas, and the high capillarity of peat will draw water to the freezing front and initiate the massive ice cores. The growing palsas may coalesce into peat plateaus. Mature palsas have mosses and lichens growing on the surface, and the resulting insulating cover, if disturbed, may initiate degradation of the permafrost in accord with the local climate. Differential thaw and settlement lead to uneven surfaces, with large blocks of thawed peat breaking off at the edges. It is not known if the process is cyclic. Palsas are similar in many ways to small pingos.

5.11 THERMOKARST

"Karst" is a term related to the development of holes or sinks in rock due to the solution of limestone or other soluble materials. Thermokarst bears a resemblance to this in external appearance, but the physical mechanisms and causes are totally different. If massive ground ice or permafrost with a high ice content is thermally disturbed and melting occurs, the volume change of the ice plus the thaw consolidation, i.e., the loss of soil volume due to loss of excess water and change of pore structure, can lead to a depression in the terrain. This depression or surface collapse is called "thermokarst." The annual melting in the active layer will not produce a thermokarst, as this occurs every year and the terrain has adjusted to these annual effects. If the insulative or upper soil layers are disturbed, damaged, or destroyed, however, the surface heat balance will be altered, and a melting of the underlying permafrost can lead to the thermokarst condition.

Often the thermokarst process is accompanied by thaw-flowing, or thaw-slumping, a process whereby the melting water forms a mud that can flow over the surface similar to a temperature climate earth flow. Thaw-flowing differs from solifluction in that it moves at a faster rate and has a breakaway scarp.

Thermokarst can be due to natural conditions or man-induced changes. Some spectacular results have occurred due to inadvertent or unavoidable changes in the active layer caused by man. A well-known case is very obvious near Inuvik, NWT. A forest fire threatened buildings, and a bulldozer was used to scrape a firebreak 8 ft wide and about 6–12 in. deep, which completely removed the vegetative layer. Massive ground ice underlay the area, and within 2 years, a thermokarst 40 ft wide and 12–15 ft deep had developed. In this case, the slumping surface was augmented by the gravity flow of excess water, mud, and sediments. Changes of this magnitude have alerted the engineer to the importance of thermokarst effects.

Natural thermokarst processes lead to a thermokarst topography that can include the following phenomena.

"Alas" is a lowland that can be many square kilometers due to the melting and disappearance of ground ice, with a lowering of the ground surface by 5–20 m (Brown 1974).

"Depression," or "thaw depression," describes any small form of an alas or simply an old thermokarst depression.

"Thaw lake," or "cave-in lake," is a lake formed by the filling of thermokarst depressions. Some of these lakes may have a preferred direction and are called "oriented lakes" (Makay 1963). The wind direction appears to be a factor in the orientation.

"Mound" refers to a residual polygonal hummock surrounded by depressions. "Cemetery mounds" are hummocks formed after the thawing of ice wedge boundaries of tundra polygons. They are also called "thermokarst mounds."

"Collapse scar" is a wet, treeless depression several meters long, adjacent to the slumping edge of a peat plateau or palsa, caused by the thawing of permafrost. The thawing edge often appears as a steep bank with leaning trees, giving rise to the expression "drunken forest."

"Beaded stream" occurs in permafrost regions and is a stream that appears to have small pools connected by the flowing stream proper. These beads are due to the melting of underlying ground ice, leading to depressions that the stream fills as small pools. Where the stream has dried up, the pools will be connected by dry stream beds.

5.12 ICING

The lower portion of an inclined aquifer that is on top of permafrost may become impermeable due to early freezing. Water then breaks out upslope of the frozen obstruction and freezes into an ice mass. This is often due to man-made obstacles, such as a road that intersects the natural drainage system and freezes early (a single set of snowshoe tracks has been blamed in Alaska). The resulting icing may make the road unuseable. Icing may also be due to a stream overflowing its banks during freezeup. The development of icing is favored by underlying impermeable permafrost, and the resulting ice masses are often buried and form new permafrost. Icings may be a few tens of square meters or they may involve thousands of acres or millions of cubic meters of ice. A road or other engineering structure that causes an icing may be impossible to use without very costly maintenance or extensive changes in the natural drainage. Icings are often referred to as *aufeis* (German) or *naleds* (Russian).

5.13 PATTERNED GROUND

The micromorphology of cold climate regions often reveals striking patterns of surface features, soil, vegetation, and rock that have regular geometrical features. Many of these patterns can be explained in terms of frost action and cryoturbation, the movement of soil due to frost action. The origin of patterned ground is polygenetic, however, with many of the processes discussed earlier acting simultaneously. Patterned ground is not confined to permafrost regions, but it has its largest scale and most spectacular form in these areas. Peatlands also display patterned ground, chiefly due to vegetative features.

The polygonal networks associated with tundra polygons have already been described. Polygonal ground may be sorted with a border of stones surrounding a fine-textured center or unsorted without a border of stones. These polygons cover tens of thousands of square miles of arctic terrain. Other prominent patterns are as follows.

5.13.1 Nonsorted Circles

Clay boils caused by frost-heave and water flow in fine mineral soils due to intergranular pressures will leave a series of circular hummocks on the order of 1–3 m high, with vegetation surrounding the hummock. The bare central area is often cracked into small dessication polygons. Hummocks, or earth hummocks, are similar patterns due to frost action, in frost-susceptible soils, with a vegetative cover. They can cover extensive areas and are a considerable impediment to walking or off-the-road vehicles.

5.13.2 Sorted Circles

These patterns are very regular in shape and consist of a circle of fines surrounded by a ring of stones or coarser material. The interior fines may gradually vary to the stones of the rim, or the change in particle size may be abrupt. The rim can be 1 ft high and 6 in. wide (Hamelin 1967). Alternate freezing and thawing is thought to be the mechanism associated with the sorting of the material, but the process is only vaguely understood. Differential forces acting on the particle volume or mass must be present to cause an outward migration and perhaps disintegration of the larger particles.

5.13.3 Nets

Nets are patterned ground intermediate between circles and polygons. They are similar to impinging circles and polygons, being of the same size as these features, and they exhibit both sorted and unsorted features. The origin of nets is probably similar to that of circles and polygons.

5.13.4 Steps

This feature occurs in groups with a steplike form and a downslope border of stones or vegetation forming a bank for upslope ground, which is relatively bare. Washburn (1973) notes that they are derived from circles, polygons, and nets, rather than arising independently. They are both sorted and non-sorted, with the nonsorted steps derived mainly from hummocks. Steps are restricted to slopes, mainly 5 to 15°.

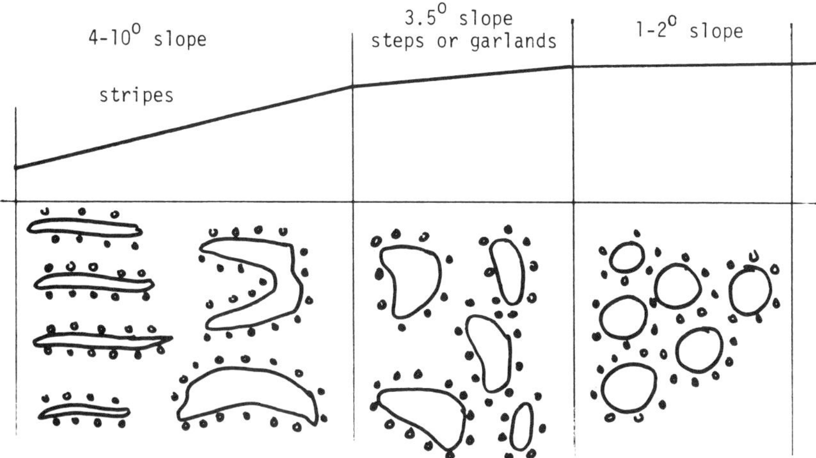

Fig. 5.8. Sorted Stone Circles Changing to Sorted Stripes on a Smooth Slope.

5.13.5 Stripes

As the name implies, stripes are patterned ground with distinct stripes that can be quite spectacular when of the sorted type. In this case, alternate rows of fines and stones orient down the steepest slope available. The stone layers may be 1 cm to 1.5 m wide, with the intervening fines several times wider. They are up to 120 m long on slopes of about 3–7°. Nonsorted stripes are similar, but are alternate rows of bare ground and vegetation. The vegetation width is usually 1 in. or less. Davies (1969) describes the formation of both steps and stripes from circles, as can be seen in Fig. 5.8.

Washburn (1973, 1979) has described the mechanisms that may give rise to patterns and the patterns associated with them, in terms of cause and effect. These relations were given in tabular form, but it should be noted that many of the causes and effects are speculative. This is emphasized by the many question marks in Washburn's tables. The mechanisms and patterns are as dicussed in this chapter.

REFERENCES

Brown, R. J. E. 1968. *Permafrost investigations in northern Ontario and northeastern Manitoba.* DBR, NRC 10465, Ottawa.

———, and Kupsch, W. O. 1974. *Permafrost terminology.* NRC Canada, Tech. Memo 111, NRCC 14274.

Corte, A. E. 1963. *Vertical migration of particles in front of a moving freezing zone.* CRREL Res. Rept. 105, Hanover, New Hampshire.

Davies, J. L. 1969. *Landforms of cold climates.* Cambridge, Massachusetts: MIT Press.

Hamelin, L.-E., and Cook, F. A. 1967. *Le periglaciaire par l'image.* Quebec: Les Presses de l'Université Laval.

Higashi, A. 1958. *Experimental study of frost heaving.* U.S.A. SIPRE Res. Rept. 45.

Lachenbruch, A. H. 1962. *Mechanics of thermal contraction cracks and ice wedge polygons in permafrost.* Geological Society of America, Special Paper No. 70.

Leffingwell, E. deK. 1915. Ground ice wedges; The dominant form of ground ice on the north coast of Alaska. *J. Geol.* 23:635–654.

Lindqvist, S., and Mattsson, J. O. 1965. *Studies on the thermal structure of a palsa.* Lund Studies on Geography, Ser. A. Physical Geography No. 34, Univ. of Lund, Sweden, pp. 38–49.

Mackay, J. R. 1963. "Origin of the pingos of the Pleistocene Mackenzie Delta area. *Proceedings first Canadian conference on permafrost.* 79–82, NRC TM 76.

———. 1963. *The Mackenzie Delta area.* NWT Geography Branch Memo No. 8, Mines and Technical Surveys, Canada.

———. 1972. *The world of underground ice.* Assoc. Am. Geog. Annals 62:1–22.

Muller, F. 1963. *Observations on pingos.* NRCC T.T. 1073.

Penner, E. 1973. *Frost heaving pressures in particulate materials.* DRB Res. Paper No. 579, NRC 13674.

Washburn, A. L. 1973. *Periglacial processes and environments.* London: Edward Arnold.

———. 1979. *Geocryology—A survey of periglacial processes and environments.* New York: Halsted-Wiley.

6
Energy Balance at Earth-Atmosphere Boundary

6.0 INTRODUCTION

The change in the thermal regime of a stratum of soil or permafrost is controlled by the energy fluxes at its boundaries. The lower boundary of the system is the lithosphere (see Chap. 4), where the geothermal gradient controls the heat flow, and the upper boundary is the surface, or contact, layer. These relations may be noted in Fig. 6.1.

The temperature of any solid body in contact with a fluid medium is governed by two kinds of processes:

(a) Energy exchange between the solid and the medium, possibly accompanied by conversion of one form of energy into another (radiant energy to internal energy, etc.).
(b) Heat propagation within the solid (conduction, convection, or radiation).

Processes of the second type take place within the body, and those of the first type occur in a layer separating the solid and the fluid. The thickness of this layer, called the "surface," or "contact," layer, depends upon the composition of the body and the fluid; it may be quite thick or essentially zero and can include vegetation, snow, etc. Within this layer, the following energy transfer processes take place:

(a) Heat transfer without change in the mechanism of heat transfer, i.e., conduction in the contact layer and to or from the surrounding fluid medium and the underlying soil.
(b) Heat transfer with change in the heat transfer mechanism, i.e., convection or radiation through the layer and then into the soil by conduction.
(c) Conversions of various forms of energy to heat (chemical, electrical, etc.).
(d) Conversion of radiant energy to chemical energy by photosynthesis.
(e) Phase transformation of water with the liberation or absorption of energy (condensation, evaporation, and fusion of water, snow, or ice).

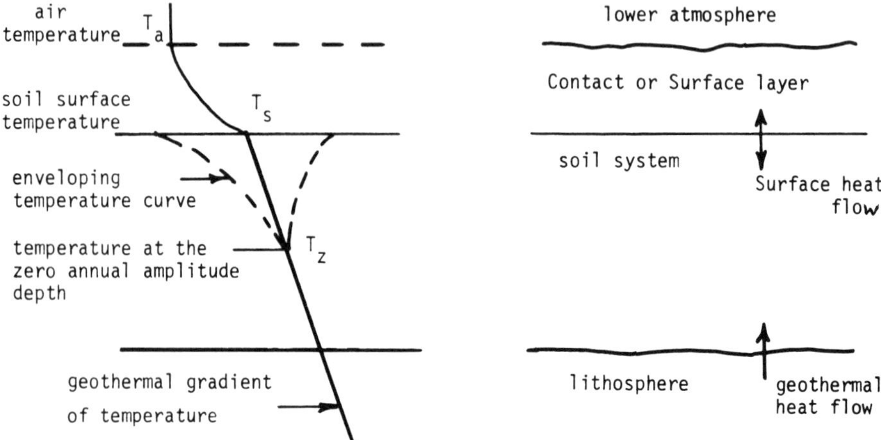

Fig. 6.1. Temperatures and Energy Flow between Earth and Atmosphere.

The energy transfers in the contact layer are extremely complex and change very rapidly with changes in the surrounding fluid medium. These changes are attenuated rapidly as the soil is penetrated downward; however, the depth of penetration of different kinds of energy is different. The lower boundary of the contact layer is defined in terms of those processes that penetrate deepest. For example, grass cover will have a lower contact layer boundary at the mineral soil surface under the sod where energy transfer is basically by conduction with some mass transfer. The soil in the sod, held by the root system, is part of the contact layer. If there is no surface cover, the contact layer is the upper layer of soil where energy conversions take place. The upper surface is assumed to be the regular surface encompassing all irregularities of the microrelief and vegetation. Thus, one may envisage a flow of energy from the medium to the contact layer, transformation to heat in this layer, and transfer of heat into the soil volume or the reverse.

6.0.1 Geophysical Processes Involved in Energy Transfer

The physical processes significant in the energy transfer can be listed as follows:

1. Surface radiation—the net sum of all the radiant heat flows occurring at the surface, including solar shortwave radiation and the longwave radiation of the surface and the atmosphere. These terms make up the most significant energy flows in the permafrost regions.

2. Atmospheric convection—turbulent heat transfer between the atmosphere and the surface layer.
3. Moisture flow—this includes the energy flows associated with the humidity of the atmosphere, i.e., evaporation and condensation, as well as the movement of moisture in the various layers.
4. Phase change—this refers to the physical change of state of moisture in any of the layers except the atmosphere, which already involves the phase change of evaporation-condensation.
5. Geothermal heat-transfer—flow of energy by conduction and convection in layers below the zone of oscillating temperature, called the "lithosphere."
6. Conduction and convection—modes of energy transfer that may exist in any of the layers.

6.0.2 Energy Balance of the Contact Layer

The energy balance of the contact layer may be written as follows for an instant of time

$$Q_N + Q_G + Q_H + Q_E + Q_B = 0 \tag{6.1}$$

where

Q_N = net radiation at the surface
Q_G = transfer of heat through the ground
Q_H = convective component at the surface
Q_E = latent heat due to evaporation, melting snow, etc. (evaporation is a negative term) and
Q_B = consumption of energy due to biochemical and other processes.

Equation 6.1 assumes that the contact layer has no mass and, thus, there will be no stored energy for this layer. This will result in errors depending upon the physical composition of the actual contact surface. Heat flow out of the contact layer is negative, and into the contact layer it is positive. Thus, a flow of energy Q_G out of the contact layer is a loss for the contact layer and a gain for the underlying soil system. Evaporation in the contact layer is always a negative heat flow, and condensation is always an energy gain.

In Eq. (6.1), some of the components may be zero and some are made up of a number of other terms. This equation allows a boundary condition to be established at the upper surface of the soil, which is used with the energy equation to obtain temperature profiles. Under natural conditions

Q_N, Q_H, and Q_E are the predominant terms, and Q_G is the unknown value of concern in the soil temperature question. Because Q_G is small compared to Q_N, etc., other small terms must be retained in Eq. (6.1) for accuracy. At present, the evaluation of Q_G from a surface energy balance is imprecise, due to the relative magnitude and uncertainty of the terms.

For example, measurements made in Russia (Budyko 1956) show $Q_N =$ 150,000 $\pm$ 15,000 kcal/m^2-yr, while Q_G is on the order of 1000 kcal/m^2 $-$ yr. Clearly, in this case, the measurements cannot yield reasonable values of the unknown Q_G, even if, at present, the net radiation can be measured to $\pm 5\%$.

6.1 RADIATION

The net radiative balance at the surface layer of the earth is

$$Q_N = Q_{TC} - Q_R - E_e + E_s\alpha_e \tag{6.2}$$

where

Q_{TC} = shortwave radiation from sun and sky, the total global radiation
Q_R = shortwave radiation reflected from the earth
E_e = longwave radiation emitted by the surface of earth
E_s = longwave radiation from the atmosphere and
α_e = longwave radiation absorptivity.

These relations are illustrated in Fig. 6.2.

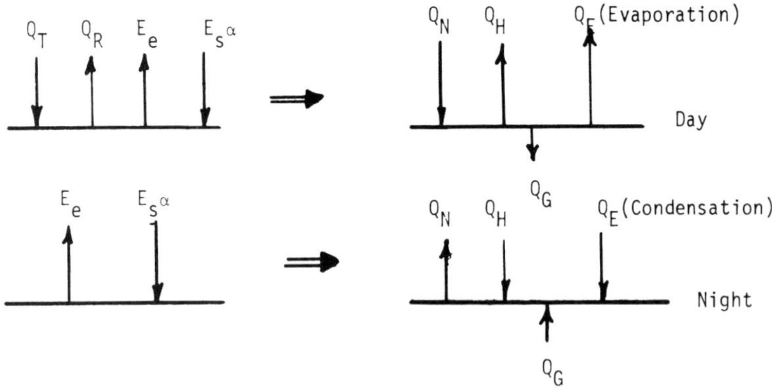

Fig. 6.2. Radiation Fluxes at Surface of Earth.

6.1.1 Annual Radiation Balance

We shall examine, in some detail, the energy transfer mechanisms, but it is useful to first look at a rough picture of the annual flow of radiant energy for the earth and its atmosphere. Consider 100 units of solar energy, which reach the top of the earth's atmosphere as shown in Fig. 6.3.

The following nomenclature may be used for the solar radiation:

Q_{TD} = direct radiation to surface of earth
Q_{TS} = diffuse radiation to surface of earth
Q_{RE} = radiation reflected from surface of earth
Q_{Rc} = radiation reflected from atmosphere
Q_{at} = radiation absorbed in troposphere and
Q_{as} = radiation absorbed in stratosphere.

Notice that about 43% of the solar energy is reflected back to space from the atmosphere and the surface combined. The albedo is defined as the ratio

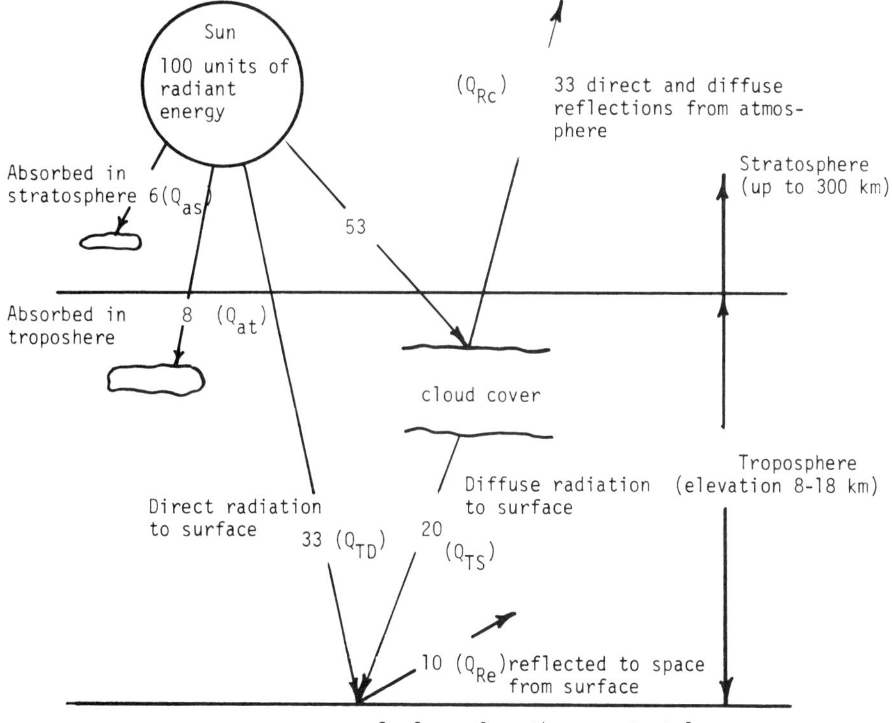

Fig. 6.3. Shortwave (Solar) Radiation Interactions for Earth-Atmosphere.

of the energy reflected to the incident energy. Thus, the albedo of the atmosphere, for solar radiation, is .33 while the albedo of the earth-atmosphere is .43. About 53% of the solar energy reaches the surface layer, and, in this example, about 10% is reflected; then the albedo of the surface layer is 0.189.

Figure 6.4 shows the longwave radiation flow for the earth-atmosphere. In this example, the surface layer emits 115 units, which is more than the solar input. This is not incorrect, but merely reflects the complicated nature of the recycling of the radiant energy in the earth-atmosphere system. The energy balances for subsystems of the total system are shown in Fig. 6.5.

Note that for the yearly energy balance at the surface layer, the heat flow into the ground is zero. This means that exactly as much energy flows out of the ground as flows in over the total year. As all the energy flows are balanced, this means that the mean temperature of the atmosphere and the earth remain constant over long periods. Of course, the temperature will fluctuate daily and seasonally about these mean values. All these values are rough estimates and cannot be used for any specific case and certainly not for short time periods.

As was mentioned before, the radiation terms involve the most significant energy flows in the energy balance. These terms will be examined in the following sections.

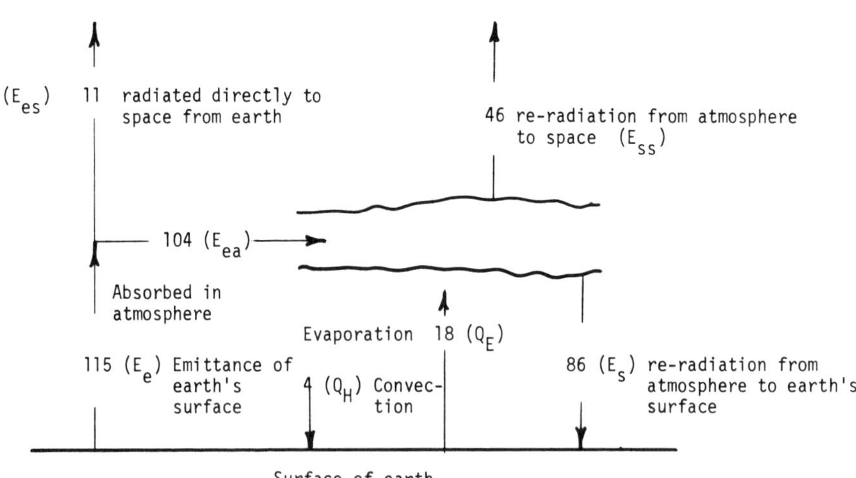

Fig. 6.4. Longwave Radiation Interactions for Earth-Atmosphere (Based on 100 Units of Solar Irradiation).

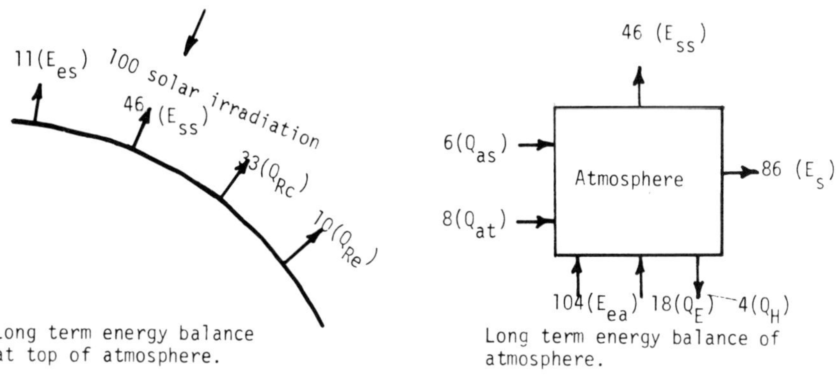

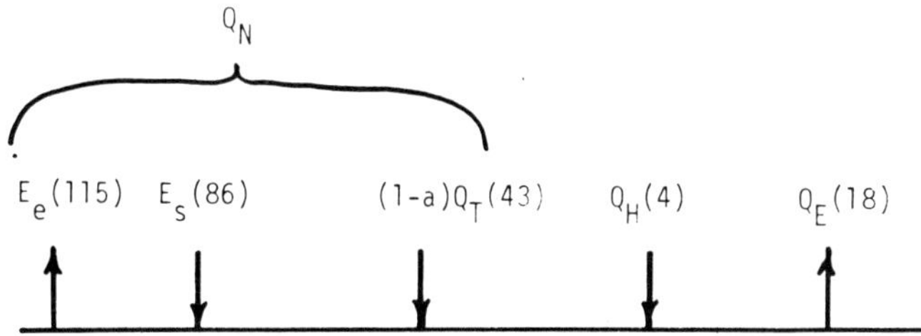

Fig. 6.5. Long-Term (Yearly) Energy Balance at Surface of Earth.

6.1.2 Shortwave Radiation

The electromagnetic spectrum is divided into a series of wavelength bands as shown below.

	Wavelength, microns[a]	Wavelength, Angstroms[b]
Cosmic Rays	$<4 \times 10^{-6}$	$<.004$
Gamma Rays	4×10^{-6}–1.4×10^{-4}	.004–1.4
X Rays	10^{-5}–2×10^{-2}	.1–200
Ultraviolet	10^{-2}–3.9×10^{-1}	100–3900
Visible light	3.9×10^{-1}–7.8×10^{-1}	3900–7800
Infrared	.78–1000	7800–10^{7}
Thermal	.1–100	1000–10^{6}
Radio and Hertzian	1000–2×10^{7}	10^{7}–2×10^{11}

[a] 1 micron = 10^{-4} cm.
[b] 1 Angstrom = 10^{-8} cm.

These bands tend to overlap, and, in fact, there is no physical difference in the radiation of each band. The names are merely historical artifacts that are convenient. The wavelength of the radiation is inversely proportional to the temperature of the source, and, thus, the higher the temperature of a source, the lower is the wavelength. The sun acts like a black body at about 5860°K or 10,460°R, and the earth behaves like a black body at 288°K. The energy spectrum of black bodies at these temperatures is such that the radiant energy of the earth and sun do not significantly overlap.

Black Body Source	99% of energy lies between
Sun	0.15–4 μ (50% in the visible spectrum)
Earth	4–120 μ (infrared or thermal)

It is customary to describe solar energy, either direct or reflected, as "shortwave" radiation and to call the radiant energy of the earth or atmosphere "longwave" radiation. As can be seen from the above values, there is very little overlap of the solar energy spectrum and that of various parts of the earth-atmosphere system.

6.2.2.1 Solar Constant

The monochromatic steradiance emitted by a black body per unit area, per unit time, and per unit solid angle normal to the surface is given by Planck's famous law

$$i_{b,\lambda} = \frac{2c_1 c^2 \lambda^{-5}}{\left[\exp\left(\frac{c\,c_1}{c_2\,\lambda T}\right) - 1\right]} \tag{6.3}$$

where

$c = 2.998 \times 10^{10}$ cm/sec, speed of light
$c_1 = 6.6254 \times 10^{-27}$ erg sec, Planck's constant and
$c_2 = 1.38049 \times 10^{-16}$ erg °K^{-1}, Boltzmann constant.

The radiation for all wavelengths from the surface can be found from

$$i_b = \int_0^\infty i_{b,\lambda}\, d\lambda = \sigma T^4 \tag{6.4}$$

where

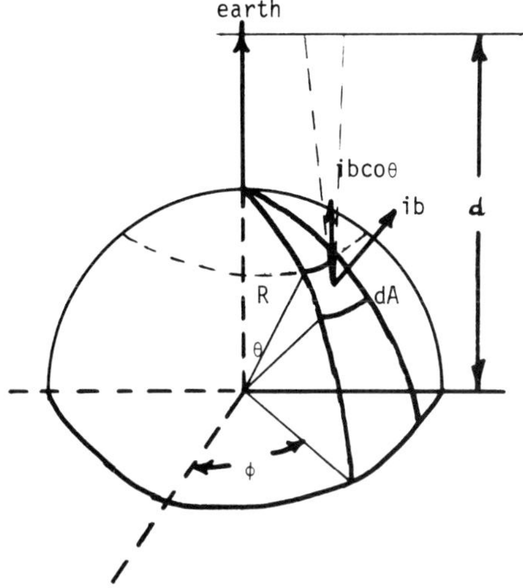

$$\sigma = \frac{2\pi^4 c_2^4}{15 c^2 c_1^3} \quad \text{is the Stefan-Boltzmann constant.}$$

$$\sigma = 5.6702 \times 10^{-12} \frac{\text{watt}}{cm^2\, K^4} = 0.1714 \times 10^{-8} \frac{\text{Btu}}{\text{hr ft}^2\, \text{R}^4}$$

$$= 1.170 \times 10^{-7} \frac{\text{ly}}{\text{day K}^4}$$

The langley is given by: 1 ly = 1 cal/cm² = 3.685 Btu/ft². Now the energy leaving an area dA of the sun, which strikes a unit area of the earth, is given by

$$ds = i_b \cos\theta \; dA \; d\omega \tag{6.5}$$

where

$dA = R^2 \sin\theta \; d\theta \; d\phi$
$d\omega = (1/r^2)$, solid angle subtended by unit area at distance of earth from sun
$r = 1.4968 \times 10^{13}$cm and
$R = 6.965 \times 10^{10}$ cm.

Integrating Eq. (6.5) over the total solar hemisphere gives

$$S = \int_o^{2\pi} \int_o^{\pi} \sigma \frac{T^4 R^2}{r^2} \cos\theta \sin\theta \; d\theta \; d\phi$$

Thus

$$S = \frac{T^4 R^2 \sigma}{r^2} \pi \tag{6.6}$$

Depending upon the solar temperature and distances used, this leads to a value of S of about 2 cal/min-cm^2. Allen (1958) notes a value of S_m = 1.99 ± .02 ly/min, and it is generally accepted that the solar constant—the radiation falling on a surface perpendicular to the sun's rays—at the top of the atmosphere is 2 ly/min. The solar constant fluctuates slightly due to geometrical changes and changes of the solar activity.

Stringer (1972) gives a relation for the solar constant at any instant

$$S = S_m \left(\frac{r_m}{r}\right)^2 \tag{6.7}$$

$$\frac{r}{r_m} = 1 - 1.6733 \times 10^{-2} \cos (.9856\ D) \tag{6.8}$$

where

r_m = mean earth-sun distance
S_m = 1.99 ly/min, mean solar constant and
D = days elapsed since December 31.

6.1.2.2 Direct Irradiation of Horizontal Surface

The motion of the sun as seen from a point on the earth is shown in Fig. 6.6. Consider a surface at the latitude ϕ as shown. The radiation falling on this surface depends on the angle between the normal to the surface and the sun, i, as well as the amount of radiant energy absorbed by the atmosphere. Consider a horizontal surface as shown in the sketch. The angle between the normal and the sun's direction is θ, the zenith angle. Suppose the change in intensity of the radiation at any point is proportional to the intensity (Beer's law). Then

$$dI = cI\ dx \tag{6.9}$$

or in terms of the zenith angle and elevation y

$$dI = -cI \sec \theta\ dy$$

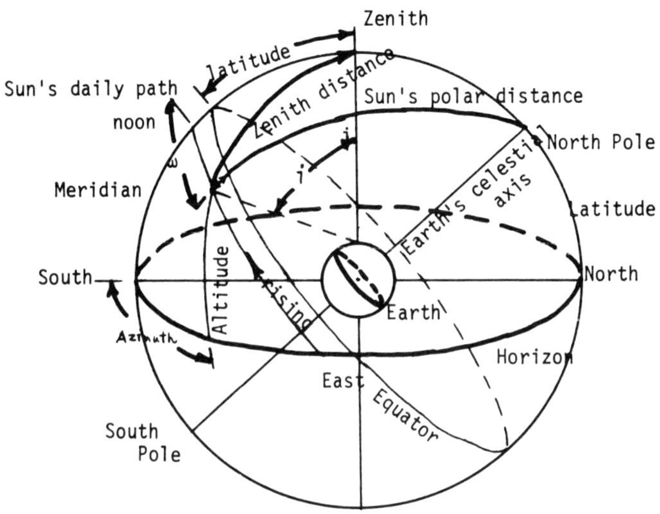

Fig. 6.6. Celestial Sphere and Sun's Coordinates Relative to Observer on Earth at Point *C*.

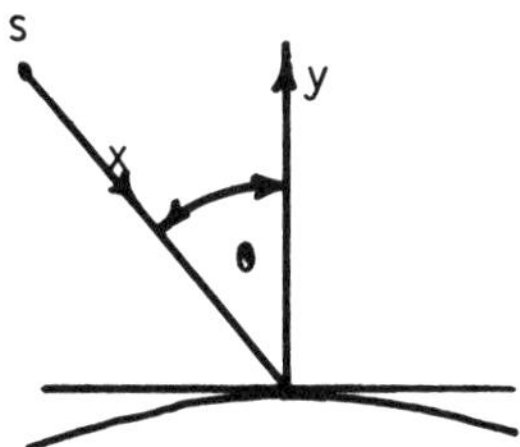

The solution to this equation is

$$I_n = S\, e^{-\sec\theta \int_o^\infty c\, dy} \equiv S \left(e^{-\int_o^\infty c\, dy} \right)^{\sec\theta}$$

Because the path of a ray does not follow this relation at all zenith angles, this equation is often written

$$I_n = S\,\tau_a^m \tag{6.10}$$

Here m is the relative air mass, defined as the ratio of the actual path traversed by the radiation to the shortest possible path. As can be seen from the above, this is sec θ if the atmospheric thickness is small compared with the radius of the earth. This will be sufficiently accurate for zenith angles between 0 and 80°.

Kasten (1964) calculated the optical air mass and obtained an approximate relation that is quite accurate

$$m = \frac{1}{\cos\theta + .15\left(\frac{\pi}{2} - \theta + 3.885\right)^{-1.253}} \tag{6.11}$$

The quantity τ_a is the transmission coefficient defined as

$$\tau_a = e^{-\int_0^\infty c\,dy}$$

The transmission coefficient depends upon the energy absorbed in the atmosphere. This is due to water vapor, carbon dioxide, ozone, and various pollutants. It ranges from about 0.81 on a clear day to 0.62 on a cloudy day, with a value of 0.7 being generally used. The energy now falling on the horizontal surface is

$$I_i = I_n \cos\theta = S\,\tau_a^m \cos\theta \tag{6.12}$$

6.1.2.3 Extraterrestrial Radiation

For the moment, let us consider the energy falling upon a horizontal surface outside the earth's atmosphere or upon the earth's surface if the atmosphere were transparent ($\tau_a = 1$)

$$I_o = S\cos\theta \tag{6.13}$$

The zenith angle is given by

$$\cos\theta = \sin\delta\sin\phi - \cos\delta\cos\phi\cos H \tag{6.14}$$

where

δ = solar declination angle
$H = (2\pi t/24)$ — hour angle and
t = hours measured from midnight. If t were measured from noon, as is often the case, the sign in Eq. (6.14) would be positive.

After Degelman (1966), the solar declination can be estimated from

$$\delta = \sin^{-1}[\sin 23.5 \cos\{.9863(D\text{-}172)\}] \tag{6.15}$$

The daily, direct solar radiation is

$$S_D = \int_{\text{sunrise}}^{\text{sunset}} S \cos\theta \, dt \tag{6.16}$$

From Eq. (6.14), since $\cos\theta = 0$, at sunrise or sunset (apparent solar time), the hour angle at sunrise is

$$H_{sr} = \cos^{-1}(\tan\delta \tan\phi) \tag{6.17}$$

Therefore

$$t_{sr} = \frac{12}{\pi} H_{sr} \tag{6.18}$$

$$t_{ss} = 24\text{-}t_{sr} \tag{6.19}$$

The length of daylight is

$$L = 24\left(1 - \frac{H_{sr}}{\pi}\right) \tag{6.20}$$

Example:

Ottawa $\quad \phi = 45.45°\text{N}$, Jan. 10, $\quad D = 10 \quad \delta = -21.96$

$H_{sr} = 114.15$, $\quad$ Longitude $= 75.62°\text{W}$

$t_{sr} = 7.61 = 7{:}37$ AM

$t_{ss} = 16.39 = 4{:}23$ PM

Standard time $= 4{:}23 + 4(75.62 - 75) + 0.08 = 4{:}33.5$

Now

$$dt = \frac{12}{\pi} \, dH \qquad \textit{and}$$

$$S_D = \int_{\text{sunrise}}^{\text{sunset}} (\sin\delta \sin\phi - \cos\delta \cos\phi \cos H) \, dt = L \sin\delta \sin\phi - \frac{24}{\pi} \cos\delta \cos\phi \int_{H_{sr}}^{\pi} \cos H \, dH$$

Thus, the daily solar flux is

$$S_D = 60 \times 24\, S\left(\frac{180 - H_{sr}}{180}\sin\delta\sin\phi + \frac{\cos\delta\cos\phi\sin H_{sr}}{\pi}\right) \tag{6.21}$$

or

$$S_D = 60 \times 24\, S\left(\frac{180 - H_{sr}}{180}\cos H_{sr} + \frac{\sin H_{sr}}{\pi}\right)\cos\delta\cos\phi \tag{6.22}$$

Above the Arctic Circle, $\phi \geq 66°33'$; then 24 hours of daylight or night can occur. Then

$$S_D = \begin{cases} 60 \times 24 S \sin\delta\sin\phi & H_{SR} = 0 \quad \text{daylight} \\ 0 & H_{SR} = \pi \quad \text{night} \end{cases}$$

6.1.2.4 Effect of Atmosphere

When the solar radiation enters the atmosphere, the flux of radiation is split into three parts. Part of the beam is scattered in all directions by air molecules, water vapor, dust, etc. Part is absorbed, chiefly by water vapor, ozone, and carbon dioxide, and the remainder passes through the atmosphere unchanged to the surface. The direct radiation to the surface is given by Eq. (6.12) as

$$Q_{TD} = \int_{\text{sunrise}}^{\text{sunset}} S\,\tau_a^m \cos\theta\, dt$$

which can be written as

$$Q_{TD} = a\int_{\text{sunrise}}^{\text{sunset}} S\cos\theta\, dt = a\, S_D \tag{6.23}$$

with reference to Fig. 6.7.
Here a is a transmission coefficient that is a function of the path length (the average daily optical air mass) and the absorption and scattering.

The indirect, scattered, or diffuse radiation reaching the surface is described in terms of a scattering coefficient β

$$Q_{TS} = \tfrac{1}{2}\,\beta\, S_D \tag{6.24}$$

The radiation reflected from the surface is

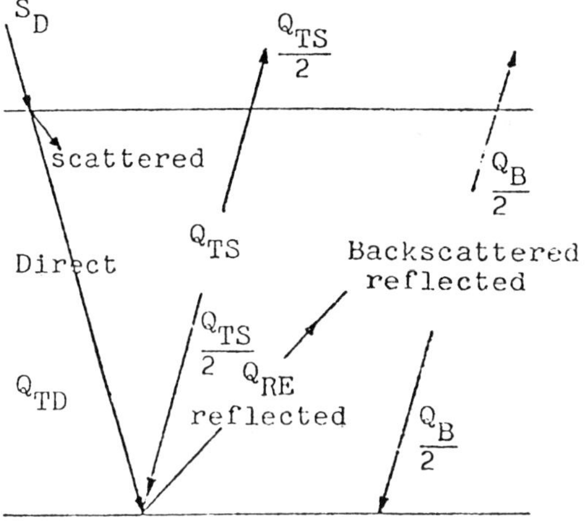

Fig. 6.7. Reflection and Scattering of Shortwave Radiation.

$$Q_{RE} = \alpha_s(a + \tfrac{1}{2}\,\beta)\,S_D \tag{6.25}$$

where α_s is the surface solar albedo or reflectivity. This quantity should not be confused with the longwave reflectivity or the emissivity.

The amount of reflected energy that is backscattered to the surface is

$$Q_B = \tfrac{1}{2}\,\beta\,Q_{RE} \tag{6.26}$$

Combining Eqs. (6.23) through (6.26) gives the total insolation (total global radiation) at the surface for clear sky conditions

$$Q_T = (a + \tfrac{1}{2}\,\beta)\,(1 + \tfrac{1}{2}\,\beta\,\alpha_s)\,S_D \tag{6.27}$$

The coefficients a and β have been obtained by a number of workers. Bolsenga (1964) notes that

$$a = a'' - d \tag{6.28}$$

$$\beta = 1 - a' + d \tag{6.29}$$

where a' is the attenuation due to scattering alone, a'' is the attenuation due to scattering plus absorption, and d is the dust attenuation. Bolsenga (1964) gives graphs for a' and a'', from which the following approximate equations were derived

$$a' = 1.041\, e^{-(.095m + .039w)} \tag{6.30}$$

$$a'' = 0.975\, e^{-(.1263m + .0513w)} \tag{6.31}$$

where m is the average daily optical air mass and w is the precipitable atmospheric water vapor in centimeters. Reitan (1963) gives an empirical equation for the water vapor as

$$\ln w = .1102 + .0614\, T_d \tag{6.32}$$

for which

w = mean monthly precipitable water vapor, cm and
T_d = mean monthly surface dew-point temperature, °C.

The average daily value of m must be evaluated from Eq. (6.11), with the average value of θ given by

$$\bar{\theta} = \frac{1}{L} \int_{\text{sunrise}}^{\text{sunset}} \cos^{-1}\left[\sin\delta \sin\theta - \cos\delta \cos\theta \cos\frac{\pi t}{12}\right] dt \tag{6.33}$$

Due to the complexity of this equation, the integration is usually done numerically, and charts have been constructed as shown in Fig. 6.8. The dust attenuation can be estimated from Table 6.1. For $\theta < 80°$, the approximation $m = 1/\cos\theta$ is acceptable; however, $m(90°) = 36.3$.

The relations presented here are quite accurate, but only approximate. If more accuracy is needed, although this seems unlikely for most engineering calculations, the details in Bolsenga (1964) can be utilized.

6.1.2.5 Effect of Clouds

The relations for the shortwave radiation striking a horizontal plane at the surface of the earth have been evaluated for clear sky conditions, i.e., the absence or near absence of clouds. Predictions of the clear sky radiation can be made within 10%. The exact coefficients one uses are not too critical, as the effects of clouds will be more uncertain than the estimates of clear sky radiation. Clouds will interpose another layer of absorbing and scattering material and must be considered in obtaining the actual radiant energy available on a surface. The ratio of the actual radiation to the clear sky radiation will be a function of the clouds

$$Q_{TC}/Q_T = f \tag{6.34}$$

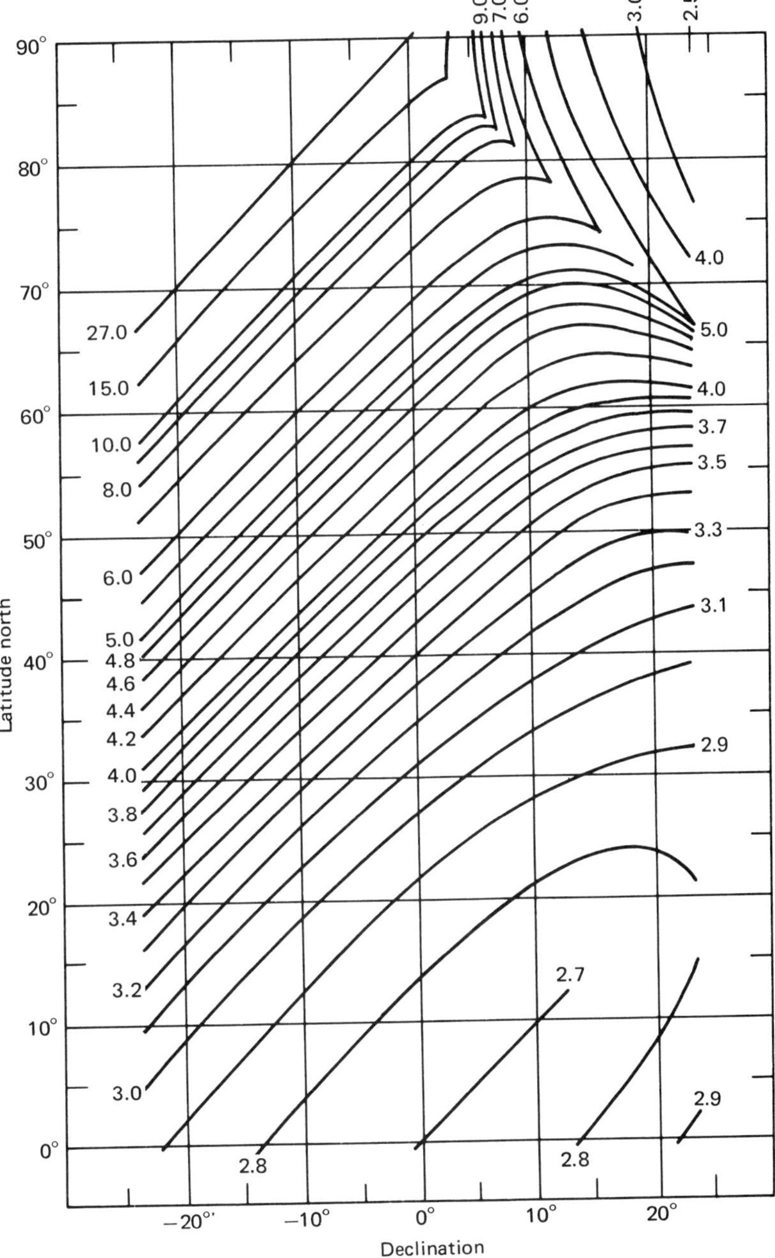

Fig. 6.8. Average Daily Optical Air Mass (after Bolsenga 1964).

Table 6.1. Mean Seasonal Values of Atmospheric Dust Attenuation, *d*, for Different Optical Air Mass, *m* (after Bolsenga 1964).

General location	Station	Season	Optical air mass, m $m=1$	$m=2$	$m=3$
U.S.A.	Lincoln, Nebraska	Winter	–	0.06	–
		Spring	0.05	0.08	–
		Summer	0.03	0.04	–
		Autumn	0.04	0.06	–
	Madison, Wisconsin	Winter	–	0.08	–
		Spring	0.06	0.10	–
		Summer	0.05	0.07	–
		Autumn	0.07	0.08	–
	Washington, D.C.	Winter	–	0.13	–
		Spring	0.09	0.13	–
		Summer	0.08	0.10	–
		Autumn	0.06	0.11	–
Atlantic Ocean	Trade-wind zone	Spring	–	–	neg.
	Equatorial zone	Spring	–	–	0.05
E. Atlantic	Cape Verde (Dunkelmeer)	Spring	–	–	0.20
	Canary Islands (Guimar)	Summer	0.00	0.01	–
Arctic Ocean	Spitzbergen (Treurenberg)	Summer	–	0.02	–
South Pacific	Samoa (Apia)	Dry	0.08	0.06	–
		Wet	0.05	0.01	–
Europe	North Sea (North Temperate Zone)	Spring	–	–	0.10
Netherlands—East Indies	Celebes Sea (Equatorial Zone)	Year	0.00	0.01	0.03

The particular form of f depends upon the type, structure, and extent of the cloud cover. In addition, one must be careful, because the comparison of cloudy sky data to clear sky data leads to f functions that depend upon the calculated values of Q_T. Different computation methods for Q_T will lead to different coefficients for f. This is not likely to be a serious error, but it should be kept in mind. The empirical coefficients of f are also dependent upon whether daily or mean monthly radiation values are being considered. One would expect that sums of daily values should lead to the monthly relations.

A. Daily Correlations. The function f has been specified as

$$f = a - b\,C_s^2 \tag{6.35}$$

or

$$f = a - b\,C_s \tag{6.36}$$

where C_s is the average fraction of sky covered by clouds during daylight hours. The value of C_s is a statistical quantity (unless calculations are made for known past conditions) and will give f value accuracies of about 20% for periods of weeks or months, but for any given day the estimate of cloud cover can be quite poor. Fortunately, most engineering work requires calculations for periods of weeks or months rather than a few days. The sunshine coefficient is sometimes used

$$f = a + b\,S \tag{6.37}$$

where S is the fraction of possible sunshine. Gorczynski (1945) gives a relation between S and C_s

$$S = (1 - C_s)(1 + .5\,C_s) \tag{6.37a}$$

This agrees fairly well with a relation derivable from the data of Scott (1964) for Camp Tuto, Greenland

$$S = (1 - C_s)(1 + C_s) \tag{6.37b}$$

The coefficients for these equations can be obtained from the following table

Reference	a	b	Eq.
Gerdel (1954)	1.0	.5	6.35
Scott (1964)	1.0	.67	6.35
Gabites (1950)	1.04	.71	6.36
Kimball (1928)	1.0	.71	6.36
Angstrom (1924)	.25	.75	6.37
Kimball (1927)	.22	.78	6.37

For Ottawa, an analysis of data by Kimura (1969) gave the following daily flux for the entire year, although there are significant seasonal fluctuations

$$f = 1.028 + .195\,C_s - .95\,C_s^2 \tag{6.38}$$

These relations can be compared as follows

C_s	Scott (1964)	Gerdel (1954)	Gabites (1950)	Kimball (1928)	Angstrom (1924)	Kimball (1927)	Kimura (1969)
0	1.0	1.0	1.04	1.0	1.0	1.0	1.03
.5	.83	.88	.69	.65	.72	.71	.89
1.0	.33	.50	.33	.29	.25	.22	.27

The preferred equation is that of Scott (1964), as it is simple and close to the average of all the results. In specific cases, local correlations will be superior. The preferred equation is then

$$f = 1 - .67\, C_s^2 \tag{6.39}$$

B. Monthly Correlations. The monthly correlations can be written as

$$f = a + b(1 - C_s)(1 + .5\, C_s) \tag{6.40}$$

where C_s is now the mean monthly cloud cover.
Fritz (1949) gives a = 0.35 and b = 0.61, though Black (1954) gives a = 0.23 and b = 0.48. Black (1956) gives a cloud cover relation that incorporates the attenuation factors

$$Q_{TC}/S_D = .803 - .34\, C_s - .458\, C_s^2 \tag{6.41}$$

This is a very simple relation, but it obviously assumes a constant clear sky transmission factor of 0.803. This is not too bad, but it cannot be as accurate as the preceding equations.

6.1.2.6 Surface Albedo for Solar Radiation

The albedo is the ratio of the flux of radiation reflected by the surface to the flux incident upon it.

$$\alpha_s \equiv \frac{Q_R}{Q_{TC}} \tag{6.42}$$

Thus, the solar energy available for energy transformations at the surface is $(1 - \alpha_s)Q_{TC}$. The quantity $\rho_s = 1 - \alpha_s$ is normally called the "reflectivity" of the surface. The albedo of a given surface will be a function of the wavelength of the radiation, and, thus, the same surface will have different albedo values for shortwave and longwave radiation.

Typical values of the solar albedo are given in Tables 6.2 through 6.6. The effect of surface modifications due to oil spilled on the vegetation can be noted in Table 6.7. The oil was on the surface for 3 years, with the vegetative cover intact, but dead. The plant cover is typical of taiga regions of the continuous permafrost zone. In this case, the oiled surfaces would absorb 10–12% more solar radiation than the unoiled, or natural, vegetation.

Table 6.2. Solar Albedo, Snow and Water.

Surface	Munn (1966)	Porkhaev (1959)	Van Wijk (1966)	Budyko (1974)
Snow, fresh	.7–.95	.85	.80–.85	.8–.95
Clean, moist snow	–	–	–	.6–.70
Snow, old	.7	–	–	–
Snow, dirty and thawing	–	.40	–	.4–.5
Compressed snow	–	–	.70	–
Melting snow	–	–	.30–.65	–
Water ($\theta = 0$ to $30°$)	–	–	.02	–
Water ($\theta = 60°$)	–	–	.06	–
Water ($\theta = 85°$)	–	–	.58	–
Ocean	.02–.07	–	–	–
Sea ice	–	–	–	.30–.40
Stable snow, high latitudes	–	–	–	.80
Stable snow, < 60°N	–	–	–	.70
Forest with stable snow cover	–	–	–	.45
Forest with unstable snow, spring	–	–	–	.25
Unstable snow cover in spring	–	–	–	.38
Unstable snow cover in autumn	–	–	–	.50
Forest, unstable snow, autumn	–	–	–	.30

Table 6.3. Solar Albedo, Vegetation.

Surface	Munn (1966)	Porkhaev (1959)	Van Wijk (1966)
Swamp with shrubs	–	.20–.25	–
Tundra	–	.15–.20	–
Brush	–	.10	–
Deciduous forest	–	.20	.16–.37
Pine forest	–	.10–.15	.10–.14
Forest, mixed	.10–.18	–	–
Wet grass in sun	.33–.37	.28	–
Wet grass, no sun	.14–.26	–	–
Dry grass	.15–.25	.19	.16–.19
Green grass	–	–	.16–.27
Prairie, wet	–	–	.22
Prairie, dry	–	–	.32
Stubble fields	–	–	.15–.17
Grain crops	–	–	.10–.25
Yellow leaves (fall)	–	–	.33–.36

Table 6.4. Solar Albedo, Soils and Minerals.

Surface	Munn (1966)	Porkhaev (1959)	Van Wijk (1966)
Dry dune sand	.37	–	.18
Moist dune sand	.24	–	.09
Light colored, bare soil	–	.35	–
Dark colored, bare soil	–	.15	–
Dark colored moist, bare soil	–	.10	–
Dark colored, freshly ploughed	–	.05	–
Concrete	–	.25	–
Gravel mixture	–	.12	–
Crushed stone	–	.14	–
Quartz sand	–	–	.35
Granite	–	–	.15
Dark clay, wet	–	–	.02–.08
Dark clay, dry	–	–	.16
Bare fields	–	–	.12–.25
Wet plowed field	–	–	.05–.14
Desert, midday	–	–	.15
Desert, low solar alt.	–	–	.35
Dry, salt cover	–	–	.50
Lime	–	–	.45
Asphalt (black top, bituminous concrete)		.1–.15 Scott (1964)	
White paint		.6 Scott (1964)	
Portland cement concrete, red brick, red tile, dark paints		.20–.35 Berg (1974)	
Yellow and buff brick or stone		.3–.5 Berg (1974)	

6.1.3 Longwave Radiation

The surface of the earth emits longwave radiation, as does the atmosphere itself. The atmosphere absorbs a large fraction of the surface radiation and then reemits this energy to the surface and to space. There is a "window" in the atmosphere for longwave radiation at about 10 μ that allows most longwave radiation that is lost from the surface to escape to space for a clear atmosphere. Cloud cover will introduce a further layer of absorbing and emitting material between the surface and space.

Table 6.5. Radiation Characteristics of Leaves (adapted from Monteith 1973). From *Principles of Environmental Physics,* J. L. Monteith, Table 5.1, p. 66.

	Solar Albedo	Emissivity
Maize	.29	.944
Tobacco	.29	.972
Cucumber	.31	–
Tomato	.28	–
Birch	.32	–
Aspen	.34	–
Oak	.30	–
Elm	.28	–
Snap Bean	–	.938
Cotton	–	.964
Sugar Cane	–	.995
Poplar	–	.927
Geranium	–	.992
Cactus	–	.927

6.1.3.1 Terrestrial Radiation

The surface of the earth will emit radiation as a gray body following the usual relation

$$E_e = \sigma \epsilon_e T_e^4 \tag{6.43}$$

where

ϵ_e = emissivity of the surface
T_e = surface temperature, absolute and
σ = Stefan-Boltzmann constant.

6.1.3.2 Longwave Absorptivity of Surfaces

The absorptivity of surfaces for longwave radiation is tabulated in many standard heat transfer texts. For longwave radiation, Kirhhoff's relation is usually adequate; that is, the emissivity of the surface is equal to the absorptivity. The same is not true for shortwave radiation. Some typical values of natural surfaces are given in Table 6.8. There is often a significant difference in the absorptivity of a surface for longwave or shortwave (solar) radiation.

Table 6.6. Solar Albedo for Vegetation (adapted from Monteith 1973). From *Principles of Environmental Physics,* J. L. Monteith, Table 5.1, p. 66.

	Solar albedo
Grass	.24
Sugar beet	.26
Barley	.23
Wheat	.26
Beans	.24
Maize	.22
Tobacco	.24
Cucumber	.26
Tomato	.23
Sorghum	.20
Sugar cane	.15
Cotton	.21
Groundnuts	.17
Heather	.14
Bracken	.24
Gorse	.18
Maquis, evergreen scrub	.21
Natural pasture	.25
Derived savanna	.15
Guinea savanna	.19
Deciduous woodland	.18
Coniferous woodland	.16
Orange orchard	.16
Aleppo pine	.17
Eucalyptus	.19
Tropical rain forest	.13
Swamp forest	.12

Table 6.7. Effect of Oil on Solar Albedo of Vegetation in Permafrost Region (Lunardini 1974).

	Solar albedo	
Surface	No oil	Oil-covered and dead
Lichens, shrubs 10–12 cm	.110–.130	.03
Gray lichens, sparse	.090–.120	.03
Light colored sand, clay	.15–.19	.07
Dead grass	–	.07
Thick green sphagnum	.140	–

Table 6.8. Longwave Absorption Coefficients (from Van Wijk 1966, with permission of Elsevier/North Holland Biomedical Press, Amsterdam).

Surface	Absorption coefficient for longwave radiation
Sand, dry-wet	.95–.98
Mineral soil, dry-wet	.95–.97
Peat, dry-wet	.97–.98
Firs	.97
Trees	.96
Grass	.96–.98
Leaves	.94–.98
Water	.95
Snow	.97

Table 6.9. Comparison of Solar and Longwave Absorptivities.

Material	Solar absorptivity	Longwave absorptivity
Red brick	.70–.77[a]	.95[c]
Clay tiles	.65–.74[a]	.93[c]
Slate	.79–.93[a]	–
Galv. iron, new	.66[a]	.23[d]
Galv. iron, dirty	.89[a]	.28[d]
Roofing paper	.88[a]	.98[c]
Asphalt	.89[a]	.93[b]
Asphalt pavement	.85[a]	–
White paper	.27[a]	.97[c]
Marble, white	.47[b]	.95[b]
Cream paints	.35[b]	.95[b]
Lampblack paint	.97[b]	.96[b]
Red paint	.74[b]	.96[b]
Yellow paint	.30[b]	.95[b]
White paint (zinc oxide)	.18[b]	.95[b]

[a] Fishenden (1932).
[b] Kreith (1973).
[c] Eckert (1959).
[d] Brown (1958).

Table 6.10. Short and Longwave Absorptivities (adapted from Wechsler 1966).

Construction materials	Emissivity, ϵ	Solar absorp-tivity, α_s
Concrete	.94	.7
Concrete cinder block	–	.85
Brick, red	.9	.55, .7
Wood, freshly planed	.9	.6
Asbestos slate (wallboard)	.96	.8
White paper	.95	.25
Black (tar) paper	.93	.95
Plaster, white	.91	.07
Bright galvanized iron	.13	.65
Oxidized (gray) galvanized iron	.28	.8
Bright aluminum foil	.04	.15
Steels (mild and stainless)	–	.5
Oxidized cast iron	–	.94
Surface Coatings		
Solid $MgCO_3$	.79	.04
White pigment (CaO)	.96	.15
Gray pigment (Co_2O_3)	.87	.97
Green pigment (Cr_2O_3)	.95	.27
White paint (.017″ on Al)	.91	.2
Black paint (.017″ on Al)	.88	.96
Aluminum paint	.4	.2
Naturally Occurring Surfaces		
Water	.95	.95
Ice	.63, .9	.31
Snow	.98	.1, .2
"Frozen soil"	.93	–
Gravel	.28	.28
Dry plowed ground	.8	.9
Clay		.4
Sand	.76	.76
Common vegetated fields and shrubs	.76	.9
Grass, high, dry	.7	.9
Forested land	.85	.9
Alfalfa (dark green)	.97	.95

This is shown in Tables 6.9 and 6.10, where it is assumed that the emissivity is equal to the absorptivity.

6.1.3.3 Clear Sky Atmospheric Radiation

The energy emitted toward the earth by the atmosphere is a function of the atmospheric temperature and water vapor pressure. It can be described by

$$\frac{E_s}{\sigma T_c^4} = \epsilon_a(T_c, e_o) \tag{6.44}$$

where e_0, T_c — atmospheric water vapor pressure and air temperature at reference level.

The effective emissivity of the atmosphere, ϵ_a, describes the effect of temperature, pressure, atmospheric gases, etc., upon the radiation. The functional relation has been given in a number of forms. One commonly used relation is that of Angstrom (1915).

$$\epsilon_a = A - Be^{-\gamma e_o} \tag{6.45}$$

Van Wijk (1966) records a number of correlations for the coefficients, with average values as follows

$$A = .78 \qquad B = .27 \qquad \gamma = 0.136\,\text{mb}^{-1}$$

Scott (1961) gives ranges of values as

$$A = .8 - 1.11 \qquad B = .24 - .41 \qquad \gamma = .017 - .122$$

but does not indicate preferred values. Scott (1964) gives a similar relation

$$\epsilon_a = .82 - (.25)10^{-.095\, e_o} \tag{6.46}$$

A simpler and preferable relation is offered by Brunt (1932)

$$\epsilon_a = a + b\sqrt{e_o} \tag{6.47}$$

There is considerable variation of the constants a and b, no doubt influenced by local conditions. These values are given below

Reference	a	$b(mb^{-1/2})$	$\epsilon_a(e_o = 10\ mb)$
Brunt (1932)	.52	.057	.70
Sellers (1965)	.60	.05	.76
Scott (1961)	.43–.68	.025–.071	.65–.76
Scott (1964)	.44	.080	.69
Yamamoto (1950)	.51	.066	.72

Brutsaert (1975) theoretically derived an equation, based upon a standard atmospheric lapse rate, and obtained a very simple equation

$$\epsilon_a = 1.24\left(\frac{e_o}{T_c}\right)^{1/7} \tag{6.48}$$

where $e_o - mb$, $T_c - {}^\circ K$.

If $T_c = 288^\circ K$, then

$$\epsilon_a = 0.553\ e_o^{\ 1/7} \tag{6.49}$$

Obviously, this equation requires no empirical constants, excluding the assumptions in its derivation, and agrees quite well with the empirical relations. An interesting relation was obtained, empirically, by Swinbank (1963) as

$$\epsilon_a = .398 \times 10^{-5} T_c^{2.148} \tag{6.50}$$

Although this relation does not contain e_o, it is not out of line with Eq. (6.48), because Deacon (1970) showed that the precipitable water is proportional to the 16.8 power of air temperature. This will lead to e_o being proportional to the 15.8 power of T_c and finally, from Eq. (6.48), ϵ_a proportional to the 2.11 power of T_c, which agrees very well with Eq. (6.50).

6.1.3.4 Net Clear Sky Longwave Radiation

From Eqs. (6.43) and 6.44), the net longwave radiation loss from the surface for a cloudless atmosphere is

$$R = E_e - \alpha_e E_s \tag{6.51}$$

or

$$\frac{R}{\sigma T_c^4} = \epsilon_e\left[\left(\frac{T_e}{T_c}\right)^4 - \epsilon_a\right] \tag{6.52}$$

where $\epsilon_e = \alpha_e$.

Simplifications often used in this equation are that the surface emissivity, which equals the absorptivity is one, and $T_e = T_c$. The latter assumption is reasonably valid for periods of 1 day if the sensible heat loss from the surface to the atmosphere is not too great. Then the net longwave radiation is given by

$$\frac{R}{\sigma T_c^4} = (1 - \epsilon_a) \tag{6.53}$$

Wexler (1941) measured the net flux for nocturnal radiation in Alaska as

$$R = 125.3 + 1.73T_c \tag{6.54}$$

where

R = ly/day and
T_c = air temperature, °C.

Scott (1964) noted that the daily summer flux of longwave radiation for cold regions (data from Camp Tuto, Greenland) is relatively constant due to the compensating effects of temperature and humidity, with $R \simeq 231$ ly/day = 850 Btu/ft² − day.

6.1.3.5 Effect of Cloud

The effect of a cloud cover will be to decrease the net flux of longwave radiation or to increase the incoming atmospheric radiation. There is no simple analytical method to take care of the absorption, reflection, and emission of clouds, and, thus, empirical relations must be used. The cloud cover can be evaluated, for the net radiation, with an equation of the form

$$\frac{R_c}{R} = a - bC_e \tag{6.55}$$

where C_e is now the 24-hour average fraction of sky covered by clouds and a and b are functions of the type and structure of the clouds. For low clouds $b = 0.9$, for cirrus clouds $b = 0.2$, and an overall average is 0.8. The coefficients have been evaluated at various locations as

Reference	a	b
Angstrom (1915)	1	.9
Berg (1974)	.992	.785
Wexler (1941)	1	.31

Phillips (1940) found $a = 1$ and b a function of cloud height

b	Cloud Base Height, km
.87	1.5
.83	2
.74	3
.62	5
.45	8

Wexler (1941) also gave a correction factor as $R_c/R = c$ where

$$c = P_c + .65\ P_p + 4\ P_o$$

where P_c, P_p, and P_o are monthly percentages of clear ($C_e < .2$), partly cloudy ($.3 < C_e < .7$), and overcast ($.8 < C_e < 1.0$) skies.

A second method involves cloud correction factors only for E_s, as E_e is not affected by clouds. Then

$$\frac{\epsilon_{ac}}{\epsilon_a} = 1 + nC_e^2 \tag{6.56}$$

Monteith (1973), gives $n = 0.2$ for low clouds (stratus, cumulus, altostratus, and altocumulus) and $n = .04$ for cirrus type clouds.

Since the terrestrial emittance is not dependent upon the cloud cover, Van Wijk (1966) suggests the following relation

$$R_c = E_e - \alpha_e\, \sigma T_c^4\, g \tag{6.57}$$

where the function g is evaluated so that Eq. (6.57) will equal Eq. (6.55) when $T_e = T_c$. This is

$$g = 1 - (1 - a)(a - bC_e) \tag{6.58}$$

Equation (6.55) is probably the best to use, being the simplest, and all give similar results. Further, Berg (1974), in comparing theory with data, found that a cloud correction on R gave better results than a correction on E_s alone. It would appear that Eq. (6.55) is preferable for cold regions and that Eq. (6.56) might be more accurate for more temperate regions; the data are very scattered, however.

Anderson (1954) proposed an equation, based on Lake Hefner studies, that incorporated the effects of cloud and humidity

$$\epsilon_a = a + be_o \tag{6.59}$$

$$a = .74 + .25\ C_e \exp(-.0584z)$$

$$b = .0037 - .0041\ C_e \exp(-.06z)$$

for $z > 1600$ ft.

The following equations are recommended for estimates of longwave radiation

$$E_e = \sigma\, \epsilon_e\, T_e^4 \tag{6.43}$$

$$E_s = \epsilon_a \, \sigma \, T_c^4 \tag{6.44}$$

$$\epsilon_a = .60 + .05 \sqrt{e_0} \tag{6.47}$$

or

$$\epsilon_a = 1.24 \left(\frac{e_o}{T_c}\right)^{1/7} \tag{6.48}$$

$$R = E_e - \epsilon_e \, E_s \tag{6.51}$$

$$\frac{R_c}{R} = 1 - .8 \, C_e \tag{6.55}$$

6.1.4 Summary of Radiation Equations for Cloudy Skies

$$Q_N = (1 - \alpha_s) \, Q_{TC} - R_c \tag{6.1}$$

$$\frac{Q_{TC}}{Q_T} = \begin{cases} 1 - .67 \, C_s^2 \text{ daily} & (6.39) \\ .35 + .61(1 - C_s)(1 + .5C_s) \text{ monthly} & 6.40) \end{cases}$$

$$Q_T = 2865.6 \, AB\left(\frac{r_m}{r}\right)^2 \cos \delta \cos \phi \left(\frac{180 - H_{sr}}{180} \cos H_{sr} + \frac{\sin H_{sr}}{\pi}\right) \tag{6.27}$$

$$A = a'' - \frac{a'}{2} + \frac{1}{2} - \frac{d}{2}$$

$$B = 1 + \frac{\alpha_s}{2} (1 - a' + d)$$

$$a' = 1.041 \, e^{-(.095m + .039w)} \tag{6.30}$$

$$a'' = .975 \, e^{-(.1263m + .051w)} \tag{6.31}$$

$$\ln w = .1102 + .0614 \, T_d \tag{6.32}$$

The value of w should be multiplied by 0.85 to adjust to daily values of precipitable water vapor

$$H_{sr} = \cos^{-1} (\tan \delta \tan \phi) \tag{6.17}$$

$$\delta = \sin^{-1} (\sin 23.5 \cos \{.9863 \, (D - 172)\}) \tag{6.15}$$

$$\frac{r}{r_m} = 1 - 1.6733 \times 10^{-3} \cos (.9856 \, D) \tag{6.8}$$

$$R_c = (1 - .8\ C_e)\ \sigma\ \epsilon_e\ T_c^4 \left[\left(\frac{T_e}{T_c} \right)^4 - \epsilon_a \right] \tag{6.55}$$

$$\epsilon_a = .6 + .05 \sqrt{e_o} \tag{6.47}$$

where

Q_T = ly/day
w = cm water
T_d = °C
T = °K
e_o = mb
d = from Table 6.1
m = from Fig. 6.1
D = days from December 31 and
ϕ = latitude.

6.2 ATMOSPHERIC FLUXES OF MOMENTUM AND ENERGY

Probably the least well understood of the energy fluxes at the earth-atmosphere interface are those of sensible and latent heat. These are, by no means, the principal energy fluxes, but they exert a critical influence upon the surface temperature. When a solid body is in contact with a fluid such that a gradient of temperature, velocity, or concentration exists between the solid and the fluid, then there can be fluxes of energy, momentum, or mass. These fluxes will also traverse the fluid from one layer to another. They are due to the movement of fluid masses, with different temperature, velocity, and concentration values, from one position to another. If the transport of energy, momentum, or mass depends upon the movement of macroscopic masses of fluid, it is called "turbulent," or "convective," transport. It is distinct from the fluxes due to the microscopic or molecular motions, which are the diffusive mechanisms relating to conductive heat transfer or vorticity transport. These molecular motions are always present, even in turbulent flows.

The molecular diffusion mechanisms are well understood, and the diffusion depends upon the transport properties of the fluid. These properties are fixed for each thermodynamic state of the fluid. In contrast, the turbulent transport depends on the state of the fluid motion as well as its properties. Thus, the turbulent flux of energy depends not only on the thermal properties of the fluid (thermal conductivity, specific heat, temperature, density), but also upon the particular type of fluid motion. It is not possible to characterize the energy flux by a property of the transport medium as is the case for conductive heat transfer. The quantities of interest (eddy coefficients) are functions of the state of motion and are not simply transport properties.

6.2.1 Turbulence

The concept of laminar flow is easy to grasp in that the layers of fluid tend to slide uniformly over one another without transverse mixing of the layers, on a macroscopic scale. The transfer of mass or energy is due to the molecular exchanges from one fluid layer to another. Turbulent motion is a distortion of this regular motion due to macroscopic, transverse, velocity components. The origin of turbulence is not completely understood, but it can be generated by a solid in contact with a fluid, different fluids in contact with each other, or thermal stratification of a fluid. In turbulent motion, the instantaneous velocity may be pictured as made up of a mean component and a rapidly fluctuating component, $u = \bar{u} + u'$. Both u and $\bar{u}$ may be functions of time, but u' is always time-dependent. Unfortunately, it is not possible to relate u' to the mean value $\bar{u}$ except in an empirical or semiempirical way. The fluctuating velocities carry masses of fluid, called "eddies," transverse to the main flow. These eddies are the mechanism for the turbulent transport of mass, momentum, and energy. They move at an eddy velocity u', v', w' relative to the mean flow of the fluid.

6.2.1.1 Thermal Turbulence

Suppose a stationary body of fluid is exposed to heating at a solid-fluid interface. The temperature of the fluid near the interface increases, and, for gases, its density decreases, and the fluid tends to rise due to buoyancy effects. This induces a convective current in the fluid, which may be thought of as thermally induced turbulence. Thus, turbulence may be introduced, augmented, or suppressed in a fluid due to thermal or temperature stratification.

6.2.1.2 Mechanical Turbulence

Now assume that a fluid is flowing isothermally over a surface. The surface, due to shear stresses, may cause the flow to become turbulent, depending upon the surface and the flow conditions. This turbulence is called "mechanical turbulence," as it is related to the shearing effects or forces at the interface. From an engineering viewpoint, this is the usual type of turbulence encountered and normally predominates even with regard to the atmosphere. In reality, mechanical and thermal turbulence are not separable and act together to generate a state of turbulence. In certain cases, the thermal effects may augment the mechanical turbulence, though for other systems they will tend to suppress the solid interface effects. No attempt will be made here to elaborate upon the complicated nature of turbulence, which, after more than 100 years of study, has not yielded to a complete analysis.

6.2.2 Equilibrium

The equilibrium of the atmosphere, and other systems, can be characterized as stable equilibrium, neutral or metastable, equilibrium, and unstable equilibrium. In all cases, the system is in equilibrium with its surroundings, but in the stable case, any disturbance from a given state will be resisted, and the system will tend to move back to the equilibrium state. In neutral equilibrium, a disturbance will not be resisted, and the system will remain in equilibrium, although it may traverse a series of different states. Unstable equilibrium is such that any small disturbance will be augmented, and the system will move to an entirely new state.

6.2.2.1 Neutral Atmosphere

For this case a part of the atmosphere may be stationary or in motion without acceleration, so that the net forces acting upon an element will be zero. The movement of a fluid particle must be along a reversible path, and, if it is also an adiabatic process, then it will be one of constant entropy. Now a perfect gas, the air, that undergoes an isentropic change of state from p_1, T_1 to p_2, T_2 will obey the following relation

$$\left(\frac{T_1}{T_2}\right)^{\gamma} = \left(\frac{p_1}{p_2}\right)^{\gamma-1} \tag{6.60}$$

or

$$T^{\gamma} = cp^{\gamma-1}$$

where $\gamma = c_p/c_v$ is the ratio of the specific heats.

Consider the vertical pressure and gravity forces acting upon a fluid element in neutral equilibrium.

$$pA - \left(p + \frac{\partial p}{\partial z}\,dz\right)A - A\rho\frac{g}{g_c}\,dz = 0$$

thus

$$\frac{\partial p}{\partial z} = \rho\frac{g}{g_c} \tag{6.61}$$

Now from Eq. (6.60)

$$\frac{\partial p}{\partial z} = -\frac{p}{\gamma - 1}\frac{\gamma}{T}\frac{\partial T}{\partial z}$$

Solving for the temperature gradient

$$\frac{\partial T}{\partial z} = -\frac{(\gamma - 1)}{\gamma R}\frac{g}{g_c}$$

For a perfect gas, $R = c_p - c_v$ and

$$\frac{\partial T}{\partial z} = -\frac{1}{c_p}\frac{g}{g_c} \tag{6.62}$$

This is the condition for neutral stability that must be met by the temperature gradient. This value of the temperature gradient is defined as the dry, adiabatic, lapse rate

$$-\frac{\partial T}{\partial z} \equiv \Gamma = \frac{1}{c_p}\frac{g}{g_c} \tag{6.63}$$

The equilibrium of the atmosphere may now be described in terms of the adiabatic lapse rate

$$-\frac{\partial T}{\partial z} = \begin{cases} \Gamma & \text{neutral} \\ <\Gamma & \text{stable} \\ >\Gamma & \text{unstable} \end{cases}$$

For a stable atmosphere, any turbulence will tend to be damped out, while for an unstable atmosphere, the turbulent effects will be augmented.

6.2.2.2 Potential Temperature

The potential temperature θ is defined as that temperature that a fluid element would have if it changes from any state p, T to a state with a reference pressure p_0 (p_0 is taken as 1000 mb) in a reversible, adiabatic process. Once again, the isentropic relation is

$$\theta = T\left(\frac{p_0}{p}\right)^{\frac{\gamma-1}{\gamma}} \tag{6.64}$$

Differentiation of Eq. (6.64) with respect to z leads to

$$\frac{1}{\theta}\frac{\partial\theta}{\partial z}=\frac{1}{T}\frac{\partial T}{\partial z}-\frac{(\gamma-1)}{\gamma}\frac{1}{p}\frac{\partial p}{\partial z}$$

Using Eq. (6.61) gives

$$\frac{1}{\theta}\frac{\partial\theta}{\partial z}=\frac{1}{T}\left(\frac{\partial T}{\partial z}+\Gamma\right) \tag{6.65}$$

Close to the ground-atmosphere interface, i.e., within the constant flux layer, the potential and actual temperatures are very nearly identical; thus, to a first approximation, from Eq. (6.65)

$$\frac{\partial\theta}{\partial z}=\frac{\partial T}{\partial z}+\Gamma$$

or

$$\theta=T+\Gamma z \tag{6.66}$$

If needed, the exact relation, Eq. (6.64), can always be used to calculate θ. The stability of the atmosphere can now be described in terms of the potential temperature gradient

$$\frac{\partial\theta}{\partial z}=\begin{cases}0 & \textit{neutral}\\ >0 & \textit{stable}\\ <0 & \textit{unstable}\end{cases}$$

6.2.3 Turbulent Equations

6.2.3.1 Momentum

We consider a system such that the only mean flow of significance is in the x direction. The Navier-Stokes equation for an incompressible fluid for the x direction is

$$\rho\frac{\partial u}{\partial t}=\frac{\partial\sigma_{xx}}{\partial x}-\rho u\frac{\partial u}{\partial x}+\frac{\partial\sigma_{xy}}{\partial y}-\rho v\frac{\partial u}{\partial y}+\frac{\partial\sigma_{xz}}{\partial z}-\rho w\frac{\partial u}{\partial z} \tag{6.67}$$

The stresses are given by

$$\sigma_{xx} = -p + \mu\left(2\frac{\partial u}{\partial x} - \frac{2}{3}\Delta\right) = -p + 2\mu\frac{\partial u}{\partial x}$$

$$\sigma_{xy} = \mu\left(\frac{\partial v}{\partial x} + \frac{\partial u}{\partial y}\right)$$

$$\sigma_{xz} = \mu\left(\frac{\partial w}{\partial x} + \frac{\partial u}{\partial z}\right)$$

where the incompressible relation $\Delta = \text{div } V = 0$ has been used. The time average of a quantity over a period t_o is defined as

$$\bar{A} = \frac{1}{t_o}\int_t^{t+t_o} A\, dt \tag{6.68}$$

The existence of a mean value requires that $\overline{A'} = 0$. The following notation is used

$$\begin{aligned} u &= \bar{u} + u' & p &= \bar{p} + p' \\ v &= \bar{v} + v' & \rho &= \rho + \rho' \\ w &= \bar{w} + w' & T &= \bar{T} + T' \end{aligned}$$

Replacing the instantaneous values in Eq. (6.67), and taking the time average gives

$$\bar{\rho}\frac{\partial \bar{u}}{\partial t} = \frac{\partial}{\partial x}(\overline{\sigma_{xx}} - \rho\bar{u}\bar{u} - \overline{\rho u'^2}) + \frac{\partial}{\partial y}(\overline{\sigma_{xy}} - \rho\bar{u}\bar{v} - \overline{\rho u'v'})$$

$$+ \frac{\partial}{\partial z}(\overline{\sigma_{xz}} - \rho\bar{u}\bar{w} - \overline{\rho u'w'}) \tag{6.69}$$

$$\overline{\sigma_{xx}} = -\bar{\rho} + 2\,\mu\frac{\partial \bar{u}}{\partial x}$$

$$\overline{\sigma_{xy}} = \mu\left(\frac{\partial \bar{u}}{\partial y} + \frac{\partial \bar{v}}{\partial x}\right)$$

$$\overline{\sigma_{xz}} = \mu\left(\frac{\partial \bar{w}}{\partial x} + \frac{\partial \bar{u}}{\partial z}\right)$$

Comparison of Eqs. (6.67) and (6.69) reveals that the turbulence has led to a new stress in the form of quantities like $-\overline{\rho u'w'}$. These are the Reynolds stresses and are due to the transport of momentum associated with the transverse velocity fluctuations u', v', w'. Physically, the Reynolds stresses can be explained as follows. Consider the laminar flow of a fluid parallel to a flat surface as shown in Fig. 6.9. The shear stress at a surface z above the solid is due to the molecular exchange of momentum and is given by

$$\tau = \mu \frac{\partial U}{\partial z}$$

where

μ = viscosity of the fluid
τ = shear stress in the plane and
U = mean velocity at z.

This equation is valid when there is an exchange of matter on the molecular level only, i.e., there is no turbulence. The above equation is also written

$$\tau = \rho \nu \frac{\partial U}{\partial z}$$

where $\nu = (\mu/\rho)$ is the kinematic viscosity.

Now assume that the fluid is in turbulent motion sketched in Fig. 6.10. Once again, the fluid at z has a mean velocity in the x direction, but it also has fluctuating components u' and w'. The instantaneous mass flux across a plane at z is given by $dm = \rho w'$; the vertical flux of x momentum is $-\rho w' u'$, which, when averaged over a finite time, is $-\overline{\rho u' w'}$. The negative sign denotes that a mass moving upward with positive w' will decrease the momentum of the above layer and vice versa. Now the shear stress associated with this momentum may be written as

$$\tau_t = -\overline{\rho u' w'}$$

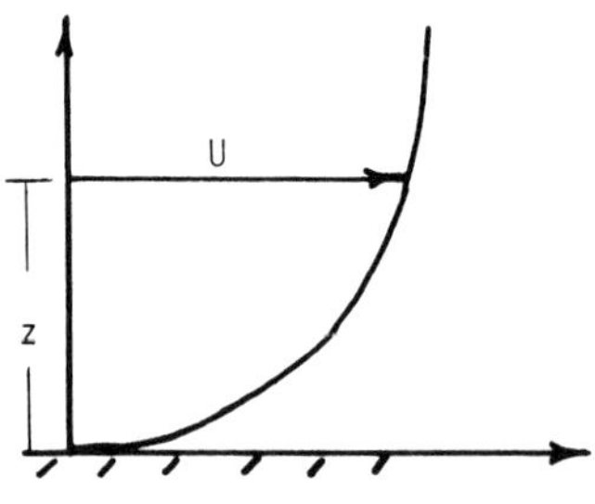

Fig. 6.9. Laminar Velocity.

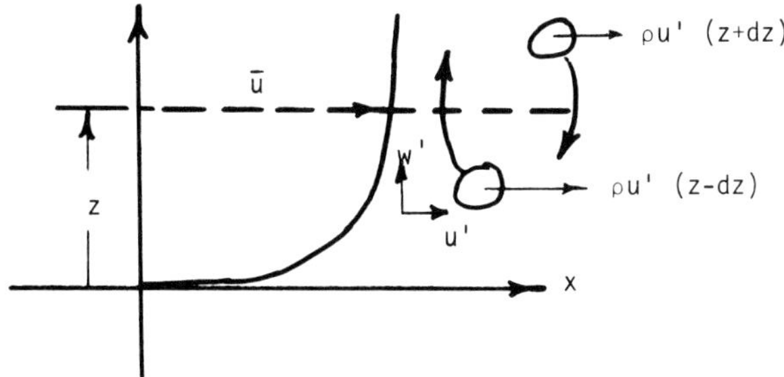

Fig. 6.10. Shear Stress and Fluctuating Velocities.

Clearly, this term is a Reynolds stress as derived from the Navier-Stokes equation. Using the stress relations and the x momentum, defined as $M_x = \bar{\rho}\,\bar{u}$, Eq. (6.69) can be written

$$\frac{DM_x}{Dt} = \frac{\partial}{\partial x}\left(-\bar{p} + \nu\frac{\partial M_x}{\partial x} - \bar{\rho}\,\overline{u'^2}\right) + \frac{\partial}{\partial y}\left(\frac{\partial M_x}{\partial y} - \bar{\rho}\,\overline{u'v'}\right) + \frac{\partial}{\partial z}\left(\nu\frac{\partial M_x}{\partial z} - \bar{\rho}\,\overline{u'w'}\right) \tag{6.70}$$

Here

$$\frac{D}{Dt} = \frac{\partial}{\partial t} + V\,\mathrm{div}$$

For the atmospheric flow, we may neglect the variations in the x and y directions, as z is the vertical. Then

$$\frac{DM_x}{Dt} = \frac{\partial}{\partial z}\left(\nu\frac{\partial M_x}{\partial z} - \bar{\rho}\,\overline{u'w'}\right) \tag{6.71}$$

Now the shear stress is

$$\tau = \nu\frac{\partial M_x}{\partial z} - \bar{\rho}\,\overline{u'w'} \tag{6.72}$$

where the Reynolds stress is often defined in terms of an eddy diffusivity ϵ_m as

$$-\bar{\rho}\,\overline{u'w'} = \bar{\rho}\,\epsilon_m\frac{\partial\bar{u}}{\partial z} = \epsilon_m\frac{\partial M_x}{\partial z} \tag{6.73}$$

A. Invariance of Shear Stress. The assumption is often made that the vertical flux of momentum, or the shear stress, is invariant. In Eq. (6.70) we retain those terms that are first-order changes in x, assuming steady-state conditions.

$$\bar{u}\frac{\partial M_x}{\partial x}+\frac{\partial \bar{p}}{\partial x}=\frac{\partial \tau}{\partial z} \tag{6.74}$$

Integrate this equation over a height h, where $\hat{F}$ is the vertical average of a function

$$\tau_h-\tau_o=h\left(\frac{1}{2}\bar{\rho}\widehat{\frac{\partial \bar{u}^2}{\partial x}}+\widehat{\frac{\partial \bar{p}}{\partial x}}\right)$$

Notice that the kinetic energy and pressure terms approach zero as h becomes small, while the shear stress approaches τ_o. Normally the kinetic energy is small compared with the pressure term. Then the shear stress variation will be small if

$$h<<\tau_o/\left(\widehat{\frac{\partial \bar{p}}{\partial x}}\right)$$

Typical atmospheric values are $\widehat{\frac{\partial \bar{p}}{\partial x}}\sim 10^{-4}$ dyne/cm^3 and $\tau_o\sim 1$ dyne/cm^2. The stress will be constant if $h<<100$ m. Thus, we conclude that, in a layer within a few tens of meters of the surface, the momentum flux or shear stress will be essentially invariant. This is the constant flux layer and allows calculations or measurements to be made of the surface flux, at an elevation above the surface. If the Coriolis force is added the results will be the same (Kraus 1972).

The following equations may then be used.

$$\tau=\nu\frac{\partial M_x}{\partial z}-\bar{\rho}\,\overline{u'w'}=(\nu+\epsilon_m)\frac{\partial M_x}{\partial x}$$

$$\frac{\partial \tau}{\partial z}=0$$

$$\frac{\partial M_x}{\partial t}=\frac{\partial}{\partial z}\left[(\nu+\epsilon_m)\frac{\partial M_x}{\partial z}\right] \tag{6.75}$$

6.2.3.2 Energy

The general energy equation can be written as

$$\rho c_p \frac{DT}{Dt} - \frac{Dp}{Dt} = -\nabla(-k \nabla T) + q''' + \mu\Phi$$

where q‴ and $\mu\Phi$ are the energy generation and the viscous dissipation terms. Once again, take the time average for the turbulent flow and neglect the energy generation (the radiation has already been neglected) and the viscous dissipation of the mean flow. Since $\rho'/\bar{\rho} << u'/\bar{u}$ and $p'/\bar{p} << u'/\bar{u}$, we drop those terms with pressure and density fluctuations

$$\bar{\rho}\, c_p \frac{D\bar{T}}{Dt} = \frac{D\bar{p}}{Dt} + \mu\,\bar{\Phi} - \frac{\partial}{\partial x}\left(-k\frac{\partial \bar{T}}{\partial x} + \bar{\rho}\, c_p \overline{u'\,T'}\right)$$
$$- \frac{\partial}{\partial y}\left(-k\frac{\partial \bar{T}}{\partial y} + \bar{\rho} c_p\, \overline{v'\,T'}\right) - \frac{\partial}{\partial z}\left(-k\frac{\partial \bar{T}}{\partial z} + \bar{\rho}\, c_p\, \overline{w'\,T'}\right) \quad (6.76)$$

The viscous dissipation is

$$\bar{\Phi} = \frac{4}{3}\left(\overline{\frac{\partial u'}{\partial x}}\right)^2 + \left(\overline{\frac{\partial v'}{\partial y}}\right)^2 + \left(\overline{\frac{\partial w'}{\partial z}}\right)^2 - \left(\overline{\frac{\partial u'}{\partial x}\frac{\partial v'}{\partial y}}\right.$$
$$\left. + \overline{\frac{\partial u'}{\partial x}\frac{\partial w'}{\partial z}} + \overline{\frac{\partial v'}{\partial y}\frac{\partial w'}{\partial z}}\right) + \left(\overline{\frac{\partial u'}{\partial y}}\right)^2 + \left(\overline{\frac{\partial v'}{\partial x}}\right)^2$$
$$+ \left(\overline{\frac{\partial w'}{\partial x}}\right)^2 + \left(\overline{\frac{\partial u'}{\partial z}}\right)^2 + \left(\overline{\frac{\partial w'}{\partial y}}\right)^2 + \left(\overline{\frac{\partial v'}{\partial z}}\right)^2$$
$$+ 2\left(\overline{\frac{\partial v'}{\partial x}\frac{\partial u'}{\partial y}} + \overline{\frac{\partial w'}{\partial x}\frac{\partial u'}{\partial z}} + \overline{\frac{\partial w'}{\partial y}\frac{\partial v'}{\partial z}}\right)$$

but for atmospheric flows we can usually drop this term. For the steady state we have

$$\bar{\rho}\, c_p\, \bar{u}\frac{\partial \bar{T}}{\partial x} = \bar{u}\frac{\partial \bar{p}}{\partial x} + \frac{\partial}{\partial z}\left(k\frac{\partial \bar{T}}{\partial z} - \bar{\rho}\, c_p\, \overline{w'\,T'}\right) \quad (6.77)$$

Note once again, the convective transport term associated with the turbulent fluctuations. The energy flux in the vertical direction is

$$-q = k\frac{\partial \overline{T}}{\partial z} - \overline{\rho}\, c_p\, \overline{w'\,T'} \tag{6.78}$$

The energy equation is

$$\overline{\rho}\, c_p\, \overline{u}\frac{\partial \overline{T}}{\partial x} = \overline{u}\frac{\partial \overline{p}}{\partial x} - \frac{\partial q}{\partial z}$$

This equation states that the horizontal flow of enthalpy (the "advection," as it is called in meteorology) equals the work of the horizontal pressure gradient plus the vertical flux of sensible heat.

A. Constancy of Heat Flux. Again, let us integrate over a height, as was done for the momentum flux

$$h\left(\overline{\rho}\, c_p\, \overline{u}\widehat{\frac{\partial \overline{T}}{\partial x}} - \overline{u}\widehat{\frac{\partial \overline{p}}{\partial x}}\right) = q_o - q_h$$

A comparison of the terms indicates that if the horizontal temperature gradient is on the order of 12°C/1000 km, the advection is 15 times the pressure term. Thus, the heat flux is constant if

$$h << \frac{q_o}{16\, \overline{u}\widehat{\frac{\partial \overline{p}}{\partial x}}} \simeq 75\ m$$

for $q_o \simeq 1$ cal/cm² − hr.
This is smaller than the constant momentum layer, but for heights less than 10 m, the constancy of fluxes is still reasonable. In this layer q given by Eq. (6.78) is constant.

B. Compressibility Effects. When the mean pressure and temperature vary, it is necessary to take into account the compressibility effects of compression and expansion work. Hinze (1975) notes that the correlation term $\overline{w'\,T'}$ can be expressed as

$$\overline{w'\,T'} = -\epsilon_h\left(\frac{\partial \overline{T}}{\partial z} - \frac{\gamma - 1}{\gamma}\frac{\overline{T}}{\overline{p}}\frac{\partial \overline{p}}{\partial z}\right) \tag{6.79}$$

Actually, ϵ_h should be expressed as a tensor, but we can consider it a zero order tensor, i.e., a scalar. Using Eq. (6.64) for the potential temperature gives

$$\overline{w' T'} = -\epsilon_h \frac{\overline{T}}{\overline{\theta}} \frac{\partial \overline{\theta}}{\partial z} \simeq -\epsilon_h \frac{\partial \overline{\theta}}{\partial z} \tag{6.80}$$

The use of the potential temperature for the heat flux is more consistent than $\overline{T}$, but the improvement in accuracy is small. For convenience, we will often use $\overline{T}$. From Eq. (6.65), one has

$$q = -\overline{\rho}\, c_p\, (\alpha + \epsilon_h) \frac{\partial \overline{T}}{\partial z} - \overline{\rho}\, c_p\, \epsilon_h\, \Gamma \tag{6.81}$$

Let the enthalpy of the fluid be defined as

$$H = \overline{\rho}\, c_p\, \overline{T} + H_o$$

Then the energy equation may be written

$$\frac{\partial H}{\partial t} = \frac{\partial}{\partial z} \left[(\alpha + \epsilon_h) \frac{\partial H}{\partial z} - \epsilon_h \frac{\partial H_a}{\partial z} \right]$$

where $(\partial H_a/\partial z) = -\overline{\rho} c_p \Gamma$ is the enthalpy flux associated with the adiabatic lapse rate. If this term is neglected, then

$$\frac{\partial H}{\partial t} = \frac{\partial}{\partial z} \left[(\alpha + \epsilon_h) \frac{\partial H}{\partial z} \right] \tag{6.82}$$

Here we note the similarity of Eq. (6.82) and that for the momentum flux, Eq. (6.75). If $(\alpha + \epsilon_h) = (\nu + \epsilon_m)$ then the momentum transport will equal that of the enthalpy for the same boundary conditions. Actually, we have neglected the momentum transport due to pressure and the energy transport due to radiation, so the analogy is not exact.

6.2.4 Stability

From this point on, we shall omit the bar over the average quantities unless noted specifically. Following Kraus (1972), we may write the turbulence energy equation for a quasi-steady, horizontally homogeneous boundary layer flow as

$$\tau \frac{\partial u}{\partial z} = \rho \epsilon + \frac{\partial}{\partial z} \overline{p' w'} + g\, \overline{\rho' w'} + \frac{1}{2} \frac{\partial}{\partial z} \overline{(u'^2 + v'^2 + w'^2) w'} \tag{6.83}$$

In this equation

$\tau\, \partial u/\partial z =$ rate of work done by the eddy stress
$\rho\, \epsilon =$ frictional dissipation of energy
$\overline{p' w'} =$ rate of work of the fluctuating pressure and
$g\, \overline{\rho' w'} =$ rate of work of the buoyancy forces; this term is zero for a hydrostatically neutral atmosphere.

The last term in Eq. (6.83) is the kinetic energy, which is usually neglected.

6.2.4.1 Richardson Numbers

The flux Richardson number is defined as the ratio of the buoyant force to that of the eddy shear stress

$$R_f = \frac{g\, \overline{\rho' w'}}{\tau \dfrac{\partial u}{\partial z}} = \frac{g\, \overline{\rho' w'}}{\rho u_*^2 \dfrac{\partial u}{\partial z}} \tag{6.84}$$

where $u_*^2 = (\tau/\rho)$ is the friction velocity. It is actually the surface shear stress and is thus a constant. The flux Richardson number can be cast into a different form as follows. A fluid at a level z will have mean values of ρ, p, and T. At a small distance from this level, $z - l$, the temperature fluctuation is

$$T' = l \frac{\partial T}{\partial z}$$

and the density is

$$\rho' = l \frac{\partial \rho}{\partial z} = -\frac{\rho}{T} T'$$

The buoyant work rate can then be written

$$g\, \overline{\rho' w'} = -g \frac{\rho}{T} T' w' = \epsilon_h \left(\frac{\partial T}{\partial z} + \Gamma \right) \frac{g\rho}{T}$$

This is the work done against gravity due to the density differences of the fluid and its surroundings at each level. This buoyant work will be zero for a neutral atmosphere. Therefore, Eq. (6.84) may be written

$$R_f = \frac{\frac{g\rho}{T}\left(\frac{\partial T}{\partial z}+\Gamma\right)\epsilon_h}{\rho\,\epsilon_m\left(\frac{\partial u}{\partial z}\right)^2} = \frac{\epsilon_h}{\epsilon_m}\,\frac{g\left(\frac{\partial T}{\partial z}+\Gamma\right)}{T\left(\frac{\partial u}{\partial z}\right)^2} \tag{6.85}$$

The gradient Richardson number is defined as

$$R_i = \frac{g\left(\frac{\partial T}{\partial z}+\Gamma\right)}{T\left(\frac{\partial u}{\partial z}\right)^2} \tag{6.86}$$

This is, again, a ratio of the buoyant work to the turbulent shear stress work, and it is used to characterize the stability of the atmosphere. R_i was originally used as a criterion for the onset of turbulence in a stably stratified flow, but it is now more useful as an indicator of the effect of thermal stratification on a turbulent flow. In general it is a function of elevation. When the Richardson number exceeds a certain value, postulated as ¼ − 1.0 (Brant 1939), the turbulence will be completely damped out by the buoyancy effect of gravity.

6.2.4.2 Monin-Obukhov Length

Another stability parameter that has gained wide acceptance is the Monin-Obukhov length. Multiply Eq. (6.83) by $k_o\ z/u_*^3$ leading to

$$\frac{k_o z}{u_*}\frac{\partial u}{\partial z} = k_o\, z\, u_*^{-3}\epsilon + z/L + \chi$$

where

$$L = \frac{\rho\, u_*^3}{k_o g\, \overline{\rho'\, w'}} \quad \text{is the Monin-Obukhov length} \tag{6.87}$$

$$\chi = \frac{z\, k_o}{\rho u_*^3}\frac{\partial\, \overline{p'\, w'}}{\partial z}$$

The dimensionless ratio z/L again relates the buoyant and friction forces and is a stability relation. This parameter may be expressed as

$$z/L = \frac{k_o z}{u_*}\frac{\partial u}{\partial z} R_f \tag{6.88}$$

We will use this equation later in relating z/L and the gradient Richardson number. Since both u_* and $g\,\overline{\rho' w'}$ are approximately constant in the constant flux layer, the value of L will not vary with height. It is also possible to express L in other forms as

$$L = \frac{\rho\, u_*^3\, T}{k_o\, \epsilon_h \left(\dfrac{\partial T}{\partial z} + \Gamma\right) g p} \tag{6.89}$$

If the molecular conduction is neglected, the sensible heat may be written

$$q = -\rho c_p\, \epsilon_h \left(\frac{\partial T}{\partial z} + \Gamma\right)$$

The sign convention used here is such that heat flow out of the atmosphere is negative, and heat flow to the atmosphere is positive. If we are looking at the ground surface, we should reverse this, so that heat flow out of the ground (into the atmosphere) will be negative. Then L is

$$L = \frac{-\rho c_p\, u_*^3\, T}{k_o\, g\, q} \tag{6.90}$$

The following values of the parameters are often used, but they are only representative and will vary from system to system. Later we shall examine further values of these parameters. Both R_i and z/L vary almost linearly with elevation

Stability	z/L	R_i
Moderately stable	0.3	0.1
Neutral	0	0
Moderately unstable	−.03	−.08

The following table, Table 6.11, notes some further physical manifestations of the stability. These relations demonstrate that z/L and R_i may be used interchangeably to denote the atmospheric stability.

6.2.5 Velocity and Temperature

6.2.5.1 Velocity, Neutral Atmosphere

Assume that the flow is isothermal, the buoyancy effects are zero, and the molecular transfer is negligible, then the turbulent shear stress of Eq. (6.72) is

Table 6.11. Buoyant and Turbulent Effects.

Relation of buoyant work to turbulent friction work	Physical effect	z/L or R_i
$\dot{W}_B > \dot{W}_f$	Buoyancy predominates, and free convection is dominant heat transfer mechanism. Very stable atmosphere.	>1
$\dot{W}_B < \dot{W}_f$	Turbulent shear work predominates, and turbulent convection is energy transfer mechanism. Unstable atmosphere.	<1
$\dot{W}_B \sim \dot{W}_f$	Free convection and turbulent exchange are of equal value, stable atmosphere.	~ 1
$\dot{W}_B = 0$	Hydrostatically neutral atmosphere.	0

$$\tau = -\rho \overline{u' w'}$$

which has already been shown to be a constant near the surface.

Prandtl's mixing length theory assumes that the velocity fluctuations are caused by fluid elements transporting momentum to various layers. A fluid particle brought from below at $z - l$ causes a fluctuation (Fig. 6.11) of

$$u' = -l \frac{\partial u}{\partial z}$$

Here, l is the mixing length, analogous to the mean free path of molecules. It is assumed that the momentum of a particle remains constant during its

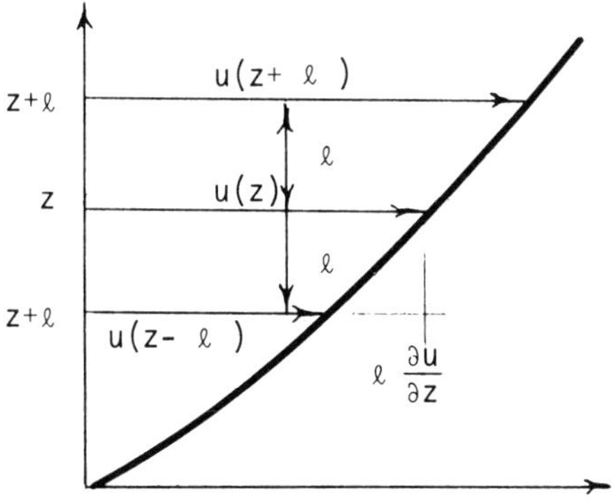

Fig. 6.11. Prandtl Mixing Length.

change from $z - l$ to z, although this is not likely. If the fluctuation in the z direction, w', is of the same order as u', then

$$\tau = \overline{\rho u' w'} = \rho l^2 \left(\frac{\partial u}{\partial z}\right)^2 = \rho l^2 \left|\frac{\partial u}{\partial z}\right| \frac{\partial u}{\partial z} \tag{6.91}$$

The mixing length is

$$\rho l^2 = -\frac{\overline{\rho u' w'}}{\left|\frac{\partial u}{\partial z}\right| \frac{\partial u}{\partial z}}$$

and since u' and w' approach zero near a solid surface, then l must also approach zero. The assumption is often made that l varies linearly with z or $l = k_o z$ where k_o is the von Karman constant. This is usually taken as 0.4, but has been estimated at a range of values. Equation (6.91) is then

$$\sqrt{\frac{\tau}{\rho}} = k_o z \frac{\partial u}{\partial z} \tag{6.92}$$

The quantity $\sqrt{\frac{\tau}{\rho}}$ is defined as the friction velocity u_*, as it has the dimensions of a velocity. The solution of Eq. (6.92) is

$$\frac{u}{u_*} = \frac{1}{k_o} \ln\left(\frac{z}{z_o}\right) \tag{6.93}$$

for a surface of roughness z_o. For a smooth surface, one has

$$\frac{u}{u_*} = 5.5 + \frac{1}{k_o} \ln\left(\frac{u_* z}{\nu}\right) \tag{6.94}$$

Smooth surface flow is unlikely at the earth-atmosphere interface. Numerous investigations have confirmed the logarithmic variation of the atmospheric wind for neutral stability. For other stabilities, however, Deacon (1949) notes that the curve departs from the logarithm as shown in Fig. 6.12.

6.2.5.2 Surface Roughness

The roughness z_o is a function of stability, but it is usually evaluated from neutral data. It can be found from Eq. (6.93) as

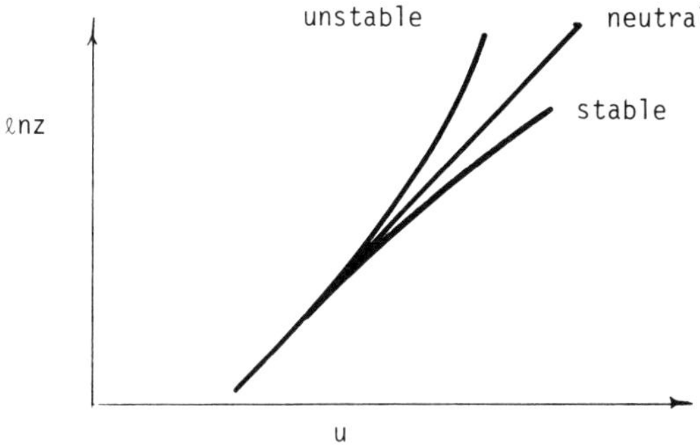

Fig. 6.12. Effect of Instability upon Velocity Profiles.

$$\frac{u_1}{u_2} = \frac{\ln(z_1/z_o)}{\ln(z_2/z_o)} \tag{6.95}$$

A universal plot of z_o versus u_1/u_2 can be drawn, and z_o can be read off the plot for data of u_1/u_2 taken during neutral conditions, $R_i = 0$. Scott (1957) used the ratio of the velocities at 80 and 40 cm. Some values for neutral stability are given in Table 6.12.

6.2.5.3 Displacement Height

If the air flows over a surface with very large roughness, such as a corn crop or a forest, then the flowing air mass is displaced upward from the original flow interface. Figure 6.13 illustrates this. In this sketch, h_1 is the actual height of the vegetative surface, and d is the displacement height—the amount by which the vegetative layer displaces the flow surfaces. The roughness is z_o, but it changes from one surface to another. Rossby (1935) suggests that the mixing length in this case should be $l = k_o(z - d)$. The solution to Eq. (6.91), with the velocity zero at $(z_o + d)$ is

$$\frac{u}{u_*} = \frac{1}{k_o} \ln\left(\frac{z-d}{z_o}\right) \tag{6.96}$$

The values of z_o and d vary in a complicated fashion with h_1, but Monteith (1973) gives the following simple relations based on many observations

$$d = .63\ h_1$$
$$z_o = .13\ h_1$$

Table 6.12. Surface Roughness Based on Neutral Stability.

Surface	Surface roughness z_o, cm		
	Van Wijk (1966)	Scott (1957)	Deacon (1953)
Very smooth (mud flats, ice)	.001	.001	.001
Lawn grass up to 1 cm	.1	–	–
Downland thin grass up to 10 cm	.7	–	–
Thick grass up to 10 cm	2.3	–	–
Thin grass up to 50 cm	5	–	–
Thick grass up to 50 cm	9	–	–
Smooth snow	.5	–	–
Smooth lawn	.5	–	–
Fallow field	2.1	–	–
Open field	3.2	–	–
Low grass	3.2	–	–
High grass	3.9	–	–
Sea surface (no breakers)	4.0	–	–
Wheat field	4.5	–	–
Turnip field	–	10	–
Grass airport, prairie surface	–	1	–
Baseball field, spinach .5–.6 cm	–	.1	–
Bare rolled soil surface	–	.01	–
Smooth snow on short grass	–	–	.005
Snow, natural prairies	–	–	.10
Desert (Pakistan)	–	–	.03
Mown grass 1.5 cm	–	–	.2
3.0 cm	–	–	.7
4.5 cm, $u = 2$ m/sec	–	–	2.4
4.5 cm, $u = 6$–8 m/sec	–	–	1.7
Long grass, 60–70 cm, 1.5 m/sec	–	–	9.0
Long grass, 60–70 cm, 3.5 m/sec	–	–	6.1
Long grass, 60–70 cm, 6.2 m/sec	–	–	3.7

These are just estimates, however, and cannot be exact for a given case. Equation (6.96) reduces to Eq. (6.93) when $d = 0$, as it should.

6.2.5.4 Temperature

The stability is related to the temperature gradient, as has been discussed. This may be noted as shown in Fig. 6.14. The thermal effects, manifested as buoyancy, cause the velocity profiles to curve away from the simple logarithmic profile.

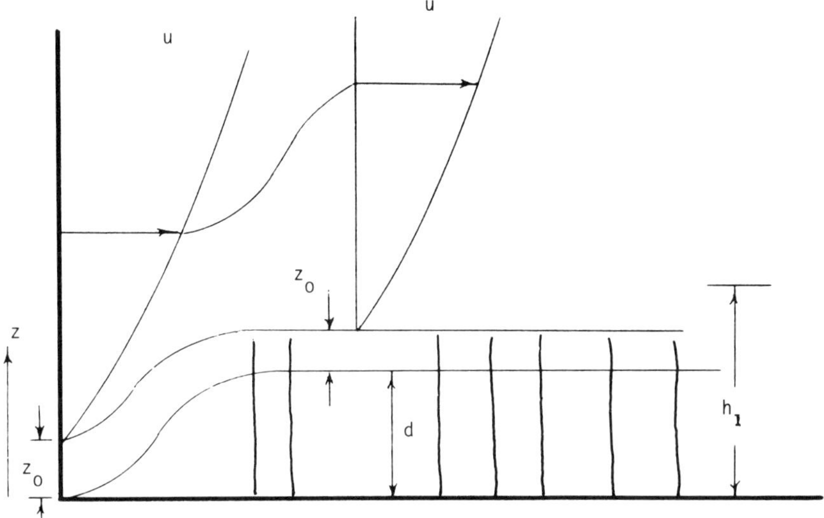

Fig. 6.13. Displacement of Velocity Profile due to Surface Roughness.

6.2.6 Turbulent Transport

We are mainly interested in the transport of sensible heat, but we must consider all the turbulent fluxes. There are three methods of obtaining a value of the sensible heat transport in the atmosphere: direct measurement, indirect measurement or the energy balance, and semiempirical calculations.

6.2.6.1 Direct Measure of Sensible Heat

The vertical flux of turbulent sensible heat has been shown to be

$$q = -\rho c_p \overline{w' T'}$$

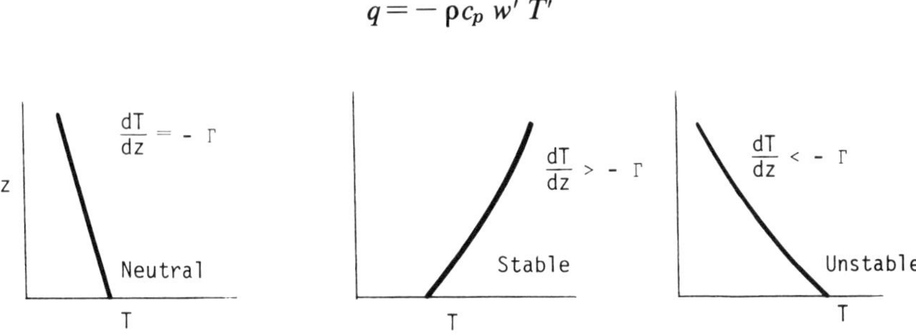

Fig. 6.14. Atmospheric Stability.

It is necessary to measure, simultaneously, the covariance of w' and T', the vertical velocity and temperature fluctuations. Unfortunately, this requires very accurate equipment and measurement techniques and is limited to short time situations. This method is the most fundamental and could be used to evaluate various theories, but there is great difficulty associated with sampling times and elevations. Measurements such as this are not standard for meteorological stations, and it is unlikely that they will be in the near future. Thus, the method is not applicable to engineering calculations at many different locations.

6.2.6.2 Energy Balance

Since the vertical flux of sensible heat cannot be conveniently measured directly for long time periods, an attempt has been made to measure the other terms in the energy balance and find q by elimination. There are two disadvantages to this method.

It is necessary to measure all the pertinent energy fluxes for a given situation. Consider Eq. (6.76) and add the energy fluxes of radiation, latent heat, and the ground conduction. Then, if the equation is integrated over a height h, with the fluxes measured at h, one has

$$-\rho c_p \overline{w' T'} = -(Q_N + Q_G + Q_E) + c_p \int_o^h \frac{\partial}{\partial x} (\rho \overline{u' T'})\, dz + c_p \int_o^h \rho u \frac{\partial T}{\partial x}\, dz + c_p \int_o^h \rho w \frac{\partial T}{\partial z}\, dz + c_p \int_o^h \frac{\partial}{\partial t} (\rho T)\, dz$$

The first three integrals on the right are the advection terms, and the last is the energy change within the volume. Now, from continuity

$$\rho w = -\int_o^z \frac{\partial \rho u}{\partial x}\, dz$$

and

$$-\rho c_p \overline{w' T'} = -(Q_N + Q_G + Q_E) + c_p \int_o^h \left[\rho u \frac{\partial T}{\partial x} - \left(\int_o^z \frac{\partial \rho u}{\partial x}\, dz \right) \frac{\partial T}{\partial z} \right] dz + c_p \int_o^h \frac{\partial}{\partial x} (\rho \overline{u' T'})\, dz + c_p \int_o^h \frac{\partial}{\partial t} (\rho T)\, dz$$

Normally, the integral terms are not measured in energy balances as they are quite small compared with Q_N or Q_E. Nevertheless, there is no assurance that they will be small compared with q, especially if obstructions to the air flow exist near a test site to cause considerable horizontal gradients.

If all the integral terms are negligible, there still is a problem. The equation is

$$-\rho c_p \overline{w' T'} = q = (Q_N + Q_G + Q_E)$$

In this balance all the errors in Q_N, Q_G, and Q_E are cumulative and will be included in the calculated value of q. As Q_N can be very large, a small error in this term can distort the value of q completely.

6.2.6.3 Aerodynamic Prediction

The third method is an attempt to calculate the heat flux from basic equations plus empirical evidence related to the stability and the structure of the velocity and temperature. This is the method we shall concentrate on, but it should be noted that the reliability of the predictions is not too good (Berg 1974).

6.2.6.4 Eddy Coefficient of Momentum

The basic equations for momentum can be written for the neutral case as

$$\frac{\partial u}{\partial z} = \frac{u_*}{k_o z}$$

$$\epsilon_{mm} = k_o z u_*$$

The diabatic case can be generalized as

$$\frac{\partial u}{\partial z} = \frac{u_*}{k_o z} \phi_m(\alpha) \tag{6.97}$$

$$\epsilon_m = \frac{\epsilon_{mm}}{\phi_m} \tag{6.98}$$

The function ϕ_m is a dimensionless relation, which has a value of 1 for neutral stability and is greater or less than 1 for stable and unstable atmospheres. The parameter α may be any of the stability parameters such as R_i, z/L, etc.

A. Aerodynamic Relation. The wind profiles show opposite curvature for lapse and inversion conditions, as has been noted. Thus, Laikhtman (1944) and independently Deacon (1949) chose an equation of the form

$$\frac{\partial u}{\partial z} = a\, z^{-\beta} \tag{6.99}$$

This amounts to a special choice of ϕ_m where a and β are functions of the stability parameter α. The solution of Eq. (6.99) is

$$u = \frac{a z_o^{(1-\beta)}}{1-\beta}\left[\left(\frac{z}{z_o}\right)^{1-\beta} - 1\right]$$

This can be expanded as

$$u = a\, z_o^{(1-\beta)}\left[\ln\frac{z}{z_o} + \frac{1-\beta}{2}\ln^2\left(\frac{z}{z_o}\right) + \ldots\right]$$

which will agree with the neutral solution, Eq. (6.93), if $a = u_*/k_o z_o(1-\beta)$. Thus

$$\frac{u}{u_*} = \frac{1}{k_o(1-\beta)}\left[\left(\frac{z}{z_o}\right)^{1-\beta} - 1\right] \tag{6.100}$$

Equation (6.100) now agrees with the neutral case at heights close to the surface, where z/z_o is small, because very close to the surface the effects of buoyancy are negligible. It also approaches the neutral case as $\beta \to 1$. Equation (6.100) can be used to find β, as z_o is known

$$u_1/u_2 = \frac{(z_1/z_o)^{1-\beta} - 1}{\left(\frac{z_2}{z_o}\right)^{1-\beta} - 1} \tag{6.101}$$

From this relation, β can be plotted versus u_1/u_2, and values can then be read off if u_1/u_2 is measured for different stabilities. Scott (1961) notes that the following values of β are in accord with the data, although these are at best estimates for small ranges of data.

Stability	β
Moderately stable	.88
Neutral	1.0
Moderately unstable	1.08

From the definition of the eddy coefficient of momentum

$$\epsilon_m = \frac{u_*^2}{\dfrac{\partial u}{\partial z}} = k_o\, u_*\, z_o \left(\frac{z}{z_o}\right)^{\beta}$$

If one uses Eq. (6.93) to find u_*, one has

$$\epsilon_m = \frac{k_o^2\, z_o \left(\dfrac{z}{z_o}\right)^{\beta} u_1}{\ln\,(z_1/z_o)} \tag{6.102}$$

where u_1 and z_1 are the reference velocity and height; the value of z_1 is often taken as 160 cm, as this is a standard meteorological height. Equation (6.102) was used by Scott (1961), although the friction velocity from Eq. (6.100) will be more consistent and leads to

$$\epsilon_m = \frac{k_o^2\, z_o\,(1-\beta) \left(\dfrac{z}{z_o}\right)^{\beta} u_1}{\left(\dfrac{z_1}{z_o}\right)^{1-\beta} - 1} \tag{6.103}$$

This relation places more emphasis upon the stability and is more consistent with the theory. The eddy coefficient can now be plotted as a function of the reference velocity, the surface roughness, and the stability. Any reference height can be used, but the value $z_1 = 160$ cm is convenient for both velocity and temperature. If data are available at some other height, then they would be used. Equation (6.103) can also be used to evaluate the mass flux, as it is usually agreed that the eddy coefficients of mass and momentum are virtually identical.

B. Similarity. Monin (1954) employed a similarity principle, at different heights, starting with Eq. (6.97). The function ϕ_m is a function of z/L and was expanded for small values of z/L as

$$\phi_m\,(z/L) = \phi_m\,(a) + \phi_m'\,(a)\,(z/L - a) + \tfrac{1}{2}\phi_m''\,(a)\,(z/L - a)^2 + \ldots$$

Now, for neutrality $z/L = 0$ from the definition of L, $\phi_m(0) = 1$, $\phi_m'(0) = \alpha$. Then Eq. (6.97) can be written

$$\frac{\partial u}{\partial z} = \frac{u_*}{k_o z}\left(1 + \alpha_1 \frac{z}{L}\right) \tag{6.104}$$

The exact form of ϕ_m has not been determined, and Eq. (6.104) implies that $\phi_m = 1$ for the neutral case. The universal constant α_1, is clearly the value of the slope of ϕ_m where z/L is zero. Monin (1954) found that $\alpha_1 \simeq 0.6$, but Priestly (1959) states that this may be considerably in error. Swinbank (1964) found

$$\phi_m = \frac{z/L}{1 - e^{-z/L}} \tag{6.105}$$

Expansion of Eq. (6.105) into a power series gives $\alpha_1 = 0.5$, although later authors found values of α_1 considerably higher, up to 7.0. Lumley (1964) found an empirical relation for unstable air in the form

$$\phi_m^4 - 14\frac{z}{L}\phi_m^3 = 1 \tag{6.106}$$

This is equivalent to a value of $\alpha_1 = 3.5$, for small values of z/L. Businger (1966) suggested, for unstable air

$$\phi_m = (1 - 16\, z/L)^{-.25} \tag{6.107}$$

This corresponds to the linear case of $\alpha_1 = 4.0$. Dyer (1974) suggested the following relations

$$\phi_m = \begin{cases} \phi_h = 1 + 5\, z/L & \text{stable} \quad (6.108) \\ \phi_h = 1 & \text{neutral} \quad (6.109) \\ \left(1 - \frac{16z}{L}\right)^{-1/4} & \text{unstable} \quad (6.110) \end{cases}$$

$$\phi_h = \left(1 - 16\frac{z}{L}\right)^{-1/2} \quad \text{for unstable} \tag{6.111}$$

The function ϕ_h is the analogue to the momentum case for the heat transfer. Méry (1969) and others have suggested that Eq. (6.108) is suitable for all stabilities where $-.03 < z/L < 0.3$. This is in accord with the linearization of Eq. (6.111), of course. Webb (1970) used a relation very similar to Eq. (6.108)

$$\epsilon_m = \frac{\epsilon_{mm}}{1 + \frac{nz}{L}} \tag{6.112}$$

where $n = 5.2$ for stable and 4.5 for unstable. These values of n are from Monteith (1973). Webb (1970) recommends the usual value of 5.

6.2.6.5 Sensible Heat

The equations for the turbulent transfer of sensible heat are assumed to be analogous to those of momentum

$$\epsilon_h = \frac{\epsilon_{hm}}{\phi_h} = \frac{k_o u_* z}{\phi_h} \tag{6.113}$$

$$\frac{\partial T}{\partial z} = \frac{-q}{\rho c_p k_o u_* z} \phi_h \tag{6.114}$$

where ϕ_h is the universal heat transfer function given by Eqs. (6.108–6.111).

A. Turbulent Prandtl Number. The molecular Prandtl number is defined as the ratio of the molecular momentum and thermal diffusivities, $P = \nu/\alpha$. Often the ratio of the turbulent diffusivities is defined as the turbulent Prandtl number

$$P_t = \frac{\epsilon_m}{\epsilon_h} \tag{6.115}$$

From Eqs. (6.98) and (6.113), we note that

$$P_t = \frac{\phi_h}{\phi_m} \tag{6.116}$$

Early attempts to estimate P_t often assumed that the turbulent ratio had the same value as in the molecular case, $P_t = \nu/\alpha$. Ogura, in Doronin (1970), theoretically predicted this relation, but numerous experiments have shown that P_t remains fairly constant for fluids with a large variation in P. Scott (1961), from limited data, noted

$$P_t = \begin{cases} 1.0 & \text{neutral, stable} \\ 0.5 - 0.67 & \text{unstable} \end{cases}$$

Since $P \sim .77$ for air, the unstable values are in accord with Ogura for air. Priestly (1959) presents a curve P_t versus R_f, which, despite considerable scatter, can be expressed

$$\ln P_t = .28 + 3.46\, R_f$$

or, approximately, for $-.5 < R_f < 0.1$

$$P_t = 2 - \sqrt{0.45 - 6.93\, R_f} \tag{6.117}$$

This relation seems reasonable for lapse and inversion, but departs from other relations in giving $P_t = 1.33$ for neutral stability. From Eqs. (6.108)–(6.111)

$$P_t = \begin{cases} 1.0 & \text{neutral, stable} \\ \left(1 - \dfrac{16z}{L}\right)^{-1/4} & \text{unstable} \end{cases} \tag{6.118}$$

Doronin (1970) relates P_t to the Richardson number

$$P_t = 0.3 + \frac{0.7}{\sqrt{1 - 28\, z_o R_i}} \tag{6.119}$$

where z_o is in centimeters.

The following relations for the turbulent Prandtl number can be used

$$\begin{array}{llll} & \textit{Similarity relations} & & \textit{Deacon relation} \\ P_t = & \begin{cases} 1.0 \\ \left(1 - \dfrac{16z}{L}\right)^{-1/4} \simeq 1 + \dfrac{4z}{L} \end{cases} & \begin{array}{l} \text{neutral, stable} \\ \text{unstable} \end{array} & \begin{array}{l} 1.0 \\ (z/z_o)^{1-\beta} \end{array} \end{array} \tag{6.120}$$

B. Stability Parameters. It is now possible to express some quantitative relations between the various stability parameters. From Eqs. (6.85) and (6.88)

$$\frac{z}{L} = \frac{k_o z}{u_*} \frac{\partial u}{\partial z} \frac{\epsilon_h}{\epsilon_m} R_i$$

Then, with Eq. (6.97)

$$R_i = \frac{\epsilon_m}{\epsilon_h} \frac{1}{\phi_m} \frac{z}{L} = \frac{P_t}{\phi_m} \frac{z}{L} = f_1\,(z/L) \tag{6.121}$$

Using Eq. (6.85)

$$R_f = \frac{1}{\phi_m}\frac{z}{L} = f_2\,(z/L) = \frac{R_i}{P_t} \tag{6.122}$$

The gradient and flux Richardson numbers are clearly related to z/L, and the exact quantitative functions f_1 and f_2 depend upon the particular similarity functions used. From Eq. (6.99), one has

$$\left(\frac{z}{z_o}\right)^{1-\beta} = \phi_m \tag{6.123}$$

The following relations can now be written

$$R_i = \begin{cases} \dfrac{z/L}{1+5\dfrac{z}{L}} & \text{neutral or stable} \quad (6.124a) \\[2ex] \dfrac{z}{L} & \text{unstable} \quad (6.124b) \end{cases}$$

$$R_f = \begin{cases} R_i & \text{neutral or stable} \quad (6.125a) \\[1ex] \dfrac{z}{L}\left(1-\dfrac{16z}{L}\right)^{1/4} & \text{unstable} \quad (6.125b) \end{cases}$$

$$\beta = \begin{cases} 1.0 & \text{neutral} \quad (6.126a) \\[1ex] 1-\dfrac{\ln\left(1+5\dfrac{z}{L}\right)}{\ln\dfrac{z}{z_o}} & \text{stable} \quad (6.126b) \\[3ex] 1+\dfrac{\ln\left(1-16\dfrac{z}{L}\right)}{4\ln\dfrac{z}{z_o}} & \text{unstable} \quad (6.126c) \end{cases}$$

Based on Eq. (6.119) Doronin (1970), the Richardson numbers are

$$R_f = \frac{R_i}{.3+\dfrac{.7}{\sqrt{1-28z_oR_i}}} \tag{6.127}$$

Finally, the following stability table can be calculated. If the values of z/L are as noted, then the other stability parameters should have the values shown.

Stability	z/L	R_i	R_f	$\beta\left(160 > \frac{z}{z_o} > 10\right)$
Neutral	0	0	0	1
Moderately stable	0.3	.12	.12	$0.60 - 0.82$
Moderately unstable	−.03	−.03	−.033	$1.02 - 1.05$

C. Surface Coefficient of Heat Transfer. The atmospheric heat flux may be written

$$q = -\rho c_p \epsilon_h \frac{\partial T}{\partial z} = -\rho c_p \frac{\epsilon_m}{P_t} \frac{\partial T}{\partial z} \tag{6.128}$$

The negative sign used for this equation is conventional in that a heat loss from the atmosphere is negative and a gain is positive. For the ground itself, however, we define the surface coefficient such that a heat loss from the ground to the atmosphere is negative. The opposite signs will be used for the atmosphere and the ground

$$q = h(T_1 - T_s) \tag{6.129}$$

where T_1 is the air temperature at the reference level and T_s is the ground surface temperature. The basic equation for h is

$$h = \rho c_p \left(\int_{z_o}^{z_1} \frac{P_t}{\epsilon_m} \, dz \right)^{-1} \tag{6.130}$$

In this relation, the assumption has been made that the surface temperature T_s exists at the level z_o. This is not exact, but there are insufficient data to further refine the assumption.

1. Aerodynamic formula. Using the relations for ϵ_m given by Eq. (6.103) and P_t given by Eq. (6.120) yields the three stability formulas for the Deacon equation

$$\frac{h}{\rho c_p k_o^2 u_1} = \begin{cases} \dfrac{1}{\ln^2(z_1/z_o)} & \text{neutral} \quad (6.131a) \\[2ex] \left(\dfrac{1-\beta}{\left(\dfrac{z_1}{z_o}\right)^{1-\beta} - 1} \right)^2 & \text{stable} \quad (6.131b) \\[2ex] \dfrac{2(1-\beta)^2}{\left[\left(\dfrac{z_1}{z_o}\right)^{1-\beta} - 1\right]^2 \left[\left(\dfrac{z_1}{z_o}\right)^{1-\beta} + 1\right]} & \text{unstable} \quad (6.131c) \end{cases}$$

2. Similarity equations. The general relation for h is given by Eqs. (6.114) and (6.129)

$$h = \rho c_p k_o \left(\int_{z_0}^{z} \frac{\phi_h \, dz}{u_* z} \right)^{-1} \tag{6.132}$$

The neutral case is identical to Eq. (6.131a), and the stable case is given by

$$u_* = \frac{k_o \, u_1}{\ln \dfrac{z_1}{z_0} + 5\left(\dfrac{z_1}{L} - \dfrac{z_0}{L}\right)}$$

$$\frac{h}{\rho c_p k_o^2 \, u_1} = \frac{1}{\left[\ln \dfrac{z_1}{z_0} + \dfrac{5}{L}(z_1 - z_0)\right]^2} \tag{6.133}$$

The unstable case is more complicated, as $(\partial u/\partial z)$ is difficult to integrate exactly. Nevertheless, an approximate relation for u_*, valid for small values of z/L is

$$u_* = \frac{k_o \, u_1}{\ln\left(\dfrac{z_1}{z_0} \dfrac{1 - 4\dfrac{z_0}{L}}{1 - 4\dfrac{z_1}{L}}\right)}$$

$$\frac{h}{\rho c_p k_o^2 u_1} = \frac{1}{\ln\left(\dfrac{z_1}{z_0} \dfrac{1 - 4\dfrac{z_0}{L}}{1 - 4\dfrac{z_1}{L}}\right) \ln\left(\dfrac{\sqrt{1 - 16\dfrac{z_1}{L}} - 1}{\sqrt{1 - 16\dfrac{z_0}{L}} - 1} \, \dfrac{\sqrt{1 - 16\dfrac{z_0}{L}} + 1}{\sqrt{1 - 16\dfrac{z_1}{L}} + 1}\right)} \tag{6.134}$$

The difficulty in using Eqs. (6.133) and (6.134) is that L is not known until the heat transfer is known when Eq. (6.90) may be used to calculate L. One can iterate a solution by guessing L, calculating h, calculating q if $(T_1 - T_s)$ is known, and then calculating L. Iteration is continued until the values of L agree. This method is probably too cumbersome for the small increase in accuracy, and an estimate of z/L is the simplest method that gives adequate accuracy.

3. Nondimensional relations. The results of the previous two sections can be nondimensionalized in terms of the Nusselt number, $N = hz_1/k$, and the Reynolds number, $R = u_1 z_1/\nu$. A j factor is defined as

$$j = \frac{N}{k_0^2 RP}$$

The following table results for the j factor

	Stable	Neutral	Unstable
Aerodynamic, Eq. (6.131)	$\left(\frac{\alpha}{z^\alpha - 1}\right)^2$	$\frac{1}{\ln^2 z}$	$\frac{2}{(z^\alpha + 1)}\left(\frac{\alpha}{z^\alpha - 1}\right)^2$
Similarity, Eqs. (6.133), (6.134)	$\left[\ln z + \frac{5}{4}(b - a)\right]^{-2}$	$\frac{1}{\ln^2 z}$	$\left[\ln \frac{z(1-a)}{1-b} \ln\left(\frac{\sqrt{1-4b}-1}{\sqrt{1-4a}-1}\frac{\sqrt{1-4a}+1}{\sqrt{1-4b}+1}\right)\right]^{-1}$

where

$$z = \frac{z_1}{z_0} \qquad \alpha = 1 - \beta \qquad a = \frac{4z_0}{L} \qquad b = \frac{4z_1}{L}$$

The case for an aerodynamically smooth surface can be written as

$$j = \frac{0.535}{P} \tag{6.135}$$

A comparison of these equations can be made for the following cases:

Unstable: $z_1 = 160$ cm	$z_0 = 1$ cm	$z_1/L = -0.03$	$\beta = 1.08$
Stable: "	"	$z_1/L = +\ .3$	$\beta = \ .88$

	j factor		
	Stable	Neutral	Unstable
Aerodynamic	.0205	0.0388	.0690
Similarity	.0232	0.0388	.0414

The difference for the unstable case is exaggerated because $\beta = 1.08$ is not equivalent to $z_1/L = -.03$. A value of $\beta = 1.04$ gives $j = .052$, which is about 24% higher than the similarity relation.

It should be emphasized that all these relations are based on the analogy between the transfer of momentum and energy. The only detailed information is that for the momentum transfer or the velocity profiles, with little data on the turbulent eddy coefficient for energy transfer. Due to the nature of the flow problem, it is unlikely that this restriction will be removed soon.

D. Two-Level Method. A further method of evaluating the heat flux is based upon the use of data at two elevations. Assume that data on the temperature and velocity are available at two levels above the actual surface. The heat transfer is

$$q = \frac{\rho c_p (T_2 - T_1)}{\int_{z_1}^{z_2} \frac{P_t}{\epsilon_m}\, dz}$$

For the neutral case, with the usual relations

$$q = \frac{\rho c_p k_o^2 (u_2 - u_1)(T_2 - T_1)}{\ln^2\left(\frac{z_1}{z_2}\right)} \tag{6.136}$$

This can be written

$$\frac{q}{(u_2 - u_1)(T_2 - T_1)} = 0.355$$

if $z_2/z_1 = 2$, u-cm/sec, T — °C.

Aerodynamic relations. If we use the relations P_t and ϵ_m from the Deacon equations

$$\frac{q}{(u_2 - u_1)(T_2 - T_1)} = \frac{\rho c_p k_o^2 (1-\beta)^2}{\left(\left(\frac{z_2}{z_o}\right)^{1-\beta} - \left(\frac{z_1}{z_o}\right)^{1-\beta}\right)^2} = 0.135, \text{ stable} \tag{6.137a}$$

$$= \frac{2\rho c_p k_o^2 (1-\beta)^2}{\left(\left(\frac{z_2}{z_o}\right)^{1-\beta} - \left(\frac{z_1}{z_o}\right)^{1-\beta}\right)^2 \left(\left(\frac{z_2}{z_o}\right)^{1-\beta} + \left(\frac{z_1}{z_o}\right)^{1-\beta}\right)} = 0.937, \text{ unstable} \tag{6.137b}$$

where $z_o = 1$ cm, $z_1 = 40$ cm, $(z_2/z_1) = 2$, $\beta = 1.08$ unstable, $\beta = .88$ stable. The equations obviously depend upon the surface roughness. This method eliminates the need to have the surface temperature, but it does require measurements at two levels, which are not routinely available.

Similarity equations. Following the same procedure we have

$$\frac{q}{(u_2 - u_1)(T_2 - T_1)} = \frac{\rho c_p k_0^2}{\left(\ln \frac{z_2}{z_1} + 5\frac{(z_2 - z_1)}{L}\right)^2} = .158, \text{ stable} \quad (6.138a)$$

$$= \frac{\rho c_p k_0^2}{\ln\left(\frac{z_2}{z_1} \frac{1 - \frac{4z_1}{L}}{1 - \frac{4z_2}{L}}\right) \ln\left(\frac{\sqrt{1 - \frac{16z_2}{L}} - 1}{\sqrt{1 - \frac{16z_1}{L}} - 1} \frac{\sqrt{1 - \frac{16z_1}{L}} + 1}{\sqrt{1 - \frac{16z_2}{L}} + 1}\right)} = 0.541, \text{ unstable} \quad (6.138b)$$

where $(z_2 - z_1)/L = 0.3$, stable; $z_2/z_1 = 2.0$.
Interestingly, these equations do not depend upon the surface roughness.

6.2.6.6 Heat Flux Calculation

Once the surface coefficient is known and the temperature difference is available, then the sensible heat flux can be calculated. The calculations are still in doubt, however, due to the daily fluctuations of ΔT and h. Normally, we will have only a mean temperature difference and a mean wind speed for a location. Use of these data can lead to significant errors in the calculated heat flux even if h is calculated quite accurately. Unfortunately, there are few data available for ΔT and wind speed over 24-hour periods. Table 6.13 lists the Johns Hopkins University (1954) data for Seabrook, N.J., August 21, 1952. The values of h were calculated according to the equations of this chapter.
The heat flux calculated on an hourly basis is

$$q_{\text{aero}} = -42.0 \text{ cal/cm}^2\text{-day}$$
$$q_{\text{sim}} = -20.6 \text{ cal/cm}^2\text{-day}$$

Table 6.14 compares the average values of the surface coefficient calculated by the aerodynamic and similarity methods.

Table 6.13. One-Day Temperature and Velocity Data (from Johns Hopkins University 1954).

Time	$\Delta T = T_{160} - T_5$, °C	Velocity u_{160}, cm/sec	Stability	h Aerodynamic	h Similarity	q Aerodynamic	q Similarity
				cal/cm²-hr-°C		cal/cm²-hr	
0–1	.23	97	S	.044	.150	.010	.035
1–2	.20	104		.047	.161	.010	.032
2–3	−.02	127		.058	.187	−.001	−.004
3–4	+.25	166		.076	.257	.019	.064
4–5	.84	198		.090	.307	.076	.258
5–6	.51	176		.080	.273	.041	.139
6–7	−.37	212		.846	.462	−.313	−.171
7–8	−.89	338		1.349	.737	−1.20	−.656
8–9	−2.66	338		1.349	.737	−3.587	−1.960
9–10	−3.76	548		2.187	1.194	−8.221	−4.490
10–11	−2.12	580	U	2.314	1.264	−4.906	−2.680
11–12	−1.89	580		2.314	1.264	−4.374	−2.389
12–13	−2.73	500		1.995	1.090	−5.446	−2.976
13–14	−1.73	548		2.187	1.194	−3.783	−2.066
14–15	–	461		1.839	1.005	–	–
15–16	−3.97	461		1.839	1.005	−7.302	−3.990
16–17	−.45	342		1.365	.745	−.614	−.335
17–18	.39	331		1.321	.721	.515	.281
18–19	.62	246		.112	.381	.069	.237
19–20	.41	210	S	.096	.326	.039	.133
20–21	.74	296		.135	.459	.100	.340
21–22	1.12	631		.287	.978	.322	1.096
22–23	.47	386		.168	.571	.079	.268
23–24	.55	311		.142	.482	.075	.256

Table 6.14. Comparison of Theoretical Equations for *h*.

	Day (unstable)	Night (stable)	Mean
Average, u, cm/sec	437	244	341
h (aero)	1.742	.111	.93
h (similarity)	.952	.379	.67
ΔT °C	−1.835	.492	−.672
$\bar{u} = 341$, $\bar{h}_2$(aero)	1.36	.155	.76
$\bar{u} = 341$, $\bar{h}_2$(similarity)	.742	.53	.64

The following equations can be used to estimate the daily heat flux

$$q_1 = 12\,(\bar{h}_s\,\Delta T_s + \bar{h}_u\,\Delta T_u)$$
$$q_2 = 12\,\bar{h}_u\,\Delta T_u$$
$$q_3 = 72\,\bar{h}_2\,\Delta T$$
$$q_4 = 24\,\bar{h}_{2_s}\,\Delta T \text{ (based on the sinusoidal variation; see below)}$$

where s and u refer to the stable and unstable values.
Note that if the values of h and ΔT fluctuate sinusoidally as

$$h = h_m + A \sin \omega t$$
$$\Delta T = \Delta T_m + B \sin \omega t$$
$$\omega = \frac{2\pi}{t_g}$$

Then the heat flux for one period ($t_g = 24$ hours) is

$$q = \int_0^{t_g} h \Delta T\, dt = 24\, h_m\, \Delta T_m + 12\, AB$$

The following values of q were calculated. The exact value is from the hourly sums using Table 6.13.

Case	q Aerodynamic	q Similarity
Exact	−42	−20.6
1	−37.7	−18.7
2	−38.4	−21
3	−36.8	−31.0
4	−21.9	−12.0

The following conclusions can be drawn:

1. Despite the closeness of h from the aerodynamic and similarity equations, the daily heat flux is considerably different due to the opposing effects of the stable heat flow period.
2. Actual daily values of q are much different from simple mean value calculations. This shows that calculation errors may be due to daily variations of h and ΔT and not necessarily errors in h.
3. The relation for q_2 is accurate if daytime values of ΔT and u are available.
4. If only mean values of ΔT and u are available, relation q_3 can be

used, but the validity of this result is not known for other than summer days in temperate climates. Arctic values may be quite different.

6.3 ATMOSPHERIC MASS TRANSFER

The transfer of mass in a fluid (the atmosphere) is analogous to the transfer of heat and momentum. One can write

$$m_v = -D\frac{\partial c}{\partial z} - \overline{v'c'} \tag{6.139}$$

where

m_v = mass flux rate
D = molecular diffusion coefficient for the mass in the fluid and
c = concentration of mass in the fluid.

An eddy diffusion can be defined as before

$$m_v = -D\frac{\partial c}{\partial z} - \epsilon_v\frac{\partial c}{\partial z} = -(D+\epsilon_v)\frac{\partial c}{\partial z} \tag{6.140}$$

If the turbulent effects predominate, then

$$m_v = -\epsilon_v\frac{\partial c}{\partial z} \tag{6.141}$$

and this equation is clearly analogous to the equations for heat and momentum. When the vapor partial pressure p_w is used, then

$$m_v = -\frac{\rho}{p}\frac{M_w}{M_a}\epsilon_v\frac{\partial p_w}{\partial z} \tag{6.142}$$

where

ρ, p are the density and pressure of the mixture and
M_w, M_a = molecular weights of water and air.

The evaporation heat flux may be written by multiplying Eq. (6.142) by the latent heat of vaporization L

$$Q_E = -\frac{\rho}{p}\frac{M_w}{M_a} L\, \epsilon_v \frac{\partial p_w}{\partial z} \tag{6.143}$$

If $\epsilon_v = \epsilon_m$, then one can use the appropriate form of ϵ_m to integrate Eq. (6.141) and find the evaporative heat flux. This is identical to the procedure discussed under sensible heat.

6.3.1 Ratio of Evaporative and Sensible Heat Fluxes

The ratio of the sensible to the evaporative heat flux is called the "Bowen ratio," and it can be written

$$b = \frac{Q_H}{Q_E} \tag{6.144}$$

Now Bowen (1926) originally derived the ratio for molecular effects, but the same details will hold for turbulent effects only

$$b = \frac{c_p p\, M_a}{L\, M_w} \frac{\epsilon_h}{\epsilon_v} \frac{\partial T}{\partial p_w} \tag{6.145}$$

If $\epsilon_h/\epsilon_v = 1$, this equation is usually written

$$b = \frac{c_p p\, M_a}{L\, M_w} \frac{\Delta T}{\Delta p_w} \tag{6.146}$$

This relation is often used to calculate either Q_H from Q_E or vice versa.

REFERENCES

Allen, S. W. 1958. Solar radiation. *Q. J. R. Meteorol. Soc.* 84:307–318.

Anderson, E. R. 1954. *Energy budget studies, water loss investigations: Lake Hefner studies.* Tech. Rept. U.S. Geol. Survey Professional Pap. 269.

Angstrom, A. 1915. *A study of radiation of the atmosphere.* Smithsonian Miscellaneous Collection No. 65, Washington, D.C.

———. 1924. *Q. J. R. Meteorol. Soc.* 50:123.

Berg, R. L. 1974. *Energy balance on a paved surface.* Hanover, New Hampshire: CRREL, TR 226.

Black, J. N. 1956. The distribution of solar radiation over the earth's surface. *Arch. Meteorol. Geophys. Bioklimatol.* 7:165–189.

———, Boynthon, C. W., and Prescott, J. A. 1954. Solar radiation and the duration of sunshine. *Q. J. R. Meteorol. Soc.* 80:231–235.

Bolsenga, S. J. 1964. *Daily sums of global radiation for cloudless skies.* CRREL Research Report 760.

Bowen, I. S. 1926. The ratio of heat losses by conduction and by evaporation from any water surface. *Phys. Rev.* 27:779–789.

Brant, D. 1939. Physical and dynamical meteorology. 2nd ed. Cambridge: the University Press.

Brown, A. I., and Marco, S. 1958. *Introduction to heat transfer.* 3rd ed. New York: McGraw-Hill.

Brunt, D. 1932. Notes on radiation in the atmosphere. *Q. J. R. Meteorol. Soc.* 58:389–420.

Brutsaert, W. 1975. On a derivable formula for long-wave radiation from clear skies. *Water Resour. Res.* 11 (5).

Budyko, M. I. 1956. *The heat balance of the earth's surface.* Translation U.S. Dept. of Commerce, Washington, D.C. 1958.

———. 1974. *Climate and life.* New York: Academic Press.

Businger, J. A. 1966. Transfer of momentum and heat in the planetary boundary layer. *Proceedings Symposium on Arctic Heat Budget and Atmospheric Circulation.* 305–32, The Rand Corp.

Deacon, E. L. 1949. Vertical diffusion in the lowest layer of the atmosphere. *Q. J. R. Meteorol. Soc.* 75:88–103.

———. 1953. *Vertical profiles of mean wind in the surface layers of the atmosphere.* Geophysical Mem. No. 91, London: Air Ministry, Met. Office.

———. 1970. The derivation of Swinbank's long-wave radiation formula. *Q. J. R. Meteorol. Soc.* 96:313–319.

Degelman, L. O. 1966. *The Development of a mathematical model for predicting solar heat gains through building walls and roofs.* Pennsylvania State Univ., Better Building Report No. 6.

Doronin, Yu. P. 1970. *Thermal interaction of the atmosphere and the hydrosphere in the Arctic.* Israel Program for Scientific Translations, Keter Press.

Dyer, A. J. 1974. A review of flux profile relationships in boundary layer. *Meteorologie* 7:363–372.

Eckert, E. R. G., and Drake, R. M. 1959. *Heat and mass transfer.* 2nd ed. New York: McGraw-Hill.

Fishenden, M., and Saunders, O. A. 1932. *Calculations of heat transmission.* London: H. M. Stationery Office.

Fritz, S., and MacDonald, T. N. 1949. Average solar radiation in the United States. *Heat. Vent.* 46:61–64.

Gabites, J. F. 1950. Seasonal variations in the atmospheric heat balance. Thesis, MIT Meteorological Dept.

Gerdel, R. W., Diamond, M., and Walsh, K. J. 1954. *Nomographs for computation of radiation heat supply.* SIPRE Research Pap. 8.

Gorczynski, W. 1945. *Comparison of climate of the United States and Europe.* Herald Square Press.

Hinze, J. O. 1975. *Turbulence.* 2nd ed. New York: McGraw-Hill, p. 476.

Johns Hopkins University. 1954. Quarterly summaries of observations made at Seabrook, New Jersey. *Publications in climatology.* Johns Hopkins Univ., Laboratory of Climatology, vol. 7, Final Report, Seabrook, N.J.

Kasten, F. 1964. *A new table and approximating formula for the relative optical air mass.* CRREL Tech. Rept. 136.

Kimball, H. H. 1927. *Mon. Weather Rev.* 55:155–169.

———. 1928. *Mon. Weather Rev.* 56:393–398.

Kimura, K., and Stephenson, D. G. 1969. Solar radiation on cloudy days. *Trans. ASHRAE* 75: part 1.

Kraus, E. B. 1972. *Atmosphere-ocean interaction.* Oxford Monographs on Meterology. Oxford: Clarendon Press, p. 135.

Kreith, F. 1973. *Principles of heat transfer.* 3rd ed. New York: Intext Educational Publishers.

Laikhtman, D. L. 1944. Profile of wind and interchange in the layer of atmosphere near the ground. *Bull. Acad. Sci. USSR, Geog. Geophys. Ser.* 8(9): 1–5.

Lumley, J. L., and Panofsky, H. A. 1964. *The structure of atmospheric turbulence.* New York: Interscience.

Lunardini, V. J. 1979. "Sensible heat transfer in the atmosphere," *Proceedings of the Seventh Canadian Congress of Applied Mechanics,* 769.

Lunardini, V. J. 1974. Unpublished data for Inuvik, NWT region.

Méry, P. 1969. Reproduction en similitude de la diffusion dans la couche limite atmosphérique. *Houille Blanche* 24(4):327–344.

Monin, A. S., and Obukhov, A. M. 1954. *Basic regularity in turbulent mixing in the surface layer.* USSR Acad. Sci. Geophys. Inst. No. 24.

Monteith, J. L. 1973. *Principles of environmental physics.* London: Edward Arnold.

Munn, R. E. 1966. *Descriptive micrometeorology.* New York: Academic Press.

Phillips, H. 1940. Zur Theorie der Warmestrahlung in Bodennake. Geol. Beitrag 56.

Porkhaev, G. V. 1959. *Principles of geocryology.* Part 2, Chapter 4. Moscow: Academy of Science of the USSR. Translation NRC TT-1249, Ottawa, 1969.

Priestly, C. H. B. 1959. *Turbulent transfer in the lower atmosphere.* Chicago: Univ. of Chicago Press, p. 27.

Reitan, C. H. 1963. Surface dewpoint and water vapor aloft. *J. Appl. Meteorol.* 2:776–779.

Rossby, C. G., and Montgomery, R. G. 1935. *The layer of frictional influence on wind and ocean currents.* Pap. Phys. Ocean Meteorology, MIT, No. 3.

Scott, R. F. 1957. Estimation of the heat transfer coefficient between air and the ground surface. *Trans. Am. Geophys. Union* 38:25–32.

———. 1961. Heat transfer at the air-ground interface with special reference to airfield pavements. MIT Dept. Civil Engr. Tech. Rep. 63.

———. 1964. *Heat exchange at the ground surface.* CRREL Monograph II-A1.

Stringer, E. T. 1972. *Techniques of climatology.* San Francisco: W. H. Freeman.

Swinbank, W. C. 1963. Long-wave radiation from clear skies. *Q. J. R. Meteorol. Soc.* 89:339–348.

———. 1964. The exponential wind profile. *Q. J. R. Meteorol. Soc.* 90(384):119.

Van Wijk, W. R. 1966. *Physics of plant environment.* Chap. 3. Amsterdam: North-Holland.

Webb, E. K. 1970. Profile relationships: The long-linear range and extension to strong stability. *Q. J. R. Meteorol. Soc.* 96:67.

Wechsler, A. E., and Glaser, P. E. 1966. *Surface characteristics effect on thermal regime, phase I.* CRREL Special Report 88.

Wexler, H. 1941. *Observations of nocturnal radiation at Fairbanks, Alaska.* U.S. Weather Bureau, Monthly Weather Review, Supp. No. 46.

Yamamoto, G. 1950. *On nocturnal radiation.* Sci. Rept. Tohoku Univ. Ser. 5,2.

7
Heat Transfer Without Phase Change

7.0 INTRODUCTION

Many natural processes occur at such a slow rate that they can be analyzed on a steady-state basis with adequate accuracy. In general, phase change is so important that it cannot be neglected, but there are a number of situations where one may safely ignore the effects of melting or thawing. In these cases, relatively simple conduction theory can give excellent results. The steady-state solutions given here are also important as they can be used as the basis for transient problems with phase change as is discussed in Chapters 10 and 11.

7.1 STEADY STATE

7.1.1 Semi-Infinite Plane

Consider a plane surface, $y = 0$, as shown in Fig. 7.1. The temperature over the half-plane $x > 0$ is T_1, and it is T_2 for $x < 0$. This might represent a shallow lake or a very large building that is at a temperature different from the ambient ground surface temperature. Let $v = T - T_1$; then the problem is to find a solution to the steady-state conduction equation

$$\frac{\partial^2 v}{\partial x^2} + \frac{\partial^2 v}{\partial y^2} = \nabla^2 v = 0 \tag{7.1}$$

with the following conditions

$$v = 0 \qquad x > 0 \qquad y = 0$$

$$v = v_o = T_2 - T_1 \qquad x < 0 \qquad y = 0$$

Problems of this type are most easily solved by using the conformal mapping technique (see Chap. 3). A function that will map the region $y > 0$ in the z plane into a strip $0 < v < \pi$ in the w plane is given by

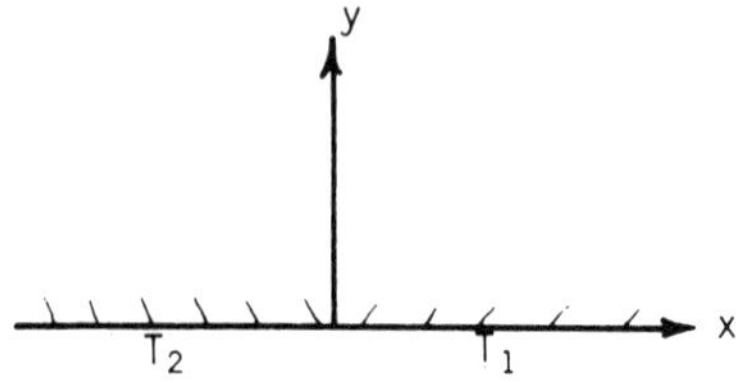

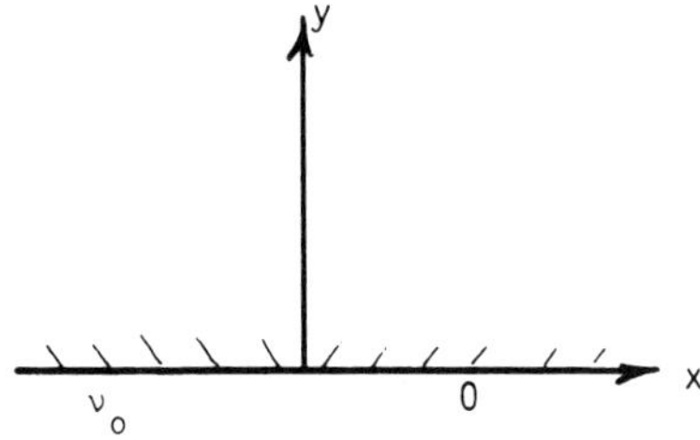

Fig. 7.1. Plane Surface with Temperature Difference.

$$w = \ln z \tag{7.2}$$

Numerous mappings of practical regions are given by Churchill (1960). The problem now is to find a solution of the conduction equation in the w plane that will satisfy the boundary conditions for the strip (Fig. 7.2).

Now, in the $u - v$ plane, a harmonic function that is 0 at $v = 0$ and ν_o at $v = \pi$ is

$$\nu = \frac{\nu_o}{\pi} v \tag{7.3}$$

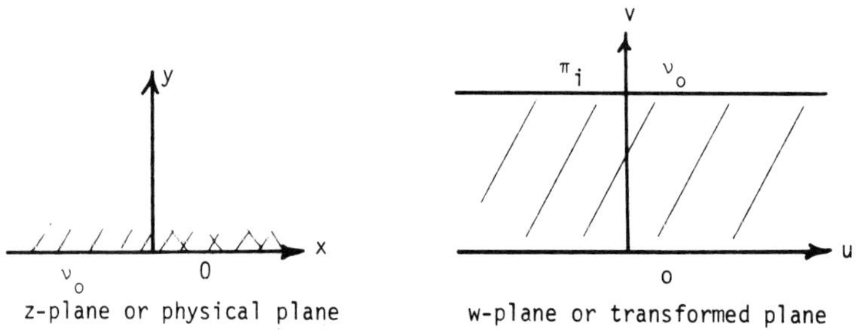

Fig. 7.2. Relation of Transformed Plane and Physical Plane.

One can use the mapping function to find a relation between u, v, and x, y

$$w = u + iv = \ln(re^{i\theta}) = \ln r + i\theta$$

where $r = (x^2 + y^2)^{1/2}$, $\theta = \tan^{-1}(y/x)$ are the usual relations between the cylindrical and cartesian coordinates.
Thus

$$v = \arctan\left(\frac{y}{x}\right) \tag{7.4}$$

and the solution in the physical plane is

$$\nu = \frac{\nu_o}{\pi}\arctan\left(\frac{y}{x}\right) \tag{7.5}$$

A quick inspection shows that Eq. (7.5) does satisfy the conditions of the problem.

Isothermal Surfaces. The lines of constant temperature can be found by setting ν equal to a constant in Eq. (7.5)

$$\nu = \nu_c = \frac{\nu_o}{\pi}\arctan\left(\frac{y}{x}\right)$$

$$\frac{\nu_c \pi}{\nu_o} = c = \arctan\left(\frac{y}{x}\right)$$

Thus

$$y = ax \tag{7.6}$$

where

$$a = \tan c$$

Also

$$\theta = \tan^{-1}(a) = \tan^{-1}(\tan c) = c$$

The isothermal lines are the radial lines as shown in Fig. 7.3. This solution does not contain the property values of the medium (it is assumed homoge-

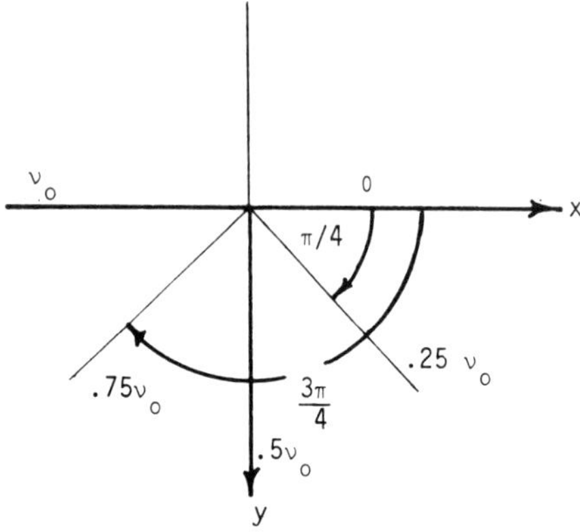

Fig. 7.3. Isothermal Surfaces for Plane Problem.

neous). This is typical of the temperature relations for steady-state conduction without internal heat generation.

Heat Flow. The lines of heat flow must always be perpendicular to the isothermal lines and should be circles in this case. This can be shown by using the Fourier relation for the heat flow

$$\frac{q}{kA} = -\left(\frac{\partial v}{\partial n}\right) \tag{7.7}$$

where n is the direction of the heat flux. This can be written at any location x, y as

$$\frac{q}{kA} = \left[\left(\frac{\partial v}{\partial x}\right)^2 + \left(\frac{\partial v}{\partial y}\right)^2\right]^{1/2} = \frac{v_o}{\pi}\left\{\frac{1}{x^4}\left(\frac{1}{1+\frac{y^2}{x^2}}\right)^2 (y^2 + x^2)\right\}^{1/2}$$

Thus the heat flow lines are the circles

$$(x^2 + y^2) = \frac{1}{\left(\frac{\pi q}{v_o kA}\right)^2} \tag{7.8}$$

Thus, heat flow lines are circles concentric at the origin (Fig. 7.4).

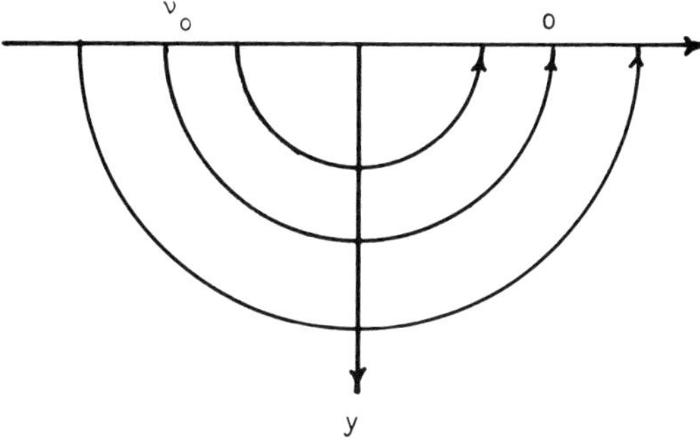

Fig. 7.4. Heat Flow Lines for Plane Surface.

Geothermal Gradient. The effect of the increase of temperature with depth, or the geothermal gradient, can be included easily. If the geothermal gradient is given by

$$\nu = Gy \tag{7.9}$$

where G is the gradient, then, by superposition of Eqs. (7.5) and (7.9)

$$\nu = \frac{\nu_o}{\pi} \arctan\left(\frac{y}{x}\right) + Gy$$

or

$$\frac{\nu}{\nu_o} = \frac{1}{\pi} \tan^{-1}\left(\frac{y}{x}\right) + \frac{G}{\nu_o} y \tag{7.10}$$

The isothermal lines can again be found by setting $\nu = \nu_c$ and solving for the equation of the lines. These are given by

$$\tan\left(\pi \frac{\nu_c}{\nu_o} - \pi \frac{G}{\nu_o} y\right) = \frac{y}{x}$$

or

$$\frac{\tan\left(\pi \frac{\nu_c}{\nu_o}\right) - \tan\left(\pi \frac{G}{\nu_o} y\right)}{1 + \tan\left(\pi \frac{\nu_c}{\nu_o}\right) \tan\left(\pi \frac{G}{\nu_o} y\right)} = \frac{y}{x} \tag{7.11}$$

For example, let $\nu_c/\nu_o = 1.0$; then Eq. (7.11) is

$$-\tan\left(\frac{\pi\, G\, y}{\nu_o}\right) = \frac{y}{x}$$

which can be written as

$$x = -\frac{\nu_o}{\pi G} \qquad \text{if } y \text{ is small}$$

If $x = 0$ then $y/x \to \infty$ and $\tan^{-1}(\infty) = \pi/2$.
Thus

$$\pi\left(1 - \frac{G}{\nu_o} y\right) = \frac{\pi}{2}$$

and

$$y = \frac{\nu_o}{2G}$$

The ratio of these results may be written

$$\left|\frac{y_o}{x_o}\right| = \frac{\nu_o\, \pi G}{2\, G\, \nu_o} = \frac{\pi}{2} = 1.572$$

The isothermal lines can also be found graphically by drawing Eqs. (7.6) and (7.9) and connecting the diagonals as shown by Fig. 7.5. The graphical ratio is

$$\left|\frac{y_o}{x_o}\right|_{\text{graph}} = \frac{33.5}{21.4} = 1.565$$

which is quite accurate even for a rough sketch.

7.1.2 Infinite Strip

The case of the infinite plane is instructive, but not too practical. Consider a long strip of width $2a$ at a temperature different from the surrounding medium. This could represent a river, a road, or a long narrow building. The temperature beneath the road can be found from conformal mapping

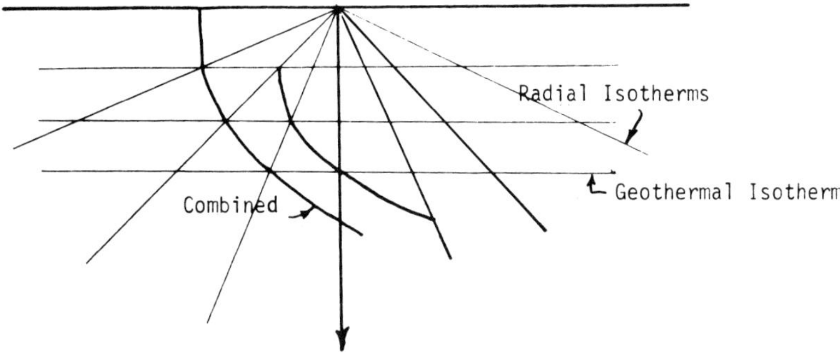

Fig. 7.5. Graphical Solution for Isothermal Lines.

because the strip will map into the same region as before, as shown in Fig. 7.6, using the mapping function

$$w = \ln \frac{z - a}{z + a} \tag{7.12}$$

Again an obvious solution in the w plane is

$$\nu = \frac{\nu_o}{\pi} v$$

The relation between u, v and x, y is found as follows

$$w = \ln \left(\frac{z - a}{z + a}\right) + i\, arg \left(\frac{z - a}{z + a}\right)$$

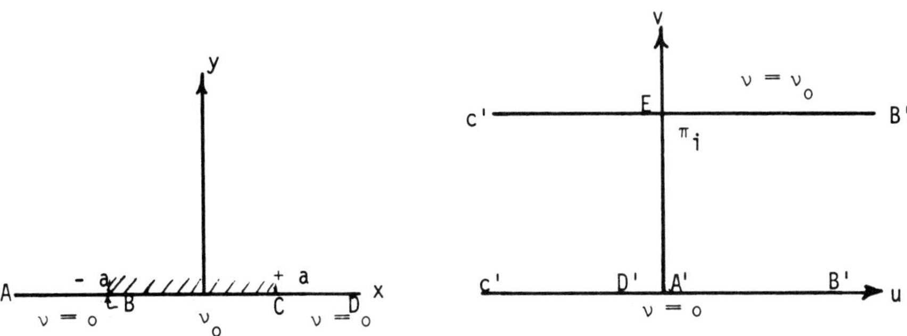

Fig. 7.6. Geometry of the Infinite Strip.

$$\frac{z-a}{z+a}=\frac{x-a+iy}{x+a+iy}=r\,e^{i\theta}=r(\cos\theta+i\sin\theta)=p+iq$$

where $\theta = \arg\dfrac{z-a}{z+a}$

$$\frac{(x-a)+iy}{(x+a)+iy}\,\frac{(x+a)-iy}{(x+a)-iy}=\frac{x^2+y^2-a^2+i\,2ay}{(x+a)^2+y^2}$$

Now

$$r\cos\theta=p$$
$$r\sin\theta=q$$

and

$$\theta=\tan^{-1}\left(\frac{q}{p}\right)$$

From above

$$q=\frac{2ay}{(x+a)^2+y^2},\qquad p=\frac{x^2+y^2-a^2}{(x+a)^2+y^2}$$

Thus

$$\nu=\frac{\nu_o}{\pi}\tan^{-1}\frac{2ay}{x^2+y^2-a^2}\tag{7.13}$$

Isothermal Lines. Again set $\nu = \nu_c$ in Eq. (7.13), and one has

$$x^2+\left(y-\frac{a}{c}\right)^2=\frac{a^2}{c^2}(c^2+1)\tag{7.14}$$

$$c=\tan\left(\frac{\pi\nu_c}{\nu_o}\right)$$

Thus, the isotherms are circles with centers along the y axis and radii

$$\frac{a}{c}\sqrt{1+c^2}$$

Note that at $y = 0$, $x = \pm a$. Thus, all circles pass through the points ($\pm$ a, 0).

If $a/c = 0$, then $c = \infty$ and $r = a$; thus

$$\tan \frac{\pi \nu_c}{\nu_o} = \infty \qquad \textit{and} \qquad \frac{\nu_c}{\nu_o} = 0.5$$

It is useful to solve the same problem by superposition. Take the solution to the semi-infinite plane

$$\nu = \frac{\nu_o}{\pi} \tan^{-1}\left(\frac{y}{x}\right)$$

Suppose we shift the origin, as shown in Fig. 7.7, such that

$$x' = (x + a)$$

and

$$x'' = -x' = (-x - a)$$

Thus we have two problems, one in the $x' - y$ system and the other for $x'' - y$ (Fig. 7.7).
The solutions, from Eq. (7.5), are

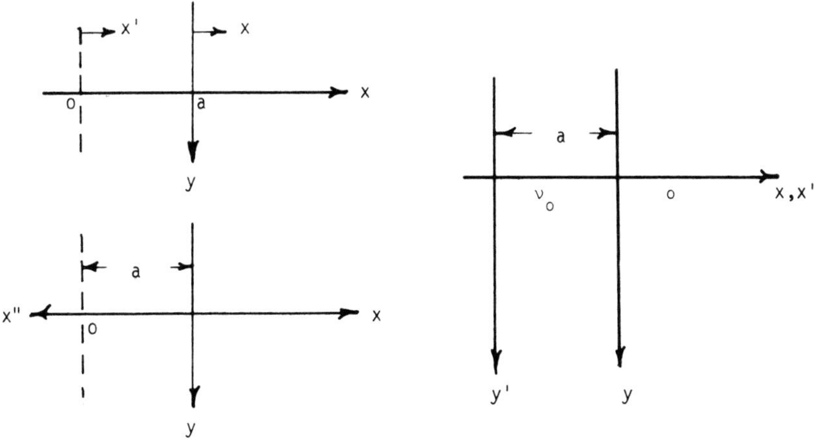

Fig. 7.7. Superposition Geometry.

$$\nu' = \frac{\nu_o}{\pi} \tan^{-1}\left(\frac{y}{x' - a}\right)$$

$$\nu'' = \frac{\nu_o}{\pi} \tan^{-1}\left(\frac{y}{-x'' - a}\right)$$

If we combine ν' and ν'', we will have the system shown in Fig. 7.8. Thus, if we subtract ν_o from this system, we will have the desired system. Then the solution is

$$\nu = \nu' + \nu'' - \nu_o = \nu_o \left\{\frac{1}{\pi}\left(\tan^{-1}\left(\frac{y}{x - a}\right) + \tan^{-1}\left(\frac{y}{-x - a}\right)\right) - 1\right\} \quad (7.15)$$

or

$$\frac{\pi\nu}{\nu_o} = \tan^{-1} p - \tan^{-1} q - \pi$$

Equation (7.15) does not appear to be the same as Eq. (7.13), but is in fact identical.

Let

$$\theta_1 = \tan^{-1} p$$
$$\theta_2 = \tan^{-1} q - \pi$$

Then

$$\tan\left(\frac{\pi\nu}{\nu_o}\right) = \tan\left(\theta_1 + \theta_2\right)$$

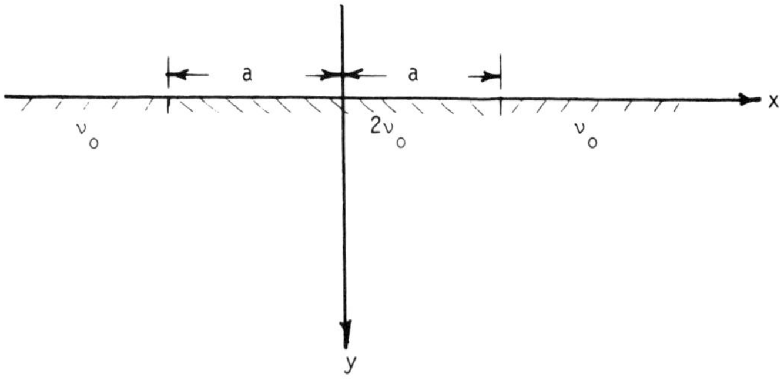

Fig. 7.8. Superposition of Infinite Planes.

$$= \frac{\tan\theta_1 + \tan\theta_2}{1 - \tan\theta_1 \tan\theta_2}$$

After some manipulation

$$\tan\left(\frac{\pi\nu}{\nu_o}\right) = \frac{p - q}{1 - pq} = \frac{2\,a\,y}{x^2 - y^2 - a^2}$$

and finally

$$\nu = \frac{\nu_o}{\pi}\tan^{-1}\left(\frac{2\,a\,y}{x^2 + y^2 - a^2}\right)$$

which is the same as Eq. (7.13) obtained from conformal mapping.

Road with Geothermal Gradient. By superposition

$$\nu = \frac{\nu_o}{\pi}\tan^{-1}\left(\frac{2ay}{x^2 + y^2 - a^2}\right) + G\,y \tag{7.16}$$

7.1.3 Temperature beneath Corners

The three-dimensional problem of the temperature beneath a corner of angle θ is of importance and has been solved by Brown (1962) for the geometry shown in Fig. 7.9.

The solution is given by

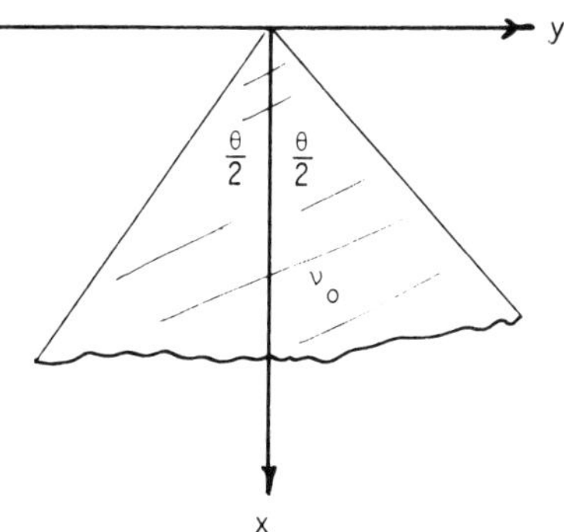

Fig. 7.9. Geometry for Plane Wedge. $\nu = 0$ everywhere outside corner; $z =$ direction downward, perpendicular to xy.

$$\nu = \frac{\nu_o}{\pi} \tan^{-1} \left\{ \frac{2z(x + \sqrt{x^2 + y^2 + z^2}) \tan \theta/4}{y^2 + z^2 - (x + \sqrt{x^2 + y^2 + z^2})^2 \tan^2 \theta/4} \right\} \tag{7.17}$$

The isotherm equation can be found as follows

$$\tan \frac{\pi \nu_c}{\nu_o} = \frac{2(p + \sqrt{p^2 + q^2 + 1}) \tan \theta/4}{q^2 + 1 - (p^2 + 2p\sqrt{p^2 + q^2 + 1} + p^2 + q^2 + 1) \tan^2 \theta/4}$$

where

$$p = \frac{x}{z} \qquad q = \frac{y}{z}$$

Then

$$p^2 + q^2 + 1 = \left[p(1 + \tan \theta/2) + \tan(\theta/2) \cot \frac{\pi \nu_c}{\nu_o} \right]^2$$

Finally the isotherm equation is

$$\left(\frac{x}{z} + \frac{\cot \frac{\pi \nu_c}{\nu_o}}{\sin \frac{\theta}{2}} \right)^2 - \left(\frac{y}{z} \right)^2 \cot^2 \frac{\theta}{2} = \cot^2 \frac{\theta}{2} \operatorname{csec}^2 \frac{\pi \nu_c}{\nu_o} \tag{7.18}$$

Now as $z \to \infty$ $x/z = y/z = 0$ and

$$\cot^2 \frac{\pi \nu_c}{\nu_o} = \cot^2 \frac{\theta}{2} \operatorname{cosec}^2 \frac{\pi \nu_c}{\nu_o} \sin^2 \frac{\theta}{2}$$

which yields

$$\frac{\nu_c}{\nu_o} = \frac{\theta}{2\pi}$$

If $\theta = \pi/2$, $\nu_c/\nu_o = 0.25$.

Example 1: Let the corner be a right angle, $\theta = \pi/2$, then Eq. (7.18) is

$$\left(\frac{x}{2} + 1.414 \cot \pi \frac{\nu_c}{\nu_o} \right)^2 - \left(\frac{y}{z} \right)^2 = \operatorname{cosec}^2 \left(\pi \frac{\nu_c}{\nu_o} \right)$$

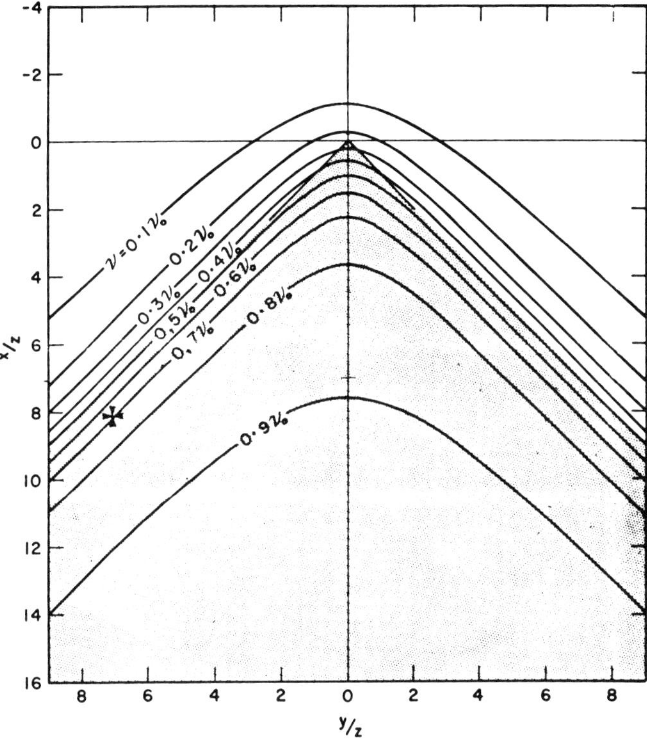

Fig. 7.10. Steady Temperature in the Ground Under Rectangular Corner. Adapted from Brown (1962), by permission of National Research Council of Canada, Ottawa.

Figure 7.10, taken from Brown (1962), shows the isotherm for this case plotted as a function of x/z and y/z.
Let the depth z be 10 ft

$$T_2 = 70°F \qquad T_1 = 20°F \qquad \nu_o = 50°F$$

(a) $y = 0 \quad x = 60 \quad x/z = 6$
From Fig. 7.10

$$\frac{\nu_1}{\nu_o} = 0.86 \qquad \text{and} \qquad \nu_1 = 43°F$$

Therefore, $T_a = 63°F$, the ground temperature at a depth of 10 ft, at position a.
(b) $y = 80 \quad x = 60 \quad y/z = 8 \quad x/z = 6$

$$\frac{\nu_2}{\nu_o} = 0.19 \qquad \nu_2 = 9.5$$

$$T_b = 29.5°F$$

The effect of the geothermal gradient can be obtained by adding Gz to each value. If G is .02°F/ft, then the temperature added would be 0.2°F.

Example 2: Suppose $\theta = \pi$. This is the case of the semi-infinite plane, solved earlier. Now

$$\sin\frac{\theta}{2} = 1 \qquad \cot\frac{\theta}{2} = 0$$

Then Eq. (7.18) is $(x/z + \cot c)^2 = 0$ and $x/z = -\cot c$ or $z/x = -\tan c$. Note that this is the same as Eq. (7.6), with different notation, which it must reduce to.

7.1.4 Finite Surface Areas

The temperature beneath a number of practical finite surface areas can be obtained by superposition of the preceding infinite cases.

Temperature beneath a Rectangle. A building will be represented by a rectangular surface area at a temperature different from the surrounding terrain. This problem can be solved by superposing right angle corners as given in Section 7.1.3. Consider two right angles superposed as shown in Fig. 7.11*a*. The intersection of these corners gives a rectangular area with temperature $2\nu_o$. Now add two more right angles, 3 and 4, as shown in Fig. 7.11*b*.
Now we have the case of a rectangle of temperature $2\nu_o$ and surroundings at ν_o.

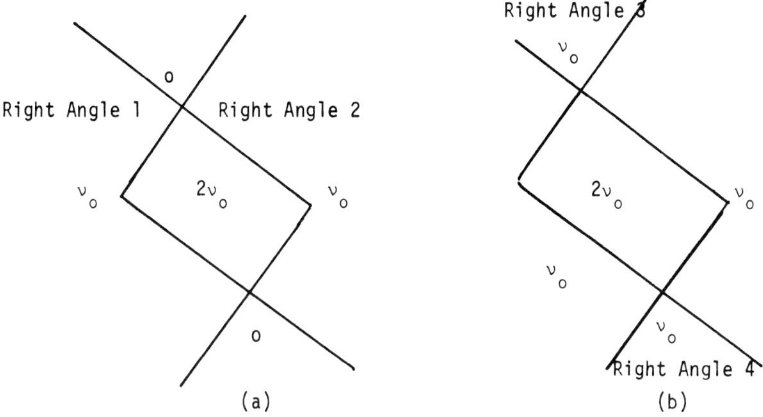

Fig. 7.11. Superposition Solution for Rectangle.

Analytic Solution. The superposition can also be done analytically, leading to the following solution

$$\frac{2\pi\nu}{\nu_o} = \tan^{-1}\frac{(x+a)(y+b)}{z\sqrt{z^2+(x+a)^2+(y+b)^2}} - \tan^{-1}\frac{(x-a)(y+b)}{z\sqrt{z^2+(x-a)^2+(y+b)^2}}$$

$$- \tan^{-1}\frac{(x+a)(y-b)}{z\sqrt{z^2+(x+a)^2+(y-b)^2}} + \tan^{-1}\frac{(x-a)(y-b)}{z\sqrt{z^2+(x-a)^2+(y-b)^2}} \quad (7.19)$$

where $2a$, $2b$ are the dimensions of the building in the x and y directions. Equation (7.19) is examined again in Chap. 11, where phase change is considered.

Example 3: Consider a square of side length 150 ft. Let us find the temperature at a depth 15 ft below the center of the square. The diagonal of the square is $150\sqrt{2}$.

For rectangles 1 and 2 (refer to Fig. 7.11)

$$\frac{x}{z} = \frac{75\sqrt{2}}{15} = 7.07 \qquad \frac{y}{z} = 0$$

From Fig. 7.10

$$\left(\frac{\nu_1}{\nu_o}\right) = 0.88$$

For rectangles 3 and 4

$$\frac{x}{z} = -7.07, \qquad \frac{y}{z} = 0$$

Again from Fig. 7.10 (read $y/z = 7.07$, $x/z = 0$), $(\nu_1/\nu_o) = .02$. Thus, the combination of all four rectangles is the sum of each one

$$\left(\frac{\nu_1}{\nu_o}\right) = 2(.88) + 2(.02) = 1.80$$

As the outside temperature is ν_o, we must subtract ν_o from ν_1 and

$$\nu_1 = .80\ \nu_o$$

Using Eq. (7.19), the value is $\nu_1/\nu_o = 0.823$, which is close enough agreement considering the inaccuracy of reading Fig. 7.10. In terms of the inside and outside temperatures T_i, T_o

$$T_i = T_o + .80(T_1 - T_o)$$

Rectangular Intersections. Figure 7.10 can also be used, with superposition, for an intersection formed by four large buildings.

Example 4: If d is the street width then

$$x = \frac{d}{2}\sqrt{2}, \quad d = 66\ ft.$$

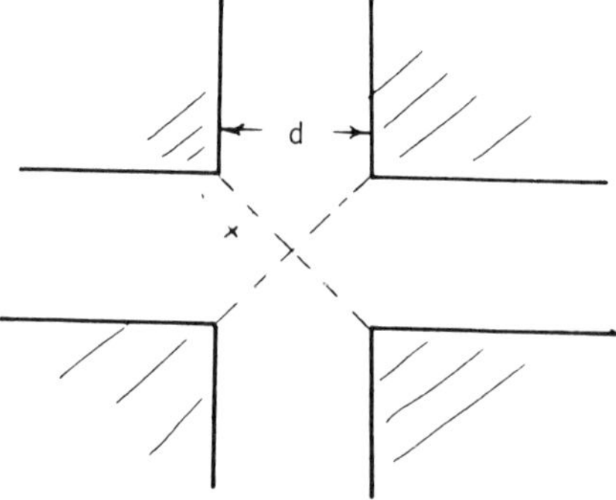

for a typical street. Let $z = 33\sqrt{2} = 46.7$ ft. The ratio is

$$\frac{x}{z} = 1.0$$

Then from Fig. 7.10, for each corner

$$\nu = 0.1\,\nu_o$$

Thus, for the four corners one has

$$\nu = 0.4\,\nu_o$$

If

$$T_1 = 70°F\ (\text{mean annual temperature of building floor})$$
$$T_2 = 50°F\ (\text{mean annual ground temperature})$$
$$\therefore \nu = 0.4\ (20) = 8°F$$

The temperature is

$$T_x = 50 + 8 + 0.47 = 58.5°F\ (\text{geothermal gradient of } .01°F/ft)$$

The temperature versus depth can be found as follows, where T_G is the temperature effect of the geothermal gradient

z	T_x'	T_G	T_x
0	50	0	50
46.7	58.0	0.5	58.5
100	66.0	1	67
∞	70.0	∞	∞

Irregular Areas. The previous methods are limited to regular areas if a reasonable amount of work is to be done. Lachenbruch (1957) has found an expression for the temperature, at any depth z, beneath the vertex of a circular segment of radius R and angle θ.
This is

$$\nu = \frac{\theta\nu_o}{2\pi}\left\{erfc\frac{z}{2\sqrt{\alpha t}} - \frac{z}{\sqrt{z^2+R^2}}\,erfc\frac{\sqrt{z^2+R^2}}{2\sqrt{\alpha t}}\right\} \tag{7.20}$$

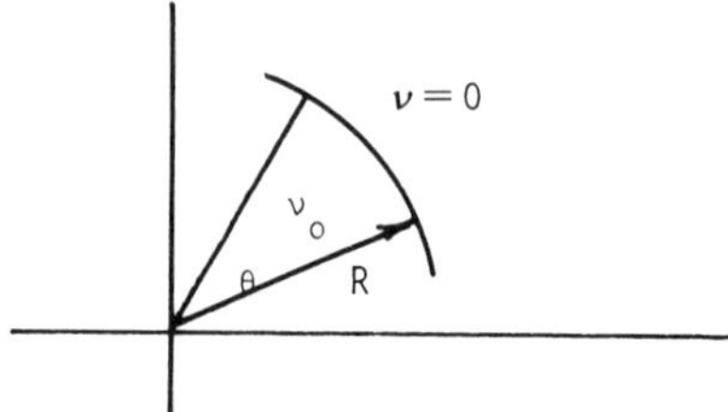

In Eq. (7.20), it is assumed that the segment temperature changes suddenly to ν_o while the remaining surface is at zero. The steady state is obtained by letting $t \to \infty$ in Eq. (7.20). Since $(z/2\sqrt{\alpha t}) \to 0$ and erfc $0 = 1$, one has

$$\nu = \frac{\theta \nu_o}{2\pi}\left\{1 - \frac{1}{\sqrt{1+(R/z)^2}}\right\} \tag{7.21}$$

Equation (7.21) is plotted as curve 0 in Fig. 7.12. It is now easy to obtain steady temperatures beneath irregular areas.

Note that if $R \to \infty$ and $\theta = \pi/2$, then curve 0 should be the same as Fig. 7.10 for the rectangular corner under the vertex.

Thus for Fig. 7.12, as $R/z \to \infty$

$$\frac{360}{\theta}\frac{\nu}{\nu_o} = 1.0$$

Therefore

$$\frac{\nu}{\nu_o} = \frac{\theta}{360} = \frac{90}{360} = 0.25$$

which is the temperature beneath the corner as $z \to \infty$ in Fig. 7.10.

Example 5: Let $R = 500$ ft, $z = 100$, and $\theta = 27°$. Then

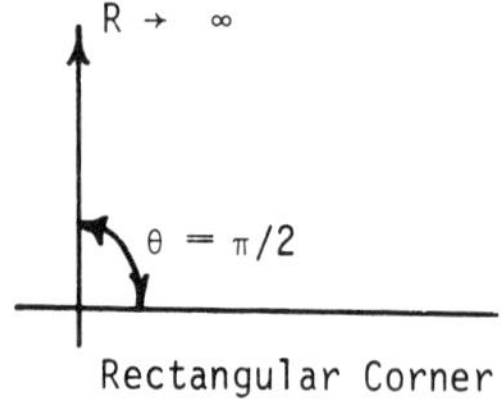

Rectangular Corner

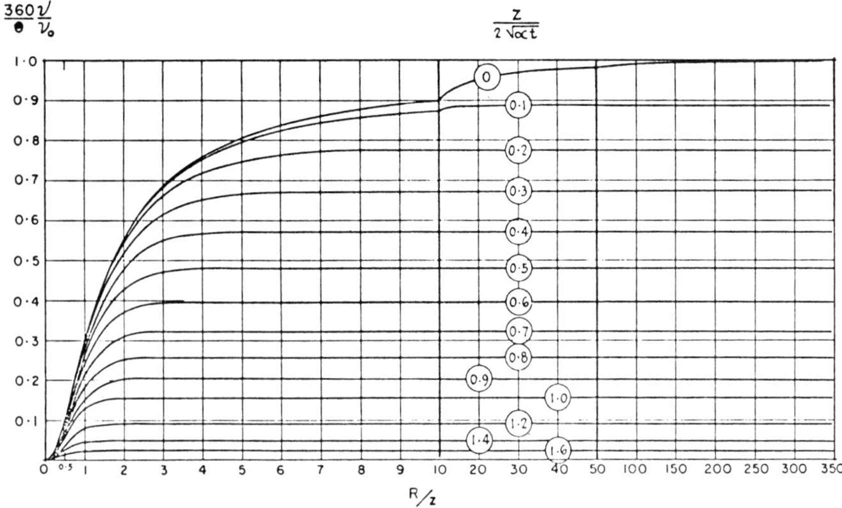

Fig. 7.12. Temperature at Depth z under the Apex of a Circular Sector of Radius R and Angle θ Degrees. Adapted from Brown (1962), by permission of National Research Council of Canada, Ottawa.

$$\frac{360}{27}\frac{\nu}{\nu_o} = .80, \text{ from Fig. 7.12}$$

$$\frac{\nu}{\nu_o} = (.80)\frac{(27)}{310} = .06$$

If $\nu_o = 20$ then

$$\nu = 1.2°F + 1 = 2.2°F \text{ above ambient}$$

Figures 7.13 and 7.14 indicate how to handle irregular areas when the point of interest lies inside the area or outside the area. For a point inside the area, sum the effects of all the sectors. For a point outside the area, subtract the temperature effects of sectors R_1 from the effects of sectors R_2 (see Fig. 7.14). Here again superposition is valid.

Example 6: Find the temperature 100 ft beneath the center of a circular lake, of radius 1000 ft. Let the temperature of the lake bottom be 50°F and the surrounding ground surface 22°F; thus ν_o =28. Consider a quadrant of the circle, $R/z = 10$; thus from Fig. 7.12

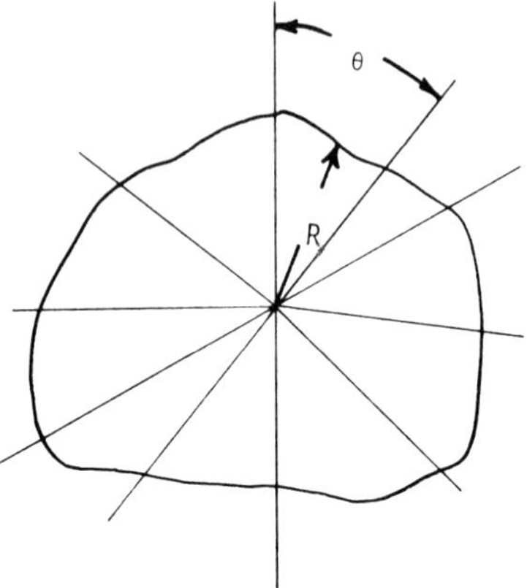

Fig. 7.13. Temperature at a Point Inside Area.

$$\frac{360}{90}\frac{\nu}{\nu_o} = 0.9$$

or

$$\frac{\nu}{\nu_o} = \frac{0.9}{4}$$

For the four quadrants

$$\nu = 0.9\nu_o = 25.2°F$$

Then

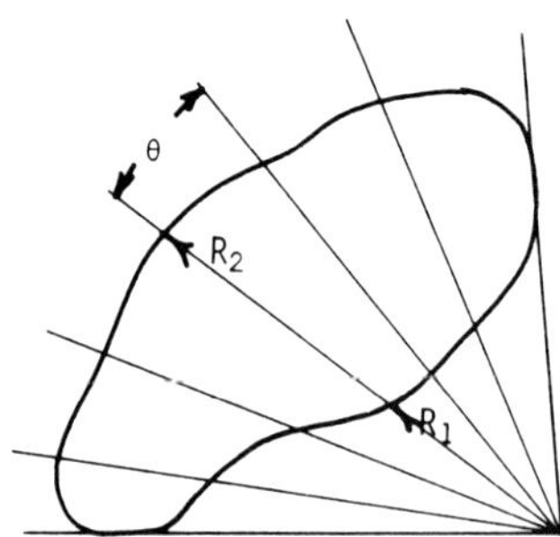

Fig. 7.14. Temperature at Point Outside Area.

$$T = 22 + 25.2 + 1 = 48.2°F$$

At $z = 1000$,
$T = 22 + 8.4 + 10 = 40.4°F$; here the geothermal gradient is significant.

7.1.5 Buried Pipelines

Here we shall investigate the temperature and heat transfer effects of a pipeline or cable buried in the ground without phase change effects. The important effects of phase change are discussed in Chap. 10. The actual system is as shown in Fig. 7.15, but let us first look at the case of a pipe buried in an infinite medium. In this case, a solution of the conduction equation, in cylindrical coordinates, must be found such that

$$\frac{d}{dr}\left(r\frac{d\theta}{dr}\right) = 0 \tag{7.22}$$

$$\theta = \theta_o = (T_p - T_G) \qquad r = r_o \tag{7.22a}$$

$$\left(\frac{d\theta}{dr}\right)_{r_o} = \frac{-Q}{2\pi kLr_o} \tag{7.22b}$$

where $\theta = T - T_G$ and Q = heat transfer rate from the pipe. The solution of this system is

$$\theta = \frac{-Q}{2\pi kL} \ln r + b$$

where b is a constant that may be dropped as it will not affect the solution to the problem of the pipe buried at a finite depth h. Thus, a solution is

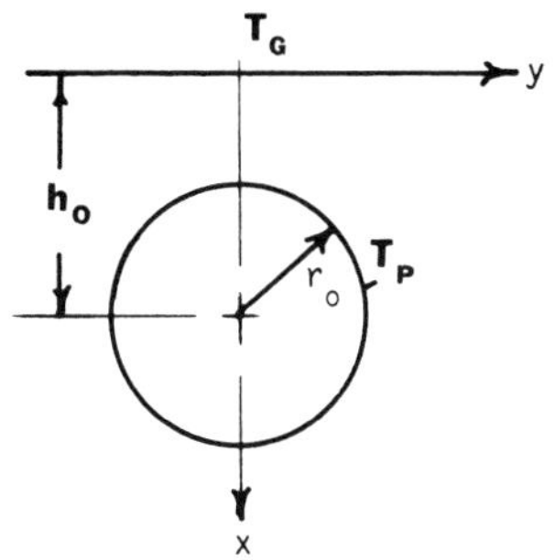

Fig. 7.15. Pipe Buried at Finite Depth.

$$\theta = \frac{Q}{2\pi kL} \ln r \tag{7.23}$$

This solution may be thought of as a line source of heat located at the origin of the coordinate system. In this sense it is analogous to the hydrodynamic case of a source at the origin.

Obviously this does not satisfy the boundary condition $\theta = 0$ along $x = 0$. Now imagine a sink of the same strength $(Q/2\pi kL)$ located at $x = -a$, as shown in Fig. 7.16, where a is related to h_o.

$$\theta_{\text{sink}} = \frac{Q}{2\pi kL} \ln r \tag{7.24}$$

Thus, the potential at any point $P(x,y)$ is the superposition of Eqs. (7.23) and (7.24).

$$\theta = \frac{Q}{2\pi kL} (\ln r_2 - \ln r_1) = \frac{Q}{2\pi kL} \ln \frac{r_2}{r_1} \tag{7.25}$$

Now, along $x = 0$, $\theta = 0$, as is required.
From the geometry of Fig. 7.16

$$r_1^2 = (a - x)^2 + y^2$$

$$r_2^2 = (a + x)^2 + y^2$$

Putting these into Eq. (7.25), one obtains the complete potential equation

$$\theta = \frac{Q}{2\pi kL} \ln \left\{\frac{(a + x)^2 + y^2}{(a - x)^2 + y^2}\right\}^{1/2} = \frac{Q}{4\pi kL} \ln \left\{\frac{(a + x)^2 + y^2}{(a - x)^2 + y^2}\right\} \tag{7.26}$$

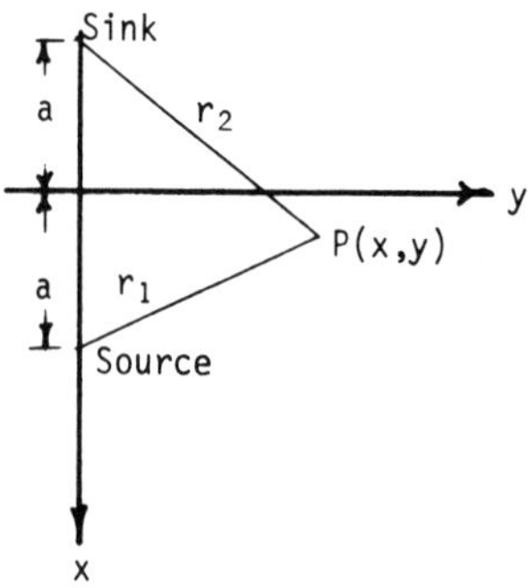

Fig. 7.16. Source and Sink for Buried Pipe.

Lines of Constant Temperature. From Eq. (7.26)

$$\frac{4\pi kL\theta}{Q} = \ln\frac{(a+x)^2+y^2}{(a-x)^2+y^2}$$

or

$$e^{\frac{4\pi kL\theta}{Q}} = c = \frac{(a+x)^2+y^2}{(a-x)^2+y^2} \tag{7.27}$$

Note that c varies for each point x, y. The isotherms will be traced by lines along which c remains constant.

From Eq. (7.27) after algebraic manipulation

$$y^2 + \left(x - \frac{c+1}{c-1}a\right)^2 = \frac{4ca^2}{(c-1)^2} \tag{7.28}$$

Thus the isotherms are circles with center and radii given by

$$\left\{\frac{c+1}{c-1}a,\ 0\right\} \tag{7.28a}$$

$$r = \frac{2\sqrt{c}}{c-1}a \tag{7.28b}$$

Now the horizontal isotherm is $\theta = 0$, as noted previously, and from Eq. (7.27) this gives $c = 1$. Then $r = \infty$, and we have a straight line as should be the case. The surface of the pipe is the isotherm θ_o with radius

$$r = r_o = \frac{D}{2} = \frac{2\sqrt{c_o}}{c_o - 1}a \tag{7.29}$$

and the depth of the centerline of the pipe is h_o which, from Eq. (7.28a), must be

$$h_o = \frac{c_o+1}{c_o-1}a \tag{7.30}$$

From Eqs. (7.29) and (7.30), a and c_o may be found. Note that since $c_o >$ 0, then $h_o > a$, and the fictitious source is always located above the pipe center line.

Now

$$a = \frac{c_o - 1}{c_o + 1} h_o \tag{7.30a}$$

and

$$c_o^2 + \left[2 - 16\left(\frac{h_o}{D}\right)^2\right] c_o + 1 = 0 \tag{7.31}$$

The solution of Eq. (7.31), using the positive root as is physically required, is

$$c_o = \left(8\left(\frac{h_o}{D}\right)^2 - 1\right) + 4\frac{h_o}{D}\sqrt{4\left(\frac{h_o}{D}\right)^2 - 1} \tag{7.32}$$

Also

$$c_o = (\gamma + \sqrt{\gamma^2 - 1})^2$$

where $\gamma = h_o/r_o$.

As

$$e^{\frac{4\pi kL\theta_o}{Q}} = c_o$$

Then

$$\frac{4\pi kL\theta_o}{Q} = \ln c_o$$

Or

$$Q = \frac{2\pi kL\theta_o}{\ln(\gamma + \sqrt{\gamma^2 - 1})} = \frac{2\pi kL\theta_o}{\cosh^{-1}\gamma} \tag{7.33}$$

This is sometimes written

$$Q = \frac{\theta_o}{R}$$

where R is the thermal resistance of the surrounding soil

$$R = \frac{\ln(\gamma + \sqrt{\gamma^2 - 1})}{2\pi kL} = \frac{\cosh^{-1}\gamma}{2\pi kL} \tag{7.34}$$

If $\gamma >> 1$, a good approximation is

$$Q = \frac{2\pi kL\theta_o}{\ln\{2\gamma\}} \tag{7.35}$$

Note that

$$c_o = e\,\frac{4\pi kL\theta_o}{Q}$$

$$c = e\,\frac{4\pi kL\theta}{Q}$$

Then

$$c = (c_o)\,\frac{\theta}{\theta_o} \tag{7.36}$$

Equation (7.36) can be used with Eqs. (7.28a) and (7.28b) to plot the isotherms for any pipe. Note that k does not enter into these expressions, as is usual for steady-state solutions.

An alternative format for the above equation is as follows

$$\frac{\theta}{\theta_o} = \frac{T - T_G}{T_p - T_G} = \frac{\operatorname{arccosh}(h/r)}{\operatorname{arccosh}\gamma} \tag{7.37}$$

$$h^2 - r^2 = h_o^2 - r_o^2 \tag{7.38}$$

These relations allow the temperature isotherms to be evaluated more quickly. The isotherm for the fusion temperature is often of interest. This is given by

$$\operatorname{arccosh}(h/r) = \left(\frac{T_f - T_G}{T_p - T_o}\right) \operatorname{arccosh}(h_o/r_o) \tag{7.39}$$

7.1.5.1 Frozen and Thawed Properties

With phase change, the properties of the frozen and thawed media will be different. If the thermal conductivity varies only with temperature, then the solutions of Section 7.1.5 will be valid with a property correction.

Let a new temperature be defined by

$$T^1 = T_f + \frac{1}{k_f} \int_{T_f}^{T} k(u)\, du \tag{7.40}$$

The thermal conductivities are related to temperature by

$$k = \begin{cases} k_t & T > T_f \\ k_f & T \leq T_f \end{cases} \tag{7.41}$$

Then the temperature to use will be

$$T^1 = \begin{cases} T_f + k_{tf}\,(T_1 - T_f) & \textit{thawed region} \\ T_2 & \textit{frozen region} \end{cases} \tag{7.42}$$

where $k_{tf} = k_t/k_f$.

Equations (7.40) and (7.41) will reduce the nonconstant property problem to one of homogeneous properties (see Chap. 8).

Equation (7.39), for the steady-state thaw interface, is now

$$\operatorname{arccosh} \frac{h_f}{r_f} = \left[\frac{T_f - T_G}{k_{tf}(T_p - T_f) + (T_f - T_G)} \right] \operatorname{arccosh} \frac{h_o}{r_o} \tag{7.43}$$

The pipe surface temperature is given by

$$T_p^1 = T_f + k_{tf}\,(T_p - T_f) \tag{7.44}$$

The heat transfer then changes to

$$Q = \frac{2\pi k_t\, L\, [T_f - T_G + k_{tf}\,(T_p - T_f)}{\ln\,(\gamma + \sqrt{\gamma^2 - 1})} \tag{7.45}$$

The temperature profiles are

$$\frac{T^1 - T_G}{T_p^1 - T_G} = \frac{\operatorname{arccosh}\,(h/r)}{\operatorname{arccosh}\,\gamma} \tag{7.46}$$

Example 7: Consider a buried water pipe with the following characteristics

$$T_p = 42°\text{F} \qquad T_G = 21°\text{F} \qquad D = 3 \text{ in.} \qquad h_o = 72 \text{ in.}$$
$$\theta_o = 21 \qquad k_t = 1.0 \text{ Btu/hr-ft-°F}$$

The freezing isotherm will be $\theta_f = 32 - 21 = 11°\text{F}$.

$$\therefore c_f = e^{(.524 \ln c_o)} = (c_o)^{.524}$$

$$c_o = 8\left(\frac{h_o}{D}\right)^2 - 1 + \frac{4h_o}{D}\sqrt{4\left(\frac{h_o}{D}\right)_2 - 1}$$

$$c_o = 16\left(\frac{h_o}{D}\right)^2 = 9220$$

$$a \sim 72 \text{ in.}$$

$$c_f = e^{(.524)(2.3)\ 3.965} = e^{4.78} = 120$$

Center of freezing isotherm

$$L_f = \frac{c_f + 1}{c_f - 1} a = 73.2''$$

$$r_f = 13.24 \text{ in.}$$

The heat transfer can be calculated from

$$q = \frac{Q}{L} = \frac{4\pi\ k\theta_o}{\ln c_o} = 28.9 \frac{\text{Btu}}{\text{hr-ft}}$$

If the frozen conductivity is $k_f = 1.5$, then with Eq. (7.43), the steady-state thaw bowl is 19.68 in. This is due to the increased conductivity of the frozen layer. The heat transfer will increase to 36.5 Btu/hr-ft.

7.1.5.2 Pipe Insulation

The effect of insulation upon the pipe heat transfer can be easily handled. The heat flow equation, Eq. (7.45) does not specify that the pipe be bare metal and thus the system shown in Figure 7.17 has heat flow

$$\frac{Q}{L} = 2\pi k_t b[(T_f - T_G) + k_{tf}(T_i - T_f)]$$

where

$$b = \frac{1}{\ln(\gamma_i + \sqrt{\gamma_i^2 - 1})}$$

$$\gamma_i = h_o/r_i$$

It is well-known that the radial heat flow through a hollow cylinder is

$$\frac{Q}{L} = \frac{2\pi\, k_i\,(T_p - T_i)}{\ln\,(r_i/r_o)}$$

If the heat flow through the ground is equated to the heat flow through the insulation, the insulation temperature T_i can be found.

$$T_i - T_f = (T_p - T_f)\left(\frac{1 - b\alpha\beta}{1 + b\alpha}\right) \tag{7.47}$$

where

$\alpha = \dfrac{k_t}{k_i}\, ln\left(\dfrac{r_i}{r_o}\right)$ is the insulation parameter and

$$\beta = k_{ft}\left(\frac{T_f - T_G}{T_p - T_f}\right)$$

This assumes that the outer surface of the insulation has a constant temperature. This is not quite true but will be satisfactory if the insulation is not very thick. The temperature in the thawed zone is

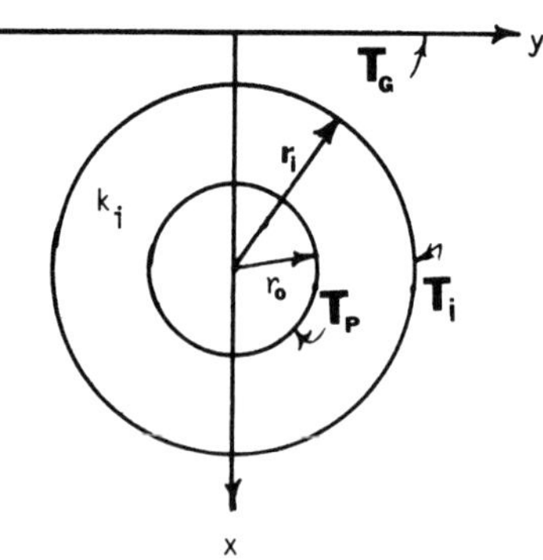

Fig. 7.17. Insulated Buried Pipe.

$$\frac{T_1 - T_f}{T_p - T_f} = \frac{1+\beta}{1+b\alpha}\,\frac{\operatorname{arccosh}(h/r)}{\operatorname{arccosh}(\gamma_i)} - \beta \tag{7.48}$$

The temperature in the frozen zone is

$$\frac{T_2 - T_G}{T_f - T_G} = \frac{1+\beta}{\beta(1+b\alpha)}\,\frac{\operatorname{arccosh}(h/r)}{\operatorname{arccosh}(\gamma_i)} \tag{7.49}$$

The heat transfer is

$$\frac{Q}{L} = 2\pi k_f \frac{(1+\beta)\,b}{\beta(1+b\alpha)}(T_f - T_G) \tag{7.50}$$

The location of the freezing isotherm is

$$\operatorname{arccosh}\left(\frac{h}{r}\right)_f = \frac{\beta}{1+\beta}(1+b\alpha)\operatorname{arccosh}(\gamma_i) \tag{7.51}$$

This can also be written in terms of the depth to the freezing isotherm, $\xi_\infty = y_\infty/r_i$,

$$\ln\left(\frac{\xi_\infty + a}{\xi_\infty - a}\right) = \frac{1}{b}\,\frac{\beta}{1+\beta}(1+b\alpha)$$

where

$$a = \sqrt{\gamma_i^2 - 1}$$

The insulation needed to keep the thaw within the insulation, i.e., $\xi_\infty \leq \gamma_i + 1$, is

$$\left(\frac{h_o}{r_o}\right)^{k_t/k_i} = \left(\gamma_i + \sqrt{\gamma_i^2 - 1}\right)^{1/\beta}$$

This equation can be solved for γ_i and then $r_i = h_o/\gamma_i$ can be found.

The heat flow is then

$$\frac{Q}{L} = \frac{2\pi k_f\, a\,(T_p - T_G)}{(a+b)\ln\left(\gamma_i + \sqrt{\gamma_i^2 - 1}\right)} \tag{7.51}$$

Example 8: Consider an insulated pipe, with the same temperatures and other conditions as Ex. 7, and

$$k_i = 0.1\ Btu/hr\text{-}ft\text{-}^\circ F \qquad k_f = 1.5\ Btu/hr\text{-}ft\text{-}^\circ F$$
$$r_i = 5.5\ in.$$

Then

$$\gamma_i = h_o/r_i = 13.1 \qquad b = 0.3064 \qquad \alpha = 13.0 \qquad \beta = 1.65$$

The heat flow is

$$\frac{Q}{L} = 6.83\ Btu/hr\text{-}ft$$

7.1.5.3 Conformal Mapping Method

It is of interest to examine the same problem from the viewpoint of conformal mapping. It has been shown by Caratheodory (1958) that the following mapping function will map the semi-infinite domain in the z plane into an annular region in the ω plane (Fig. 7.18)

$$\omega = R_o \frac{(B - r_o)(R_o + 1) - Z(R_o - 1)}{(B - r_o)(R_o + 1) + Z(R_o - 1)} \tag{7.52}$$

Thus, the problem is reduced to the solution of the Laplace equation in the ω plane with constant temperatures on the cylindrical surfaces. Equation

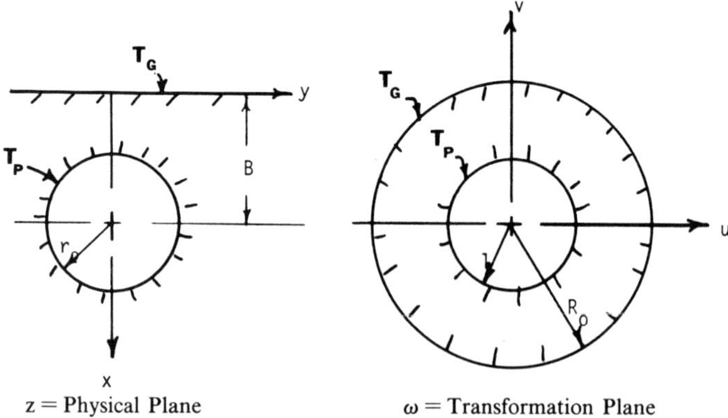

Fig. 7.18. Transformation Plane for Buried Pipe.

(7.52) is a case of the general Mobius transformation called the "bilinear" or "linear fractional" transformation.

The following equations can be shown to be valid

$$R_o = \left(\frac{B}{r_0}\right) + \sqrt{\left(\frac{B}{r_0}\right)^2 - 1} \tag{7.53}$$

$$c_o = (R_o)^2 = 2\frac{B}{r_o} R_o - 1 \tag{7.54}$$

Now the temperature solution in the ω plane is

$$\frac{\theta}{\theta_o} = \frac{\ln (R_o/r)}{\ln R_o} \tag{7.55}$$

where

$$\left(\frac{r}{R_o}\right)^2 = \frac{(B - r_o)^4(R_o + 1)^4 + (x^2 + y^2)^2(R_o - 1)^4}{\{[(B - r_o)(R_o + 1) + x(R_o - 1)]^2 + y^2(R_o - 1)^2\}^2} - \frac{2(x^2 - y^2)(B - r_o)^2(R_o^2 - 1)^2}{\{[(B - r_o)(R_o + 1) + x(R_o - 1)]^2 + y^2(R_o - 1)^2\}^2}$$

After algebraic manipulation, it can be shown that

$$\left(\frac{R_o}{r}\right)^2 = \frac{(a + x)^2 + y^2}{(a - x)^2 + y^2} \tag{7.56}$$

And thus Eq. (7.55), using Eq. (7.56), may be written

$$\frac{\theta}{\theta_o} = \frac{1}{\ln c_o} \ln \left\{\frac{(a + x)^2 + y^2}{(a - x)^2 + y^2}\right\} \tag{7.57}$$

Equation (7.57) is identical to the solution obtained by the image method. The conformal mapping technique is more powerful and can be used to study other problems associated with buried pipelines.

7.1.5.4 Constant Surface Heat Flux

The previous results are for pipes with a constant surface temperature, but for some conditions the pipe may have a constant surface heat flux. Examples are buried cables or heat traces with constant joulean heating. In some cases the actual problem may lie between the constant temperature and constant heat flux conditions.

Thiyagarajan and Yovanovich (1974) have evaluated the conductive resistance for a constant heat flux. The resistance is defined such that

$$Q = \frac{T_p - T_G}{R_q} \tag{7.58}$$

where

Q = heat flow per unit length of pipe
T_p = average pipe surface temperature and
R_q = conductive resistance between the pipe and the horizontal ground surface.

The resistance is

$$\pi k R_q = \frac{\operatorname{arccosh} \gamma}{2} + \sum_{n=1}^{\infty} \frac{e^{-2n \cosh^{-1} \gamma}}{n} \tanh\left(n \cosh^{-1} \gamma\right) \tag{7.59}$$

The ratio of the constant heat flux resistance to the constant temperature resistance R_q/R_T, rapidly approaches 1 as γ exceeds 1.5. For example at $\gamma = 1.5$, $R_q/R_T = 1.25$, though at $\gamma = 2.0$ it is 1.1. Thus, the constant temperature resistance is acceptable for all cases unless the burial depth ratio is considerably less than 2.0.

7.1.6 Conductive Resistances

For steady-state conduction problems, certain geometrical combinations arise frequently. It is assumed that the inner and outer bounding surfaces are isothermal and that the medium contained within the boundaries is homogeneous. The heat transfer rate per unit length is given by

$$\frac{q}{L} = \frac{\Delta T}{R} \tag{7.60}$$

where ΔT is the temperature difference of the surfaces and R is the conductive resistance.

Some practical shapes are listed in Table 7.1.

Table 7.1. Conductive Resistance (selected from Andrews, (1965), Rudenberg (1925), and Thornton (1977).

Physical System	Geometry	Resistance
Hollow Cylinder	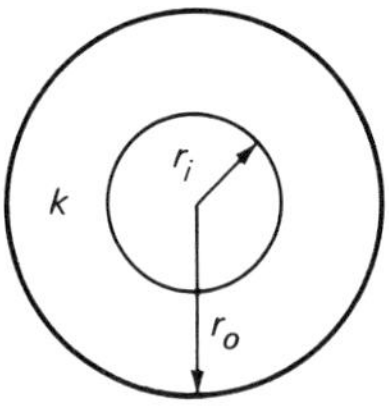	$\frac{\ln (r_o/r_i)}{2\pi k}$
Hollow Sphere		$\frac{r_o - r_i}{4\pi k r_o r_i}$
Buried Sphere	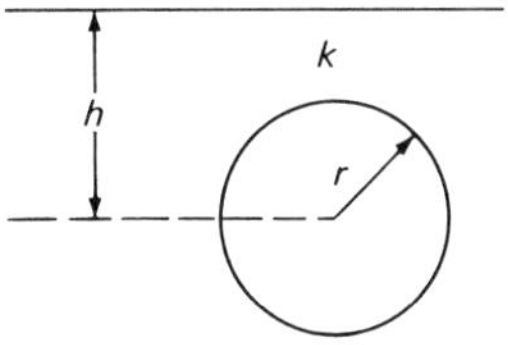	$\frac{1 - \frac{r}{2h}}{4\pi k r}$
Sphere Buried at Infinite Depth	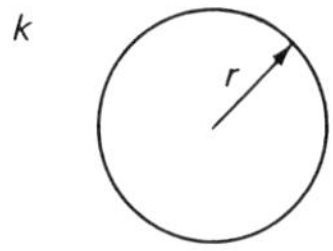	$4\pi k r$
Square Insulation Around Cylinder	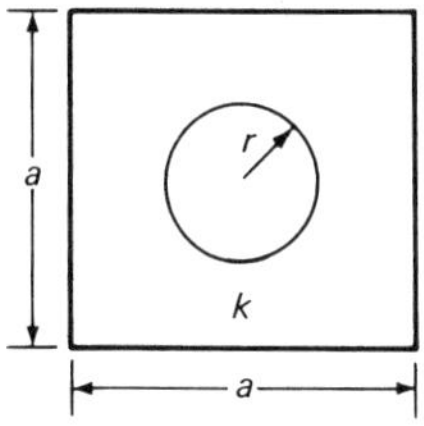	$\frac{1}{2\pi k} \ln \left(\frac{1.08a}{2r}\right)$
Eccentric Cylinders	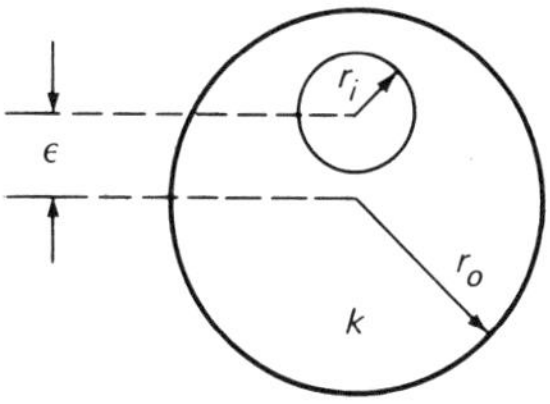	$\frac{1}{2\pi k} \operatorname{arccosh} \frac{1 + \left(\frac{r_o}{r_i}\right)^2 - \left(\frac{\epsilon}{r_i}\right)^2}{2\frac{r_o}{r_i}}$

Table 7.1. (Continued)

Physical System	Geometry	Resistance
Two Buried Cylinder	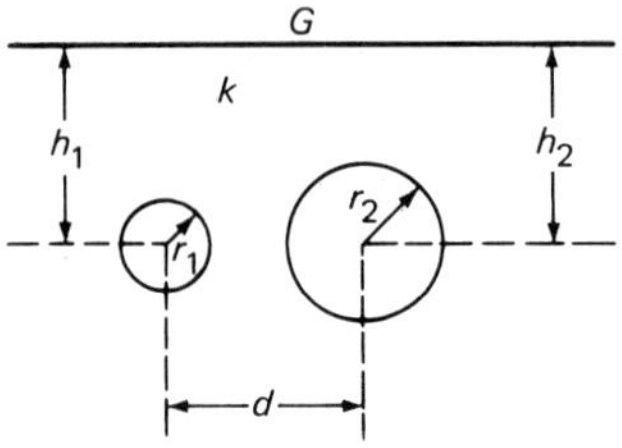 $\mu_1 = h_1/r_1 > 3$ $\mu_2 = h_2/r_2 > 3$ $d > 3(r_1 + r_2)$	$R_{1-2} = \frac{1}{2\pi k} \frac{\ln 2\mu_1 \cdot \ln 2\mu_2 - (\ln p)^2}{\ln p}$ $R_{1-G} = \frac{1}{2\pi k} \frac{\ln 2\mu_1 \cdot \ln 2\mu_2 - (\ln p)^2}{\ln 2\mu_2 - \ln p}$ $R_{2-G} = \frac{1}{2\pi k} \frac{\ln 2\mu_1 \cdot \ln 2\mu_2 - (\ln p)^2}{\ln 2\mu_1 - \ln p}$ $p = \sqrt{\frac{(h_1 + h_2)^2 + d^2}{(h_1 - h_2)^2 + d^2}}$
Two Cylinders Buried at Infinite Depth	See above	$R_{1-2} = \frac{1}{2\pi k} \operatorname{arccosh} \left(\frac{d^2 - r_1^2 - r_2^2}{2 r_1 r_2} \right)$
Rectangular Buried Duct	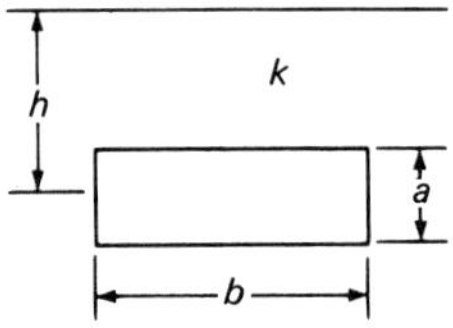	$\frac{1}{k\left(5.7 + \frac{b}{2a}\right)} \ln \left(\frac{3.5\, h}{b^{.25}\, a^{.75}} \right)$
Cylinder Buried Vertically	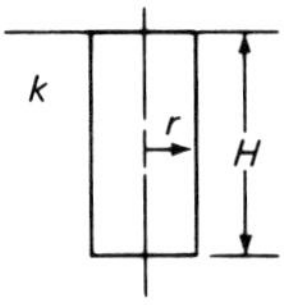	$\frac{1}{2\pi k H} \ln (2\, H/r)$

7.1.6.1 Convective Resistance for Enclosed Cylinder

A common geometry is a cylinder enclosed within a rectangular (sometimes circular) duct. Figure 7.19 shows a cylindrical enclosure that has the same surface area as the rectangular duct.

The heat flow from the inner to the outer cylinder can be calculated from Eq. (7.60), with

$$R = \frac{\ln (D_o/D_i)}{2\pi k_e} \tag{7.61}$$

The equivalent thermal conductivity, k_e, is found from simple relations. Calculate the Grashoff-Prandtl number product

$$N = \frac{g(T_1 - T_2)\,\delta^3}{T_f \alpha \nu} \tag{7.62}$$

where

g = acceleration of gravity
$\delta = (D_o - D_i)/2$
$T_f = (T_1 + T_2)/2$, temperature to evaluate the properties of the gas between the ducts
α = thermal diffusivity of gas
ν = kinematic viscosity of gas and
k = thermal conductivity of gas.

Then

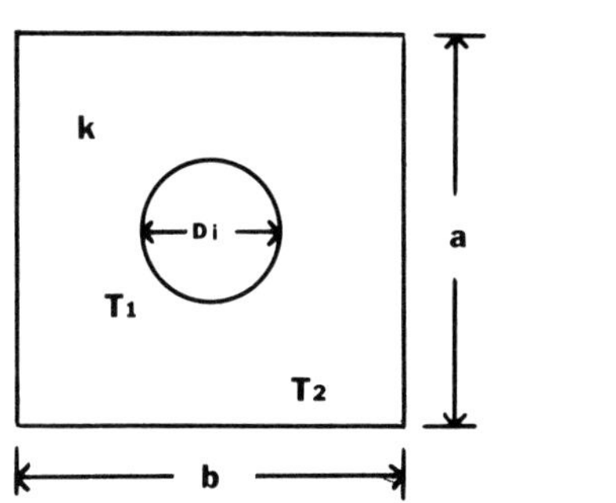

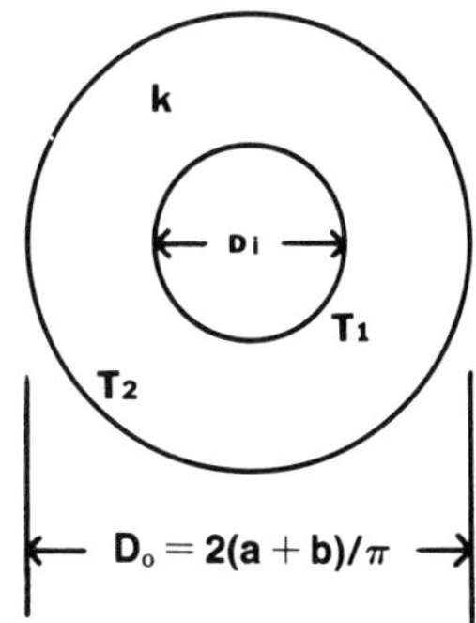

Fig. 7.19. Equivalent Circular and Rectangular Ducts.

Table 7.2. Comparison of Insulation Properties of Various Materials.

Material	Thermal conductivity		Density,	Effect of moisture content, w on thermal	Relative volumetric
	Cal/h·m·°C	Btu/hr-ft-°F	kg/m³	conductivity, k	cost
Polyurethane					
New	0.014	.009	30	Negligible	4–6.5
Aged	0.02	.013	30	Negligible	
High Strength	0.07	.047	65	Nil	
Expanded Polystyrene	0.025–0.030	.017–.02	15–45	Negligible	3.5
Molded Polystyrene	0.026–0.033	.017–.022	15–30	Absorption, $w = 3\%$ max.	2
Sulfur Foam	0.036	.024	175	Absorption, $w = 2\%$ max.	
Insulating Concrete	0.09–0.52	.06–.349	300–1500	Nonrepellent	
Glass Fiber	0.03–0.045	.02–.03	25–55	Deterioration at $w = 8\%$, $k = 0.06$	1
Glass Foam	0.05	.034	150–200	Nil	
Vermiculite	0.06	.040	200	Deteriorates	
Sawdust, Dry	0.05–.08	.034–.054	150–250	Deteriorates	
Water	0.50	.336	1000	–	
Ice	1.90	1.277	900	–	

$$k_e/k \begin{cases} .11\ N^{.29} & 6000 < N < 10^6 \\ .4\ N^{.2} & 10^6 < N < 10^9 \end{cases} \quad \begin{matrix} (7.63a) \\ (7.63b) \end{matrix}$$

7.1.6.2 Thermal Conductivity

Some representative values for the thermal conductivity of some materials are given in Tables 7.2 and 7.3. The thermal conductivity varies with the thermodynamic and physical states, and these values are merely illustrative. The reader should refer to primary sources for more specific design information.

Table 7.3. Thermal Conductivities of Common Materials.

Material	Unit weight (dry), kg/m³	Specific heat capacity, cal/g°C or Btu/lb-°F	Thermal conductivity	
			cal/m·h·°C	Btu/ ft·h·°F
Air, No Convection (0°C)	–	0.24	0.020	0.014
Polyurethane Foam	32	0.4	0.021	0.014
Polystyrene Foam	30	0.3	0.031	0.020
Rock Wool, Glass Wool	55	0.2	0.034	0.023
Snow, New, Loose	85	0.5	0.07	0.05
Snow, on Ground	300	0.5	0.20	0.13
Snow, Drifted and Compacted	500	0.5	0.6	0.4
Ice at −40°C	900	0.5	2.29	1.54
Ice at 0°C	900	0.5	1.90	1.28
Water (0°C)	1000	1.0	0.50	0.34
Rock, Typical	2500	0.20	1.9	1.3
Wood, Plywood, Dry	600	0.65	0.15	0.10
Wood, Fir or Pine, Dry	500	0.6	0.10	0.07
Wood, Maple or Oak, Dry	700	0.5	0.15	0.10
Insulating Concrete (varies)	200 to 1500		0.16 to 0.50	0.04 to 0.35
Concrete	2500	0.16	1.5	1.0
Asphalt	2000		0.62	0.42
Polyethelene, High Density	950	0.54	0.31	0.21
PVC	1400	0.25	0.16	0.11
Asbestos Cement	1900		0.56	0.38
Wood Stave (varies)			0.22	0.15
Steel	7500	0.12	37	25
Ductile Iron	7500		45	30
Aluminum	2700	0.21	175	115
Copper	8800	0.1	325	220

7.1.7 Utilidors

A utilidor is an enclosed box that contains utility lines carrying water, sewage, steam, electricity, etc. The utilidor may be above or below ground, depending upon the thermal conditions of the system. Consider a simple utilidor, with a single pipe, as shown in Fig. 7.20. The steady-state heat flow from this system is

$$\frac{q}{L} = \frac{T_g - T_\infty}{R_T} \tag{7.64}$$

$$R_T = \sum_{i=1}^{7} R_i \tag{7.65}$$

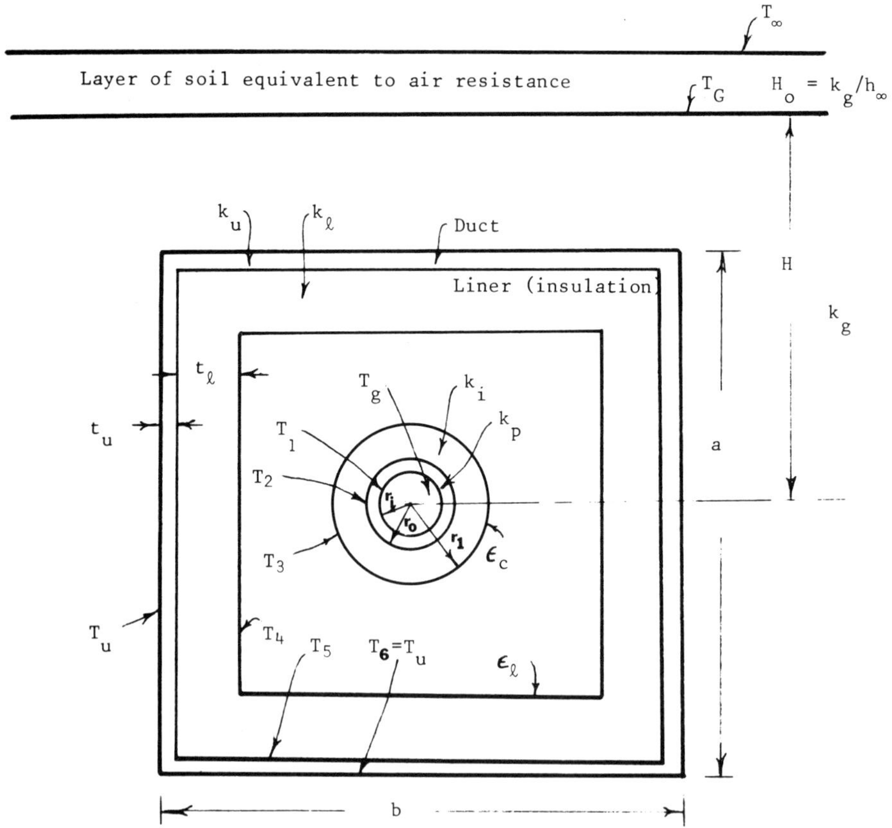

Fig. 7.20. Buried Utilidor System.

The individual resistances of this series system are calculated using the formulae in Table 7.1.

The convective resistance of the fluid flowing inside the pipe is

$$R_1 = \frac{1}{2\pi r_i h_i} \tag{7.66}$$

where h_i is calculated with standard correlations. This resistance can usually be safely neglected.

The conductive resistances of the pipe wall and the insulation are

$$R_2 = \frac{\ln (r_o/r_i)}{2\pi k_p} \tag{7.67}$$

$$R_3 = \frac{\ln (r_1/r_o)}{2\pi k_i} \tag{7.68}$$

The resistance, R_4, across the air gap in the utilidor is due to convection and radiation. The convective resistance is given by Eq. (7.61)

$$R_c = \frac{\ln (D_o/D_i)}{2\pi k_e}$$

The radiative resistance is

$$R_r = \frac{1}{2\pi r\, h_r} \tag{7.69}$$

$$h_r = \frac{4\,\overline{T}^3}{p} \tag{7.70}$$

where

$\overline{T}$ is the average absolute temperature of the outer surface of the insulation layer and the inner surface of the duct liner

$\sigma = .1718 \times 10^{-8}$ Btu/hr-ft^2 $-R^4$, constant

$$p = \frac{1}{\epsilon_c} + \frac{A_1}{A_c}\left(\frac{1}{\epsilon_l} - 1\right)$$

$A_1 = 2\pi r_1$, outside area of inner cylinder

$A_c = 2(b + a) - 4(t_l + t_u)$, inside area of duct liner and

ϵ_c, ϵ_l = emissivities of the insulation and liner.

Then

$$R_4 = \frac{1}{\dfrac{1}{R_c} + \dfrac{1}{R_r}} \tag{7.71}$$

The resistance of the duct liner and the duct are

$$R_5 = \frac{t_l}{P_l k_l} \tag{7.72}$$

$$R_6 = \frac{t_u}{P_u k_u} \tag{7.73}$$

where P_l, P_u are the mean perimeters of the duct liner and the duct. The last resistance, R_7, is

$$R_7 = \begin{cases} \dfrac{1}{k_g\left(5.7 + \dfrac{b}{2a}\right)}\left[\ln \dfrac{3.5(H + H_o)}{a^{.75}\, b^{.25}}\right] & \text{if the utilidor is buried} \quad (7.74a) \\[2ex] \dfrac{1}{2(a+b)\, h_\infty} & \text{if above ground} \quad (7.74b) \end{cases}$$

The outside convective coefficient can be estimated, see also Chapter 6, with

$$h_\infty = .31\left(\frac{T_u - T_\infty}{a+b}\right)^{.25} \sqrt{12.5\, u_\infty + 1} \tag{7.75}$$

where

u_∞ = wind velocity, mph
h_∞ = Btu/hr-ft-°F
a, b = ft and
T = °F.

The calculation is an iterative one, with an initial guess made for the unknown temperatures and then a check of the estimated temperatures, with repetition until the temperatures no longer change. For most problems the solution will converge rapidly.

Example 9: Estimate the heat loss from the above ground utilidor shown. A bare asbestos cement pipe is enclosed in an air-filled duct. The solution starts by guessing the following temperatures

$$T_1 = 45°F$$
$$T_2 = 27°F$$
$$T_3 = -49°F$$

For this problem

$$R_1 \approx .033°\text{F-hr-ft/Btu (could be neglected)}$$

$$R_2 = \frac{\ln\left(\frac{.299}{.244}\right)}{2\pi(.375)} = .086$$

$$R_3 = 0 \text{ (no insulation on pipe)}$$

The Grashof-Prandtl parameter is

$$N = \frac{32.2(45 - 27)(.3375)^3}{496(12.44 \times 10^{-5})^2} = 2.081 \times 10^6$$

$r_i = .2438$ ft
$r_o = .299$ ft
$T_g = 50°F$ $\quad T_\infty = -50°F$ $\quad \epsilon_c = .93$

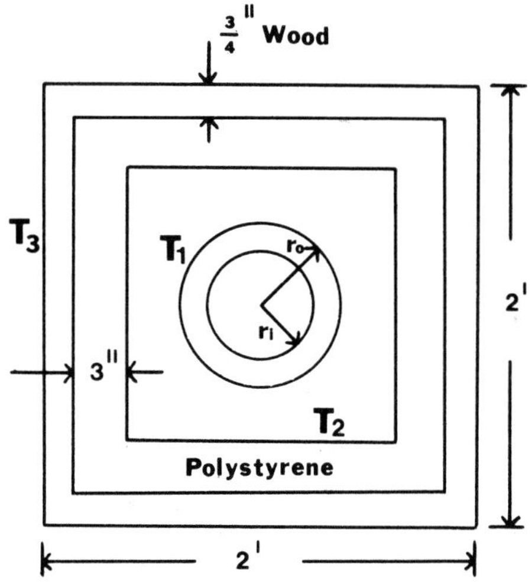

$u_\infty = 5$ mph $\qquad \epsilon_l = .91$
$k_u = .075$ Btu/hr-ft-°F
$k_l = .02$
$k_p = .375$
$P_u = 8.25$ ft $\qquad P_l = 7.0$ ft
$D_o = 1.273$ ft $\qquad D_i = .598$ ft
$\delta = .3375$

The properties of the air in the duct

$$\nu = 12.44 \times 10^{-5} \text{ft}^2/\text{sec}$$
$$k = .0136 \text{ Btu/hr-ft-°F}$$
$$P = .717$$

Then from Eq. (7.63*b*), $k_e = .100$ and

$$R_c = \frac{\ln\left(\dfrac{1.273}{.598}\right)}{2\pi(.1)} = 1.205$$

Now

$$p = \frac{1}{.93} + \frac{1.879}{5.5}\left(\frac{1}{.91} - 1\right) = 1.109$$

$$\overline{T} = \left(\frac{45 + 27}{2}\right) + 460 = 496°R$$

$$h_r = \frac{4(.1714 \times 10^{-8})(496)^3}{1.109} = .754$$

$$R_r = .7056$$

$$R_4 = \frac{1}{\dfrac{1}{1.205} + \dfrac{1}{.7056}} = .445$$

$$R_5 = 1.786$$
$$R_6 = .101$$

$$h_\infty = .31\left(\frac{1}{4}\right)^{.25}\sqrt{12.5(5) + 1} = 1.75$$

$$R_7 = .072$$

Finally

$$R_T = 2.523$$

$$q/L = \frac{100}{2.523} = 3.96 \text{ Btu/hr-ft}$$

The temperature guesses may be checked by calculating the temperature drop across each resistance. The calculation gives

$$T_1 = 45.3\,°\text{F}, \qquad T_2 = 27.7, \qquad T_3 = -47.2$$

In this case, the initial guess was chosen to avoid iteration, but the calculations are simply repeated using the new temperatures each time until there is no significant change.

If there is more than one pipe in the utilidor, the calculation is essentially the same, except for the simplifying assumption that each pipe acts independently of the others. With some experience it will be noted that some of the resistances predominate and the heat transfer calculation can be simplified by considering only the major resistances.

7.1.7.1 Multiple Pipe Utilidors

With more than one pipe in the utilidor, it is convenient to define an air temperature in the utilidor as T_A. The interior air temperature can be evaluated as

$$T_A = \frac{T_\infty/R_s + \sum_{i=1}^{N} T_i/R_i}{1/R_s + \sum_{i=1}^{N} 1/R_i} \tag{7.76}$$

where

T_i = internal fluid temperature of each pipe
R_i = total resistance of each pipe, including surface convection and
R_s = resistance of the utilidor from the inside air to the exterior.

The heat flow is

$$q/L = \frac{T_A - T_\infty}{R_s} = (T_A - T_\infty)/R_s \tag{7.77}$$

$$q/L = \sum_{i=1}^{N} \left(\frac{T_i - T_A}{R_i} \right) \tag{7.78}$$

Example 10: A high-temperature water pipe is added to the utilidor of Example 9. The second pipe has an outside radius of 1 in. and 1.5 in. of insulation, with $k = .04$ Btu/hr-ft-°F. The water temperature is 250°F. Calculate the heat loss.

Assume $T_A = 45$, $T_{2i} = 80$°F (temperature of pipe insulation) and $R_s = 1.91$ from Example 9. The convection coefficient for the original pipe is

$$h_1 = .23 \left(\frac{50 - 45}{.299} \right)^{1/4} = .47$$

Then

$$R_1 = .033 + .086 + \frac{1}{2\pi(.299)\,(.47)} = 1.263$$

$$R_2 = \frac{\ln\left(\frac{.2083}{.0833}\right)}{2\pi(.04)} + \frac{1}{2\pi(.2083).23\left(\frac{80 - 45}{.2083}\right)^{.25}} = 4.57$$

Then

$$T_A = \frac{\frac{-50}{1.91} + \frac{50}{1.263} + \frac{250}{4.57}}{\frac{1}{1.91} + \frac{1}{1.265} + \frac{1}{4.57}} = 44.4°F$$

$$q/L = 49.4 \text{ Btu/hr-ft}$$

Check the insulation temperature of the hot pipe

$$T_{2i} = 250 - (3.65)(49.4) = 69.6°F$$

Repeat with $T_A = 44.4$, $T_{2i} = 69.6$; the results will be essentially the same.

7.2 STEADY PERIODIC PROBLEMS

Many significant heat transfer problems dealing with soil systems or permafrost systems deal with boundary conditions that vary periodically or nearly so with time. If the surface disturbance continues indefinitely, such problems are called "steady periodic," because the temperature at a point repeats itself continuously with time. The main simplification for such problems is that the initial conditions are not important. A given system will eventually attain the same steady periodic solution regardless of the initial state of the temperature.

7.2.1 Homogeneous, Infinite Soil System

One of the classic problems of geotechnical heat transfer deals with the case of a homogeneous semi-infinite solid without phase change, for which the surface temperature is varying periodically. This problem may be stated as

$$\frac{\partial v}{\partial t} = \alpha \frac{\partial^2 v}{\partial x^2} \tag{7.79}$$

$$v = v_s \qquad t = 0$$

$$v = v_o = f(t) \qquad x = 0$$

$$v \rightarrow const. \qquad x \rightarrow \infty$$

where $v = (T - T_s)$, $v_s = T_o - T_s$, and T_o is the initial temperature of the soil. For a periodic system, the value of T_o, or even if it is a constant, is not important as the initial effect will drop out; T_s = mean annual ground surface temperature.

Separation of Variables. The system of Eq. (7.79) may be solved by simple separation of variables, although this is not the most efficient method. Let

$$v = F(t)\, G(x) \tag{7.80}$$

where as usual F is a function only of time and G is a function only of x. Setting Eq. (7.80) in Eq. (7.79) gives

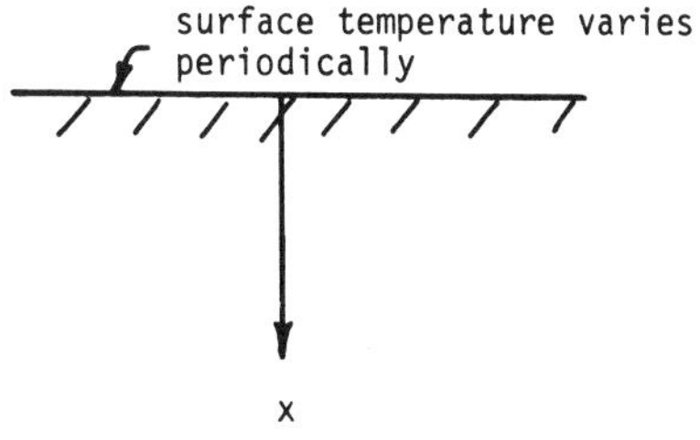

$$\frac{F'(t)}{\alpha F(t)} = \frac{G''(x)}{G(x)} = \pm i\lambda^2$$

Choose $i\lambda^2$ to make time appear as the periodic variable

$$F'(t) \pm i\lambda^2 \alpha F(t) = 0$$

$$G''(x) \pm i\lambda^2 G(x) = 0$$

The solutions to these equations are

$$F = D_1 e^{\pm i\lambda^2 \alpha t} \tag{7.81}$$

$$G = D_2 e^{-\lambda\sqrt{\pm i}\, x} \tag{7.82}$$

Expanding $\sqrt{\pm i}$ and eliminating those solutions that approach infinity as x approaches infinity, one can write the complete solution as

$$v = e^{-\frac{\lambda x}{\sqrt{2}}}\left\{C_1 e^{i\left(\lambda^2 \alpha t - \frac{\lambda x}{\sqrt{2}}\right)} + C_2 e^{-i\left(\lambda^2 \alpha t - \frac{\lambda x}{\sqrt{2}}\right)}\right\}$$

or

$$v = e^{-\frac{\lambda x}{\sqrt{2}}}\left\{A \cos\left(\lambda^2 \alpha t - \frac{\lambda x}{\sqrt{2}}\right) + B \sin\left(\lambda^2 \alpha t - \frac{\lambda x}{\sqrt{2}}\right)\right\}$$

$$v = C e^{-\frac{\lambda x}{\sqrt{2}}} \cos\left(\lambda^2 \alpha t - \frac{\lambda x}{\sqrt{2}} - \delta\right) \tag{7.83}$$

where

$$\delta = \tan^{-1}\frac{B}{A} \qquad C = \sqrt{A^2 + B^2}$$

Surface Temperature a Simple Harmonic Function. Let the surface temperature be given by

$$v = v_0 = T_a \cos\left(\frac{2\pi t}{p}\right) \tag{7.84}$$

where $T_a = T_{s\,\max} - T_s$, amplitude of the surface temperature change, and $p =$ period of the temperature oscillation.
Now using Eqs. (7.84) and (7.83)

$$C = T_a \qquad \lambda = \sqrt{\frac{2\pi}{p\alpha}}$$

Finally the temperature of the solid is

$$T = T_s + T_a\, e^{-\sqrt{\frac{\pi}{p\alpha}}\,x} \cos\left(\frac{2\pi t}{p} - \sqrt{\frac{\pi}{p\alpha}}\,x\right) \tag{7.85}$$

This result has been obtained by a different method in Chap. 3. Notice that Eq. (7.85) will not satisfy the initial condition unless

$$v_s = T_a\, e^{-\sqrt{\frac{\pi}{p\alpha}}\,x} \cos \sqrt{\frac{\pi}{p\alpha}}\,x$$

which is unlikely. This is not a problem because the solution is expected to be valid only a long time after the temperature fluctuation has started. A solution valid immediately after $t = 0$ will be discussed later.
The temperature at any depth x will have a maximum value when

$$\frac{2\pi t}{p} - \sqrt{\frac{\pi}{p\alpha}}\,x = 2\,m\pi \qquad \textit{for } m = 0, 1, 2, \ldots$$

The time at which the temperature will be maximum is

$$t_m = mp + \frac{1}{2}\sqrt{\frac{p}{\alpha\pi}}\,x \tag{7.86}$$

The temperature oscillations at any depth x will have the same period or frequency as those of the surface, but will lag the surface fluctuations by $\frac{1}{2}\sqrt{\frac{p}{\alpha\pi}}\,x$. The amplitude of the temperature will diminish by the factor $e^{-\sqrt{\pi/p\alpha}\;x}$ at any depth x. These relations can be visualized in Fig. 7.21.

Equation (7.85) may be written as

$$\frac{T - T_s}{T_a} = e^{-2\pi\left(\frac{x}{2\sqrt{\pi p\alpha}}\right)} \cos 2\pi\left(\frac{t}{p} - \frac{x}{2\sqrt{\pi p\alpha}}\right)$$

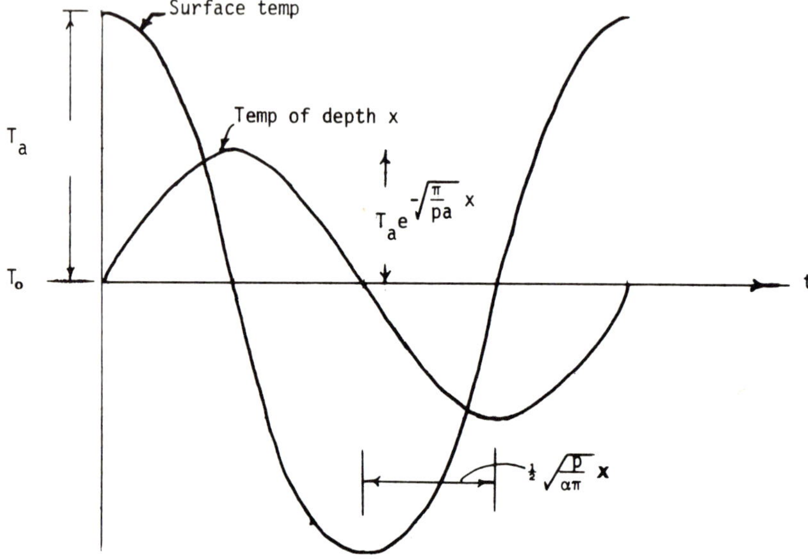

Fig. 7.21. Comparison of Surface-Temperature Variation and Temperature Variation at Depth.

The amplitude of this curve will always be less than the enveloping curve

$$\frac{T - T_s}{T_a} = e^{-2\pi\left(\frac{x}{2\sqrt{\pi p \alpha}}\right)} \tag{7.87}$$

For this equation it can be seen that if $x/2\sqrt{\pi p\alpha}$ is greater than 0.8, then the amplitude of the curve will be less than 1% of the surface fluctuation. The depth at this amplitude is called the "depth of the zero temperature amplitude." If one is referring to a yearly annual temperature fluctuation at the surface, this is the depth of the zero annual amplitude of the temperature fluctuation. The temperature of the medium will be disturbed down to this depth, which may be expressed as

$$\frac{x_z}{2\sqrt{\pi p\alpha}} = 0.8 \tag{7.88}$$

The depth at which the temperature wave is damped out will depend upon the period of the surface fluctuation and the thermal diffusivity of the medium (Fig. 7.22).

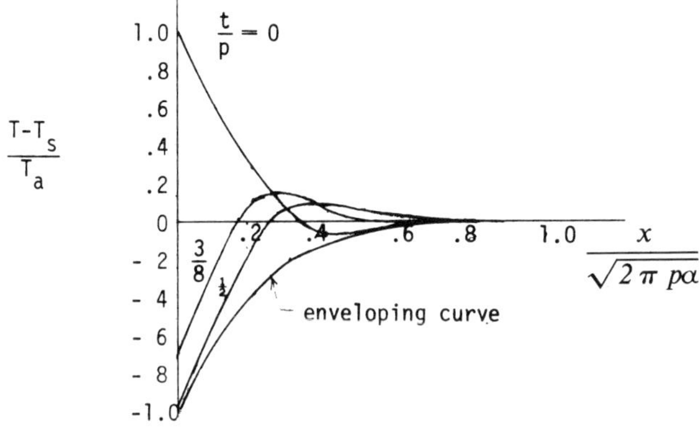

Fig. 7.22. Penetration of a Temperature Oscillation into an Infinitely Thick Solid.

It is of interest to look at the temperature versus depth for two different times. These are shown in Fig. (7.23). The enveloping curve is the well-known trumpet-shaped region, which will include all of the possible temperature variations.

Example 11: Calculate the depth to which the diurnal and the yearly temperature fluctuations will penetrate into a soil with $\alpha = .039$ ft^2/hr.

The depth is

$$x_z = .8(2\sqrt{\pi p \alpha})$$

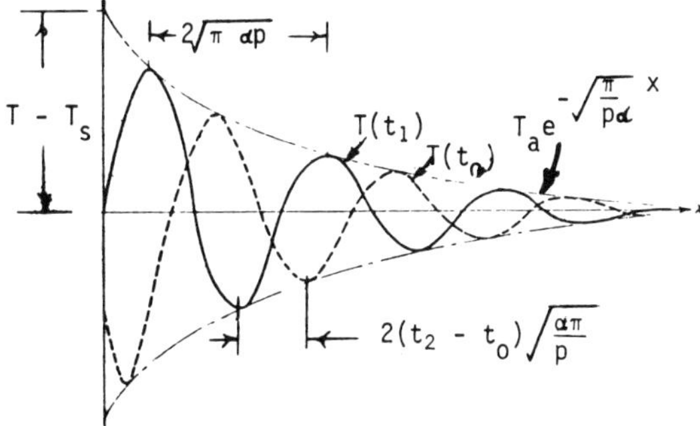

Fig. 7.23. Characteristics of a Temperature Oscillation Penetrating an Infinitely Thick Medium.

Thus the daily fluctuation will penetrate to

$$x_z = 2.73 \text{ ft}$$

and the annual will penetrate to

$$x_z = 52.2 \text{ ft}$$

This example shows that the yearly temperature fluctuation will control the maximum depth at which temperature changes may occur. It is also clear that very long period temperature fluctuation, on the order of hundreds of years, can penetrate to rather extreme depths. These effects may be masked, however, by geothermal effects.

Heat Transfer. The instantaneous heat transfer at the ground surface, $x = 0$, is given by

$$dQ = -kA\left(\frac{\partial T}{\partial x}\right)_{x=o} dt \tag{7.89}$$

Now from Eq. (7.85)

$$\left(\frac{\partial T}{\partial x}\right)_{x=o} = T_a\sqrt{\frac{\pi}{p\alpha}}\left\{\sin\left(\frac{2\pi t}{p}\right) - \cos\left(\frac{2\pi t}{p}\right)\right\}$$

and the heat transfer is

$$dQ = -kA\sqrt{\frac{\pi}{2p\alpha}}\, T_a \sin\left(\frac{2\pi t}{p} - \frac{\pi}{4}\right) dt \tag{7.90}$$

The instantaneous heat transfer rate is given by

$$q = \frac{dQ}{dt}$$

and it can be seen that q will be zero, over one cycle, when

$$\sin\left(\frac{2\pi t}{p} - \frac{\pi}{4}\right)$$

is zero or when

$$\frac{2\pi t}{p} - \frac{\pi}{4} = 0, \pi$$

Thus, the heat flow will be in one direction, in this case out of the soil system for

$$\frac{p}{8} < t < \frac{5}{8}p$$

During this time the total heat flowing out of the soil will be given by

$$Q_t = \int_{p/8}^{5p/8} q\, dt$$

which reduces to

$$Q_t = -\, kA\, T_a \sqrt{\frac{p}{2\alpha\pi}} \tag{7.91}$$

where the negative sign indicates a heat flow out of the system. During the remaining half-cycle, there will be a heat flow into the soil that will be identical to Eq. (7.91), except for the sign, as the heat flow is steady, periodic. These relations can be examined in Fig. (7.24).

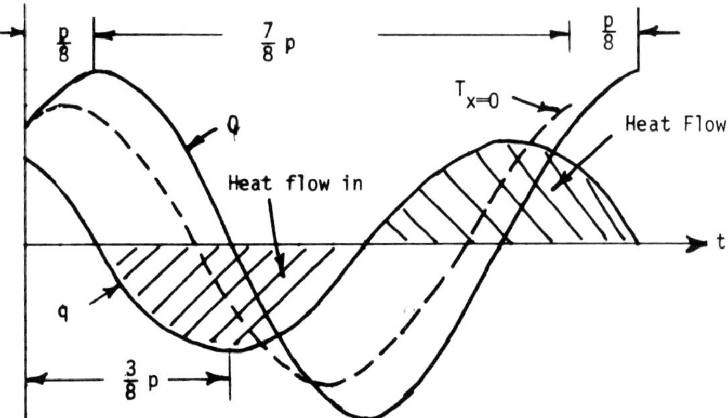

Fig. 7.24. Phase Relationships in Periodic Heat Flow.

7.2.1.1 Maximum Soil Temperature

A question of some interest is the evaluation of the properties (mechanical or thermal) of soil layers. These properties usually are highly dependent upon the temperature of the soil. The maximum temperatures of a soil, throughout the year, can be evaluated from Eq. (7.87). Consider the case of permafrost, which has a mean annual temperature T_o. This is the temperature at the depth of zero annual amplitude. The permafrost maximum temperature will be T_f, and we will calculate the temperature starting at this depth (Fig. 7.25). The temperature is

$$T = T_o + (T_f - T_o)\, e^{-z\sqrt{\frac{\pi}{p\alpha}}} \tag{7.92}$$

Equation (7.92) gives the maximum temperatures during the entire year. These maximums will be reached at different times for each depth. If the soil varies considerably, each layer can be converted to an equivalent thickness of the first layer with

$$\frac{z_2}{z_1} = \sqrt{\frac{\alpha_2}{\alpha_1}} \tag{7.93}$$

Thus, a layer of soil with diffusivity α_2 will be equivalent to a thickness $z_1\sqrt{\frac{\alpha_2}{\alpha_1}}$ of a layer with α_2, for transient temperature estimates.

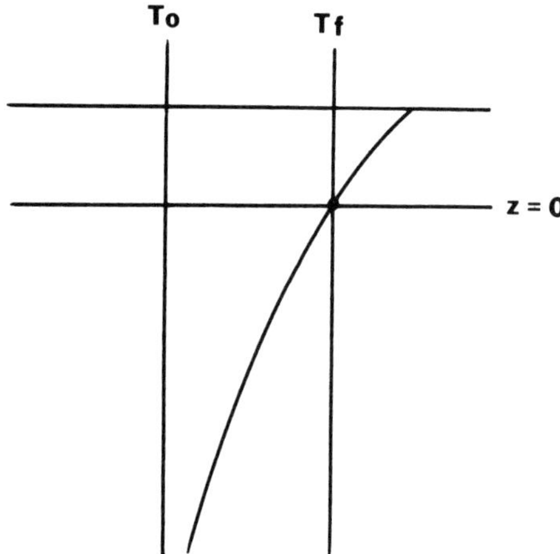

Fig. 7.25. Maximum Temperature Envelope.

Example 12: A permafrost soil has two layers of material, $d_1 = 2$ ft thickness below the permafrost table with $\alpha_1 = .039$ ft²/hr, and a second layer with $\alpha_2 = .025$ ft²/hr. Estimate the maximum temperatures at 1 ft and 3 ft below the permafrost table.

Write Eq. (7.92) for the first layer, with a period of 1 year

$$T = 27 + 5\,e^{-.0959z}$$

Then the following table can be calculated

Actual depth, ft	Equivalent depth, ft	Temperature, °F
0	0	32
1	1	31.54
3	2.80	30.82

7.2.2 Periodic Ambient Fluid

Consider the same problem except that the surrounding fluid is fluctuating with time instead of the ground surface temperature.

Let the ambient fluid temperature vary as follows

$$T_f = \overline{T}_f + T_{fa} \cos\left(\frac{2\pi t}{p}\right) \tag{7.94}$$

Then the boundary condition of the surface of the soil is

$$\left(\frac{\partial T}{\partial x}\right)_{x=0} = \frac{h}{k}(T - T_f)_{x=0}$$

where

$\overline{T}_f$ = mean temperature of the fluid
T_{fa} = amplitude of the fluid temperature oscillation and
h = surface coefficient of heat transfer.

The solution for the temperature follows in the same way as before and is

$$T = \bar{T}_f + \frac{T_{fa}\, e^{-\sqrt{\frac{\pi}{\alpha p}}x} \cos\left\{\frac{2\pi t}{p} - \sqrt{\frac{\pi}{\alpha p}}\, x - \tan^{-1}\left(1 + \sqrt{\frac{h^2 \alpha p}{k^2 \pi}}\right)^{-1}\right\}}{\sqrt{1 + 2\sqrt{\frac{\pi k^2}{\alpha p h^2}} + \frac{2\pi k^2}{\alpha p h^2}}} \qquad (7.95)$$

The difference between the fluid temperature and the soil surface temperature at any instant of time is

$$\Delta T = \frac{T_{fa} \cos\left\{\frac{2\pi t}{p} - \tan^{-1}(1 + r)^{-1}\right\}}{\sqrt{1 + \frac{2}{r} + \frac{2}{r^2}}} - T_{fa} \cos\frac{2\pi t}{p} \qquad (7.96)$$

where

$$r = \sqrt{\frac{h^2 \alpha p}{k^2 \pi}}$$

This relation does not give any further useful information because both the mean annual ground temperature and the mean annual fluid temperature are $\bar{T}_f$. This is due to the mathematical nature of the solution.

7.2.3 Half-Plane with Harmonic Surface Temperature

Let one half-plane be at zero and the other have a harmonic temperature of amplitude μ_o and period p; the mean temperature is zero. Then from Brown (1963)

$$\mu = \frac{\mu_o}{\pi}\left\{\beta \sin \omega t + 2\omega \sum_{n=1}^{\infty} (-1)^n \sin (n\beta) \times \right.$$

$$\left. \int_0^{\infty} \frac{\omega \sin \omega t + \alpha u^2 [\cos \omega t - e^{-\alpha u^2 t} J_n(ur)]}{u(\alpha^2 u^4 + \omega^2)}\, du \right\} \qquad (7.97)$$

The temperature beneath a semi-infinite strip with surrounding temperature fluctuating can be found by the superposition method discussed in Section 7.3.3, but

$$\mu' = \mu_o\, e^{-\left(z\sqrt{\frac{\pi}{p\alpha}}\right)} \sin\left(\frac{2\pi t}{p} - z\sqrt{\frac{\pi}{p\alpha}}\right)$$

must be subtracted from the final result.

7.2.4 Homogeneous System, Initial Transient

Let us consider again the problem of the sinusoidal surface variation. Using separation of variables, we found a solution valid for long times, Eq. (7.85). The complete solution can be found using Duhamel's integral; see Section 3.5.2 for the details of this method.

First consider a simpler case

$$\frac{\partial v}{\partial t} = \alpha \frac{\partial^2 v}{\partial x^2} \tag{7.98}$$

$$v(0,t) = 0$$

$$v(x,0) = f(x)$$

The solution to this problem is well known, using Fourier integrals

$$v = \frac{2}{\pi} \int_0^\infty dx' \int_0^\infty e^{-\alpha\beta^2 t} f(x') \sin \alpha x' \sin \alpha x \, d\beta \tag{7.99}$$

If the initial temperature distribution is constant, $f(x) = V$, then Eq. (7.99) can be integrated to obtain

$$v = V \, erf\left(\frac{x}{2\sqrt{\alpha t}}\right) \tag{7.100}$$

If the surface temperature is held at V and the initial temperature is zero, then the solution for this case is

$$v = V\left\{1 - erf\left(\frac{x}{2\sqrt{\alpha t}}\right)\right\} = V \, erfc \frac{x}{2\sqrt{\alpha t}} \tag{7.101}$$

Equation (7.101) is the solution for a step change at the surface of V and a uniform initial temperature of zero and is the fundamental step change solution to be used with Duhamel's integral.

Now pose the actual problem

$$\frac{\partial v}{\partial t} = \alpha \frac{\partial^2 v}{\partial x^2}$$

$$v(x,0) = 0$$

$$v(0,t) = \phi(t)$$

Here $v = T - T_o$ where T_o is a constant initial temperature of the soil. In these problems the implied second boundary condition in x is that the temperature remains finite as x approaches infinity. One may immediately write Duhamel's integral as

$$v = \int_0^t \phi(\lambda) \frac{\partial F(x, t-\lambda)}{\partial t} \, d\lambda \tag{7.102}$$

Now, in this case, from Eq. (7.101), one has

$$F(x,t) = erfc\left(\frac{x}{2\sqrt{\alpha t}}\right) \tag{7.103}$$

Here a few details on the properties of the error function may be noted

$$erf\, x = \frac{2}{\sqrt{\pi}} \int_0^x e^{-\beta^2} d\beta$$

It can be shown that

$$\int_0^\infty e^{-\beta^2} \, d\beta = \frac{\sqrt{\pi}}{2}$$

and thus

$$erf\, \infty = \frac{2}{\sqrt{\pi}} \int_0^\infty e^{-\beta^2} \, d\beta = \frac{2}{\sqrt{\pi}} \frac{\sqrt{\pi}}{2} = 1$$

Now

$$\begin{aligned} erfc\, x \equiv 1 - erf\, x = 1 - \frac{2}{\sqrt{\pi}} \int_0^x e^{-\beta^2} \, d\beta &= \frac{2}{\sqrt{\pi}} \int_0^\infty e^{-\beta^2} \, d\beta - \frac{2}{\sqrt{\pi}} \int_0^x e^{-\beta^2} \, d\beta \\ &= \frac{2}{\sqrt{\pi}} \int_0^x e^{-\beta^2} \, d\beta + \frac{2}{\sqrt{\pi}} \int_x^\infty e^{-\beta^2} \, d\beta - \frac{2}{\sqrt{\pi}} \int_0^x e^{-\beta^2} \, d\beta \end{aligned}$$

Thus

$$erfc\, x = \frac{2}{\sqrt{\pi}} \int_x^\infty e^{-\beta^2} \, d\beta$$

Then Eq. (7.103) may be written as

$$F(x,t) = \frac{2}{\sqrt{\pi}} \int_{x/2\sqrt{\alpha t}}^{\infty} e^{-\beta^2} d\beta$$

and

$$F(x,t-\lambda) = \frac{2}{\sqrt{\pi}} \int_{x/2\sqrt{\alpha(t-\lambda)}}^{\infty} e^{-\beta^2} d\beta \tag{7.104}$$

Using Leibnitz's rule

$$\frac{\partial F}{\partial t} = \frac{x e^{-\frac{x^2}{4k(t-\lambda)}}}{2\sqrt{\pi k(t-\lambda)^3}}$$

Thus the general solution of Eq. (7.102) is

$$v = \frac{x}{2\sqrt{\pi\alpha}} \int_o^t \frac{\phi(\lambda)\, e^{\overline{4\alpha(t-\lambda)}^{\,x^2}}}{(t-\lambda)^{3/2}} d\lambda \tag{7.105}$$

Equation (7.105) may be rewritten as follows. Let

$$\mu = \frac{x}{2\sqrt{\alpha(t-\lambda)}}$$

Equation (7.105) becomes

$$v = \frac{2}{\sqrt{\pi}} \int_{\frac{x}{2\sqrt{\alpha t}}}^{\infty} \phi\left(t - \frac{x^2}{4\alpha\mu^2}\right) e^{-\mu^2} d\mu \tag{7.106}$$

The solution for an arbitrary surface temperature variation may be reduced to the sinusoidal case by letting

$$\phi(t) = A \cos(\omega t - \epsilon) \tag{7.107}$$

where A, ϵ are constants.
Then the sinusoidal solution is

$$v = \frac{2A}{\sqrt{\pi}} \int_{\frac{x}{2\sqrt{\alpha t}}}^{\infty} \cos\left(\omega t - \frac{\omega x^2}{4\alpha\mu^2} - \epsilon\right) e^{-\mu^2} d\mu \tag{7.108}$$

It is known that

$$\frac{2}{\sqrt{\pi}}\int_0^\infty \cos\left(\omega t - \frac{\omega x^2}{4\alpha\mu^2} - \epsilon\right) e^{-\mu^2}\, d\mu = e^{-x\sqrt{\frac{\omega}{2\alpha}}} \cos\left(\omega t - x\sqrt{\frac{\omega}{2x}} - \epsilon\right)$$

Then the solution is

$$v = Ae^{-x\sqrt{\frac{\omega}{2\alpha}}} \cos\left(\omega t - x\sqrt{\frac{\omega}{2\alpha}} - \epsilon\right)$$

$$- \frac{2A}{\sqrt{\pi}}\int_0^{\frac{x}{2\sqrt{\alpha t}}} \cos\left(\omega t - \frac{\omega x^2}{4\alpha\mu^2} - \epsilon\right) e^{-\mu^2}\, d\mu \quad (7.109)$$

In Eq. (7.109), we may note that the second term is a transient that decays as the time increases, leaving the steady periodic oscillation of period $2\pi/\omega$, as before. Equation (7.109) is the complete solution valid for all times. One can find the time required to reach the steady state by numerically evaluating the second term if this is significant.

If $\epsilon = 0$ in Eq. (7.107) and $A = T_a$, then the first term of Eq. (7.109), the long-time solution, is the same as Eq. (7.85).

7.2.5 Nonhomogeneous System

It is very rare to find soil systems that are homogeneous under actual conditions. Solutions can be carried out for a system made up of layers with different thermal properties with a sinusoidal boundary condition. If there are more than two layers, the algebraic work becomes very tedious. Lachenbruch (1959) has obtained results for two- and three-layer systems using the Laplace transform method.

7.2.5.1 The Two-Layer Problem

Consider a semi-infinite medium made up of two layers. The problem may be formulated as follows (Fig. 7.26).

Let $\theta = T - T_o$ where T_o is the mean annual ground temperature.

$$\frac{\partial^2\theta_1}{\partial x^2} = \frac{1}{\alpha_1}\frac{\partial\theta_1}{\partial t} \qquad 0 < x < X \qquad (7.110)$$

$$\theta_1(0,t) = A_o \sin\omega t \qquad t > 0$$

$$\theta_1(x,0) = 0$$

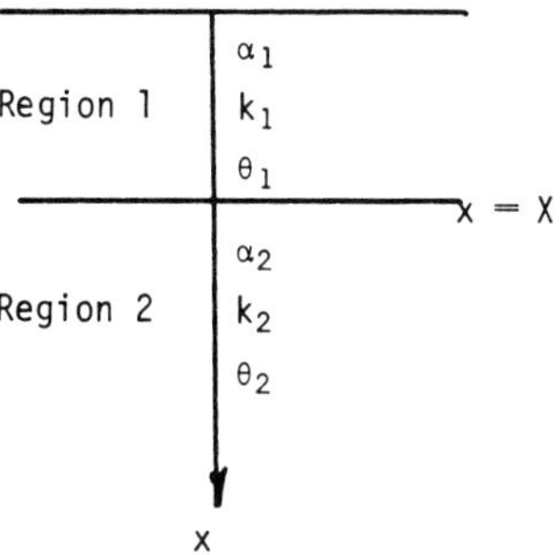

Fig. 7.26. Two-Layer Problem.

$$\frac{\partial^2\theta_2}{\partial x^2}=\frac{1}{\alpha_2}\frac{\partial\theta_2}{\partial t} \qquad X<x<\infty \tag{7.111}$$

$$\theta_2\,(x,0)=0$$

$$\theta_2\,(\infty,t)\rightarrow 0$$

$$\theta_1\,(X,t)=\theta_2\,(X,t) \tag{7.112}$$

$$k_1\frac{\partial\theta_1\,(X,t)}{\partial x}=k_2\frac{\partial\theta_2\,(X,t)}{\partial x} \tag{7.113}$$

Equations (7.112) and (7.113) are the temperature and energy equalities at the interface $x = X$.

The Laplace transform of the temperature function is defined as

$$\bar{\theta}(x,p)=\int_0^\infty e^{-pt}\,\theta(x,t)\,dt$$

Applying this to the system of Eqs. (7.91) and (7.92) leads to

$$\frac{d^2\bar{\theta}_1}{dx^2}=\frac{1}{\alpha_1}\left[p\,\bar{\theta}_1-\theta_1\,(x,0)\right]=\frac{p}{\alpha_1}\bar{\theta}_1 \tag{7.114}$$

$$\bar{\theta}_1\,(0,p)=A_o\frac{\omega}{p^2+\omega^2} \tag{7.114a}$$

$$\frac{d^2\bar{\theta}_2}{dx^2}=\frac{p}{\alpha_2}\bar{\theta}_2 \tag{7.115}$$

$$\bar{\theta}_2\,(\infty,p)\rightarrow 0$$

$$\bar{\theta}_1(X,p)=\bar{\theta}_2(X,p) \tag{7.116}$$

$$k_1 \frac{d\bar{\theta}_1(X,p)}{dx} = k_2 \frac{d\bar{\theta}_2(X,p)}{dx} \tag{7.117}$$

The solutions are

$$\bar{\theta}_1 = C e^{\sqrt{\frac{p}{\alpha_1}}x} + D e^{-\sqrt{\frac{p}{\alpha_1}}x} \qquad 0 < x < X \tag{7.118}$$

$$\bar{\theta}_2 = E d^{-\sqrt{\frac{p}{\alpha_2}}x} \tag{7.119}$$

Now use Eqs. (7.114*a*), (7.116), and (7.117) to determine the coefficients *C, D, E.* After some algebra

$$\bar{\theta}_1 = A_o \frac{\omega}{p^2 + \omega^2} e^{\sqrt{p}\frac{x}{\sqrt{\alpha_1}}} \frac{1 + M e^{-2\sqrt{p}\frac{X-x}{\sqrt{\alpha_1}}}}{1 + M e^{-2x\sqrt{p/\alpha_1}}} \tag{7.120}$$

and

$$\bar{\theta}_2 = A_o \frac{\omega}{p^2 + \omega^2} e^{-\sqrt{p}\left[\frac{X}{\sqrt{\alpha_1}} + \frac{x-X}{\sqrt{\alpha_2}}\right]} \frac{1 + M}{1 + M e^{-2\sqrt{p}\frac{x}{\sqrt{\alpha_1}}}} \tag{7.121}$$

where $M = (\beta_1 - \beta_2)/(\beta_1 + \beta_2)$ and $\beta = \sqrt{\rho\ c\ k}$
Equations (7.120) and (7.121) may be inverted using the inversion theorem. The functions are of Class II with simple poles at $p = \pm\ i\omega$ and a branch point at $p = 0$.

From Chap. 3, the inversion theorem is

$$\theta(x,t) = \frac{1}{2\pi i} \int_{\gamma - i\infty}^{\gamma + i\infty} \bar{\theta}(x,p)\, e^{-pt}\, dp \tag{7.122}$$

For a Class II function the inverse is

$$\theta(x,t) = \sum_{n=1} Res(a_n) + \frac{1}{2\pi i} \lim_{\rho \to o} \int_{C_p} e^{pt} \bar{\theta}(x,t) dp$$
$$+ \frac{1}{2\pi i} \int_o^\infty e^{-tp} [\bar{\theta}(x, \rho e^{-i\pi}) - \bar{\theta}(x, \rho e^{i\pi})]\, dp \tag{7.123}$$

It can be shown that

$$\lim_{\rho \to o} \int_{C_p} e^{pt} \overline{\theta}(x,t)\, dp \to 0$$

and the last term of Eq. (7.123) vanishes as $t \to \infty$, due to the exponential terms. They represent the initial transient terms, which do not concern us here. Thus we must evaluate the residues at the poles $p = \pm i\omega$

$$\theta(x,t) = Res\,(a_1) + Res\,(a_2)$$

For θ_1 the residues are as follows

$$Res\,(i\omega) = \frac{A_o}{2i} e^{-\frac{x}{\sqrt{\alpha_1}}\sqrt{i\omega}} \frac{\left(1 + Me^{-2\frac{(X-x)}{\sqrt{\alpha_1}}\sqrt{i\omega}}\right)}{\left(1 + Me^{\frac{-2X}{\sqrt{\alpha_1}}\sqrt{+i\omega}}\right)}$$

$$Res\,(-i\omega) = \frac{A_o}{2i} e^{-\frac{x}{\sqrt{\alpha_1}}\sqrt{-i\omega}} \frac{\left(1 + Me^{-2\frac{(X-x)}{\sqrt{\alpha_1}}\sqrt{-i\omega}}\right)}{\left(1 + Me^{\frac{-2X}{\sqrt{\alpha_1}}\sqrt{-i\omega}}\right)}$$

Now use

$$i^{1/2} = \frac{\sqrt{2}}{2}(1+i) \text{ and } (-i)^{-1/2} = \frac{\sqrt{2}}{2}(1-i)$$

Then after a great deal of algebraic manipulation

$$\begin{aligned}
\theta_1(x,t) = A_o\, e^{-x\sqrt{\frac{\omega}{2\alpha_1}}} \frac{1}{S} \Bigg\{ & \sin\left(\omega t - x\sqrt{\frac{\omega}{2\alpha_1}}\right) \\
& + Me^{-2(X-x)\sqrt{\frac{\omega}{2\alpha_1}}} \sin\left(\omega t - (2X-x)\sqrt{\frac{\omega}{2\alpha_1}}\right) \\
& + Me^{-2X\sqrt{\frac{\omega}{2\alpha_1}}} \sin\left(\omega t + (2X-x)\sqrt{\frac{\omega}{2\alpha_1}}\right) \\
& + M^2 e^{-2(2X-x)\sqrt{\frac{\omega}{2\alpha_1}}} \sin\left(\omega t + x\sqrt{\frac{\omega}{2\alpha_1}}\right) \Bigg\}
\end{aligned} \quad (7.124)$$

$$\theta_2(x,t) = A_o \frac{(1+M)\, e^{\left(-X\sqrt{\frac{\omega}{2\alpha_1}} - (x-X)\sqrt{\frac{\omega}{2\alpha_2}}\right)}}{S} \left\{ \sin \omega t \right.$$

$$- \left(X\sqrt{\frac{\omega}{2\alpha_1}} + (x-X)\sqrt{\frac{\omega}{2\alpha_2}}\right) + Me^{-2X\sqrt{\frac{\omega}{2\alpha_1}}} \sin \omega t$$

$$\left. + \left(X\sqrt{\frac{\omega}{2\alpha_1}} - (x-X)\sqrt{\frac{\omega}{2\alpha_2}}\right)\right\} \tag{7.125}$$

where

$$S = 1 + 2\,Me^{-2X\sqrt{\frac{\omega}{2\alpha_1}}} \cos 2X\sqrt{\frac{\omega}{2\alpha_1}} + M^2\, e^{-4X\sqrt{\frac{\omega}{2\alpha_1}}}$$

Note that if $\beta_1 = \beta_2$ and $\alpha_1 = \alpha_2$

$$\theta_1(x,t) = \theta_2(x,t) = A_o\, e^{-x\sqrt{\frac{\omega}{2\alpha}}} \sin\left(\omega t - x\sqrt{\frac{\omega}{2\alpha}}\right)$$

as was seen before for the homogeneous system.

7.2.5.2 Depth of Fill Problem

Now let us consider the depth of fill needed to reduce the amplitude of θ at $x = X$ to $F = T_f - T_o$ where T_f is the phase change temperature.

From Eqs. (7.124) or (7.125)

$$\theta(X,t) = A_o \frac{(1+M)\, e^{-X\sqrt{\frac{\omega}{2\alpha_1}}}}{S} \left\{ \sin\left(\omega t - X\sqrt{\frac{\omega}{2\alpha_1}}\right) \right.$$

$$\left. + Me^{-2X\sqrt{\frac{\omega}{2\alpha_1}}} \sin\left(\omega t + X\sqrt{\frac{\omega}{2\alpha_1}}\right)\right\}$$

This can be written as

$$\theta(X,t) = A(X) \sin\{\omega t - \phi(X)\} \tag{7.126}$$

where

$$A(X) = A_o e^{-X\sqrt{\frac{\omega}{2\alpha_1}}} \left(\frac{1+M}{\sqrt{S}}\right)$$

$$\tan\phi = \frac{1 - Me^{-2X\sqrt{\frac{\omega}{2\alpha_1}}}}{1 + Me^{-2X\sqrt{\frac{\omega}{2\alpha_1}}}} \tan\left(X\sqrt{\frac{\omega}{2\alpha_1}}\right)$$

Note that for the amplitude $A(x)$ the only parameter of the second layer involved is β_2. The damping is greater or lesser than in the homogeneous case as β_2 is greater or less than β_1. The ratio $A(X)/A_o$ is plotted in Fig. 7.27. At large depths the amplitude differs from the homogeneous case by a factor

$$1 + M = \frac{2\beta_1}{\beta_1 + \beta_2}$$

Note that since

$$A(X) = A_o e^{-X\sqrt{\frac{\omega}{2\alpha_1}}} \frac{(1+M)}{\sqrt{S}}$$

the damping effect is increased by a low diffusivity in layer 1 and a low ratio $1 + M = 2\beta_1/(\beta_1 + \beta_2)$. Thus, add a thin layer with a low β between the layers for greatest effect.

Layer	Value of β
1	Low
2	Very low (organic material)
3	High

Skaven-Haug (1959) notes that this effect has been used on Norwegian railroads to reduce the amount of frost heave. The principle is shown by Fig. 7.28.

Some typical thermal parameters taken in the field are listed in Table 7.4.

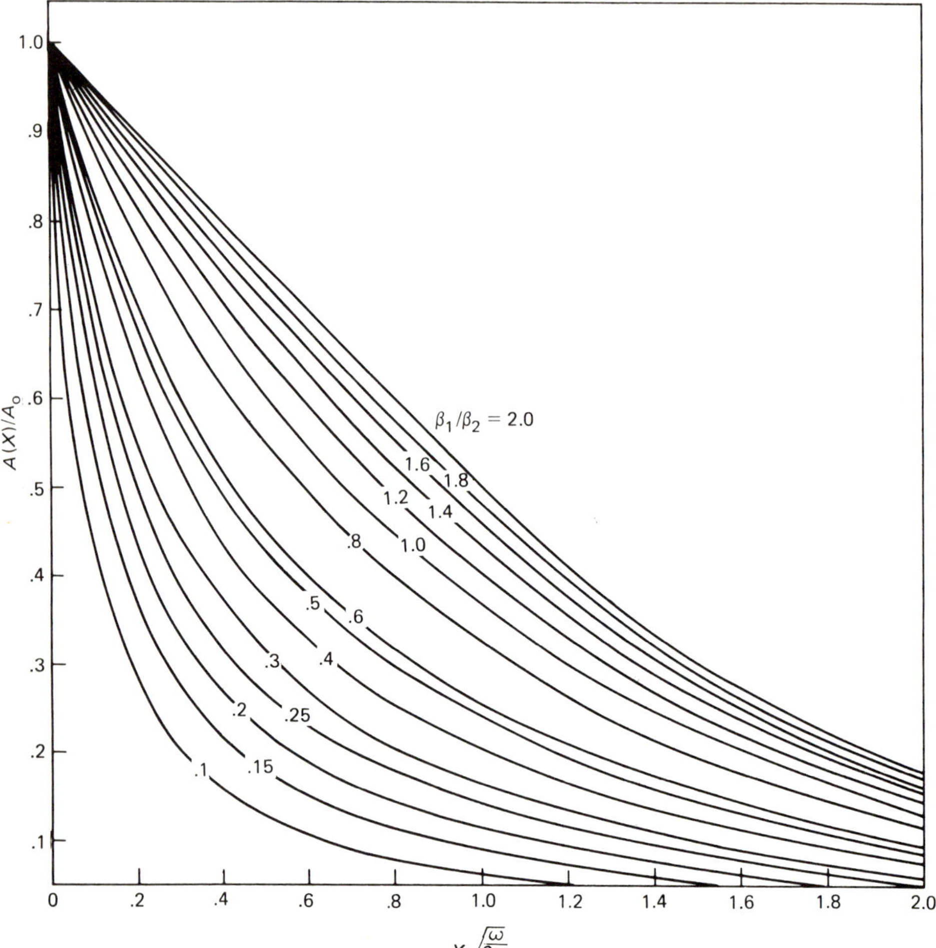

Fig. 7.27. Temperature Amplitude at the Interface of Two Materials (after Lachenbruch 1959).

Example 13: Consider a layer of gravel overlaying a layer of icy silt at Inuvik, NWT. What depth of gravel will yield a temperature of T_f at the interface? How does this compare with the homogeneous case? For Inuvik: $T_o = -9°C$; $A_o = 18°C$; therefore $F = 9°C$. Now $A(X)/A_o = 9/18$ and $\beta_1/\beta_2 = 0.6$. From Fig. 7.27, $X/\gamma_1 = 0.46$ and $X = 4.4$ ft of gravel needed to keep the bottom of the gravel frozen. For the homogeneous case, $\beta_1/\beta_2 = 1.0$, $X/\gamma = 0.68$, and $X = 6.5$ ft. Thus, the difference between the two calculations shows a saving of 2.1 ft with the more accurate two-layer method.

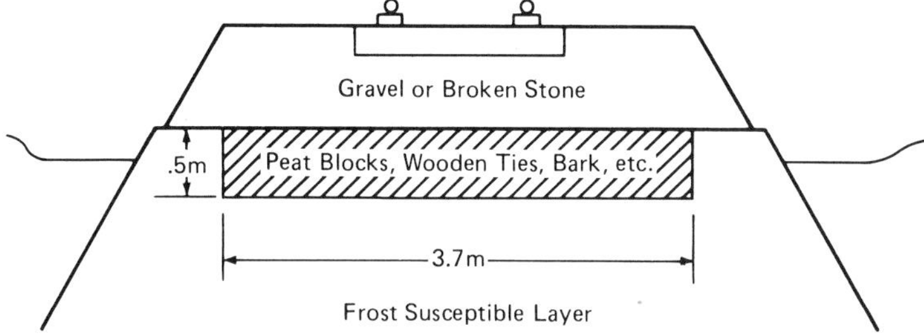

Fig. 7.28. Practical Two-Layer System.

Table 7.4. Thermal Properties of Soils in Situ (after Lachenbruch 1959).

	k, cal/sec-cm°C	ρ, g/cm³	c, cal/g-°C	β, cal/cm²-°Csec$^{1/2}$	α, cm²/sec	γ^a cm
Gravel	.003	2.0	.18	.033	.0083	290
Sandy Gravel	.006	2.1	.20	.050	.014	375
Icy Silt	.006	1.6	.31	.055	.012	345
Frozen Organic Silty Clay	.003	1.35	.32	.036	.007	265
Spruce Logs	.0004	.5	.4	.009	.002	140
Dry Peat	.0004	.4	.5	.009	.002	140
Icy Peat	.0045	.9	.4	.040	.0125	355
Wet Peat—Thawed	.0013	1.0	1.0	.036	.0013	115
Snow, Fresh Drift	.0002	.2	.45	.0042	.0022	150
Snow, Packed Drift	.0006	.35	.45	.0097	.0038	195

$^a\gamma = \sqrt{2\alpha/\omega}$.

The following calculations show the effect of an intermediate layer of spruce logs 1 ft thick, on the depth of gravel needed. Consider the three cases shown below

1	2	3
Gravel	Gravel	Gravel
		Spruce Logs
	Icy Silt	Icy Silt

The depth of gravel needed for different temperatures is shown in the following table. Notice the very significant effect of the layer of organic material, or other material with a very low β value.

	Depth of gravel needed, ft		
$A(X)/A_o$	1	2	3
0.38	9.10	6.4	.43
0.20	15.2	12.2	8.0
0.40	9.0	6.2	0

It is also possible for the homogeneous case to underestimate the depth of gravel needed.

7.2.6 Periodic Heat Flow—Finite Areas

An important problem in periodic heat transfer deals with the case of finite surface areas. In general these cases can be simplified by superposition, as will be shown.

7.2.6.1 Temperature beneath Apex of Circular Segment

Consider a finite circular segment of radius R and angle θ whose surface temperature oscillates harmonically while the surrounding area is at zero. This problem has been solved by Lachenbruch (1957) and the relation is

$$\frac{360}{\theta}\frac{\mu}{\mu_o} = e^{-z\sqrt{\frac{\pi}{p\alpha}}}\sin\left(\frac{2\pi t}{p} - z\sqrt{\frac{\pi}{p\alpha}}\right) - \frac{1}{\sqrt{1+\left(\frac{R}{z}\right)^2}}\, e^{-z\sqrt{\frac{\pi}{p\alpha}}\sqrt{1+\left(\frac{R}{z}\right)^2}}\sin\left(\frac{2\pi t}{p} - z\sqrt{\frac{\pi}{p\alpha}}\sqrt{1+\left(\frac{R}{z}\right)^2}\right) \quad (7.127)$$

where

μ = temperature beneath the vertex of the segment
p = period

$t =$ time
$\alpha =$ thermal diffusivity
$R =$ radius of segment and
$\theta =$ angle of segment.

The time when the temperature μ is a maximum or minimum is given by $t = p/4$ and $3p/4$.
Thus

$$\frac{360}{\theta}\frac{\mu}{\mu_o} = e^{-q}\cos q - \frac{1}{\sqrt{1+\left(\frac{R}{z}\right)^2}} e^{-q\sqrt{1+\left(\frac{R}{z}\right)^2}} \cos q\sqrt{1+\left(\frac{R}{z}\right)^2} \tag{7.128}$$

where $q = z\sqrt{\dfrac{\pi}{p\alpha}}$

Equation (7.128) is plotted as Fig. 7.29. Thus, one may use Fig. 7.29 in the same way as Fig. 7.12.

7.2.6.2 Rectangular Area

As was discussed, after a time the initial transient dies out, and only the periodic effects are significant. Here it is assumed that outside the surface of interest the temperature varies periodically, either daily or yearly, with the yearly effect being most significant. Suppose we have the case where the area is assumed to be at ν_o compared with the surrounding terrain, which undergoes a sinusoidal oscillation of period P and amplitude μ_o about zero

$$\mu_o \sin \omega t = \mu_o \sin \frac{2\pi}{p} t$$

This can be broken into three simpler cases by superposition (Fig. 7.30).

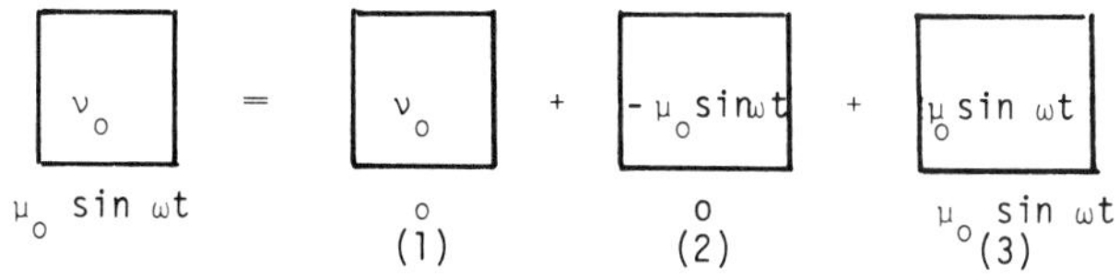

Fig. 7.30. Superposition for Rectangular Area.

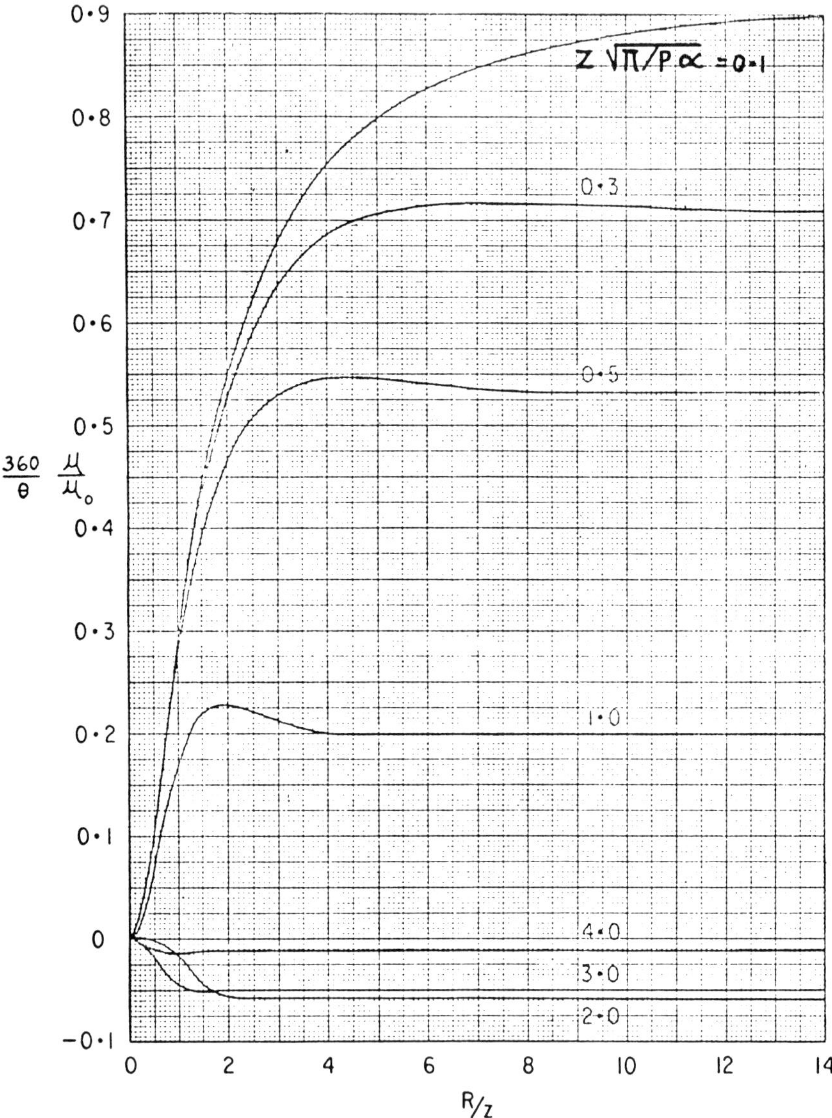

Fig. 7.29. Temperature μ at Depth z under Apex of a Circular Sector at Time when Sinusoidal Surface Temperature has Maximum Value of μ_o Higher than Surrounding Ground Surface at Zero. At minimum μ_o, use negative of values (from Brown, 1962, by permission of National Research Council of Canada, Ottawa).

ν_o	$=$	ν_o	$+$	$-\mu_o \sin \omega t$	$+$	$\mu_o \sin \omega t$
$\mu_o \sin \omega t$		0		0		$\mu_o \sin \omega t$
		Case (1)		Case (2)		Case (3)

Case 1: Step change of the temperature at time zero. The problem of the sudden change of temperature of a circular segment has been discussed in Section 7.1.4, and Fig. 7.12 may be used with the appropriate value of the time parameter $z/2\sqrt{\alpha t}$. In this case the time is infinite and the time parameter value is zero.

Case 2: Region temperature fluctuates sinusoidally. This part of the problem is the case discussed in Section 7.2.6.1. Cases 1 and 2 may be solved by breaking up the region into the segments of a circle.

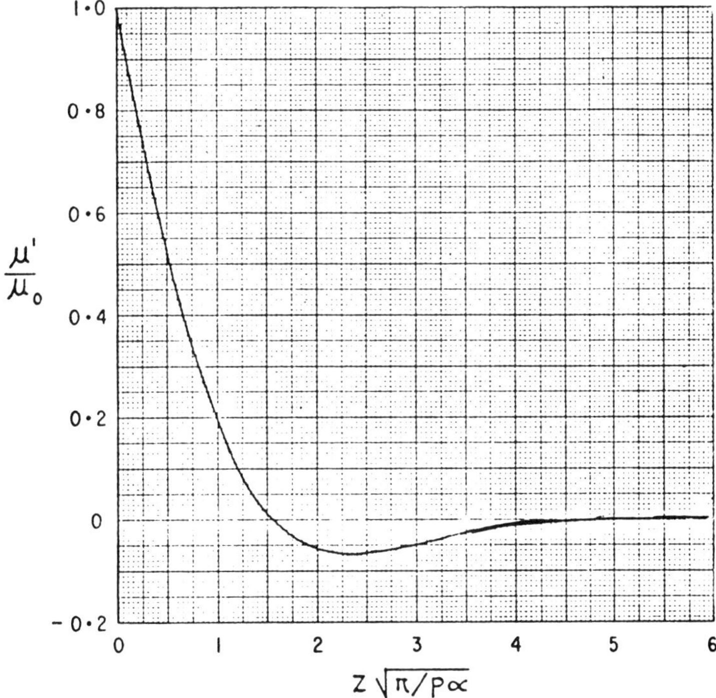

Fig. 7.31. Temperature Depth z due to a Sinusoidal Temperature of Period p at Ground Surface When the Surface Temperature has Maximum Value of μ_o above Mean Ground Temperature (after Brown 1962, by permission of National Research Council of Canada, Ottawa).

Case 3: Periodic surface temperature of infinite region. This is the familiar periodic oscillation of Section 7.2.1, which is given by

$$\frac{\mu'}{\mu_o} = e^{-z\sqrt{\frac{\pi}{p\alpha}}} \sin\left(\frac{2\pi t}{p} - z\sqrt{\frac{\pi}{p\alpha}}\right)$$

This is plotted as Fig. 7.31 for the time when μ' is a maximum or a minimum.

Example 14: Find the winter temperature 20 ft beneath the center of a 40 × 100 ft building in Hanover, N.H. The mean annual ground surface temperature is 49°F, and the floor of the building has a temperature of 70°F and $G = .01$°F/ft. The soil beneath the building has $\alpha = .036$ ft²/hr. Since $p = 1$ year = 8766 hours

$$z\sqrt{\frac{\pi}{p\alpha}} = 2.0$$

The rectangle is divided into four equal areas. Using Figs. 7.12 and 7.29, we can make a table as follows

R/Z	Steady temp. $\frac{360}{\theta}\frac{\nu}{\nu_0}$ (Fig. 7.12)	Periodic temp. $\frac{360}{\theta}\frac{\nu}{\nu_0}$ (Fig. 7.27)	
2.50	.628	+.057	
2.59	.640	+.056	
2.39	.612	+.056	Note that positive values
1.76	.507	+.050	are used because winter
1.42	.430	+.039	is being considered.
1.23	.372	+.028	
1.10	.330	+.022	
1.04	.310	+.019	
1.01	.292	+.016	

Thus for cases 1 and 2, respectively, summing the temperature columns and multiplying by 4

$$\frac{\nu}{\nu_o} = 0.458$$

$$\frac{\mu}{\mu_o} = 0.038$$

Now from Fig. 7.31 at $z\sqrt{\frac{\pi}{p\alpha}} = 2.0$, read

$$\frac{\mu'}{\mu_o} = +.057$$

Therefore

$$\frac{\mu}{\mu_o} = -.038 + (.057) = +.019$$

Then

$$T = T_o + .458\nu_o + .019\mu_o + 20G$$

Now $\mu_o = (49\text{–}28) = 21\,°$F, since the minimum monthly average temperature of ground surface in January is 28°F for Hanover and $\nu_o = 70\text{–}49 = 21°F$. Finally

$$T = 49 + (.458 + .019)\ 21 + 0.2 = 59.2°F$$

Note that the periodic effect is quite small beneath the building.

7.3 TRANSIENT TEMPERATURES

In transient problems, the time is a variable, and one must consider the variation of temperature with time and position. Unlike the steady periodic problems, the temperature will change continuously with time. In many cases the calculation details do not differ from the steady-state cases, which are in fact simply the transient solutions as time approaches infinity.

7.3.1 Sudden Change of Surface Temperature

The transient temperature beneath the apex of a circular sector was given by Lachenbruch (1957), as discussed in Section 7.1.4.

If the temperature of the area in question suddenly increases or decreases with regard to the surrounding temperature, then Fig. 7.12 may be used with the appropriate values of $z/2\sqrt{\alpha t}$. The calculation procedure is identical with the previous cases for irregular or regular areas. The following examples, taken from Brown (1963), will help to fix the ideas in mind.

Example 15: A building 40 × 100, in Ottawa, Canada, has its surface temperature increased to 70°F at time zero. The soil under the building has a thermal diffusivity $\alpha = .036$ ft²/hr. For Ottawa, the mean annual ground surface temperature is about 50°F. Calculate the temperature at the center of the building at a depth of 20 ft after 1.7 years.

The rectangle may be broken into four quadrants and each quadrant subdivided as shown. Then using Fig. 7.12, the table and calculations follow

R	R/Z	$\frac{360}{\theta} \cdot \frac{\nu}{\nu_o}$
50.1	2.51	.500
51.8	2.59	.505
47.8	2.39	.495
35.2	1.76	.430
28.4	1.42	.370
24.6	1.23	.325
22.1	1.11	.295
20.8	1.04	.280
20.1	1.005	.265

Now these total to $\frac{360}{\theta} \frac{\nu}{\nu_o} = 3.465$. Thus $\frac{\nu}{\nu_o} = .0963$ for each quadrant.

$$\frac{\nu}{\nu_o} = 4 \times .0963 = 0.385$$

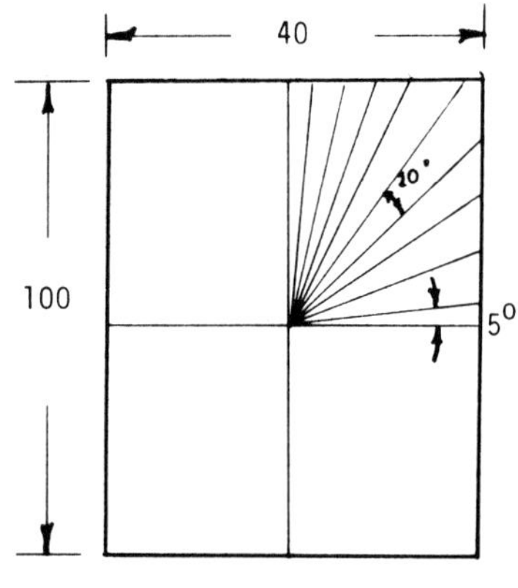

$z = 20$ ft
$G = .01$°F/ft
$\alpha = .036$ ft²/hr
$t = 1.7$ yr $= 14892$ hr
$\therefore z/2\sqrt{\alpha t} = 0.43$

The temperature is

$$T = 50 + 20\,(.385) = 50 + 7.7 = 57.7°F + .2 = 57.9°F$$

7.3.2 Step Change of Surface Temperature

Let the entire surface $z = 0$ be at zero temperature and then one-half of the plane is suddenly raised to v_o indefinitely. The solution to this problem is given by Carslaw and Jaeger (1959) as

$$v = \frac{v_o}{\pi}\left\{\beta + 2\sum_{n=1}^{\infty} (-1)^n \sin(n\beta) \int_0^{\infty} e^{-\alpha u^2 t} \frac{J_n(ur)\,du}{u}\right\} \tag{7.129}$$

where

$$\beta = \tan^{-1}\left(\frac{z}{x}\right)$$

$r^2 = x^2 + z^2$
J_n = Bessel function of first kind of order n and
u = dummy variable.

Carslaw and Jaeger use a wedge of angle θ_o. In Eq. (7.129) the angle is limited to $\theta_o = \pi$.

7.3.2.1 Temperature under a Strip

By replacing x by $(x - a)$ and $(-x - a)$ in Eq. (7.129), the temperature under a strip of width $2a$ may be found. Since $v = 2v_o$ on the strip and v_o on the surrounding, we must subtract the effect of suddenly increasing the surface by v_o.

This is

$$v = v_o erfc\left(\frac{z}{2\sqrt{\alpha t}}\right) \tag{7.130}$$

as noted in Section 7.2.4.
Thus the solution is

$$v = v_1 + v_2 - v_o erfc\left(\frac{z}{2\sqrt{\alpha t}}\right) \tag{7.131}$$

where

ν_1 is Eq. (7.128) with $x = x - a$ and
ν_2 is Eq. (7.128) with $x = -x - a$.

7.3.2.2 Thermal Effects of the Ocean

The problem of 7.3.2 can represent the effect of a sudden inundation of a land surface by the ocean or other large body of water. Initially let the temperature be the mean annual temperature of the ground at a given location. Thus the temperature is initially constant at T_e (Fig. 7.32). At time $t = 0$, let the positive half of the x plane jump to T_s and remain there. Let $\theta = T - T_e$.

Then the problem can be stated as follows

$$\frac{\partial^2\theta}{\partial x^2} + \frac{\partial^2\theta}{\partial z^2} = \frac{1}{\alpha}\frac{\partial\theta}{\partial t} \tag{7.132}$$

$$\begin{aligned}
&\theta(x,z,o) = 0 \\
&\theta(x,o,t) = 0 \qquad x < 0 \\
&\theta_o(x,o,t) = (T_s - T_e) \qquad x > 0 \\
&\lim_{z\to\infty} \theta(x,z,t) = 0
\end{aligned}$$

Lachenbruch (1957*b*) obtained a solution by using Green's function

$$\theta = \theta_o \, \psi\left(\frac{x}{z}, m\right) \tag{7.133}$$

$$\psi = \frac{1}{2}\, erfc\sqrt{m} + \frac{1}{\pi}\int_o^{x/z} \frac{e^{-m(1+v^2)}dv}{1 + v^2} \tag{7.134}$$

where $m = z^2/4\alpha t$.

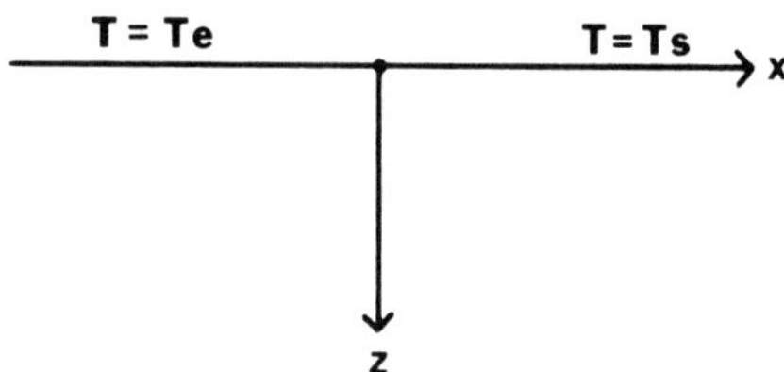

Fig. 7.32. Surface Temperatures for Land-Ocean.

Table 7.5. The Function $\psi\left(\frac{x}{z}, m\right), \frac{x}{z} \leq 0$ (after Lachenbruch 1957*b*).

$\frac{x}{z}$ \ m	0.0000	0.0001	0.0004	0.001	0.002	0.005	0.01	0.02	0.03	0.04	0.05	0.06	0.07	0.08	0.09	0.10
0.0	0.500	0.494	0.489	0.482	0.475	0.460	0.444	0.421	0.403	0.389	0.376	0.365	0.354	0.345	0.336	0.327
−0.2	0.437	0.432	0.426	0.419	0.412	0.398	0.382	0.359	0.342	0.328	0.316	0.305	0.296	0.287	0.278	0.271
−0.4	0.379	0.373	0.368	0.361	0.354	0.340	0.324	0.302	0.286	0.273	0.261	0.251	0.242	0.233	0.225	0.218
−0.6	0.328	0.322	0.317	0.310	0.303	0.289	0.274	0.252	0.237	0.224	0.213	0.204	0.195	0.187	0.180	0.173
−0.8	0.285	0.279	0.274	0.267	0.260	0.246	0.231	0.211	0.196	0.184	0.173	0.164	0.156	0.149	0.142	0.137
−1.0	0.250	0.244	0.239	0.232	0.225	0.212	0.197	0.177	0.163	0.151	0.141	0.133	0.125	0.119	0.113	0.107
−1.2	0.221	0.216	0.210	0.204	0.197	0.183	0.169	0.149	0.136	0.125	0.115	0.108	0.101	0.095	0.089	0.084
−1.4	0.197	0.192	0.186	0.180	0.173	0.160	0.146	0.127	0.114	0.103	0.095	0.087	0.081	0.075	0.070	0.066
−1.6	0.178	0.172	0.167	0.160	0.154	0.140	0.127	0.109	0.096	0.086	0.078	0.071	0.065	0.060	0.056	0.052
−1.8	0.161	0.156	0.150	0.144	0.137	0.124	0.111	0.093	0.081	0.072	0.065	0.058	0.053	0.048	0.044	0.041
−2.0	0.148	0.142	0.137	0.130	0.124	0.111	0.098	0.081	0.069	0.061	0.054	0.048	0.043	0.039	0.035	0.032
−2.5	0.121	0.116	0.110	0.104	0.097	0.085	0.072	0.057	0.047	0.040	0.034	0.029	0.026	0.022	0.020	0.017
−3.0	0.102	0.097	0.092	0.086	0.079	0.067	0.056	0.042	0.033	0.027	0.022	0.018	0.015	0.013	0.011	0.009
−4.0	0.078	0.072	0.067	0.061	0.055	0.044	0.034	0.023	0.016	0.012	0.009	0.007	0.006	0.004	0.003	0.003
−5.0	0.063	0.057	0.052	0.047	0.041	0.031	0.022	0.013	0.008	0.005	0.004	0.002	0.002	0.001	0.001	0.001
−6.0	0.053	0.047	0.042	0.037	0.031	0.022	0.014	0.007	0.004	0.002	0.001	0.001	0.001	0.000	0.000	0.000
−8.0	0.040	0.034	0.029	0.024	0.019	0.012	0.006	0.002	0.001	0.000	0.000	0.000	0.000	0.000	0.000	0.000
−10.0	0.032	0.026	0.022	0.017	0.013	0.007	0.003	0.001	0.000	0.000	0.000	0.000	0.000	0.000	0.000	0.000
− ∞	0.000	0.000	0.000	0.000	0.000	0.000	0.000	0.000	0.000	0.000	0.000	0.000	0.000	0.000	0.000	0.000

Table 7.6. The Function $\psi\left(\frac{x}{z}, m\right), \frac{x}{z} \geq 0$ (after Lachenbruch 1957*b*).

$\frac{x}{z}$ \ m	0.0000	0.0001	0.0004	0.001	0.002	0.005	0.01	0.02	0.03	0.04	0.05	0.06	0.07	0.08	0.09	0.10
0.0	0.500	0.494	0.489	0.482	0.475	0.460	0.444	0.421	0.403	0.389	0.376	0.365	0.354	0.345	0.336	0.327
0.2	0.563	0.557	0.552	0.545	0.537	0.523	0.506	0.482	0.464	0.449	0.436	0.424	0.413	0.402	0.393	0.384
0.4	0.621	0.615	0.610	0.603	0.596	0.581	0.564	0.539	0.521	0.505	0.491	0.478	0.467	0.456	0.446	0.436
0.6	0.672	0.666	0.661	0.654	0.646	0.631	0.614	0.589	0.570	0.553	0.539	0.526	0.513	0.502	0.491	0.481
0.8	0.715	0.709	0.704	0.697	0.689	0.674	0.656	0.630	0.611	0.594	0.579	0.565	0.552	0.540	0.529	0.518
1.0	0.750	0.744	0.739	0.732	0.724	0.709	0.691	0.664	0.644	0.626	0.611	0.596	0.583	0.570	0.559	0.548
1.2	0.779	0.773	0.767	0.761	0.753	0.737	0.719	0.692	0.671	0.653	0.636	0.621	0.608	0.595	0.582	0.571
1.4	0.803	0.797	0.791	0.784	0.776	0.761	0.742	0.715	0.693	0.674	0.657	0.642	0.627	0.614	0.601	0.589
1.6	0.822	0.816	0.811	0.804	0.796	0.780	0.761	0.733	0.711	0.691	0.674	0.658	0.643	0.629	0.616	0.603
1.8	0.839	0.833	0.827	0.820	0.812	0.796	0.777	0.748	0.725	0.705	0.687	0.671	0.655	0.641	0.627	0.614
2.0	0.852	0.847	0.841	0.834	0.826	0.809	0.790	0.761	0.737	0.717	0.698	0.681	0.665	0.650	0.636	0.623
2.5	0.879	0.873	0.867	0.860	0.852	0.835	0.815	0.784	0.759	0.738	0.718	0.700	0.683	0.667	0.652	0.637
3.0	0.898	0.892	0.886	0.879	0.870	0.853	0.832	0.800	0.774	0.751	0.730	0.711	0.693	0.676	0.660	0.645
4.0	0.922	0.916	0.910	0.903	0.894	0.876	0.853	0.819	0.790	0.765	0.743	0.722	0.703	0.685	0.668	0.652
5.0	0.937	0.931	0.925	0.918	0.909	0.890	0.866	0.829	0.798	0.772	0.748	0.727	0.707	0.688	0.670	0.654
6.0	0.947	0.942	0.935	0.928	0.918	0.898	0.873	0.834	0.803	0.775	0.751	0.728	0.708	0.689	0.671	0.655
8.0	0.960	0.955	0.948	0.940	0.930	0.908	0.881	0.839	0.806	0.777	0.752	0.729	0.708	0.689	0.671	0.655
10.0	0.968	0.962	0.956	0.947	0.937	0.914	0.884	0.841	0.806	0.777	0.752	0.729	0.708	0.689	0.671	0.655
∞	1.000	0.989	0.977	0.964	0.950	0.920	0.888	0.842	0.807	0.777	0.752	0.279	0.708	0.689	0.671	0.655

The function ψ has been tabulated for values of m, of interest in geological systems, as given in Tables 7.5 and 7.6.
The steady state value of Eq. (7.133) is

$$\theta(x,z) = \theta_o \left[\frac{1}{2} + \frac{1}{\pi} \tan^{-1}(x/z)\right]$$

which is equivalent to Eq. (7.5).

Equation (7.133) can be modified to include the effect of the geothermal gradient as follows

$$\theta(x,z,t) = \theta_o\psi - \frac{T_e}{z_p} z \qquad (7.135)$$

where

T_e = initial mean ground surface temperature and
z_p = permafrost thickness associated with T_e.

This relation assumes that the initial ground temperature is a linear function of depth.

The temperature disturbance beneath a semi-infinite strip of width a, with a temperature θ_o relative to the surroundings can be found from Eq. (7.133) by superposition

$$\theta(x,z,t) = \theta_o \left\{\psi\left(\frac{x}{z}, m\right) - \psi\left(\frac{x-a}{z}, m\right)\right. \qquad (7.136)$$

The application of Eq. (7.135) to the effect of ocean inundation is discussed in Chap. 4.

REFERENCES

Andrews, R. V. 1955. Solving conductive heat transfer problems with electrical analogue shape factors. *Chem. Eng. Prog.* 51 (2):67.

Brown, W. G. 1963. *Graphical determination of temperature under heated or cooled areas on the ground surface.* NRC-DBR Tech. Paper No. 163.

———. 1962. The steady temperature under a corner of a plate and under polygonal areas on the surface of a semi-infinite solid. *Q. J. Mech. Appl. Math.* 14 (part 5):393–399.

Caratheodory, C. 1958. *Conformal representation.* New York: Cambridge Univ. Press.

Carslaw, H. S., and Jaeger, J. C. 1959. *Conduction of heat in solids.* 2nd ed. Oxford: Clarendon Press, pp. 419–420.

Churchill, R. V. 1960. *Complex variables and applications.* 2nd ed. New York: McGraw-Hill.

Lachenbruch, A. H. 1959. *Periodic heat flow in a stratified medium with application to permafrost problems.* U.S. Geological Survey Bulletin 1083-A.

———. 1957. *Three-dimensional heat conduction in permafrost beneath heated buildings.* Geol. Survey Bull. No. 1052-B. Washington: U.S. Govt. Printing Office.

———. 1957*b*. Thermal effects of the ocean on permafrost. *Bull. Geol. Soc. Am.* 68:1515–1530.

Rudenberg, R. 1925. Die Ausbreitung der Luft-und Erdfelder um Hochspannungsleitungen besonders bei Erd-und Kurzschlussen. *Elektrotech Z.* 46:1342.

Sv. Skaven-Haug. 1959. Protection against frost heaving on the Norwegian railways. *Geotechnique* 9:87–106.

Thryagarajan, R., and Yovanovich, M. M. 1974. Thermal resistance of a buried cylinder with constant flux boundary condition. *J. Heat Transfer* (2):249–250.

Thornton, D. E. 1977. *Calculation of heat loss from pipes,* in *Utilities delivery in arctic regions.* EPS, Environment Canada, Rept. EPS 3—WP-77-1:131–150.

PROBLEMS

1. Consider an infinite strip with the following details

$$\nu_o = 10$$
$$a = 500$$
$$G = .01\,°\text{F/ft}$$

   Plot the isotherms for ν/ν_o = .3, .5, .7, .8, .96, 1.0. Compare this plot with the graphical method.
2. A building has a surface area of 40 × 100 ft, with a temperature of 65°F and an outside temperature of 40°F. Calculate the temperature at a depth of 20 ft at the midpoint of the long side.
3. Calculate the temperature at z = 10 ft on the circumference of the circular lake of example problem six.
4. Find the radius of the thaw bowl for a 3-in. diameter pipe buried 30 in., with a pipe temperature of 42°F and a ground surface temperature of 21°F. Plot the heat loss for a coarse-grained soil as a function of density if the water content is 10%.
5. Repeat problem 4 for T_o = 55°F.
6. Repeat problem 4 for h_o = 15 in.
7. Four inches of insulation with k_i = 0.1 Btu/hr-ft-°F are used on a pipe with D = 3 in., T_o = 42°F, T_s =21°F. Calculate the heat loss for h_o = 72, 30, and 15 in.
8. Find the temperature 20 ft beneath the center of a 40 × 100 ft building in Ottawa with an inside temperature of 70°F if the soil is,
   (a) Peat, γ_d = 0.6 g/cm³, W = 150%
   (b) Sand, γ_d = 2.0, W = 8%
   (c) Clay, γ_d = 1.3, W = 25%
9. A layer of silt underlies a layer of gravel at Baker Lake, NWT. Find the gravel depth needed to keep the bottom of the gravel frozen. Compare this with the homogeneous case.
10. Repeat problem 9 for Churchill, Manitoba.

11. Assume that at Churchill, Manitoba, a layer of sandy gravel overlies a frozen organic silty clay. Calculate the gravel depth to keep the bottom of the gravel at the freezing point, and compare this to the homogeneous case.
12. Consider a 1-ft layer of spruce logs between a frozen organic silty clay below and a sandy gravel above. Find the gravel depth needed at Baker Lake and at Hay River, NWT. Compare the three cases of one, two, and three layers.
13. Find the midsummer temperature 20 ft beneath the center of a 40 × 100 ft building in Hanover, N.H. The inside floor temperature is 70°F.
14. Repeat problem 13 for a cold storage building with a floor temperature of 0°F.
15. Consider the annual surface temperature of Hanover, N.H., to be a sinusoidal curve as shown. Calculate the temperature of a homogeneous gravel with $\gamma_d = 1.6\ g/cm^3$, $W = 15\%$, at a depth of 2 ft in January and July. Consider the system to be without phase change effects.

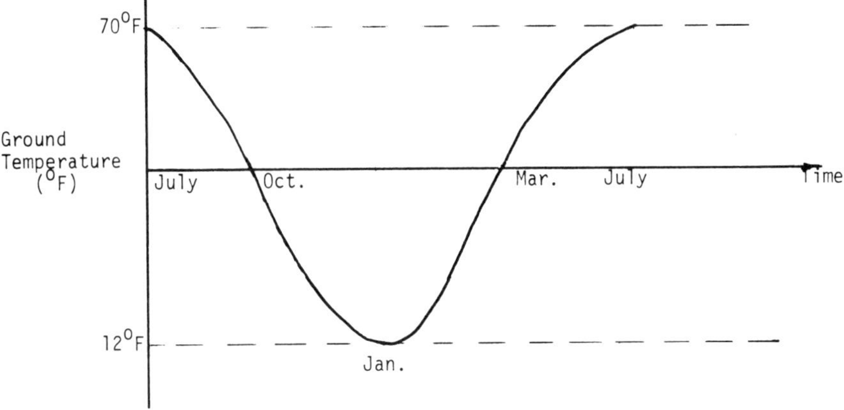

16. In geothermal heat transfer problems, discuss the effect of phase change on the solution methods detailed in this chapter. What important solution technique is now excluded and why?
17. Do a two-layer problem for heat transfer by conduction through a semi-infinite solid, using the complex temperature method. Find the final expression for the temperature distribution θ_1 in the upper layer.
18. What is the mean temperature of a circular lake of 1000-ft radius if the temperature of the ground beneath the center of the lake, at the 100-ft depth, is 32°F. The mean ground surface temperature is 22°F.
19. For problem 3, find the temperature at $z = 100$ ft.
20. For problem 3 a point is located 2000 ft from the center of the lake. Find the temperature at a depth of 100 ft.
21. Repeat problem 13 for a location on the midpoint of the long side.
22. Repeat problem 14 for a location on the midpoint of the long side.
23. Find the limits of the temperature penetration for daily and yearly periods for

fine and coarse-grained soils. Consider frozen and unfrozen cases without phase change.

24. Calculate the rate of heat loss (per unit length) from a plastic pipe of outside diameter 6.54 in. and inside diameter 5.4 in. whose thermal conductivity is .7 Btu/hr-ft-°F and is encased in 2 in. of polyurethane foam of thermal conductivity .013 when the pipe contains water at 40°F and the aboveground pipe is exposed to an air temperature of −40°F and the wind speed is 15 mph.
25. Calculate the rate of heat loss (per unit length) for the pipe in problem 24 if the pipe is buried at a depth of 4 ft in soil with a thermal conductivity of 1.16 Btu/hr-ft-°F when the fluid temperature is 40°F and the (steady-state) ground surface temperature is −40°F.
26. In example 9, suppose the same volume of insulation was used directly on the pipe, instead of as a duct liner. Calculate the heat loss.
27. In example 10, the water in the cold pipe closes to flow. Estimate the rate of heat gain of the cold pipe and the final utilidor air temperature.
28. A metal pipe of outside diameter of 6 in. is buried 4 ft below grade in a clay soil with thawed and frozen thermal conductivities of 0.60 and 1.0 Btu/hr-ft-°F, respectively. Calculate the mean size of the thawed zone and the average rate of heat loss if the mean ground temperature is 27.5°F and water at 45°F is circulated through the pipe.
29. A metal pipe of external diameter 6 in. is buried in permafrost (k = 1 Btu/hr-ft-°F) with its axis 4 ft below the ground surface whose mean temperature is 25°F. What is the minimum thickness of polyurethane foam insulation (k_1 = .014 Btu/hr-ft-°F) that will maintain the soil in a frozen state (on average) if water is flowing through the pipe at a mean temperature of 45°F? What is the average rate of heat loss if this insulation thickness is used?
30. For the buried insulated pipe in problem 29, estimate the maximum thawing under the pipe (in the fall) if the maximum thaw depth (active layer) is 4 ft and the volumetric heat capacity of the frozen soil is 30 Btu/ft³°F.

8 Analytical Methods for Conduction with Phase Change

8.1 BASIC EQUATIONS

The question of phase change occurs quite regularly in engineering problems dealing with permafrost. Because the system is frozen, or near the phase change temperature, a small change in the thermal regime can cause melting or solidification. Other important applications can be found in thermal storage systems for solar energy, the freezing of food or biological material, and the solidification or freezing of metals. During a melting or freezing process, the system will be divided into regions separated by a phase change interface, or a phase change region, that is usually at the phase change (fusion) temperature. In general, and certainly for permafrost, the thermal properties of the frozen and unfrozen regions are different, but are not strong functions of temperature. The transient nature of certain quantities is of particular interest; among these are the surface temperature, the boundary heat fluxes, and the location of the phase change interface. Figure 8.1 shows a schematic of the processes of freezing and thawing for a homogenous, semi-infinite body. The concepts of solid and liquid relate to the thermodynamic state of the water contained in the pores of a soil system. The basic concept and definitions of permafrost have been discussed in Chap. 4.

The usual problem is that of a region initially liquid at a temperature T_o, which has its surface temperature suddenly dropped to T_s for $t > 0$, but other boundary conditions are possible at $x = 0$. The material freezes at T_f, and the solidification front is denoted by $X(t)$. The process of thawing is obviously similar. Consider the case of melting. For the liquid region, convection is possible and

$$\frac{\partial}{\partial x}\left(k_l \frac{\partial T_l}{\partial x}\right) = \rho_l c_l \frac{\partial T_l}{\partial t} + \rho_l c_l\, v \frac{\partial T_l}{\partial x} \tag{8.1}$$

The solid phase may be assumed stationary and impervious to liquid motion

$$\frac{\partial}{\partial x}\left(k_2 \frac{\partial T_2}{\partial x}\right) = \rho_2 C_2 \frac{\partial T_2}{\partial t} \tag{8.2}$$

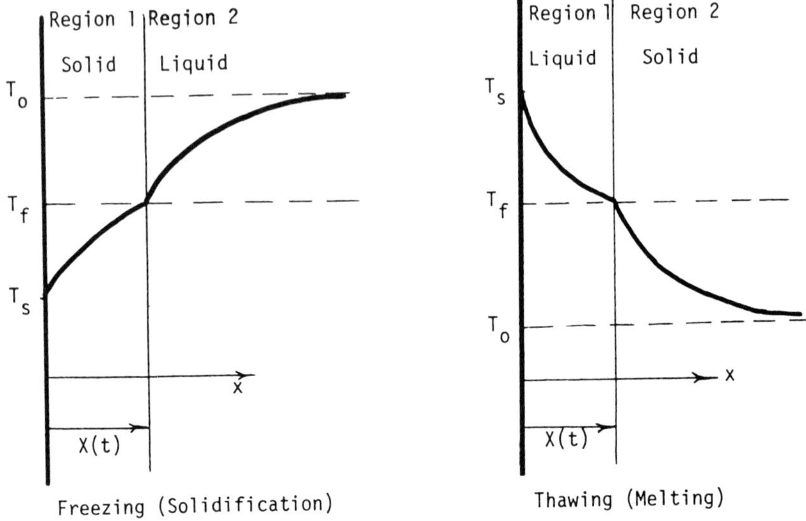

Fig. 8.1. Freezing and Thawing Relations.

The bulk velocity, v, in the liquid is due to the volume change during melting caused by the density differences of the two phases. Consider a volume of solid, $A\Delta X$, which melts in the time Δt. The mass of the material in the volume that changes phase is constant and

$$A\,v\,\Delta T = V_2 - V_1 = m\left(\frac{1}{\rho_2} - \frac{1}{\rho_1}\right)$$

Then

$$v = \left(1 - \frac{\rho_2}{\rho_1}\right)\frac{dX}{dt} \tag{8.3}$$

The energy equation for the growing phase is then

$$\frac{\partial}{\partial x}\left(k_1\frac{\partial T_1}{\partial x}\right) = \rho_1 c_1\left[\frac{\partial T_1}{\partial t} + \left(1 - \frac{\rho_2}{\rho_1}\right)\frac{dX}{dt}\frac{\partial T_1}{\partial x}\right] \tag{8.4}$$

If the thermal conductivity of each phase is constant, then Eqs. (8.1) and (8.2) are

$$\frac{\partial^2 T_1}{\partial x^2} = \frac{1}{\alpha_1}\left[\frac{\partial T_1}{\partial t} + \left(1 - \frac{\rho_2}{\rho_1}\right)\frac{dX}{dt}\frac{\partial T_1}{\partial x}\right] \tag{8.5}$$

$$\frac{\partial^2 T_2}{\partial x^2} = \frac{1}{\alpha_2}\frac{\partial T_2}{\partial t} \tag{8.6}$$

These may be thought of as the constant property forms of the energy equation. Though the constant conductivity assumption is not unreasonable, there are two cases of variable conductivity that can be handled without too much difficulty.

Case I: Temperature variation of thermal conductivity.

A new temperature can be defined as

$$\phi = \int_{T_0}^{T} k(T')\, dT'$$

Equation (8.2) can be transformed to

$$\frac{\partial^2 \phi_2}{\partial x^2} = \frac{\rho_2 c_2}{k_2(T)}\frac{\partial \phi_2}{\partial t} = \frac{1}{\alpha_2(\phi_2)}\frac{\partial \phi_2}{\partial t}$$

This is of the same form as the constant property equation and can be handled in the same way.

Case II: Spatial variation of thermal conductivity.

For this case a new space variable is defined

$$x_1 = \int_0^x \frac{dx}{k(x)}$$

The transformation of Eq. (8.2) is now

$$\frac{\partial^2 T_2}{\partial x_1^2} = k_2(x)\,\rho_2 c_2 \frac{\partial T_2}{\partial t} = \frac{1}{\alpha(x_1)}\frac{\partial T_2}{\partial t}$$

This equation can also be considered in the form of the constant property equation, with a new thermal diffusivity. In either of these cases the "thermal diffusivity" may be a function of the independent variables.

Normally, it is assumed that the density of the frozen and unfrozen phases are identical, and thus there is no convection and the problem is one of pure conduction. The basic equations are then

$$\frac{\partial^2 T_1}{\partial x^2} = \frac{1}{\alpha_1}\frac{\partial T_1}{\partial t} \qquad 0 \le x < X(t) \tag{8.7}$$

$$\frac{\partial^2 T_2}{\partial x^2} = \frac{1}{\alpha_2}\frac{\partial T_2}{\partial t} \qquad x > X(t) \tag{8.8}$$

where the thermal diffusivity may be a variable.

$$T_1(X,t) = T_2(X,t) = T_f \tag{8.9a}$$

$$T_1(x,0) = T_2(x,0) = T_o \tag{8.9b}$$

Equation (8.9*a*) describes the natural condition that the frozen and unfrozen layers equal the fusion temperature at the phase change boundary where phase change is occurring.

8.1.1 Energy Balance at the Phase Change Interface

At the interface between the phases, energy will be released or absorbed as the material freezes or thaws, respectively (Fig. 8.2). The conservation of energy applied to the mass m of the volume $A\Delta X$ that undergoes phase change during Δt is

$$(q_1 - q_2)\,\Delta t + q\Delta t - W = \Delta E$$

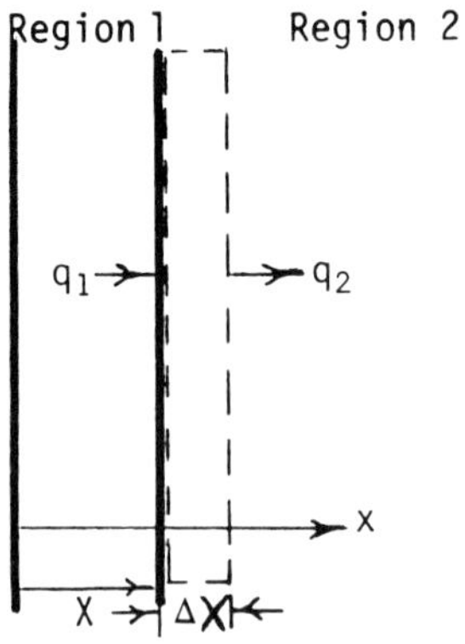

Fig. 8.2. Energy Flow at Phase Change Interface.

In this relation q_1, q_2 are conduction heat transfers, and q may represent other heat flows. The work associated with the volume change of the mass is W. The energy balance is

$$\left(-k_1\frac{\partial T_1}{\partial x}+k_2\frac{\partial T_2}{\partial x}\right)A\Delta t+p\,A\Delta X\left(1-\frac{\rho_2}{\rho_1}\right)+qA\Delta t=m\,(h^f-h^i)$$

The enthalpy change from the solid to the liquid phase is the latent heat of phase change l

$$-k_1\frac{\partial T_1}{\partial x}+k_2\frac{\partial T_2}{\partial x}+p\left(1-\frac{\rho_2}{\rho_1}\right)\frac{dX}{dt}+q=\rho_2 l\frac{dX}{dt} \qquad (8.10a)$$

A similar relation holds for the solidification process

$$-k_1\frac{\partial T_1}{\partial x}+k_2\frac{\partial T_2}{\partial x}+p\left(1-\frac{\rho_1}{\rho_2}\right)\frac{dX}{dt}+q=-\rho_1 l\frac{dX}{dt} \qquad (8.10b)$$

The work term in Eqs. (8.10) may be neglected if the system pressure p is much less than 30,000 atm. Thus this term is normally not considered, and Eqs. (8.10a) and (8.10b) may be written

$$-k_1\frac{\partial T_1}{\partial x}+k_2\frac{\partial T_2}{\partial x}=\pm\rho l\frac{dX}{dt} \qquad x=X \qquad (8.11)$$

where the upper sign is for melting and the lower sign is for freezing.

The derivation assumes that the interface motion is in the positive direction of the space coordinate. If this is reversed, as for solidification of a cylinder of water, then the signs for the latent heat are reversed in Eq. (8.10).

8.1.2 Nonlinearity of the Problems

The phase change introduces a basic nonlinearity into the boundary conditions of the problem. Consider Eq. (8.9a), the temperature at the phase change interface

$$T(x,t)_{x=X}=T_f$$

The differential of this equation is zero as T_f is a constant

$$dT=\frac{\partial T}{\partial x}dx+\frac{\partial T}{\partial t}dt=0$$

Now evaluate this at $x = X$ and

$$\left(\frac{\partial T}{\partial x}\right)\frac{dX}{dt} = -\frac{\partial T}{dt}$$

Combine this result with Eqs. (8.7) and (8.11) to obtain a boundary condition in the form

$$-k_1\left(\frac{\partial T_1}{\partial x}\right)^2 + k_2\left(\frac{\partial T_2}{\partial x}\right)\frac{\partial T_1}{\partial x} = \mp \rho l \alpha_1 \frac{\partial^2 T_1}{\partial x^2}, \qquad x = X \tag{8.12}$$

where the upper sign is again for melting. The nonlinear nature of the problem is quite obvious in this form. This inherent nonlinearity, along with the unknown motion of the phase change interface, has limited the exact solutions of phase change problems to a mere handful of cases with particularly simple geometries and boundary conditions. It has also motivated considerable interest in reliable, approximate procedures. After examining the exact solutions (these don't include numerical solutions), we shall study the approximate techniques available.

8.1.3 General Interface Conditions

Following Patel (1968) for a three-dimensional system, the interface energy balance, Eq. (8.11), can be generalized to

$$k_s\frac{\partial T_s}{\partial n} - k_L\frac{\partial T_L}{\partial n} = \rho_s v_n l - Q_n \tag{8.12a}$$

$$T_i(x,y,z,t) = T_f \qquad i = s \text{ and } L \tag{8.12b}$$

$$\text{on } f(x,y,z,t) = 0$$

where

n = outward normal to the interface, into the liquid
v_n = velocity of the interface in the normal direction
Q_n = heat flux, such as radiation, normal to interface and
$f(x,y,x,t)$ = function describing the interface separating the solid and liquid regions.

The interface position function is defined as

$$f(x,y,z,t) = 0 \tag{8.13}$$

For computational purposes, it is desirable to write Eq. (8.12a) in terms of the independent variables x,y,z. The following relations are valid on the interface

$$\bar{n} = \nabla f/|\nabla f| = \nabla T_i/|\nabla T_i|$$

$$\frac{\partial T_i}{\partial n} = \nabla T_i \cdot \bar{n} = \nabla T_i \cdot \nabla f/|\nabla f| = |\nabla T_i|$$

$$v_n = \bar{v} \cdot \bar{n} = \bar{v} \cdot \nabla f/|\nabla f| = \bar{v} \cdot \nabla T_i/|\nabla T_i|$$

$$Q_n = \bar{Q} \cdot \bar{n}/|\nabla f| = \bar{Q} \cdot \nabla T_i/|\nabla T_i|$$

Using the relations

$$dT_i = \frac{\partial T_i}{\partial x}\,dx + \frac{\partial T_i}{\partial y}\,dy + \frac{\partial T_i}{\partial z}\,dz + \frac{\partial T_i}{\partial t}\,dt$$

$$df = \frac{\partial f}{\partial z}\,dx + \frac{\partial f}{\partial y}\,dy + \frac{\partial f}{\partial z}\,dz + \frac{\partial f}{\partial t}\,dt$$

and Eq. (8.12), it is possible to obtain

$$v_n = -\frac{\partial f}{\partial t}/|\nabla f| = -\frac{\partial T_i}{\partial t}/|\nabla T|_i$$

Then Eq. (8.12a) can be written

$$k_s \nabla T_s \cdot \nabla f - k_L \nabla T_L \cdot \nabla f = -\rho_s l \frac{\partial f}{\partial t} - \bar{Q} \cdot \nabla f$$

or

$$(k_s \operatorname{grad} T_s - k_L \operatorname{grad} T_L) \cdot \operatorname{grad} f = -\rho_s l \frac{\partial f}{\partial t} - \bar{Q} \cdot \nabla f$$

or

$$k \operatorname{grad} T\Big]_L^S \cdot \operatorname{grad} f = -\rho_s l \frac{\partial f}{\partial t} - \bar{Q} \cdot \nabla f$$

It is often convenient to express all the derivatives on the interface in terms of one direction. Differentiate Eqs. (8.12b) and (8.13) to obtain

$$\frac{\partial T_i}{\partial x} = \frac{\partial f/\partial x}{\partial f/\partial z}\frac{\partial T_i}{\partial z} \tag{8.14a}$$

$$\frac{\partial T_i}{\partial y} = \frac{\partial f/\partial y}{\partial f/\partial z}\frac{\partial T_i}{\partial z} \tag{8.14b}$$

Using Eq. (8.14) the general interface condition is then

$$\left[1+\left(\frac{\partial f/\partial x}{\partial f/\partial z}\right)^2+\left(\frac{\partial f/\partial y}{\partial f/\partial z}\right)^2\right]\left(k_s\frac{\partial T_s}{\partial z}-k_L\frac{\partial T_L}{\partial z}\right)=-\rho_s l\frac{\partial f/\partial t}{\partial f/\partial z} - \left[\frac{\partial f/\partial x}{\partial f/\partial z}Q_z+\frac{\partial f/\partial y}{\partial f/\partial z}Q_y+Q_z\right] \tag{8.15}$$

If $\bar{Q}=0$ and $f(x,y,z,t) = z - s(x,y,t) = 0$, then Eq. (8.15) is

$$\left[1+\left(\frac{\partial s}{\partial x}\right)^2+\left(\frac{\partial s}{\partial y}\right)^2\right]\left(k_s\frac{\partial T_s}{\partial z}-k_L\frac{\partial T_L}{\partial z}\right)=-\rho_s l\frac{\partial s}{\partial t}$$

The two-dimensional form of Eq. (8.15) is

$$\left[1+\left(\frac{\partial f/\partial x}{\partial f/\partial y}\right)^2\right]\left(k_s\frac{\partial T_s}{\partial y}-k_L\frac{\partial T_L}{\partial y}\right)=-\rho_s l\frac{\partial f/\partial t}{\partial f/\partial y}-\left[\frac{\partial f/\partial x}{\partial f/\partial y}Q_x+Q_y\right]$$

This equation is in a convenient form for the isotherm migration method of Chap. 9 and may be expressed as

$$\left[1+\left(\frac{\partial y}{\partial x}\right)^2\right]\left(k_s\frac{\partial T_s}{\partial y}-k_L\frac{\partial T_L}{\partial y}\right)=-\rho_s l\frac{\partial y}{\partial t}-\left[\frac{\partial y}{\partial t}Q_x+Q_y\right]$$

or

$$-\frac{\partial y}{\partial t}(\rho_s l+Q_x)=\left[1+\left(\frac{\partial y}{\partial x}\right)^2\right]\left[k_s\left(\frac{\partial y}{\partial T_s}\right)^{-1}-k_L\left(\frac{\partial y}{\partial T_L}\right)^{-1}\right]+Q_y \tag{8.16}$$

8.2 AN EXACT SOLUTION

Due to the nonlinearity of the phase change system, there are very few complete, analytic solutions. The first and still the most comprehensive solution is due to Neumann (*ca.* 1860), generalized in Carslaw and Jaeger (1959). It is instructive to examine, in detail, this classic problem and its solution.

8.2.1 The Neumann Problem

Let us look at the exact solution of a classical problem in phase change often referred to as Neumann's solution (Fig. 8.3). Initially, a semi-infinite region is at a constant temperature T_o, and the temperature of the surface is suddenly dropped to T_s and held constant (step change at surface). It is assumed that initially the medium is in a liquid state, i.e., $T_o > T_f$, where

T_o = initial temperature
T_f = fusion temperature of medium and
T_s = surface temperature.

The problem can then be formulated as

$$\frac{\partial^2 T_1}{\partial x^2} = \frac{1}{\alpha_1}\frac{\partial T_1}{\partial t} \tag{8.17}$$

$$\frac{\partial^2 T_2}{\partial x_2} = \frac{1}{\alpha_2}\frac{\partial T_2}{\partial t} \tag{8.18}$$

$$\lim_{x\to\infty} T_2 = T_o \tag{8.18a}$$

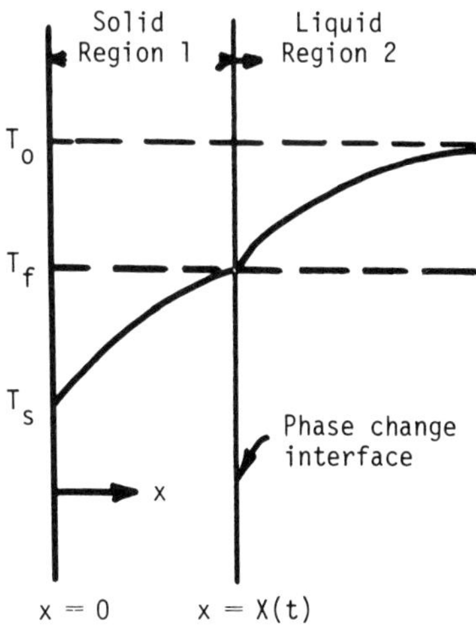

Fig. 8.3 Temperature Distribution in Partially Frozen Medium.

$$T_1(0,t) = T_s \tag{8.17a}$$

The location of the freezing interface is a function of time, and the temperatures at this location of the liquid and solid are both equal to the fusion temperature.

$$T_1(X,t) = T_2(X,t) = T_f \tag{8.17b}$$

Equation (8.11) becomes

$$k_1 \frac{\partial T_1}{\partial x} - k_2 \frac{\partial T_2}{\partial x} = \rho l \frac{dX}{dt} \tag{8.17c}$$

where l = latent heat of fusion, mass basis.

A solution to this problem is obtainable by using a well-known similarity transformation. Let

$$\eta = \frac{x}{2\sqrt{\alpha_1 t}} \tag{8.19}$$

Equation (8.17) then becomes

$$\frac{d^2T_1}{d\eta^2} + 2\eta \frac{dT_1}{d\eta} = 0 \tag{8.20}$$

The first integral of Eq. (8.20) can be obtained immediately

$$\frac{dT_1}{d\eta} = ce^{-\eta^2}$$

A second integration gives a formal solution for the temperature

$$T_1 = c\int_0^\eta e^{-\beta^2}\, d\beta \tag{8.21}$$

The error function is defined as

$$erf\eta \equiv \frac{2}{\sqrt{\pi}} \int_0^\eta e^{-\beta^2}\, d\beta$$

The error function has been numerically evaluated and tabulated (Carslaw and Jaeger 1959). Thus, a formal solution to the conduction equation is available, if the similarity transformation is valid. This will be the case if the differential equation and all the boundary and initial conditions can be expressed in terms of the single, independent variable η. This is true for Eqs. (8.17) to (8.18).

A solution of Eqs. (8.17) and (8.17*a*) is then

$$T_l = T_s + A\, erf \frac{x}{2\sqrt{\alpha_l t}} \tag{8.22}$$

It also follows that

$$T_2 = T_o - B\left[1 - erf \frac{x}{2\sqrt{\alpha_2 t}}\right] = T_o - B\, erfc \frac{x}{2\sqrt{\alpha_2 t}} \tag{8.23}$$

The notation used in Eq. (8.23) is $erfc\ \eta \equiv 1 - erf\ \eta$. Then, using Eqs. (8.24) and (8.23) in Eq. (8.17*b*) leads to

$$A\, erf \frac{X}{2\sqrt{\alpha_l t}} = T_o - B\, erfc \frac{X}{2\sqrt{\alpha_2 t}} = T_f \tag{8.24}$$

Now, as Eq. (8.24) must be satisfied for all values of time, then X must be proportional to $\sqrt{t}$, for then

$$T_s + A\, erf \frac{a}{2\sqrt{\alpha_l}} = T_o - B\, erfc \frac{a}{2\sqrt{\alpha_2}} = T_f = \text{constant}$$

where it is assumed that $X \sim a\sqrt{t}$. Thus, we may write

$$X = 2\gamma\sqrt{\alpha_l t} \tag{8.25}$$

where γ is a constant. Now, from Eqs. (8.22) and (8.23)

$$\frac{\partial T_l}{\partial x} = A \frac{1}{\sqrt{\pi \alpha_l t}} \exp\left(-\frac{x^2}{4\alpha_l t}\right)$$

$$\frac{\partial T_2}{\partial x} = B \frac{1}{\sqrt{\pi \alpha_l t}} \exp\left(-\frac{x^2}{4\alpha_2 t}\right)$$

Thus from Eqs. (8.17*c*) and (8.25)

$$k_l A e^{-\gamma^2} - k_2 B\sqrt{\alpha_{12}}\, e^{-\alpha_{12}\gamma^2} = \alpha_l\, \rho l \gamma \sqrt{\pi}$$

where α_{12} denotes the ratio of α_1 to α_2. Then using Eqs. (8.24) and (8.25)

$$T_s + A\, erf\gamma = T_o - B\, erfc\, \gamma\sqrt{\alpha_{12}} = T_f$$

$$A = \frac{T_f - T_s}{erf\gamma}$$

$$B = \frac{(T_o - T_f)}{erfc\,(\gamma\sqrt{\alpha_{12}})}$$

and finally, the equation for the constant γ is

$$\frac{e^{-\gamma^2}}{erf\gamma} - \frac{k_{12}\sqrt{\alpha_{12}}\,(T_o - T_f)e^{-\alpha_{12}\gamma^2}}{(T_f - T_s)\; erfc\,\gamma\sqrt{\alpha_{12}}} = \frac{l\gamma\sqrt{\pi}}{c_l(T_f - T_s)} \tag{8.26}$$

With Eq. (8.26), for various properties and conditions, the value of γ may be numerically evaluated; this has been done, as given in Table 8.1, for water with the following properties:

Ice	Liquid Water
$k_1 = .0053$ cal/sec-cm-°C	$k_2 = .00144$
$\alpha_1 = .0115$ cm²/sec	$\alpha_2 = .00144$
$c_1 = 0.465$ cal/g-°C	$c_2 = 1.0$
$\rho l = 73.6$ cal/cm³	

The temperatures are now given by

$$T_l = T_s + \frac{(T_f - T_s)}{erf\gamma}\, erf\frac{x}{2\sqrt{\alpha_l t}} \tag{8.27}$$

Table 8.1. Value of γ, Eq. (8.26), for Water (adapted from Carslaw and Jaeger 1959).

$(T_f - T_s)$ °C	$(T_o - T_f)$°C 0	1	2	3	4	5
1	.056	.054	.053	.051	.050	.049
2	.079	.077	.076	.074	.073	.071
3	.097	.095	.093	.091	.090	.088
4	.111	.110	.108	.106	.104	.103
5	.124	.123	.121	.119	.117	.115

$$T_2 = T_o - \frac{(T_o - T_f)}{erfc\,\gamma\sqrt{\alpha_{12}}}\, erfc\frac{x}{2\sqrt{\alpha_2 t}} \tag{8.28}$$

A solution of Eq. (8.26) valid for moist soil systems is given in Chap. 10.

8.2.1.1 Variable Thermal Properties

Cho and Sunderland (1974) have extended the Neumann problem when the thermal conductivity varies linearly with temperature. The equations are

$$C_1\frac{\partial T_1}{\partial t} = \frac{\partial}{\partial x}\left(k_1\frac{\partial T_1}{\partial x}\right), \qquad 0 < x < X(t) \tag{8.29}$$

$$C_2\frac{\partial T_2}{\partial t} + (\rho_2 - \rho_1)\, c_2\frac{dX}{dt}\frac{\partial T_2}{\partial x} = \frac{\partial}{\partial x}\left(k_2\frac{\partial T_2}{\partial x}\right), \qquad x > X(t) \tag{8.30}$$

where $C = \rho c$
The thermal diffusivities are

$$\alpha_1 = \alpha_{01}\left(1 + \beta_1\frac{T_1 - T_S}{T_o - T_S}\right) \tag{8.31}$$

$$\alpha_2 = \alpha_{02}\left(1 + \beta_2\frac{T_2 - T_S}{T_o - T_S}\right) \tag{8.32}$$

where α_{01} and α_{02} are the values of the thermal diffusivity of the solid and the liquid phases respectively, at T_S. The liquid phase does not exist at T_S, and α_{02} is obtained by extrapolation. The initial conditions are

$$T_2 = T_o > T_f, \qquad x > 0 \tag{8.33a}$$

and

$$X(0) = 0 \tag{8.33b}$$

The boundary condition at the free surface, $x = 0$, is

$$T_1(0,t) = T_S < T_f, \qquad t > 0 \tag{8.33c}$$

At the moving fusion front, $x = X(t)$, two additional conditions must be satisfied

$$T_1 = T_2 = T_f \tag{8.33d}$$

and

$$k_1 \frac{\partial T_1}{\partial x} - k_2 \frac{\partial T_2}{\partial x} = \rho_1 l \frac{dX}{dt} \tag{8.33e}$$

A modified error function is defined as follows. Consider a second-order, nonlinear ordinary differential equation.

$$\frac{d}{d\eta}\left\{(1+\beta\theta)\frac{d\theta}{d\eta}\right\} + 2\eta\frac{d\theta}{d\eta} = 0 \tag{8.34}$$

with the boundary conditions

$$\theta(0) = 0$$
$$\theta(\infty) = 1$$

If the solution is designated as $\phi_\beta(\eta)$, which can be called the "modified error function," then by definition

$$\frac{d}{d\eta}\left\{(1+\beta\phi_\beta)\frac{d\phi_\beta}{d\eta}\right\} + 2\eta\frac{d\phi_\beta}{d\eta} = 0 \tag{8.35}$$

$$\phi_\beta(0) = 0$$
$$\phi_\beta(\infty) = 1$$

Note that when $\beta = 0$, Eq. (8.34) becomes linear, and

$$\phi_o(\eta) = erf(\eta) = \frac{2}{\sqrt{\pi}}\int_o^\eta e^{-z^2}\,dz$$

When the thermal conductivity varies linearly with temperature, using the property of the modified error function $\phi_\beta(\eta)$, the solutions of Eqs. (8.29) to (8.33) are found to be

$$T_1 = T_S + (T_f - T_S)\frac{\phi_\delta(x/2\sqrt{\alpha_{o1}t})}{\phi_\delta(\lambda)} \tag{8.36}$$

and

$$T_2 = T_o - (T_o - T_f)\frac{1 - \phi_\epsilon\dfrac{x}{2\sqrt{\alpha_{o1}\alpha_{21}(1+\beta_2 a)}} + \dfrac{\left(\dfrac{1}{\rho_{12}} - 1\right)\lambda}{\sqrt{\alpha_{12}(1+\beta_2 a)}}}{1 - \phi_\epsilon(\zeta)} \tag{8.37}$$

The constants λ, δ, ϵ, and a are determined from

$$\delta\phi_\delta(\lambda) = \beta_1\theta_f$$
$$\epsilon = \frac{\beta_2(1-a)}{1+\beta_2 a}$$
$$\phi_\epsilon(\zeta) = \frac{\theta_f - a}{1-a}$$

and

$$\frac{\phi_\delta'(\lambda)}{\lambda\phi_\delta(\lambda)} - \frac{1-\theta_f}{\theta_f}\frac{C_{21}}{1+\beta_2 a}\frac{\phi_\epsilon'(\zeta)}{\zeta\{1-\phi_\epsilon(\zeta)\}} = 2h/\theta_f$$

where

$$C_{21} = C_2/C_1$$
$$\zeta = \lambda/\rho_{21}\sqrt{(1+\beta_2 a)\alpha_{21}}$$
$$h = \frac{\rho_1 l\alpha_{01}}{k_{1f}(T_o - T_S)}$$
$$k_{1f} = \text{thermal conductivity of solid at } T_f$$
$$\theta_f = \frac{T_f - T_S}{T_o - T_S}$$

The phase change location is given by

$$X = 2\lambda\sqrt{\alpha_{01}t}$$

The functions can be evaluated with the use of Figs. 8.4 through 8.7. The effect of variable conductivity on the phase change rate is small unless β is large and h is small (the Stefan number is large).

8.2.2 The Stefan Problem

If the initial temperature of the liquid is T_f, then Eq. (8.26) is

$$\gamma e^{\gamma^2} erf\,\gamma = \frac{c_l(T_f - T_s)}{l\sqrt{\pi}} \tag{8.38}$$

Now, when x is small, the error function may be approximated as

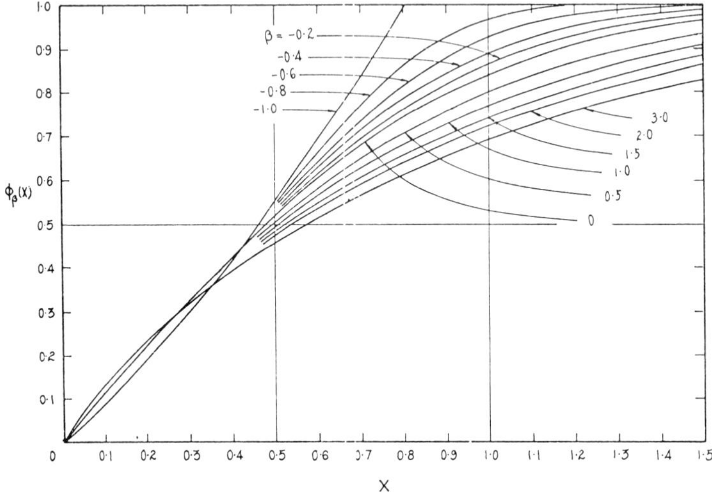

Fig. 8.4. Graph of ϕ_β (x) (from Cho and Sunderland 1974).

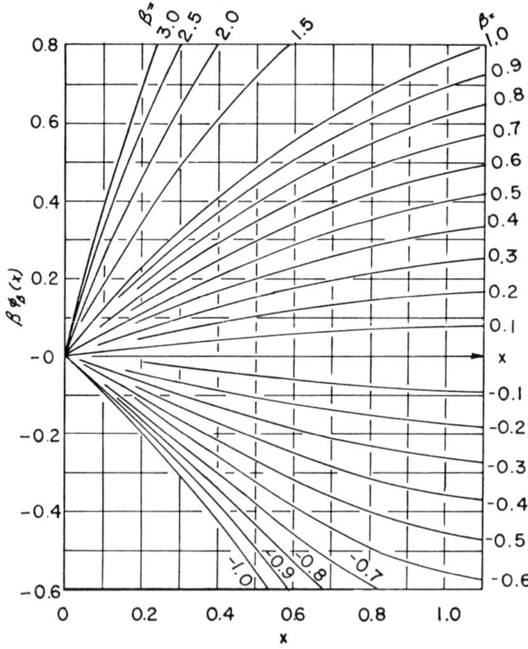

Fig. 8.5. Graph of $\beta\phi_\beta$ (x) (from Cho and Sunderland 1974).

$$erf\,x = \frac{2}{\sqrt{\pi}} \sum_{n=0}^{\infty} \frac{(-1)^n x^{2n+1}}{(2n+1)\,n!} = \frac{2}{\sqrt{\pi}} \left(x - \frac{x^3}{3} + \frac{x^5}{10} + \ldots \right)$$

Thus, if γ is small

$$erf\,\gamma \simeq \frac{2}{\sqrt{\pi}} \gamma \tag{8.39}$$

and Eq. (8.38) is

$$\gamma \frac{2}{\sqrt{\pi}} \gamma = \frac{c_l(T_f - T_s)}{l\sqrt{\pi}}$$

or

$$\gamma^2 = \frac{c_l(T_f - T_s)}{2l} \tag{8.40}$$

and

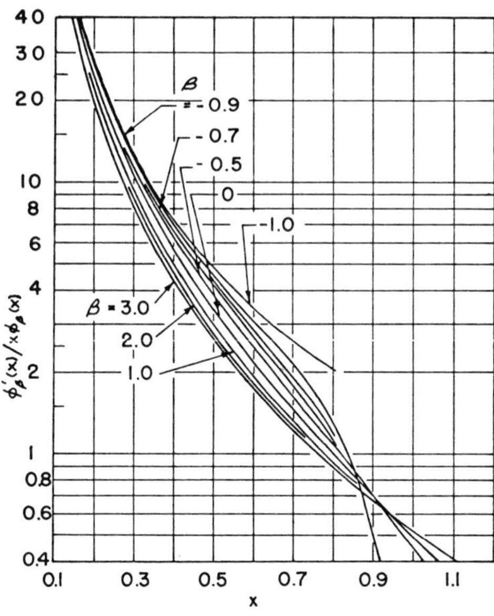

Fig. 8.6. Graph of $\phi'_\beta(x)/[x\{\phi_\beta(x)\}]$ (from Cho and Sunderland 1974).

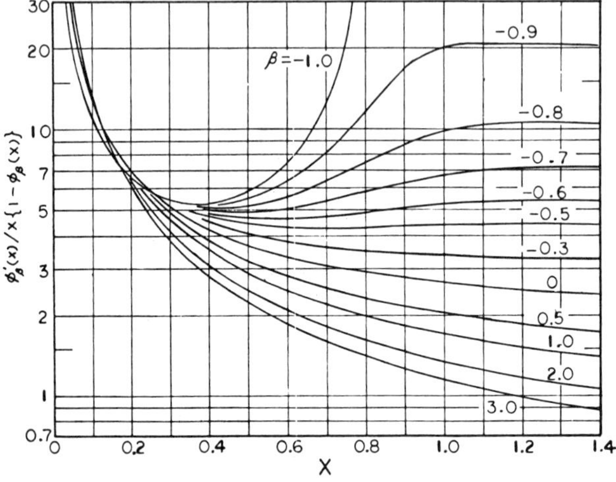

Fig. 8.7. Graph of $\phi'_\beta(x)/[x\{1-\phi_\beta(x)\}]$ (from Cho and Sunderland 1974).

$$X = \sqrt{\frac{2k_l(T_f - T_s)t}{\rho l}} \tag{8.41}$$

Equations (8.40) and (8.41) were presented by Stefan (1891), who used an approximate technique that will be discussed later. The Stefan solution is not an exact solution in the sense of the Neumann solution. For rocks or metals $[c_l(T_f - T_s)/2l] \cong 1.0$, but for water this value is very small, about 0.02, if $T_f - T_s = 5°C$. Therefore, Eq. (8.39) is an acceptable approximation for water. These values of γ compare favorably with the exact values of Table 8.1, for water and $(T_o - T_f) = 0$, as can be noted below

$T_f - T_s$ °C	1	2	3	4	5
γ (Eq. 8.40)	.056	.079	.097	.112	.126

8.2.2.1 Variable Properties

Pedroso and Domoto (1973*a*) have presented a perturbation solution for the Stefan problem if the thermal properties are variable. The original equations are

$$\frac{\partial}{\partial x}\left(k\frac{\partial T}{\partial x}\right) = \rho c \frac{\partial T}{\partial t} \tag{8.42}$$

$$T(0,t) = T_S \tag{8.42a}$$

$$T(X,t) = T_f \tag{8.42b}$$

$$T(x,0) = T_f \tag{8.42c}$$

$$\rho l \frac{dX}{dt} = k \frac{\partial\, T(X,t)}{\partial x} \tag{8.42d}$$

The following transformations are defined

$$\bar{k}(T_f - T_S) = \int_{T_S}^{T_f} k\, dy \qquad \bar{c}(T_f - T_S) = \int_{T_S}^{T_f} c\, dy$$

$$\tau = \frac{\bar{k}(T_f - T_S)}{\rho l}\, t$$

$$\theta\, \bar{k}(T_f - T_S) = \int_{T_S}^{T_f} k\, dy$$

$$f(\theta) = \frac{c(\theta)}{k(\theta)} \frac{\bar{k}}{\bar{c}}$$

$$\eta = \frac{x}{X}$$

The perturbation parameter is the Stefan number

$$S_T = \frac{\bar{c}(T_f - T_S)}{l}$$

The equations are then

$$\theta'' + S_T\, f\theta'(1)\, \eta\theta' = 0 \tag{8.43}$$

$$\theta(0) = 0 \qquad \theta(1) = 1 \qquad \frac{X^2}{2\tau} = \theta'(1)$$

The temperature and function f are expanded as

$$\theta = \sum_{i=0}^{N} S_T^i\, \theta_i \tag{8.44}$$

$$f = f(\theta_o) + S_T f'(\theta_o)\theta_1 + \frac{S_T^2}{2}\,[f''(\theta_o)\theta_1^2 + 2f'(\theta_o)\theta_2] + \ldots \tag{8.45}$$

General integral solutions are given for arbitrary functions f (Pedroso and Domoto 1973*a*). For the case where

$$f = a + b\theta$$

the interface position is given, for the first four terms of the expansion

$$X^2 = \frac{2(\tau_2 - S_T\tau_3)}{\tau_0\tau_2 + S_T(\tau_1\tau_2 - \tau_0\tau_3) + S_T^2(\tau_2^2 - \tau_1\tau_3)}\,\tau \tag{8.46}$$

where

$$\tau_0 = 1$$

$$\tau_1 = \frac{a}{3} + \frac{b}{4}$$

$$\tau_2 = -\left(\frac{2}{45}a^2 + \frac{ab}{18} + \frac{b^2}{56}\right)$$

$$\tau_3 = \frac{16}{95}a^3 + \frac{67}{2160}a^2b + \frac{29}{1512}ab^2 + \frac{63}{15680}b^3$$

This solution agrees very well with a numerical solution of Eq. (8.43).

8.3 APPROXIMATE SOLUTION METHODS

The Neumann problem has shown that error functions can lead to an exact solution, suitable for certain boundary conditions. Because of the nonlinearity of the systems, however, the method cannot be applied in general. Such a similarity solution required that the differential equation and all the initial and boundary conditions could be expressed with a single independent variable. Similarity solutions will not exist for finite domains, two phases present initially, nonuniform initial temperatures, and boundary temperatures that are arbitrary functions of time. Thus there are very few other exact solutions existing. This has prompted considerable interest in approximate methods. Aside from the usual numerical procedures, two analytical methods have been of great value: the quasi-static approximation and the heat balance integral method.

8.3.1 Quasi-Static Approximation

If the phase change interface moves relatively slowly, then an assumption can be made that the moving interface will not exert a major influence upon the temperature field during short time periods. Two approximations have been used.

The quasi-stationary assumption neglects any convection in the diffusion equation and neglects the moving interface in evaluating the temperature field and consequently the diffusive flux anywhere in the volume of interest. The procedure can handle initial conditions, but is not generally valid for all times if the temperature ahead of the interface is nonuniform or is changing due to the interface motion. The problem reduces to one of transient conduction with no phase change. The actual phase change is then solved through the interface boundary condition.

The quasi-steady approximation further simplifies the quasi-stationary problem by dropping the transient term in the energy equation. The justification for the method is somewhat tenuous because it cannot satisfy the initial conditions; the problem is mathematically so simple, however, that the idea has been used more than the quasi-stationary method.

8.3.1.1 Quasi-Stationary Approximation

The melting system will be examined again to illustrate the quasi-stationary idea. Neglect the temperature variations in the solid region, and examine only the liquid equations. This is done so that a convection term can be maintained. If the solid region were studied there could be no convective term. From Section 8.1

$$\frac{\partial^2 T}{\partial x^2}=\frac{1}{\alpha}\left[\frac{\partial T}{\partial t}+\left(1-\frac{\rho_2}{\rho}\right)\frac{dX}{dt}\frac{\partial T}{\partial x}\right] \tag{8.47}$$

$$T(x,0)=T_o$$

$$T(X,t)=T_f$$

$$T(0,t)=T_s$$

$$-k\frac{\partial T(X,t)}{dx}=\rho l\frac{dX}{dt}$$

$$X(0)=X_o$$

Here ρ_2/ρ is the ratio of the solid to liquid phase densities. Introduce the following nondimensionalizing variables

$$y=\frac{x}{X_o}\qquad \theta=\frac{T-T_o}{T_s-T_f}\qquad \phi=\frac{T_f-T_o}{T_s-T_f}\qquad \theta_s=\frac{T_s-T_o}{T_s-T_f}$$

$$S_T=c(T_s-T_f)/l\qquad \xi=\frac{X}{X_o}\qquad \tau=\frac{\alpha t}{X_o^2}$$

The dimensionless time, τ, uses a characteristic time X_o^2/α. This is the diffusion time for the liquid region, and thus the interface will move only a small distance in the characteristic time. The time domain will be relatively short, in τ, with a small movement of X.

The equations are

$$\frac{\partial^2\theta}{\partial y^2}=\frac{\partial\theta}{\partial\tau}+(1-\rho_2/\rho)\frac{d\xi}{d\tau}\frac{\partial\theta}{\partial y}$$

or

$$\frac{\partial^2\theta}{\partial y^2}=\frac{\partial\theta}{\partial\tau}+\left(\frac{\rho_2}{\rho}-1\right)S_T\frac{\partial\theta}{\partial y}\left(\frac{\partial\theta}{\partial y}\right)_\xi \tag{8.48}$$

$$\theta(y,0)=0$$
$$\theta(o,\tau)=\theta_s$$
$$\theta(\xi,\tau)=\phi$$
$$\frac{d\xi}{d\tau}=-S_T\left(\frac{\partial\theta}{\partial y}\right)_\xi \tag{8.48a}$$
$$\xi(0)=1$$

The quasi-stationary approximation, which tends to be valid if $S_T \ll 1$, is

$$\frac{\partial^2\theta}{\partial y^2}=\frac{\partial\theta}{\partial t} \tag{8.49}$$

$$\theta(y,0)=0$$
$$\theta(0,\tau)=\theta_s$$
$$\theta(\xi,\tau)=\phi$$

After solving Eqs. (8.49), the interface location is evaluated with Eq. (8.48*a*). Because of the limitations already mentioned, the method is best suited to single-phase problems. Duda and Vrentas (1969) showed that the quasi-stationary solution is the first term of an asymptotic series.

Expand the temperature and interface position as follows, with the Stefan number as the perturbation parameter

$$\theta=\theta_o+S_T\theta_1+S_T^2\theta_2+S_T^3\theta_3+\ldots \tag{8.50}$$
$$\xi=1+S_T\xi_1+S_T^2\xi_2+S_T^3\xi_3+\ldots \tag{8.51}$$

The relations at the interface position can be expanded by Taylor series, about the initial interface location

$$\left(\frac{\partial \theta_i}{\partial y}\right)_\xi = \left(\frac{\partial \theta_i}{\partial y}\right)_1 + \left(\frac{\partial^2 \theta_i}{\partial y^2}\right)_1 (S_T\xi_1 + S_T^2\xi_2 + S_T^3\xi_3 + \ldots) \qquad (8.52)$$

$$\theta_i(\xi,\tau) = \theta_i(1,\tau) + \left(\frac{\partial \theta_i}{\partial y}\right)_1 (S_T\xi_1 + S_T^2\xi_2 + S_T^3\xi_3 + \ldots) \qquad (8.53)$$

Let $\epsilon \equiv NS_T = (\rho_2/\rho - 1)S_T$. The constant ϵ will be of order S_T^2 if $\rho_2/\rho \sim 1.0$.

The procedure is called a "surface-volume perturbation," since the differential equation and the boundary conditions both contain nonlinearities. Applying Eqs. (8.50) through (8.53) to Eqs. (8.48) leads to the following system of equations up to order 2.

$$\frac{\partial^2 \theta_o}{\partial y^2} = \frac{\partial \theta_o}{\partial \tau} \qquad (8.54)$$

$$\theta_o(y,0) = 0$$

$$\theta_o(0,\tau) = \theta_s$$

$$\theta_o(1,\tau) = \phi$$

$$\frac{d\xi_o}{d\tau} = -S_T\left(\frac{\partial \theta_o}{\partial y}\right)_{\xi_o}$$

$$\xi_o(0) = 1.0$$

$$\frac{\partial^2 \theta_1}{\partial y^2} = \frac{\partial \theta_1}{\partial \tau} \qquad (8.55)$$

$$\theta_1(y,0) = \theta_1(0,\tau) = 0$$

$$\theta_1(1,\tau) = -\xi_1\left(\frac{\partial \theta_o}{\partial y}\right)_1$$

$$\frac{d\xi_1}{d\tau} = -\left(\frac{\partial \theta_o}{\partial y}\right)_1$$

$$\xi_1(0) = 0$$

$$\frac{\partial^2 \theta_2}{\partial y^2} = \frac{\partial \theta_2}{\partial \tau} + \frac{\left(\frac{\rho_2}{\rho} - 1\right)}{S_T}\frac{\partial \theta_o}{\partial y}\left(\frac{\partial \theta_o}{\partial y}\right)_1 \qquad (8.56)$$

$$\theta_2(y,0) = \theta_2(0,\tau) = 0$$

$$\theta_2(1,\tau) = -\xi_2 \left(\frac{\partial \theta_o}{\partial y}\right)_1 - \xi_1 \left(\frac{\partial \theta_1}{\partial y}\right)_1$$

$$\frac{d\xi_2}{d\tau} = -\xi_1 \left(\frac{\partial^2 \theta_o}{\partial y}\right)_1 - \left(\frac{\partial \theta_1}{\partial y}\right)_1$$

$$\xi_2(0) = 0$$

It is clear that the zeroth-order solution, Eq. (8.54), is essentially the quasi-stationary approximation, as defined earlier by Eq. (8.49).

A regular perturbation (volume) can also be used if the nonlinearities are concentrated in the differential equation. This can be done by immobilizing the interface with $\eta = y/\xi$.

Equations (8.48) are then

$$\frac{\partial^2 \theta}{\partial \eta^2} + \left(\frac{1}{\eta} + \frac{\eta}{2}\frac{d\xi^2}{d\tau}\right)\frac{\partial \theta}{\partial \eta} = \xi^2 \frac{\partial \theta}{\partial \tau} + \epsilon \frac{\partial \theta}{\partial \eta}\left(\frac{\partial \theta}{\partial \eta}\right)_1 \tag{8.57}$$

$$\theta(\eta,0) = 0$$
$$\theta(0,\tau) = \theta_s$$
$$\theta(1,\tau) = \phi$$
$$\xi(0) = 1$$

$$\frac{d\xi}{d\tau} = -\frac{S_T}{\xi}\left(\frac{\partial \theta}{\partial \eta}\right)_1$$

or

$$\xi^2 - 1 = -2S_T \int_o^\tau \frac{\partial \theta}{\partial \eta}(1,\tau')d\tau'$$

A regular perturbation will yield a set of equations equivalent to Eqs. (8.54) to (8.56).

Since the quasi-stationary method involves the solution of a transient conduction problem, the result will usually be in the form of an infinite series solution.

8.3.1.2 Quasi-Steady Approximation

The quasi-stationary method can be further simplified if the unsteady terms in the diffusion equation are also neglected. To justify this, a new characteristic time will be used. Let

$$\tau = \frac{\alpha t}{X_o^2} S_T \tag{8.58}$$

Notice that the characteristic time is now long compared with the diffusion time X_o^2/α, if $S_T < 1.0$. The new time domain is ideally suited to long-time movement of the interface. Jiji and Weinbaum (1978) used two time domains, the quasi-stationary for initial growth and Eq. (8.58) for later growth. The two times were joined at a suitable intermediate time. In this way a two-phase problem could be handled.

The nondimensional equations are

$$\frac{\partial^2 \theta}{\partial y^2} = S_T \frac{\partial \theta}{\partial \tau} - \epsilon \frac{d\xi}{d\tau} \frac{\partial \theta}{\partial y} \tag{8.59}$$

$$\theta(y,0) = 0$$
$$\theta(0,\tau) = \theta_s$$
$$\theta(\xi,\tau) = \phi$$

$$\frac{d\xi}{d\tau} = -\left(\frac{\partial \theta}{\partial y}\right)_\xi$$

$$\xi(0) = 1$$

It is now obvious that for small Stefan numbers, the diffusion equation is

$$\frac{\partial^2 \theta}{\partial y^2} = 0 \tag{8.60}$$

Thus, no transient term need be considered, and the solution is extremely simple. Solutions are far easier to obtain, compared to the quasi-stationary equations, but the validity of the solution is limited, as the initial conditions cannot be met. Nevertheless, this concept is very widely used for freezing and thawing problems.

The quasi-steady method can also be examined from the viewpoint of perturbation solutions. Use the following series expansions

$$\theta = \theta_o + S_T\theta_1 + S_T^2\theta_2 + \ldots \tag{8.61}$$

$$\xi = \xi_o + S_T\xi_1 + S_T^2\xi_2 + \ldots \tag{8.62}$$

$$\left(\frac{\partial \theta_i}{\partial y}\right)_\xi = \left(\frac{\partial \theta_i}{\partial y}\right)_{\xi_o} + \left(\frac{\partial^2 \theta_i}{\partial y^2}\right)_{\xi_o} (S_T\xi_1 + S_T^2\xi_2 + \ldots) \tag{8.63}$$

$$\theta_1(\xi,\tau) = \theta_i(\xi_o,\tau) + \left(\frac{\partial \theta_i}{\partial y}\right)_{\xi_o} (S_T\xi_1 + S_T^2\xi_2 + \ldots) \tag{8.64}$$

The following systems of equations are generated

$$\frac{\partial^2 \theta_o}{\partial y^2} = 0 \tag{8.65}$$

$$\theta_o(0,\tau) = \theta_s$$

$$\theta_o(\epsilon_o,\tau) = \phi$$

$$\frac{d\xi_o}{d\tau} = -\left(\frac{\partial \theta_o}{\partial y}\right)_{\xi_o}$$

$$\xi_o(0) = 1$$

$$\frac{\partial \theta_1}{\partial y^2} = \frac{\partial \theta_o}{\partial \tau} \tag{8.66}$$

$$\theta_1(0,\tau) = 0$$

$$\theta_1(\xi_o,\tau) = -\xi_1 \left(\frac{\partial \theta_o}{\partial y}\right)_{\xi_o}$$

$$\frac{d\xi_1}{d\tau} = -\left[\xi_1 \left(\frac{\partial^2 \theta_o}{\partial y^2}\right)_{\xi_o} + \left(\frac{\partial \theta_1}{\partial y}\right)_{\xi_o}\right]$$

$$\xi_1(0) = 0$$

$$\frac{\partial^2 \theta_2}{\partial y^2} = \frac{\partial \theta_1}{\partial \tau} + \frac{N}{S_T}\left(\frac{\partial \theta_o}{\partial y}\right)\left(\frac{\partial \theta_o}{\partial y}\right)_{\xi_o} \tag{8.67}$$

$$\theta_2(0,\tau) = 0$$

$$\theta_2(\xi_o,\tau) = -\xi_1 \left(\frac{\partial \theta_1}{\partial y}\right)_{\xi_o} - \xi_2 \left(\frac{\partial \theta_o}{\partial y}\right)_{\xi_o}$$

$$\frac{d\xi_2}{d\tau} = -\xi_2 \left(\frac{\partial^2 \theta_o}{\partial y^2}\right)_{\xi_o} - \xi_1 \left(\frac{\partial^2 \theta_1}{\partial y^2}\right)_{\xi_o} - \left(\frac{\partial \theta_2}{\partial y}\right)_{\xi_o}$$

$$\xi_2(0) = 0$$

The zeroth solution is the quasi-steady approximation. Pedroso and Domoto (1973) demonstrated this for a spherical system. Lock (1969*b*) derived the zeroth- and first-order sytems above for the solidification of the semi-infinite medium.

The quasi-steady method is so simple that solutions for the above system can be written down

$$\theta_o = \frac{-y}{\xi_o} + \theta_s$$

$$\xi_o = \sqrt{1 + 2\tau}$$

$$\theta_1 = \left(\frac{\xi_1}{\xi_o} - \frac{1}{6}\right)\frac{y}{\xi_o} + \frac{1}{6\xi_o^3}y^3$$

$$\xi_1 = \frac{1}{6}\left(\frac{1}{\xi_o} - \xi_o\right)$$

$$\theta_2 = dy + cy^2 + ay^3 + by^5$$

$$\xi_o d = -\frac{1}{36}\left(\frac{1}{\xi_o^4} - 1\right) + \frac{\xi_2}{\xi_o} - \frac{1}{6}\left(-\frac{1}{2\xi_o^2} + \frac{1}{3}\right) - \left(\frac{1}{40} + \frac{N}{2S_T}\right)$$

$$a\xi_o^3 = \frac{1}{6}\left(-\frac{1}{2\xi_o^2} + \frac{1}{3}\right)$$

$$\xi_o^5 b = \frac{1}{40}$$

$$c\xi_o^2 = \frac{N}{2S_T}$$

$$\xi_2 = \frac{1}{4}\left(\frac{1}{5} + \frac{N}{S_T}\right)\frac{1}{\xi_o} + \frac{1}{4}\left(\frac{1}{18} - \frac{1}{5} - \frac{N}{S_T}\right)\xi_o - \frac{1}{72}\xi_o^{-3}$$

A. The Stefan Problem. It is of interest to return to the solution of Section 8.2.2 for the case of a constant surface temperature. The solution was based on the assumption that $c_l(T_f - T_s)/l$ was small, which is equivalent to the quasi-steady approximation. The problem reduces to

$$\frac{\partial^2 T_l}{\partial x^2} = 0 \tag{8.68}$$

$$T_l(0,t) = T_s$$

$$T_l(X,t) = T_f$$

$$k_l \frac{\partial T_l(X,t)}{\partial x} = \rho_l l \frac{dX}{dt}$$

The solution to this system is

$$T_l = (T_f - T_s)\frac{x}{X} + T_s \tag{8.69}$$

$$X = \sqrt{\frac{2k_l}{\rho_l l}(T_f - T_s)t} \tag{8.70}$$

The phase change depth X given by Eq. (8.70) is identical to that of Eq. (8.41). Thus Stefan, in effect, seems to have been the first to use the quasi-steady method.

The results are also valid if the surface temperature is a function of time. Then

$$T_l = [T_f - T_s(t)]\frac{x}{X} + T_s(t) \tag{8.71}$$

$$X = \sqrt{\frac{2k_l}{\rho_l l}\int_0^t [T_f - T_s(t')]dt'} \tag{8.72}$$

A major limitation of the quasi-steady approximation is the failure to account for the sensible heat during the phase change. The heat flow from the surface is given by

$$q = -k_l A\left(\frac{\partial T}{\partial x}\right)_{x=0}$$

Using Eq. (8.69), this is

$$q = -k_l A\frac{(T_f - T_s)}{X}$$

The total heat flow at the surface during a given time is

$$Q = \int_0^t q\,dt$$

This can be evaluated using Eq. (8.70) as

$$Q = -k_l A(T_f - T_s)\int_0^t \frac{dt}{\sqrt{\dfrac{2k_l(T_f - T_s)t}{\rho_l l}}}$$

or

$$Q = -A X L$$

The total surface energy flow, during the time that a layer of thickness X freezes, is simply the latent heat. Thus, the method does not take into account any sensible heat, although the temperature of the frozen layer does decrease with time. This is in contrast to the exact solution, with $T_o = T_f$, where it can be shown that the heat removed equals the latent heat plus the sensible heat involved in lowering the temperature of the frozen layer. This limitation is directly associated with the assumption of a Stefan number of zero.

8.3.2 The Heat Balance Integral Method

A second approximate method that has been used with good results for phase change problems involves the concept of the temperature penetration depth. Consider the semi-infinite solid shown in Fig. 8.8. At a time t, after the surface temperature has dropped to T_s, the temperature in the solid will be disturbed to a depth $\delta(t)$. Beyond this depth, the temperature of the solid remains at the initial temperature T_o, and no energy is transferred beyond this point. The penetration distance δ is analogous to the boundary layer thickness in fluid mechanics. The solution method is analogous to the momentum integral method, in that the basic equations are satisfied on average over the volume of thickness $\delta(t)$. The conduction equation is

$$\alpha \frac{\partial^2 T}{\partial x^2} = \frac{\partial T}{\partial t} \tag{8.73}$$

Now, integrate this equation over the distance $\delta(t)$. Thus

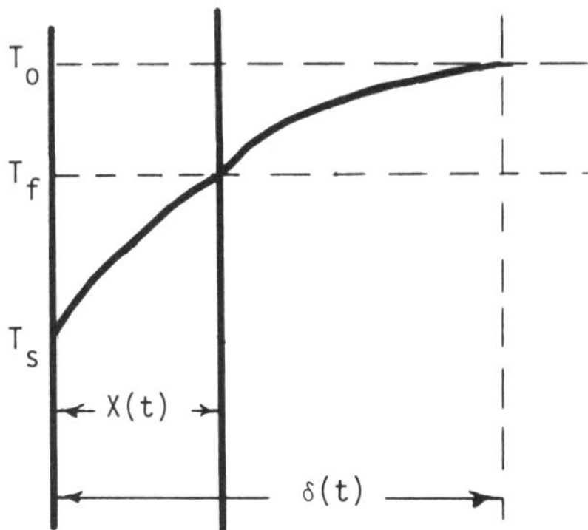

Fig. 8.8. Temperature Penetration Depth.

$$\int_0^{\delta(t)} \alpha \frac{\partial^2 T}{\partial x^2} dx = \int_0^{\delta(t)} \frac{\partial T}{\partial t} dx$$

The properties are assumed constant and

$$\alpha \int_0^{\delta} \frac{\partial^2 T}{\partial x^2} dx = \alpha \left[\frac{\partial T(\delta,t)}{\partial x} - \frac{\partial T(0,t)}{\partial x} \right] \tag{8.74}$$

Leibnitz's rule for a general function is

$$\frac{d}{dt}\int_{a(t)}^{b(t)} f(x,t)dx = f(b,t)\frac{db}{dt} - f(a,t)\frac{da}{dt} + \int_a^b \frac{\partial f(x,t)}{\partial t} dx$$

Then

$$\int_0^{\delta(t)} \frac{\partial T}{\partial t} dx = \frac{d}{dt}\int_0^{\delta(t)} T(x,t)dx - T(\delta,t)\frac{d\delta}{dt}$$

Let

$$\theta = \int_0^{\delta} T(x,t)\, dx \tag{8.75}$$

Then the integral equation is

$$\frac{d\theta}{dt} + \alpha \frac{\partial T(0,t)}{\partial x} - T_o \frac{d\delta}{dt} = 0 \tag{8.76}$$

This equation is valid if there is no phase change. Consider the case of phase change where the properties of the frozen region differ from those of the thawed region. There will then be two integral equations as follows

$$\frac{d\theta_1}{dt} - T_f \frac{dX}{dt} - \alpha_1 \left[\frac{\partial T_1(X,t)}{\partial x} - \frac{\partial T_1(0,t)}{\partial x} \right] = 0 \tag{8.77}$$

$$\frac{d\theta_2}{dt} - T_o \frac{d\delta}{dt} + T_f \frac{dX}{dt} + \alpha_2 \frac{\partial T_2(X,t)}{\partial x} = 0 \tag{8.78}$$

where

$$\theta_1 = \int_0^X T_1(x,t)\, dx$$

$$\theta_2 = \int_X^\delta T_2(X,t)\, dx$$

and $T_l(X,t) = T_2(X,t) = T_f$ and $T_2(\delta,t) = T_o$ have been used.

The solution of a general problem with superheat or subcooling will involve two coupled parameters X and δ. The solution will normally be tedious. However, assume that the initial temperature is T_f. Then the problem reduces to only one differential equation because the penetration distance δ is now identical to phase change depth X

$$\frac{d\theta_l}{dt} - T_f \frac{dX}{dt} - \alpha_l \left[\frac{\partial T_l(X,t)}{\partial x} - \frac{\partial T_l(0,t)}{\partial x} \right] = 0 \tag{8.79}$$

$$\theta_l = \int_0^X T_l(x,t)\, dx \tag{8.80}$$

8.3.2.1 The Stefan Problem

As an example of the method, the solution of the Stefan problem will be given. Goodman (1958) has solved this problem, and other problems also. The heat balance integral is

$$\frac{d\theta_l}{dt} = \alpha_l \left[\frac{\partial T_l(X,t)}{\partial x} - \frac{\partial T_l(0,t)}{\partial x} \right] + T_f \frac{dX}{dt} \tag{8.81}$$

$$\theta_l = \int_0^X T_l\, dx \tag{8.80}$$

The boundary conditions are

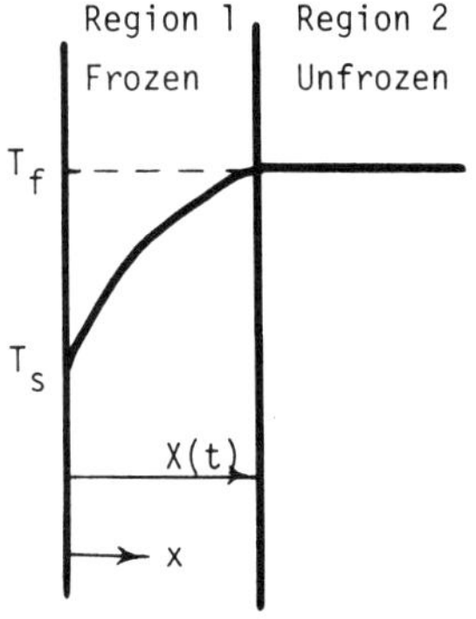

$$T_l(X,t) = T_f \tag{8.81a}$$

$$T_l(0,t) = T_s \tag{8.82b}$$

$$k_l \frac{\partial T_l(X,t)}{\partial x} = \rho_l l \frac{dX}{dt} \tag{8.81c}$$

Assume that T_l is a quadratic function of x

$$T_l = T_f + a_l(x - X) + b(x - X)^2 \tag{8.82}$$

To avoid a second-order differential equation for X, boundary condition (8.81c) must be written in the form of Eq. (8.12)

$$-k_l \left[\frac{\partial T_l(X,t)}{\partial x}\right]^2 = \rho_l l \alpha_l \frac{\partial^2 T_l(X,t)}{\partial x^2} \tag{8.83}$$

The temperature is then given by

$$T_l = T_f + a(x - X)\frac{x}{X} - \frac{lS_T}{c_l}\frac{(x - X)^2}{X^2} \tag{8.84}$$

where

$$S_T = \frac{c_l(T_f - T_S)}{l}$$

$$\frac{\partial T_l(X,t)}{\partial x} = a \tag{8.85}$$

$$\frac{\partial T_l(0,t)}{\partial x} = -a + \frac{2lS_T}{c_l X} \tag{8.86}$$

The coefficient a is given by the solution to

$$a^2 + \frac{2la}{c_l X} - 2\left(\frac{l}{c_l}\right)^2 \frac{S_T}{X^2} = 0 \tag{8.87}$$

The positive root of this equation is chosen as $a > 0$, for $[\partial T(X,t)/\partial x] > 0$. Then

$$a = \frac{l}{c_l X}[-1 + \sqrt{1 + 2S_T}] \tag{8.88}$$

The equation for X is

$$X\frac{dX}{dt} = 6\alpha_l \left(\frac{1-\sqrt{1+2\,S_T}+2\,S_T}{5+2\,S_T+\sqrt{1+2\,S_T}}\right) \tag{8.89}$$

The solution is then

$$X=\sqrt{3}\left(\frac{1-\sqrt{1+2\,S_T}+2\,S_T}{5+2\,S_T+\sqrt{1+2\,S_T}}\right)^{1/2} 2\sqrt{\alpha_l t} \tag{8.90}$$

The solution is in the same form as the exact solution for the Neumann problem, Eq. (8.25). The Stefan solution Eq. (8.41) can also be put into this form

$$X= 1/2\sqrt{2S_T} \quad 2\sqrt{\alpha_l t} \tag{8.91}$$

For $T_f - T_S = 5°\text{C}$, and using the properties of ice, $c_1 = 0.465$ cal/g-°C, $l = 79.71$ cal/g, then $S_T = .0292$.

Solution	Exact	Quasi-static	Integral method
Value of γ	.124	.1208	.1207

The integral solution is virtually identical to the much simpler quasi-steady solution, and both are −2.7% in error. Of course, the integral solution is not limited to small Stefan numbers.

This example illustrates a problem with the integral method for a quadratic temperature approximation. By using different combinations of the equations, different solutions can be obtained. Equation (8.90) was obtained by using Eqs. (8.79), (8.80), (8.81*c*), and (8.86). When Eqs. (8.81*c*) and (8.85) are used, then

$$X=\left(\frac{\sqrt{1+2S_T}-1}{2}\right)^{1/2} 2\sqrt{\alpha_l t} \tag{8.92}$$

This solution is a generalization of the quasi-steady solution, with a quadratic temperature in place of the linear temperature profile. If Eqs. (8.79), (8.80), (8.85), and (8.86) are used, then

$$X=\sqrt{6}\left(\frac{1+S_T-\sqrt{1+2\,S_T}}{2S_T-1+\sqrt{1+2\,S_T}}\right)^{1/2} \tag{8.93}$$

Let us compare these equations with $S_T = 0.9205$

Solution	Exact	Eq. (8.91)	Eq. (8.90)	Eq. (8.92)	Eq. (8.93)
Value of γ	0.600	.678	.638	.586	.747

Apparently, Eq. (8.92) is the most accurate, with Eq. (8.90) close behind. This quandary can be eliminated if a cubic temperature profile is used, but the computational work is greatly increased; refer to Goodman (1958) for further discussion.

8.3.2.2 Approximate Neumann Solution

A simple, but useful, approximate solution to the Neumann problem can be obtained by the use of the heat balance integral method. The system of equations, derived earlier, for the freeze problem, is

$$\frac{d\theta_1}{dt} - T_f\frac{dX}{dt} - \alpha_1\left[\frac{\partial T_1(X,t)}{\partial x} - \frac{\partial T_1(0,t)}{\partial x}\right] = 0 \tag{8.77}$$

$$\frac{d\theta_2}{dt} - T_o\frac{d\delta}{dt} + T_f\frac{dX}{dt} + \alpha_2\frac{\partial T_2(X,t)}{\partial x} = 0 \tag{8.78}$$

$$\theta_1 = \int_o^X T_1(x,t)\,dx \tag{8.80}$$

$$\theta_2 = \int_X^\delta T_2(x,t)\,dx \tag{8.94}$$

$$T_1(o,t) = T_s \tag{8.95a}$$

$$T_1(X,t) = T_f \tag{8.95b}$$

$$T_2(\delta,t) = T_o \tag{8.95c}$$

$$\frac{\partial T_2(\delta,t)}{\partial x} = 0 \tag{8.95d}$$

$$T_2(X,t) = T_f \tag{8.95e}$$

$$k_1\frac{\partial T_1(X,t)}{\partial x} - k_2\frac{\partial T_2(X,t)}{\partial x} = \rho_1 l\frac{dX}{dt} \tag{8.11}$$

$$-k_1\left[\frac{\partial T_1(X,t)}{\partial x}\right]^2 + k_2\frac{\partial T_1(X,t)}{\partial x}\frac{\partial T_2(X,t)}{\partial x} = \rho_1 l\alpha_1\frac{\partial^2 T_1(X,t)}{\partial x^2} \tag{8.12}$$

Assume that the solutions for the phase change interface, X, and the thermal penetration depth, δ, are

$$X = 2\gamma\sqrt{\alpha_1 t} \tag{8.25}$$

$$\delta = 2\lambda\sqrt{\alpha_2 t} \tag{8.96}$$

A linear approximation for the temperture in region 1, satisfying Eq. (8.95a,b), is

$$T_1 = T_s + (T_f - T_s)\frac{x}{X} \tag{8.97}$$

A quadratic approximation for T_2, which satisfies Eqs. (8.95c,d,e) is

$$T_2 = T_f + 2\Delta T\frac{x - X}{\delta - X} - \Delta T\frac{(x - X)^2}{(\delta - X)^2} \tag{8.98}$$

where $\Delta T = T_o - T_f$.

The key to the solution is to recognize, from the exact solution, that δ/X should not vary with time, although it is a function of ϕ, S_T, and the property values. Thus, let $\delta/X = b$, where b is $b(\phi$, S_T, properties). Equation (8.98) is

$$T_2 = T_f + \frac{2\Delta T}{(b-1)}\frac{x - X}{X} - \frac{\Delta T(x - X)^2}{(b-1)^2X^2} \tag{8.99}$$

Then, with Eq. (8.99), Eq. (8.78) is

$$\left(T_f + \frac{2}{3}\Delta T\right)(b-1)\frac{dX}{dt} - T_o b\frac{dX}{dt} + \frac{2\alpha_2\Delta T}{(b-1)X} + T_f\frac{dX}{dt} = 0 \tag{8.100}$$

Using Eq. (8.25), the value of b can be obtained from Eq. (8.100)

$$b = \sqrt{2.25 + \frac{3\alpha_{21}}{\gamma^2}} - \frac{1}{2} \tag{8.101}$$

Equation (8.80) for θ_1 is

$$\theta_1 = \frac{T_S + T_f}{2}X \tag{8.102}$$

In Eq. (8.77), the term $[\partial T_1(X,t)/\partial x]$ is evaluated using Eq. (8.11) and Eqs. (8.97) and (8.99). This leads to the following differential equation for X

$$\left(\frac{1}{2}+\frac{1}{S_T}\right)\frac{dX}{dt}=\frac{\alpha_1}{X}\left(1-\frac{2k_{21}\phi}{b-1}\right) \tag{8.103}$$

The solution to Eq. (8.103) is straightforward and can be written as

$$X=2\gamma\sqrt{\alpha_1 t} \tag{8.25}$$

$$\gamma^2=\frac{-b_1-\sqrt{b_1^2-4a\,S_T^2}}{2a} \tag{8.104}$$

$$a=\left(S_T+2+\frac{2k_{21}\,\phi S_T}{\alpha_{21}}\right)(S_T+2)$$

$$b_1=-2\,S_T\left(S_T+2+\frac{k_{21}\,\phi S_T}{\alpha_{21}}\right)-\frac{4}{3}\frac{(k_{21}\,\phi S_T)^2}{\alpha_{21}}$$

As often occurs with the heat balance integral method, a more complicated approximation for T_1 does not significantly improve the accuracy. A quadratic approximation, T_1, that satisfies Eqs. (8.95a,b) and Eq. (8.12) is

$$T_1=T_f+\frac{px(x-X)}{X^2}-\frac{lS_T}{c_1}\frac{(x-X)^2}{X^2} \tag{8.105}$$

$$p=-\frac{l}{c_1}+\frac{k_{21}\Delta T}{(b-1)}+\sqrt{\left(\frac{l}{c_1}-\frac{k_{21}\Delta T}{b-1}\right)^2+2\left(\frac{l}{c_1}\right)^2 S_T}$$

The equation for X is now

$$\left(\frac{p}{6}+\frac{l\,S_T}{3\,c_1}+\frac{l}{c_1}\right)\frac{dX}{dt}=\frac{\alpha_1}{X}\left[-p+\frac{2l}{c_1}S_T-\frac{2k_{21}\Delta T}{b-1}\right]$$

The solution to this equation is

$$\gamma^2=\frac{3(1+2S_T-3\nu-q)}{5+2\,S_T+\nu+q} \tag{8.106}$$

$$q=\sqrt{(1-\nu)^2+2S_T}$$

$$\nu=\frac{k_{21}\,\phi\,S_T}{b-1}$$

The approximate equations for γ, Eq. (8.106) and Eq. (8.104), can be compared to exact values, Eq. (8.26), given by Carlsaw and Jaeger (1959), for water-ice systems. The values are shown in Table 8.2.

Clearly, both approximate equations yield good results for small values of S_T. Sparrow (1978) has calculated exact values of γ for $k_{21} = \alpha_{21} = 1$. These comparisons are shown in Table 8.3. Typical soil property ratios are also compared in Table 8.4.

These calculations indicate that Eq. (8.105) is a very good approximate equation for γ over a wide range of S_T, ϕ, and property values.

It should be within 5% for all practical cases for soil systems. Nixon and McRoberts (1973) made a parametric study of Eq. (8.26), but presented a graph only for $\alpha_{21} = 0.49$, which was said to represent most soil conditions. A semiempirical relation was found for γ with $\phi = 0$ as

Table 8.2. Accuracy of Approximate γ-Values for Ice-Water System (from Lunardini and Varotta 1980).

		γ				
S_T	ϕ	Exact, Eq. (8.26)	Eq. (8.105)	% Error, Eq. (8.105)	Eq. (8.106)	% Error, Eq. (8.106)
.0063	0	.056*	.0559	− .2	—	—
.0063	5	.049*	.0486	− .8	—	—
.0292	0	.124*	.1241	+ .08	—	—
.0292	1.0	.115*	.1155	.4	—	—
.1	0	.2200	.2181	− .90	.2232	1.5
.1	5	.1176	.1190	1.2	.1200	2.0
1	0	.6201	.5774	−6.9	.6600	6.4
1	5	.1449	.1479	2.1	.1497	3.3

($k_{21} = .2717$, $\alpha_{21} = .1252$)
* Values from Carslaw and Jaeger (1959).

Table 8.3. Comparison of γ-Values, $\alpha_{21} = k_{21} = 1.0$ (from Lunardini and Varotta 1980).

		γ				
S_T	ϕ	Exact, Eq. (8.26)	Eq. (8.105)	% Error, Eq. (8.105)	Eq. (8.106)	% Error, Eq. (8.106)
.0058	0	.0538	.0538	0	.0539	0.2
.0058	5	.0459	.0458	− .2	.0458	−0.2
1	4	.1278	.1276	− .2	.1290	0.9
1	0	.6201	.5774	−6.9	.6600	6.4
1	4	.1704	.1706	.1	.1737	1.9

Table 8.4. Accuracy of Approximate γ-Values, $k_{21} = 0.51$, $\alpha_{21} = .3355$ (from Lunardini and Varotta 1980).

		γ				
S_T	ϕ	Exact, Eq. (8.26)	Eq. (8.105)	% Error, Eq. (8.105)	Eq. (8.106)	% Error, Eq. (8.106)
.0058	0	.0538	.0538	0	.0539	0.2
.0058	5	.0468	.0465	0	.0465	0
.03	0	.1219	.1215	− .30	.1225	0.5
.03	5	.0860	.0861	.1	.0865	.6
.1	0	.2200	.2181	− .9	.2232	1.5
.1	5	.1168	.1174	.5	.1185	1.5
1	0	.6201	.5774	−6.9	.6600	6.4
1	5	.1460	.1477	1.2	.1496	2.5

$$\gamma = \sqrt{\frac{S_T}{2}}\left(1 - \frac{S_T}{8}\right) \tag{8.107}$$

In the limit as $\phi \to 0$, Eq. (8.104) reduces to

$$\gamma = \sqrt{\frac{S_T}{2 + S_T}} \tag{8.108}$$

and Eq. (8.106) is

$$\gamma = \sqrt{\frac{3(1 + 2\,S_T - \sqrt{1 + 2\,S_t})}{5 + 2\,S_T + \sqrt{1 + 2\,S_T}}} \tag{8.109}$$

Notice that Eq. (8.106) reduces to Eq. (8.90) of the previous section for the Stefan problem, as it should. The quasi-steady Stefan solution has already been shown to be

$$\gamma = \sqrt{\frac{S_T}{2}} \tag{8.110}$$

All the above equations are quite close and accurate for small S_T values, with Eqs. (8.108) and (8.90) showing better accuracy as the Stefan number exceeds 1. Equation (8.90) gives the best overall accuracy.

8.3.2.3 Refinement of Heat Balance Integral Method

The problem of the proper approximation to use for the assumed temperature profile can be eased with a refinement to the heat balance integral method discussed by Bell (1978). Instead of following only the phase change and 0°C penetration depths, any number of isotherms can be followed by writing the heat balance integral for an arbitrary number of regions. Let us again look at the Stefan problem described in Section 8.3.2.1. The frozen region will be divided into two parts where $T = T_a = (T_f + T_S)/2$ at $x = X_1$. The accuracy increases with an increasing number of subdivisions, but the computational work also increases. The heat balance integral for region 1 is

$$\alpha\left[\frac{\partial T_1(X_1,t)}{\partial x} - \frac{\partial T_1(0,t)}{\partial x}\right] = \frac{d}{dt}\int_0^{X_1} T_1 dx - T_a\frac{dX_1}{dt} \tag{8.111}$$

and that for region 2 is

$$\alpha\left[\frac{\rho l}{k}\frac{dX}{dt} - \frac{\partial T_2(X_1,t)}{\partial x}\right] = \frac{d}{dx}\int_{X_1}^{X} T_2\, dx - T_f\frac{dX}{dt} + T_a\frac{dX_1}{dt} \tag{8.112}$$

Quadratic temperature approximations are used in each region with

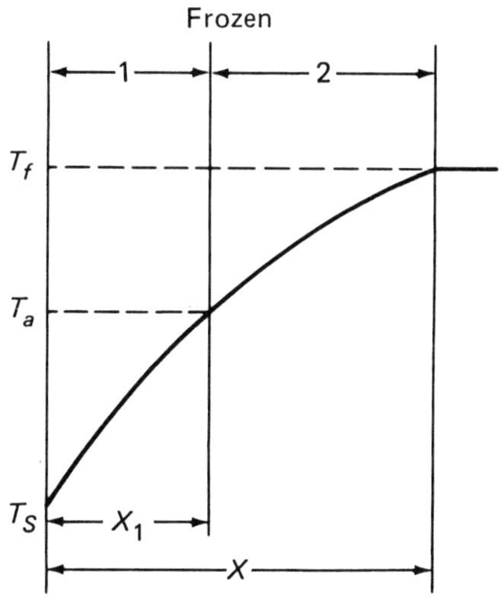

$$T_l = ax^2 + bx + c \tag{8.113}$$
$$T_2 = dx^2 + ex + f \tag{8.114}$$

The conditions to be satisfied by T_1 are

$$T_l(X_l,t) = T_a \tag{8.115a}$$
$$T_l(0,t) = T_S \tag{8.115b}$$

and the energy flux continuity relation

$$\frac{\partial T_l(X_l,t)}{\partial x} = \frac{\partial T_2(X_l,t)}{\partial x} \tag{8.115c}$$

The conditions to be satisfied by T_2 are

$$T_2(X,t) = T_f \tag{8.115d}$$
$$T_2(X_l,t) = T_a \tag{8.115e}$$
$$\left(\frac{\partial T_2}{\partial x}\right)^2 = -\frac{l}{c}\frac{\partial^2 T_2}{\partial x^2} \tag{8.115f}$$

Equations (8.115) are solved simultaneously for the six unknowns in Eqs. (8.113) and (8.114). With these relations, Eqs. (8.111) and (8.112) can then be solved for X_1 and X.

Bell (1978) showed that the error for the Stefan problem can be reduced from 6.5% with one region to 1.2% for two subdivisions. The refinement described here can reduce the errors on the heat balance integral method, but it also tends to negate the simplicity of the method. For more than two subdivisions or for more complicated boundary conditions, it is likely that numerical solutions will be required of the set of simultaneous differential equation.

8.4 PROBLEMS IN THE SEMI-INFINITE REGION

We will examine a number of problems for the semi-infinite solid, both exact and approximate.

8.4.1 Neumann Problem with Melting Temperature Range

Not all materials exhibit a fixed phase change temperature. For soils, rocks, metal alloys, etc., melting will occur over a temperature range. Cho and Sunderland (1969) give an exact solution for the case of a binary eutectic mixture. The method is outlined here. Tien and Geiger (1967) give an approximate solution when the mixture is initially at the liquidus temperature. The system is described in terms of three regions: a completely solidified region, an initially liquid region, and a region with both solid and liquid phases formed by isothermal planes at the liquidus (T_f) and solidus (T_{fs}) temperatures (Fig. 8.8a). The liquid begins to freeze at T_f, and freezing is complete when the solid front reaches the volume of interest. Equilibrium freezing occurs if the element is completely frozen just as the solid front reaches it. Normal nonequilibrium freezing is such that the element still has a liquid fraction, which then freezes isothermally at T_{fs} before the solid front moves on. It is assumed that the properties of the solid and liquid do not vary with temperature, and volume changes are negligible. The solid fraction distribution, in the freezing zone, is linear with distance. It has a value of zero at the liquidus front and f at the solidus front. For binary eutectic mixtures, f is the solid fraction at the eutectic composition. Tien and Geiger (1967) have shown that the linear assumption is not important for the phase change process.

The solid fraction is then given by

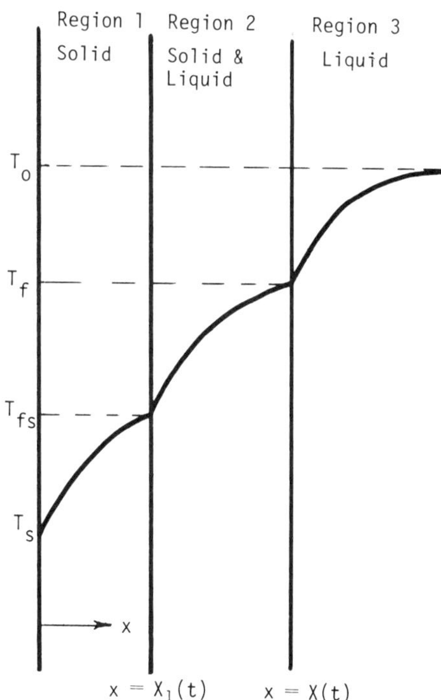

Fig. 8.8*a*. Solidification with a Mixed Phase Region.

$$f_s = f\left(1 - \frac{x - X_1}{X - X_1}\right) \qquad X_1 < x < X \tag{8.116}$$

The energy liberated per unit volume when a volume of liquid changes to a solid is

$$Q = \frac{l\,dm_s}{V} = \frac{l}{V}d(m\,f_s) = \frac{l}{V}\left[m\frac{df_s}{dt} + f_s\frac{dm}{dt}\right]$$

$$Q = \rho_2 l\frac{df_s}{dt} \tag{8.117}$$

This will represent a uniformly distributed heat source in the freezing zone

$$Q = \frac{\rho_2 l f}{(X - X_1)^2}\left[(X - x)\frac{dX_1}{dt} + (x - X_1)\frac{dX}{dt}\right]$$

The equations for this problem are

$$\frac{\partial^2 T_1}{\partial x^2} = \frac{1}{\alpha_1}\frac{\partial T_1}{\partial t} \qquad 0 < x < X_1 \tag{8.118}$$

$$\frac{\partial^2 T_2}{\partial x^2} + \frac{\rho_2 l}{k_2}\frac{df_s}{dt} = \frac{1}{\alpha_2}\frac{dT_2}{dt} \qquad X_1 < x < X \tag{8.119}$$

$$\frac{\partial^2 T_3}{\partial x^2} = \frac{1}{\alpha_3}\frac{\partial T_3}{\partial t} \qquad x > X \tag{8.120}$$

$$T_3(x,0) = T_o \tag{8.121a}$$

$$T_3(\infty,t) = T_o \tag{8.121b}$$

$$T_1(0,t) = T_s \tag{8.121c}$$

$$T_1 = T_2 = T_{fs} \qquad x = X_1 \tag{8.121d}$$

$$T_2 = T_3 = T_f \qquad x = X \tag{8.121e}$$

At the solidus front, the remaining liquid will solidify, and the energy balance is

$$k_1\frac{\partial T_1}{\partial x} = k_2\frac{\partial T_2}{\partial x} + \rho_1 l(1 - f)\frac{dX_1}{dt} \qquad x = X_1 \tag{8.121f}$$

At the liquidus front, the solid fraction is zero, and the phase change for the freezing zone has been accounted for by the energy generation term. Then, only equality of heat fluxes is required

$$k_2 \frac{\partial T_2}{\partial x} = k_3 \frac{\partial T_3}{\partial x} \qquad x = X \tag{8.121g}$$

The solution proceeds in the same fashion as for Section 8.2.1 except that there are two parameters, defined as

$$X_l = 2\lambda \sqrt{\alpha_l t} \tag{8.122}$$

$$X = 2\eta \sqrt{\alpha_l t} \tag{8.123}$$

The parameters λ and η must be found by the simultaneous solution of the following equations

$$\frac{e^{-\lambda_2}}{erf\lambda} + \frac{\sqrt{\pi}\, k_{21} fl}{2\, c_{p_2}(T_{fs} - T_s)(\eta - \lambda)} - \frac{k_{21}}{\sqrt{\alpha_{21}}} \frac{\left(T_f - T_{fs} + \frac{lf}{c_{p_2}}\right) e^{-\frac{\lambda^2}{\alpha_{21}}}}{(T_f - T_s)\left(erf\frac{\eta}{\sqrt{\alpha_{21}}} - erf\frac{\lambda}{\sqrt{\alpha_{21}}}\right)} = \frac{\sqrt{\pi}(1-f)l\lambda}{c_{p_l}(T_{fs} - T_s)} \tag{8.124}$$

$$\frac{(T_f - T_{fs} + lf/c_{p_2})e^{-\frac{\eta^2}{\alpha_{21}}}}{\left(erf\frac{\eta}{\sqrt{\alpha_{21}}} - erf\frac{\lambda}{\sqrt{\alpha_{21}}}\right)} - \frac{\sqrt{\pi\alpha_{21}}\, lf}{2\, c_{p_2}(\eta - \lambda)} = (T_o - T_f)k_{32} \frac{\sqrt{\alpha_{23}}\, e^{-\frac{\eta^2}{\alpha_{31}}}}{erfc\frac{\eta}{\sqrt{\alpha_{31}}}} \tag{8.125}$$

The temperatures are given by

$$\frac{T_l - T_s}{T_{fs} - T_s} = \frac{erf\left(\frac{x}{2\sqrt{\alpha_l t}}\right)}{erf\lambda} \tag{8.126}$$

$$\frac{T_2 - T_s}{T_{fs} - T_s} = 1 - \frac{lf\left(\frac{x}{2\sqrt{\alpha_1 t}} - \lambda\right)}{c_{p2}(T_{fs} - T_s)(\eta - \lambda)}$$

$$+ \frac{(T_f - T_{fs} + lf/c_{p2})}{(T_{fs} - T_s)} \frac{\left[erf\left(\frac{x}{2\sqrt{\alpha_2 t}}\right) - erf\left(\frac{\lambda}{\sqrt{\alpha_{21}}}\right)\right]}{\left(erf\frac{\eta}{\sqrt{\alpha_{21}}} - erf\frac{\lambda}{\sqrt{\alpha_{21}}}\right)} \quad (8.127)$$

$$T_3 = T_o - (T_o - T_f)\frac{erfc\left(\frac{x}{2\sqrt{\alpha_3 t}}\right)}{erfc\frac{\eta}{\sqrt{\alpha_{31}}}} \quad (8.128)$$

8.4.1.1 Melting Temperature Range, Approximate Solution

Tien and Geiger (1967) presented an interesting approximate solution to the preceeding problem. The initial liquid temperature is at the liquidus temperature; thus there are only two regions to consider.

$$\frac{\partial^2 T_1}{\partial x^2} = \frac{1}{\alpha_1}\frac{\partial T_1}{\partial t} \qquad 0 < x < X_1 \quad (8.129)$$

$$\frac{\partial^2 T_2}{\partial x^2} + \frac{\rho_2 l}{k_2}\frac{df_s}{dt} = \frac{1}{\alpha_2}\frac{\partial T_2}{\partial t} \qquad X_1 < x < X \quad (8.130)$$

$$T_1(0,t) = T_s \quad (8.131a)$$

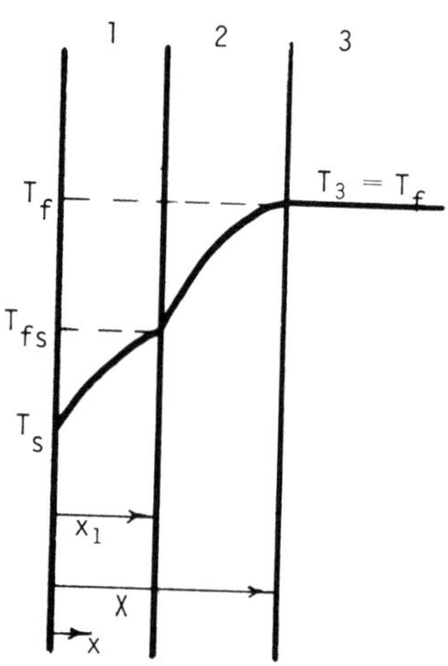

$$T_1(X_1,t) = T_2(X_1,t) = T_{fs} \tag{8.131b}$$

$$T_2(X,t) = T_f \tag{8.131c}$$

$$k_1 \frac{\partial T_1}{\partial x} = k_2 \frac{\partial T_2}{\partial x} + \rho_1 l(1-f)\frac{dX_1}{dt};\ x = X_1 \tag{8.131d}$$

$$\frac{\partial T_2}{\partial x} = 0 \qquad x = X \tag{8.131e}$$

$$\frac{df_s}{dt} = f\frac{d}{dt}\left(1 - \frac{(x - X_1)}{\Delta X}\right) \tag{8.131f}$$

where

$$\Delta X = X - X_1 \tag{8.131g}$$

The solution in the solid region is given by Eq. (8.126)

$$\frac{T_1 - T_s}{T_{fs} - T_s} = \frac{erf\left(\dfrac{x}{2\sqrt{\alpha_1 t}}\right)}{erf\lambda}$$

where, as previously

$$X_1 = 2\lambda\sqrt{\alpha_1 t} \tag{8.122}$$

In the mixed phase region, the heat balance integral method is used. The heat balance equation is given by

$$\frac{1}{\alpha_2}\left[\frac{d\theta_2}{dt} - T_2(X,t)\frac{dX}{dt} + T_2(X_1,t)\frac{dX_1}{dt}\right] = \frac{\partial T_2(X,t)}{\partial x}$$
$$- \frac{\partial T_2(X_1,t)}{\partial x} + \frac{\rho_2 lf}{k_2}\left(\frac{1}{2}\frac{d\Delta X}{dt} + \frac{dX_1}{dt}\right) \tag{8.132}$$

with

$$\theta_2 = \int_{X_1}^{X} T_2(x,t)\, dx \tag{8.133}$$

A quadratic temperature profile is assumed, which satisfies Eqs. (8.131*b,c,e*).

$$T_2 = T_{fs} + \frac{2\Delta T}{\Delta X}(x - X_1) - \frac{\Delta T}{(\Delta X)^2}(x - X_1)^2$$

where $\Delta T = (T_f - T_{fs})$. Equation (8.132) is then

$$\left(1 + \frac{l}{c_2}\frac{f}{\Delta T}\right)\frac{dX_1}{dt} + \left(\frac{1}{3} + \frac{1}{2}\frac{lf}{c_2 \Delta T}\right)\frac{d\Delta X}{dt} = \frac{2\alpha_2}{\Delta X} \tag{8.134}$$

With Eq. (8.126), the boundary condition (8.131d) yields a second differential equation

$$\frac{(T_{fs} - T_s)}{\sqrt{\pi\,\alpha_1 t}} \frac{e^{-\left(\frac{X_1}{2\sqrt{\alpha_1 t}}\right)^2}}{erf\lambda} = \frac{2k_2 \Delta T}{\Delta X} + \rho_2 l(1 - f)\frac{dX_1}{dt} \tag{8.135}$$

The following equations will be used

$$X_1 = 2\lambda\sqrt{\alpha_1 t} \tag{8.122}$$

$$\Delta X = 2\lambda_1\sqrt{\alpha_2 t} \tag{8.136}$$

Inserting Eqs. (8.122) and (8.136) into Eqs. (8.134) and (8.135) yields the equations for λ, λ_1, the solutions of which are

$$\lambda_1 = \frac{\sqrt{\alpha_{12}}}{k_{12}\dfrac{(T_{fs} - T_s)}{\Delta T}\dfrac{e^{-\lambda^2}}{\sqrt{\pi}\,erf\lambda} - \dfrac{\alpha_{12} l(1-f)\lambda}{c_2 \Delta T}} \tag{8.137}$$

$$B_1\lambda + \frac{B_4}{B_2\dfrac{e^{-\lambda^2}}{erf\lambda} - B_3\lambda} = B_2\frac{e^{-\lambda^2}}{erf\lambda} - B_3\lambda \tag{8.138}$$

where

$$B_1 = 1 + \frac{lf}{c_2 \Delta T}$$

$$B_2 = k_{12}\left(\frac{T_{sf} - T_s}{\Delta T}\right)\frac{\alpha_{21}}{\sqrt{\pi}}$$

$$B_3 = \frac{l(1-f)}{c_2 \Delta T}$$

$$B_4 = \left(\frac{1}{3} + \frac{lf}{2c_2\Delta T}\right)\alpha_{21}$$

8.4.2. Sinusoidal Surface Temperature

The Neumann solution is for a surface temperature that instantaneously changes to a fixed value at the beginning of phase change. Practical surface temperatures will rarely be of this form, and it is of some interest to examine the relation between the Neumann solution and a sinusoidal surface temperature. This variable surface temperature problem cannot be solved exactly, as has been previously discussed, but an acceptable quasi-steady solution can be found.

Consider a system that is initially at the fusion value and then undergoes a sinusoidal surface temperature variation

$$\frac{\partial^2 T}{\partial x^2} = \frac{1}{\alpha}\frac{\partial T}{\partial t} \tag{8.139}$$

$$T(X,t) = T_f \tag{8.139a}$$

$$T(o,t) = T_s(t) = T_f + \theta_c \sin\left(\frac{2\pi t}{P}\right) \tag{8.139b}$$

$$k\frac{\partial T(X,t)}{\partial x} = \pm \rho l \frac{dX}{dt} \tag{8.139c}$$

The sign in Eq. (8.139*c*) denotes the thawing (−) or the freezing situation. Nondimensionalize by using

$$\theta = \frac{T - T_f}{\theta_c} \qquad \tau = \frac{t}{t_c} = \frac{2\pi t}{P} \qquad S_T = \frac{c\theta_c}{l}$$

$$y = \left(\frac{\rho l}{k\theta_c t_c}\right)^{1/2} x \qquad \xi = \left(\frac{\rho l}{k\theta_c t_c}\right)^{1/2} X$$

The equations to solve are

$$\frac{\partial^2 \theta}{\partial y^2} = S_T \frac{\partial \theta}{\partial T} \tag{8.140}$$

$$\theta(\xi,\tau) = 0 \tag{8.140a}$$

$$\theta(o,\tau) = \theta_s(\tau) = -\sin\tau \tag{8.140b}$$

$$\frac{d\xi}{d\tau} = \pm \frac{\partial \theta(\xi,\tau)}{\partial y} \tag{8.140c}$$

$$\xi(o) = 0 \tag{8.140d}$$

The solution will be obtained for the first-order quasi-steady equations. These equations are described in Section 8.3.1.2. The solution here will start with freezing an initially thawed medium and then thawing the frozen system. The equations for each of the phase changes are found, using the properties of the frozen and thawed material. The effect due to the difference between the maximum thaw and freeze depths will not be considered. This will not be a serious error if the Stefan numbers are less than 1.

The solutions are straightforward and are

$$\theta = \theta(\tau)\left(1 - \frac{y}{2\sin\frac{\tau}{2}}\right) + \frac{S_T}{2} y\left(y - 2\sin\frac{\tau}{2}\right)\left[2\sin^2\frac{\tau}{2} - 1 - \frac{1}{6}\sin\frac{\tau}{2}\left(y + 2\sin\frac{\tau}{2}\right)\right] \tag{8.141}$$

$$\xi = 2\sin\frac{\tau}{2} - \frac{S_T}{3}\left[\frac{2}{3} + 2\cos\frac{\tau}{2} - \frac{8}{3}\cos^3\frac{\tau}{2}\right] \tag{8.142}$$

Equations (8.141) and (8.142) can then be used for $0 \leq \tau \leq \pi$ with the properties of the frozen and thawed material.

Lock et al. (1969) give a similar solution for the phase change

$$\xi = 2\sin\frac{\tau}{2} - \frac{S_T}{3}\sin\frac{\tau}{2}\sin\tau \tag{8.143}$$

This relation does not appear to be as accurate as Eq. (8.142), although the values are similar.

Seban (1971) noted that the even simpler Stefan equation gave acceptable results and also that the effect of convection due to density variation, in water, is small but may be important for some systems (see also Yen 1968).

The same procedure can also be used for other surface temperature variations. If the surface temperature is not symmetric about the fusion temperature, the results are more complicated, but follow in the same way.

Equation (8.143) can be compared to the step change of surface temperature, the Neumann solution. Using the approximation for γ given by Eq. (8.108), Section 8.3.2.2, with a constant surface temperature equivalent to the sinusoidal temperature, gives the step change solution as

$$\xi = 2\sqrt{\frac{\tau}{\pi + S_T}} \tag{8.144}$$

Figure 8.9 shows the freeze depths and the freezing rates for $S_T = 0.1$. Notice the total freeze depths, for the two surface temperature cases, are within 1%, but the freeze depths, and especially the freezing rates, differ considerably at intermediate times. Thus, the Neumann solution will model a variable surface temperature if the total phase change depth is desired, but will not give comparable intermediate values for the phase change depth or rate.

8.4.3. Variable Latent Heat

For many materials the latent heat is a fixed quantity or varies weakly with the thermodynamic state. For soil systems, however, the latent heat is directly proportional to the water content, and, since the water content can vary spatially, it follows that the latent heat will also. We will ignore the effects due to the fact that all the water in a soil system need not change phase.

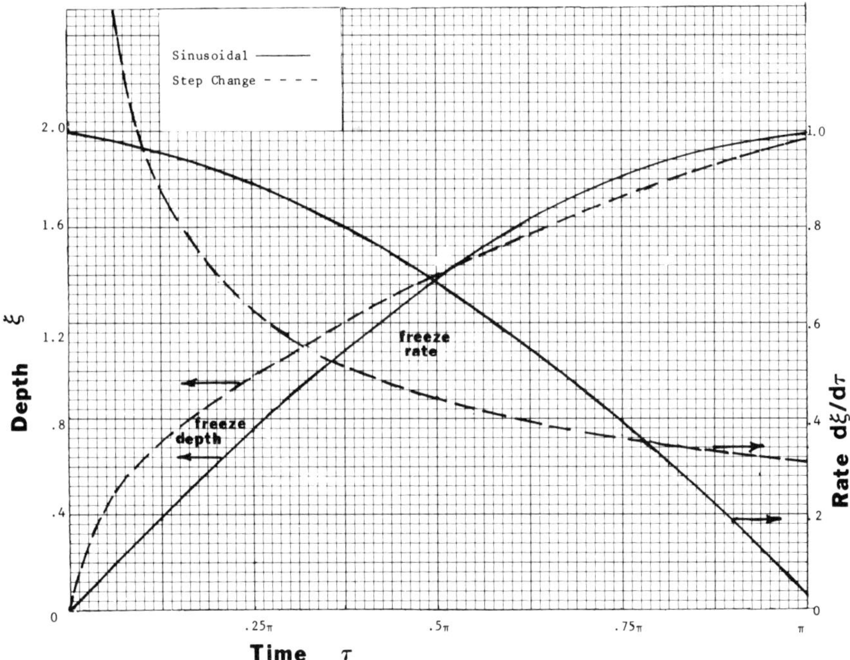

Fig. 8.9. Freezing of Semi-Infinite Medium with Sinusoidal and Step Change Surface Temperatures.

For simplicity, based on data for clay soils, Lock 1969b assumed that the water content decays exponentially from the surface value. The equations for a freezing system, initially at the freezing temperature, are

$$\frac{\partial^2 \theta}{\partial y^2} = S_T \frac{\partial \theta}{\partial \tau} \tag{8.145}$$

$$\theta(o,\tau) = -1$$

$$\theta(\xi,\tau) = 0$$

$$\frac{\partial \theta}{\partial y}(\xi,\tau) = f(\xi)\frac{d\xi}{d\tau}$$

$$\xi(o) = 0$$

$$l = l_o f(y) = l_o e^{-y}$$

The notation is that of Section 8.4.2, except that $S_T = c\theta_c/l_o$.

The zeroth-order, quasi-steady equation for the phase change interface is

$$\frac{d\xi}{d\tau} = \frac{1}{\xi f(\xi)} \tag{8.146}$$

The solution to this equation depends upon the water content function f.

If the water content is constant, then the Stefan solution results

$$\tau_1 = \frac{\xi^2}{2} \tag{8.147}$$

With an exponential water content function

$$\tau_2 = 1 - (1+\xi)e^{-\xi} \tag{8.148}$$

Finally, if the average value of the latent heat between the surface and the freezing depth is used

$$\tau_3 = \frac{\xi}{2}(1 - e^{-\xi}) \tag{8.149}$$

From these results, it can be noted that the time to freeze a layer ξ = 1, for an exponential decay will be about 50% that of the constant latent heat solution and 85% that for an average latent heat assumption. Some caution should therefore be used in applying the constant or average water content solutions.

8.4.4 Convective Boundary Conditions, Solid Initially at Fusion Temperature

8.4.4.1 Analogue Solution

There are no exact solutions for this problem when the initial temperature is different from the fusion temperature. Kreith and Romie (1955) used an electrical analogue to obtain a solution when the initial temperature is at the fusion value.

The problem is

$$\frac{\partial^2 T_l}{\partial x^2} = \frac{1}{\alpha_l}\frac{\partial T_l}{\partial x} \tag{8.150}$$

$$T_2 = T_f$$

$$T_l(X,t) = T_f \tag{8.150a}$$

$$T_l(x,0) = T_f \tag{8.150b}$$

$$X(0) = 0 \tag{8.150c}$$

$$k_l\frac{\partial T_l}{\partial x} = \rho_l l \frac{dX}{dt} \qquad x = X \tag{8.150d}$$

At the surface of the solid, the conduction will equal the convection heat transfer

$$k_l\frac{\partial T_l}{\partial x} = h(T - T_a) \qquad x = 0 \tag{8.150e}$$

The analogue solutions for the surface temperature and the depth of freeze are given in Figs. 8.10 and 8.11. These figures can also be used for melting if the properties of the thawed material are used and if the latent heat is taken as negative so that $l/c_1(T_f - T_a)$ will be positive.

The depth of freeze for this problem agrees within 10% with the Neumann solution ($T_o = T_f$, $T_s = T_a$) if Xh/k is greater than 1. The heat flow to the surface "sees" two thermal resistances, a conductive resistance, which increases with time, and a convective resistance, which is constant. Thus, after a certain time, the effect of the surface resistance is approaching zero, and the solidification proceeds essentially as in the constant surface temperature case.

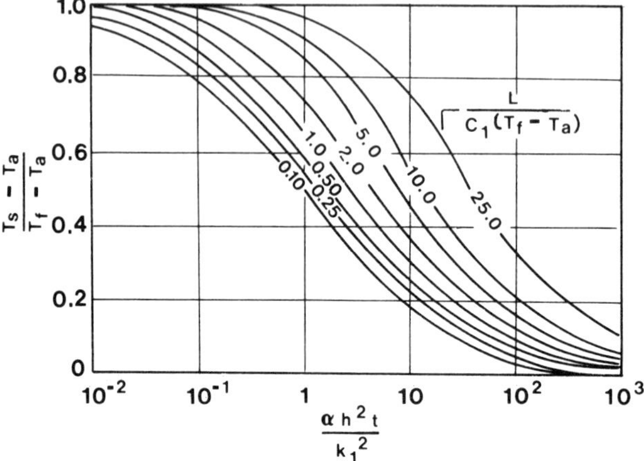

Fig. 8.10. Temperature-Time History of the Surface $x = 0$ during Solidification of a Semi-Infinite Region Subjected to Convection Boundary Condition at the Surface $x = 0$ (from Krieth and Romie 1955).

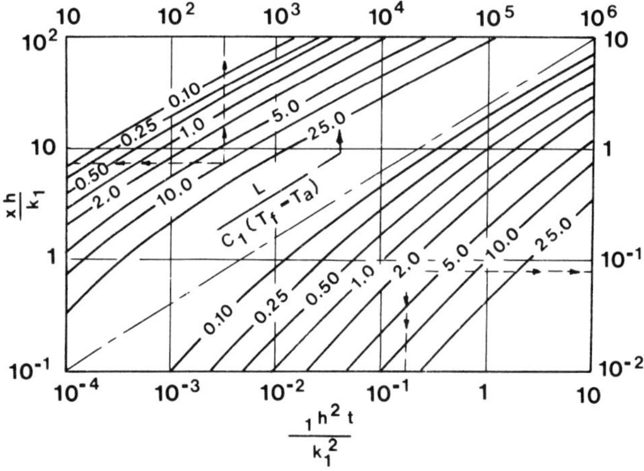

Fig. 8.11. Depth of Solidification as a Function of Time for a Semi-Infinite Region Subjected to Boundary Condition at the Surface $x = 0$ (from Krieth and Romie 1955).

8.4.4.2 Quasi-Steady Approximations

Examine the problem studied under Section 8.4.4.1. No exact solution exists, but the graphs of Kreith and Romie (1955) can be examined. The problem reduces to the following for the zeroth, quasi-steady approximation

$$\frac{\partial^2 T_1}{\partial x^2} = 0 \tag{8.151}$$

$$T_2 = T_f$$

$$T_1(X,t) = T_f \tag{8.151a}$$

$$T_1(x,0) = T_f \tag{8.151b}$$

$$X(0) = 0 \tag{8.151c}$$

$$k_1 \frac{\partial T_1}{\partial x} = \rho_1 l \frac{dX}{dt} \qquad x = X \tag{8.151d}$$

$$k_1 \frac{\partial T_1}{\partial x} = h(T - T_a(t)] \qquad x = 0 \tag{8.151e}$$

The solution to (8.151) and (8.151*e*) is

$$T = a\left(x + \frac{k_1}{h}\right) + T_a(t) \tag{8.152}$$

Then from Eqs. (8.151*a*) and (8.151*d*)

$$\frac{dX}{dt} = \frac{k_1}{\rho_1 l} \frac{(T_f - T_a)}{\left(X + \dfrac{k_1}{h}\right)}$$

The solution to this equation that satisfies Eq. (8.151*c*) is

$$\left(X + \frac{k_1}{h}\right)^2 \equiv p^2 = \frac{2k_1}{\rho_1 l}\int_o^t (T_f - T_a(t))\, dt + \left(\frac{k_1}{h}\right)^2 \tag{8.153}$$

If the surface coefficient h is a function of time, then the equation for the phase change interface is

$$\frac{dp^2}{dt} = \frac{2k_1}{\rho_1 l}(T_f - T_a) - 2p\frac{k_1}{h^2}\frac{dh}{dt}$$

Unfortunately, this equation cannot be simply integrated. With $T_a(t)$, a constant, Eq. (8.153) was first reported by London and Seban (1943). Equation (8.153) was given in its present form by Foss and Fan (1972) and was independently derived and used for surface temperature calculations during freezing and thawing by Lunardini (1978). If T_a is constant, then Eq. (8.153) is

$$\left(X+\frac{k_l}{h}\right)^2=\frac{2k_l}{\rho_l l}(T_f-T_a)t+\left(\frac{k_l}{h}\right)^2 \tag{8.154}$$

and the surface temperature (valid also for $T_a = T_a(t)$) is

$$T_s = T_a + \frac{T_f - T_a}{p}\frac{k_l}{h} \tag{8.155}$$

where

$$p^2 = 2\alpha_l S_T t + \left(\frac{k_l}{h}\right)^2$$

$$S_T = \frac{c_l(T_f - T_a)}{l}$$

Equation (8.151*b*) cannot be strictly satisfied, as there is no time coordinate in Eq. (8.151). Nevertheless, Eq. (8.153) does give $T_S(0) = T_f$, and the temperature in region 1 is valid only for $t > 0$. Equations (8.154) and (8.105) agree very well with Figs. (8.10) and (8.11), the analogue solutions of Kreith and Romie (1955), particularly for small Stefan numbers, as is expected.

8.4.4.3 Heat Balance Integral Approximation

Goodman (1958) has solved this problem using the integral method. The labor involved in the integral method tends to far exceed that of the quasi-steady approximation. The equations for the freeze problem are

$$\frac{d\theta}{dt} - T_f\frac{dX}{dt} = \alpha_l\left[\frac{\rho_l l}{k_l}\frac{dX}{dt} - \frac{\partial T(0,t)}{\partial x}\right] \tag{8.156}$$

$$\theta = \int_0^X T(x,t)\,dx \tag{8.157}$$

$$T(X,t) = T_f \tag{8.158a}$$

$$k_l\frac{\partial T(0,t)}{\partial x} = h[T(0,t) - T_a] \tag{8.158b}$$

$$-k_l\left[\frac{\partial T(X,t)}{\partial x}\right]^2 = \rho_l l \alpha_l \frac{\partial^2 T_l(X,t)}{\partial x^2} \tag{8.158c}$$

Boundary conditions (8.158*a, b, c*) are used to find the coefficients of the quadratic temperature approximation.

$$T = T_f + a(x - X) + b(x - X)^2$$

$$a = \frac{-(1+S) + \sqrt{(1+S)^2 + (\beta - 1)\,S(S+2)}}{\dfrac{k}{2h\Delta T}(\beta - 1)\,S(S+2)}$$

$$b = -\frac{1}{4}\frac{(\beta - 1)}{\Delta T}\,a^2$$

where

$$\theta = \frac{h^2(T_f - T_a)t}{k_l \rho_l l} = \frac{h^2 \Delta T\,t}{k_l \rho_l l}$$

$$S = \frac{h}{k_l}X$$

$$\beta = 1 + 2\,S_T = 1 + 2\frac{c_l \Delta T}{l}$$

$$\theta = \frac{1}{12\beta}\Bigg\{[(1+2\beta) + (2+\beta)S][1 + \beta S(2+S)]^{1/2}$$

$$- \frac{2(\beta - 1)}{\sqrt{\beta}}\ln\frac{[1 + \beta S(2+S)]^{1/2} + [(1+S)\beta]^{1/2}}{1+\sqrt{\beta}}$$

$$-4\beta(\beta - 1)\ln\frac{-1 + \beta(2+S) + [1 + \beta S(2+S)]^{1/2}}{2\beta}$$

$$+ (\beta^2 + 5\beta)\frac{S^2}{2} + 2(\beta^2 + 4\beta - 2)S - (1 + 2\beta)\Bigg\} \quad (8.159)$$

The surface temperature is given by

$$\frac{T_f - T(0,t)}{\Delta T} = \frac{(\beta - 1)S^2 + 2(\beta - 2)S - 2 + 2[1 + \beta S(2+S)]^{1/2}}{(\beta - 1)(2+S)^2} \quad (8.160)$$

Equations (8.159) and (8.160) are plotted as Figs. 8.12 and 8.13. Comparison of these figures shows that the integral approximation is quite close to the analogue solution. It is interesting to note that if the Stefan number is zero ($\beta = 1$), then the integral solutions for the phase change interface and the surface temperature are

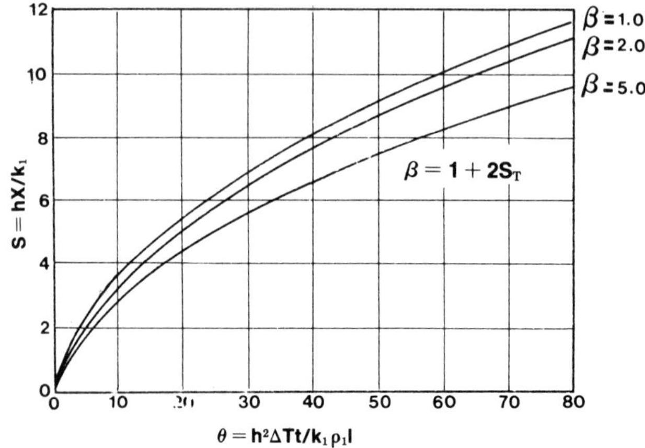

Fig. 8.12. Thickness of Melt versus Time for Aerodynamic Heating or Radiation Boundary Condition, Eq. (8.159).

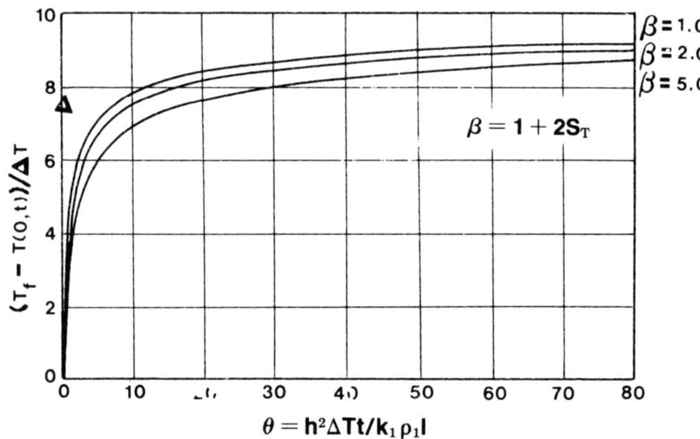

Fig. 8.13. Surface Temperature versus Time for Aerodynamic Heating or Radiation Boundary condition, Eq. (8.160).

$$\theta = \frac{S^2}{2} + S \tag{8.161}$$

$$\frac{T_f - T(0,t)}{\Delta T} = \frac{S}{1+S} \tag{8.162}$$

Replacing θ and S by the usual variables t and X leads to

$$\left(X+\frac{k_1}{h}\right)^2=\frac{2k_1\Delta T}{\rho_1 l}t+\left(\frac{h}{k_1}\right)^2$$

$$T(0,t)=T_a+\frac{\Delta T}{\left(X+\frac{k_1}{h}\right)}\frac{k_1}{h}$$

These are identical to Eqs. (8.154) and (8.155) of the quasi-steady approximation.

8.4.4.4 Constant Heat Flux at Lower Boundary

A problem with significance in terms of the freezing or melting of ice layers over bodies of water can be formulated if the heat flux from the melting liquid is assumed to be constant. Foss and Fan (1974) solve the problem using the quasi-steady method. The initial temperature distribution is not known, but the initial air temperature and the surface temperature of the water are at the freezing temperature. At $t = 0$ the air temperature drops below freezing and may then vary with time (Fig. 8.14). The equations are

$$\frac{\partial^2 T_1}{\partial x^2}=\frac{1}{\alpha_1}\frac{\partial T_1}{\partial t}=0 \tag{8.163}$$

$$k_1\frac{\partial T_1(0,t)}{\partial x}=h_a[T_1(0,t)-T_a(t)] \tag{8.163a}$$

$$T_1(0,0)=T_f \tag{8.163b}$$

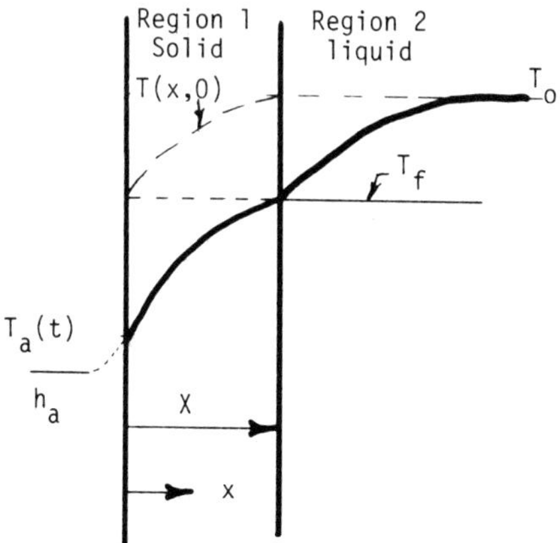

Fig. 8.14. Freezing with Surface Heat Flow.

$$T_a(0) = T_f \tag{8.163c}$$

$$T_1(X,t) = T_f \tag{8.163d}$$

$$k_1 \frac{\partial T_1(X,t)}{\partial x} - k_2 \frac{\partial T_2(X,t)}{\partial x} = \rho_1 l \frac{dX}{dt} \tag{8.163e}$$

At the solid-liquid interface, the heat flux from the liquid is assumed to be constant; therefore

$$k_2 \frac{\partial T_2(X,t)}{\partial x} = Q \tag{8.163f}$$

The solution to Eq. (8.163) with boundary conditions (8.163*a*) and (8.163*d*) is

$$T_1 = \frac{\left(T_f + \dfrac{h\, T_a\, X}{k_1}\right)}{1 + \dfrac{h\, X}{k_1}} \left(1 + \frac{h\, x}{k_1}\right) - \frac{h}{k_1} T_a x \tag{8.164}$$

The differential equation for X is obtained from Eq. (8.163*e*) using Eq. (8.164)

$$\rho_1 l \frac{dX}{dt} = \frac{h_a(T_f - T_a)}{1 + \dfrac{hX}{k_1}} - Q \tag{8.165}$$

The integration of Eq. (8.165) depends upon the functional form of the time variation of the ambient temperature.

(i) Constant Ambient Temperature. For a constant ambient temperature, Eq. (8.165) can be easily integrated. The solution is

$$\frac{h_a Q}{k_1 \rho_1 l} t = \frac{h_a(T_f - T_a)}{Q} \ln \left\{ \frac{h_a(T_f - T_a) - Q}{h_a(T_f - T_a) - Q\left(1 + \dfrac{h_a X}{k_1}\right)} \right\} - \frac{h_a X}{k_1} \tag{8.166}$$

or

$$\frac{\theta^2}{4} = \frac{S_t}{Q_x^2} \ln \left[\frac{S_t - Q_x}{S_t - Q_x(1 + S)} \right] - \frac{S}{Q_x} \tag{8.167}$$

where

$$\theta^2 = \frac{4h_a{}^2 t}{k_l \rho_l c_l}$$

$$S_t = \frac{c_l(T_f - T_a)}{l}$$

$$Q_x = \frac{Qc_l}{h_a l}$$

$$S = \frac{h_a X}{k_l}$$

The steady-state value of X can be found by letting $dX/dt = 0$ in Eq. (8.165). Then

$$X_{\max} = k_l \left[\frac{T_f - T_a}{Q} - \frac{1}{h_a} \right] \tag{8.168}$$

$$S_{\max} = \frac{S_t}{Q_x} - 1 \tag{8.169}$$

These relations are plotted in Fig. 8.15 for some values of S_t, T_a, and Q_x.

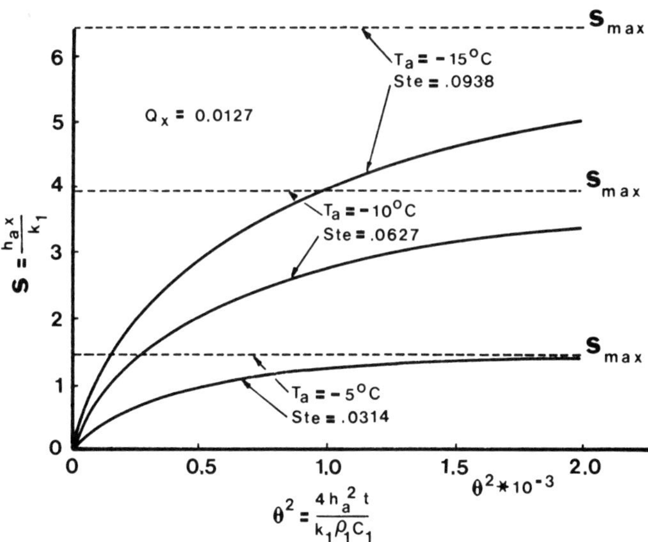

Fig. 8.15. Phase Change Depth for a Constant-Surface Temperature, Eq. (8.167).

(ii) Sinusoidal Ambient Temperature. If the ambient temperature varies with time, a numerical solution of Eq. (8.160) will normally be necessary. Foss and Fan (1974) present a solution for a particular yearly sinusoidal, with daily fluctuations included, and a particular value of Q_x, as shown in Fig. 8.16.

8.4.5 Specified Surface Heat Flux

8.4.5.1 Constant Surface Heat Flux

Kreith and Romie (1955) present an analogue solution for this case with the initial temperature at the fusion temperature. The problem is the same as in Section 8.4.4.1, except that condition (8.150e) is

$$\frac{\partial T_l}{\partial x} = G = \text{constant} \tag{8.170}$$

The solutions, in graphical form, are given for the freeze depth and the surface temperature in Fig. 8.17. Evans et al. (1950) presented an exact solution for this problem by assuming Taylor series expansions for $X(t)$ about $t = 0$. The solution is a series

$$\sigma = \tau - \frac{1}{2}\tau^2 + \frac{5}{6}\tau^3 - \frac{17}{8}\tau^4 + \frac{827}{51}\tau^5 - \ldots \tag{8.171}$$

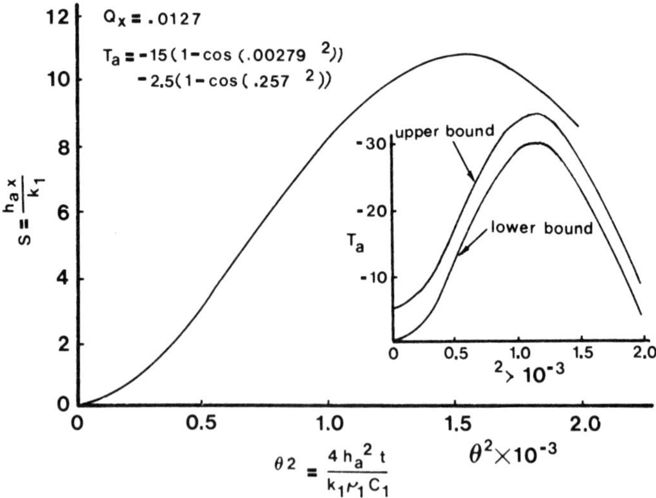

Fig. 8.16. Phase Change Depth for a Sinusoidal Ambient Temperature.

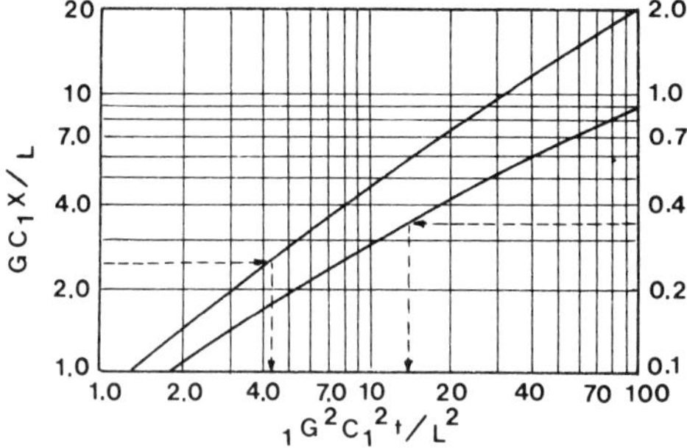

Fig. 8.17. Depth of Solidification and Temperature-Time History of the Surface of a Semi-Infinite Solid from the Surface of which Heat is Transferred at a Constant Rate (adapted from Kreith and Romie 1955).

where

$$\sigma = \frac{Gc_l X}{l} \qquad \tau = \frac{\alpha_l G^2 c_l^2 t}{l^2}$$

The equation is valid only for values of $\tau < 1$, unless many more terms are included.

8.4.5.2 Variable Surface Heat Flux

The initial temperature is again at T_f. The problem is formulated as

$$\frac{\partial^2 T_l}{\partial x^2} = \frac{1}{\alpha_l} \frac{\partial T_l}{\partial t} \tag{8.172}$$

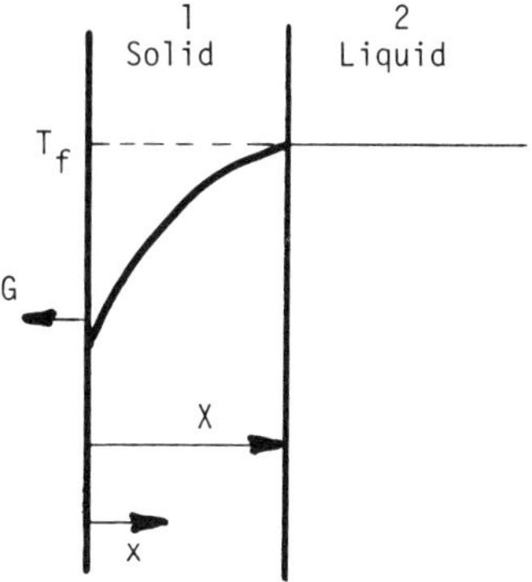

$$T_2 = T_f$$
$$T_l(X,t) = T_f$$
$$T_l(x,0) = T_f$$
$$X(0) = 0$$
$$k_l \frac{\partial T_l(X,t)}{\partial x} = \rho_l l \frac{dX}{dt}$$
$$\frac{\partial T_l(0,t)}{\partial x} = G(t)$$

A heat-balance integral solution for the problem can be found following Goodman (1958). The heat balance integral is

$$\frac{d}{dt}\left[\theta_l - \left(T_f + \frac{\alpha_l \rho_l l}{k_l}\right) X\right] = -\alpha_l G \qquad (8.173)$$

where

$$\theta_l = \int_0^X T_l(x,t)\, dx$$

Equation (8.173) can be integrated immediately to give

$$\theta(X) - \theta(0) - T_f X - \frac{l}{c_l} X = -\alpha_l \int_0^t G(t)\, dt \qquad (8.174)$$

The temperature is again assumed quadratic and with the boundary conditions is

$$T = T_f + a(x - X) + \frac{a - G}{2X}(x - X)^2 \qquad (8.175)$$

where a is given by

$$a = \frac{l}{2c_l X}\left(-1 + \sqrt{1 + 4\frac{G\, c_l X}{l}}\right) \qquad (8.176)$$

Using Eqs. (8.175) and (8.176) in Eq. (8.174) gives the solutions for X and $T_1(0,t)$ as

$$\tau = \frac{\sigma}{6}(5 + 6 + \sqrt{1 + 4\sigma}) \tag{8.177}$$

$$4c_l(T_f - T(0,t)) = 2\sigma - 1 + \sqrt{1 + 4\sigma} \tag{8.178}$$

where

$$\tau = \frac{\alpha_l G c_l^2}{l^2} \int_o^t G(t)\, dt \qquad \sigma = \frac{c_l G X}{l}$$

These solutions are plotted on Figs. 8.18 and 8.19. The results agree very well with the analogue solution of Kreith and Romie (1955), Fig. 8.17, and also with the exact solution of Eq. (8.171) for small values of time (τ). Solutions of this type are not valid for a pulse-type function for G, which vanishes after some finite time. Two-parameter integral heat balance solutions can be used in these cases; see Goodman (1958).

8.4.6 Constant Phase Change Rate

Stefan (1891) gives a solution for the case of a constant heat flux at the phase change interface. The equations are

$$\frac{\partial^2 T_l}{\partial x^2} = \frac{1}{\alpha_l}\frac{\partial T_l}{\partial t} \tag{8.179}$$

$$T(x,0) = T(X,t) = T_f$$

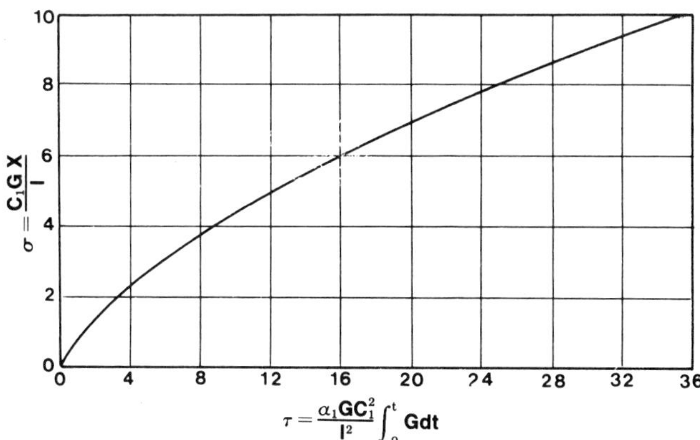

Fig. 8.18. Thickness of Melt Versus Time, for a Given Heat Flux at Boundary, Eq. (8.177).

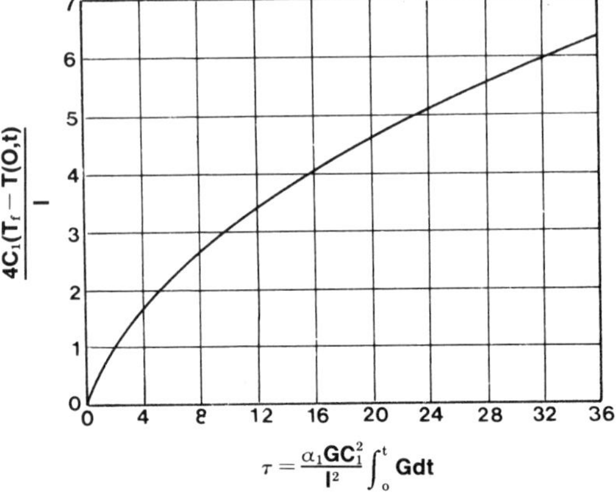

Fig. 8.19. Temperature-Time History on Boundary for a Given Heat Flux at Boundary, Eq. (8.178).

$$k_l \frac{\partial T_l}{\partial x} = \rho_l l \frac{dX}{dt}$$

$$X(0) = 0$$

$$\frac{\partial T_l}{\partial x} = G \qquad x = X$$

The solution to this system is straightforward, and the equations for the temperature of the solid and the phase change location are

$$(T_l - T_f)\frac{c_l}{l} = 1 - e^{\left(\frac{G^2 c_l k_l t}{\rho_l l^2} - \frac{G c_l x}{l}\right)} \tag{8.180}$$

$$X = \frac{k_l G t}{\rho_l l} \tag{8.181}$$

The problem does not have significant practical value, as it requires that a variable surface temperature be imposed upon the solid to maintain melting at a constant rate.

8.4.7 Ablation with Complete Removal of Melt

Phase change problems for which the melting (or vaporized) material is removed from the system are not significant for permafrost systems. Nevertheless, the solutions might be useful for ice melting from vertical surfaces where the water can run off the surface due to gravity.

8.4.7.1 Constant Surface Heat Flux

For this problem, there are two time domains to consider: the time before the surface temperature reaches the fusion value, during which no phase change occurs, and the phase change with removal of melt, during which the surface temperature remains at T_f. These are shown in Fig. 8.20.

Premelt Solution. Initially, the solid is at T_o and at $t = 0$, a heat flux, q, is applied. The thermal penetration depth is $\delta(t)$; when the surface temperature is T_f, $\delta = \delta_m$, and melting begins. This problem has been solved exactly (Carslaw and Jaeger 1959), but the integral method approximation will be used, as this method is used for the melting problem. The problem can be formulated as

$$\frac{\partial^2 T}{\partial x^2} = \frac{1}{\alpha}\frac{\partial t}{\partial t} \tag{8.182}$$

$$T(x,0) = T_o \tag{8.182a}$$

$$k\frac{\partial T(0,t)}{\partial x} = -q \tag{8.182b}$$

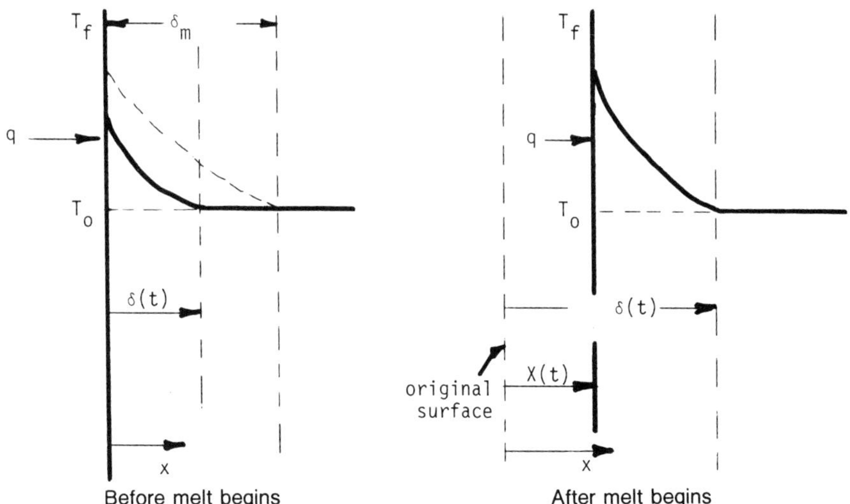

Fig. 8.20. Complete Removal of Melt From Surface.

At the thermal penetration depth, the temperature is T_o, and the heat flux is 0; then

$$T(\delta,t) = 0 \tag{8.182c}$$

$$\frac{\partial T(\delta,t)}{\partial x} = 0 \tag{8.182d}$$

The heat balance integral is given by Eq. (8.76)

$$\frac{d\theta}{dt} + \alpha\frac{\partial T(0,t)}{\partial x} - T_o\frac{d\delta}{dt} = 0 \tag{8.183}$$

with

$$\theta = \int_o^\delta T\,dx \tag{8.184}$$

The temperature is assumed to have a quadratic profile, and with Eqs. (8.182*a,b,c*)

$$T = T_o + \frac{q}{2\delta k}(x-\delta)^2 \tag{8.185}$$

Using Eqs. (8.183–8.185), the equation for δ is

$$\frac{d\delta^2}{dt} = 6\alpha$$

with the solution

$$\delta = \sqrt{6\alpha t} \tag{8.186}$$

The surface temperature is

$$T(0,t) = T_o + \frac{q\delta}{2k}$$

The value of δ and t, when the surface temperature reaches the melt value, are given by

$$\delta_m = \frac{2k(T_f - T_o)}{q} \tag{8.187}$$

$$t_m = \frac{2}{3}\frac{k^2(T_f - T_o)^2}{\alpha q^2} \tag{8.188}$$

The exact value for the time t_m is

$$t_m = 0.785\frac{k^2(T_f - T_o)^2}{\alpha q^2}$$

If a quartic temperature profile is used, the additional smoothness relations at δ can be used

$$\frac{\partial^2 T(\delta,t)}{\partial x^2} = \frac{\partial^3 T(\delta,t)}{\partial x^3} = 0 \tag{8.189}$$

Then

$$T = T_o + \frac{q}{4k\delta^3}(x-\delta)^4$$

$$\delta_m = \frac{4k(T_f - T_o)}{q}$$

$$t_m = 0.800\frac{k^2(T_f - T_o)^2}{\alpha q^2}$$

Melt Solution. Once the surface temperature reaches T_f, at $t = t_m$, melting begins, and the problem is

$$\frac{\partial^2 T}{\partial x^2} = \frac{1}{\alpha}\frac{\partial t}{\partial t} \tag{8.190}$$

$$T(X,t) = T_f \tag{8.190a}$$

$$T(\delta,t) = T_o \tag{8.190b}$$

$$\frac{\partial T(\delta,t)}{\partial x} = 0 \tag{8.190c}$$

$$q + k\frac{\partial T(X,t)}{\partial x} = \rho l\frac{dX}{dt} \tag{8.190d}$$

$$\delta(0) = \delta_m \tag{8.190e}$$

$$X(0) = 0 \tag{8.190f}$$

The heat balance integral equation is now, from Eq. (8.78)

$$\frac{d\theta}{dt} - T_o \frac{d\delta}{dt} + T_f \frac{dX}{dt} + \frac{\alpha \rho l}{k} \frac{dX}{dt} = \frac{\alpha q}{k} \tag{8.191}$$

with

$$\theta = \int_X^\delta T(x,t)\, dx \tag{8.192}$$

The quadratic temperature assumption satisfying Eqs. (8.190*a,b,c*) yields

$$T = T_f - 2\frac{(T_f - T_o)}{(\delta - X)}(x - X) + \frac{(T_f - T_o)}{(\delta - X)^2}(x - X)^2 \tag{8.193}$$

Equations (8.191–8.193) yield

$$\frac{1}{3}\frac{dP}{d\Omega} + (1 + \nu)\frac{dS}{d\Omega} = \nu \tag{8.194}$$

Equations (8.190*d*) and (8.193) lead to a second differential equation

$$\frac{dS}{d\Omega} = 1 - \frac{2}{P} \tag{8.195}$$

where

$$P = \frac{q(\delta - X)}{(T_f - T_o)k} \qquad \nu = \frac{\alpha \rho l}{k(T_f - T_o)}$$

$$\Omega = \frac{q^2 t}{\rho l k(T_f - T_o)} \qquad S = \frac{X q}{k(T_f - T_o)}$$

Eliminating $dS/d\Omega$ from Eqs. (8.194) and (8.195) gives

$$\frac{dP}{d\Omega} - \frac{6(1 + \nu)}{P} = -3$$

The solution to this equation is

$$\Omega = -\frac{1}{3}\left[P - 2 + 2(1 + \nu)\ln\left(\frac{2(1 + \nu) - P}{2\nu}\right)\right] \tag{8.196}$$

From Eq. (8.196)

$$d\Omega = \frac{P\,dP}{3[2(1+\nu) - P]}$$

and then

$$dS = \left(1 - \frac{2}{P}\right)\frac{P\,dP}{3[2(1+\nu) - P]}$$

The solution to this equation is

$$S = -\frac{1}{3}\left[P - 2 + 2\nu \ln\left(\frac{2(1+\nu) - P}{2\nu}\right)\right] \tag{8.197}$$

Equations (8.196) and (8.197) are the parametric equations for S as a function of Ω. Goodman (1958) plotted these equations for some values of ν, as shown on Fig. 8.21. Landau (1950) solved this problem numerically, and the results agree quite well with Eqs. (8.196) and (8.197), but as $\nu \to 0$ the values tend to diverge as Ω becomes small. This can be seen on Fig. 8.22.

A steady-state solution for S can be easily found. If $dS/d\Omega = S_\infty$ then, from Eq. (8.195), $P =$ constant and then

$$S_\infty = \frac{\nu}{1+\nu}$$

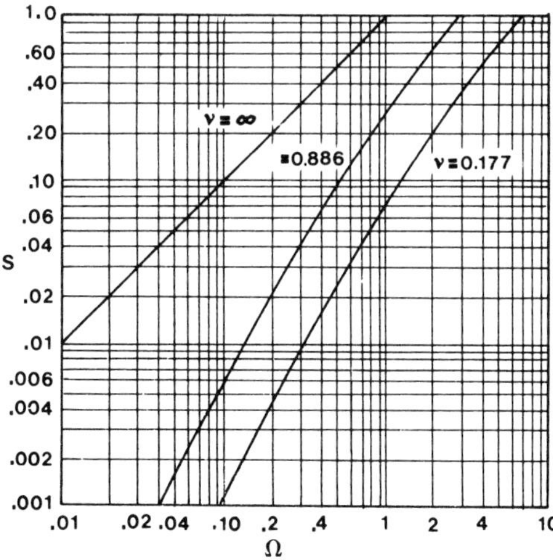

Fig. 8.21. Melt-Line Location Versus Time, with Complete Removal of Melt, Eq. (8.197) (adapted from Goodman 1958).

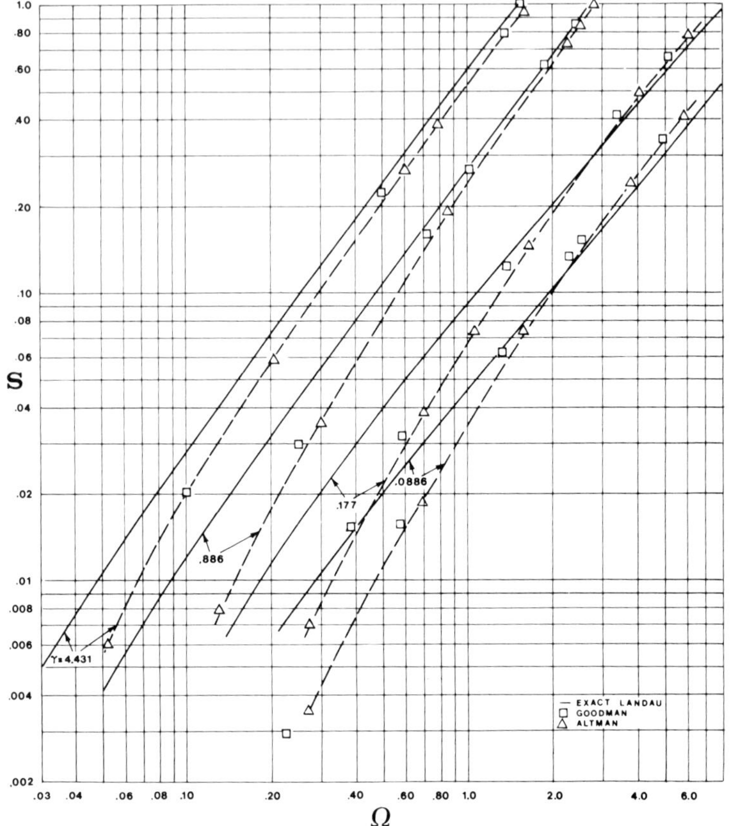

Fig. 8.22. Melt-Line Location versus Time, Constant Surface Heat Flux.

and, for large values of time

$$S = \frac{\nu}{1+\nu}\,\Omega$$

This equation is identical to the exact solution of Landau (1950).

8.4.7.2 Variable Surface Heat Flux

The procedure detailed in the previous section can be used for the case of variable surface heat flux. Following Altman (1961), a fourth-order polynomial for the temperature will be used, with Eq. (8.189). Then

$$T = T_o + \frac{q}{4k\delta^3}(x-\delta)^4$$

The surface temperature is

$$T(0,t) = T_o + \frac{q\delta}{4k}$$

The equation for δ and the time when the surface temperature is T_f are

$$\delta(t) = \sqrt{20\alpha}\left[\frac{1}{q}\int_o^t q\,dt\right]^{1/2} \tag{8.198}$$

$$\delta_m = \frac{4k(T_f - T_o)}{q(t)_m} \tag{8.199}$$

$$\int_o^{t_m} q\,dt = \frac{4k^2(T_f - T_o)^2}{5\alpha\, q(t_m)} \tag{8.200}$$

Equations (8.199) and (8.200) can be used to evaluate δ_m and t_m if the transient function $q(t)$ is specified.

The temperature profile after melting begins is

$$\frac{T - T_f}{T_f - T_o} = -4\left(\frac{x-X}{\delta - X}\right) + 6\left(\frac{x-X}{\delta - X}\right)^2 - 4\left(\frac{x-X}{\delta - X}\right)^3 + \left(\frac{x-X}{\delta - X}\right)^4 \tag{8.201}$$

Using Eqs. (8.190*d*), (8.191), (8.192), and (8.201) again leads to two differential equations

$$\frac{d}{dt}(\delta - X) + 5(1+\nu)\frac{dX}{dt} = \frac{5\alpha q}{k(T_f - T_o)} \tag{8.202}$$

$$\frac{dX}{dt} = \frac{1}{\rho l}\left[q - \frac{4k(T_f - T_o)}{\delta - X}\right] \tag{8.203}$$

Using Eq. (8.202) in (8.203) produces an equation in $(\delta - X)$

$$\frac{d}{dt}(\delta - X) - \frac{20k(T_f - T_o)}{\rho l}(1+\nu)\frac{1}{(\delta - X)} = \frac{-5q(t)}{\rho l}$$

The initial value is again $\delta(0) - X(0) = \delta_m$. This equation can be solved numerically, given $q(t)$, and then

$$X = \frac{1}{\rho l}\int_{o}^{t}\left[q(t) - \frac{4k(T_f - T_o)}{(\delta - X)}\right] dt$$

For $q(t)$ a constant, Altman (1961) presented the solution

$$\Omega = -\frac{1}{5}\left[P - 4 + 4(1+\nu)\ln\left(\frac{4(1+\nu) - P}{4\nu}\right)\right]$$

$$S = -\frac{1}{5}\left[P - 4 + 4\nu\ln\left(\frac{4(1+\nu) - P}{4\nu}\right)\right]$$

As often occurs in integral solutions, this fourth-power approximation is not superior to Goodman's (1958) second-power solution except that t_m is much closer to the exact value.

8.5 PROBLEMS FOR THE FINITE SLAB

Most of the solutions associated with a finite thickness are similar to those of the semi-infinite region, with minor adjustments. The problems may be adapted to the freezing of water of finite depth, ice formation on solid surfaces, etc.

8.5.1 Constant Surface Temperature

8.5.1.1 Exact Solution, Initial Temperature at Fusion Point

Consider a slab of thickness L with the temperature of the liquid initially at the fusion temperature T_f (Fig. 8.23). At $t = 0$, the surface temperature drops to T_s and is held there. The surface at $x = L$ is effectively insulated, as the liquid temperature has a constant value T_f. The equations are

$$\frac{\partial^2 T_1}{\partial x^2} = \frac{1}{\alpha_1}\frac{\partial T_1}{\partial t} \qquad 0 \leq x \leq X \tag{8.204}$$

$$T(0,t) = T_s$$

$$T(X,t) = T_f$$

$$k_1\frac{\partial T_1}{\partial x} = \rho_1 l\frac{dX}{dt} \qquad x = X$$

Ruoff (1958) presented an exact solution for this problem by means of a similarity transformation. The results will be seen to be identical to the equivalent Neumann solution for the semi-infinite solid. Let

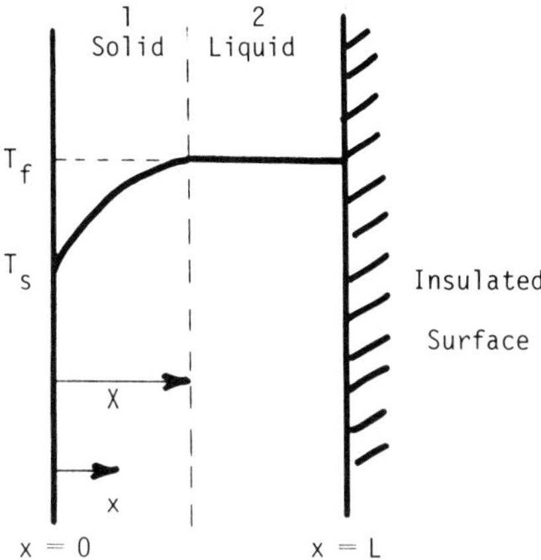

Fig. 8.23. Freezing a Finite Slab.

$$\eta = \frac{x}{2\sqrt{\alpha_l t}}$$

Then Eqs. (8.204) become

$$\frac{d^2T}{d\eta^2} + 2\eta \frac{dT}{d\eta} = 0 \qquad 0 \leq \eta \leq p \tag{8.205}$$

$$T(0) = T_S \tag{8.205a}$$

$$T(p) = T_f \tag{8.205b}$$

$$\frac{dT(p)}{d\eta} = \frac{2l}{c_l}\left[p + 2\,t\frac{dp}{dt}\right] \tag{8.205c}$$

where

$$p = \frac{X}{2\sqrt{\alpha_l t}}$$

Equation (8.205) is integrated in a straightforward manner, and with Eqs. (8.205*a*) and (8.205*b*)

$$(T_f - T_S) = \beta \int_0^p e^{-\eta^2}\, d\eta \tag{8.206}$$

Since the left-hand side of this equation is constant, then p must also be a constant

$$p = \gamma \tag{8.207}$$

Boundary condition (8.205*c*), with Eq. (8.207), leads to

$$\beta = \frac{2l}{c_1} \gamma e^{\gamma^2} \tag{8.208}$$

The equation for γ is obtained with Eq. (8.206) and Eq. (8.208) as

$$\gamma e^{\gamma^2} erf\, \gamma = \frac{c_1(T_f - T_S)}{l\sqrt{\pi}}$$

The temperature in the solid region is then

$$T = T_S + \frac{(T_f - T_S)}{erf\, \gamma} erf\, \eta$$

The phase change interface is

$$X = 2\gamma\sqrt{\alpha_1 t}$$

Notice that these equations are identical to Eqs. (8.26), (8.27), and (8.25), respectively, for the semi-infinite solid. Thus, the solutions are identical until $X = L$, when the slab is completely frozen. The method of this section is, in fact, simply the concept underlying the definition of the error function.

8.5.1.2 Insulated Surface at $x = L$

There are no exact solutions for the finite slab when the initial temperature is different from the fusion temperature. Cho and Sunderland (1969) present an approximate solution for the case of a slab insulated at the nonisothermal surface (Fig. 8.24). The problem is described as

$$\frac{\partial^2 T_1}{\partial x^2} = \frac{1}{\alpha_1}\frac{\partial T_1}{\partial t} \tag{8.209}$$

$$\frac{\partial^2 T_2}{\partial x^2} = \frac{1}{\alpha_2}\frac{\partial T_2}{\partial t} \tag{8.210}$$

$$T_2(x,0) = T_o \tag{8.211a}$$

$$T_1(0,t) = T_s \tag{8.211b}$$

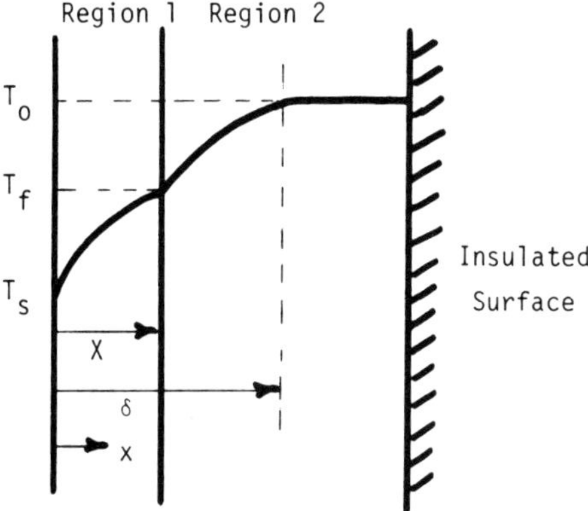

Fig. 8.24. Two-Phase Finite Slab.

$$\frac{\partial T_2(L,t)}{\partial x} = 0 \tag{8.211c}$$

$$T_1(X,t) = T_2(X,t) = T_f \tag{8.211d}$$

$$-k_1\frac{\partial T_1(X,t)}{\partial x} + k_2\frac{\partial T_2(X,t)}{\partial x} = -\rho_1 l\frac{dX}{dt} \tag{8.211e}$$

The assumption of a thermal penetration depth δ, gives rise to two new boundary conditions

$$\frac{\partial T_2(\delta,t)}{\partial x} = 0 \tag{8.211f}$$

$$T_2(\delta,t) = T_o \tag{8.211g}$$

For the solid region, the exact solution for the semi-infinite solid, Eq. (8.27), is applicable

$$\frac{T_1 - T_S}{T_f - T_S} = \frac{erf\left(\dfrac{x}{2\sqrt{\alpha_1 t}}\right)}{erf\lambda} \tag{8.212}$$

$$X = 2\lambda\sqrt{\alpha_1 t} \tag{8.213}$$

In the liquid region, the heat balance integral will be used; see Eq. (8.78)

$$\frac{d\theta_2}{dt} - T_o\frac{d\delta}{dt} + T_f\frac{dX}{dt} + \alpha_2\frac{\partial T_2(X,t)}{\partial x} = 0 \tag{8.214}$$

where

$$\theta_2 = \int_X^{\delta} T_2 \, dx \tag{8.215}$$

Assume a quartic temperature for the liquid region. Then, with Eqs. (8.211*d,f,g*)

$$T_2 = T_o - (T_o - T_f)\left(\frac{\delta - x}{\delta - X}\right)^4 \tag{8.216}$$

Assume that the thermal penetration depth is

$$\delta = 2\beta\sqrt{\alpha_l t} \tag{8.217}$$

Equations (8.213–8.217) then lead to

$$\beta - \lambda = \frac{5}{2}\left[-\lambda + \sqrt{\lambda^2 + \frac{8}{5}\frac{\alpha_2}{\alpha_l}}\right] \tag{8.218}$$

Then using Eq. (8.211*e*)

$$\frac{e^{-\lambda^2}}{erf\lambda} - \frac{2(T_o - T_f)k_2\sqrt{\pi}}{(T_f - T_S)\, k_l(\beta - \lambda)} = \frac{l\lambda\sqrt{\pi}}{c_l(T_f - T_S)} \tag{8.219}$$

Equations (8.218) and (8.219) may be easily solved for λ and β. When the temperature penetration reaches the far wall, $\delta = L$, then the insulated wall temperature will start to decrease. The times when $\delta = L$ and when $X = L$ are given by

$$t_l = \frac{L^2}{4\alpha_l\beta^2} \tag{8.220}$$

$$t_2 = \frac{L^2}{4\alpha_l\lambda^2} \tag{8.221}$$

The insulated wall temperature for $t_l < t < t_2$ may be estimated by assuming

$$T_2(x,t) = T_w - (T_w - T_f)\left(\frac{L - x}{L - X}\right)^4$$

where $T_w = T_2(L,t)$ is the temperature of the insulated wall when $t_1 < t < t_2$. Using the heat balance integral in region 2 again, a differential equation for the transient temperature T_w may be generated

$$(L-X)\frac{d}{dt}(T_w - T_f) + \left(\frac{5\alpha_2}{L-X} - \frac{dX}{dt}\right)(T_w - T_f) = 0$$

The solution to this equation is

$$\frac{T_2(L,t) - T_f}{T_o - T_f} = \left(\frac{a}{b}\right)^{-(1+c)} e^{c\left(\frac{1}{b} - \frac{1}{a}\right)} \qquad t_1 < t < t_2 \qquad (8.222)$$

where

$$a = 1 - \frac{2\lambda}{L}\sqrt{\alpha_1 t}$$

$$b = 1 - \frac{\lambda}{\beta}$$

$$c = \frac{5}{2}\lambda^2 \frac{\alpha_2}{\alpha_1}$$

When the time is greater than t_2, the entire liquid region has frozen, and the problem reduces to that of a single phase. Cho and Sunderland (1969) also give a solution to the above problem if two distinct phase change temperatures exist, with two different latent heats.

8.5.1.3 Extended Phase Change Temperature Range, Insulated Surface

An approximate solution to this problem is given by Cho and Sunderland (1969). The method follows closely the results of Section 8.4.1. The equations, following the notation of Section 8.4.1, are (Fig. 8.25)

$$\frac{\partial^2 T_1}{\partial x^2} = \frac{1}{\alpha_1}\frac{\partial T_1}{\partial t} \qquad 0 < x < X_1 \qquad (8.223)$$

$$\frac{\partial^2 T_2}{\partial x^2} + \frac{\rho_2 l}{k_2}\frac{df_s}{dt} = \frac{1}{\alpha_2}\frac{\partial T_2}{\partial t} \qquad X_1 < x < X \qquad (8.224)$$

$$\frac{\partial^2 T_3}{\partial x^2} = \frac{1}{\alpha_3}\frac{\partial T_3}{\partial t} \qquad X < x < L \qquad (8.225)$$

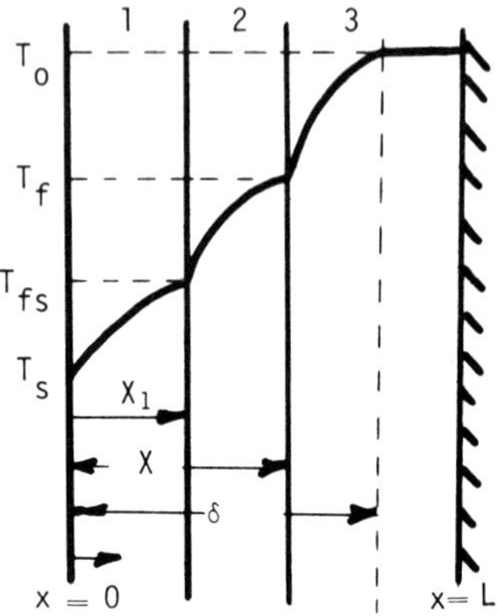

Fig. 8.25. Phase Change Regions for Finite Slab.

$$T_3(x,0) = T_0 \tag{8.226a}$$

$$\frac{\partial T_3(L,t)}{\partial x} = 0 \tag{8.226b}$$

$$T_1(0,t) = T_S \tag{8.226c}$$

$$T_1(X_1,t) = T_2(X_1,t) = T_{fS} \tag{8.226d}$$

$$T_2(X,t) = T_3(X,t) = T_f \tag{8.226e}$$

$$k_1 \frac{\partial T_1(X_1,t)}{\partial x} = k_2 \frac{\partial T_2(X_1,t)}{\partial x} + \rho_1 l (1-f) \frac{dX_1}{dt} \tag{8.226f}$$

$$k_2 \frac{\partial T_2(X,t)}{\partial x} = k_3 \frac{\partial T_3(X,t)}{\partial x} \tag{8.226g}$$

$$\frac{\partial T_3(\delta,t)}{\partial x} = 0 \tag{8.226h}$$

$$T_3(\delta,t) = T_o \tag{8.226i}$$

The usual assumptions, as follows, are used

$$X_1 = 2\lambda\sqrt{\alpha_1 t}$$
$$X = 2\eta\sqrt{\alpha_1 t}$$
$$\delta = 2\beta\sqrt{\alpha_1 t}$$

The exact solutions for regions 1 and 2 are given by Eqs. (8.27) and (8.127)

$$\frac{T_1 - T_S}{T_{fS} - T_S} = \frac{erf\dfrac{x}{2\sqrt{\alpha_1 t}}}{erf\lambda} \tag{8.227}$$

$$\frac{T_2 - T_S}{T_{fS} - T_S} = 1 - \frac{lf\left(\dfrac{x}{2\sqrt{\alpha_1 t}} - \lambda\right)}{c_{p2}(T_{fS} - T_S)(\eta - \lambda)} + \frac{T_f - T_{fS} + lf/c_{p2}}{(T_{fS} - T_S)} \frac{\left(erf\dfrac{x}{2\sqrt{\alpha_2 t}} - erf\dfrac{\lambda}{\sqrt{\alpha_{21}}}\right)}{erf\dfrac{\eta}{\sqrt{\alpha_{21}}} - erf\dfrac{\lambda}{\sqrt{\alpha_{21}}}} \tag{8.228}$$

The heat balance integral for region 3 is again

$$\frac{d\theta_3}{dt} - T_o\frac{d\delta}{dt} + T_f\frac{dX}{dt} + \alpha_3\frac{\partial T_3(X,t)}{\partial x} = 0 \tag{8.229}$$

$$\theta_3 = \int_X^\delta T_3\,dx$$

The fourth-power temperature for region 3 is

$$T_3 = T_o - (T_o - T_f)\left(\frac{\delta - x}{\delta - X}\right)^4$$

Then from Eq. (8.229)

$$\beta - \eta = \frac{5}{2}\left[-\eta + \sqrt{\eta^2 + \frac{8}{5}\frac{\alpha_3}{\alpha_1}}\right] \tag{8.230}$$

The equation for λ follows directly from Eq. (8.124)

$$\frac{e^{-\lambda^2}}{erf\lambda} + \frac{\sqrt{\pi}\,k_{21}\,fl}{2\,c_{p2}(T_{fS} - T_S)(\eta - \lambda)} - \frac{k_{21}\left(T_f - T_{fS} + \dfrac{lf}{c_{p2}}\right)e^{-\frac{\lambda^2}{\alpha_{21}}}}{\sqrt{\alpha_{21}}\,(T_f - T_S)\left(erf\dfrac{\eta}{\sqrt{\alpha_{21}}} - erf\dfrac{\lambda}{\sqrt{\alpha_{21}}}\right)} = \frac{\sqrt{\pi}\,(1-f)l\lambda}{c_{p1}(T_{fS} - T_S)} \tag{8.231}$$

Equation (8.226g) leads to

$$\frac{-lf}{c_{p2}(\eta-\lambda)}+\frac{2\left(T_f-T_{fS}+\frac{lf}{c_{p2}}\right)\sqrt{\alpha_{12}}\,e^{-\eta^2\alpha_{12}}}{\sqrt{\pi}\left(erf\frac{\eta}{\sqrt{\alpha_{21}}}-erf\frac{\lambda}{\sqrt{\alpha_{21}}}\right)}=\frac{4k_{32}(T_o-T_f)}{\beta-\eta} \tag{8.232}$$

The simultaneous solution of Eqs. (8.230–8.232) will give λ, η, β. The temperature of the slab at $x = L$ follows exactly, as in Section 8.5.1.2

$$\frac{T_3(L,t)-T_f}{T_o-T_f}=\left(\frac{a}{b}\right)^{-(1+c)}e^{c\left(\frac{1}{b}-\frac{1}{a}\right)} \qquad t_1<t<t_2$$

where

$$a=1-\frac{2\eta}{L}\sqrt{\alpha_1 t}$$

$$b=1-\frac{\eta}{\beta}$$

$$c=\frac{5}{2}\frac{\alpha_3}{\alpha_1\eta^2}$$

8.6 CYLINDRICAL PROBLEMS, INFINITE DOMAIN, GROWTH OUTWARD

In contrast to the Neumann problem, there are no general, exact solutions for phase change in cylindrical coordinates. Because this is an important geometry for practical applications, a significant effort has been expended upon analytical solutions—limited to certain domains, or approximate solutions. The geometry of the system, shown in Fig. 8.26, is that of a circular cylinder surrounded by an infinite, homogeneous medium.

The energy condition at the phase change interface is given by the same relation as in the cartesian case

$$-k_1\frac{\partial T_1}{\partial r}+k_2\frac{\partial T_2}{\partial r}=\pm\rho l\frac{dR}{dt} \tag{8.233}$$

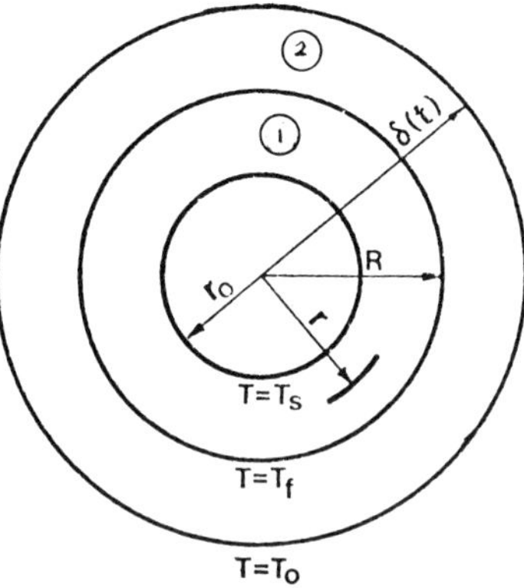

Fig. 8.26. Cylindrical Phase Change Notation.

where the upper sign is for melting and the lower is for freezing. This assumes that the interface motion is in the positive direction of the coordinate r. The superheat or subcooling parameter, ϕ, is defined as

$$\phi = \begin{cases} \dfrac{T_o - T_f}{T_f - T_s} & \text{freeze} \\ \dfrac{T_f - T_o}{T_s - T_f} & \text{thaw} \end{cases}$$

8.6.1 Zero Superheat

A number of solutions are available for the case when the temperature of the medium is initially uniform at the fusion temperature.

8.6.1.1 Constant Phase Change Rate

A simple system that has an exact solution requires that the temperature gradient at the phase change interface be a constant. This case has been discussed by Kreith and Romie (1955) for inward solidification. The problem for outward solidification can be written as follows

$$\frac{1}{r}\frac{\partial}{\partial r}\left(r\frac{\partial T}{\partial r}\right) = \frac{1}{\alpha}\frac{\partial T}{\partial t} \tag{8.234}$$

$$T(R,t) = T_f \tag{8.234a}$$

$$k\frac{\partial T(R,t)}{\partial r} = \rho l \frac{dR}{dt} \tag{8.234b}$$

$$T(r_o,t) = T_s \tag{8.234c}$$

The temperature gradient at the phase change interface is specified as

$$\frac{\partial T(R,t)}{\partial r} = G \tag{8.234d}$$

It is not necessary to solve Eq. (8.234) to find the location of the freezing front. From Eqs. (8.234*b*) and (8.234*d*)

$$\frac{dR}{dt} = \frac{k}{\rho l} G$$

Thus, the rate of movement of the phase change interface is a constant. The location is

$$R = \frac{kG}{\rho l} t + r_o$$

or

$$\beta = 1 + \frac{G\, r_o}{(T_f - T_s)} \tau \tag{8.235}$$

where

$$\tau = \frac{\alpha_l S_T t}{r_o^2}$$

$$\beta = \frac{R}{r_o}$$

Kreith and Romie (1955) used an iterative type series solution for the temperature of the solid. This will not be given here, as the problem is not too practical due to the necessity of imposing a complicated transient temperature at the cylinder surface.

8.6.1.2 Zero Sensible Heat, $S_T = 0$

If the sensible heat of the material is also neglected, then particularly simple solutions are available. Since the Stefan number is a measure of the sensible heat, this situation is equivalent to assuming that the Stefan number is 0. Carslaw and Jaeger (1959) presented a thaw solution when the surrounding medium is at the phase change temperature T_f, using the quasi-steady approximation. The problem is as follows

$$\frac{d}{dr}\left(r\frac{dT}{dr}\right)=0 \tag{8.236}$$

$$T(R)=T_f \tag{8.236a}$$

$$T(r_o)=T_s \tag{8.236b}$$

Equation (8.233), the interface energy balance is now

$$-\left(k\frac{\partial T}{\partial r}\right)_t+\left(k\frac{\partial T}{\partial r}\right)_f=\rho l\frac{dR}{dt} \tag{8.236c}$$

The solution for the phase change location is straightforward and is

$$\frac{1}{2}\beta^2\ln\beta-\frac{\beta^2}{4}+\frac{1}{4}=\tau \tag{8.237}$$

Equation (8.237) will overestimate the thaw beneath the pipe because all the energy transfer from the pipe will go into thawing. A correction could be made by using an effective latent heat in place of l, but better relations are available for this problem. The heat transfer at the cylinder surface is

$$q^*=\frac{qc_l}{2\pi k_l l}=\frac{S_T}{\ln\beta}$$

8.6.1.3 Heat Balance Integral Approximation

The heat balance integral can also be conveniently used for the no superheat problem. The equations are

$$\frac{\partial}{\partial r}\left(r\frac{\partial T}{\partial r}\right)=\frac{r}{\alpha}\frac{\partial T}{\partial t} \tag{8.238}$$

$$T(R,t)=T_f \tag{8.238a}$$

$$T(r_o,t)=T_s \tag{8.238b}$$

$$T(r,o)=T_f \tag{8.238c}$$

$$-k\frac{\partial T(R,t)}{\partial r} = \rho l \frac{dR}{dt} \tag{8.238d}$$

Integration of Eq. (8.238) once, over the space coordinate, yields the heat balance integral equation

$$\alpha\left[R\frac{\partial T(R,t)}{\partial r} - r_o\frac{\partial T(r_o,t)}{\partial r}\right] = \frac{d\theta}{dt} - RT_f\frac{dR}{dt} \tag{8.239}$$

where the integrated temperature is

$$\theta = \int_{r_o}^{R} rT\,dr \tag{8.240}$$

Lardner and Pohle (1961) noted that the logarithmic approximation is more appropriate than a polynomal in r as the area is varying with r. They suggest that $T = f(r)\ln r$ be used as an approximation. The temperature is, thus, assumed to be a logarithmic relation satisfying Eqs. (8.238a) and (8.238b).

$$T = T_s - (T_s - T_f)\frac{\ln\dfrac{r}{r_o}}{\ln\dfrac{R}{r_o}} \tag{8.241}$$

Equations (8.239–8.241) then yield a differential equation for the phase change location

$$\left[\left(\frac{\beta}{2} - \frac{\beta^2 - 1}{4\beta\ln\beta}\right)S_T + \beta\ln\beta\right]d\beta = d\tau$$

This can be integrated to give

$$\frac{S_T}{4}\left[\beta^2 - 1 - 2(\ln\beta) - (\ln\beta)^2 - \ldots\frac{2^n(\ln\beta)^n}{nn!}\ldots\right] + \frac{\beta^2}{2}\ln\beta - \frac{\beta^2}{4} + \frac{1}{4} = \tau \tag{8.242}$$

This equation reduces to the zero sensible heat solution, Eq. (8.237), when the Stefan number is 0.

8.6.2 Finite Superheat

The more practical problems for which the initial temperature is not at the fusion value, present significantly more difficult analyses. Several methods have been utilized. All the problems assume that the pipe is buried at an infinite depth. A finite burial depth will present very severe restraints on the solution methods.

8.6.2.1 Quasi-Steady Solution

A simple solution can again be obtained with the quasi-steady state approximation. Khakimov (1957) investigated the problem and introduced the concept of a thermal layer of influence around the buried pipe. Let us consider the case of thawing of a medium initially frozen at $T = T_o$. A hot cylinder with a surface temperature T_s is inserted in the medium at time zero. The thaw effect is assumed to extend, at any time t, to a finite distance δ, as shown in Fig. 8.26. That is, the temperature of the medium will be T_o at some location sufficiently far from the hot cylinder. This concept of a temperature penetration is also used in the heat balance integral method and in boundary layer theory.

Khakimov (1957) assumed, based on experimental evidence, that $b = \delta/R$ was a constant equal to 4.5. The assumption that b is a constant, for a given S_T and ϕ, is correct for the Neumann problem (see Lunardini and Varotta 1980 and Section 8.3.2.2), but it is invalid for a cylindrical system. Nevertheless, it does simplify the equations and yields a reasonable solution.

The temperature will be the solution of

$$\frac{d}{dr}\left(r\frac{du}{dr}\right) = 0 \tag{8.243}$$

for each region with the boundary conditions

$$u_1 = u_2 = u_f \qquad r = R \tag{8.243a}$$

$$u_1 = u_s \qquad r = r_o \tag{8.243b}$$

$$u_2 = 0 \qquad r = \delta \tag{8.243c}$$

where

$$u = T - T_o$$

The solutions for the temperatures are

$$u_1 = \frac{(T_f - T_s)}{\ln\left(\frac{R}{r_o}\right)} \ln\left(\frac{r}{r_o}\right) + u_s$$

$$u_2 = \frac{u_f}{\ln\left(\frac{R}{\delta}\right)} \ln\left(\frac{r}{\delta}\right)$$

The total energy added to the thawing medium will be the latent heat to thaw the layer between r_o and R and the sensible heat to increase the temperature of all the layers up to $r = \delta$. Thus

$$Q = \pi(R^2 - r_o^2)L + 2\pi\, C_1 \int_{r_o}^{R} r\, u_1\, dr + 2\pi\, C_2 \int_{R}^{\delta} r\, u_2\, dr$$

Carrying out the integration yields

$$Q = \pi(R^2 - r_o^2)L + 2\pi\, C_1 \left[(T_f - T_s)\left\{\frac{R^2}{2} - \frac{(R^2 - r_o^2)}{4\ln\frac{R}{r_o}}\right\} + \frac{u_s}{2}(R^2 - r_o^2)\right] - \pi R^2 p \quad (8.244)$$

where

L = volumetric latent heat
C_1, C_2 = volumetric specific heat of thawed and frozen layers

$$p = C_2\, u_f\left(\frac{1 - b^2}{2\ln b} + 1\right)$$

During a small time increment, the change in the energy absorbed by the system must equal the energy added to the system at the cylinder surface

$$q_p = -k_1\, 2\pi r_o \left(\frac{du_1}{dr}\right)_{r=r_o} = \frac{dQ}{dt} \quad (8.245)$$

Using Eqs. (8.244) and (8.245) yields

$$\frac{k_1}{r_o^2 C_1}\int_0^t dt = \int_1^{\beta}\left\{mx\ln x + \frac{x}{2} - \frac{1}{4}\frac{x^2 - 1}{x\ln x}\right\} dx$$

where

$$m = \frac{L - p + u_f C_l}{C_l(T_s - T_f)}$$

The solution to this equation is

$$4\tau = 2mS_T\beta^2 \ln\beta + S_T(1 - m)(\beta^2 - 1) - S_T\psi \tag{8.246}$$

where

$$m = \frac{1}{S_T} + \phi\left\{1 - k_{21}\alpha_{12}\left(\frac{1 - b^2}{2\ln b} + 1\right)\right\}$$

$$\psi = \int_1^\beta \frac{x^2 - 1}{x \ln x}\, dx$$

The function ψ is plotted as Fig. 8.27, and thus the thaw for any soil can be approximated. The method is simple but limited in application to first estimates of thaw depth.

8.6.2.2 Heat Balance Integral Solution

The heat balance integral method may also be applied to the problem of finite superheat, but the labor is considerably increased. The problem may be stated as follows, for a melting system

$$\frac{\partial}{\partial r}\left(r\frac{\partial T_l}{\partial r}\right) = \frac{r}{\alpha_l}\frac{\partial T_l}{\partial t} \tag{8.247}$$

$$T_l(R,t) = T_f \tag{8.247a}$$

$$T_l(r_o,t) = T_s \tag{8.247b}$$

$$T_l(r,o) = T_o \tag{8.247c}$$

$$-k_l\frac{\partial T_l(R,t)}{\partial r} + k_2\frac{\partial T_2(R,t)}{\partial r} = \rho_l l\frac{dR}{dt} \tag{8.247d}$$

$$\frac{\partial}{\partial r}\left(r\frac{\partial T_2}{\partial r}\right) = \frac{r}{\alpha_2}\frac{\partial T_2}{\partial r} \tag{8.248}$$

$$T_2(R,t) = T_f \tag{8.248a}$$

$$T_2(\delta,t) = T_o \tag{8.248b}$$

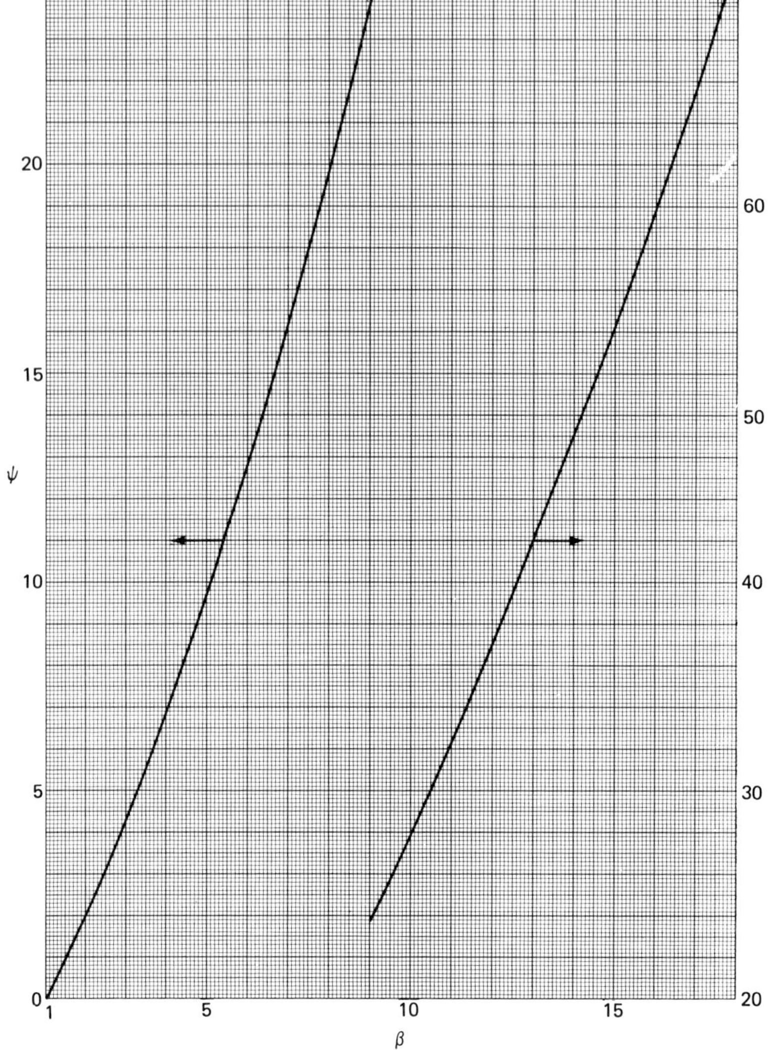

Fig. 8.27. Function for Phase Change around Cylinder.

$$\frac{\partial T_2(\delta,t)}{\partial r} = 0 \tag{8.248c}$$

The integrated temperatures are defined as usual

$$\theta_1 = \int_{r_0}^{R} r\, T_1\, dr$$

$$\theta_2 = \int_R^{\delta} r\, T_2\, dr$$

The heat balance integral equations are

$$\alpha_1 \left[R \frac{\partial T_1(R,t)}{\partial r} - r_o \frac{\partial T_1(r_o,t)}{\partial r} \right] = \frac{d\theta_1}{dt} - R T_f \frac{dR}{dt}$$

$$-\alpha_2 R \frac{\partial T_2(R,t)}{\partial r} = \frac{d\theta_2}{dt} - T_o \delta \frac{d\delta}{dt} + T_f R \frac{dR}{dt}$$

An approximation for T_1 that satisfies Eqs. (8.247*a–c*) is

$$T_1 = T_s - \frac{(T_s - T_f) \ln \left(\dfrac{r}{r_o} \right)}{\ln \dfrac{R}{r_o}}$$

A similar approximation for T_2 that satisfies Eqs. (8.248*a–c*) is

$$T_2 = T_o + (T_f - T_o) \frac{\delta - r}{\delta - R} \frac{\ln \left(\dfrac{r}{\delta} \right)}{\ln \dfrac{R}{\delta}}$$

The previous relations now lead to the following system of equations to solve

$$\alpha_{12} S_T \frac{dF}{d\tau} = \frac{1}{\Omega - 1} + \frac{1}{\ln \Omega} \tag{8.249}$$

$$F(\beta,\Omega) = \frac{\beta^2}{2} + \frac{\beta^2}{(1-\Omega)\ln\Omega} \left[-\frac{\Omega}{4}(\Omega^2 - 1) + \left(\frac{\Omega}{2} - \frac{1}{3} \right) \ln \Omega + \frac{1}{9}(\Omega^3 - 1) \right] \tag{8.250}$$

$$\left[S_T \left(\frac{\beta}{2 \ln \beta} - \frac{\beta^2 - 1}{4\beta (\ln \beta)^2} \right) + \beta \right] \frac{d\beta}{d\tau} = \frac{1}{\ln \beta} - k_{21}\phi \left(\frac{1}{\Omega - 1} + \frac{1}{\ln \Omega} \right) \tag{8.251}$$

where $\Omega = \delta / R$.

Equations (8.249–8.251) were solved numerically with a fourth-order Runge-Kutta, predictor-corrector technique. Because the problem, as specified, is initially singular at the origin, the Neumann solution was used to start the solution. Sparrow (1978) solved this problem numerically, with $\alpha_{12} = k_{12} = 1$, for a range of S_T and ϕ, also using the Neumann solution for starting. The much simpler method presented here is within 5% of Sparrow's results.

The calculations have been generalized for a range of α_{12}, k_{12}, pertinent for soil systems, and are presented in Chap. 10. Tien and Churchill (1965) also numerically evaluated freezing outside a circular cylinder. Their calculations are more extensive than Sparrow's (1978) and also agree very well with the solution presented here. Their numerical method was totally different from that of Sparrow (1978).

The numerical solutions, though very good, are not as convenient as analytical solutions. Thus, it is valuable to examine further methods with which to obtain approximate solutions that yield acceptable accuracy.

8.6.2.3 Coordinate Transformation

A method suggested by Lin (1971) uses a coordinate transformation to reduce a problem with a variable phase change area, such as a cylinder, to one of constant phase change area, the semi-infinite solid. Since the relations for the semi-infinite solid are well known, this is a useful procedure, but it is limited in accuracy. The following transformation will reduce the cylindrical system to the constant area case

$$y = r_o \ln\left(\frac{r}{r_o}\right) \tag{8.252}$$

The phase change interface, which is the value of y when $r = R$, is related by

$$\eta = r_o \ln\left(\frac{R}{r_o}\right) \tag{8.253}$$

The governing equations for the cylindrical system then transform into the following system, valid near the phase change interface where $r \approx R$.

$$\frac{\partial^2 T}{\partial y^2} = \frac{k}{\rho l \alpha} \frac{\partial T(\eta,\eta)}{\partial y} \frac{\partial T}{\partial \eta}$$

$$T(y,0) = T_f$$
$$T(\eta,\eta) = T_f$$
$$T(0,\eta) = T_S$$

Solutions of this sytem of equations are universal functions for all cross-sectional areas. The solutions are valid, however, only near the phase change interface. The system of equations need not be solved to use the method. The phase change interface rate of movement is given by

$$\frac{d\eta}{dt} = \frac{k}{\rho l}\left[\frac{A(r_o)}{A(R)}\right]^2 \frac{\partial T(\eta,\eta)}{\partial y}$$

For the constant area case $A(r_o) = A(R)$ and

$$\frac{d\eta}{dt} = \frac{k}{\rho l}\frac{\partial T(\eta,\eta)}{\partial y} = g(\eta)$$

Then the generalized case is

$$\frac{dR}{dt} = \frac{r_o}{R}g(\eta)$$

Thus, if the velocity of the phase change interface for the constant area case is given by

$$\frac{dX}{dt} = g(X)$$

then, the phase change interface velocity for the cylindrical system is

$$\frac{dR}{dt} = \frac{r_o}{R}g\left(r_o \ln\frac{R}{r_o}\right)$$

The plane solution is given by Eq. (8.25). From this

$$g(X) = \frac{2\gamma^2\alpha_l}{X}$$

Finally, the velocity of the cylindrical interface is

$$\frac{dR}{dt} = \frac{2\gamma^2\alpha_l}{R\ln\dfrac{R}{r_o}} \tag{8.254}$$

Equation (8.254) may be integrated to give

$$2\beta^2 \ln \beta - \beta^2 + 1 = \frac{8\gamma^2}{S_T} \tau \tag{8.255}$$

This solution may be compared to the quasi-steady solution, Eq. (8.237), for the case of no superheat, i.e., $S_T = 0$, $\phi = 0$. In the limit as $\phi \to 0$, the parameter γ is given by Eq. (8.108) and is $\gamma^2 = S_T/(2 + S_T)$ and in this case, Eq. (8.255) is identical to Eq. (8.237).

Although Eq. (8.255) is in an extremely convenient form for cylindrical systems, its accuracy is limited to certain ranges of S_T, ϕ, and τ. In comparison with the numerical solutions, Eq. (8.255) is accurate if $\tau/S_T < 1.0$, when $\phi = 4$. For lesser values of ϕ, the time limit when Eq. (8.255) is accurate will increase. The relation is so simple that it may be of value for quick, more or less crude, estimates, especially when $\phi \approx 0.0$.

8.6.2.4 Effective Thermal Diffusivity

Churchill and Gupta (1977) have introduced another method that allows the Neumann solution to be used for more complex geometries. The procedure involves replacing the nonlinear phase change problem with its linear analogue, which does not include phase change. The thermal diffusivity of the latter problem is then replaced by an effective diffusivity that includes the latent heat. Since many solutions are available for nonphase change problems, the method has potential for application to numerous freeze-thaw problems.

The effective diffusivity to use is based upon the fact that the solution to the zero latent heat analogue of the Neumann problem (simply transient conduction in the semi-infinite medium) can be forced to agree with the Neumann solution if an effective diffusivity replaces the actual diffusivity. If the location of the freezing surface is important, then one value of effective diffusivity is used, while if the surface heat flux density is desired, a second thermal diffusivity is used.

The following relations can be used for the heat flux density and the phase change location, respectively

$$\alpha_{eq} = \alpha_1 \, [(\phi + 1) \, erf\gamma]^2 \tag{8.256}$$

$$\alpha_{eq} = \alpha_1 \left[\frac{\gamma}{erf^{-1}\left\{\dfrac{1}{\phi + 1}\right\}} \right]^2 \tag{8.257}$$

where α_1 is the thawed value for thawing and the frozen value for freezing, and erf is the error function.

The method consists of first solving the linear conduction problem and then finding the relation between the space coordinate and the freezing (or simply the 32°F) isotherm. In this relation, one replaces the thermal diffusivity by Eq. (8.257) to obtain the phase change interface location for the phase change problem. It is likely that numerical evaluation will be necessary due to the complex inverse relation between the space coordinate and the freezing isotherm in the linear problem. The method has considerable power to handle complicated problems, but the range of validity is not known. The method cannot take into account the difference in thermal properties of the regions for which the temperature is above or below 32°F, when finding the no-phase-change solution.

Churchill and Gupta (1977) applied the method to cylinders and to freezing in a corner with good results. They used the exact solution for the cylindrical, no-phase-change problem given by Carslaw and Jaeger (1959). This required the use of tabulated, numerical values. The following section will derive an approximate solution to the problem that yields good results.

The first step in the solution of the phase change problem is to solve for the temperature of the pure conduction problem with zero latent heat. For the case of a cylinder surrounded by an infinite medium, an exact solution for the surface step change situation is given by Carslaw and Jaeger (1959). The solution involves an infinite series of Bessel functions, however, which were approximated by error functions for small values of time. The final results are in graphical form. An approximate solution to this problem will be found by using the heat balance integral method. The equations to solve are given below, referring again to Fig. 8.26. The melting case will be examined, but the results will apply to freezing also

$$\frac{\partial}{\partial r}\left(r\frac{\partial T}{\partial r}\right)=\frac{r}{\alpha}\frac{\partial T}{\partial t} \tag{8.258}$$

$$T(\delta,t)=T_o \tag{8.258a}$$

$$T(r_o,t)=T_s \tag{8.258b}$$

$$\frac{\partial T}{\partial r}(\delta,t)=0 \tag{8.258c}$$

The heat balance integral of Eq. (8.258) is a single integration over space and reduces to

$$-\alpha r_o\frac{\partial v(r_o,t)}{\partial r}=\frac{d\theta}{dt}$$

$$\theta=\int_{r_o}^{\delta} r\,v\,dr$$

where

$$v = \frac{T - T_o}{T_s - T_o}$$

For simplicity, a polynomial in r is assumed for the temperature

$$v = \left(\frac{\delta - r}{\delta - r_o}\right)^n \qquad r_o \leq r \leq \delta \tag{8.259}$$

which satisfies Eqs. (8.258*a–c*) and also the smoothness relations

$$\frac{\partial^{n-1} v(\delta, t)}{\partial v^{n-1}} = 0 \tag{8.260}$$

Integer values of n are used. Boundary layer theory has shown that the additional smoothness relations of Eq. (8.260) improve the accuracy of integral methods to a certain extent. The initial condition $v(r,0) = 0$ cannot be met, but this will not seriously affect the final results.

The equations lead to a differential equation for δ as follows

$$\left(\frac{2\Delta^2}{n+2} + \Delta\right)\frac{d\Delta}{d\tau} = \frac{n(n+1)}{S_T} \tag{8.261}$$

where

$$\Delta = \frac{\delta - r_o}{r_o}$$

Equation (8.261) is easily integrated to give

$$\frac{4}{3(n+2)}\Delta^3 + \Delta^2 = 2n(n+2)\frac{\tau}{S_T} \tag{8.262}$$

Volkov and Li-Orlov (1970) noted that the accuracy of the integral method could be improved by integrating Eq. (8.258) twice over the space coordinate. El-Genk and Cronenberg (1979) applied this idea to the Neumann problem with apparent success. For the cylindrical system, however, this resulted in considerably poorer results than the integral heat balance for any given value of $n \geq 2$.

Equations (8.259) and (8.262) complete the solution of the no-phase-change problem. These equations are acceptable when compared to the exact results of Carslaw and Jaeger (1959). The location of the phase change interface is found by using the movement of the isotherm with the fusion value. Thus, Eq. (8.259) gives

$$v_f = \frac{T_f - T_o}{T_s - T_o} = \left(\frac{\delta - r_f}{\delta - r_o}\right)^n$$

The location of the fusion value isotherm is then

$$\Delta_1 = \Delta\left[1 - \left(\frac{\phi}{\phi + 1}\right)^{1/n}\right] \tag{8.263}$$

where

$$\Delta_1 = \frac{r_f - r_o}{r_o}$$

Finally, the effective thermal diffusivity, Eq. (8.257), will replace the thermal diffusivity in Eq. (8.262). The solution for the actual phase change interface location is then

$$\beta = \left[1 - \left(\frac{\phi}{\phi + 1}\right)^{.05}\right]\{(a + d)^{1/3} + (a - d)^{1/3} - 5.5\} + 1 \tag{8.264}$$

$$a = 6930\frac{\alpha_{eq}}{\alpha}\frac{\tau}{S_T} - 166.375$$

$$d = [a^2 - 27680.6406]^{1/2}$$

The accuracy of the solution increases as n increases. Above $n = 20$, the improvement is slight, and thus $n = 20$ was used to obtain Eq. (8.264). This equation, combined with Eq. (8.257), for the effective thermal diffusivity, and Eq. (8.104), for γ, is a simple relation for freezing or thawing about a cylinder, which is accurate enough for most engineering purposes. A comparison of Eq. (8.264), with numerical results showed that for most cases it will be accurate to within 10% or less of the numerical results. This is felt to be satisfactory for engineering calculations.

The instantaneous heat flow from the cylinder and the total heat loss or gain during a given time are also quantities of interest. The surface heat flux is given by

$$q^* = \frac{qc_l}{2\pi\ k_l l} = \frac{20\ S_T(1+\phi)}{e} \tag{8.265}$$

where

$$e = (a+d)^{1/3} + (a-d)^{1/3} - 5.5$$

as before, and Eq. (8.256) is used with a and d.

The integrated heat transfer is

$$Q_t = \int_o^t q\ dt$$

This can be written in nondimensional form as

$$Q_t^* = \frac{Q_t}{2\pi r_o^2 \pi \rho_l l} = (1+\phi)\frac{S_T}{21}\left(\frac{e^2}{22} + e\right) \tag{8.266}$$

Figures 8.28 and 8.29 show a comparison of the effective diffusivity method with the numerical calculations. The equations of this section are quite good. The effective diffusivity for heat transfer does not seem as accurate as that

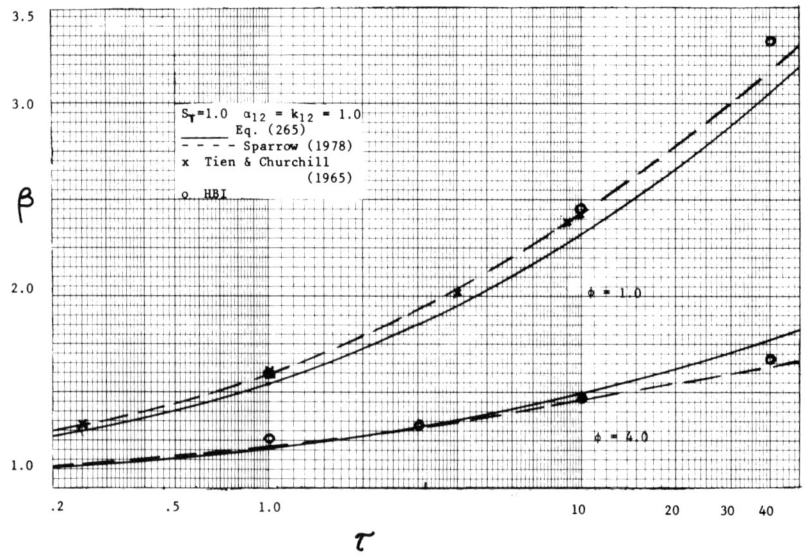

Fig. 8.28. Thawing around a Circular Pipe.

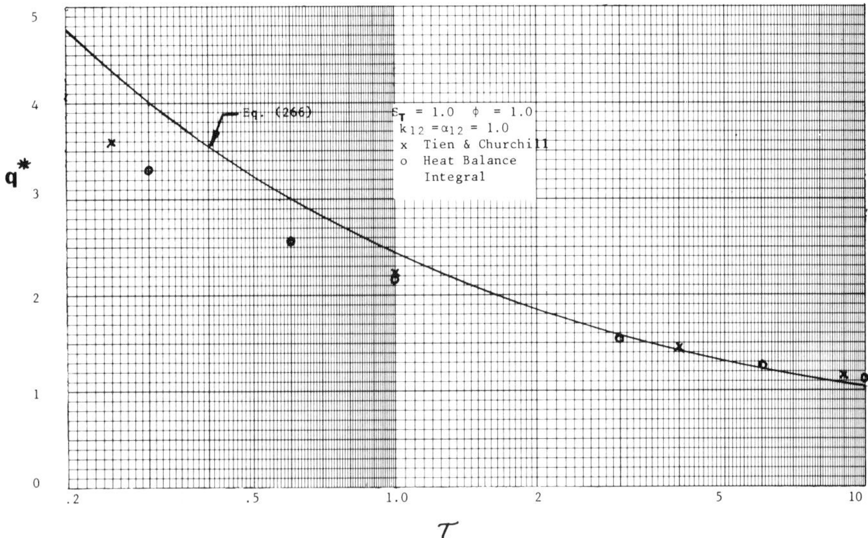

Fig. 8.29. Heat Flow from a Circular Pipe with Thaw.

for phase location, but it still gives adequate results. The equations can be applied to any values of α_{12} and k_{12}, of course.

The method of this section cannot be applied when the superheat parameter ϕ is zero, but for this case there are sufficiently accurate solutions available, as given earlier. The effective thermal diffusivity method, used here, should be adaptable to many problems of freezing and thawing.

8.6.2.5 Constant Surface Heat Flux

The problems discussed so far have used a constant surface temperature for the cylinder. Goodling and Khader (1975) discuss numerical solutions for a constant heat flux, q″, at the cylinder surface. The heat flux from the liquid region is accounted for by a surface coefficient of heat transfer as shown in Fig. 8.30. Thus, the conduction equation need be handled only in the solid region. Nondimensional equations are

$$\frac{\partial^2\theta}{\partial y^2} + \frac{1}{y}\frac{\partial\theta}{\partial y} = S_l\,\frac{\partial\theta}{\partial\tau} \qquad (8.267)$$

$$\theta(\beta,\tau) = 0$$

$$\frac{\partial\theta(1,\tau)}{\partial y} = -1$$

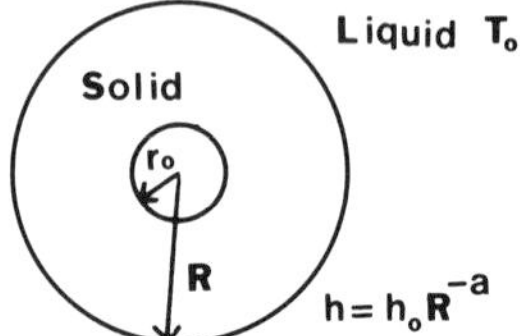

Fig. 8.30. Cylindrical Phase Change with Convection.

$$\frac{d\beta}{d\tau} = \frac{\partial\theta(\beta,\tau)}{\partial y} - p\beta^a$$

$$\beta(0) = 1$$

where

$$\theta = \frac{(T - T_f)k}{q'' r_o}$$

$$y = \frac{r}{r_o}$$

$$\tau = q'' \frac{t}{\rho l r_o}$$

$$S_l = q'' \frac{c r_o}{lk}$$

$$p = \frac{h_o(T_f - T_o)}{q'' \, r_o^a}$$

Numerical solutions are given for $a = 0.25$, $p = 0.6$, and $S_l = 0.2\text{–}10$.

An integral heat balance solution is given where the approximate temperature is defined as a logarithmic function. The integral solution is

$$\tau = \int_1^\beta \frac{2d\beta}{\sqrt{\left(\frac{1}{S_l\beta \ln \beta} - \frac{p}{\beta^a}\right) + \frac{4}{S_l\beta^2 \ln \beta}} - \left(\frac{1}{S_l\beta \ln \beta} + \frac{p}{\beta^a}\right)}$$

8.7 CYLINDRICAL PROBLEMS, FINITE GEOMETRY

The solutions for an infinite medium, discussed in Section 8.6, give results that indicate too fast a growth rate for the thaw around a buried pipe when compared with the numerical results of Lachenbruch (1970) for the transient thaw depth around a hot pipe buried in permafrost. The solution of Lachenbruch assumes two-dimensional heat transfer, with no heat transfer along the axis of the pipe (Fig. 8.31). The problem is formulated as follows

$$\frac{\partial^2 T_1}{\partial x^2}+\frac{\partial^2 T_1}{\partial y^2}=\frac{1}{\alpha_1}\frac{\partial T_1}{\partial t} \qquad y>0 \qquad T>T_f \tag{8.268}$$

$$\frac{\partial^2 T_2}{\partial x^2}+\frac{\partial^2 T_2}{\partial y^2}=\frac{1}{\alpha_2}\frac{\partial T_2}{\partial t} \qquad T<T_f \tag{8.269}$$

The ground is initially frozen, with the following initial temperature distribution

$$T(x,y,0)=T_G+A\exp\left(-y\sqrt{\frac{\omega}{2\alpha_2}}\right)\sin\left(\omega t-y\sqrt{\frac{\omega}{2\alpha_2}}\right)$$

where

T_G = mean ground or air temperature
A = amplitude of yearly air or ground temperature fluctuation
ω = frequency of temperature fluctuation (1 year)

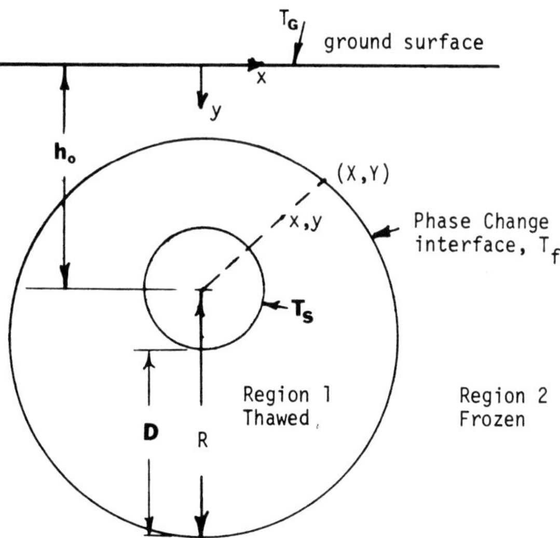

Fig. 8.31. Thaw around a Buried Pipe in Permafrost.

$$T(x,0,t,) = T_G + A \sin \omega t$$
$$T = T_G \qquad y \rightarrow \infty$$

$$\frac{\partial T}{\partial x} = 0 \qquad x = 0 \text{ or } x \rightarrow \infty$$

$$T = T_s \text{ on the pipe surface}$$
$$T_1(X,Y) = T_2(X,Y) = T_f$$

and X, Y are the coordinates of the phase change interface.

The energy balance at the phase change interface (X, Y) is

$$\left.\begin{aligned} k_1 \frac{\partial T_1}{\partial x} - k_2 \frac{\partial T_2}{\partial x} = L \frac{\partial X}{\partial t} \\ k_1 \frac{\partial T_1}{\partial y} - k_2 \frac{\partial T_2}{\partial y} = L \frac{\partial Y}{\partial t} \end{aligned}\right\} \quad (x,y) = (X,Y) \tag{8.270}$$

An exact solution to this problem is beyond present mathematical techniques. Numerical solutions for the thaw depth beneath the pipe, as a function of soil properties and temperatures, have been given by Lachenbruch (1970), Hwang (1972), and Gold (1972). These numerical results are discussed in Chap. 10. Although numerical techniques allow great flexibility with respect to properties, temperature conditions, etc., they are usually cumbersome and often expensive. In many cases preliminary estimates will suffice, especially where site conditions and data do not warrant the use of more exact analyses.

8.7.1 Quasi-Static Approximations

There are no solutions, even approximate, for the complete problem as described above, in the semi-infinite domain. Nevertheless, the quasi-static approximation has been used when the original ground temperature has a constant value of T_G, which is also the ground surface temperature. The problem is as follows. With Eqs. (8.268) and (8.269)

$$T(x,y,0) = T(x,0,t) = T_G$$
$$T(x,y,t) = T_s \quad \text{on the pipe surface}$$

Porkhayev (1963) assumed that the phase change energy balance at the centerline below the pipe was the controlling mechanism for the phase change. Equation (8.270) then is

$$k_1 \frac{\partial T_1}{\partial y} - k_2 \frac{\partial T_2}{\partial y} = L \frac{dY}{dt} \tag{8.271}$$

where the phase change is at $Y(t)$.

Using the well-known steady-state solution (see Chap. 7), it is possible to integrate Eq. (8.271)

$$\int_{1+\mu}^{Y/r_o} \frac{f(\xi) - 1}{1 - \alpha\left[1 - \dfrac{1}{f(\xi)}\right]} \frac{d\xi}{f'(\xi)} = k_1 \frac{(T_s - T_f)\, t}{L} \tag{8.272}$$

where $\xi = y/r_o$

$$\alpha = \frac{k_2}{k_1} \frac{(T_G - T_f)}{(T_s - T_f)}$$

$$f(\xi) = \frac{\ln\left[\dfrac{\xi + \sqrt{\mu^2 - 1}}{\xi - \sqrt{\mu^2 - 1}}\right]}{\ln\left(\mu + \sqrt{\mu^2 - 1}\right)}$$

$$\mu = h_o/r_o$$

Hwang (1977) evaluated Eq. (8.272) numerically for parametric values of α and μ related to pipelines in the arctic. The results are discussed in Chap. 10. Porkhayev's solution yields an underestimate of the thaw depth, and Hwang (1977) suggests that the mean of the Porkhayev solution and the infinite domain solution, Eq. (8.237) or Eq. (8.264), will give a value adequate for engineering purposes but this will not be generally true.

Thornton (1976) also used the steady-state solution, but with the average value of the heat flux around the circular thaw interface in satisfying the energy condition of Eq. (8.270). This method led to the thawbowl decreasing as the pipe burial increases, which is contrary to expectations. Thornton's results should, thus, be used with some caution.

8.8 CYLINDRICAL PROBLEMS, INWARD PHASE CHANGE

The question of phase change inwardly from a cylindrical surface is important in dealing with the freezing of water in pipes and allied problems (Fig. 8.32). A number of solutions are available for this finite domain problem.

8.8.1 Zero Superheat or Subcooling

As expected, when the initial temperature of the system is at the fusion value the analysis is considerably simplified.

8.8.1.1 Constant Phase Change Rate

The solution to this problem, as given by Kreith and Romie (1955), has been discussed in Section 8.6.1.1. The problem is expressed as

$$\frac{1}{r}\frac{\partial}{\partial r}\left(r\frac{\partial T}{\partial r}\right)=\frac{1}{\alpha}\frac{\partial T}{\partial t} \tag{8.273}$$

$$T = T_f \tag{8.273a}$$

$$k\frac{\partial T}{\partial r} = -\rho l\frac{dR}{dt} \qquad r = r_o - R \tag{8.273b}$$

$$\frac{\partial T}{\partial r} = G \tag{8.273c}$$

$$T(r_o,0) = T_f \tag{8.273d}$$

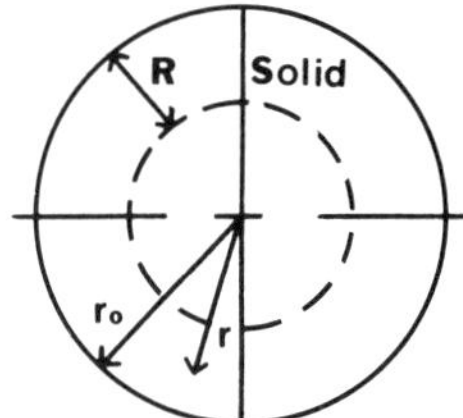

Fig. 8.32. Inward Solidification in a Cylinder.

The interface moves at a constant rate given by

$$R=\frac{-kGt}{\rho l} \tag{8.274}$$

The time to attain complete solidification is

$$t_f=\frac{-\rho l r_o}{kG} \tag{8.275}$$

8.8.1.2 Zero Sensible Heat

Once again the quasi-static procedure will lead to simple and useful results. The problem may be described as follows

$$\frac{d}{dr}\left(r\frac{dT}{dr}\right) = 0 \tag{8.276}$$

$$\left.\begin{aligned} T &= T_f \\ k\frac{\partial T}{\partial r} &= -\rho l\frac{dR}{dt} \end{aligned}\right\} r = r_o - R$$

$$T(r_o,0) = T_f$$
$$T(r_o,t) = T_s$$

Equation (8.276) is easily solved for the temperature, which is given by

$$T = T_s + \frac{(T_f - T_s)}{\ln\left(\dfrac{r_o - R}{r_o}\right)} \ln\left(\frac{r}{r_o}\right)$$

The solidification interface can now be evaluated

$$2(1-\beta)^2 \ln(1-\beta) - (1-\beta)^2 + 1 = 4\tau \tag{8.277}$$

where

$$\beta = R/r_o$$

$$\tau = \frac{\alpha_s t}{r_o^2} S_T$$

$$S_T = \frac{c_s(T_f - T_s)}{l}$$

The solidification time is

$$\tau_f = \frac{1}{4} \tag{8.278}$$

8.8.1.3 Finite Sensible Heat

A solution for the constant surface temperature problem was given by Riley et al. (1974), using a perturbation method valid for small values of the Stefan number. The first two terms of the series for the solidification time gave good results. The relation is

$$\tau_f = \frac{1}{4}(1 + S_T) \tag{8.279}$$

It is clear that this result will be close to Eq. (8.278) for small values of the Stefan number.

8.8.1.4 Integral Method

An integral method can be used to examine this problem. The basic equations are those of Section 8.8.1.1, with Eq. (8.273*c*) replaced by

$$T(r_o,t) = T_s \tag{8.280}$$

An equivalent form of Eq. (8.273*b*) has been discussed previously and is

$$\frac{\partial T}{\partial t} = -\frac{k}{\rho l}\left(\frac{\partial T}{\partial r}\right)^2 \qquad r = r_o - R \tag{8.280a}$$

The integrated energy equation is arrived at by integrating Eq. (8.273) over the solidified region and using Eq. (8.273*b*)

$$(r_o - R)\frac{\rho l}{k}\frac{dR}{dt} + r_o\frac{\partial T(r_o,t)}{\partial r} = -\frac{1}{\alpha}\int_{r_o}^{r_o - R} r\frac{\partial T}{\partial t}\,dr$$

A second equation is obtained by multiplying Eq. (8.273) by $r(\partial T/\partial r)$, integrating and using Eq. (8.280*a*) to obtain

$$\frac{\rho l}{k}(r_o - R)^2\frac{\partial T(r_o - R,t)}{\partial t} + r_o^2\left[\frac{\partial T(r_o,t)}{\partial r}\right]^2 = -\frac{2}{\alpha}\int_{r_o}^{r_o - R} r^2\frac{\partial T}{\partial r}\frac{\partial T}{\partial t}\,dr$$

A single-parameter relation for the temperature was used

$$\frac{T - T_s}{T_f - T_s} = \frac{r_o - r}{R}$$

which then led to a closure time of

$$\tau_f = \frac{1}{4}(2 + S_T) - \frac{1}{9}(3 + S_T) \tag{8.281}$$

This result was not too accurate, and Poots (1962) used a two-parameter relation

$$\frac{T - T_s}{T_f - T_s} = \frac{r_o - r}{R} + \frac{r_o - r}{R}\left(1 - \frac{r_o - r}{R}\right) g$$

Unfortunately, the resulting equations for R and g had to be numerically integrated, but the closure time was much more accurate.

The results of these calculations are summarized in Table 8.5. Equation (8.279) is probably the best relation to use, but the simple quasi-steady analysis is surprisingly good. The average of Eqs. (8.279) and (8.269) gives

$$\frac{\alpha t_f}{r_o^2} = \frac{1}{4S_T} + \frac{1}{8} \tag{8.282}$$

and this relation is very good.

8.8.1.5 Surface Heat Flux Specified

If the heat flux at the surface of the cylinder is specified, approximate solutions have been given by Shamsundar and Sparrow (1974) for two cases.

a) Constant Heat Flux. At the cylinder the heat flux is specified as

$$-k\frac{\partial T}{\partial r} = q_o \tag{8.283}$$

The basic equations are as given in Section 8.8.1.1. An approximate solution is obtained by expanding the equations in a truncated series such that the energy equation and interface boundary condition are satisfied exactly only

Table 8.5. Complete Solidification of a Circular Cylinder.

	$\tau_f / S_T = \alpha t_f / r_o^2$				
S_T	Quasi-Steady	Riley (1974)	Poots (1962*a*)	Beckett (1971)[b]	Allen & Severn (1962)[b]
.05	5.00	5.25	3.47	5.30	–
.1	2.50	2.75	1.81	2.69	–
.2	1.25	1.50	.97	–	–
.25	1.00	1.25	.81	1.19	–
.641	.39	.64	.40(.52)[a]	–	.47
1.0	.25	.50	.31	–	–

[a] Two-parameter method.
[b] Numerical solutions.

at the phase change interface. The method is due to Megerlin (1968). The phase change interface is given by

$$2\frac{S\tau}{S_T}=\frac{\beta}{2}(2-\beta)+\frac{1}{2}[1-(1-\beta)^2\sqrt{1-4S\ln(1-\beta)}]$$

$$+\frac{1}{4}\sqrt{\pi S}\,e^{1/2S}\frac{1}{2S}\left\{erf\sqrt{\frac{1}{2S}-2\ln(1-\beta)}-erf\sqrt{\frac{1}{2S}}\right\} \quad (8.284)$$

where $S = cq_o r_o/kl$ is a form of Stefan number and $S_T = c(T_f - T_s)/l$ is the usual Stefan number. The solidification time is given by

$$\frac{S\tau_f}{S_T}=\frac{1}{2}+\frac{\sqrt{2\pi S}}{8}\exp\left(\frac{1}{2S}\right)\text{erfc}\sqrt{\frac{1}{2S}} \quad (8.285)$$

(b) Constant Ambient Temperature. The second case involves convection at the surface into an ambient at T_∞

$$-k\frac{\partial T}{\partial r}=h(T-T_\infty) \qquad r=r_o \quad (8.286)$$

In this case, numerical evaluation is needed to find the location of β. For small specific heats (Stefan number), however, an approximation for the closure time is given by the quasi-steady method.
The solidification interface is

$$2\tau=\frac{2\beta-\beta^2}{B_i}+\left[\ln(1-\beta)+\frac{1}{2}\right](1-\beta)^2+\frac{1}{2} \quad (8.287)$$

where $B_i = hr_o/k$ is the Biot number, and $S_T = c(T_f - T_\infty)/l$.
The solidification time is then

$$\tau_f=\frac{1}{2}\left(\frac{1}{B_i}+\frac{1}{2}\right) \quad (8.288)$$

When the Biot number is very large, the ambient temperature will be the same as the surface temperature. Then

$$\tau_f=\frac{1}{4}$$

This is exactly the result of the simple quasi-steady procedure, Eq. (8.278), for a constant surface temperature. See Section 10.7 for further details.

Goodling and Khader (1974) numerically solved this problem with the addition of radiation at the cylinder surface. Unfortunately, their calculations were such that only one value of the ambient to fusion temperature ratio was studied, which was not the right range for water solidification. One can roughly estimate that if $\sigma \epsilon T_f^3 r_o/k < 1/3 \ B_i^2$, the solidification time with radiation will be at least 90% of the value for convection only. This estimate will vary with the ratio of the ambient to fusion, absolute temperatures.

Figure 8.33 is a plot of the numerical evaluation of the approximate solution for the closure time of Shamsundar and Sparrow (1974). The results are in good agreement (less than 10% difference) with Goodling and Khadar (1974) if the Stefan number is less than 1. Above $S_T = 1.0$ the results are less accurate.

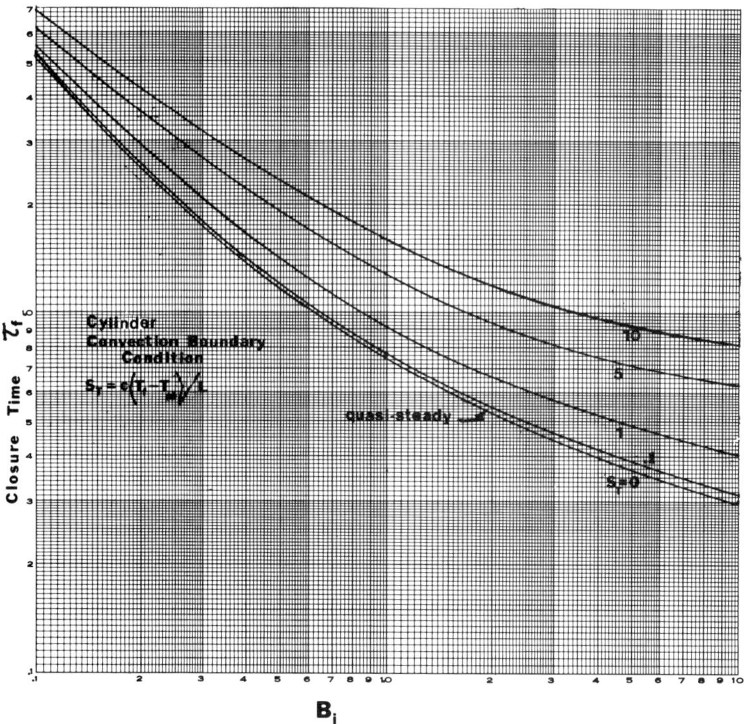

Fig. 8.33. Complete Freeze of a Circular Cylinder.

8.8.2 Finite Superheat

There are very few analytical results available for the inward solidification of cylinders if the initial temperature is not at the fusion value. Jiji and Weinbaum (1978) used a perturbation method to arrive at approximate solutions for annular regions. They considered the cases of an insulated inner cylinder and an isothermal inner cylinder. For the insulated case at small Stefan numbers, which was used as the perturbation parameter, the complete solidification time of a tube was the same as the quasi-steady solution. They concluded that the initial temperature was not too significant for the insulated system; thus, the zero sensible heat solutions presented in Section 8.1 should be reasonable at small Stefan numbers. For isothermal systems, the superheat is quite significant even at small S_T.

8.9 SPHERICAL PROBLEMS, OUTWARD GROWTH

The techniques associated with spherical systems follow closely those of cylindrical coordinates. Once again, no complete solutions are available.

8.9.1 Zero Superheat

The quasi-steady approximation will yield a simple solution. The freeze problem will be given (Fig. 8.34)

$$\frac{d}{dr}\left(r^2\frac{dT}{dr}\right)=0 \tag{8.289}$$

$$T(r_o,t)=T_s$$

$$T(R,t)=T_f$$

$$\frac{dR}{dt}=\frac{k}{\rho l}\frac{\partial T(R,t)}{dr}$$

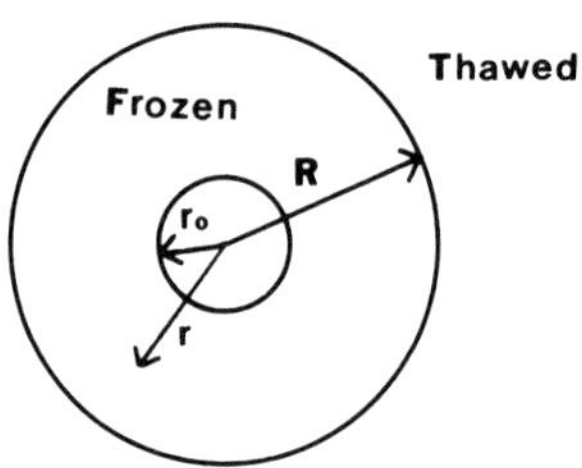

Fig. 8.34. Phase Change for Spherical Systems.

The solution for the temperature is

$$\frac{T-T_s}{T_f-T_s}=\frac{1-\dfrac{r_o}{r}}{1-\dfrac{r_o}{R}}=\frac{1-\dfrac{r_o}{r}}{1-\dfrac{1}{\beta}}$$

The phase change interface is given by

$$\tau=\frac{1}{3}\beta^2-\frac{\beta^2}{2}+\frac{1}{6} \tag{8.290}$$

The heat transfer into the sphere, at the surface, is

$$q^*=\frac{q}{4\pi(T_f-T_s)kr_o}=\frac{\beta}{\beta-1} \tag{8.291}$$

These relations are, as usual, for negligible sensible heat or $S_T = 0$.

Pedroso and Domoto (1973) used a perturbation method to obtain a solution valid for small Stefan numbers. The problem is

$$\frac{1}{r}\frac{\partial^2(rT)}{\partial r^2}=\frac{1}{\alpha}\frac{\partial T}{\partial t} \tag{8.292}$$

$$T(r_o,t)=T_s$$

$$T(R,t)=T_f$$

$$\frac{dR}{dt}=\frac{k}{\rho l}\frac{\partial T(R,t)}{\partial r}$$

The temperature and the phase interface velocity were expanded in Taylor's series about $S_T = 0$

$$u=u_o+u_1S_T+u_2S_T^2+\ldots$$

$$g=g_o+g_1S_T+g_2S_T^2+\ldots$$

where

$$u=\frac{T-T_s}{T_f-T_s}$$

$$g=\frac{d\beta}{d\tau}$$

The zeroth-order solutions are exactly the quasi-steady solutions just discussed. The first three terms yield

$$\frac{u}{u_o} = 1 + \frac{1}{6}\left[1 - \left(\frac{r}{R}\right)^2 u_o^2\right]\frac{S_T}{\beta^2} - \left\{\frac{1}{36}\left[1 - \left(\frac{r}{R}\right)^2 u_o^2\right] + \frac{4\beta - 1}{120}\left[1 - \left(\frac{r}{R}\right)^4 u_o^4\right]\right\}\left(\frac{S_T}{\beta^2}\right)^2 \tag{8.293}$$

$$\frac{g}{g_o} = 1 - \frac{1}{3}\frac{S_T}{\beta} + \frac{1 - 6\beta}{45}\frac{S_T^2}{\beta^3} \tag{8.294}$$

where

$$u_o = \frac{1 - \dfrac{r_o}{r}}{1 - \dfrac{1}{\beta}} \qquad g_o = \frac{1}{\beta(\beta - 1)}$$

Equation (8.294) can be integrated to give the phase change position

$$\tau = \frac{3(\beta - 1)^2 + 2(\beta - 1)^3}{6} + \frac{(\beta - 1)^2}{6} S_T - \frac{1}{45}\frac{(\beta - 1)^2}{\beta} S_T^2 \tag{8.295}$$

The first term of this series is the quasi-steady solution, Eq. (8.290).

8.9.2 Finite Superheat

Gupta (1973) used the quasi-steady method of Khakimov (1957) to include the effects of the surrounding medium at temperatures other than the fusion value (Fig. 8.35).

The equations are

$$\frac{d}{dr}\left(r^2 \frac{dT}{dr}\right) = 0 \tag{8.296}$$

for each region and

$$T_1(r_o,t) = T_S$$
$$T_1(R,t) = T_2(R,t) = T_f$$
$$T_2(\delta,t) = T_o$$

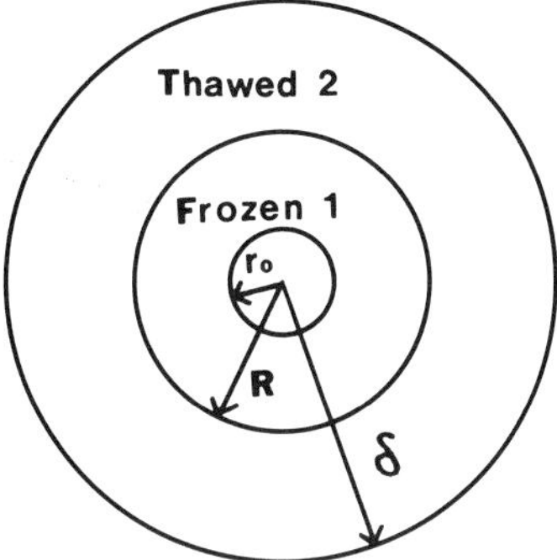

Fig. 8.35. Temperature Penetration for Freeze outside a Sphere.

The temperatures are

$$\frac{T_1 - T_S}{T_f - T_S} = \frac{1 - \dfrac{r_o}{r}}{1 - \dfrac{r_o}{R}}$$

$$\frac{T_2 - T_o}{T_f - T_o} = \frac{1 - \dfrac{\delta}{r}}{1 - \dfrac{\delta}{R}}$$

As discussed in Section 8.6.2.1, the heat extracted at the surface of the sphere, in a small time, must equal the change in the latent and sensible heats of the surrounding medium. The phase change interface position is then

$$\tau = \left(\frac{\beta^3}{3} - \frac{\beta^2}{2} + \frac{1}{6}\right)\left[1 - \phi S_T\left\{1 + k_{21}\alpha_{12}\left(\frac{b^2}{2} + \frac{b}{2} - 1\right)\right\} + \frac{S_T}{6}(\beta^2 - \beta - \ln b)\right] \quad (8.297)$$

where $b = \delta/R$. A value of $b = 4.5$ was suggested by Khakimov (1957). The surface heat transfer is again given by Eq. (8.291).

Equation (8.297) reduces to the quasi-steady solution when $S_T = \phi = 0$. Equations (8.297) and (8.295) are quite close for small values of S_T. As the Stefan number increases, Eq. (8.295) increasingly differs from Eq. (8.297), at values of β less than 10. Because Eq. (8.297) is not limited to small values of S_T, it should be more accurate at large Stefan numbers.

The method of Lin (1971) (see Section 8.6.2.3) can also be used. This will give a solution for the phase change interface as follows

$$\frac{\beta^3}{3} - \frac{\beta^2}{2} + \frac{1}{6} = \frac{2\gamma^2}{S_T}\tau$$

The parameter γ is given by the exact solution to the Neumann problem, as discussed in Sections 8.2 and 8.3. This will likely be acceptable for small superheat parameters ϕ and small times τ, but should be used with caution.

8.9.3 Constant Surface Heat Flux

The case of a constant surface heat flux has been numerically examined by Goodling and Khadar (1975). The equations are as given in Section 8.6.2.5, except that the diffusion equation is

$$\frac{\partial^2\theta}{\partial y^2} + \frac{2}{y}\frac{\partial\theta}{\partial y} = S_l \frac{\partial\theta}{\partial\tau}$$

Very limited ranges of the parameters are given, as noted in the above-mentioned section.

An integral solution for the phase change interface is

$$\tau = \int_1^\beta \frac{2d\beta}{\sqrt{\left(\dfrac{1}{S_l\beta(\beta-1)} - \dfrac{p}{\beta^a}\right)^2 + \dfrac{4}{S_l\beta^3(\beta-1)}} - \left(\dfrac{1}{S_l(\beta-1)\beta} + \dfrac{p}{\beta^a}\right)} \tag{8.298}$$

The time to reach a given solidification location with Eq. (8.298) is about 10–30% greater than the numerical solution of Goodling and Khadar (1975), with $S_1 \sim 10$, $a = .25$, and $p = .6$. For lesser values of S_1, the agreement is much better.

8.10 SPHERICAL PROBLEMS, INWARD GROWTH

Several approximate solutions are available for the inward solidification of spheres. No solution exists for problems with finite superheat.

8.10.1 Constant Interface Temperature Gradient

Kreith and Romie (1955) have examined the linear problem of a constant temperature gradient applied at the interface location

$$\frac{1}{r}\frac{\partial}{\partial r}\left(r^2\frac{\partial T}{\partial r}\right)=\frac{1}{\alpha}\frac{\partial T}{\partial t} \tag{8.299}$$

$$\left.\begin{aligned} T &= T_f && (8.299a)\\ k\frac{\partial T}{\partial r} &= -\rho l\frac{dR}{dt} && (8.299b)\\ \frac{\partial T}{\partial r} &= G && (8.299c)\end{aligned}\right\} r = r_o - R$$

$$T(r_o,0)=T_f \tag{8.299d}$$

The solution for the interface position is given by

$$\beta=\frac{G\tau}{(T_f-T_S)} \tag{8.300}$$

The closure time occurs when $\beta = 1$; therefore

$$\tau_f=\frac{(T_f-T_S)}{G} \tag{8.301}$$

This relation, for the closure time, is not exact, as a solution for the surface temperature showed that at $\beta \approx 0.9$, the surface temperature will exceed T_f if G is a constant. Thus, realistically, when T_S reaches a value of T_f, it will be held constant, and closure will occur with a variable temperature gradient, if surface melting is to be avoided. Solutions for the transient surface temperature are given by Kreith and Romie (1955).

8.10.2 Constant Surface Temperature

A quasi-steady solution can be obtained for this case. The equations are

$$\frac{d}{dr}\left(r^2\frac{dT}{dr}\right)=0$$

Boundary conditions, Eqs. (8.299a,b) apply. The temperature is given by

$$\frac{T-T_S}{T_f-T_S}=\frac{-(r_o-R)}{R}\left(1-\frac{r_o}{r}\right)$$

The interface position is

$$\tau = \frac{\beta^2}{2} - \frac{\beta^3}{3} \tag{8.302}$$

This yields a closure time of

$$\tau_f = \frac{1}{6} \tag{8.303}$$

Pedroso and Domoto (1973b) used a perturbation solution to obtain the following relation for the phase change interface

$$\tau = \frac{S_T}{9K}[3+(3-S_T)K]\left[1-\beta_1-\left(\frac{3-S_TK}{3K}\right)\ln\frac{3+(3-S_T)K}{3+(3\beta_1-S_T)K}\right] + \frac{3-S_T}{6}(1-\beta_1^2) - \frac{1-\beta_1^3}{3} \tag{8.304}$$

where K is a function of S_T, given below, and $\beta_1 = 1 - \beta$.

K	14.96	9.099	6.699	5.350	4.474	3.854
S_T	.1	.2	.3	.4	.5	.6

3.391	3.030	2.791	2.503	1.351	.9287
.7	.8	.9	1	2	3

Thus, the closure time is

$$\tau_f = \frac{S_T}{9K}[3+(3-S_T)K]\left[1-\left(\frac{3-S_TK}{3K}\right)\ln\frac{3+(3-S_T)K}{3-S_TK}\right] + \frac{3-S_T}{6} - \frac{1}{3} \tag{8.305}$$

This relation was shown to be accurate for $S_T \leq 1.0$.

Riley et al. (1974) also used a perturbation method to give a closure solution of

$$\tau_f = \frac{1}{6} + \frac{1}{6}S_T - \frac{S_T^{3/2}}{3\sqrt{2\pi}} + 0(S_T^2) \tag{8.306}$$

This relation is also valid at small Stefan numbers, but is less accurate than Eq. (8.305) at $S_T > 0.5$.

Table 8.6. Closure Time for Constant Surface Temperature Spheres Initially at Freezing Temperatures.

	Closure Time τ_f				
S_T	Numerical solution, Tao (1967)	Quasi-steady Eq. (8.303)	Eq. (8.305)	Eq. (8.306)	Eq. (8.307)
0	–	.1667	.1667	.1667	.0833
.1	.192	.1667	.1805	.1791	.0924
.3	–	.1667	.2168	.1948	.1104
.5	.237	.1667	.2302	.2030	.1285
.8	–	.1667	.2663	.2048	.1556
1	.284	.1667	.2901	.2004	.1736
2	.360	.1667	.408	.1239	.2639

The heat balance integral technique was used by Poots (1962). This method, following Goodman (1967), did not give accurate results for the closure time, which was

$$\tau_f = \frac{1}{4}(S_T + 2) - \frac{2}{9}(S_T + 3) + \frac{1}{16}(S_T + 4.0) \tag{8.307}$$

A two-parameter method led to much better results, but required numerical evaluation, which tends to lessen the value of the heat balance integral method.

The results for the closure time of spheres are shown in Table 8.6. Equation (8.305) clearly gives superior results.

8.10.3 Convection and Radiation Boundary Condition

Goodling and Khadar (1974) numerically evaluated the freezing of a sphere with convection and radiation at the surface of the sphere. The numerical results were too limited to reproduce, but, if $\sigma \epsilon\ T_f^3 r_o/k < 1/3\ B_i^2$, the solidification time with radiation will exceed 90% of the time with only convection. Thus, for soil systems or atmospheric conditions, it can be anticipated that surface radiation will not be significant.

REFERENCES

Allen, D. N. de G., and Severn, R. T. 1962. The application of the relaxation method to the solution of non-elliptic partial differential equations. *Q. J. Mech. Appl. Math.* 15:53.

Altman, M. 1961. Some aspects of the melting solution for a semi-infinite slab. *Chem. Eng. Prog. Symp. Ser.* 57:16–23.

Beckett, P. M. 1971. Ph.D. thesis. Hull Univ., England.

Bell, G. E. 1978. A refinement of the heat balance integral method applied to a melting problem. *Int. J. Heat Mass Transfer* 21:1357–1362.

Carslaw, H. S., and Jaeger, J. C. 1959. *Conduction of heat in solids.* Oxford: Clarendon Press.

Cho, S. H., and Sunderland, J. E. 1969. Heat conduction problems with melting or freezing. *J. Heat Transfer* 91: 421–426.

______. 1974. Phase change problems with temperature dependent thermal conductivity. *J. Heat Transfer* 96:214–217.

Churchill, S. W., and Gupta, J. P. 1977. Approximations for conduction with freezing or melting. *Int. J. Heat Mass Transfer* 20:1251–1253.

Duda, J. C., and Vrentas, J. S. 1969. Perturbation solutions of diffusion-controlled moving boundary problems. *Chem. Eng. Sci.* 24:461–470.

El-Genk, M. S., and Cronenberg, A. W. 1979. Some improvements to the solution of Stefan-like problems. *Int. J. Heat Mass Transfer* 22:167–170.

Evans, G. W., Isaacson, E., and MacDonald, J. K. L. 1950. Stefan-like problems. *Q. Appl. Math.* 8 (3):312–319.

Foss, S. D., and Fan, S. S. T. 1972. Approximate solution to the freezing of the ice-water system. *J. Water Resour. Res.* 8 (4):1083–1086.

______. 1974. Approximate solution to the freezing of the ice-water system with constant heat flux in the water phase. *J. Water Resour. Res.* 10 (3):511–513.

Gold, L. W., Johnston, G. N., Slusarchuk, W. A., and Goodrich, L. E. 1972. Thermal effects in permafrost. *Proceedings Canadian Northern Pipeline Research Conference.* Ottawa, Canada, pp. 25–45.

Goodling, J. S., and Khader, M. S. 1974. Inward solidification with radiation—convection boundary condition. *J. Heat Transfer* 96 (1):114–115.

______. 1975. Results of the numerical solution for outward solidification with flux boundary conditions. *J. Heat Transfer* 97 (2):307–309.

Goodman, T. R. 1958. The heat-balance integral and its application to problems involving a change of phase. *Tran. ASME* 80:335–342.

______. 1964. Application of integral methods to transient nonlinear heat transfer. *Advances in heat transfer,* vol. 1, eds. T. F. Irvine and J. P. Hartnett, New York: Academic Press, pp. 52–122.

______, and Shea, J. J. 1960. The melting of finite slabs. *J. Appl. Mech.* 27:16–24.

Gupta, J. P. 1973. An approximate method for calculating the freezing outside spheres and cylinders. *Chem. Eng. Sci.* 28:1629–1633.

Hwang, C. T. 1977. On quasi-static solutions for buried pipes in permafrost. *Can. Geotech. J.* 14(2):180–192.

______, Murray, D. W., and Brooker, E. W. 1972. A thermal analysis for structures on permafrost. *Can. Geotech. J.* 9(2):33–46.

Jiji, L. M., and Weinbaum, S. 1978. Perturbation solutions for melting or freezing in annular regions initially not at the fusion temperature. *Int. J. Heat Mass Transfer* 21:581–592.

Khakimov, K. R. 1957. Voprosy teorii i praktiki iskusstvennogo zamorazhivaniya grantov. Moskva: *Isdat el'stvo Akademii Nauk SSSR* (TT 66–51051 U.S. Dept. of Commerce).

Kreith, F., and Romie, F. E. 1955. A study of the thermal diffusion equation with boundary conditions corresponding to solidification or melting of materials initially at the fusion temperature. *Proc. Phys. Soc. London Sect. B* 68:277–291.

Lachenbruch, A. H. 1970. *Some estimates of the thermal effect of a heated pipeline in permafrost.* U.S. Geological Survey Circular No. 632, Washington, D.C.

Landau, H. G. 1950. Heat conduction in a melting solid. *Q. Appl. Math.* 8:81–94.

Lardner, T. J., and Pohle, F. V. 1961. Application of the heat balance integral to problems of cylindrical geometry. *Trans. ASME Ser. E.* 83: (2):310–312.

Lin, S. 1971. One-dimensional freezing or melting process in a body with variable cross-sectional area. *Int. J. Heat Mass Transfer* 14:153–156.

Lock, G. S. H. 1969. On the use of asymptotic solutions to plane ice-water problems. *J. Glaciology* 8 (53):285–300.

———, Gunderson, J. R., Quon, D., and Donnelly, J. K. 1969. A study of one-dimensional ice formation with particular reference to periodic growth and decay. *Int. J. Heat Mass Transfer* 12:1343–1352.

London, A. L., and Seban, R. A. 1943. Rate of ice formation. *Trans. ASME* 65 (7):771–779.

Lunardini, V. J. 1977. "Thawing of permafrost beneath a buried pipe." *J. Can. Pet. Technol.* 16(4):34–37.

Lunardini, V. J. 1980. "Phase change around a circular pipe," U.S. Army Cold Regions Research and Engineering Lab, Report 149, Hanover, N.H.

———, and Varotta, R. 1980. *Approximate solution to Neumann problem for soil systems.* ASME Paper 80-PET-14, presented at the Energy Sources Technology Conference, ASME, New Orleans, Louisiana.

Megerlin, F. 1968. Geometrisch eindimensionale Warmeleitung beim Schmelzen und Erstarren. *Forsch. Ingenieurwes.* 34:40.

Neumann, F. ca. 1860. Lectures given in 1860s. See Riemann-Weber. Die partiellen Differentialgleichungen. *Physik* (5th ed., 1912), 2:121.

Nixon, J. F., and McRoberts, E. C. 1973. A study of some factors affecting the thawing of frozen soils. *Can. Geotech. J.* 10:439–452.

Ozisik, M. N. 1968. *Boundary value problems of heat conduction.* Scranton, Pennsylvania: International Textbook.

Patel, R. D. 1968. Interface conditions in heat conduction problems with change of phase. *AIAA Journal* 6 (12):2454.

Pedroso, R. I., and Domoto, G. A. 1973*a*. Planar solidification with fixed wall temperature and variable thermal properties. *J. Heat Transfer* 95 (4):533–535.

———. 1973*b*. Perturbation solutions for spherical solidification of saturated liquids. *J. Heat Transfer* 95 (1):42–46.

Poots, G. 1962. On the application of integral methods to the solution of problems involving the solidification of liquids initially at fusion temperature. *Int. J. Heat Mass Transfer* 5:525–531.

———. 1962*a*. An approximate treatment of heat conduction problem involving a two-dimensional solidification front. *Int. J. Heat Mass Transfer* 5:339–348.

Porkhayev, G. V. 1963. Temperature fields in foundations. *Proceedings, First International Conference on Permafrost.* NRC No. 1287, pp. 285–291.

Riley, D. S., Smith, F. I., and Poots, G. 1974. The inward solidification of spheres and circular cylinders. *Int. J. Heat Mass Transfer* 17:1507–1516.

Ruoff, A. L. 1958. An alternate solution to Stefan's problem. *Q. Appl. Math.* 16:197–201.

Seban, R. A. 1971. A comment on the periodic freezing and melting of water. *Int. J. Heat Mass Transfer* 14:1862–1864.

Shamsundar, N., and Sparrow, E. M. 1974. Storage of thermal energy by solid-liquid phase change—Temperature drop and heat flux. *J. Heat Transfer* 96 (4):541–543.

Sparrow, E. M., Ramadhyani, S., and Patankar, S. V. 1978. Effect of subcooling on cylindrical melting. *J. Heat Transfer* 100 (3):395–402.

Stefan, J. 1891. Uber die Theorie des Eisbildung, insbesonder uber die Eisbildung im Polarmere. *Ann. Phys. u Chem. Neue Folge* 42 (2):269–286.

Tao, L. C. 1967. Generalized numerical solutions of freezing a saturated liquid in cylinders and spheres. *AICHE Journal* 13 (1):165.

Thornton, D. E. 1976. Steady-state and quasi-static thermal results for bare and insulated pipes in permafrost. *Can. Geotech. J.* 13(2):161–170.

Tien, L. C., and Churchill, S. W. 1965. Freezing front motion and heat transfer outside an infinite, isothermal cylinder. *AICHE Journal* 11 (5):790–793.

———, and Geiger, G. E. 1967. A heat transfer analysis of the solidification of a binary eutectic system. *J. Heat Transfer* 89:230–234.

Volkov, V. N., and Li-Orlov, V. K. 1970. A refinement of the integral method in solving the heat conduction equation. *Heat Transfer Sov. Res.* (2):41–47.

Yen, Y. C. 1968. On the effect of density variation on natural convection in a melted water layer. *Chem. Engr. Prog. Symp. Ser.* No. 29. American Institute of Chemical Engineers, p. 245.

9

Finite Difference Methods for Freezing and Thawing

9.0 INTRODUCTION

It is not our intention here to present a comprehensive review of numerical methods in heat transfer, as excellent treatises are already available: Carnahan (1969), Welty (1978), Dusinberre (1961), Schenck (1963), Ockendon (1975), Forsythe (1960). The mathematical difficulties associated with heat conduction problems with moving boundaries are so great, however, that numerical methods are very frequently used. Hence, a certain familiarity with numerical techniques for freezing and thawing is necessary. In general, no attempt will be made to present rigorous proofs; rather, the operational aspects of the problem solution will be concentrated on. There are no computer programs presented because we believe that the range of input-output devices is so varied that any specific instructions used in a program will have to be modified. The reader should be able to write a program fitting given hardware with the information presented here.

Linear Partial Differential Equations. We will be concerned, for the most part, with the partial differential equations of heat conduction. These equations are often referred to in the mathematical literature in terms of certain classifications. Let us assume we are dealing with a function of four variables, $T(x,y,z,t)$. Then a general partial differential equation may be written as

$$A_x \frac{\partial^2 T}{\partial x^2} + A_y \frac{\partial^2 T}{\partial y^2} + A_z \frac{\partial^2 T}{\partial z^2} + A_t \frac{\partial^2 T}{\partial t^2} + B_x \frac{\partial T}{\partial x} + B_y \frac{\partial T}{\partial y} + B_z \frac{\partial T}{\partial z} + B_t \frac{\partial T}{\partial t} + CT - D = 0 \quad (9.1)$$

An elliptic equation is one for which all the A_i are nonzero, with the same sign. Consider Eq. (9.2)

$$\frac{\partial^2 T}{\partial x^2} + \frac{\partial^2 T}{\partial y^2} + \frac{\partial^2 T}{\partial z^2} = 0 \quad (9.2)$$

This is the steady-state heat conduction equation and is an elliptic equation. A hyperbolic equation has all A_i nonzero with the same sign, except for one of the A_i. The telephone equation, Eq. (9.3), is an example of a hyperbolic equation

$$LC\frac{\partial^2 V}{\partial t^2} - \frac{\partial^2 V}{\partial x^2} + (RC + GL)\frac{\partial V}{\partial t} + RGV = 0 \tag{9.3}$$

If A_k is zero, the other A_i are nonzero with the same sign, and B_k is nonzero, then the equation is parabolic. The transient conduction equation in two dimensions, written as Eq. (9.4), is an example of a parabolic equation

$$\frac{\partial^2 T}{\partial x^2} + \frac{\partial^2 T}{\partial y^2} - \frac{1}{\alpha}\frac{\partial T}{\partial t} = 0 \tag{9.4}$$

9.1 GENERAL NUMERICAL METHODS

9.1.1 Nodal System and Development of Equations

There are two methods to generate a system of finite difference equations for heat transfer problems: start with a given partial differential equation, and express it in finite difference form or write an energy balance for specified regions of interest.

The basic spatial network, or nodal system, is as shown in Fig. 9.1. In this example, a two-dimensional space network is used, but the two dimensions could be space and time. If time is used, then the superscript n will be used to denote the time step. The primary assumption made is that the

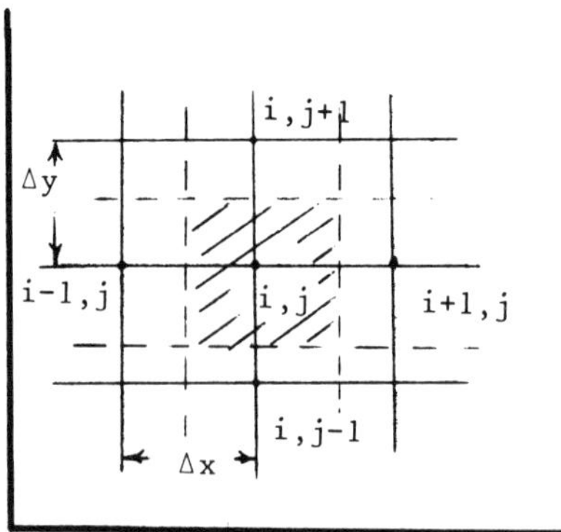

Fig. 9.1. Two-Dimensional Space Grid.

temperature or other properties at nodal point *i,j* represents the temperature over the entire finite volume $\Delta x \Delta y$ centered at *i,j,* the cross-hatched region shown. It is assumed that a space dimension of 1 exists normal to $\Delta x \Delta y$.

9.1.2 Errors and Related Concepts

Any finite difference solution will vary somewhat from the exact solution of the partial differential equation. These errors are due to several factors.

Truncation Error. This is the difference between the finite difference solution and the exact solution of the differential equation due to replacing derivatives at a point by finite differences over a region. These errors are related to the finite size of the mesh in space and time. A sketch of these concepts is shown by Fig. 9.2.

Roundoff Error. The error due to the inability of the computation to retain sufficient significant figures is called the "roundoff error." With modern computing systems this error is usually minor.

Convergence. This is the condition where the solution of the finite difference problem, excluding roundoff error, approaches the exact solution of the differential equation as the grid size approaches zero.

Consistency. This is the quality wherein the solution of the finite difference equation approaches that of the given partial differential equations and not some other partial differential equations. Consistency is usually assumed to

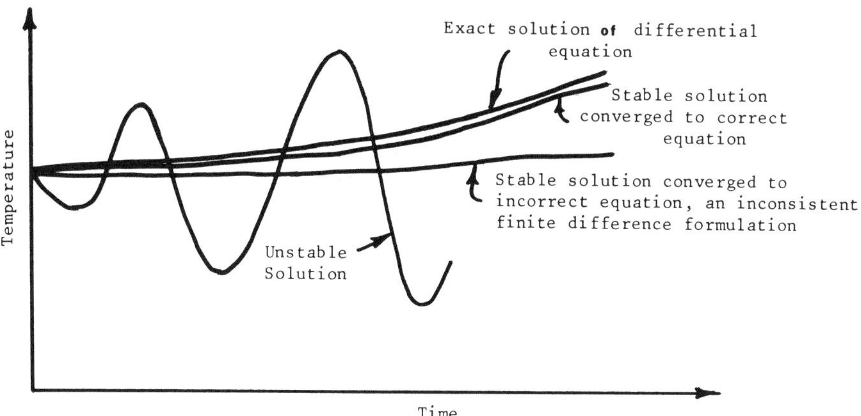

Fig. 9.2. Finite Difference Errors.

exist, without verification, but it is possible to set up a finite difference formulation that is inconsistent with the given differential equation, as exemplified in Fig. 9.2. The solution converges stably, but to the wrong differential equation.

Stability. This concept implies the boundedness of the finite difference result as the solution proceeds. Instability may also result if errors grow more rapidly than the rate of convergence as the mesh size approaches zero. For linear partial differential equations, given a consistency criterion, stability is a necessary and sufficient condition for convergence (Carnahan 1969). This conclusion may not follow for nonlinear equations.

9.2 FINITE DIFFERENCE APPROXIMATIONS FOR DERIVATIVES

This section will investigate methods of replacing a derivative of a function at a point by values of the function near the point of interest.

For ease of presentation, let us examine one-dimensional relations; the extension to multidimensions is straightforward. The function $T(x)$ can be expanded in a Taylor series about x as

$$T(x+\Delta x) = T(x) + \Delta x \frac{dT(x)}{dx} + \frac{\Delta x^2}{2}\frac{d^2T(x)}{dx^2} + \frac{\Delta x^3}{6}\frac{d^3T(x)}{dx^3} + \ldots \tag{9.5}$$

and

$$T(x-\Delta x) = T(x) - \Delta x \frac{dT(x)}{dx} + \frac{\Delta x^2}{2}\frac{d^2T(x)}{dx^2} - \frac{\Delta x^3}{6}\frac{d^3T(x)}{dx^3} + \ldots \tag{9.6}$$

Truncate Eq. (9.5) after the first two terms, and solve for the first derivative

$$\frac{dT(x)}{dx} = \frac{T(x+\Delta x) - T(x)}{\Delta x} + 0(\Delta x) \tag{9.7}$$

The symbol $0(\Delta x)$ indicates the error due to truncation is on the order of Δx. This is the first forward difference approximation of the first derivative. In a similar fashion, from Eq. (9. 6)

$$\frac{dT(x)}{dx} = \frac{T(x) - T(x-\Delta x)}{\Delta x} + 0(\Delta x) \tag{9.8}$$

which is the backward difference approximation. Now subtract Eq. (9.6) from Eq. (9.5), and neglect terms of order $(\Delta x)^3$ and higher. Then

$$\frac{dT(x)}{dx} = \frac{T(x+\Delta x) - T(x-\Delta x)}{2\Delta x} + O(\Delta x^2) \tag{9.9}$$

This represents the first central difference approximation and is the approximation generally used, particularly in heat transfer problems, for spatial derivatives, though the forward or backward forms are used for time derivatives. The central difference truncation error is considerably less than that of Eqs. (9.7) or (9.8).

The second derivative can be found by adding Eqs. (9.5) and (9.6) and truncating after the Δx^2 terms

$$\frac{d^2T(x)}{dx^2} = \frac{T(x+\Delta x) + T(x-\Delta x) - 2T(x)}{\Delta x^2} + O(\Delta x^2) \tag{9.10}$$

This is also a central difference relation. If the central difference operand is defined as

$$\delta_x T(x) = \frac{T\left(x+\frac{\Delta x}{2}\right) - T\left(x-\frac{\Delta x}{2}\right)}{\Delta x} \tag{9.11}$$

then

$$\delta_x^2 T(x) = \frac{T(x+\Delta x) + T(x-\Delta x) - 2T(x)}{\Delta x^2} \tag{9.12}$$

The previous equations can now be expressed in terms of the notation defined for the finite difference grid of Fig. 9.1.

$$\text{(forward)} \qquad \left.\frac{dT}{dx}\right|_i = \frac{T_{i+1} - T_i}{\Delta x} \tag{9.7a}$$

$$\text{(backward)} \qquad \left.\frac{dT}{dx}\right|_i = \frac{T_i - T_{i-1}}{\Delta x} \tag{9.8a}$$

$$\text{(central)} \qquad \left.\frac{dT}{dx}\right|_i = \frac{T_{i+1} - T_{i-1}}{2\Delta x} \tag{9.9a}$$

$$\text{(central)} \qquad \left.\frac{d^2T}{dx^2}\right|_i = \frac{T_{i+1} + T_{i-1} - 2T_i}{\Delta x^2} \tag{9.10a}$$

Salvadori and Baron (1952) list finite difference formulae for derivatives up to fourth order and with accuracy on the order of $(\Delta x)^4$. As we are concerned with the heat conduction equation, let us write the finite difference form of the transient conduction equation with energy generation. If the properties can be considered constant, then the transient, one-dimensional, conduction equation is

$$\frac{\partial^2 T}{\partial x^2}+\frac{q_g}{k}=\frac{1}{\alpha}\frac{\partial T}{\partial t} \tag{9.13}$$

Using Eq. (9.10*a*), this can be expressed as

$$T_{i+1}+T_{i-1}-2T_i+q_{gi}\frac{\Delta x^2}{k_i}=\frac{\Delta x^2}{\alpha_1}\left(\frac{\partial T}{\partial t}\right)_i \tag{9.14}$$

The time derivative approximation will be examined shortly.

9.3 ENERGY BALANCE METHOD FOR FINITE DIFFERENCES

In problems with variable properties or unusual boundaries, it is often preferable to obtain the finite difference equation from an energy balance. Consider a simple one-dimensional model with only conduction and heat generation considered, governed by Eq. (9.13), if the properties are constant. The general energy balance is given by the first law of thermodynamics, which, for this system, can be expressed as

$$\Sigma Q=\Delta H \tag{9.15}$$

For a constant pressure process, the summation of all of the energy flows, as heat, is equal to the change of enthalpy of the mass of the system. Referring to Fig. 9.3, this is

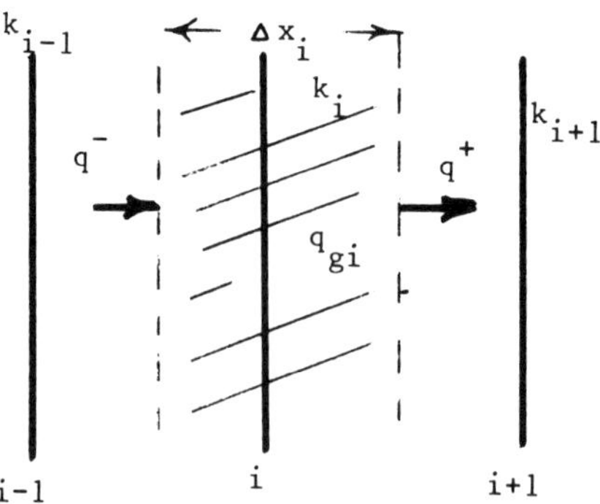

Fig. 9.3. One-Dimensional Heat Flow.

$$q_g + q^- + q^+ = \Delta H \tag{9.16}$$

The heat flow due to conduction from element $i - 1$ to element i, is

$$q^- = \frac{-(T_i - T_{i-1})\Delta t}{\dfrac{\Delta x_{i-1}}{2Ak_{i-1}} + \dfrac{\Delta x_i}{2Ak_{i+1}}}$$

The heat flow from i to $i + 1$ is

$$q^+ = \frac{-(T_i - T_{i+1})\Delta t}{\dfrac{\Delta x_i}{2Ak_i} + \dfrac{\Delta x_{i+1}}{2Ak_{i+1}}}$$

The energy generated is

$$q_{gi}\,\Delta x_i\,A\,\Delta t$$

Finally, the enthalpy change is

$$\Delta H = A\,\Delta x_i\,[(\rho c)_i^{n+1} T_i^{n+1} - (\rho c)_i^n T_i^n]$$

where n denotes the time levels.
Equation (9.16) then becomes

$$\frac{-2(T_i - T_{i-1})}{\dfrac{\Delta x_{i-1}}{k_{i-1}} + \dfrac{\Delta x_i}{k_i}} + \frac{-2(T_i - T_{i+1})}{\dfrac{\Delta x_i}{k_i} + \dfrac{\Delta x_{i+1}}{k_{i+1}}}$$

$$+ q_{gi}\,\Delta x_i = \frac{\Delta x_i}{\Delta t}[(\rho c)_i^{n+1} T_i^{n+1} - (\rho c)_i^n T_i^n] \tag{9.17}$$

The time level for the spatial derivatives has been omitted for the present. An energy balance may also include radiation and convection terms, as well as conduction. If the properties are constant and $\Delta x_i = \Delta x$, then Eq. (9.17) is

$$T_{i+1} + T_{i-1} - 2T_i + q_{gi}\frac{\Delta x^2}{k_i} = \frac{\Delta x^2}{\alpha_i}\frac{T_i^{n+1} - T_i^n}{\Delta t} = \frac{\Delta x^2}{\alpha_i}\left(\frac{\partial T}{\partial t}\right)_i$$

This is identical to Eq. (9.14) and shows that the heat balance method is equivalent to the central difference approximation method. Complete heat balance equations can be expressed more conveniently in terms of thermal conductances and resistances, as described by Welty (1978).

9.4 TIME DOMAIN

The form of the finite difference equation is influenced by the method used in carrying out the time derivative approximation.

9.4.1 Explicit Method

Return to Eq. (9.14), and exclude the energy generation term for simplicity

$$T_{i+1} + T_{i-1} - 2T_i = \frac{\Delta x^2}{\alpha}\left(\frac{\partial T}{\partial t}\right)_i \tag{9.18}$$

Use the forward difference relation, Eq. (9.7*a*), for the time derivative, and express the temperatures on the left-hand side of Eq. (9.18) in terms of their values at the beginning of the time step. Thus

$$T^n_{i+1} + T^n_{i-1} - 2\,T^n_i = \frac{\Delta x^2}{\alpha\Delta t}(T^{n+1}_i - T^n_i)$$

or

$$T^{n+1}_i = \left(1 - \frac{2\alpha\Delta t}{\Delta x^2}\right) T^n_i + \frac{\alpha\Delta t}{\Delta x^2}\left(T^n_{i+1} + T^n_{i-1}\right) \tag{9.19}$$

Simplifying slightly, this is

$$T^{n+1}_i = (1 - 2M)\,T^n_i + M(T^n_{i+1} + T^n_{i-1}) \tag{9.20}$$

where

$$M = \frac{\alpha\,\Delta t}{\Delta x^2}$$

It is clear from Eq. (9.20) that the temperature T^{n+1}_i at the end of each time step can be evaluated completely in terms of the quantities known at the beginning of the time step. Stated differently, the temperatures for each time step are evaluated in terms of the parameters of the previous time step. Further, each nodal point equation has only one unknown quantity, the temperature at the nodal point for the new time increment of interest. Then each unknown can be solved for explicitly without recourse to systems of simultaneous equations. This simplicity is purchased at a certain cost, however. Equation (9.20) will converge to the actual solution of the differential equation, only if $0 < M \leq ½$ (Carnahan 1969). Physically, this restriction can be examined as follows. At the beginning of a time interval let

$$T_{i+1}^{n} = T_{i-1}^{n} = T_{1}$$

and let $T_i^n > T_1$. Then $T_i^n = T_1 + \delta$, where δ is any positive number. At the end of the time step, from Eq. (9.20)

$$T_i^{n+1} = T_1 + (1 - 2M)\delta$$

Now let $1 - 2M < 0$ or $M > ½$; then $T_i^{n+1} < T_1$. But this would require a heat flow from the lower temperature T_1 to the higher temperature T_i^n during the time interval, violating the second law of thermodynamics and directly leading to the above criterion for M. Physical arguments of this type lead to Dusinberre's (1961) stability guide that all the coefficients of Eq. (9.19) should be positive to guarantee stability. This means that if the truncation error is to be reduced, by reducing Δx, then the time step must also be reduced, so that $\Delta t \leq (\Delta x^2/2\alpha)$. This can be a severe restriction in terms of computational time. In addition to this limitation, the explicit method does not fully reflect the influence of all the surrounding nodes, at the previous time step, upon the value of temperature at the present time step. Nevertheless, the simplicity of the explicit method is a great advantage computationally.

9.4.2 Implicit Method

The limitations of the explicit method can be avoided by considering a different time step scheme. Write the finite difference form of Eq. (9.18) using the space values at the present time and the backward difference form, Eq. (9.8*a*), for the time derivative

$$T_{i+1}^{n} + T_{i-1}^{n} - 2T_i^n = \frac{\Delta x^2}{\alpha \Delta t}(T_i^n - T_i^{n-1}) \tag{9.21}$$

For convenience, replace the time step n by $n + 1$, and then

$$T_{i+1}^{n+1} + T_{i-1}^{n+1} - 2T_i^{n+1} = \frac{\Delta x^2}{\alpha \Delta t}(T_i^{n+1} - T_i^{n}) \tag{9.22}$$

The basic form then

$$-MT_{i-1}^{n+1}+(1+2M)T_i^{n+1}-M\,T_{i+1}^{n+1}=T_i^n \tag{9.23}$$

Notice that in this formulation the temperature T_i^{n+1} is influenced fully by all the nodal points in contact with it, at the prior time (Fig. 9.4). It can be shown that Eq. (9.23) is stable (it converges to the partial differential equation) for all values of $M > 0$ (see Carnahan 1969). The disadvantage of the implicit formulation is that more than one unknown is involved at each time step, and thus a system of simultaneous equations must be solved at each time step. For certain systems of equations, however, this is not a severe restriction, and since the time step is independent of the space increment, the implicit method may yield comparable accuracy with reduced computer time, in comparison with an explicit formulation.

Example 1: Consider the solution to the transient heating of a semi-infinite slab

$$\frac{\partial^2 T}{\partial x^2}=\frac{1}{\alpha}\frac{\partial T}{\partial t} \tag{9.24}$$

$$T(x,0)=f(x) \tag{9.24a}$$

$$T(0,t)=p(t) \tag{9.24b}$$

$$T(L,t)=S(t) \tag{9.24c}$$

Use only the four nodal points, as shown in Fig. 9.5, where the temperatures at $i=0$, and $i=4$ are known. Then, using Eq. (9.23), the system of equations for each time step is

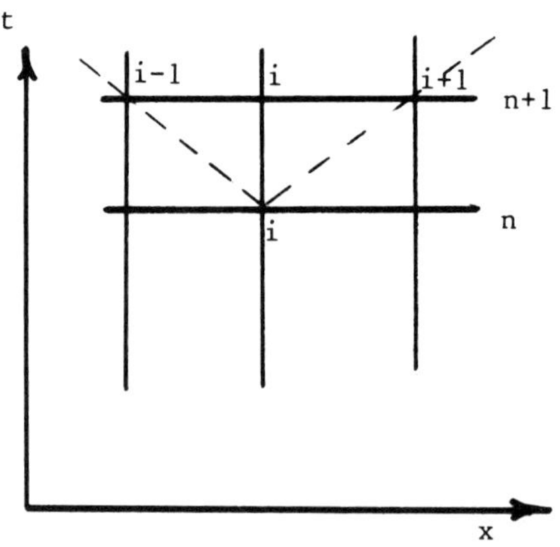

Fig. 9.4. Implicit Finite Differences.

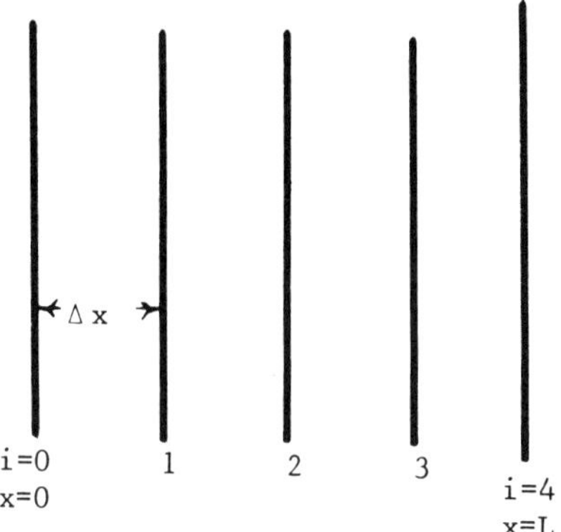

Fig. 9.5. Nodal System for One Space Dimension.

$$(1+2M)T_1^{n+1} - MT_2^{n+1} = T_1^n + Mp(n+1) \tag{9.25a}$$

$$-M\,T_1^{n+1} + (1+2M)T_2^{n+1} - MT_3^{n+1} = T_2^n \tag{9.25b}$$

$$-M\,T_2^{n+1} + (1+2M)T_3^{n+1} = T_3^n + MS(n+1) \tag{9.25c}$$

A set of three simultaneous equations for the unknown temperatures is, thus, developed.

Equation (9.25) is written in this form to illustrate the tridiagonal nature of the matrix for the set of equations. This is shown more clearly in the matrix below for six equations

$$\begin{bmatrix} a_{11} & a_{12} & 0 & 0 & 0 & 0 \\ a_{21} & a_{22} & a_{23} & 0 & 0 & 0 \\ 0 & a_{32} & a_{33} & a_{34} & 0 & 0 \\ 0 & 0 & a_{43} & a_{44} & a_{45} & 0 \\ 0 & 0 & 0 & a_{54} & a_{55} & a_{56} \\ 0 & 0 & 0 & 0 & a_{65} & a_{66} \end{bmatrix} \begin{bmatrix} T_1 \\ T_2 \\ T_3 \\ T_4 \\ T_5 \\ T_6 \end{bmatrix} = \begin{bmatrix} b_1 \\ b_2 \\ b_3 \\ b_4 \\ b_5 \\ b_6 \end{bmatrix} \tag{9.26}$$

It is well worth the effort to formulate the difference equations in such a way that a tridiagonal system is to be evaluated. In certain cases, the solution can be effected quite efficiently by the Gaussian elimination method. This method is preferable to Gauss-Seidel iteration if the coefficients of the matrix are the same for each time step, as the computation time is significantly reduced. Complete and efficient computer programs for solving tridiagonal matrix systems are given by Carnahan (1969) and Welty (1978). Thus, the solutions to problems of this type can be considered routine.

9.4.3 Crank-Nicolson Method

The accuracy of Eqs. (9.20) and (9.23) are limited to $0(\Delta t + \Delta x^2)$ due to the first difference approximation for the time change of temperature. The accuracy can be improved by multiplying Eq. (9.23) by β, a coefficient between 0 and 1, Eq. (9.20) by $(1 - \beta)$, and adding the resulting equations. Then

$$\beta(T_{i+1}^{n+1} - T_{i-1}^{n+1}) - \left(2\beta + \frac{1}{M}\right) T_i^{n+1}$$

$$= -(1-\beta)(T_{i+1}^{n} + T_{i-1}^{n}) + \left[2(1-\beta) - \frac{1}{M}\right] T_i^n \quad (9.27)$$

The Crank-Nicolson method (1977), a widely used procedure, uses $\beta = ½$. Thus

$$-M(T_{i+1}^{n+1} - T_{i-1}^{n+1}) + 2(M+1)\, T_i^{n+1} = M(T_{i+1}^{n} + T_{i-1}^{n})$$

$$+ 2(1-M)T_i^n \quad (9.28)$$

This equation is not much more complicated than Eq. (9.23), but there is a significant increase in accuracy. The increased accuracy is due to using a central difference operator for the time based upon

$$\left.\frac{dT}{dt}\right|_i^{n+1/2} = \frac{T_i^{n+1} - T_i^n}{\Delta t} + 0(\Delta t^2) \quad (9.29)$$

The general, multidimensional, conduction equation is

$$\nabla^2 T = \frac{1}{\alpha}\frac{\partial T}{\partial t} \quad (9.30)$$

Then the generalized form of the Crank-Nicolson formulation is

$$T^{n+1} - T^n = \frac{\alpha \Delta t}{2}(\nabla^2 T^{n+1} + \nabla^2 T^n) \quad (9.31)$$

In one-space dimension, this reduces to Eq. (9.28).

Forsythe and Wasow (1960) note that the accuracy can be further improved to $0(\Delta x^4)$ if $\beta = (6M + 1)/12M$, and this accuracy will apply to heat transfer problems of the type discussed here. The parameter M is, naturally, not restricted for this implicit method.

9.5 UNCONDITIONALLY STABLE EXPLICIT METHODS

The advantage of unconditionally stable systems has prompted the suggestion of various ingenious formulations. These methods remove the restriction on M and retain the explicit nature of the equations.

9.5.1 The DuFort-Frankel Method

This scheme is described by DuFort and Frankel (1953). Write the time and space derivatives as

$$\left.\frac{dT}{dt}\right|_i^n = \frac{T_i^{n+1} - T_i^{n-1}}{2\Delta t} \tag{9.32}$$

$$\left.\frac{d^2T}{dx^2}\right|_i = \frac{T_{i-1}^n - T_i^{n-1} + T_{i+1}^n - T_i^{n+1}}{\Delta x^2} \tag{9.33}$$

Then the difference equation, for Eq. (9.30), is

$$(1 + 2M)\, T_i^{n+1} = 2M(T_{i-1}^n - T_i^{n-1} + T_{i+1}^n) + T_i^{n-1} \tag{9.34}$$

The equations are clearly explicit for the unknown temperature at each time step. As a cautionary note, the consistency of the method depends upon the way in which Δt and Δx tend to zero (see Carnahan 1969).

9.5.2 The Saul'yev Method

This calculation method alternates the direction of the evaluation of the heat balance at each nodal point, at each time step, from left to right and right to left. This is an example of an alternating direction explicit procedure (Saul'yev 1964). For the time step from n to $n + 1$, evaluate the temperature from the left boundary to the right boundary with

$$(1 + M)T_i^{n+1} = (1 - M)T_i^n + M(T_{i-1}^{n+1} + T_{i+1}^n) \tag{9.35}$$

As the calculation moves from left to right, T_{i-1}^{n+1} is known from the previous nodal point evaluation (at $n + 1$) or from the left-hand boundary condition, and all the T_i^n are known. The next time step from $n + 1$ to $n + 2$ is evaluated from right to left with

$$(1 + M)T_i^{n+2} = (1 - M)T_i^{n+1} + M(T_{i-1}^{n+1} + T_{i+1}^{n+2}) \tag{9.36}$$

Again T_{i+1}^{n+2} is known at the right-hand boundary.

Other schemes are also available that have the simplicity of the explicit method and do not have the stability restriction on M.

Example 2: Consider the problem of Example 1, but use the Saul'yev method. It is assumed that T_o, T_L, T_i^0, and M are known. Again use four space divisions; then for the first time increment ($n = 0$) moving from left to right, Eq. (9.35) gives

$$(1+M)T_1^1=(1-M)T_1^0+M(T_0+T_2^0)$$
$$(1+M)T_2^1=(1-M)T_2^0+M(T_1^1+T_3^0)$$
$$(1+M)T_3^1=(1-M)T_3^0+M(T_2^1+T_L)$$

For the second time increment (n is still 0), we use Eq. (9.36) and move from the right-hand boundary to the left-hand boundary

$$(1+M)T_3^2=(1-M)T_3^1+M(T_2^1+T_L)$$
$$(1+M)T_2^2=(1-M)T_2^1+M(T_1^1+T_3^2)$$
$$(1+M)T_1^2=(1-M)T_1^1+M(T_o+T_2^2)$$

Thus, for $n = 0$, the first sweep to the right allows the three unknown temperatures to be explicitly evaluated at $n = 1$; then, with n still 0, the sweep to the left gives the three unknown temperatures at $n = 2$, still with an explicit calculation. The advantage here is that M is not restricted in value. Of course, for actual problems, the space would be broken into many increments to minimize truncation error.

9.6 ALTERNATING DIRECTION IMPLICIT METHOD

For problems with only one space dimension, the implicit equation can lead to the efficient tridiagonal matrix, as given by Eq. (9.23). With more than one space dimension, the method of Section 9.4.2 will yield matrices more complicated than the tridiagonal type, but a scheme can be devised for multidimensional problems that will also have tridiagonal matrices. Three time levels are used. Consider the equation

$$\frac{\partial^2 T}{\partial x^2}+\frac{\partial^2 T}{\partial y^2}=\frac{1}{\alpha}\frac{\partial T}{\partial t} \tag{9.37}$$

Two sets of equations are used over each half-time interval $\Delta t/2$. In the first half-time increment, the x derivatives are in the implicit form, and the y derivatives are explicit. During the second half of the time interval, the order is reversed.

Thus, for the interval n to $n + ½$, with $\Delta x = \Delta y$ for simplicity

$$(T_{i-1,j}^{n+1/2} - 2\,T_{i,j}^{n+1/2} + T_{i+1,j}^{n+1/2}) + (T_{i,j-1}^{n} - 2T_{i,j}^{n} + T_{i,j+1}^{n}) = \frac{2}{M}(T_{i,j}^{n+1/2} - T_{i,j}^{n})$$

or

$$T_{i-1,j}^{n+1/2} - 2\left(1 + \frac{1}{M}\right) T_{i,j}^{n+1/2} + T_{i+1,j}^{n+1/2} = 2\left(1 - \frac{1}{M}\right) T_{i,j}^{n} - T_{i,j-1}^{n} - T_{i,j+1}^{n} \quad (9.38)$$

This tridiagonal matrix can be solved for the intermediate time step values $T_{i,j}^{n+1/2}$.

To complete the calculation, reverse the implicit-explicit order for the x and y derivatives, giving

$$T_{i,j-1}^{n+1} - 2\left(1 + \frac{1}{M}\right) T_{i,j}^{n+1} + T_{i,j+1}^{n+1} = 2\left(1 - \frac{1}{M}\right) T_{i,j}^{n+1/2} - T_{i-1,j}^{n+1/2} - T_{i+1,j}^{n+1/2} \quad (9.39)$$

The solution of this tridiagonal system completes the time step. Note that the intermediate values $T_{i,j}^{n+1/2}$ are known from the first half-time step calculation. The method is unconditionally stable. For further details, see Douglas and Gunn (1964).

9.7 GENERALIZED FORMULATION

The conduction equation, Eq. (9.30), can be written in a general finite difference form as

$$\begin{aligned}(T_{i,j}^{n+1} - T_{i,j}^{n})/M = {} & f_1(T_{i+1,j}^{n+1} - T_{i,j}^{n+1}) + f_2(T_{i+1,j}^{n} - T_{i,j}^{n}) \\ & + f_3(T_{i-1,j}^{n+1} - T_{i,j}^{n+1}) + f_4(T_{i-1,j}^{n} - T_{i,j}^{n}) + f_5(T_{i,j+1}^{n+1} - T_{i,j}^{n+1}) \\ & + f_6(T_{i,j+1}^{n} - T_{i,j}^{n}) + f_7(T_{i,j-1}^{n+1} - T_{i,j}^{n+1}) + f_8(T_{i,j-1}^{n} - T_{i,j}^{n}) \quad (9.40)\end{aligned}$$

All the previous formulations depend upon how the f_i are chosen. In one space dimension, Eq. (9.40) can be rearranged as

$$[1 + (f_1 + f_2)M]\, T_i^{n+1} - Mf_1 T_{i+1}^{n+1} - Mf_3 T_{i-1}^{n+1}$$
$$= [1 - (f_2 + f_4)M] T_i^n + f_2 M T_{i+1}^n + f_4\, T_{i-1}^n \quad (9.41)$$

If $f_i = ½$, then the Crank-Nicolson formula, Eq. (9.28), is again obtained. The Saul'yev formula for the explicit, alternating, direction, Eq. (9.35), occurs when $f_1 = f_4 = 0$ and $f_2 = f_3 = 1$.

The methods described here can generally be extended to multidimensional problems. This introduction should allow the reader to follow the many specific, finite difference schemes that have been recently applied to practical problems of freezing and thawing.

A second, numerical approach is that of finite elements. This method has been recently applied to a number of phase change problems, but the concept will not be developed here. Computer implementation of most finite element methods tend to be time consuming. The reader is referred to Wilson et al. (1978), Comini et al. (1974), and O'Neill and Lynch (1979).

9.8 PHASE CHANGE AND LATENT HEAT

The numerical procedures described are general and can be applied to any heat transfer problem. For cold climate regions, the major complication to heat transfer problems is due to the freezing and thawing of water in the soil system. The phase change effect can lead to singularities in certain functions, such as temperature gradients, at the phase change interface. These singularities adversely affect the accuracy of the numerical formulations. The enthalpy change relations during the phase change processes will be associated with the water within the solid volume. The soil solids will exert little influence upon the enthalpy changes except as sources of impurities for the water or in terms of surface effects upon the phase change of water (see Section 4.6).

The removal (or addition) of energy from a substance, at constant pressure, will result in a decrease (or increase) of the enthalpy of the material. This effect is manifested by the temperature variation as the process continues. The cooling curve for a pure substance, such as water, is sketched in Fig. 9.6. The temperature of the water will decrease as heat is extracted, until the fusion temperature, T_f, is reached. Heat removal or addition, with a temperature change, is described as a sensible heat effect or process, i.e., a change, during which, the temperature is varying. The specific heat at constant pressure, is defined as the enthalpy change per unit of temperature change

$$C_p \equiv \left(\frac{\partial H}{\partial T}\right)_p \quad (9.42)$$

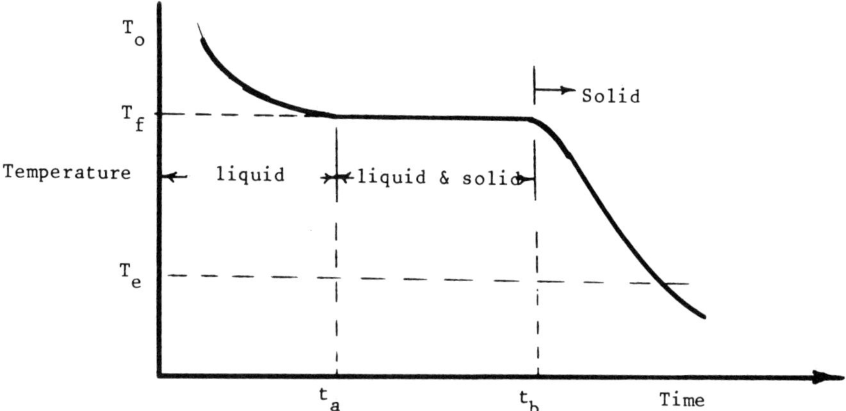

Fig. 9.6. Cooling Curve for a Pure Substance.

where H is the enthalpy of the substance per unit volume. Because the specific heat of solids or liquids, for constant pressure and constant volume, are nearly equal, no further subscript will be used for the specific heat at constant pressure.

The enthalpy change during a sensible heat process is then

$$H_o - H_f = \int_{T_0}^{T_f} C\,dT \tag{9.43}$$

It is well known that the enthalpy may decrease (increase) without a change of temperture. This may occur during phase change, as for water, or during chemical reaction and molecular rearrangement without phase change. For water, a phase change will occur at T_f, and the liquid will be changed to a solid (ice) at constant temperature. Under this condition, the specific heat of the water would be infinite, according to Eq. (9.42). Thus, the specific heat of a substance is not defined during the constant temperature enthalpy change. The change in enthalpy of the material, during the constant temperature process, is called the "latent" (hidden) heat. The enthalpy of a liquid will be greater than that of its solid form, and thus the freezing process will result in a liberation of energy and the thawing process will be one of energy absorption. This absorption or release of energy will occur at the interface where phase change is occurring.

In Fig. 9.6, the enthalpy decreases at constant temperature T_f, from time t_a to t_b. A mixed phase exists, during the time interval $(t_b - t_a)$, of saturated water and saturated ice (or different chemical or molecular structures for other substances) with only saturated liquid at t_a and only saturated solid at t_b. If cooling continues, the enthalpy will continue to decrease, but during a sensible heat process. The complete enthalpy change, going from T_o to T_e, can then be written as

$$H_o - H_e = \int_{T_o}^{T_f} C\,dT + L + \int_{T_f}^{T_e} C\,dT \tag{9.44}$$

The enthalpy can be expressed as a function of temperature, as shown in Fig. 9.7.

Here, the enthalpy change L is the latent heat of fusion, where H_f is the enthalpy of a saturated liquid (water) at a given pressure, and H_i is the enthalpy of a saturated solid (ice) at the same pressure. The specific heats are generally functions of temperature and can be quite different for liquid and solid phases. The mass specific latent heat, l, for water is 144 Btu/lbm or 79.6 cal/g. The total energy given up during the phase change process will, of course, be a function of the total mass of water contained in a given volume of soil and the percentage of this water that actually changes phase. Thus, in Eq. (9.44), for a soil

$$L = \frac{\gamma_d l(W - W_u)}{100} \tag{9.45}$$

where

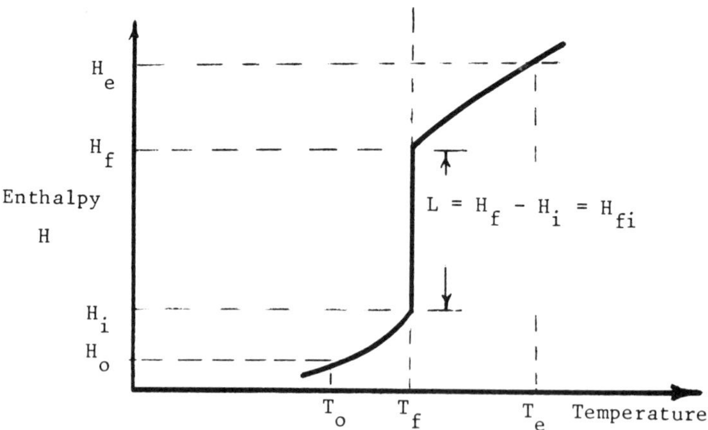

Fig. 9.7. Enthalpy vs. Temperature for Pure Water.

γ_d = dry unit weight of the soil system
W = original water content percent based on the dry unit weight and
W_u = unfrozen water at the end of the phase change.

The unfrozen water content is discussed in Section 4.7.

If the specific heats of the two phases are not functions of temperature, then a simple form of the enthalpy curve exists, as shown in Fig. 9.8.

It should be noted that the quantity of energy released during freezing is identical to that absorbed during thawing if the same mass of water undergoes phase change, although for soils there is a hysteresis effect. The phase change can also occur over a finite temperature range ΔT. This type of enthalpy curve is more realistic for the water in soil systems. This suggests an approximation to the actual enthalpy versus temperature curve for pure substances, as noted by Fig. 9.9. The idealization of the discontinuous enthalpy function by a function that is piecewise continuous will be discussed later.

9.9. PROBLEM FORMULATION FOR PHASE CHANGE

The moving boundary of the phase change interface presents special difficulties for numerical procedures. Since the position of the boundary is dependent upon the unknown temperature field, the location of the phase boundaries are not known a priori and must be monitored carefully during the calculation procedure. Coupled with the possibility of rapidly changing (or even discontinuous) properties near the phase front, this leads to exceptional difficulties for the numerical calculation.

A heat transfer problem for which a moving interface exists between two

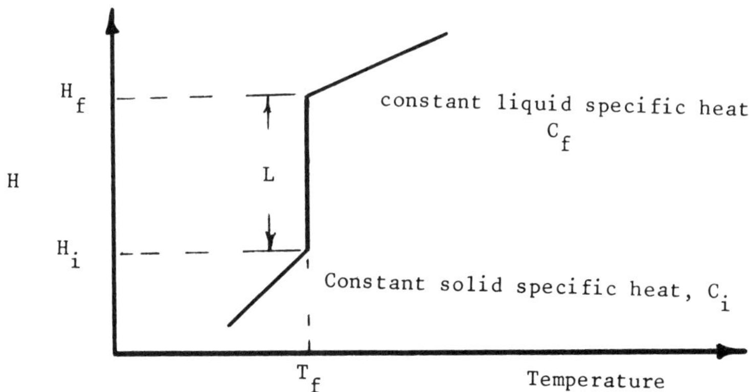

Fig. 9.8. Ideal Enthalpy Curve.

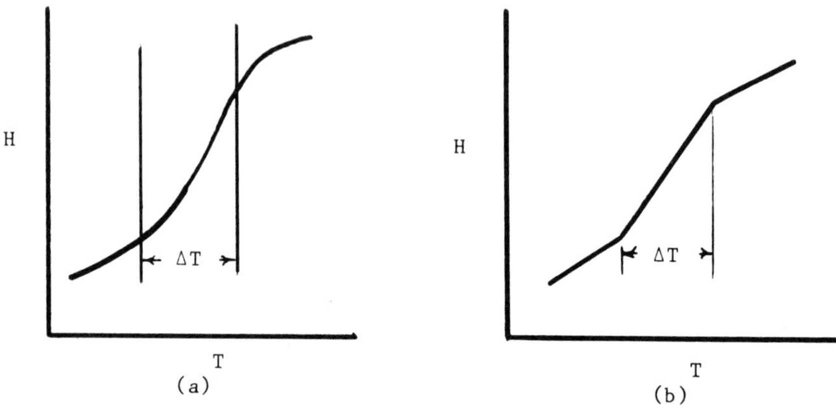

Fig. 9.9. Enthalpy of a Soil System (a) and Idealized Soil Enthalpy (b).

phases, and for which the unknown location of the interface must be evaluated as part of the temperature field, is defined as a Stefan problem. Such problems, however, may exist for other physical systems, such as concentration changes, although the thermal problem originally gave rise to the term "Stefan problem."

9.9.1 Boundary Value Formulation

The physical system is, as usual, modeled mathematically by a more or less simplified set of equations. Several approaches are available, of which the boundary value description is the most popular. Examine the simple system with phase change involved in a melting case (Fig. 9.10). The general conduction equation can be written for regions 1 and 2 as

$$\frac{\partial}{\partial x}\left(k_1 \frac{\partial T_1}{\partial x}\right) = \frac{\partial H_1}{\partial t} \tag{9.46}$$

$$T_1(x,o) = \phi_1(x) \tag{9.46a}$$

$$T_1(o,t) = f_1(t) < T_f \tag{9.46b}$$

$$\frac{\partial}{\partial x}\left(k_2 \frac{\partial T_2}{\partial x}\right) = \frac{\partial H_2}{\partial t} \tag{9.47}$$

$$T_2(x,o) = \phi_2(x) \tag{9.47a}$$

$$T_2(o,t) = f_2(t) > T_f \tag{9.47b}$$

$$T_1(X,t) = T_2(X,t) = T_f \tag{9.48}$$

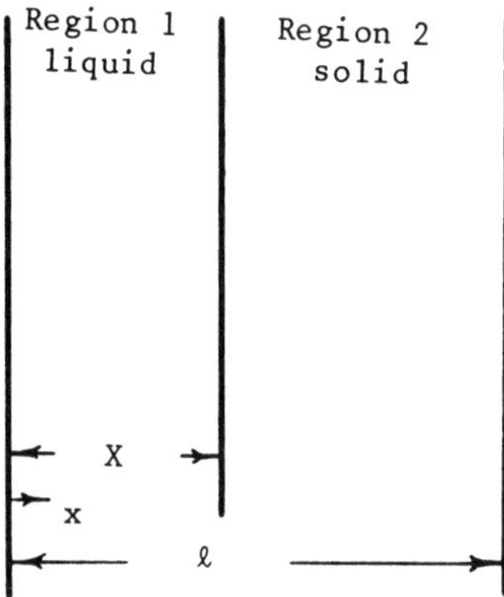

Fig. 9.10. Melting System.

The enthalpies of regions 1 and 2 are H_1, H_2. The enthalpy versus temperature relations are needed to complete the formulation. If the enthalpy relations are described in terms of the fusion temperature and the specific heats as discussed previously, then the enthalpies are

$$dH = \begin{cases} C_t\,dT & T > T_f \quad x > X \\ L & T = T_f \quad x = X \\ C_f\,dT & T < T_f \quad x < X \end{cases} \tag{9.49}$$

The Eqs. (9.46) and (9.47) are described in terms of the specific heats as usual with

$$\frac{\partial}{\partial x}\left(k_1 \frac{\partial T_1}{\partial x}\right) = C_1 \frac{\partial T_1}{\partial t} \qquad x < X \tag{9.50}$$

$$\frac{\partial}{\partial x}\left(k_2 \frac{\partial T_2}{\partial x}\right) = C_2 \frac{\partial T_2}{\partial t} \qquad x > X \tag{9.51}$$

The enthalpy jump is accommodated by the energy balance at the phase change interface

$$k_2 \frac{\partial T_2}{\partial x} - k_1 \frac{\partial T_1}{\partial x} = L \frac{dX}{dt} \qquad x = X \tag{9.52}$$

Equation (9.52) is the energy balance equation for the node or nodes undergoing phase change. The fact that this boundary moves from node to node, in a fashion not known a priori, is part of the complication of the numerical procedures.

9.9.1.1 Types of Boundary Conditions

There are three or four kinds of boundary conditions that are important for conduction problems. These will be listed for future reference.

A. First Kind. The first kind of boundary condition relates to surface temperature, $T(0,t)$ = constant. Actually, this is a special case of type C or D given below. This boundary condition is used for most exact solutions such as that of Neumann (see Section 8.2.1).

B. Second Kind. The heat flux or the temperature gradient at the surface is specified, $Q = -k(\partial T(0,t)/\partial x)$. This type is often used in ablation problems (Section 8.4.7).

C. Third Kind. A heat balance at the surface is set up. This is often referred to as a "radiative boundary condition" (see Section 8.4.4).

$$h[T_\infty - T(0,t)] = -k\frac{\partial T(0,t)}{\partial x}$$

If the surface coefficient h is very large, then this reduces to the first kind of boundary condition. The surface energy balance is of particular importance for engineering systems, such as roads and airfields, in permafrost regions.

D. Fourth Kind. Again, the surface temperature is specified, but now it is a transient, $T(0,t) = f(\text{time})$. Clearly type A is a special case of this condition. This boundary condition is often used as an approximation to the surface energy balance.

9.9.1.2 Weak Solution

The boundary value formulation, as given above, may preclude certain interesting problems. In particular, when the latent heat is released over a finite temperature range, as is the case for many soil systems, the energy balance equation, Eq. (9.52), must be modified. Although this can be done for simple problems, it may not be convenient for many practical situations. A weak version of the boundary value problem may be posed that offers a generalization of the problem good enough to obtain physically acceptable solutions. This concept is described by various names, such as the "enthalpy method," "apparent specific heat," "energy source method," etc. A complete description is given in the section dealing with the apparent heat capacity, Section 9.11. The weak solution is an integral formulation, but the solution method of the apparent heat capacity is in the usual boundary value form.

9.9.1.3 Boundary Immobilization

A second approach to the numerical solution of the boundary value problem involves a change of independent variables that will fix or immobilize the moving boundary. This complicates the governing differential equations, but it greatly simplifies the energy balance at the phase change interface, allowing standard numerical procedures to be utilized.

9.9.1.4 Integral Equations

It is possible to solve the energy equations on average over a given volume rather than at each point within the volume. The heat balance integral method is the most popular example of this procedure (see Section 8.3.2). The energy balance boundary condition equations are accounted for, but the mathematics is greatly simplified. The concept is widely used in fluid mechanics in the form of the momentum integral method. It is also possible to rewrite the equations in the form of Volterra integrals or integrodifferential equations for the phase boundary. The concept of embedding the equations for the moving boundary problem into a general conduction problem with fixed boundaries has been used in this sense. This usually requires, however, that certain parts of the problem can be solved analytically, and the approach is not as universal as finite differences.

9.9.2 Variational Formulation

Quite distinct from the boundary value method of posing the Stefan problem is that of the variational concept. No true variational principle holds for heat conduction problems with moving bounaries, but the temperature can be shown to satisfy a variational inequality. This idea has been used to discuss the existence and uniqueness of solutions. Biot's principle connects the variation in thermal potential and dissipation functions, which are defined concepts, to a generalized work that is the product of temperature and normal heat flux at a boundary. The Stefan problem is then reduced to the integration of the associated Lagrangian equation. Biot's principle is a minimum dissipation principle and is based upon physical concepts, rather than simply mathematical manipulations and analogies. A number of other concepts have also been used, such as perturbation and asymptotic solutions. The various formulations often suggest different, practical, numerical schemes, and these will be examined in the following sections.

9.10. NUMERICAL PROCEDURES FOR BOUNDARY VALUE PROBLEMS

The basic equations of heat conduction can be rewritten for the one-dimensional system as follows

$$\frac{\partial}{\partial x}\left(k_1 \frac{\partial T_1}{\partial x}\right) = C_1 \frac{\partial T_1}{\partial t} \qquad x < X \tag{9.53}$$

$$\frac{\partial}{\partial x}\left(k_2 \frac{\partial T_2}{\partial x}\right) = C_2 \frac{\partial T_2}{\partial x} \qquad x > X \tag{9.54}$$

$$T_1(X,t) = T_2(X,t) = T_f \tag{9.55}$$

$$k_2 \frac{\partial T_2(X,t)}{\partial x} - k_1 \frac{\partial T_1(X,t)}{\partial x} = L \frac{dX}{dt} \tag{9.56}$$

along with appropriate initial and boundary conditions.

Relatively few attempts have been made to fully incorporate the coupling equations, Eqs. (9.55) and (9.56), with the conduction equations, Eqs. (9.53) and (9.54), in numerical solutions. This is due to the discontinuities in temperature gradient and enthalpy at the phase change interface, as has been discussed previously. These singularities are severe enough to discourage complete numerical solutions of the phase change problem.

The complete system of equations described above can be numerically solved with the nodes at the phase change interface handled by equations of the type of Eq. (9.56). This is probably the most satisfying procedure physically, as it attempts to follow the actual latent heat and specific heat variations and to account for the specific heat jump at a fixed or very narrow band of fusion temperature. Because of the difficulty of locating the phase interchange nodes with a reasonably efficient calculation scheme, this method has not been used as extensively as one might think.

9.10.1 Elementary Schemes

The most straightforward procedure that can be used is as follows, referring to Fig. 9.11.

(a) Write the usual difference equations for the two phases.
(b) For the node(s) at which phase change is occurring:

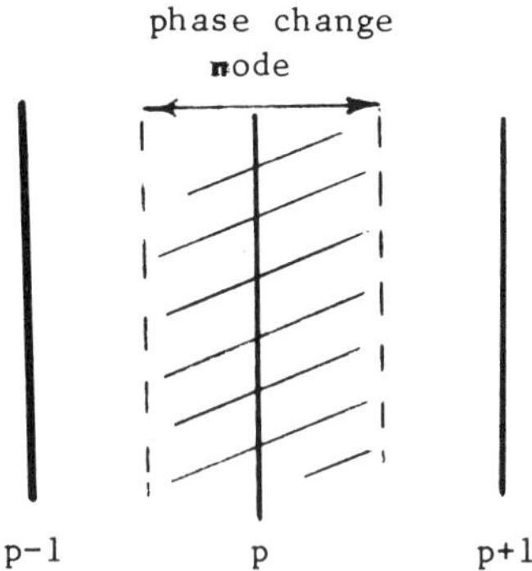

Fig. 9.11. Simple Nodal System, Fixed Grid.

(i) hold the temperature of the node constant at T_f to satisfy Eq. (9.55).
(ii) Evaluate the energy balance at node p, to find the heat gain or loss; this will satisfy Eq. (9.56).
(iii) Maintain the node at T_f until the heat gain or loss equals the latent heat capacity (or the enthalpy change) of the nodal volume. The phase change interface then moves to the next mode.

This scheme, exemplified by Dusinberre (1975), is simple, but is not widely used due to the following limitations:

(i) The phase front is not continuously followed. It jumps from one node to the next.
(ii) The location of each phase change node must be carefully monitored.
(iii) The temperatures near the phase change nodes are not accurate due to the imprecise location of the spatial coordinate with the temperature T_f. It can be anywhere within the Δx of the node.
(iv) The latent heat effect can be missed entirely if the phase front traverses more than one Δx in a single time step.
(v) An enthalpy budget or its equivalent is required for each node.

Doherty (1970) and Lachenbruch (1970) used the method to examine a two-dimensional problem of heat flow from a pipe buried in permafrost. The procedure is as follows.

(i) Melting begins when the node temperature exceeds the melting temperature after a time step Δt, i.e., $T_{i,j}^{n+1} > T_f > T_{i,j}^{n}$.

(ii) An energy budget is set up for the node (i,j) holding $T_{i,j}^{n}$ at T_f, and the energy flow is calculated as

$$E_{i,j}^{n+m} = \sum_{k=1}^{m} (T_{i,j}^{n+k} - T_f)\, C_f\, \rho \Delta x \Delta y$$

where $T_{i,j}^{n+k}$ is the excess temperature at (i,j), calculated as if no melting occurred, but with $T_{i,j}$ held at T_f. The concept is that of the excess temperature scheme of Dusinberre (1961).

(iii) Melting is complete when the total latent heat of the node is used up

$$E_{i,j}^{n+m} \geq \rho l P \Delta x \Delta y$$

where P denotes the fraction of water changing phase.

(iv) Finally, the temperature of the node at the end of time step $n + m$ is

$$T_{i,j}^{n+m} = T_f + \frac{E_{i,j}^{n+m} - \rho l P \Delta x \Delta y}{\rho C_t \Delta x \Delta y}$$

A simpler procedure would seem to be to simply calculate the enthalpy change of the node without using values of specific heat until the melting is complete.

9.10.2 Advanced Schemes

The problems of the elementary method have been successfully avoided with various calculation schemes.

9.10.2.1 Variable Space Nodes

Murray and Landis (1959) devised a scheme to track the moving interface by allowing the size of the space mesh to vary continuously with time. The concept is shown in Fig. 9.12.

Let $\theta = (T - T_f)$

r = number of increments in the solid zone and
N = total number of increments

The substantial derivative at x_i is used

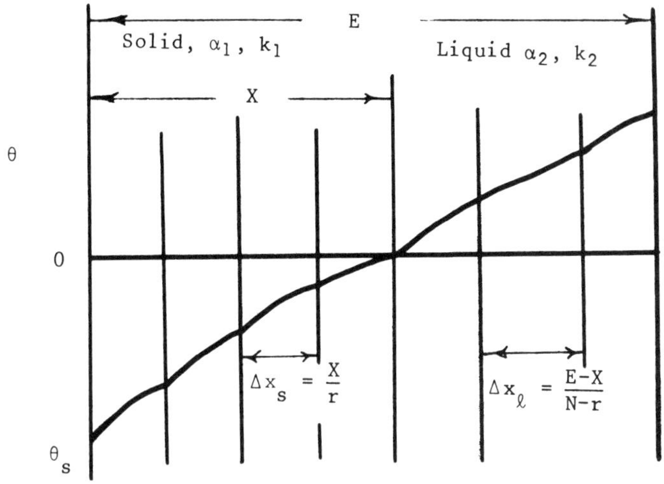

Fig. 9.12. Variable Space Mesh.

$$\left.\frac{d\theta}{dt}\right|_i = \left.\frac{\partial\theta}{\phi x}\right|_i \frac{dx}{dt} + \left.\frac{\partial\theta}{\partial t}\right|_i$$

The solid and liquid nodal equations are

$$\frac{\theta_i^{n+1} - \theta_i^n}{\Delta t} = \frac{i}{X} \frac{\theta_{i+1}^n - \theta_{i-1}^n}{2} \frac{dX}{dt} + \alpha_1 r^2 \frac{\theta_{i-1}^n - 2\theta_i^n + \theta_{i+1}^n}{X^2} \qquad i = 1, \ldots r-1 \quad (9.57)$$

$$\frac{\theta_i^{n+1} - \theta_i^n}{\Delta t} = \frac{N-i}{E-X} \frac{\theta_{i+1}^n - \theta_{i-1}^n}{2} \frac{dX}{dt} + \alpha_2 (N-r)^2 \frac{\theta_{i-1}^n - 2\theta_i^n + \theta_{i+1}^n}{(E-X)^2} \qquad i = r-1, \ldots N-1 \quad (9.58)$$

The energy balance equation is

$$X^{n+1} - X^n = \frac{\Delta t}{\rho l} \left\{ k_1 r \frac{\theta_{r-2}^n - 4\theta_{r-1}^n}{2X^n} + k_2 (N-r) \frac{\theta_{r+2}^n - 4\theta_{r+1}^n}{2(E-X^n)} \right\} \quad (9.59)$$

Nodal temperature interpolation is required, and the method is restricted to homogeneous media with a finite material thickness. A constant space increment is also discussed to allow the temperatures near the phase change region to be evaluated more accurately, but nodal iteration is required. Heitz and Westwater (1970) extended the method to improve the starting solution when only one phase is present and also considered variable phase density and the effects of free convection. It is preferable to use a system that has a fixed spatial network for the nodal points.

9.10.2.2 Time Discretization Only

If only the time functions are replaced by finite differences, a set of differential equation in the spatial variables results. This is particularly useful for one-dimensional problems. The method will be illustrated for a single phase, one-dimensional Stefan problem (Fig. 9.13), following Meyer (1970).

The standard equations for a single-phase melting problem, are

$$\frac{\partial^2 u}{\partial x^2} - \frac{1}{\alpha}\frac{\partial u}{\partial t} = 0 \tag{9.60}$$

$$u(o,t) = \psi(t) \tag{9.60a}$$

$$u(x,o) = u_o \qquad 0 \leq x \leq S(o) \tag{9.60b}$$

$$u(x,o) = o \qquad if\, S(o) = o$$

$$u(s,t) = o \tag{9.60c}$$

$$\frac{dS}{dt} + \gamma \frac{\partial u(S,t)}{\partial x} = o \tag{9.60d}$$

where

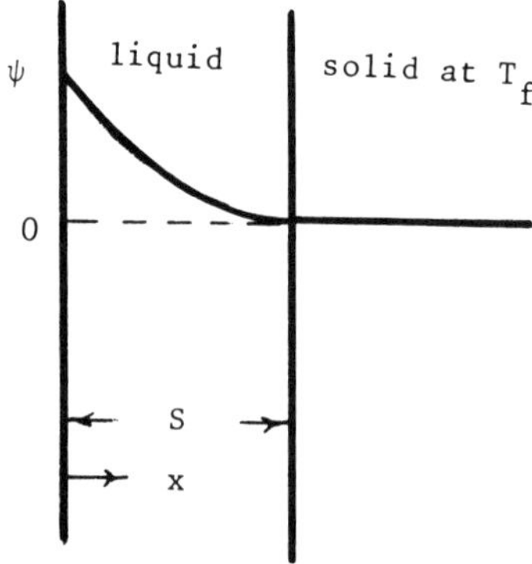

Fig. 9.13. Single-Phase Melting Problem.

$$u = T - T_f$$

$$\gamma = \frac{k}{\rho l}$$

Discretize the time derivatives with

$$\frac{\partial u}{\partial t} = \frac{u_n - u_{n-1}}{\Delta t}$$

Then for each time step n

$$u_n'' - p(u_n - u_{n-1}) = 0 \tag{9.61}$$

$$u_n(o) = \psi_n = \psi(n\Delta t) \tag{9.61a}$$

$$u_n(S_n) = 0 \tag{9.61b}$$

$$S_n - S_{n-1} + \gamma \Delta t\, u_n'(S_n) = 0 \tag{9.61c}$$

where $p = 1/\alpha\Delta t$.

The problem has now been reduced to a two-point boundary value problem with the extra condition, Eq. (9.61*c*), used to evaluate the unknown interface. The solution of this formulation of the problem is not particularly advantageous, and the system will be further reduced to an initial value problem.

Let

$$u_n = W_n v_n + \omega_n \tag{9.62}$$

$$u_n' = v_n \tag{9.63}$$

This substitution will split the second-order operator of Eq. (9.61) into two first-order operators. The new equations are

$$W_n' = 1 - p\, W_n^2 \tag{9.64}$$

$$W_n(o) = 0 \tag{9.64a}$$

$$\omega_n' = -p\, W_n(\omega_n - u_{n-1}) \tag{9.65}$$

$$\omega_n(o) = \alpha_n \tag{9.65a}$$

$$S_n - S_{n-1} + \Delta t\, \gamma\, v_n(S_n) = 0 \tag{9.66}$$

$$v_n' = p(W_n v_n + \omega_n - u_{n-1}) \tag{9.67}$$

$$v_n(S_n) = \frac{-\omega_n(S_n)}{W_n(S_n)} \tag{9.67a}$$

The solution algorithm is as follows.

The functions u_{n-1}, S_{n-1} are given from the previous time step.

(i) Solve Eqs. (9.64) and (9.65) for W_n, ω_n.
(ii) Find S_n using Eq. (9.66) as

$$S_n - S_{n-1} = -\gamma \Delta t\, v_n(S_n) = \gamma \Delta t \frac{\omega_n(S_n)}{W_n(S_n)}$$

This is the root of the following equation

$$\phi(x) = x - S_{n-1} - \gamma \Delta t \frac{\omega_n(x)}{W_n(x)} = 0 \tag{9.68}$$

This equation can be solved by Newton's iteration method, starting with S_{n-1}.

(iii) Solve Eq. (9.67) for v_n.
(iv) Finally evaluate u_n with Eq. (9.62).

For this example problem, Eq. (9.64) is a Riccati equation and an exact solution is possible

$$W(x) = \frac{1}{\sqrt{p}} \tanh \sqrt{p}\, x$$

$$\omega_n(x) = \operatorname{sech} \sqrt{p}\, x \left[\psi_n + \int_0^x \sqrt{p} \sinh (\sqrt{p} r)\, u_{n-1}(r) dr \right]$$

$$\phi_n = (x - S_{n-1}) \sinh \sqrt{p}\, x$$

$$- \gamma \sqrt{\frac{\Delta t}{\alpha}} \left[\psi_n + \sqrt{p} \int_0^{S_n = x} \sinh (\sqrt{p} r) u_{n-1}(r) dr \right] = 0$$

$$v_n = \cosh \sqrt{p}\, x \left[v_n(S_n) \operatorname{sech} (\sqrt{p} S_n) + p \int_{S_n}^{x} \operatorname{sech} (\sqrt{p} r)[\omega_n(r) - u_{n-1}(r)] dr \right]$$

with

$$v_n(S_n) = \frac{-\omega_n(S_n)}{W_n(S_n)}$$

In general, analytic solutions will not be possible, and numerical schemes for evaluating Eqs. (9.64) to (9.67) will be necessary. The method can be extended to two phase problems and multidimension problems, but the mathematical work is greatly increased. Meyer et al. (1972) applied the method to the design of foundations in permafrost regions for one-dimensional heat flow. The system had an arbitrary number of layers of different thicknesses and thermal properties. The method is flexible and seems accurate, but the computational work is great and probably too cumbersome for routine calculations.

9.10.2.3 Preferred One-Dimensional Method

Following Goodrich (1978), a clever scheme, eliminating most of the above problems, such as nodal iteration and excessive computer time, for one-dimensional problems, will be examined. The spatial grid is as sketched, in Fig. 9.14, with the thermal properties within a volume element constant, but different—if necessary, for each nodal volume.

The basic equations are, as usual

$$\frac{\partial}{\partial x}\left(k\frac{\partial T}{\partial x}\right)=\frac{\partial(CT)}{\partial t} \tag{9.69}$$

$$k_f\frac{\partial T}{\partial x}\bigg)_f - k_u\frac{\partial T}{dx}\bigg)_u = \pm L\frac{dX}{dt} \tag{9.70}$$

Noting that the nodal volume is identified at the beginning of a region Δx, and using the Crank-Nicolson formula, Eq. (9.28), with

$$T^n_{i-1/2}=\frac{T^n_{i-1}+T^n_i}{2}$$

Equation (9.69) can be written

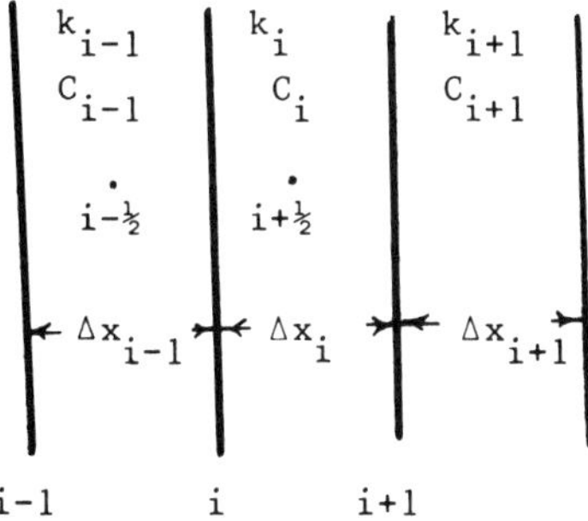

Fig. 9.14. One-Dimensional Grid.

$$\Delta_{i-1} T_{i-1}^{n+1} + (k_i + \Delta_{i-1} + \Delta_i) T_i^{n+1} - \Delta_i T_{i+1}^{n+1}$$
$$= \Delta_{i-1} T_{i-1}^{n} + (K_i - \Delta_{i-1} - \Delta_i) T_i^{n} + \Delta_i T_{i-1}^{n} \quad (9.71)$$

where

$$\Delta_i = \frac{k_i}{\Delta x_i} \qquad K_i = \frac{(C_{i-1}\Delta x_{i-1} + C_i \Delta x_i)}{\Delta t}$$

If $i = p$ denotes the node in which phase change takes place, then Eq. (9.71) is used for all nodes except p and $p + 1$. The phase change moves through node p as shown in Fig. 9.15. Using central differences, and letting $T_f = 0$ for simplicity, nodes p and $p + 1$ are written

$$\frac{\frac{1}{2}(C_{p-1}\Delta x_{p-1} + C_a \overline{G})(T_p^{n+1} - T_p^n)}{\Delta t} = \frac{1}{2}\Delta_{p-1}(T_{p-1}^{n+1} - T_p^{n+1}$$
$$+ T_{p-1}^{n} - T_p^n) + \frac{1}{2}\frac{k_a}{\overline{G}}(T_p^{n+1} + T_p^n) \quad (9.72)$$

$$\frac{\frac{1}{2}(C_{p+1}\, x_{p+1} + C_b \overline{H})(T_{p+1}^{n+1} - T_{p+1}^n)}{\Delta t} = \frac{1}{2}\Delta_{p+1}(T_{p+2}^{n+1} - T_{p+1}^{n+1} + T_{p+2}^{n} - T_{p+1}^{n})$$
$$- \frac{1}{2}\frac{k_b}{\overline{H}}(T_{p+1}^{n+1} + T_{p+1}^n) \quad (9.73)$$

where

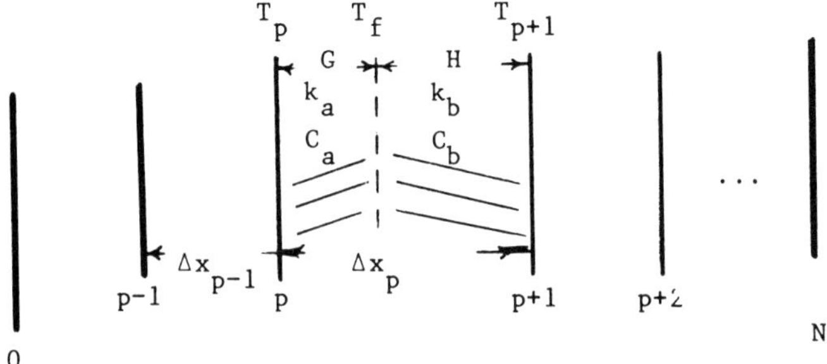

Fig. 9.15. Phase Change Nodal Points.

$$\overline{G} = \frac{G^{n+1} + G^n}{2} \qquad \overline{H} = \Delta x_p - \overline{G}$$

Equation (9.70) is now written

$$\pm 2L \frac{G^{n+1} - G^n}{\Delta t} + \frac{1}{2} C_a(T_{p-1}^{n+1} + T_{p-1}^n) - \frac{1}{2} C_b(T_{p+1}^{n+1} + T_{p+1}^n)$$

$$= \frac{k_a}{\overline{G}}(T_p^{n+1} + T_p^n) + \frac{k_b}{\overline{H}}(T_{p+1}^{n+1} + T_p^n) \quad (9.74)$$

The procedure is as follows:

(i) Use Eq. (9.71) and evaluate in the forward direction, starting at $i = 1$, which will allow all temperatures T_i^{n+1} up to $i = p - 1$ to be evaluated in the tridiagonal matrix formulation.

(ii) Again, using Eq. (9.71), starting at $i = N - 1$, evaluate T_i^{n+1} back to $i = p + 2$, as in step (i).

(iii) Solve Eqs. (9.72) to (9.74) simultaneously for T_p^{n+1}, T_{p+1}^{n+1} and G^{n+1}.

(iv) When $G^{n+1} > \Delta x_p$, let the time increment be divided into two steps, $\Delta t = \Delta t_1 + \Delta t_2$, where Δt_1 is the time required for T_f to move to the nodal boundary $p + 1$. The time increment Δt_1 can be approximated by

$$\Delta t_1 = \frac{\Delta x_p - G^n}{G^{n+1} - G^n} \Delta t$$

which assumes that the interface moves at a constant velocity during Δt.

(v) Finally, the nodal temperatures are updated using Δt_1, after which the phase change displacement is calculated into the element $p + 1$ during Δt_2. The temperatures are then updated to T_i^{n+1}. The procedure is accurate and relatively simple to program. It would not appear easy to adapt the method to multidimensional problems, however. Goodrich (1978) claims accuracy much improved over the apparent enthalpy method.

9.11 APPARENT HEAT CAPACITY

The most common way to handle the latent heat effect numerically, is to include it as an energy source or sink in the energy equation. An apparent specific heat can be defined to account for the entire enthalpy change, including latent heat. The apparent specific heat idea assumes or approximates the latent heat effect as taking place over a finite temperature range.

9.11.1 Property Changes during Phase Change

From Eq. (9.43), it can be noted that the enthalpy change during a sensible heat change is the area beneath the specific heat curve when plotted against temperature. Let C_a be the apparent specific heat, which will include any latent heat effects. In the actual case, the specific heat may change as a step function at T_f, as shown by the dashed line in Fig. 9.16. Now assume that the latent heat effect takes place over some finite ΔT. The area beneath the curve for C_a^* must equal the latent heat; thus

$$C_a^* = \frac{L}{\Delta T} \tag{9.75}$$

The apparent specific heat is assumed constant during the phase change. As $\Delta T \rightarrow 0$, the apparent specific heat during phase change approaches infinity, as we have noted previously for the actual specific heat. It is possible to use various smooth curves for the apparent specific heat; an example is sketched in Fig. 9.17.

The curve assumed must satisfy the actual latent heat effect

$$L = \int_{T_f - \Delta T}^{T_f} C_a dT \tag{9.76}$$

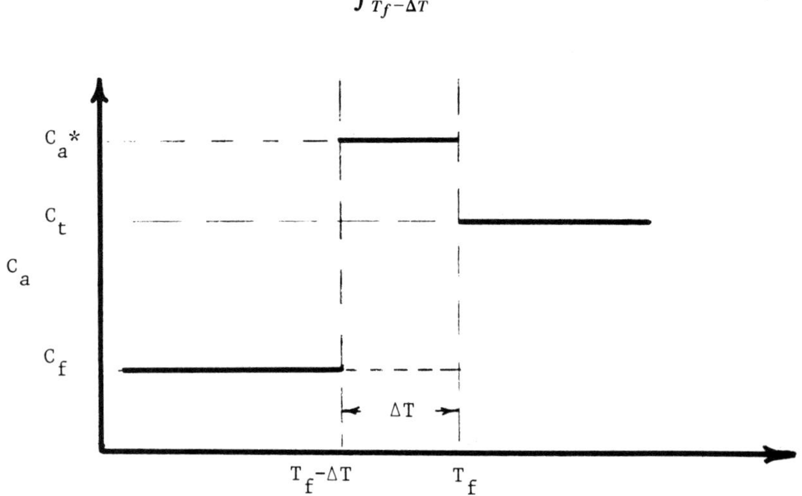

Fig. 9.16. Apparent Heat Capacity, Step Changes.

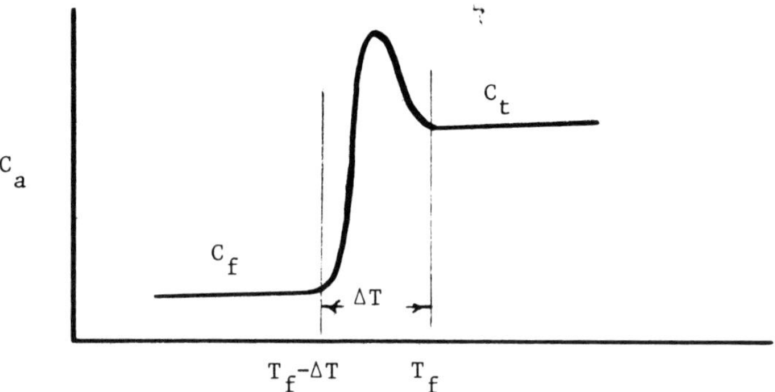

Fig. 9.17. Smoothly Varying Apparent Specific Heat.

The area beneath the assumed specific heat curve must equal the actual latent heat effect in the temperature increment ΔT, but Bonacina et al. (1974) note that the shape of the assumed curve did not seem important with regard to the temperature fields calculated. Bonacina et al. (1973) also suggested that the latent heat effect can be assumed to occur over a temperature range of $T_f \pm \Delta T$. Then the apparent specific heat can be written as

$$2C_a\Delta T = \int_{T_f-\Delta T}^{T_f} C_f\, dT + L + \int_{T_f}^{T_f+\Delta T} C_t\, dT \tag{9.77}$$

If the specific heats are constant with temperature, the apparent specific heat is

$$C_a = \begin{cases} C_t & T > T_f + \Delta T \\ \dfrac{C_f + C_t}{2} + \dfrac{L}{2\Delta T} & T_f - \Delta T < T < T_f + \Delta T \\ C_f & T < T_f - \Delta T \end{cases} \tag{9.78}$$

Bonacina et al. (1973) also suggest that the thermal conductivity can be approximated as

$$k_a = \begin{cases} k_t & T > T_f + \Delta T \\ k_f + \dfrac{k_t - k_f}{2\Delta T}[T - (T_f - \Delta T)] & T_f - \Delta T < T < T_f + \Delta T \\ k_f & T < T_f - \Delta T \end{cases} \tag{9.79}$$

These relations have the advantage of being defined in terms of the temperature, rather than the position of the node.

9.11.1.1 Unfrozen Water Relations

For soil systems, the water in the soil will usually not all freeze, even at temperatures considerably lower than the usual freezing point for pure water. In this case, particularly valid for soils with fine solid sizes, such as clays and silts, the latent heat effect will be spread over a finite freezing range (see Section 4.7). Where the latent heat effect does occur over a temperature range, the quantities can be related to the unfrozen water content W_u. The enthalpy change can be written for the phase change region below T_f, as

$$dH = C_f\,dT + L\,dW_u \tag{9.80}$$

and the apparent specific heat is

$$C_a\,dT = C_f\,dT + L\,dW_u \tag{9.81}$$

or

$$C_a = C_f + \frac{L\,dW_u}{dT} \tag{9.82}$$

Thus, one approximation is

$$C_a = \begin{cases} C_t & T > T_f \\ C_f + L\dfrac{\partial W_u}{\partial T} & T \leq T_f \end{cases} \tag{9.83}$$

The unfrozen water content may be of a form shown in Fig. 9.18. The step change from the initial water content W_i to zero is often used for simplicity, shown as the dashed curve in Fig. 9.18. For many soils, especially fine-grained soils, a certain amount of water will remain unfrozen even at temperatures far below the normal freezing point. A typical curve is shown in Fig. 9.18. If the temperature range over which phase change occurs is ΔT, then

$$C_a = C_f + L\frac{\Delta W_u}{\Delta T} = C_f + L\frac{W_i - W_a}{\Delta T} = C_f + \frac{L'}{\Delta T} \tag{9.84}$$

Here L' is the latent heat associated with the soil water that actually changes phase. An apparent specific heat is then

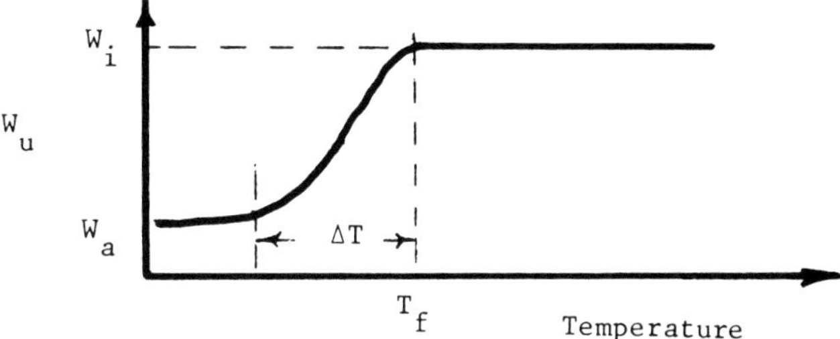

Fig. 9.18. Unfrozen Water in Fine-Grained Soil.

$$C_a = \begin{cases} C_t & T > T_f \\ C_f + \dfrac{L'}{\Delta T} & T_f - \Delta T \leq T \leq T_f \\ C_f & T < T_f - \Delta T \end{cases} \tag{9.85}$$

This formulation has the advantage that the apparent specific heat will automatically equal C_f when $T < T_f - \Delta T$, because for this temperature range $L' \equiv 0$. Equations (9.85) or (9.78) can be used, depending upon the particular problem at hand.

9.11.2 The Weak Solution

Consider again the basic equations for a one-dimensional, two-phase, melting problem, written explicitly with the enthalpy (see Fig. 9.19).

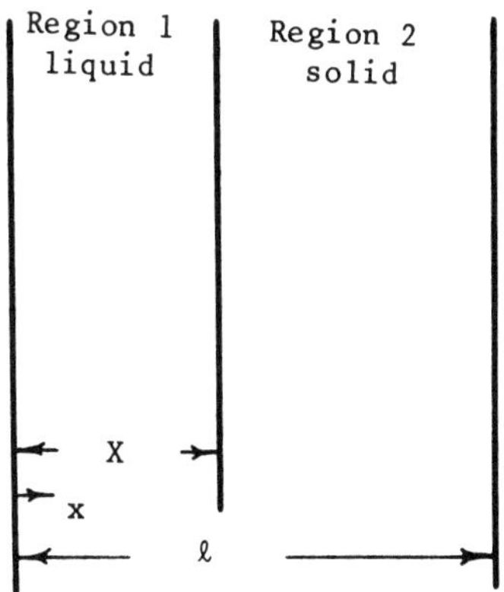

Fig. 9.19. Melting System.

$$\frac{\partial}{\partial x}\left(k_1 \frac{\partial T_1}{\partial x}\right) = \frac{\partial H_1}{\partial t} \tag{9.86}$$

$$T_1(x,o) = h_1(x) \tag{9.86a}$$

$$T_1(o,t) = g_1(t) \tag{9.86b}$$

$$\frac{\partial}{\partial x}\left(k_2 \frac{\partial T_2}{\partial x}\right) = \frac{\partial H_2}{\partial t} \tag{9.87}$$

$$T_2(x,o) = h_2(x) \tag{9.87a}$$

$$T_2(o,t) = g_3(t) \tag{9.87b}$$

$$T_2(l,t) = g_2(t) \tag{9.87c}$$

$$T_1(X,t) = T_2(X,t) = T_f \tag{9.88}$$

$$k_2 \frac{\partial T_2}{\partial x} - k_1 \frac{\partial T_1}{\partial x} = L \frac{dX}{dt} \tag{9.89}$$

Let us look at the analogous problem with no phase change, assuming constant properties

$$k \frac{\partial^2 T}{\partial x^2} - \rho \frac{\partial H}{\partial t} = 0 \qquad \begin{array}{l} 0 \le x \le l \\ 0 \le t \le t_o \end{array} \tag{9.90}$$

$$T(o,t) = g_1(t) \tag{9.90a}$$

$$T(l,t) = g_2(t) \tag{9.90b}$$

$$H(x,o) = h_o(x) = Ch_1(x) \tag{9.90c}$$

A test function, $\phi(x,t)$, has the following properties

$$\phi(o,t) = \phi(l,t) = 0 \tag{9.91a}$$

$$\phi(x,t_o) = 0 \tag{9.91b}$$

Multiply Eq. (9.90) by the test function ϕ, and integrate over space and time

$$\int_o^{t_o}\int_o^l \left(k \frac{\partial^2 T}{\partial x^2} - \rho \frac{\partial H}{\partial t}\right) \phi \, dx \, dt = 0 \tag{9.92}$$

Now integrate Eq. (9.92) by parts, and use the conditions (9.90*a*), (9.90*b*), (9.90*c*), (9.91*a,b*) to give

$$\int_o^{t_o}\left\{\int_o^l k\,\phi\frac{\partial}{\partial x}\left(\frac{\partial T}{\partial x}\right)dx\right\}dt=\int_o^{t_o}\left\{\left[k\,\phi\frac{\partial T}{\partial x}\right]_o^l-\int_o^l k\frac{\partial \phi}{\partial x}\frac{\partial T}{\partial x}dx\right\}dt$$

$$=-\int_o^{t_o}\int_o^l k\frac{\partial \phi}{\partial x}\frac{\partial T}{\partial x}dx\,dt=-\int_o^{t_o}\left\{\left[k\frac{\partial \phi}{\partial x}\,T\right]_o^l-\int_o^l k\,T\frac{\partial^2\phi}{\partial x^2}dx\right\}dt$$

$$=\int_o^{t_o}k\,g_2\frac{\partial\phi(l,t)}{\partial x}dt+\int_o^{t_o}k\,g_1\frac{\partial\phi(o,t)}{\partial x}dt+\int_o^{t_o}\int_o^l k\,T\frac{\partial^2\phi}{\partial x^2}dx\,dt$$

Also

$$\int_o^l\int_o^{t_o}\rho\phi\frac{\partial H}{\partial t}dt\,dx=\int_o^l\left.\rho\phi H\right]_o^{t_o}-\int_o^{t_o}\rho H\frac{\partial\phi}{\partial t}dt\Big\}dx$$

$$=-\int_o^l\rho\phi(x,o)h_o(x)dx-\int_o^t\int_o^l\rho H\frac{\partial\phi}{\partial t}dx\,dt$$

Then the system of Eqs. (9.90), (9.91) is

$$\int_o^{t_o}\int_o^l\left(k\,T\frac{\partial^2\phi}{\partial x^2}+\rho H\frac{\partial\phi}{\partial t}\right)dx\,dt=-\int_o^l\rho\phi(x,o)\,h_o(x)\,dx$$
$$+\int_o^{t_o}\left[kg_2(t)\frac{\partial\phi(l,o)}{\partial x}-k\,g_1(t)\frac{\partial\phi(o,t)}{\partial x}\right]dt \quad (9.93)$$

Assume that the material undergoes a phase change at T_f, where the enthalpy jump is given by the usual relations

$$H=\begin{cases}CT & T<T_f\\ \mathrm{CT}+\mathrm{L} & T\geq T_f\end{cases} \quad (9.94)$$

This gives rise to the usual temperature gradient discontinuities at T_f.

A weak solution of system Eq. (9.90) is defined to be a pair of functions H and T, related by Eq. (9.94), for which Eq. (9.93) is satisfied for all test functions $\phi(x,t)$. The weak solution is a formulation in which it is not necessary to describe explicitly the behavior of the solution near any of its singularities. The derivatives of temperature do not appear in the weak formulation, Eq. (9.93), and it is no longer necessary to consider separate liquid and solid regions. The entire region of interest is considered, and the shape of the phases is later determined from the solution. Any classical solution of the original boundary value problem—Eq. (9.86) to (9.89) is a solution of Eq. (9.93) and is, thus, also a weak solution. The importance of the weak solution is that certain finite difference schemes will converge to the weak solution of the boundary value problem despite discontinuities of temperature gradient across phase boundaries; see Kamenomostskaja (1961) and Atthey (1974), for explicit finite differences, and Jerome (1977), for implicit finite difference formulations. The method is also powerful, in that it can be easily extended to multidimensional problems, unlike the prior schemes considered that directly solved the energy balance equation at the phase change interface.

The weak solution functions, H and T, will also automatically satisfy the energy balance at the phase change interface, Eq. (9.89). This can be shown by integrating Eq. (9.92) over the solid and liquid regions separately and combining the resulting relations with Eq. (9.93).

Consider now the following system of equations with the enthalpy written explicitly, as given previously

$$k\frac{\partial^2 T}{\partial x^2}-\frac{\partial H}{\partial t}=0 \tag{9.90}$$

$$T(\mathrm{o,t})=g_1(t) \tag{9.90a}$$

$$T(l,t)=g_2(t) \tag{9.90b}$$

$$H(x,o)=h_o(x) \tag{9.90c}$$

$$H=\begin{cases} CT & T<T_f \\ CT+L & T\geq T_f \end{cases} \tag{9.94}$$

Alternatively, one could use the apparent heat capacity with the equations given by

$$\frac{\partial}{\partial x}\left(k_a\frac{\partial T}{\partial x}\right)=C_a\frac{\partial T}{\partial t} \tag{9.95}$$

$$T(x,o)=h_0(x) \tag{9.95a}$$

$$T(o,t)=g_1(t) \tag{9.95b}$$

$$T(l,t)=g_2(t) \tag{9.95c}$$

$$C_a=\begin{cases} C_t & T>T_f+\Delta T \\ \dfrac{C_f+C_t}{2}+\dfrac{L}{2\Delta T} & T_f-\Delta T\leq T\leq T_f+\Delta T \\ C_f & T<T_f-\Delta T \end{cases} \tag{9.78}$$

A finite difference solution of the systems, Eqs. (9.90), (9.94) or Eqs. (9.95), (9.78), will then converge to the weak solution, which in turn approximates the complete solution, Eqs. (9.86) to (9.89). Thus, the approximate solution has been reduced to the same type of equations we encountered without phase change. The energy balance at the phase change interface and the discontinuities are thus avoided. The interface disappears, and no special equations are needed in any time or space regions. This is a step in the direction of the weak solution. The replacement of the smooth enthalpy function by a piecewise continuous function, Eq. (9.94), gives a solution that tends to the weak solution of the partial differential equation. This may be the best computational scheme for multidimensional problems (Meyer 1978).

9.11.3 Latent Heat as a Heat Source

Using the enthalpy concept, an energy balance on a volume element for a small time Δt is

$$Q\Delta t = \Delta H = C_a \Delta T \tag{9.96}$$

The apparent heat capacity, rewriting Eq. (9.84), is

$$C_a = C_o + L\gamma_d\, W \frac{\Delta W_u}{\Delta T} = C_o + L_o \frac{\Delta W_u}{\Delta T} \tag{9.97}$$

$$C_o = \begin{cases} C_t & T > T_f \\ C_f & T \leq T_f \end{cases} \tag{9.98}$$

where

L = volumetric latent heat of water
L_o = volumetric latent heat of soil volume
W = water content based on soil mass
γ_d = dry unit weight of soil
W_u = fraction of unfrozen water at a given temperature
Q = net heat flow to volume element in time Δt and
ΔT = temperature change of the volume in time Δt.

Equation (9.96) can be rewritten as follows, using the apparent heat capacity

$$Q\Delta t - L_o \Delta W_u = C_o \Delta T \tag{9.99}$$

In this form, the latent heat effect appears as a source or sink term in the energy equation. Ho et al. (1970) used this concept to investigate the temperatures beneath insulated roads. An explicit finite difference formulation leads to the following nodal equation

$$T_i^{n+1} = A\,T_{i+1}^n - (A + B - 1)T_i^n + B\,T_{i-1}^n - L_o\,[W_u(T_i^{n+1}) - W_u(T_i^n)] \quad (9.100)$$

where A and B are parameters dealing with the geometry and thermal properties of the nodes. If a functional relation between the unfrozen water fraction and the temperature is known, then Eq. (9.100) can be solved easily (see Section 4.7). A Crank-Nicolson formulation might be preferable if the iteration at each time step can be handled easily. The following W_u functions were used, without reference to validity.

$$W_u = \frac{C + e^{(a\,T - b)}}{100} \quad (9.101)$$

	C	$A(°F)^{-1}$	b
Gravel-sand	0	.17	.73
Silt-clay	38	.25	4.0

These curves have a form similar to those of Fig. 9.18, as expected.

The method requires iteration, but is easily adaptable to multidimensional systems. It is necessary to have accurate data for the soil system in regard to the unfrozen water, as exemplified by Eq. (9.101). In the explicit format, the calculation time may become excessive. Nakano and Brown (1971) used an implicit formulation to investigate the effect of a freezing zone of finite width on the Neumann problem. They found that a finite phase change zone can have a significant effect upon the soil temperatures, but the effect upon the actual phase change depth is not serious if the unfrozen water is considered in the Neumann solution (Nixon and McRoberts 1973).

9.11.4 Finite Difference Schemes

As has been mentioned previously, finite difference schemes for the apparent heat capacity method should converge to the weak solution if minimal care is taken. A number of these schemes will be examined for guidance in the solution of practical problems.

9.11.4.1 Minimal Iteration

The apparent heat capacity method has been used by Couch et al. (1970) to investigate the thawing about an oil well located in permafrost. The system is two-dimensional in space (radius and depth) and transient. The basic equation is

$$\frac{1}{r}\frac{\partial}{\partial r}\left(r\,k\frac{\partial T}{\partial r}\right)+\frac{\partial}{\partial z}\left(k\frac{\partial T}{\partial z}\right)=\rho\,\frac{\partial H}{\partial t} \tag{9.102}$$

The enthalpy term is expanded as usual with

$$\frac{\partial H}{\partial t}=\frac{\partial H}{\partial T}\frac{\partial T}{\partial t}=C(t)\frac{\partial T}{\partial t} \tag{9.103}$$

Thus the finite difference, implicit form, of the following equation is to be examined

$$\frac{1}{r}\frac{\partial}{\partial r}\left(r\,k\frac{\partial T}{\partial r}\right)+\frac{\partial}{\partial z}\left(k\frac{\partial T}{\partial z}\right)=\rho\,C(t)\frac{\partial T}{\partial t} \tag{9.104}$$

Due to chemical impurities, capillary suppression of the melting point, and interface effects, the water in a soil system will change phase over a temperature range, rather than at a specific fusion temperature, as has been discussed. Then the enthalpy curve for water is approximated, as usual, by the relation shown in Fig. 9.20. The apparent specific heat of the water also follows the usual approximations, where

$$C_w=\begin{cases} C_f & T< T_1 \quad \text{frozen zone} \\ C^*=\dfrac{L}{\Delta T} & T_1\leq T\leq T_2 \quad \text{mixed phase region} \\ C_t & T> T_2 \quad \text{thawed zone}\end{cases} \tag{9.85}$$

The apparent specific heat of the water is sketched in Fig. 9.21. The choice of ΔT is somewhat arbitrary, but if ΔT is too small, numerical roundoff

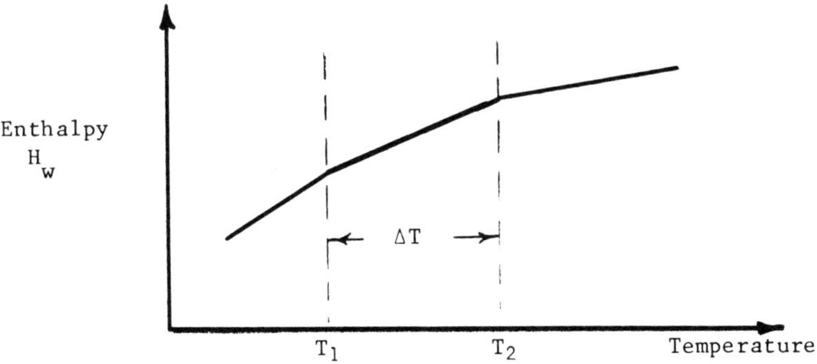

Fig. 9.20. Enthalpy of Water in Soil System.

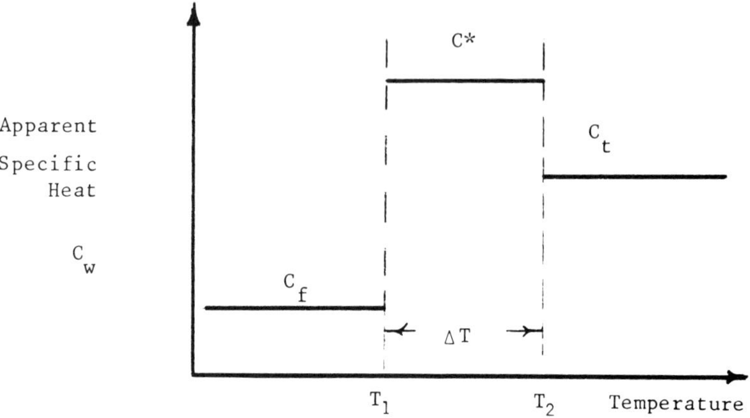

Fig. 9.21 Apparent Specific Heat of Water in Soil.

errors may be important. A value of ΔT of several degrees Fahrenheit was used. The apparent specific heat of the soil system is defined, for the midpoint of a time step, as follows

$$C_a = C(T_{i,j}^{n+1/2}) = \int_{T_{i,j}^{n}}^{T_{i,j}^{n+1}} C(T)\,dT \tag{9.105}$$

$$= (1-n)\frac{\rho_r C_r}{\rho_{i,j}} + \frac{n\rho_w S_w}{\rho_{i,j}(T_{i,j}^{n+1} - T_{i,j}^{n})}\int_{T_{i,j}^{n}}^{T_{i,j}^{n+1}} C_w(T)\,dT$$

where

ρ_r, C_r = density and specific heat of the solids
n = porosity of the medium
S_w = water saturation, fraction of pore space and
ρ_w = density of water

From Equation (9.85), the apparent specific heat relation is

$$\frac{1}{(T_{i,j}^{n+1} - T_{i,j}^{n})}\int_{T_{i,j}^{n}}^{T_{i,j}^{n+1}} C_w\,dT = \begin{cases} C_f & T_{i,j}^{n} < T_1 \\ C^* & T_1 \le T_{i,j}^{n} \le T_2 \\ C_t & T_{i,j}^{n} > T_2 \end{cases} \tag{9.106}$$

If a nodal point changes from one phase to another, in a given time step, the apparent specific heat is adjusted appropriately. For example, if a node changes from the frozen to the mixed phase, i.e., $T_{i,j}^{n} < T_1$, and $T_1 < T_{i,j}^{n+1} < T_2$, then

$$\int_{T_{i,j}^{n}}^{T_{i,j}^{n+1}} C_w(T)\,dT = C_f(T_1 - T_{i,j}^{n}) + C^*(T_{i,j}^{n+1} - T_1) \tag{9.107}$$

The calculation scheme proceeds as follows:

Given $T_{i,j}^{n}$, properties, geometry, boundary conditions,

(i) Initially assume that no nodal point changes phase. Check $T_{i,j}^{n}$ for phase and then evaluate C_a from Eqs. (9.105) and (9.106).
(ii) Calculate $T_{i,j}^{n+1}$, which is the solution of a system of linear simultaneous equations.
(iii) Check to see if $T_{i,j}$ has changed phase after step (ii),
(iv) If any $T_{i,j}$ has changed phase, recalculate C_a, as in step (i), and recalculate $T_{i,j}^{n+1}$. Continue to iterate until no node changes phase.

The advantage of this method is that each calculation involves only a system of linear equations if the correct range has been used for each C_a. This avoids the iteration required for each time step when a continuous function for $C(T)$ is assumed, rather than the piecewise continuous function used here.

There is the problem that a node may entirely cross the melting range in one iteration and recross it during the next. Thus, a cycle may be set up with no convergence for that node. To avoid this, if $T_{i,j}$ has passed from the solid to the liquid during an iteration, it is placed in the melting range to find C_a for the next iteration. It will require at least three iterations for the node to go from the frozen to the completely thawed state in one time step.

9.11.4.2 Three Time Levels

The equations can be put into a tridiagonal matrix format by using three time levels, as exemplified by Bonacina and Comini (1973). The three time level temperature is described by Bonacina et al. (1973*b*) as

$$T_i^* = \frac{1}{3}(T_i^{n+1} + T_i^n + T_i^{n-1}) \tag{9.108}$$

The finite difference form of Eq. (9.95) is

$$C_a(T_i^n)\,\gamma_t T_i^n = \frac{\Delta t}{\Delta x^2}\gamma_x\,[k_a(T_i^n)\,\gamma_x T_i^*] \tag{9.109}$$

with

$$\gamma_t T_i^n = \frac{T_i^{n+1} - T_i^{n-1}}{2}$$

$$\gamma_x T_i^n = T_{i+1/2}^n - T_{i-1/2}^n$$

The final form of Eq. (9.109) is

$$\begin{aligned}[C_a(T_i^n) + M(k^+ + k^-)]\, T_i^{n+1} - Mk^+\, T_{i+1}^{n+1} - k^-\, M\, T_{i-1}^{n+1} \\ = C_a(T_i^n) - M(k^+ + k^-)\,(T_i^n + T_i^{n-1}) + Mk^+(T_{i+1}^n + T_{i+1}^{n-1}) \\ + Mk^-(T_{i-1}^n + T_{i-1}^{n-1}) \quad (9.110)\end{aligned}$$

where

$$k^+ = k_a\left(\frac{T_i^n + T_{i+1}^n}{2}\right)$$

$$k^- = k_a\left(\frac{T_i^n + T_{i-1}^n}{2}\right)$$

$$M = \frac{2}{3}\frac{\Delta t}{\Delta x^2}$$

This equation is now in the tridiagonal form and can be solved as usual. Bonacina and Comini (1973) examined the case of boundary conditions of the first kind with good results, but care had to be exercised in the choice of ΔT for the phase change. Best results occurred if two or three nodes were simultaneously involved in phase change. Cleland and Earle (1977) used the same method to study boundary conditions of the third kind. The results were less accurate than for the first kind, with an estimated error of 8%, but the boundary conditions of the third kind are more practical for many situations.

Talwar and Dilpare (1977) used the method of this section, with an explicit formulation, to study the inward solification of water in a cylinder. The explicit formulation required considerable computational care with respect to stability. The freezing range chosen (ΔT in Eqs. (9.78) and (9.79)) did not affect the freeze time if ΔT was less than 1.1°C.

9.11.4.3 Enthalpy Formulation

An alternative but equivalent method is to formulate the equations in terms of the enthalpy. The following one-dimensional problems will be examined to present the idea. The equations, as given before, are

$$k\frac{\partial^2 T}{\partial x^2} - \frac{\partial H}{\partial t} = 0 \quad (9.90)$$

$$T(o,t) = g_1(t) = T_o \quad (9.90a)$$

$$T(l,t) = g_2(t) = T_l \quad (9.90b)$$

$$H(x,o) = h_o(x) \quad (9.90c)$$

The temperature may then be defined in terms of the enthalpy. The following relations assume, for simplicity, that the enthalpy is arbitrarily assigned a value of zero as $(T - T_f)$ approaches zero from the negative side. Inverting Eq. (9.94)

$$T = \begin{cases} T_f + \dfrac{H}{C_f} & H \leq 0 \\ T_f & 0 < H < L \\ T_f + \dfrac{H - L}{C_t} & \mathrm{H} \geq L \end{cases} \quad (9.111)$$

As the temperature is usually given initially, the condition (9.90*c*), can also be described in terms of Eq. (9.111). Note that no energy condition need be posed because it is incorporated into the equations. A finite difference scheme should then converge to the weak solution, as defined earlier, which in turn will approximate the original boundary value problem (Atthey 1974).

Write Eq. (9.90) in an explicit formulation

$$H_i^{n+1} = H_i^n + N(T_{i+1}^n - 2T_i^n + T_{i-1}^n) \quad (9.112)$$

where

$$N = \frac{k\,\Delta T}{\Delta x^2}$$

For simplicity, take a very elementary case with only two unknown nodal temperatures. Then the system of equations to be solved is

$$H_1^{n+1} = H_1^n + N(T_2^n - 2T_1^n + T_o) \quad (9.113a)$$

$$H_2^{n+1} = H_2^n + N(T_l - 2T_2^n + T_1^n) \quad (9.113b)$$

The boundary temperatures T_o, T_l are given.

The solution proceeds as follows:

(i) The initial values of H_1^0, H_2^0 are given by Eq. (9.90*c*). T_1^0, T_2^0 are then obtained from H_1^0, H_2^0, and Eq. (9.111).

(ii) Calculate H_1^1, H_2^1 from Eqs. (9.113*a,b*).

(iii) Calculate T_1^1, T_2^1 from

$$T_i^n = \begin{cases} T_f + \dfrac{H_i^n}{C_f} & H_i^n \leq 0 \\ T_f & 0 < H_i^n < L \\ T_f + \dfrac{H_i^n - L}{C_t} & H_i^n \geq L \end{cases}$$

(iv) Repeat for the next time step.

The interface has not been followed explicitly, but it may be found by interpolation. If an implicit format is used, then the enthalpy equations will lead to a system of nonlinear algebraic equations.

9.11.4.4 Multidimensional Problems

The apparent heat capacity method, described earlier, seems to be the most adaptable to numerical solutions of phase change problems in more than one dimension. The equations are straightforward and do not require special formulations. There may, however, be questions involving the oscillatory movement of the phase change interface if computational care is not taken.

Crowley (1978) has extended the one-dimensional enthalpy method of the last section to the two-dimensional solidification of square cylinders when the surface temperature is lowered in a transient manner. Initially, the cylinders are liquid with no superheat. An explicit formulation was used with no computational difficulties. The results agreed well with data and with other numerical methods.

Meyer (1973) approximated the discontinuous enthalpy function by a piecewise, continuous linear function of temperature, with the latent heat effect spread over a finite temperature range $2\Delta T$. The enthalpy function is as noted in Fig. 9.9, except that the fusion temperature is in the center of a freezing range $2\Delta T$. An implicit formulation leads to a nonlinear set of equations that was solved iteratively without complications. It was speculated that for numerical work it may be preferable to smooth the enthalpy function, which is a step function, rather than the specific heat, which is a Dirac delta function. The method is applied to the problem of melting of a rectangular cylinder of ice due to heating through a rectangular duct in the ice block.

There are no other results to compare this particular problem to, but the method is equivalent to the enthalpy formulation.

9.12 MOVING BOUNDARY IMMOBILIZATION

As has been noted, there are considerable computational difficulties associated with Stefan problems. These are mainly due to the necessity of treating a linear differential equation (heat conduction) in a region with space boundaries that vary with time (the moving interface). In many cases, it may be much simpler to numerically solve a nonlinear differential equation with fixed spatial boundaries. Thus, the use of a change in one or more space variables, in order to immobilize the position of the phase change interface, may be advantageous. The basic idea was mentioned by Landau (1950) but not utilized. Ferriss (1975) examined a simple case for oxygen diffusion problems; the analogous temperature problem will be examined here.

Consider the Neumann problem, for solidification, with only one phase. The solution to this problem is known exactly, of course, but it is a useful problem to study the technique of phase change interface immobilization (Fig. 9.22).

$$\frac{\partial^2 T}{\partial x^2}=\frac{1}{\alpha}\frac{\partial T}{\partial t} \tag{9.114}$$

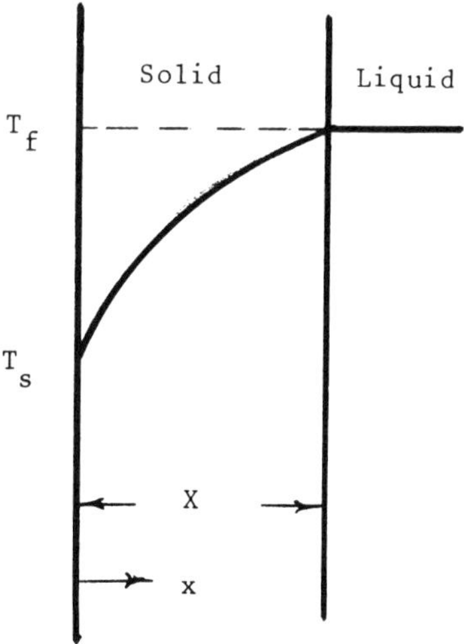

Fig. 9.22. Semi-Infinite Slab.

$$\left.\begin{aligned} -k\frac{\partial T}{\partial x} &= L\frac{dX}{dt} \\ T &= T_f \end{aligned}\right\} x = X \qquad \begin{aligned}(9.114a)\\(9.114b)\end{aligned}$$

$$T(x,o) = \phi_1(x) = T_f \qquad (9.114c)$$

$$T(o,t) = T_s(t) \qquad (9.114d)$$

$$X(o) = 0 \qquad (9.114e)$$

Now let

$$\eta = \frac{x}{X(t)} \qquad (9.115)$$

The following transformations can be used

$$\frac{\partial^2 T}{\partial x^2} = \frac{1}{X^2}\frac{\partial^2 T}{\partial \eta^2} \qquad (9.116)$$

$$\frac{\partial T}{\partial t} = \frac{\partial T}{\partial t} - \frac{\eta}{X}\frac{\partial T}{\partial \eta}\frac{dX}{dt} \qquad (9.117)$$

The system of Eq. (9.114) becomes

$$\alpha\frac{\partial^2 T}{\partial \eta^2} = S\frac{\partial T}{\partial t} - \frac{\eta}{2}\frac{dS}{dt}\frac{\partial T}{\partial \eta} \qquad (9.118)$$

$$\left.\begin{aligned} \frac{\partial T}{\partial \eta} &= -\frac{L}{2k}\frac{dS}{dt} \\ T &= T_f \end{aligned}\right\} \eta = 1 \qquad \begin{aligned}(9.118a)\\(9.118b)\end{aligned}$$

$$T(\eta,o) = T_f \qquad (9.118c)$$

$$T(o,t) = T_s(t) \qquad (9.118d)$$

$$S(o) = 0 \qquad (9.118e)$$

where $S = X^2$ has been used.

Thus the problem has been transformed into a region where $0 < \eta \leq 1$, and the interface is always at $\eta = 1$. The governing equation is now nonlinear, but this can be handled without undue difficulty.

Use the Crank-Nicolson scheme, Section 9.4.3, to write

$$\left(\frac{dT}{d\eta}\right)_i^{n+1/2}=\frac{1}{4\Delta\eta}[T_{i+1}^{n+1}-T_{i-1}^{n+1}+T_{i+1}^{n}-T_{i-1}^{n}]$$

$$\left.\frac{d^2T}{d\eta^2}\right|_i^{n+1/2}=\frac{1}{2\Delta\eta^2}[T_{i+1}^{n+1}-2T_i^{n+1}+T_{i-1}^{n+1}+T_{i+1}^{n}-2T_i^{n}+T_{i-1}^{n}]$$

$$\left.\frac{dT}{dt}\right|_i=\frac{T_i^{n+1}-T_i^{n}}{\Delta t}$$

$$\left.\frac{dS}{dt}\right|_i=\frac{S^{n+1}-S^{n}}{\Delta t}$$

The value of S to use in Eq. (9.118) can be chosen as

$$S=\frac{S^{n+1}+S^{n}}{2}$$

The finite difference form of Eq. (9.118) is

$$(M+a)T_{i+1}^{n+1}+(M-a)T_{i-1}^{n+1}-(2M+b)T_i^{n+1}$$
$$=-(M+a)T_{i+1}^{n}-(M-a)T_{i-1}^{n}+(2M-b)T_i^{n} \quad (9.119)$$

where

$M=\alpha\Delta t/\Delta\eta^2$
$a=(S^{n+1}-S^{n})i/2$ and
$b=S^{n+1}+S^{n}$

Equation (9.119) can be combined with the boundary conditions, Eqs. (9.118*b,c,d*), to yield a tridiagonal matrix

$$\tilde{P}(S^{n+1})\tilde{T}=\tilde{R} \quad (9.120)$$

where the elements of matrix $\tilde{P}$ are functions of S^{n+1}. The remaining boundary condition, Eq. (9.118*a*), can be written in backward difference formulation as

$$F(S^{n+1})\equiv\frac{1}{2\Delta\eta}(3T_N^{n+1}-4T_{N-1}^{n+1}+T_{N-2}^{n+1})+\frac{L}{2k\Delta t}(S^{n+1}-S^{n})=0 \quad (9.121)$$

where N is the boundary node.
An iterative solution of Eq. (9.120) and Eq. (9.121) can then be carried out at each time step.

Example 3: Use boundary immobilization for the system of Fig. 9.22, assuming the space is divided into three increments for simplicity. The region, with the new independent variable η is as shown in Fig. 9.22a. Only two unknown temperatures, T_1 and T_2, are involved. Then Eq. (9.120) reduces to the set of two equations for the first time step

$$(M + a_1)T_2^1 - (2M + b)T_1^1 = b\,T_f - (M - a_1)T_s^1 \qquad (9.122a)$$

$$(M - 2a_1)T_1^1 - (2M + b)T_2^1 = -(M + 2a_1 + b)T_f \qquad (9.122b)$$

where

$$a_1 = \frac{1}{2}[S^1 - S(o)] = \frac{1}{2}S^1$$

Equation (9.121) is

$$F(S^1) = \frac{1}{2\Delta\eta}[3T_f - 4T_2^1 + T_1^1] + \frac{LS^1}{2k\Delta t} = 0 \qquad (9.123)$$

The solution method is to guess a value of S^1, solve Eqs. (9.122) for T_1^1, T_2^1, using any convenient tridiagonal algorithm, and then check the guess with Eq. (9.123). Iterate until $|(F(S^1)| < \epsilon$, and then repeat the process for the next time step.

Duda et al. (1975) used the method for a multidimensional cylindrical system with a boundary transformation for each of the two phases present. The procedure involved three iterative schemes embedded within an overall iteration loop, but it was not conceptually difficult. Sparrow et al. (1978) applied the concept to the melting of a solid about a cylinder with phase boundary immobilization for the solid/liquid interface and also for the temperature wave in the solid. An implicit finite difference scheme was used.

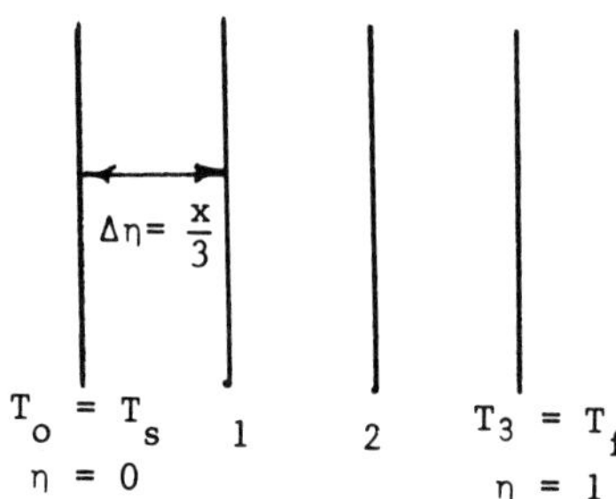

Fig. 9.22a. Simple grid system for boundary immobilization.

9.13 ISOTHERM MIGRATION METHOD

The previous methods sought to find a solution of the temperature as a function of the space variables and time. It is also possible to change the roles of the dependent variable, temperature, and one of the independent variables. The position of an isotherm will be examined as a function of time, hence the name, "isotherm migration." A one-dimensional problem will be recast so that position is the dependent variable and time and temperature are the independent variables. The study of the way in which a fixed temperature moves through the medium is of particular advantage for phase change problems because the phase change interface is almost always at a fixed temperature. The motion of the boundary arises naturally from the solution, without special treatment. The method eliminates the usual property reevaluation at each time step, as the properties simply move along with the particular isotherm. There are disadvantages, however, as will be noted.

The equations governing the isotherm motion can be derived by transformation of the usual equations, as was done by Rose (1967), where numerical methods were not used in the solution. Dix and Cizek (1970) used an energy balance to arrive at the same equations and discussed certain numerical details.

It is instructive to examine first an energy balance for a control volume in order to gain insight into the concept. The method also allows effects, such as energy generation, to be easily incorporated into the equations. Consider an energy balance on a control volume contained between two isotherms, T and $T + dT$, as shown in Fig. 9.23. The system is one-dimensional in space, with energy generation and variable properties.

At each time, the isotherms are moving at an unknown velocity, and

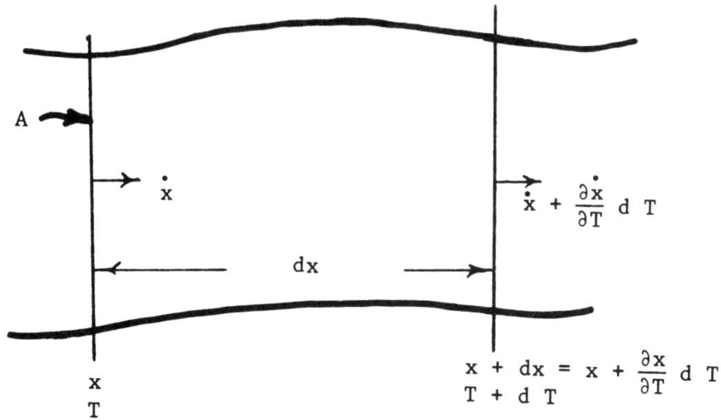

Fig. 9.23. Control Volume for Isotherm Migration.

thus there is a flux of energy at the surfaces x and $x + dx$ due to conduction and convection. The conduction at x (or at isotherm T) is

$$q_{c,x} = -k\,A\frac{\partial T}{\partial x} = -k\,A\left(\frac{\partial x}{\partial T}\right)^{-1}$$

Then the net rate of conduction for the element is

$$q_c = A\frac{\partial}{\partial t}\left[k\left(\frac{\partial x}{\partial T}\right)^{-1}\right]dT \tag{9.124}$$

Even though the material may be solid, the fact that the boundary is moving means there is an enthalpy flux (convection) that, at boundary T, is

$$q_{v,s} = -\rho c A T\dot{x}$$

where $\dot{x} = \dfrac{\partial x}{\partial t}$.

Again, the net rate of convection for the element is

$$q_v = -\frac{\partial}{\partial T}(-\rho c A\dot{x}T) = \rho c A\left[\dot{x} + T\frac{\partial \dot{x}}{\partial T}\right]dT \tag{9.125}$$

The energy generation is given by

$$q_G = q_g A\frac{\partial x}{\partial T}dT \tag{9.126}$$

where $dx = (\partial x/\partial T)\,dT$ has been used and q_g is the energy generation rate per unit volume.

The energy change of the mass in the volume is not due to a temperature change, as the average temperature of the mass between the isotherms is always constant. Thus, the stored energy changes are simply due to volume changes in a time increment. The stored energy change is then given by

$$\Delta q_s = \frac{q_{s,t+dt} - q_{s,t}}{\Delta t} = \rho A c T\frac{\partial(dx)}{\partial t} = \rho A c T\frac{\partial \dot{x}}{\partial T}dT \tag{9.127}$$

Combining Eqs. (9.124) to (9.127) yields

$$A\frac{\partial}{\partial T}\left(k\left[\frac{\partial x}{\partial T}\right]^{-1}\right)dT + \rho c A\left[\dot{x} + T\frac{\partial \dot{x}}{\partial T}\right]dT + q_g A\frac{\partial x}{\partial T}dT = \rho A c T\frac{\partial \dot{x}}{\partial T}dT$$

or

$$\rho c A\dot{x} = -A\frac{\partial}{\partial T}\left[k\left(\frac{\partial x}{\partial T}\right)^{-1}\right] - q_g A\frac{\partial x}{\partial T} \tag{9.128}$$

Carrying out the differentiation finally gives the differential equation for the motion of an isotherm

$$\dot{x} = \alpha\left(\frac{\partial x}{\partial T}\right)^{-2}\frac{\partial^2 x}{\partial T^2} - \frac{1}{\rho c}\left(\frac{\partial x}{\partial T}\right)^{-1}\frac{\partial k}{\partial T} - \frac{q_g}{\rho c}\frac{\partial x}{\partial T} \tag{9.129}$$

This equation will hold for all isotherms except that for the case of phase change. For phase change, the usual energy balance will lead to the following isotherm

$$\pm L\,\dot{x} = k_s\left(\frac{\partial T_s}{\partial x}\right) - k_l\left(\frac{\partial T_l}{\partial x}\right) \tag{9.130}$$

where the positive sign is for melting, and the negative is for freezing. Equation (9.129) is used for all isotherms without phase change, and Eq. (9.130) is used for the phase change isotherm.

The same result can be obtained by transforming the conduction equation. The differential equation governing the heat flow problem has been used previously and is

$$\frac{\partial}{\partial x}\left(k\frac{\partial T}{\partial x}\right) + q_g = \rho c\frac{\partial T}{\partial t} \tag{9.131}$$

Consider an isotherm temperature $T(x,t)$. The differential of this temperature is zero, as any isotherm value remains fixed

$$dT = \frac{\partial T}{\partial x}dx + \frac{\partial T}{\partial t}dt = 0$$

Then the relation on the isotherm is

$$\frac{\partial x}{\partial t} = -\frac{\partial T}{\partial t}\frac{\partial x}{\partial T} \tag{9.132}$$

Replacing $\partial T/\partial t$ with the aid of Eq. (9.131) yields

$$\frac{\partial x}{\partial t} = -\frac{1}{\rho c}\frac{\partial x}{\partial T}\left[\frac{\partial}{\partial x}\left(k\frac{\partial T}{\partial x}\right) + q_g\right]$$

Using

$$\frac{\partial\left(k\dfrac{\partial T}{\partial x}\right)}{\partial x} = \frac{\partial\left(k\dfrac{\partial T}{\partial x}\right)}{\partial T}\frac{\partial T}{\partial x}$$

leads to

$$\frac{\partial x}{\partial t} = -\frac{1}{\rho c}\left\{\frac{\partial}{\partial T}\left[k\left(\frac{\partial x}{\partial T}\right)^{-1}\right] + q_g\frac{\partial x}{\partial T}\right\}$$

This equation is identical to Eq. (9.128), derived previously.

Example 4: Let us clarify the numerical procedure by examining a simple case. Consider the melting of a semi-infinite slab with no initial subcooling and constant properties with a step change at the surface; this is the old, familiar, Stefan approximation. The problem can be written as

$$\frac{\partial^2 T}{\partial x^2} = \frac{1}{\alpha}\frac{\partial T}{\partial t} \tag{9.133}$$

$$T(o,t) = T_s \qquad t > 0 \tag{9.133a}$$

$$T = T_f \qquad t = 0 \qquad 0 < x < \infty \tag{9.133b}$$

$$\left.\begin{aligned} &T = T_f \\ &L\frac{dS}{dt} = -k\frac{\partial T}{\partial x} \end{aligned}\right\} x = S(t) \tag{9.133c}$$

The isotherm migration method was used by Crank and Phahle (1973) for the melting of an ice slab, described by Eq. (9.133). The transformed system of equations is

$$\dot{x} = \alpha\left(\frac{\partial x}{\partial T}\right)^{-2}\frac{\partial^2 x}{\partial T^2} \tag{9.134}$$

$$x = 0 \qquad T = T_s \tag{9.134a}$$

$$\left.\begin{aligned} &x = S \\ &\frac{dS}{dt} = -\frac{k}{L}\left(\frac{\partial x}{\partial T}\right)^{-1} \end{aligned}\right\} T = T_f \tag{9.134b}$$

Basically, a solution is sought for $x = S(t)$ on the isotherm $T = T_f$. Consider the (t, T) grid as shown in Fig. 9.24. At $t = 0$, the initial condition is $T = T_f$, and the $T = T_s$ isotherm is always associated with $x = 0$.
An explicit formulation of Eq. (9.134) leads to

$$x_i^{n+1} - x_i^n = 4\alpha\Delta t \frac{(x_{i+1}^n - 2x_i^n + x_{i-1}^n)}{(x_{i+1}^n - x_{i-1}^n)^2} \qquad i = 1, 2, \ldots N-1 \tag{9.135a}$$

$$S^{n+1} - S^n = \frac{k}{L}\Delta t \frac{(T_s - T_f)}{N(S^n - x_1^n)} \tag{9.135b}$$

This example points out one of the disadvantages of the isotherm migration method: initial values of x_i are required. It is necessary to start the solution at some small time, after which the calculation is routine. To illustrate the procedure, take the Stefan solution for this problem given by

$$T = T_s - (T_s - T_f)\frac{x}{S} \tag{9.136}$$

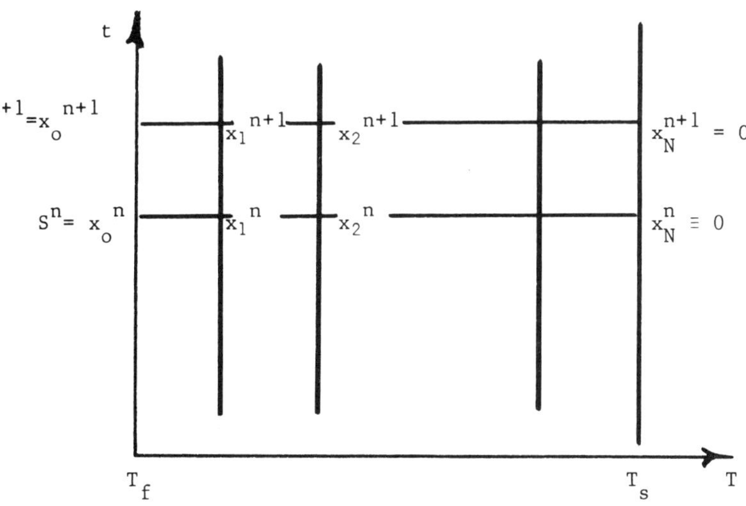

Fig. 9.24. $T - t$ Grid for Isotherm Migration.

$$S = \sqrt{2\frac{k}{L}(T_s - T_f)\,t} \tag{9.137}$$

For this example, assume that the phase change region is split into only two parts, $N = 2$, and start the solution at some arbitrary small time t^*. Then the problem reduces to

$$x_1^{n+1} - x_1^n = \frac{4\alpha\Delta t\,(S^n - 2x_1^n)}{(S^n)^2}$$

$$S^{n+1} - S^n = \frac{\dfrac{k}{L}\dfrac{(T_s - T_f)}{2}\Delta t}{S^n - x_1^n}$$

with

$$x_1^0 = \frac{T_s - T_f}{2p}$$

$$s^0 = \frac{(T_s - T_f)}{p}$$

$$p = \frac{(T_s - T_f)}{\sqrt{2\dfrac{k}{L}(T_s - T_f)t^*}}$$

Stability relations require that

$$\Delta t \leq \frac{1}{8\alpha}(x_{i+1}^n - x_{i-1}^n)^2 \tag{9.138}$$

It is not possible to choose Δt a priori, due to the nonlinearity of Eq. (9.138). This can be done by trial and error after values of x_i have been calculated.

For this trivial case, x_1 is the $(T_s + T_f)/2$ isotherm. Clearly, any value for N could be chosen. Stability relations must, of course, be observed for the explicit formulation, which may well limit the time step seriously. The Crank-Nicolson implicit formulation would obviate this, even though the algebraic equations would be nonlinear. A more elaborate forward difference scheme could also be used in place of Eq. (9.135*b*) to improve the accuracy.

Crank and Gupta (1975) have extended the method to the solidification of a square prism of fluid, initially without superheat. Crank and Crowley

(1978) used the method in conjunction with orthogonal flow lines to solve multidimensional freezing problems.

The isotherm migration method has the advantages that the region represented is the only region in which a temperature change has occurred and no interpolation of isotherms is needed; property reevaluation at each time step for temperature dependent properties is eliminated, as the same properties are carried along with each isotherm; finally, the phase change interface is followed exactly and directly calculated. The following disadvantages must be considered, however: the governing equation becomes nonlinear, this is not significant for explicit methods, but can be a factor in implicit schemes; a starting solution is often required, necessitating a small time solution; the value of the isotherm position may be multivalued for some boundary and initial conditions, and if the temperature rises and then falls on a boundary, the same temperature may occur at two different positions at the same time.

REFERENCES

Atthey, D. R. 1974. A finite difference scheme for melting problems. *J. Inst. Math. Its Appl.* 13:353–366.

Bonacina, C., and Comini, G. 1973. On the solution of the nonlinear heat conduction equations by numerical methods. *Int. J. Heat Mass Transfer* 16:581–589.

———, ———, Fasano, A., and Primicerio, M. 1973. Numerical solution of phase change problems. *Int. J. Heat Mass Transfer* 16:1825–1832.

———, ———, ———, and ———. 1974. On the estimation of thermophysical properties in nonlinear heat conduction problems. *Int. J. Heat Mass Transfer* 17:861–867.

Carnahan, B., Luther, H. A., and Wilkes, J. O. 1969. *Applied numerical methods.* New York: Wiley.

Cleland, A. C., and Earle, R. L. 1977. The third kind of boundary condition in numerical freezing calculations. *Int. J. Heat Mass Transfer* 20:1029–1034.

Comini, G., Del Giudice, S., Lewis, R. W., and Zienkiewicz, O. C. 1974. Finite element solution of nonlinear heat conduction problems with special reference to phase change. *Int. J. Num. Meth. Engr.* 8:613.

Couch, E. J., Keller, H. H., and Watts, J. W. 1970. Permafrost thawing around producing oil wells. *J. Can. Pet. Technol.* April–June:107–111.

Crank, J., and Crowley, A. B. 1978. Isotherm migration along orthogonal flow lines in two dimensions. *Int. J. Heat Mass Transfer* 21:393–398.

———, and Gupta, R. S. 1975. Isotherm migration method in two dimensions. *Int. J. Heat Mass Transfer* 18:1101–1107.

———, and Nicolson, P. 1947. A practical method for numerical evaluation of solutions of partial differential equations of the heat conduction type. *Proc. Cambridge Philos. Soc.* 43:50–67.

———, and Phahle, R. D. 1973. Melting ice by the isotherm migration method. *J. Inst. Math. Its Appl.* 9:12–14.

Crowley, A. B. 1978. Numerical solution of Stefan problems. *Int. J. Heat Mass Transfer* 21:215–219.

Dix, R. C., and Cizek, J. 1970. The isotherm migration method for transient heat conduction analysis. *Proc. 4th Int. Conf. Heat Transfer* 1:1–10.

Doherty, P. C. 1970. *Hot pipe.* U.S. Geological Survey Computer Contribution No. 4.

Douglas, J., and Gunn, J. E. 1964. A general formulation of alternating direction methods: Part 1. Parabolic and hyperbolic problems. *Numerische Mathematik* 6:428–453.

Duda, J. L., Malone, M. F., and Notter, R. N. 1975. Analysis of two-dimensional diffusion controlled moving boundary problems. *Int. J. Heat Mass Transfer* 18:901–910.

Du Fort, E. C., and Frankel, S. P. 1953. Stability conditions in the numerical treatment of parabolic differential equations. *Math. Tables Aids Comput.* 7:135–152.

Dusinberre, G. M. 1961. *Heat transfer calculations by finite difference.* Scranton, Pennsylvania: International Textbook.

Dusinberre, G. M. 1945. Numerical methods for transient heat flow. *Trans. ASME* 67:703.

Ferriss, D. H. 1975. Fixation of a moving boundary by means of a change of independent variable. In *Moving boundary problems in heat flow and diffusion,* eds., J. R. Ockendon, and N. R. Hodgkins, pp. 251–255. Oxford: Clarendon Press.

Forsythe, G. E., and Wasow, W. R. 1960. *Finite difference methods for partial differential equations.* New York: Wiley.

Goodrich, L. E. 1978. Efficient numerical technique for one-dimensional thermal problems with phase change. *Int. J. Heat Mass Transfer* 21:615–621.

Heitz, W. L., and Westwater, J. W. 1970. Extension of the numerical methods for melting and freezing problems. *Int. J. Heat Mass Transfer* 13:1371–1375.

Ho, D. M., Harr, M. E., and Leonards, G. A. 1970. Transient temperature distribution in insulated pavements—Predictions vs. observations. *Can. Geotech. J.* 7:275–284.

Jerome, J. W. 1977. Nonlinear equations of evolution and a generalized Stefan problem. *J. Diff. Eqn.* 26:240–261.

Kamenomostskaja, S. L. 1961. On the Stefan problem. *Mat. Sb.* 53(95):489–514.

Lachenbruch, A. H. 1970. *Some estimates of the thermal effect of a heated pipeline in permafrost.* U.S. Geological Survey Circular No. 632, Washington, D.C.

Landau, H. G. 1950. Heat conduction in a melting solid. *Q. J. Appl. Mech.* 8:81–94.

Meyer, G. H. 1970. On a free interface problem for linear ordinary differential equations and the one-phase Stefan problem. *Num. Math.* 16:248–267.

———. 1973. Multidimensional Stefan problems. *SIAM J. Num. Anal.* 10(3):522–538.

———. 1978. The numerical solution of multidimensional Stefan problems—A survey. In *Moving boundary problems,* eds. D. G. Wilson, A. D. Solomon, and P. T. Boggs, pp. 73–89. New York: Academic Press.

———, Keller, H. H., and Couch, E. J. 1972. Thermal model for roads, airstrips, and building foundations in permafrost regions. *J. Can. Pet. Technol.* 11(12):13–25.

Murray, W. D., and Landis, F. 1959. Numerical and machine solutions of transient heat conduction problems involving melting or freezing. Part 1—Method of analysis and sample solutions. *J. Heat Transfer* 81:106–112.

Nakano, Y., and Brown, J. 1971. Effect of a freezing zone of finite width on the thermal regime of soils. *J. Water Resour. Res.* 7:1226–1233.

Nixon, J. F., and McRoberts, E. C. 1973. A study of some factors affecting the thawing of frozen soils. *Can. Geotech. J.* 10:439–452.

Ockendon, J. R., and Hodgkins, W. R. eds. 1975. *Moving boundary problems in heat flow and diffusion.* Oxford: Clarendon Press.

O'Neill, K., and Lynch, D. R. 1979. A finite element method for porous medium freezing, using hermite basis functions and a continuously deforming coordinate system. In *Numerical Methods in Thermal Problems,* eds. R. W. Lewis and K. Morgan, pp. 548–562. *Proc. 1st Int. Conf.,* Swansea, Wales.

Rose, M. E. 1967. On the melting of a slab. *SIAM J. Appl. Math.* 15(3):495–504.

Salvadori, M. G., and Baron, M. L. 1952. *Numerical methods in engineering.* Englewood Cliffs, New Jersey: Prentice-Hall.

Saul'yev, V. K. 1964. *Integration of equations of parabolic type by the method of nets.* New York: Macmillan.

Schenck, H. 1963. *Fortran methods in heat flow.* New York: Ronald Press.

Sparrow, E. M., Ramadhyani, S., and Patankar, S. 1978. Effect of subcooling on cylindrical melting. *J. Heat Transfer* 100:395–402.

Talwar, R., and Dilpare, A. L. 1977. *A two-dimensional numerical solution to freezing/melting in cylindrical coordinates.* ASME Paper 77-WA/HT-11.

Welty, J. R. 1978. *Engineering heat transfer.* New York: Wiley.

BIBLIOGRAPHY

Numerical Methods

Altman, M. 1960. Some aspects of the melting solution for a semi-infinite slab. *Chem. Eng. Prog. Symp. Ser.* 53:16.

Boley, B. A. 1961. A method of heat conduction analysis of melting and solidification problems. *J. Math. Phys.* 40:300.

Booth, F. 1948. A note on the theory of surface diffusion reactions. *Trans. Faraday Soc.* 44:796.

Citron, S. J. 1960. Heat conduction in a melting slab. *J. Aerosp. Sci.* 27:219.

Dewey, C. F., Schlesinger, S. I., and Sashkin, L. 1960. Temperature profiles in a finite solid with moving boundary. *J. Aerosp. Sci.* 27:59.

Douglas, J. 1954. *On the numerical integration of a quasi-linear differential equation.* Humble Production Research Report, Humble Oil Co., Houston, Texas.

———, and Gallie, T. M. 1955. On the numerical integration of a parabolic differential equation subject to a moving boundary condition. *Duke Math. J.* 22:557.

Eyres, N. R., Hartree, D. R., Ingham, J., Jackson, R., Sarjant, R. J., and Wagstaff, J. B. 1946. The calculation of variable heat flow in solids. *Proc. R. Soc. London Ser. A* 240.

Forster, C. A. 1954. *Finite difference approach to some heat conduction problems involving change of state.* Report of the English Electric Co., Luton, England.

Goodman, T. R. 1959. The heating of slabs with arbitrary heat inputs. *J. Aerosp. Sci.* 26:187.

Kreith, F., and Romie, F. E. 1955. A study of the thermal diffusion equation with boundary conditions corresponding to solidification or melting of materials initially at the fusion temperature. *Proc. Phys. Soc.* 68:277.

Liebmann, G. 1956. Solution of transient heat transfer problems by the resistance-network analog method. *Trans. ASME* 78:1267.

Lotkin, M. 1958. The numerical integration of the heat conduction equation. *J. Math. Phys.* 37:178.

———. 1960. The calculation of heat flow in melting solids. *Q. J. Appl. Math.* 18:79.

Master, J. I. 1956. Problem of intense surface heating of a slab accompanied by change of phase. *J. Appl. Phys.* 27:477.

Miller, M. L. 1959. *Transient one-dimensional heat-conduction analysis for heterogeneous structures including an ablating surface.* ASME Paper No. 59-HT-22.

Poots, G. 1962. An approximate treatment of a heat conduction problem involving a two-dimensional solidification front. *Int. J. Heat Mass Transfer* 5:339.

Price, P. H., and Slack, M. R. 1952. Stability and accuracy of numerical solutions of the heat flow equation. *Br. J. Appl. Phys.* 3:379.

———. 1954. The effect of latent heat on numerical solutions of the heat flow equation. *Br. J. Appl. Phys.* 5:285.

Sarjant, R. J., and Slack, M. R. 1954. Numerical analysis of transient heat flow with variable properties. *J. Iron Steel Inst.* 177:428.

Vasilev, F. P., and Uspenskii, A. B. 1963. A finite difference method for the solution of a two phase Stefan problem with a quasilinear equation. *Sov. Math. Dokl.* 4:1475.

Variational Technique

Biot, M. A. 1955. Variational principles in irreversible thermodynamics with application to viscoelasticity. *Phys. Rev.* 97:1463.

———. 1956. Thermoelasticity and irreversible thermodynamics. *J. Appl. Phys.* 27:240.

———. 1957. New methods in heat flow analysis with application to flight structures. *J. Aeronaut. Sci.* 24:857.

———. 1958. *Further developments of new methods in heat flow analysis.* Cornell Aeronautical Labs. Inc., Report #SA-987—S-5, Buffalo, N.Y.

———, and Agrawal, H. G. 1964. Variational analysis of ablation for variable properties. *J. Heat Transfer* 86:437.

———, and Daughaday, H. 1962. Variational analysis of ablation. *J. Aerosp. Sci.* 29:227.

Chambers, L. G. 1956. A variational principle for the conduction of heat. *Q. J. Mech. Appl. Math.* 9:234.

Citron, S. J. 1960. A note on the relation of Biot's method in heat conduction to a least-squares procedure. *J. Aerosp. Sci.* 27:317.

Herivel, J. W. 1954. A general variational principle for dissipative systems. *Proc. R. Ir. Acad. Sec. A* 56:37.

Lardner, T. J. 1963. Biot's variational principle in heat conduction. *AIAA J.* 1:196.

Rosen, P. 1953. On variational principles for irreversible processes. *J. Chem. Phys.* 21:1220.

Contour Integral Technique

Grinbert, G. A. 1951. Letters to the editor. *J. Tech. Phys.* USSR 21:382.

Liubov, B. Ya. 1947. Computation of the rate of solidification of an ingot taking into account the temperature dependence of the thermophysical parameters of the metal. *Doklady Akademiia Nauk, SSSR* 57:763.

Redozubov, D. V. 1957. On linear heat problems with one moving boundary. *Sov. Phys. Tech. Phys.* 2:1993.

———. 1959. A method of applying contour integrals to the solutions of one-dimensional problems of the heat conductivity theory. *Inzh. Fiz. Zh.* 2:76.

———. 1960. The solution of linear thermal problems with a uniformly moving boundary in a semi-infinite region. *Sov. Phys. Tech. Phys.* 5:570.

———. 1962. *The Stefan's problem for a linear initial temperature distribution in a semi-infinite region.* Bulletin, Izvestiia Akademiia Nauk SSSR, Geophysics Series, 364.

Experimental Results

Adams, C. M., and Taylor, F. 1957. Flow of heat from sand castings by conduction, radiation, and convection. *Trans. Am. Foundrymen's Soc.* 65:170.

Bishop, H. F., and Pellini, W. S. 1952. Solidification of metals. *Foundry* 8(1):86.

Chalmers, B. 1954. Melting and freezing. *J. Met.* 519.

———. 1964. *Principles of solidification.* New York: Wiley.

Chvorinov, N. 1940. Control of solidification of castings by calculations. *Giesserei* 27:177.

Clark, K. L. 1945. Methods employed to obtain rates of solidification. *Trans. Am. Foundrymen's Soc.* 53:88.

Cole, G. S., and Winegard, W. C. 1962. Thermal convection ahead of a solid-liquid interface. *Can. Metall. Q.* 1:29.

———. 1965. Thermal convection during horizontal solidification of pure metals and alloys. *J. Inst. Met.* 93:153.

Elmer, S. L. 1932. Ice formation on pipe surfaces. *Refrig. Eng.* 24:17.

Moon, J. S., and Keeler, R. N. 1962. A theoretical consideration of directional effects in heat flow at the interface of dissimilar metals. *Int. J. Heat and Mass Transfer* 5:967.

Robinson, A. T., McAlexander, R. L., Ransdell, J. D., and Wolfson, M. R. 1963. Transpiration cooling with liquid metals. *AIAA J.* 1:89.

Ruddle, R. W. 1957. *The solidification of castings.* Institute of Metals, London, p. 79.

Thomas, L. J., and Westwater, J. W. 1963. Microscopic study of solid-liquid interfaces during melting and freezing. *Chem. Eng. Prog. Symp. Ser.* 59:155.

Tiller, W. A., Jackson, K. A., Rutter, J. W., and Chalmers, B. 1953. The redistribution of solute atoms during the solidification of metals. *Acta Metall.* 1:428.

Wagner, C. 1954. Theoretical analysis of diffusion of solutes during the solidification of alloys. *J. Met.* 6:154.

Winegard, W. C. 1961. Fundamentals of the solidification of metals. *Metall. Rev.* 6:57.

10
Phase Change Heat Transfer Applications

10.0 INTRODUCTION

Our previous discussions have dealt with heat transfer problems in which there was no change of phase. If one is investigating a system that contains water or another liquid, there will be a possibility of freezing or thawing, if the temperature of the system attains a value at or near the fusion temperature. Water is an important constituent of permafrost, or any soil system, and phase changes will often be a significant physical effect. Thus, the problems of freezing and thawing are extremely important from an engineering viewpoint and have received considerable attention, theoretically and especially from a practical viewpoint. Despite this importance, investigators often attempt to ignore the effects of phase change due to the formidable analytical problems introduced by melting or freezing. From a mathematical viewpoint, phase change will usually change the basic problem from a linear to a nonlinear form, either through the equations themselves or through the boundary conditions. This means that most of the standard mathematical techniques for solving the problem are no longer valid.

The solution for phase change in a homogeneous, semi-infinite medium (Neumann problem) has been given in Chap. 8. The phase change depth is

$$X = 2\gamma\sqrt{\alpha_1 t} \tag{10.1}$$

The equation for the constant γ is

$$\frac{e^{-\gamma^2}}{erf\,\gamma} - \frac{k_2}{k_1}\sqrt{\frac{\alpha_1}{\alpha_2}}\,\frac{(T_o - T_f)\,e^{-\frac{\alpha_1}{\alpha_2}\gamma^2}}{(T_f - T_s)\,erfc\,\gamma\sqrt{\frac{\alpha_1}{\alpha_2}}} = \frac{l\gamma\sqrt{\pi}}{c_1\,(T_f - T_s)} \tag{10.2}$$

Graphical solutions for γ are given in this chapter.

The quasi-steady approximation to the Neumann problem, the Stefan solution (Fig. 10.1), has also been discussed in Chap. 8. The basic relation is

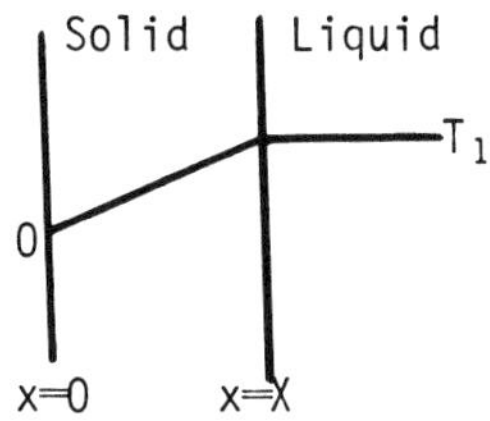

Fig. 10.1. Single-Phase Freezing.

$$X = \sqrt{\frac{2k_f(T_f - T_s)}{L}} \tag{10.3}$$

The quasi-steady assumption will be good if the latent heat is much larger than the sensible heat effects or if the sensible heat effects are taken as zero. This would be true for water and also soils with a high water content, but may not be valid if the water content of a soil is low.

Equation (10.3) was reported by Stefan (1889, 1891), in a study of the growth of polar ice.

The Stefan equation is often generalized to

$$X = \sqrt{\frac{2k_f}{L}\int_o^t (T_f - T_s)\, dt} \tag{10.4}$$

10.1 FORMATION AND GROWTH OF AN ICE LAYER

Consider the growth of a sheet of ice over a river or lake as shown in Fig. 10.2. The ice can have a snow layer upon it, and the temperature of the water can be above the freezing point. The interface between the ice and the water is the phase transformation plane. The difference between the heat flow from the interface to the air and the heat flow from the water to the interface is the latent heat removed in the formation of a layer of ice. The energy balance, neglecting radiative effects, leads to

$$\frac{(T_f - T_a)}{\left(\frac{1}{h_a} + \frac{X}{k_i} + \frac{d}{k_s}\right)} - h_w\,(T_w - T_f) = L\frac{dX}{dt} \tag{10.5}$$

The surface coefficient, h_a, can be found by the methods of Chap. 6. Following Zarling (1978), dimensionless relations can be used

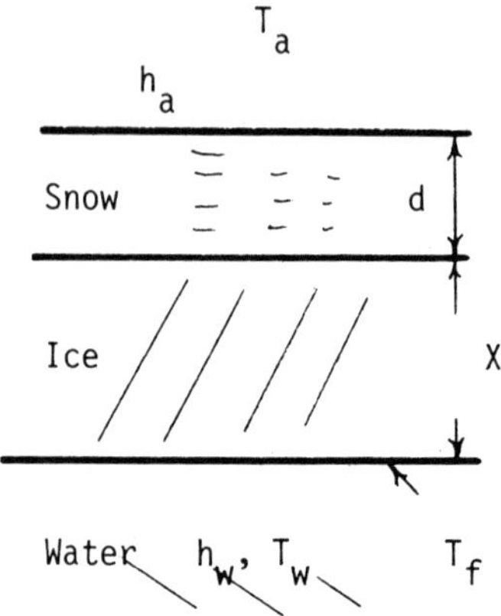

Fig. 10.2 Ice Growth over Water.

$$\epsilon = \frac{h_a}{k_i} X \qquad D = \frac{dh_a}{k_s} \qquad P = \frac{h_w (T_w - T_f)}{h_a (T_f - T_a)} \qquad \theta = \frac{(T_f - T_a)}{L \, k_i} h_a^2 \, t$$

Then Eq. (10.5) is

$$d\theta = \frac{1 + \epsilon + D}{1 - P(1 + \epsilon + D)} d\epsilon \tag{10.6}$$

This equation can be integrated if $\epsilon = 0$ at $t = 0$. The solution is

$$\theta = -\frac{1}{P^2} \ln \left[1 - \frac{P\epsilon}{1 - P(1 + D)} \right] - \frac{\epsilon}{P} \tag{10.7}$$

10.1.1 Zero Snow Depth

The assumption that $\epsilon = 0$, initially, means the snow layer must be zero at least to start the solution. In this case $D = 0$, and Eq. (10.7) is

$$\theta = \frac{-1}{P^2} \ln \left(1 - \frac{P\epsilon}{1 - P} \right) - \frac{\epsilon}{P} \tag{10.8}$$

10.1.2 Water at Freezing Point

For this case $P = 0$ and

$$\epsilon = -(1 + D) + \sqrt{(1 + D)^2 + 2\theta} \tag{10.9}$$

10.1.2.1 Zero Snow Depth

Equation (10.9) now reduces to

$$\epsilon = -1 + \sqrt{1 + 2\theta} \tag{10.10}$$

10.1.2.2 Zero Snow Depth, Ice Surface at T_a

This very special case will reduce to the Stefan equation. Equation (10.6) reduces to

$$d\theta = (1 + \epsilon)\, d\epsilon$$

or

$$\left(1 + \frac{h_a X}{k_i}\right) dX = (T_f - T_a) \frac{h_a}{L}\, dt \tag{10.11}$$

Now if the ice surface temperature equals the air temperature, then this implies that $h_a \rightarrow \infty$.

Thus

$$X\, dX = \frac{k_i}{L} (T_f - T_a)\, dt$$

The solution is

$$X = \sqrt{\frac{2\, k_i}{L} \int_o^t (T_f - T_a)\, dt}$$

which is identical to Eq. (10.4) with $k_i = k_f$ and $T_a = T_s$.

10.1.3 Frazil Ice

Under certain conditions, ice may form at an accelerated rate due to the presence of frazil or dendritic ice. Water moving down a rapids may be in contact with very cold air but fail to freeze. It will then subcool and nucleate into a dense mixture of platelike ice crystals (frazil ice) and water, essentially at the fusion temperature. The situation leading to Eq. (10.6) now changes due to the layer of frazil ice and water (Fig. 10.3). Since this layer is at T_f, there will be no heat flow into the growing ice layer from the warmer water

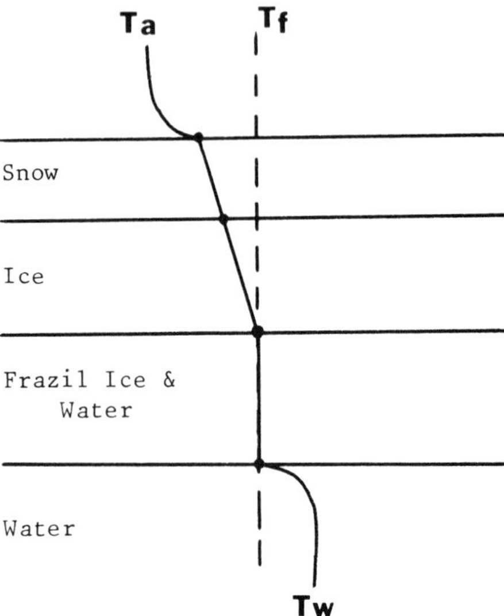

Fig. 10.3. Ice Growth with Frazil Ice.

below. The convective heat flow $h_w\ (T_w - T_f)$ will simply melt the frazil ice at its lower boundary. The energy balance will then be

$$\frac{(T_f - T_a)}{\dfrac{1}{h_a} + \dfrac{X}{k_i} + \dfrac{d}{k_s}} = L_f \frac{dX}{dt} \tag{10.12}$$

The latent heat will be reduced due to the frazil ice in the water layer at the solid ice interface. If the volume fraction of water in the frazil ice and water mixture is x_w, then the volumetric latent heat of the mixture is

$$L_f = Lx_w$$

The frazil ice layer accelerates the rate of ice formation due to its insulating effect, at the lower boundary, and due to the reduced latent heat of the ice and water mixture. The solutions of Section 10.1.2 are valid if L is replaced by Lx_w. The numerical integration of Eq. (10.12), for variable conditions, is discussed by Calkins (1979).

10.1.4 Empirical Relation

For ice growth, an empirical coefficient is often introduced to account for the neglect of snow cover, sensible heat, and the meteorological effects

Table 10.1. Empirical Relations for Ice Growth over Water.

a	Conditions	Reference
.9–.95	Practical maximum for snow free ice	Rhodes (1974)
.8	Windy lakes with no snow	Michel (1971)
.7–.8	Medium size lakes with moderate snow cover	"
.58–.65	Rivers with moderate flow	Rhodes (1974)
.4–.5	Rivers with snow	Michel (1971)
.2–.4	Small river with rapid flow	"

$$X = a\sqrt{\int_o^t (T_f - T_a)\, dt} \tag{10.13}$$

where

X = ice thickness, in. and

$\int_o^t (T_f - T_a)\, dt$ = air freezing index, °F − days.

Values of the coefficient a are given in Table 10.1.

10.2 MODIFIED BERGGREN EQUATION

The Neumann equation is used as the basis for many phase change studies, and it is usable for engineering purposes. Berggren (1943) was apparently the first to actually apply the Neumann solution to soil phase change problems. Aldrich and Paynter (1953) later used the Stefan form of the phase change equation to arrive at the modified Berggren equation.

10.2.1 Freeze Case

Let us rewrite Eq. (10.1) in the modified form

$$X = \lambda \sqrt{\frac{2k_f v_s t}{L}} \tag{10.14}$$

where λ is a function determined from the Neumann solution, with assumptions on the thermal properties of the frozen and unfrozen media.

Equation (10.14) is simply another form of the Neumann equation (Eq. (10.1)) and is called the "modified Berggren equation." Various forms of this equation have been shown to yield excellent results when applied to the in situ phase change of soils (McRoberts 1975).

Now the freezing index at the surface of the solid may be defined as

$$I_s = \int_o^t (T_f - T_s)\, dt = \int_o^t v_s\, dt = v_s t$$

This is not the same as the air freezing index, but analysis of data (see Sanger 1963 indicates that $(v_s t)$ is related to the air freezing index, denoted by I_f. This assumes that the surface temperature of a soil during freezing is a constant determined by I_f and the length of the freezing season. In fact, the surface temperature varies continuously, but the average value obtained from the freezing index is suitable for use in the Neumann solution (see Section 8.4.2).

Thus Eq. (10.14) is

$$X = \lambda \sqrt{\frac{2k_f I_f n}{L}} \tag{10.15}$$

The parameter n is discussed in Section 10.5

For calculation purposes one can use

$$X = 16.33\ \lambda \sqrt{\frac{k_f I_f}{L}}$$

where

$|X| = \text{cm}$
$|k| = \text{kcal/(hr-m-°C)}$
$|I| = \text{°F-day}$ and
$|L| = \text{cal/cm}^3$.

The coefficient λ can be found by equating Eqs. (10.1) and (10.15) and using Eq. (10.2) for the freezing case. This leads to the following equation for λ

$$\frac{e^{-\lambda^2\mu}}{erf\,\lambda\sqrt{\mu}} - \frac{p\alpha e^{-q\lambda^2\mu}}{erfc\ r\lambda\sqrt{\mu}} = \frac{\lambda}{2}\sqrt{\frac{\pi}{\mu}} \tag{10.16}$$

The parameters α and μ take into account the soil temperatures, specific heat, and latent heat and are

$$\alpha = \frac{T_o - T_f}{T_f - T_s}$$

$$\mu = \frac{C_f}{2L}(T_f - T_s)$$

The quantities p, q, and r describe the thermal properties of the soil system

$$p = \frac{k_u}{k_f}\sqrt{\frac{\alpha_f}{\alpha_u}} \qquad q = \frac{\alpha_f}{\alpha_u} \qquad r = \sqrt{\frac{\alpha_f}{\alpha_u}}$$

These relations are all for the freezing case. The thawing case will require further discussion. Aldrich and Paynter (1953) used the relations

$$\alpha = \frac{T_o - T_f}{T_f - T_s}\frac{C_u}{C_f}$$

$$\mu = \frac{C_f}{L}(T_f - T_s)$$

and noted that calculations with typical soil properties indicated

$$\frac{\alpha_f}{\alpha_u} \simeq 1.0, \qquad \frac{C_u}{C_f} \simeq 1, \qquad \text{and thus } k_u/k_f = 1.0$$

Actually this is valid only for the case where the water content is zero.

Clearly p, q, and r will vary for different soil systems, but fortunately a relatively simple procedure can be used to generate these functions for any soil system.

In Chap. 4, Section 4.12.1.1, it was noted that the thermal conductivity of a mixture could be expressed as

$$k = k_s^{x_s} k_l^{x_l}$$

where k_s and k_l are the thermal conductivities of the solid and liquid phases and x_s and x_l are the volume fractions of the solid and liquid phases. For soil systems, the thermal conductivity of the solids will not vary significantly as phase change occurs (Kersten 1949), and thus the ratio of the frozen to unfrozen conductivity will be related to the thermal conductivity of ice and water as follows

$$\frac{k_u}{k_f} = \left(\frac{k_w}{k_i}\right)^{x_l}$$

where k_w and k_i are the thermal conductivities of water and ice. The volumetric specific heat for the system may be expressed as follows, for the thawed and frozen states

$$C_u = C_{su}\,(1 - x_l) + C_w\, x_l$$
$$C_f = C_{sf}\,(1 - x_l) + C_i\, x_l$$

where

C_{su}, C_{sf} = specific heats of unfrozen and frozen solids and
C_w, C_i = specific heats of water and ice.

It is fortunate that the specific heats of soil solids and ice are all about the same. For example, the specific heat of organic solids is 0.461 cal/cc-°C, for mineral solids it is 0.420, and for ice it is 0.459 (Lunardini 1971). If one assumes that the values for the solids, except for ice, change little through the phase change then

$$\frac{C_u}{C_f} = 1 + \left(\frac{C_w}{C_i} - 1\right) x_l$$

or

$$\frac{C_u}{C_f} = 1 + 1.189\, x_l$$

The property values to use in Eq. (10.16) can then be expressed as functions of the soil water content

$$q = R^{x_l}(1 + 1.189\, x_l)$$
$$r = \sqrt{q}$$
$$p = \frac{r}{R^{x_l}}$$

where

$$R = \frac{k_i}{k_w}$$

The ratio of the thermal conductivity of ice to water, R, is 3.89. Figures 10.4–10.9 give the values of λ to use in Eq. (10.15) for the freezing case.

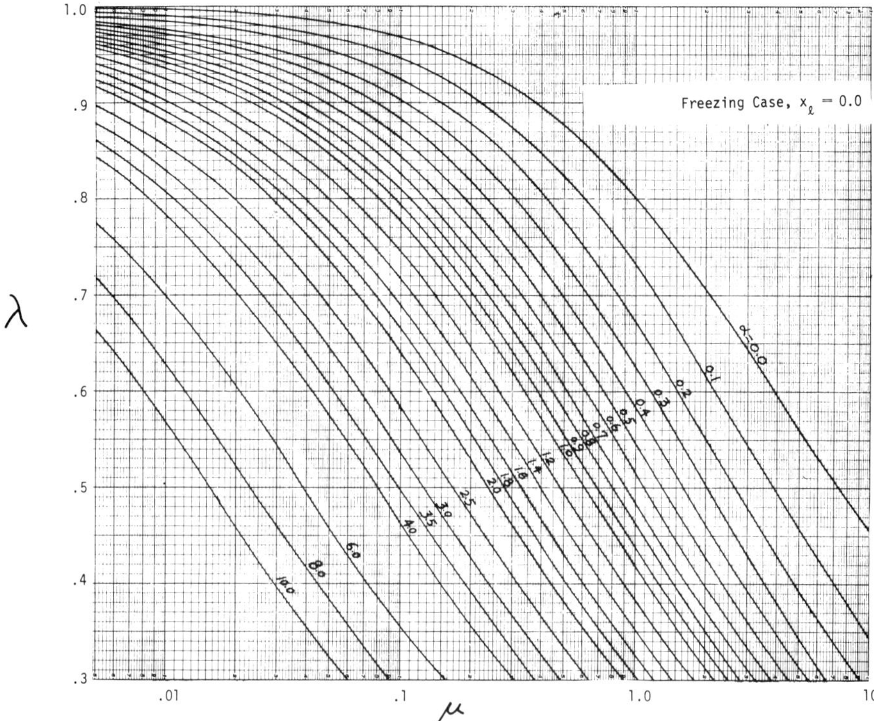

Fig. 10.4. Freezing Case, $x_l = 0.0$.

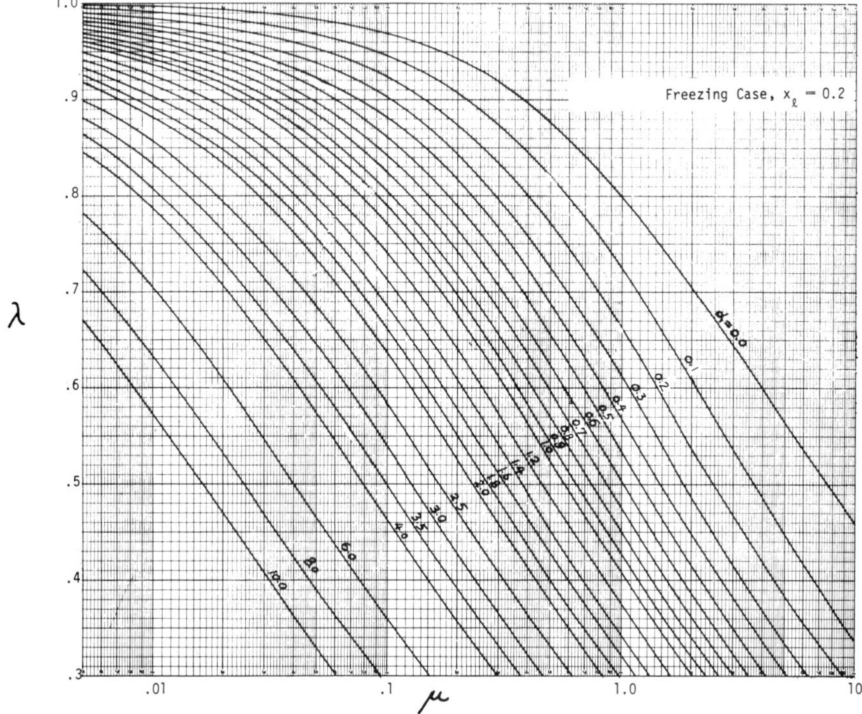

Fig. 10.5. Freezing Case, $x_l = 0.2$.

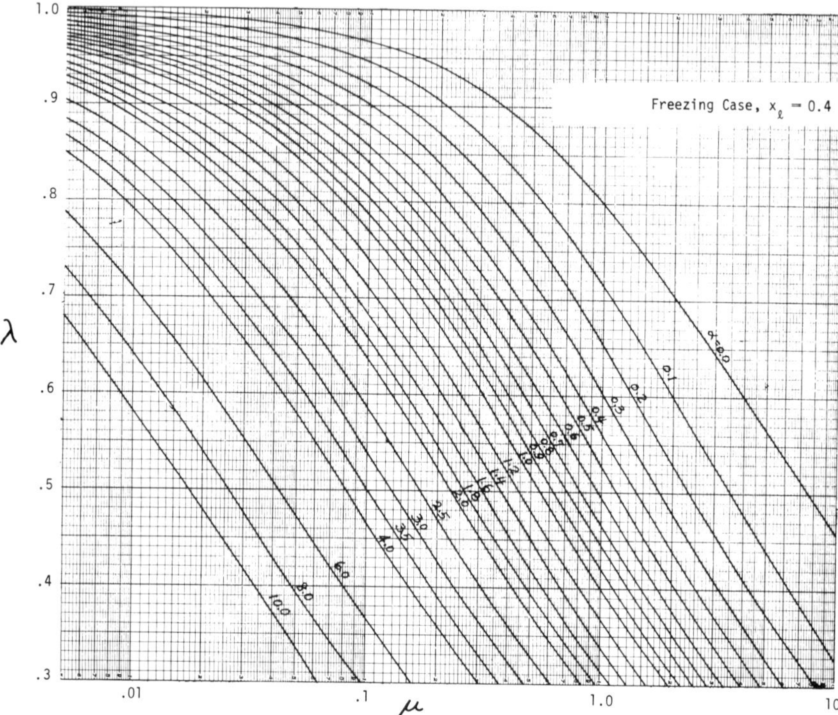

Fig. 10.6. Freezing Case, $x_l = 0.4$.

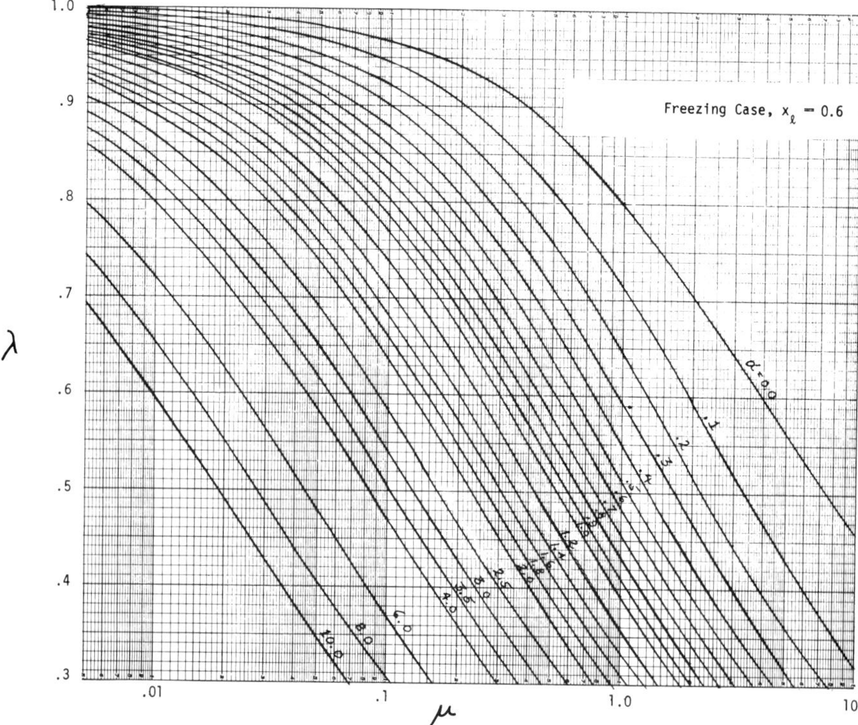

Fig. 10.7. Freezing Case, $x_l = 0.6$.

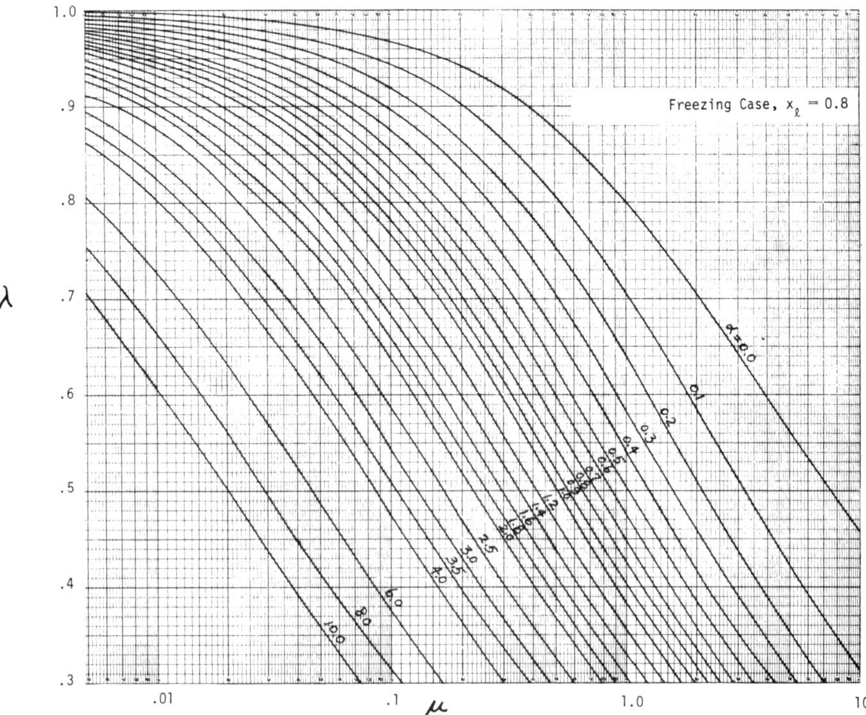

Fig. 10.8. Freezing Case, $x_l = 0.8$.

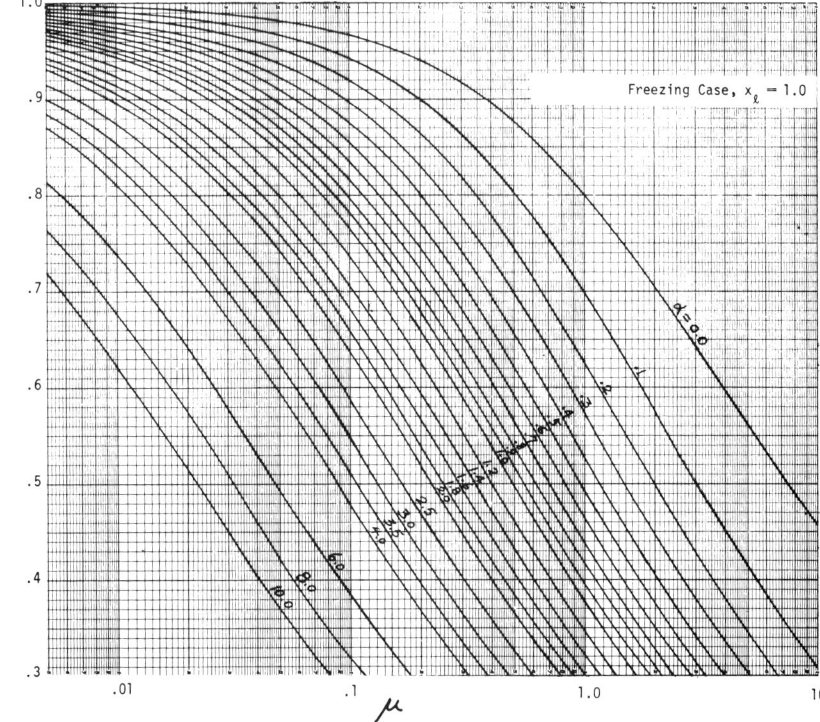

Fig. 10.9. Freezing Case, $x_l = 1.0$.

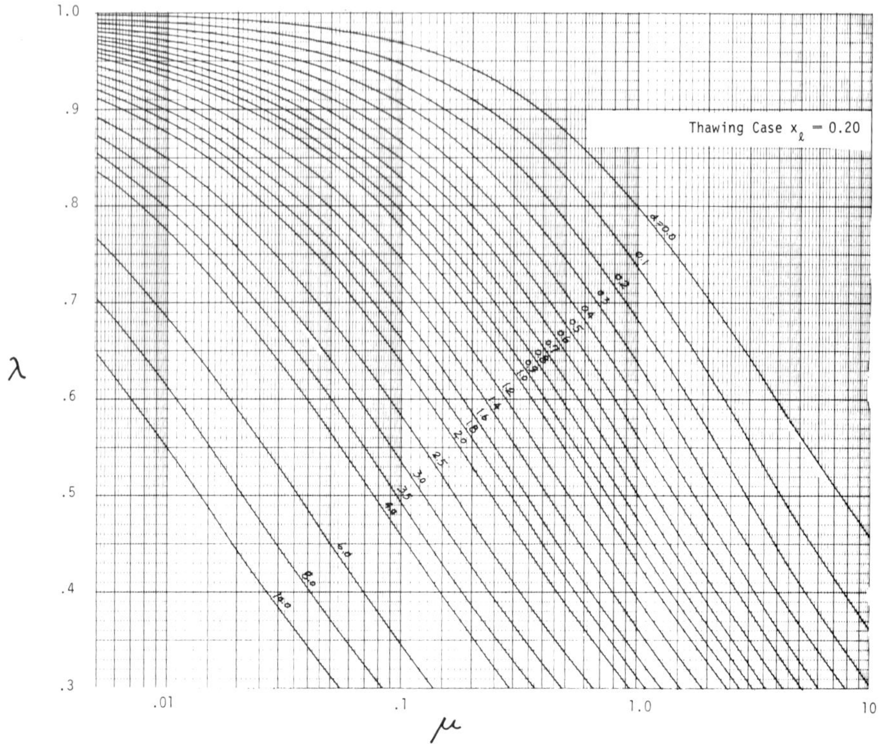

Thawing Case $x_\ell = 0.20$

Fig. 10.10. Thawing Case, $x_l = 0.20$.

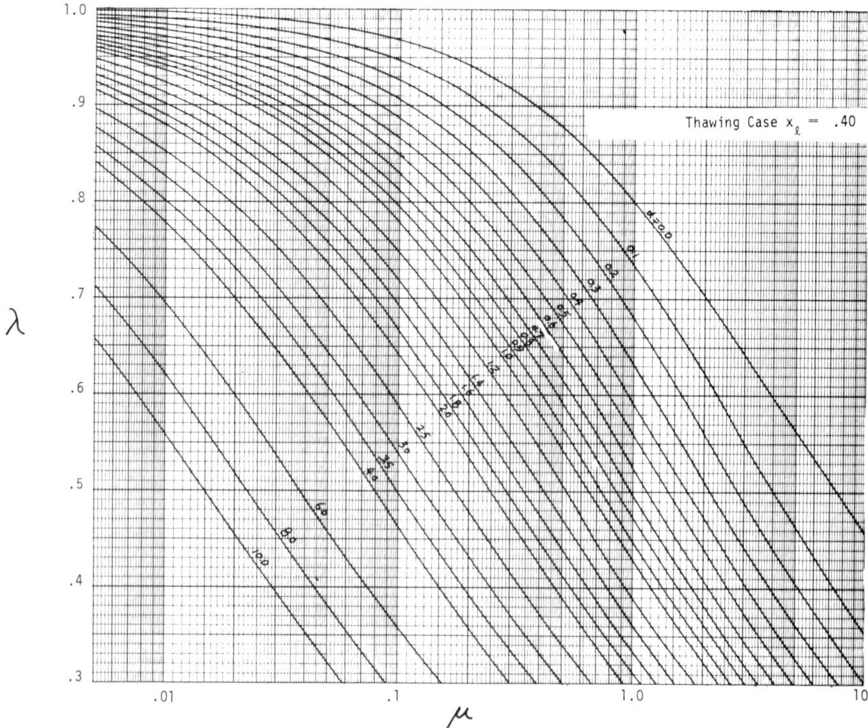

Fig. 10.11. Thawing Case, $x_l = .40$.

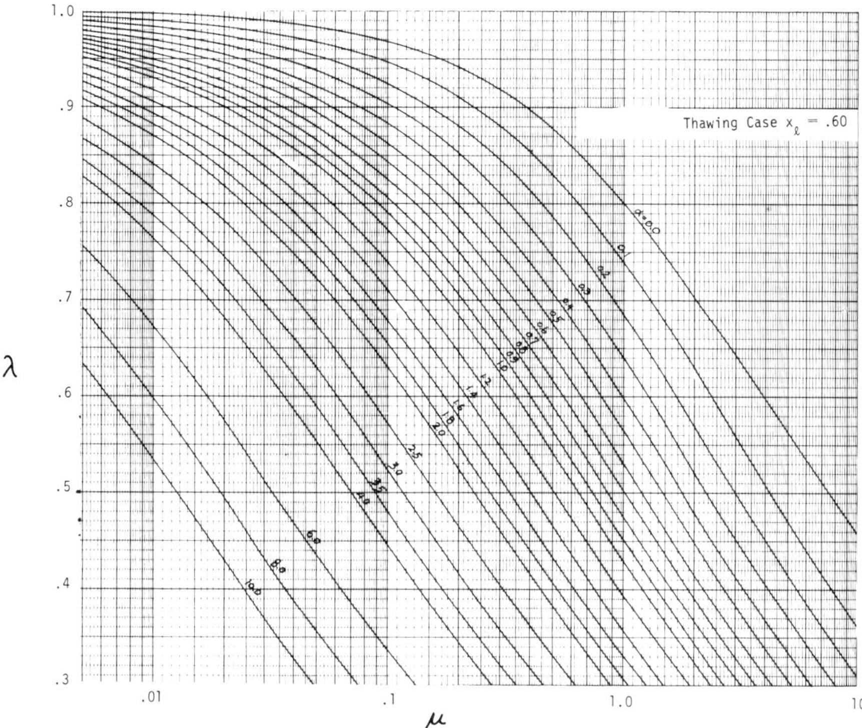

Fig. 10.12. Thawing Case, $x_l = .60$.

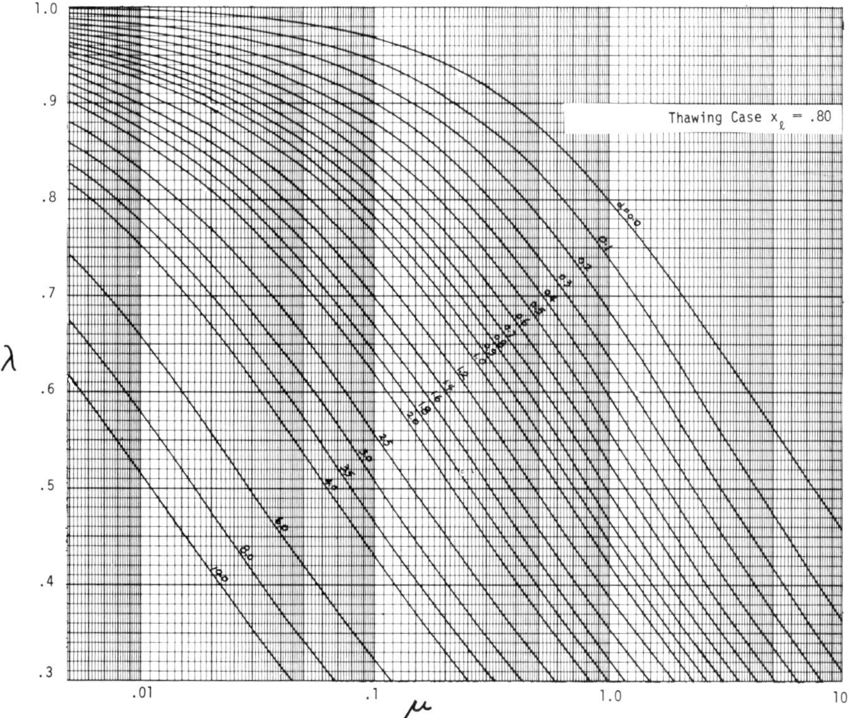

Fig. 10.13. Thawing Case, $x_l = .80$.

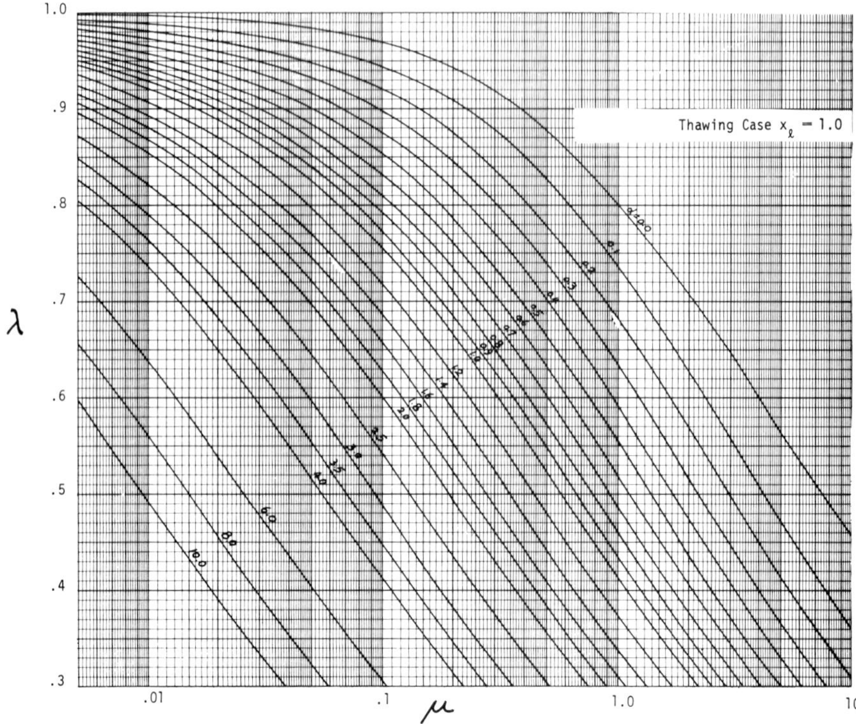

Fig. 10.14. Thawing Case, $x_l = 1.0$.

10.2.2 Thaw Case

At first glance, it might appear that the same relations could be used for the thawing or the freezing case. This, however, is not true. In the thawing case, the medium is initially frozen at T_0, and energy must be conducted through the thawed layer from the phase change interface. Because the thermal conductivity of the thawed region is considerably less than that of the frozen, the heat flow will be reduced even with the same temperature gradient. The general form of the equation will be the same, however, with appropriate changes for the property values. The thaw depth is again expressed as

$$X = \lambda \sqrt{\frac{2k_u I_t}{L}} \tag{10.17}$$

and λ is again given by Eq. (10.16), but the parameters are now

$$\alpha = \frac{T_f - T_o}{T_s - T_f} \qquad \mu = \frac{C_u}{2L}(T_s - T_f)$$

The functions p, q, and r all change due to the property changes of the thawed and frozen states. The parameter R is still 3.89

$$q = \alpha_{uf} = \frac{1}{R^{x_l}(1 + 1.189\, x_l)}$$

$$r = \sqrt{q}$$

$$p = k_{fu}\sqrt{\alpha_{uf}} = r\, R^{x_l}$$

The λ values for thawing are now given in Figs. 10.10 through 10.14. Notice that when $x_l = 0$, the λ values are the same for freezing or thawing, the Aldrich (1953) case. Unless further information is available, one can assume that the initial temperature of the soil system is given by T_a, the mean annual air temperature, and the surface temperature is given by

$$T_s - T_f = \frac{I_t}{\theta_s}$$

where

I_t = thawing index and
θ_s = length of the thawing season.

It should be clear that the charts for λ, Figs. 10.4–10.14, are for the exact solution of the Neumann problem with the property values typical of soil systems. An approximate equation for γ, to use in Eq. (10.1), is also given in Section 8.3.2.2.

Example 1: Thawing of a homogeneous system.

Consider the case of thawing at Chesterfield, NWT. The soil is homogeneous, fine-grained, with $\gamma_d = 1.3$ g/cm³, $W = 30.8\%$. The thaw season is 112 days, and the thawing index is 1320°F-day. The mean annual air temperature is 11°F.

The soil properties, from the charts of Chap. 4, are

$$k_u = 0.92\,\frac{kcal}{hr\text{-}m\text{-}^\circ C} \qquad C_u = 0.611\,\frac{cal}{cm^3 - {}^\circ C}$$

$$L = \frac{\gamma_d W h_{if}}{100} = 1.3\,(.308)(79.7) = 31.9\ \text{cal/cm}^3$$

Then

$$T_s - T_f = \frac{I_t}{\theta_s} = 11.8°\text{F}$$

$$\alpha = \frac{T_f - T_a}{T_s - T_f} = \frac{21}{11.8} = 1.78$$

$$\mu = \frac{C_u}{2L}\,(T_s - T_f) = .063$$

$$x_l = \frac{W\gamma_d}{100\rho_w} = (.308)(1.3) = 0.4$$

From Fig. 10.11, $\lambda = 0.73$. The thaw depth is

$$X = 16.33\ \lambda\ \sqrt{\frac{k_u I_t}{L}}$$

$$X = 73.6\ \text{cm}$$

Example 2: Freezing of a homogeneous system.

Take the same case as Example 1, but let $T_0 = 53°$F, i.e., T_0 is 21°F above freezing, whereas in Example 1 it was 21°F below freezing. The freezing index is 8750°F-day and $\theta_s = 253$ days.

The frozen soil properties are

$$k_f = 1.49 \qquad C_f = 0.41$$

Then

$$T_f - T_s = \frac{8750}{253} = 34.6°\,F$$

i.e., the average surface temperature, T_s, during the freezing is −2.6°F

$$\alpha = \frac{53 - 32}{34.6} = 0.61$$

$$\mu = \frac{0.41}{2(31.9)}\,\frac{(34.6)}{(1.8)} = 0.124$$

Now from Fig. 10.6, $\lambda = 0.816$, and the depth of freezing is

$$X = 16.33\ \lambda \sqrt{\frac{k_f I_f}{L}} = 269.4 \text{ cm}$$

Actually the freeze at Chesterfield can only be as deep as the depth of the thawed layer, as the underlying soil is permanently frozen. This illustrates the significantly different effects of thawing and freezing at a given location. Note also that the λ value is different for freezing and thawing.

10.2.3 Existence of Permafrost

Examples 1 and 2 show that the freeze depth exceeded the thaw depth, and thus permafrost is likely to exist and be maintained under these conditions. A simple criterion for the presence of permafrost is that the depth of freeze should exceed the depth of thaw. That is

$$X_f > X_t$$

Now using Eqs. (10.14) and (10.17), this leads to

$$I_f > \left(\frac{\lambda_t}{\lambda_f}\right)^2 \left(\frac{k_t}{k_f}\right) \left(\frac{n_t}{n_f}\right) I_t \tag{10.18}$$

Thus the freeze index I_f must exceed some multiple of the thaw index I_t.

Example 3: Consider the data of Chesterfield from Examples 1 and 2.

$$\frac{k_t}{k_f} = 0.62 \qquad \left(\frac{\lambda_t}{\lambda_f}\right)^2 = 0.79$$

Now n_t/n_f can vary from 1.5 to 3 for natural surfaces (see Table 10.2). Thus, for permafrost to exist here, $I_f > 1.47\ I_t \cong 2000$. As I_f exceeds this value, permafrost is to be expected and does exist.

10.2.4 Sinusoidal Surface Temperature and Permafrost

The formation and growth of permafrost are interesting and complex phenomena. A sinusoidal air or surface temperature with a period of 1 year can approximate natural conditions to allow us to examine the origin of permafrost. Section 8.4.1.2 gave an approximate solution for phase change with a sinusoidally varying surface temperature. Let us use the conditions of Example 1 of this chapter, except that the ground will initially be thawed at the fusion temperature. This will create a distortion of the real case situation, but the results will still adequately model a permafrost system. A zeroth-order quasi-steady approximation will suffice.

The equations to solve are

$$\frac{\partial^2 \theta}{\partial y^2} = 0 \tag{10.19}$$

$$\theta(\xi,\tau) = 0 \tag{10.19a}$$

$$\theta(0,\tau) = \phi(\tau) \tag{10.19b}$$

$$\frac{d\epsilon}{d\tau} = \pm \frac{\partial \theta(\xi,\tau)}{\partial y} \tag{10.19c}$$

$$\xi(0) = 0 \tag{10.19d}$$

The positive sign is for freezing, and the negative sign is for the thaw part of the cycle in Eq. (10.19*c*). The following variables have been used

$$\theta = \frac{T - T_f}{\theta_c} \qquad \tau = \frac{t}{t_c} = \frac{2\pi t}{P} \qquad S_T = \frac{c\theta_c}{l}$$

$$y = \frac{x}{\sqrt{\alpha S_T t_c}} \qquad \xi = \frac{X}{\sqrt{\alpha S_T t_c}}$$

The air temperature at Chesterfield, NWT, which will be taken as the surface temperature, is closely represented by

$$\phi(\tau) = \sin\tau + .68(\cos\tau - 1)$$

This equation is shown in Fig. 10.15. A value of $\phi = 0$ refers to the fusion temperature. Thawing occurs for $0 \leq \tau < \tau_1 = 1.9478$ radians and freezing for the remainder of the period P of 1 year. The thaw season is about 113 days. The solution for the phase change depth is given by

$$\frac{\xi^2}{2} = \begin{cases} 1 - \cos\tau + .68(\sin\tau - \tau) & \textit{thawing} \quad (10.20) \\ \cos\tau - .68(\sin\tau - \tau) - \cos\tau_1 + .68(\sin\tau_1 - \tau_1) & \textit{freezing} \quad (10.21) \end{cases}$$

The solutions assume that the ground is initially unfrozen at the fusion temperature and freezes to a maximum depth. At the time of maximum freeze, the surface temperature and that of the entire frozen layer will have returned

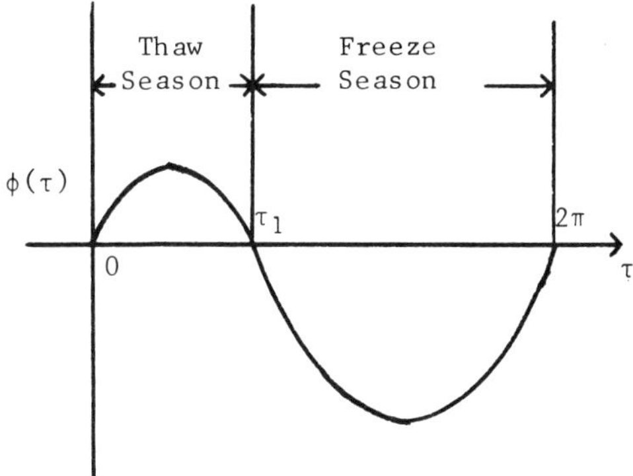

Fig. 10.15. Sinusoidal Air Temperature.

to the fusion value. Thawing will then occur down to a maximum value. If the freeze and thaw depths were equal, the solution would be exact for the repeat cycles, and the ground would change phase periodically. Due to temperature asymmetry and differing property values, however, this will not be the case, and the cycles will not simply repeat. There will be an excess of freezing, for this case, and the lower boundary of the frozen layer will grow in a complicated fashion, not covered by the simple solution used here. Additionally, finite superheat and subcooling will further complicate the actual growth. Nevertheless, plotting Eqs. (10.20) and (10.21) in a piecewise fashion, with frozen and thawed properties, will demonstrate the processes of permafrost generation and growth.

Figure 10.16 shows these results for a few yearly cycles, using the thermal properties of Example 1. Notice that the initial freeze depth is over three times the subsequent thaw depth. In this figure, the lower boundary of the "permafrost" is simply sketched in and does not represent actual calculations. If the surface temperature remained uniformly periodic, there would be an eventual equilibrium of the thickness of permafrost with a constant active layer as shown. In reality, the surface temperature and the lengths of the thaw and freeze seasons will oscillate about mean values, and the active layer will also oscillate about some long-term value, as shown in Fig. 10.17. This simple example reasonably illustrates the processes of permafrost birth and growth, although it is highly idealized.

If a surface coefficient of heat transfer h is used with $\phi(\tau)$ as the air temperature, then the surface boundary condition is

$$-\frac{1}{B_i}\frac{\partial\theta(0,\tau)}{\partial y} = \phi(\tau) - \theta(0,\tau)$$

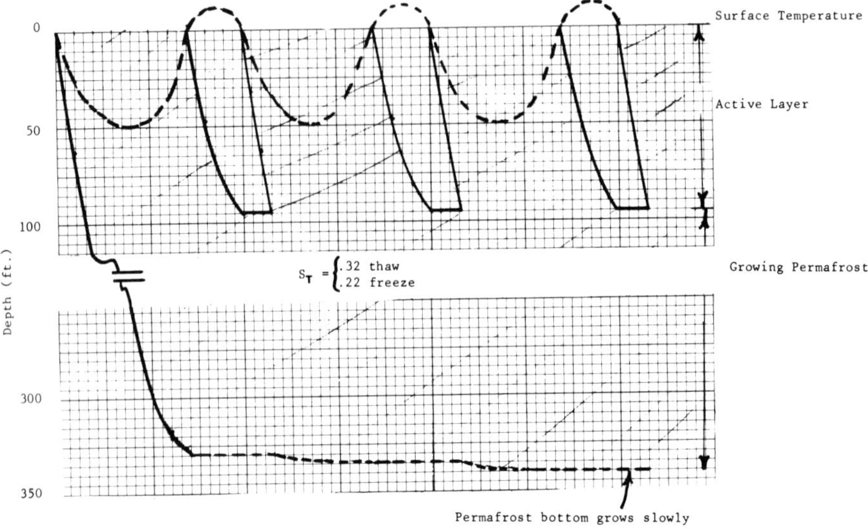

Fig. 10.16. Active Layer Formation and Permafrost Growth.

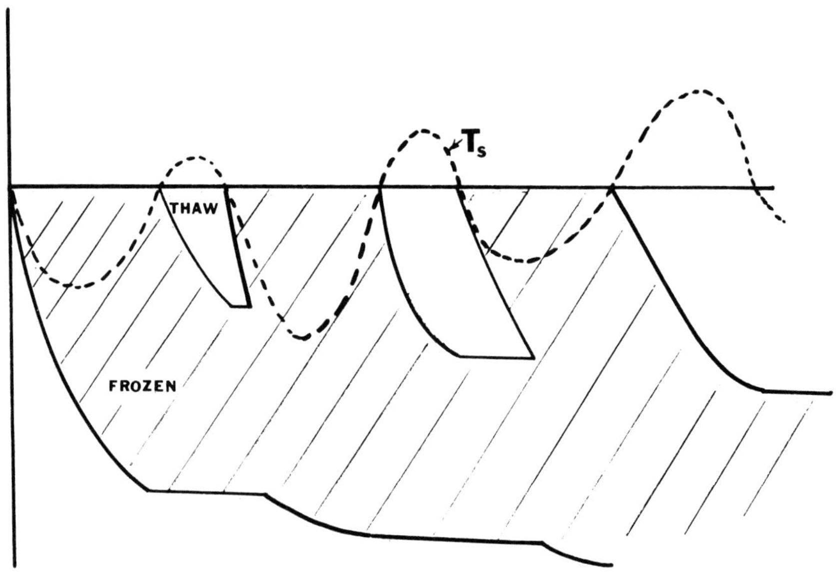

Fig. 10.17. Active Layer Varying from Year to Year.

and

$$\frac{\xi^2}{2}+\frac{1}{B_i}\xi=\begin{cases}1-\cos\tau+.68(\sin\tau-\tau) & \textit{thaw} \quad (10.22)\\ \cos\tau-.68(\sin\tau-\tau)-.32413 & \textit{freeze} \quad (10.23)\end{cases}$$

Assuming that the surface coefficient is

$$h=\begin{cases}5\ \ kcal/(hr-m^2-{}^\circ C) & \textit{freeze}\\ 20 & \textit{thaw}\end{cases}$$

then the Biot number is

$$B_i=\begin{cases}3.54 & \textit{freeze}\\ 17.8 & \textit{thaw}\end{cases}$$

These values will have very little effect upon the phase change depths; the maximum freeze is now 304 cm, and the thaw is 91 cm. Of course, this assumes that h is a constant for a given season, which is not correct, and that phase change starts when the air temperature is at the freezing value, also incorrect.

10.3. EFFECTIVE LATENT HEAT

The quasi-steady method (Stefan type equations) will always yield a rate of freeze that is too large due to the following three assumptions:

(a) neglect of sensible heat,
(b) the temperature gradient in the frozen region at the interface is not as steep as the straight line would indicate,
(c) neglect of the heat flow from the unfrozen to the frozen region.

One can approximately include some of these effects while maintaining the simplicity of the Stefan approach, as has been pointed out by Janson (1963)

The temperature in the frozen layer, T_1, is (see Fig. 10.18)

$$T_1=\frac{(T_f-T_s)x}{X(t)}+T_s \tag{10.24}$$

The energy balance at the interface is

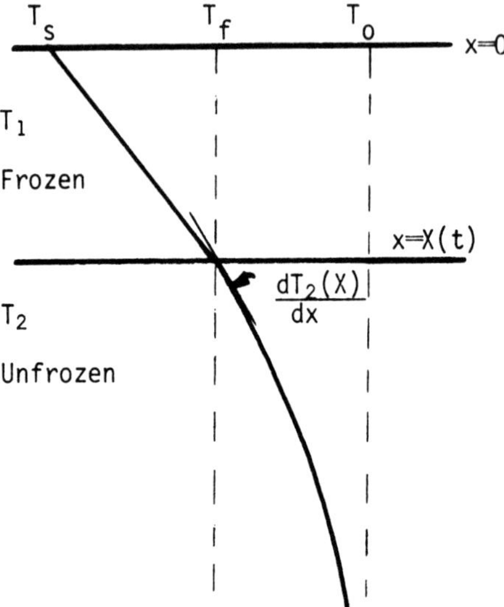

Fig. 10.18. Two-Phase Freezing.

$$A\left(k_f\frac{\partial T_1}{\partial x} - k_u\frac{\partial T_2}{\partial x}\right)_{x=X} = \frac{dE}{dT} \tag{10.25}$$

Here the energy E is considered to be made up of three components

E_1 = energy removal to cool unfrozen layer from T_o to T_f
E_2 = energy removal of the latent heat and
E_3 = energy removal to cool frozen layer (of thickness X) from T_f to $\frac{(T_s + T_f)}{2}$

Thus

$$E = \rho AXc_u(T_o - T_f) + \rho AXl + \rho AXc_f\frac{(T_f - T_s)}{2}$$

where c_f, c_u are the specific heats of the frozen and unfrozen layer. We can define an effective latent heat by

$$E = AXL_e \tag{10.26}$$

where

$$L_e = L + C_u(T_o - T_f) + C_f \frac{(T_f - T_s)}{2} \tag{10.27}$$

The equation governing the motion of the interface is then found by using Eqs. (10.24) and (10.27) in Eq. (10.25)

$$\frac{dX}{dt} = \frac{1}{L_e}\left[k_f \frac{(T_f - T_s)}{X} - k_u S_2\right] \tag{10.28}$$

where the temperature gradient in the unfrozen region at the interface is

$$S_2 = \frac{\partial T_2(X)}{\partial x}$$

The ratio of the temperature gradient at $x = X$ in the unfrozen region to the gradient in the frozen region can be expressed as

$$\left(\frac{\partial T_2/\partial x}{\partial T_1/\partial x}\right)_{x=X} = \delta \tag{10.29}$$

An estimate of the ratio of the temperature gradients on each side of the phase change interface can be obtained from the Neumann solution. Thus

$$\delta(x) = \frac{(T_o - T_f)}{(T_f - T_s)} \frac{erf\gamma}{erfc\left(\gamma\sqrt{\frac{\alpha_f}{\alpha_u}}\right)} \sqrt{\frac{\alpha_f}{\alpha_u}}\, e^{\frac{x^2}{4t}\left(\frac{1}{\alpha_f} - \frac{1}{\alpha_u}\right)} \tag{10.30}$$

At the interface, $x = X$ and $(X^2/4t) = \gamma^2\alpha_f$, then

$$\delta(X) = \frac{(T_o - T_f)}{(T_f - T_s)} \frac{erf\gamma}{erfc\left(\gamma\sqrt{\frac{\alpha_f}{\alpha_u}}\right)} \sqrt{\frac{\alpha_f}{\alpha_u}}\, e^{-\gamma^2\left(\frac{\alpha_f}{\alpha_u} - 1\right)} \tag{10.31}$$

Now if one can assume that γ is small, then

$$erf\gamma \cong \frac{2}{\sqrt{\pi}}\gamma$$

$$erfc\left(\gamma\sqrt{\frac{\alpha_f}{\alpha_u}}\right) \approx 1 - \frac{2}{\sqrt{\pi}}\gamma\sqrt{\frac{\alpha_f}{\alpha_u}}$$

and

$$\gamma = \sqrt{\frac{c_f(T_f - T_s)}{2l}} \tag{10.32}$$

These would be good approximations if $\gamma < .25$.

Using these relations, it follows that

$$\delta = \frac{(T_o - T_f)}{(T_f - T_s)} \frac{p}{(1-p)} e^{-\gamma^2\left(\frac{\alpha_f}{\alpha_u} - 1\right)} \tag{10.33}$$

where

$$p = \frac{2}{\sqrt{\pi}} \sqrt{\frac{\alpha_f}{\alpha_u}}\, \gamma$$

Equation (10.28) then becomes

$$\frac{dX}{dt} = \frac{(T_f - T_s)}{L_e X} (k_f - k_u \delta)$$

The solution to this equation is

$$X = \sqrt{\frac{2(k_f - k_u\delta)(T_f - T_s)t}{L_e}}$$

or

$$X = \sqrt{\frac{2(k_f - k_u\delta)\, I_f}{L_e}} \tag{10.34}$$

This equation now includes corrections for the sensible heat and the heat flow from the lower layer into the frozen layer, while maintaining the simplicity of the Stefan equation.

If γ, as given by Eq. (10.32), is not small, then the exact solution for γ may be obtained from Figs. 10.2–10.12 with $\gamma = \lambda\sqrt{\mu}$.

Ruckli (1950) assumed a linear temperature in the frozen soil and a parabolic distribution in the unfrozen soil and arrived at the effective latent heat of Eq. (10.27).

The U.S. Army Corps of Engineers (1949) considered the sensible heat needed to reduce the soil to the freezing level and arrived at

$$L_e = L + C_u (T_o - T_f) \tag{10.35}$$

called the "Minneapolis formula." Jumikis (1955) and Beskow (1935) used only the sensible heat below the freezing level to obtain

$$L_e = L + C_f \frac{(T_f - T_s)}{2} \tag{10.36}$$

Of these studies, only Ruckli (1950) considered the effect of conductive heat flow from the unfrozen layer.

Example 4: Take the case of freezing at Chesterfield, NWT, with the same data as Example 2. From Eq. (10.32), $\gamma = 0.352$ (the exact value is 0.287). From Eq. (10.33), $\delta = 0.343$

$$L_e = \frac{3.413(21)}{(1.8)} + 31.9 + \frac{.611(34.6)}{2(1.8)} = 42.56$$

and from Eq. (10.34)

$$X = 253.7\,\text{cm} \qquad (X = 269.4,\ \text{exact})$$

which compares well with the value in Example 2. If the exact value of γ is used in Eq. (10.34), then $X = 263.6$ cm.

Since the exact solutions for the Neumann problem have been graphed the principal value of the quasi-steady, or Stefan, solution is its application to more complex problems for which solutions are unavailable.

10.4 SURFACE *n* FACTOR

10.4.1 Definition

For a homogeneous system, we have noted that the total phase change depth can be calculated from the Stefan equation, Eq. (10.3) or, more accurately, the modified Berggren equation

$$X_B = \lambda \sqrt{\frac{2k}{L} I_s} \tag{10.14}$$

Berg (1973) showed that the modified Berggren equation gave good results in general, but was not applicable to residual thaw layers, and Sanger (1959) also noted that Eq. (10.14) would not yield good results for insulation layers.

The accurate use of Eq. (10.14) depends upon the correct value of I_s, the surface index of thawing or freezing, being used. Unfortunately, this quantity is not a standard meteorological measurement and is rarely available for a given case. It is, in fact, a variable for a given location and surface that depends upon many interrelated factors. Normally, the air index, I_a, for a location is the only available datum, and this is used with a defined quantity, the n factor, as

$$X_B = \lambda \sqrt{\frac{2k_f}{L} n\, I_a} \tag{10.15}$$

Thus the n factor is defined as

$$n = \frac{I_s}{I_a} = \frac{\int_0^{\theta_s} (T_f - T_s)\, dt}{\int_0^{\theta_a} (T_f - T_a)\, dt} \tag{10.37}$$

The definition of n is clear enough, but often there is confusion due to attempts to calculate n from predicted and measured depths of phase change. Calculated n factors for the Stefan and Berggren equations are related by

$$n_S = \lambda^2 n_B \tag{10.38}$$

These values are confusing and should be avoided. Equation (10.15) should be used with an appropriate value of n based on data or surface temperature calculations.

10.4.2 Surface Temperature

The n factor is another way of expressing the surface temperature, as the mean surface temperature, for the thaw season, can be found from

$$\overline{T}_s = T_f - n(T_f - \overline{T}_a)\frac{\theta_a}{\theta_s} \tag{10.39}$$

where θ_a, θ_s are the lengths of the air and surface temperature thaw seasons. The value of n depends directly upon the surface temperature, and any calculation of n will be a prediction of the surface temperature, which is a function

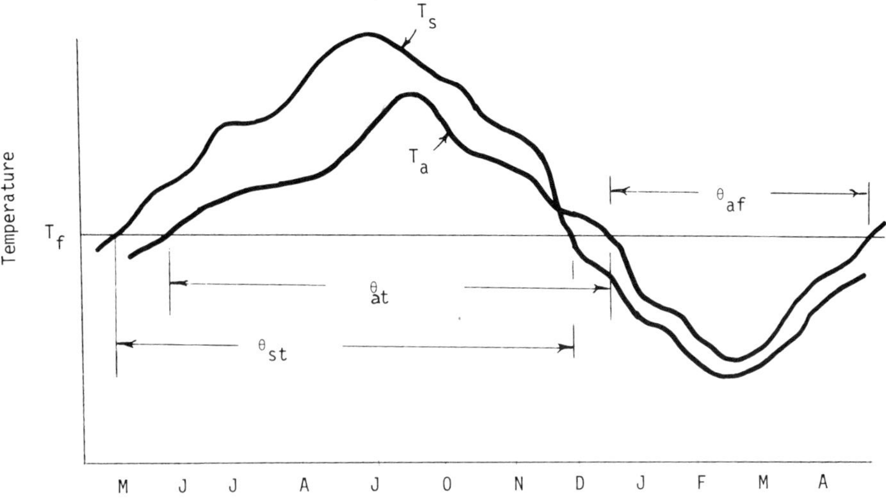

Fig. 10.19. Phase Change Seasons for Air and Surface Temperatures.

of the total heat transfer at the surface. The relations between the air and surface temperatures and the season lengths are illustrated in Fig. 10.19. The surface heat transfer can be idealized as shown in Fig. 10.20; the details are given in Chap. 6

Q_N = net radiation
Q_L = evapotranspiration
Q_H = sensible heat transfer and
Q_G = heat flow to or from soil system.

Any calculation of T_s will thus depend upon:

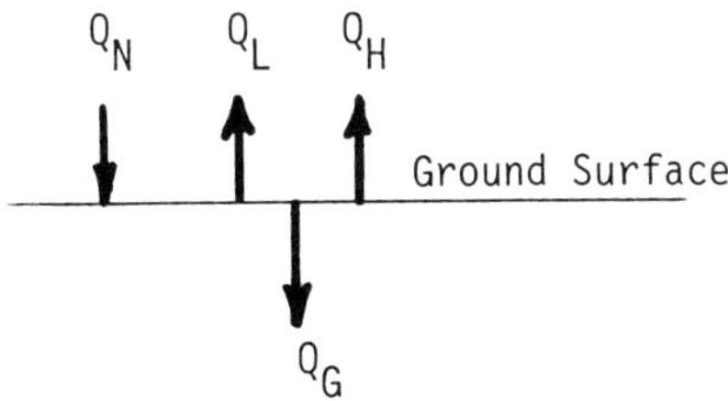

Fig. 10.20. Energy Transfer at Surface.

(i) Net shortwave radiation—latitude, season, cloud cover data.
(ii) Net longwave radiation—air temperature, relative humidity, surface temperature.
(iii) Surface layer characteristics—albedo, reflectivity, roughness, thermal resistance.
(iv) Evapotranspiration-precipitation, soil moisture.
(v) Sensible heat—wind speed, stability of atmosphere.
(vi) Soil system—thermal properties, permeability.

The average surface temperature and n factor may be expected to change significantly, even for a given surface and location, from year to year. The same can be said for different sites, surfaces, and soil systems.

10.4.3 *n* Factor Data

The majority of the experimental studies of the n factor have been for paved surfaces, which are of most significance from an engineering viewpoint. Tables 10.2 through 10.5 list the n values for general surfaces, concrete pavements, asphalt pavements, and surfaces that have been insulated at depth. Not all the values listed in these tables are experimental; some of them are calculated, but others are simply "best estimates" that have been used for design purposes. A description of the underlying soils and the meteorological data is given in Lunardini (1977) for each case.

An examination of Table 10.4 for asphalt shows that the freezing n values vary from 0.25–2.48, and the thaw values vary from 1.39–2.28. These are remarkable ranges and show that the value of n to use for a given site cannot be chosen simply from the tabulated data.

10.4.3.1 Time Variation of *n* Factor

The n factor for a given site may be expected to vary with time, from month to month and from year to year. Table 10.6 shows the variation of the thawing n value with the month for a site in Alaska. It is obvious that a seasonal n value cannot be obtained from measurements made over a period of time that is short compared to the phase change season. From data of Esch (1973), the monthly n value was calculated for Chitina, Alaska, and is shown in Table 10.7. The variation of n for different months is striking and is closely related to the net radiation values. Table 10.8 shows the freezing n values as a function of the month and different freeze seasons.

Table 10.9 shows yearly changes of the freezing n factor for Canaan Valley (asphalt) and Grafton, West Virginia (PCC). Table 10.10 shows data for

Table 10.2. *n* Factor Data, General Surfaces (from Lunardini 1978*a*). Reprinted from *Proceedings Third International Conference on Permafrost*, vol. 1.

Case	Surface type	Freeze		Thaw		Location	Symbol for figures	Reference
		n	I_a	n	I_a			
1	Spruce trees, brush, moss over peat soil	.29	5042	.37	3055	Fairbanks, Alaska	● freeze	U.S. Army (1950*a*)
2	Brush & trees cleared, moss in place, peat soil	.25	5042	.73	3055	" "	○ thaw	" "
3	Vegetation and 16 in. of soil stripped clean	.33	5042	1.22	3055	" "		" "
4	Turf	.5	–	1.0	–	Alaska and Greenland	NA	U.S. Army (1966)
5	Snow	1.0	–	–	–	" "		" "
6	Sand & gravel	.9	–	2.0	–	" "		" "
7	Gravel	.76	5042	1.99	3055	Fairbanks	∅ freeze	U.S. Army (1950*a*)
8	Gravel	.63	5042	2.01	3055	" "	⊘ thaw	" "
9	Gravel	.6	5042	1.4	3055	Fairbanks Alaska	NA	Carlson (1953)
10	Elevated building	–	–	1.0	–	" "	NA	U.S. Army (1966)
11	Pavement without snow	.9	–	–	–	" "		" "
12	Pavement north of 45°N	.9	–	–	–	General	NA	Sanger (1959)
63	Gravel	1.0	5400	1.47	2440	Chitina, Alaska	○ freeze ● thaw	Esch (1973)
64	Gravel colored dark	–	–	1.40	3320	Fairbanks	⊗	Berg (1973)
74	Sandy soil, with snow	.49	1908	–	–	Lakselv	NA	Heiersted (1975) (Norway)
75		.02	2034	–	–	Os		
76		.53	342	–	–	Amli, 1974		" "
77		1.39	234	–	–	Amli, 1975		" "
78	General	.8	–	–	–	Southern Canada $\phi < 50°N$	NA	McCormick (1971)
79	General	.9	–	–	–	Northern Canada $\Phi > 60°N$		" "
80	Gravel	–	–	1.5	2680	Fairbanks	⊘	U.S. Army (1972)
81	Gravel colored dark	–	–	1.27	2720	" "	⊗	" "

Table 10.3. *n* Factor Data, Concrete Pavements (from Lunardini 1978*a*). Reprinted from *Proceedings Third International Conference on Permafrost,* vol. 1.

Case	Freeze		Thaw		Location	Symbol for Figures	Reference
	n	I_a	n	I_a			
13	.74	5042	2.07	3055	Fairbanks	● freeze	U.S. Army (1950*a*)
14	.75	5042	1.85	3055	" "	-●- thaw	
15	.81	5042	2.02	3055	" "		
16	.85	5042	2.13	3055	" "		
17	.69	5042	2.08	3055	" "		
18	.6	5042	1.4	3055	" "	NA	Carlson (1953)
19			1.3–2.2	–	Alaska, Greenland	NA	U.S. Army (1966)
20	.87	965	–	–	Dow AFB, Maine	■	U.S. Army (1950*b*)
21	.95	2304	–	–	Presqu'Isle, Maine	△	
22	.92	1314	–	–	Sioux Falls, S. Dakota	▼	
23	.65–.67	1461	–	–	West Union, Iowa	◫	Oosterbaan (1965)
24	.66	663	–	–	Waltham, Mass	⊟	Quinn (1962)
25	.71	663	–	–	" "		
26	.35	259	–	–	1963/64 Beckeley	▶◁	Moulton (1968)
	.27	125	–	–	1964/65		
	.53	239	–	–	1965/66		(West Virginia) ↓
27	.54	206	–	–	1963/64 Benwood		
	.36	880	–	–	1964/65		
	.42	296	–	–	1965/66		
28	.27	241	–	–	1963/64 Grafton		
	.34	215	–	–	1964/65		
	.71	174	–	–	1965/66		
29	.16	86	–	–	1963/64 North		
	.12	165	–	–	1964/65 Charleston		
30	.87	278	–	–	1963/64 Pike		
	.56	146	–	–	1964/65		
	.29	295	–	–	1965/66		
31	.45	284	–	–	1964/65 Cave		
	.33	308	–	–	1965/66		
68	.79	–	1.71	–	Minnesota	NA	Braun (1957)
85	.62	1363	–	–	Hanover, New Hampshire	▲	Berg (1977)

Table 10.4. *n* Factors, Asphalt Pavements (from Lunardini 1978*a*). Reprinted from *Proceedings Third International Conference on Permafrost*, vol. 1.

Case	Freeze n	Freeze I_a	Thaw n	Thaw I_a	Location	Symbol for Figures	Reference
	.96	620	–	–	1963/64 Caanan Valley	×	Moulton (1968) ↓
32	.74	292	–	–	1964/65		
	.63	537	–	–	1965/66		
	.44	234	–	–	1963/64 Fairmont		
33	.39	136	–	–	1964/65		
	.39	235	–	–	1965/66		
34	.36	181	–	–	1963/64 Flatwoods		
	.29	166	–	–	1964/65		
	.41	277	–	–	1965/66		
35	.25	82	–	–	1963/64 Nitro		
	.27	83	–	–	1964/65		
36	.49	153	–	–	1964/65 Point Mt.		
	.49	423	–	–	1965/66		
37	.79	282	–	–	1963/64 Slatyfork		
38	.35	514	–	–	1963/64 Sophia		
	.29	165	–	–	1964/65		
	.41	462	–	–	1965/66		
39			1.6–2.7	–	Alaska, Greenland	NA	U.S. Army (1966)
40	.55	1314	–	–	Sioux Falls, S.D.	▽	U.S. Army (1950*b*)
41	.75	965	–	–	Dow, AFB, Maine	□	
42	.78	2304	–	–	Presqu'Isle, Maine	+	
43	.78	5042	2.11	3055	Fairbanks	○ freeze	U.S. Army (1950*a*)
44	.65	5042	2.28	3055	" "	○ thaw	
45	.72	5042	2.15	3055	" "		
46	.8	923	–	–	Minnesota	NA	Kersten (1955)
47	.6	5042	1.4	3055	Alaska	NA	Carlson (1953)
48	.55	580	–	–	Indiana	NA	Yoder (1952)
49	.58	2600	–	–	Sudbury, Ont.		Penner (1966)
62	1.0	5400	1.73	2440	Chitina, Alaska	⊖ freeze ⊖ thaw	Esch (1973)

Table 10.4. Cont.

Case	Freeze		Thaw		Location	Symbol for Figures	Reference
	n	I_a	n	I_a			
66	–	–	1.96	3320	Fairbanks	⦵	Berg (1973)
67[a]			.98	3320	Fairbanks	○	Berg (1973)
69	.74	–	1.91	–	Minnesota	NA	Braun (1957)
70	1.26	1908	–	–	Lakselv	◐	Heiersted (1975)
71	1.02	2034	–	–	Os	⦶	
72	2.48	342	–	–	Amli, 1974	⊗	(Norway)
73	1.90	234	–	–	Amli, 1975		
82	–	–	1.58	2720	Fairbanks	○	U.S. Army (1972)
83	–	–	1.39	2720	" "		
84*	–	–	1.15	2720	" "	○	

[a] Asphalt painted white.

Table 10.5. *n* Factor Data, Insulated Systems (from Lunardini 1978*a*). Reprinted from *Proceedings Third International Conference on Permafrost,* vol. 1.

Case	Surface Type	Freeze		Thaw		Location	Symbols for Figures	Reference
		n	I_a	n	I_a			
50	Asphaltic concrete	.68	2600	–	–	Sudbury, Ont.	△	Penner (1966)
51	PCC	.97	1302	–	–	Midland, Mich.	⬓	Oosterbaan (1965)
52	PCC	1.0	1261	–	–	" "		" "
53	PCC	.78	1461	–	–	West Union, Iowa	◨	" "
54	PCC	.98	663	–	–	Waltham, Mass.	⬒	Quinn (1962)
55	Asphalt	.96	7279	1.74	2052	Kotzebue	◑ freeze	Rhode (1976)
56	" "	1.04	6549	1.66	2386	" "	◑ thaw	" "
57	" "	.99	5677	–	–	" "		" "
58	" "	.95	7279	2.00	2052	" "		" "
59	" "	1.01	6549	1.94	2386	" "		" "
60	Asphalt	1.0	5400	1.84	2440	Chitina, Alaska	◒ freeze ◒ thaw	Esch (1973)
61	Gravel	.7	5400	1.38	2440	" "	⊙ freeze ⊗ thaw	" "
65	Asphalt			1.72	3320	Fairbanks	⊗	Berg (1973)
86	PCC	.65	1363	–	–	Hanover, N.H.	△	Berg (1977)

Table 10.6. Thawing *n* Values, Fairbanks, Alaska, 1971 (Berg 1973). Reproduced from *Proceedings Second International Conference on Permafrost,* 1973, p. 585, with permission of National Academy of Sciences.

Surface	May	June	July	Aug.	Sept.	Wt. Average
Asphalt painted white	–	1.25[a]	1.11	1.03	.76	.98
Asphalt control	2.7[a]	1.88[a]	1.85	1.79	1.74	1.96
Asphalt with peat insulation	2.28[a]	1.75[a]	1.73	1.61	1.44	1.72
Dark gravel	–	1.79[a]	1.52	1.39	1.15	1.40

[a] Only part of month.

Table 10.7. Thawing *n* Values, Chitina, 1972.

Surface	May	June	July	Aug.	Sept.
Asphalt, insulated	2.0	2.05	1.55	1.69	2.04
Gravel	–	1.53	1.20	1.52	1.49

Table 10.8. Freezing *n* Factor, site near Hanover, New Hampshire, U.S.A. (Lobacz 1977).

Surface	Dec. 1973	Jan. 1974	Feb. 1974	1973–74	Jan. 1975	Feb. 1975	1974–75
Conventional CE asphaltic concrete	.34	.84	.53	.57	.86	.58	.67
Fill depth asphaltic concrete	.23	.60	.44	.43	.67	.48	.52
"	.23	.61	.48	.43	–	–	–

Kotzebue, Alaska. The longer the phase change season, the smaller the yearly variation in the value of *n*.

10.4.4 Theoretical *n* Factors

10.4.4.1 Thaw Case

Actually there is no significant difference between thawing and freezing, but the thaw case will be presented here. Lunardini (1978*a*) has shown that an approximate equation can be derived for the *n* factor, as follows. The surface

Table 10.9. Freezing n Values, West Virginia (Moulton 1968).

Surface	1963–64	1964–65	1965–66
Asphalt	.96	.74	.63
PCC	.27	.34	.71

Table 10.10. Variation of n Factors with Time, Kotzebue, Alaska (Rhode 1976).

	Freeze			Thaw	
Surface	70–71	71–72	72–73	1971	1972
Asphalt, insulated	.96	1.04	.99	1.74	1.66
Asphalt, insulated	.95	1.01	–	2.00	1.94

temperature and rate of thaw have been derived in Section 8.4.4.2 and are given by the following for a homogeneous soil system

$$T_s = T_e - \frac{(T_e - T_f)}{p}\frac{k_t}{h} \tag{10.40}$$

$$p^2 = \frac{2k_t}{L}\int_o^{\theta}(T_e - T_f)\,d\theta + \left[\frac{k_t}{h}\right]^2 \tag{10.41}$$

$$T_e = \frac{q}{h} + T_a \tag{10.42}$$

where

T_e = equivalent temperature related to the net, nonconvective heat flux at surface

q = net surface heat flux excluding convection and

θ = time.

The quantities T_s, T_e, p, h, and q are all functions of time and must be known if the n factor is to be evaluated.

Over the thaw season, a seasonal average value of these variables will be used according to the formula

$$\bar{f} = \frac{1}{\theta_s}\int_o^{\theta_s} f\,d\theta \tag{10.43}$$

where θ_s is the length of the thaw season. The averages are computed over the length of the surface thaw season, θ_s, as this is the significant time for the heat transfer to occur. Then

$$T_s = \overline{T}_e - \frac{(\overline{T}_e - T_f)}{p}\, r \tag{10.44}$$

$$p = \sqrt{b\theta + r^2} \tag{10.45}$$

where

$$r = \frac{k_t}{h}$$

$$b = \frac{2k_t}{L}(\overline{T}_e - T_f)$$

The surface index I_s may be calculated as

$$I_s = \int_0^{\theta_s} (T_s - T_f)\, d\theta \tag{10.46}$$

Using Eqs. (10.44) and (10.45), this leads to

$$I_s = (\overline{T}_e - T_f)\,\theta_s - \frac{rL}{k_t}\left(\sqrt{r^2 + b\theta_s} - r\right) \tag{10.47}$$

Now

$$I_e = \int_0^{\theta_s} (T_e - T_f)\, d\theta = (\overline{T}_e - T_f)\,\theta_s$$

and

$$I_s = I_e - m\left[\sqrt{1 + \frac{2I_e}{m}} - 1\right] \tag{10.48}$$

where

$$m = \frac{Lk_t}{h^2}$$

Now divide Eq. (10.48) by m, giving

$$\frac{I_s}{m} = \frac{I_e}{m} - \left[\sqrt{1 + \frac{2I_e}{m}} - 1\right] \tag{10.49}$$

noting that

$$\frac{I_s}{m} = \frac{nI_a}{m}$$

Thus, a nondimensional equation for n is

$$\pi_1 = \pi_2 - (\sqrt{1 + 2\pi_2} - 1) \tag{10.50}$$

where

$$\pi_1 = \frac{nI_a}{m}$$

$$\pi_2 = \frac{I_e}{m}$$

The index I_e can also be expressed as

$$I_e = \left[\left(\frac{\bar{q}}{\bar{h}}\right) + \bar{T}_{as} - T_f\right]\theta_s = \left(\frac{\bar{q}}{\bar{h}}\right)\theta_s + I_{as} \tag{10.51}$$

Note that I_{as} is used, not the air thawing index I_a. It is very likely that I_{as} is approximately equal to I_a, and thus I_a could be used if no information is available on I_{as}. Now $(\bar{q}\theta_s)$ is the total heat transfer at the surface, exclusive of convection, during the thaw season and will be denoted as Q. Then π_2 can also be written as

$$\pi_2 = \frac{Q}{m\bar{h}} + \frac{I_{as}}{m}$$

Using the fact that the net heat flow into the ground must equal the energy required to thaw the soil to a depth d, and the sensible heat associated with the soil temperature increase, it is possible to obtain the following relation between π_1 and π_2

$$\pi_2 = \pi_1 + \frac{d\bar{h}}{k_t} \tag{10.52}$$

This gives a remarkably simple relation between π_1 and π_2, which avoids the difficulty of calculating $\bar{q}$. It is necessary, however, to know the thaw depth beforehand in order to use Eq. (10.52). The quantity $d\bar{h}/k_t$ has the form of the well-known Biot number, but it lacks the same thermal significance.

10.4.4.2 Freeze Case

The details on freezing are similar to thawing, with a few changes in definitions. We use the same definitions of $\overline{T}_e$, $\overline{T}_a$, $\bar{h}$, $\bar{q}$, but

$$I_e = (T_f - \overline{T}_e)\theta_s$$
$$I_{as} = (T_f - \overline{T}_a)\theta_s$$
$$I_s = (T_f - \overline{T}_s)\theta_s$$

We also have

$$T_s = \overline{T}_e + \frac{T_f - \overline{T}_e}{p}\frac{k_f}{h}$$

$$p = \sqrt{b\theta + r^2}$$

$$r = \frac{k_f}{h} \qquad b = \frac{2k_f}{L}(T_f - \overline{T}_e)$$

$$m = \frac{k_f L}{h^2}$$

With these relations, we again arrive at Eq. (10.50) and the same definitions of π_1 and π_2, although

$$\pi_2 = \frac{-\frac{\bar{q}}{h}\theta_s + I_{as}}{m}$$

Equation (10.52) remains unchanged for the freezing case.

10.4.4.3 Effect of Wind Speed on *n* Factor

Equations (10.49) and (10.51) can be combined to give

$$n = \frac{\delta}{I_a} - \frac{m}{I_a}\left(\sqrt{1 + \frac{2\delta}{m}} - 1\right) \tag{10.53}$$

where

$$\delta = \begin{cases} I_a - \dfrac{Q_f}{h} & \text{freeze} \\ I_a + \dfrac{Q_t}{h} & \text{thaw} \end{cases}$$

The variation of n with $\bar{u}$, the average wind speed, can be evaluated from this equation. Sanger (1963) noted that for arctic and subarctic regions the thaw n factor decreases with increasing wind speed, and the U.S. Army (1966) results noted the inverse for the freezing n factor. The following values, typical for the northern locations listed, were used to evaluate these trends:

Surface and Location	Q Btu/ft²-season	I_a °F-day
Asphalt, freeze, Fairbanks	−8330	5042
Asphalt, freeze, Chitina	12,500	5400
Asphalt, thaw, Fairbanks	100,500	2985
PCC, thaw, Fairbanks	68,000	3055

The surface coefficient is given by the following relations for northern regions, taken from Chap. 6

$$h = \begin{cases} 12.63\,\bar{u} & \text{thaw season} \\ 3.51\,\bar{u} & \text{freeze season} \end{cases}$$

and Lk_f was taken as 70,000 Btu²/ft⁴-day-°F. Based on these values and Eq. (10.53), the variation of n with $\bar{u}$ is shown in Table 10.11.

Table 10.11. Variation of n Factor with Wind Speed.

$\bar{u}$ Wind speed, mph	Thaw: Asphalt	Thaw: PCC	Freeze: Asphalt, Fairbanks	Freeze: Asphalt, Chitina
2	1.96	1.55	.64	.28
6	1.34	1.20	.85	.69
10	1.21	1.12	.90	.80
16	1.13	1.08	.94	.87

Berg and Quinn (1976) have plotted the difference in surface and air temperatures versus the absorbed shortwave radiation, with wind speed as a parameter, for Fairbanks, Alaska. The data show that the temperature difference for thawing will decrease with increasing wind speed. Information of this kind can then be used to evaluate ΔT or n for different sites if the absorbed shortwave radiation and wind speed are known. Their method is an approximation to the more comprehensive method discussed here.

The simple theory tends to explain the general trends for n, based on data or field experience. One general observation (U.S. Army 1966), that n_f increases with latitude, is not necessarily correct, however. Both the theory and data confirm this.

10.4.5 Correlation of n Factor Data

Equation (10.50) indicates that it should be possible to correlate the n factor data with the use of only two parameters. This has been done for the data in Tables 10.2 to 10.5, plotted as Fig. 10.21, taken from Lunardini (1978*b*). The plot tends to corroborate the theoretical prediction.

Figure 10.22 is a plot of the data on the I_s, δ coordinates. The data

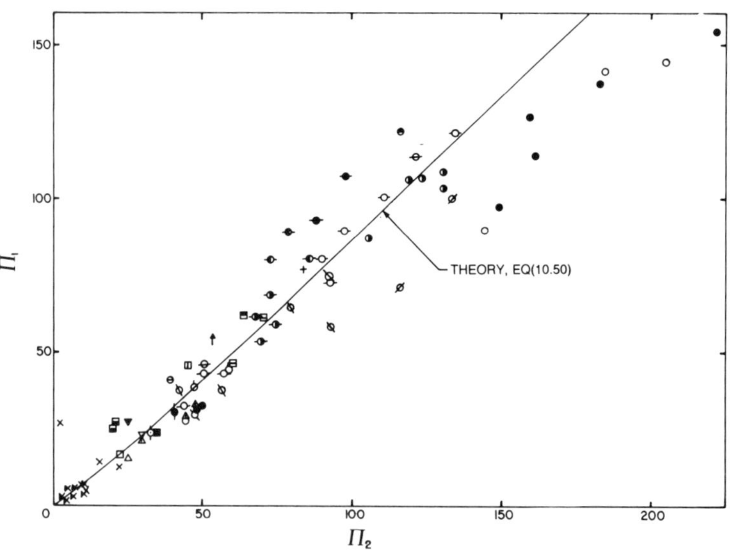

Fig. 10.21. Dimensionless Correlation of Data (from Lunardini 1978*a*). Reprinted from *Proceedings Third International Conference on Permafrost,* vol. 1.

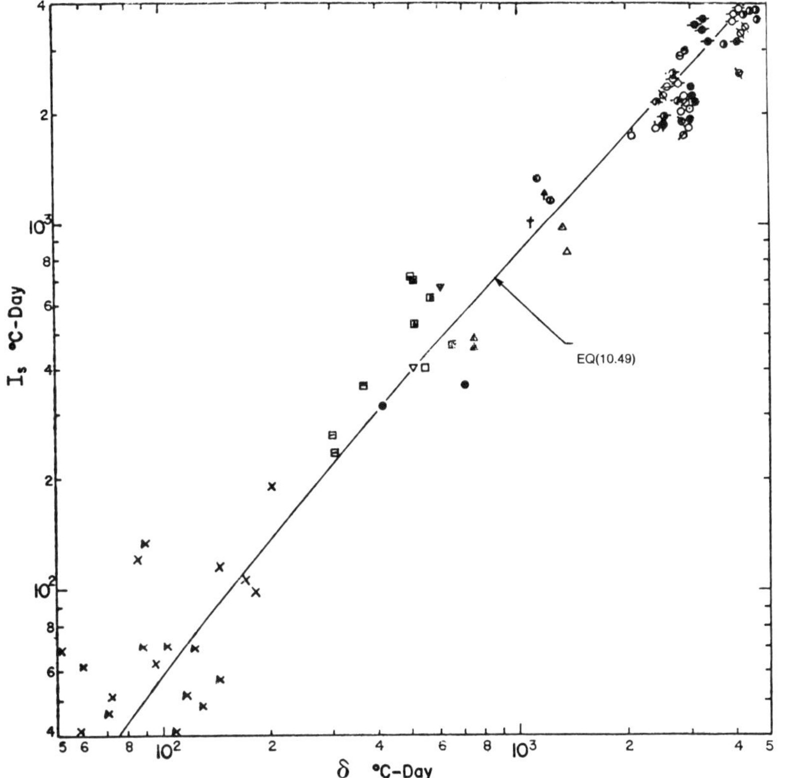

Fig. 10.22. Surface Index versus δ, All Surfaces (from Lunardini 1978*a*). Reprinted from *Proceedings Third International Conference on Permafrost,* vol. 1.

now exhibit a manageable trend, and values of I_s or n can be obtained for a site with some confidence. The surface coefficient will vary considerably from site to site, but a value of $h = 50$ Btu/day-ft^{2}-°F was chosen as a representative value.

The value of m is not known for many of the cases, but a value of 28 is quite reasonable, using an average value of kL from available data. These values were used for the theoretical curve plotted on Fig. 10.22.

Finally, Fig. 10.23 shows another plot of the data with I_s versus I_a. This correlation is surprisingly good, but the data tend to scatter more as I_a increases. Of course this plot is suggested from the theory, where I_s is certainly a function of I_a, but not a simple function.

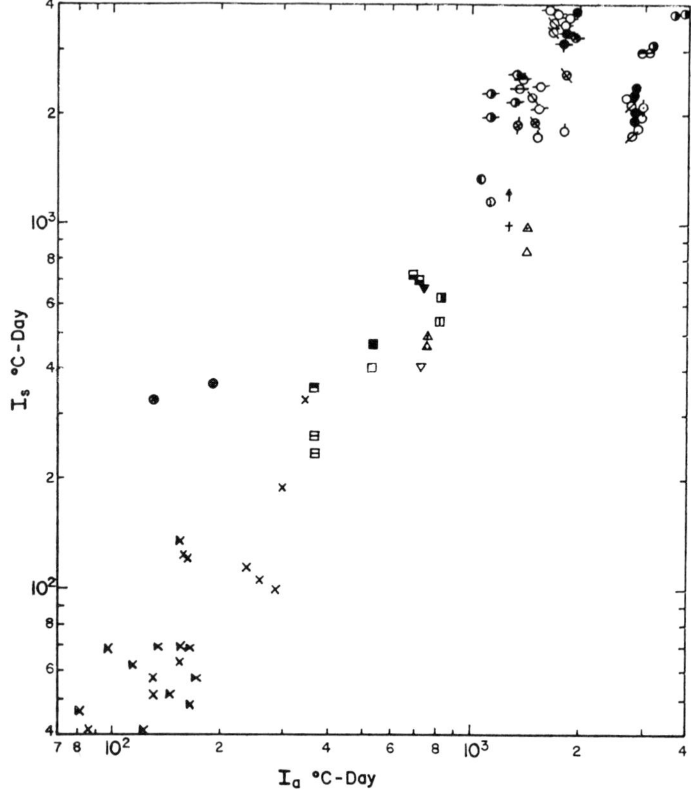

Fig. 10.23. Surface Index versus Air Index (from Lunardini 1978*a*). Reprinted from *Proceedings Third International Conference on Permafrost,* vol. 1.

10.5 FREEZING AND THAWING IN MULTILAYER SYSTEMS, DEGREE-DAY METHOD

10.5.1 Modified Berggren Equation Concept

The quasi-steady approach can be used for more complicated systems, such as multilayer soils. The quasi-steady method will not include sensible heat or the effect of the thawed layer, but they can be included in the same way as in the modified Berggren equation. The calculation is simplest with the use of λ, as has been seen already. Consider the system shown in Fig. 10.24, which consists of two layers. Layer 1 is already frozen (it may contain no water), and layer 2 is partially frozen. The location of the phase change

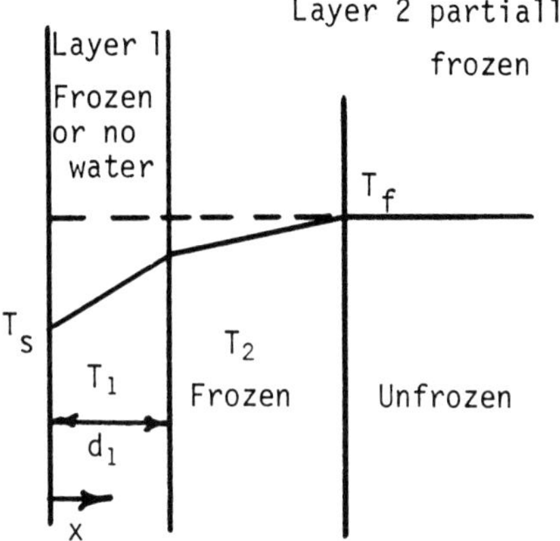

Fig. 10.24. Multilayer Freeze.

surface is given as usual by $x = X$. Let $v = (T_f - T)$. The simplified equations governing this system are as follows

$$\frac{\partial^2 v_1}{\partial x^2} = 0 \tag{10.54}$$

$$\frac{\partial^2 v_2}{\partial x^2} = 0 \tag{10.55}$$

The boundary conditions are

$$v_1 = v_s \qquad x = 0 \tag{10.56}$$

$$v_2 = 0 \qquad x = X \tag{10.57}$$

At the interface between layers 1 and 2 ($x = d_1$), the temperatures must be equal, and the heat fluxes must be equal, as there is no phase change here

$$k_1 \frac{\partial v_1}{\partial x} = k_{2f} \frac{\partial v_2}{\partial x} \tag{10.58}$$

$$v_1 = v_2 \tag{10.59}$$

Subscript f on the properties of region 2 refers to the frozen condition. The solutions satisfying Eqs. (10.54) to (10.56) are

$$v_1 = Ax + v_s \tag{10.60}$$

$$v_2 = Cx + D \tag{10.61}$$

These equations contain three unknowns A, C, and D which can be found using Eqs. (10.57) through (10.59). One has

$$A = \frac{k_{2f}}{k_1} C$$

$$C = \frac{-v_s}{\left(\frac{k_{2f}}{k_1} - 1\right) d_1 + X}$$

$$D = \frac{v_s X}{\left(\frac{k_{2f}}{k_1} - 1\right) d_1 + X}$$

It should be noted that A, C, and D are not constants, but functions of time. The temperature in region 2 is then

$$v_2 = \frac{-v_s}{\left(\frac{k_{2f}}{k_1} - 1\right) d_1 + X} (x - X) \tag{10.62}$$

or

$$T_2 = T_f - \frac{(T_f - T_s)(X - x)}{\left(\frac{k_{2f}}{k_1} - 1\right) d_1 + X} \tag{10.63}$$

where $d_1 < x < X$.

To proceed further, we must find a relation for X. The energy balance at the phase change interface, $x = X$, is

$$-k_{2f} \frac{\partial v_2}{\partial x} = \rho_2 l_2 \frac{dX}{dt} \tag{10.64}$$

Now substituting Eq. (10.62) into Eq. (10.64) yields

$$\frac{dX}{dt} = \frac{-a}{b + X} \tag{10.65}$$

where

$$a = \frac{k_{2f} v_s}{\rho_2 l_2}$$

$$b = \left(\frac{k_{2f}}{k_1} - 1\right) d_1$$

Equation (10.65) may be integrated over the frozen layer only

$$\int_{d_1}^{X} (b + X)\, dX = -\int_{o}^{t} a\, dt \tag{10.66}$$

One has

$$b(X - d_1) + \frac{1}{2}(X + d_1)(X - d_1) = \frac{k_{2f}}{\rho_2 l_2}(v_s t) \tag{10.67}$$

Earlier it was noted that $v_s t$ is related to the freezing (or thawing) index I by $v_s t = I_2 \lambda^2$. This use of λ allows a correction factor for sensible heat, etc., not taken into account by the quasi-steady calculations. Using the relation $(X - d_1) = d_2$

$$b\, d_2 + \frac{1}{2} d_2(2d_1 + d_2) = \frac{k_{2f}\lambda^2 I_2}{\rho_2 l_2} \tag{10.68}$$

Here we have assumed that region 2 has a thickness of d_2 when it is completely frozen.

Let us now introduce the thermal resistance of a layer

$$R = \frac{d}{k} \tag{10.69}$$

Then, $b = k_{2f} R_1 - d_1$, and Eq. (10.68) may be written as

$$I_2 = \frac{L_2\, d_2}{\bar{\lambda}^2}\left(R_1 + \frac{R_2}{2}\right) \tag{10.70}$$

where $\bar{\lambda}$ = average value of λ for all layers including the frozen part of 2. Equation (10.70) gives the degree-days required to move the freezing interface through the second layer only. This may be generalized for any layer n as

$$I_n = \frac{L_n d_n}{\bar{\lambda}^2} \sum_{i=1}^{n-1} \left(R_i + \frac{R_n}{2}\right) \tag{10.71}$$

where

$$\bar{\lambda} = \text{weighted average for layers through } n \text{ and}$$

$$\sum_{i=1}^{n-1} R_i = \text{thermal resistance of the layers above } n.$$

Equation (10.71) may be written for specific units as

$$I_n = \frac{0.75\, L_n\, d_n}{\bar{\lambda}^2} \sum_{i-1}^{n-1} \left(R_i + \frac{R_n}{2} \right)$$

where

$I =$ °F-day
$L =$ cal/cm^3
$d =$ cm and
$R =$ hr-m^2-°C/kcal

Equation (10.71) is a modification of the Neumann equation for layered systems, which may be called the modified Berggren equation for layered systems. Aldrich and Paynter (1953) recommended a similar set of equations

$$X = \lambda \sqrt{\frac{2I}{\left(\frac{L}{k}\right)_e}} \qquad (10.72)$$

where

$$\left(\frac{L}{k}\right)_e = \frac{2}{X^2} \left[R_1 \left(\frac{L_1 d_1}{2} + L_2 d_2 + \ldots L_n\, d_n \right) + R_2 \left(\frac{L_2 d_2}{2} + L_3 d_3 + \ldots L_n\, d_n \right) + \ldots + R_n \frac{L_n d_n}{2} \right] \qquad (10.73)$$

$$X = d_1 + d_2 + \ldots d_n \qquad (10.74)$$

The coefficient λ is calculated using

$$\bar{C} = \frac{\Sigma\, C_i d_i}{X}, \qquad \bar{L} = \frac{\Sigma\, L_i d_i}{X}$$

with all properties at the mean of the frozen and unfrozen values.

In practice a guess is made of X; $\left(\frac{L}{k}\right)_e$ is calculated from Eq. (10.73), λ is found, and X is calculated from Eq. (10.72). Iteration is used until the initial and final values of X are sufficiently close.

The U.S. Army Corps of Engineers (1966) uses a simpler version, which is essentially identical to Eq. (10.71) with the average of the frozen and unfrozen properties and λ calculated from

$$\overline{C} = \frac{\Sigma\, C_i d_i}{\Sigma\, d_i} \qquad \text{etc.}$$

10.5.2 Stefan Equation Basis

If the sensible heat effect is ignored, then $\lambda \equiv 1$, and Eq. (10.71) reduces to

$$I_1 = \frac{L_1 d_1 R_1}{2}$$

$$I_n = L_n\, d_n \sum_{i=1}^{n-1} \left(R_i + \frac{R_n}{2} \right) \tag{10.75}$$

These results are the Stefan equation modifications for layered systems and are referred to in the literature as the "St. Paul (Minnesota) equations" (Kersten 1959).

10.5.3 Layer of Soil Overlain by Insulating Layer

Consider the depth of penetration of the freezing isotherm through a layer of soil overlain by layers of material (soil, building materials, etc.).

From Eq. (10.70)

$$I = \frac{L\,X}{\overline{\lambda}^2} \left(R_1 + \frac{R}{2} \right) \tag{10.76}$$

where

I = degree-days required to move through layer X
L, k, R = refer to properties of soil layer and
R_1 = thermal resistance of overlayer.

Note from Fig. 10.25 that X is measured from the bottom of the upper layer, not from the surface itself. Now the thermal resistance may be expressed as

$$R=\frac{X}{k}$$

Substituting this into Eq. (10.76) and solving for X, one has

$$X=k\,R_1\left(\sqrt{1+\frac{2\bar{\lambda}^2 I}{kL\,R_1^2}}-1\right) \tag{10.77}$$

or

$$X=\frac{k}{k_1}\,d_1\left(\sqrt{1+\frac{267\bar{\lambda}^2 I\,k_1}{\frac{k}{k_1}\,L\,d_1^2}}-1\right)$$

where

$k=$ kcal/(hr-m-°C)
$d, X=$ cm
$I=$ °F-day and
$L=$ cal/cm³.

This equation is useful in calculating the depth of thaw or freeze beneath buildings, roads, etc.

The temperature in the first layer is

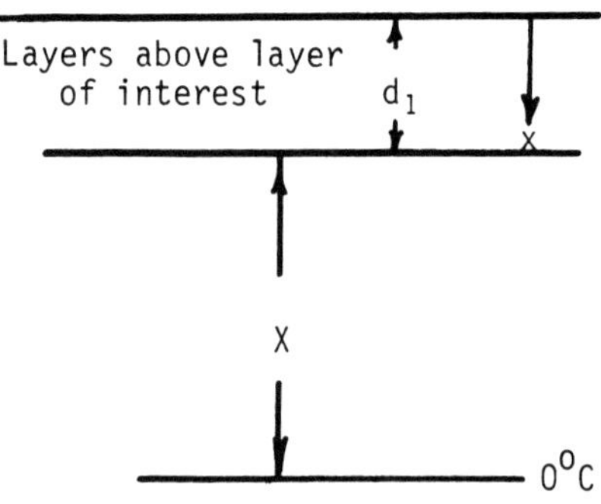

Fig. 10.25. Insulation Layer.

$$T_1 = T_s + \frac{(T_f - T_s)\,x}{d_1 + \dfrac{k_1}{k_{2f}} X} \tag{10.78}$$

and the temperature in the soil is

$$T_2 = T_f - \frac{(T_f - T_s)(d_1 + X - x)}{\dfrac{k_{2f}}{k_1} d_1 + X} \tag{10.79}$$

These equations are also valid for the thaw case, except that the subscripts refer to the properties of the unfrozen material.

Example 5: Consider an airstrip at Thule Air Base, Greenland, having these characteristics: mean annual air temperature T_m = 12°F; I = thawing index = 1560°F-days; length of thaw season = 105 days. Find the depth of thaw expected in the layered system below:

Depth, ft	Field data Material	γ_d, lb_m/ft^3	W, %	Layer
0	AC[a]	–	–	
.4	AC	–	–	a
1	GW–GP	155	2.4	b
2	GP	157	1.8	b
3	GW	151	3.2	c
4	GW	151	3.2	c
5	GP	152	2.1	c
6	SM	136	6.5	d
7	SM	144	4.6	e
8	SC	143	4.6	e
9	SM–GM	140	2.8	e

[a] Asphalt.

The following table gives the calculation procedure:

Layer	d, cm	γ_d	W	k_u	C_u	L	ΣCd	Σd	$\bar{C}$	ΣdL
a	12.2	–	0	1.3	.45	0	5.49	12.2	.45	0
b	48.8	2.5	2.1	3	.46	4.18	27.94	61	.46	204
c	91.4	2.42	2.83	3	.45	5.46	69.1	152.4	.45	703.1
d	30.5	2.2	6.5	2.8	.50	11.4	84.32	182.0	.46	1050.7
e_1	23	2.3	4.6	2.9	.48	8.43	95.36	205.9	.46	1244.6
e_2	25			2.9	.48	8.43	–	–	.46	–

Layer	$\bar{L}$	μ	x_l	λ	R	ΣR	$\Sigma\left(R_i + \frac{R_m}{2}\right)$	d	I_n	ΣI_n
a	0	–	0	–	.094	.094	–	12.2	0	0
b	3.34	.57	.053	.465	.1627	.2565	.1754	48.8	124.1	124.1
c	4.61	.403	.07	.52	.3047	.5612	.4089	91.4	566.0	690.1
d	5.75	.330	.14	.56	.1089	.6701	.6157	30.5	512.0	1202.1
e_1	6.05	.314	.11	.56	.0793	.7494	.7098	23	329.1	1531.2 Low
e_2	6.05	.314	.11	.56	.0862	–	.7132	25	359.5	1561.6 OK

∴ Thaw depth = 207.9 cm = 6.82 ft. The observed value is 6.5 ft.

$[k] = \dfrac{\text{kcal}}{\text{hr-m-°C}}$

$[C] = \text{cal/cm}^3\text{-°C}$

$[L] = \text{cal/cm}^3$

$[R] = \dfrac{\text{hr-m}^2\ \text{°C}}{\text{kcal}}$

$[I] = \text{°F-day}$

$[\gamma_d] = \text{g/cm}^3$

The initial temperature of the ground may be taken as the mean annual air temperature

$$T_o = T_a = -11.1\text{°C}$$

$$T_s - T_f = \frac{1560}{105} = 14.86\text{°F} \therefore T_s = 8.26\text{°C}$$

$$\alpha = \frac{T_f - T_o}{T_s - T_f} = 1.35$$

$$\bar{\mu} = \frac{\bar{C}}{2\bar{L}}(T_s - T_f) = 4.13\frac{\bar{C}}{\bar{L}}$$

$$I_n = .75\frac{L_n d_n}{\bar{\lambda}^2}\Sigma\left(R_i + \frac{R_n}{2}\right)$$

$$\bar{C} = \frac{\Sigma Cd}{\Sigma d} \qquad \bar{L} = \frac{\Sigma Ld}{\Sigma d}$$

$$x_l = \frac{W\gamma_d}{100}$$

10.5.4 Freezing (Thawing) of a Two-Layer System

Assume there are two soil layers, both of which have latent heat. A solution can be obtained with the quasi-steady approximation (Fig. 10.26). The solution has already been given for the time until layer 1 freezes (see Section 8.3.1.2.).

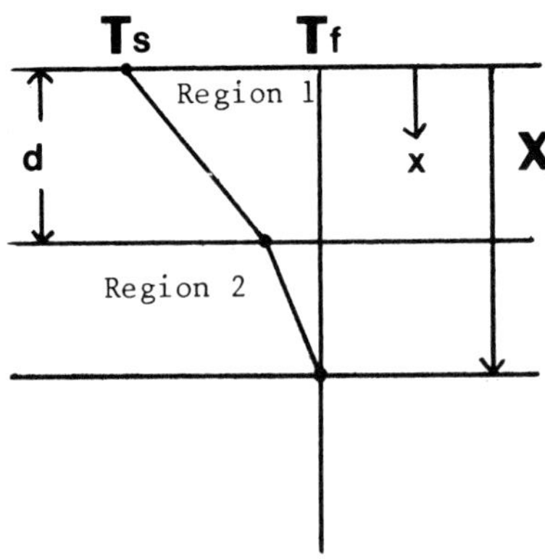

Fig. 10.26. Freezing of Two-Layer System.

$$\left.\begin{aligned} T_1 &= (T_f - T_s)\frac{x}{X}\,T_s \\ X &= \sqrt{\frac{2k_1}{\rho_1 l_1}(T_f - T_s)t} \end{aligned}\right\} t \leq t_o$$

The time t_o when layer 1 is frozen is

$$t_o = \frac{\rho_1 l_1\, d^2}{2k_1\,(T_f - T_s)} \tag{10.80}$$

For time $t > t_o$, two regions must be considered

$$\frac{d^2T_1}{dx^2} = 0 \tag{10.81}$$

$$T_1(0) = T_s \tag{10.81a}$$

$$\left.\begin{aligned} k_1\frac{dT_1}{dx} &= k_2\frac{dT_2}{dx} \\ T_1 &= T_2 \end{aligned}\right\} x = d \tag{10.81b}$$

$$\frac{d^2T_2}{dx^2} = 0 \tag{10.82}$$

$$\left.\begin{aligned} k_2\frac{dT_2}{dx} &= \rho_2 l_2\frac{dx}{dt} \\ T_2 &= T_f \end{aligned}\right\} x = X \tag{10.82a}$$

The solution to these equations is straightforward and

$$X = \sqrt{d^2k_{21}^2 + 2\,k_2\frac{(T_f - T_s)(t - t_o)}{\rho_2 l_2}} - d(k_{21} - 1) \tag{10.83}$$

where

$$k_{21} = k_2/k_1$$

$$T_1 = \frac{k_{21}\,(T_f - T_s)}{(k_{21} - 1)\,d + X}\,x + T_s \tag{10.84}$$

$$T_2 = \frac{(T_f - T_s)\,(x - X)}{(k_{21} - 1)\,d + X} + T_f \tag{10.85}$$

Notice that Eq. (10.83) reduces to Eq. (10.77), with $\lambda \equiv 1$, if layer 1 has zero latent heat ($t_0 \equiv 0$), taking account that X is measured from the bottom of the insulating layer in Eq. (10.77).

10.6. BURIED PIPES

With the discovery of significant fossil fuels in the north came the immediate problem of the best means of moving these vital fuels to the southern population. No method of transporting oil and, especially, gas is as convenient as pipelines. The introduction of a hot pipeline into permafrost will lead to considerable thawing around the pipe, however, with a distinct possibility of pipeline failure due to excess settlement or flow of the thawed soil. The rate and magnitude of the thaw beneath a hot pipe have been studied extensively, but, unfortunately, the nonlinear nature of the problem precludes an exact solution at this time. Fortunately, the quasi-steady method can be used with reasonable accuracy. There is always the possibility of using numerical techniques, as will be discussed later. The available theoretical solutions have been given in Sections 8.6 and 8.7.

10.6.1 Infinite System

No exact solutions are available for cylindrical coordinates, even when the pipe is buried at an infinite depth. Valuable approximations are obtainable, however, if the quasi-steady approximation is used. All the solutions use the following energy balance at the phase change interface

$$\int_{\substack{\text{phase change}\\ \text{interface}}} \left[\left(k\frac{\partial T}{\partial n}\right)_t - \left(k\frac{\partial T}{\partial n}\right)_f\right] ds = L\frac{dA}{dt} \tag{10.86}$$

where the derivative $\partial/\partial n$ is with respect to the outward normal of the boundary, ds is an element of length along the phase change boundary, and A is the area within the boundary where phase change has occurred.

In addition some useful approximations derived from the heat balance integral method will be presented.

10.6.1.1 Zero Sensible Heat

Carslaw and Jaeger (1959) present a solution when the surrounding soil is at the phase change temperature T_f (Fig. 10.27).

The solution is given by Eq. (8.237)

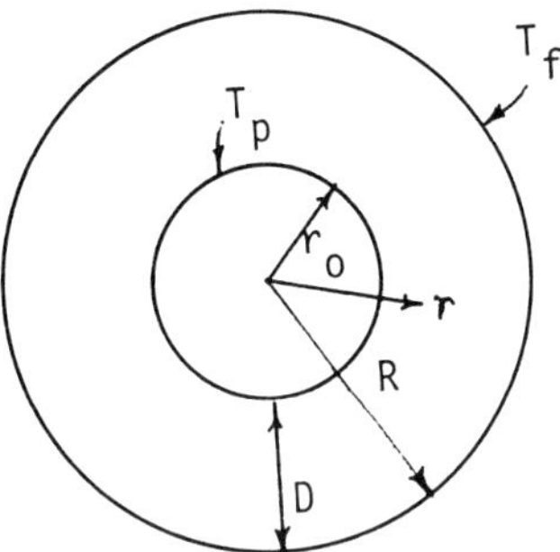

Fig. 10.27. Thaw around Buried Pipe.

$$2R^2 \ln\left(\frac{R}{r_o}\right) - R^2 + r_o^2 = 4\frac{k_t\,(T_p - T_f)}{L}\,t \tag{10.87}$$

The thaw depth beneath the pipe is shown in Fig. 10.28. Equation (10.87) is the curve for $S_T = 0$. The solution, taking into account the sensible heat, is given in Section 8.6.1.3 and was used to calculate the thaw depth for the values of the Stefan number on Fig. 10.28. Equation (10.87) will overestimate the thaw beneath the pipe, as all the energy transfer from the pipe will go into thawing. A correction can be made by using the effective latent heat in place of L, as noted in Section 10.3.

10.6.1.2 Effect of Superheat

A simple quasi-steady approximation was derived in Section 8.6.2.1. The equations are

$$4\tau = 2mS_T\beta^2 \ln \beta + S_T(1 - m)(\beta^2 - 1) - S_T\,\psi \tag{10.88}$$

where

$$m = \frac{1}{S_T} + \phi\left\{1 - k_{21}\alpha_{12}\left(\frac{1 - b^2}{2\ln b} + 1\right)\right\}$$

$$\psi = \int_1^\beta \frac{x^2 - 1}{x \ln x}\,dx$$

The function ψ is plotted and given in Chap. 8.

Equation (10.88) compares reasonably well with numerical solutions, and thus the quasi-steady method is suitable for an infinite domain or for short

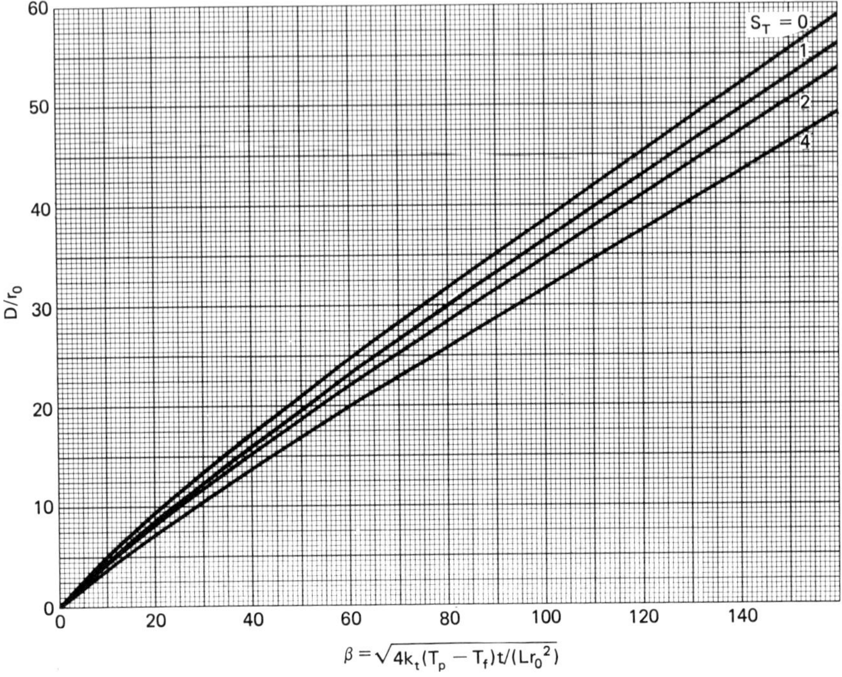

$\sqrt{4k_t(T_p - T_f)t/(L\, r_o^2)}$

D/r_o

Fig. 10.28. Thaw beneath a Pipe, Infinite System.

growth times. If the system is a pipe buried at a finite depth or if the growth time is long, these solutions will tend to give too rapid a growth rate.

The heat balance integral solution (Section 8.6.2.2) was shown to be very close to complete numerical solutions. This solution has been evaluated for phase change of pipes buried in soil, and the phase change depth and pipe surface heat transfer are given in Figs. 10.29 to 10.37. These figures are for certain values of the Stefan number and superheat parameter; interpolation can be used for other values. The water content parameter, x_l, adjusts for the soil properties, as discussed in Chap. 4.

An approximate solution, which is fairly accurate and simple to use, was discussed in Section 8.6.2.4. This can be used for cases not covered by Figs. 10.29–10.37.

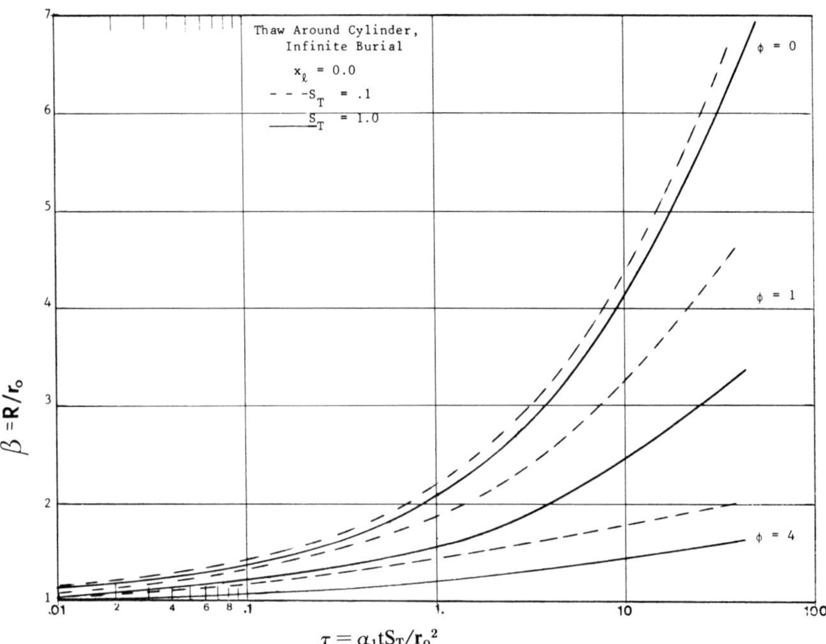

Fig. 10.29. Radius versus Time, $k_{12} = 1.0$, $\alpha_{12} = 1.0$.

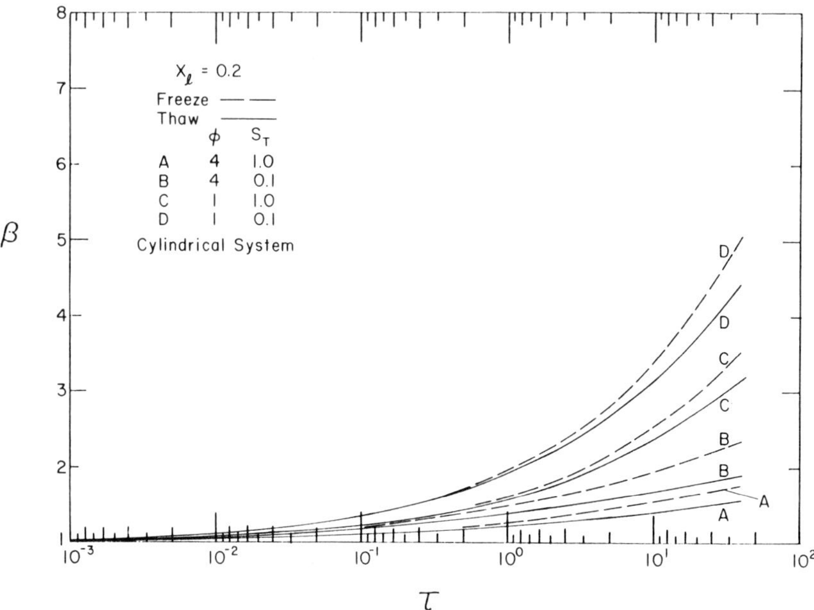

Fig. 10.30. Radius versus Time ($k_{12} = .7621$, $\alpha_{12} = .6326$, thaw).

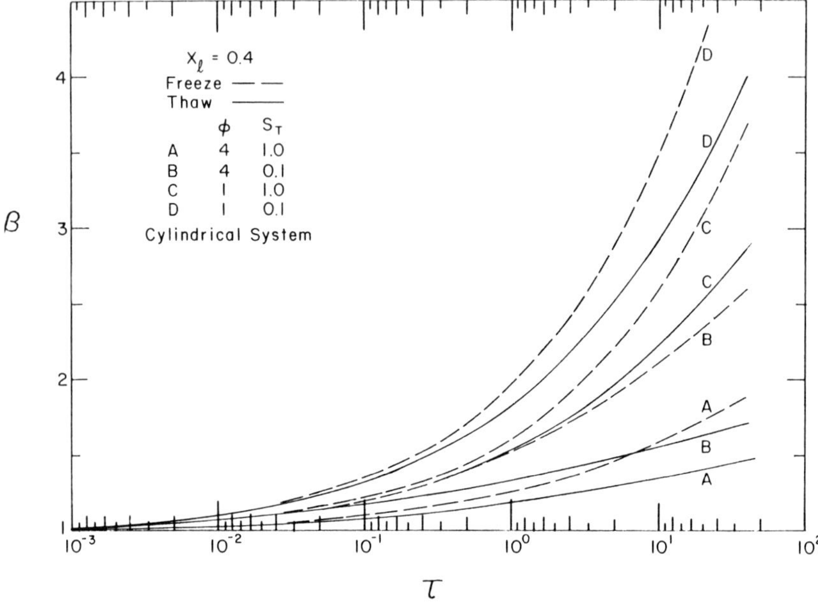

Fig. 10.31. Radius versus Time (k_{12} = .5808, α_{12} = .4121, thaw).

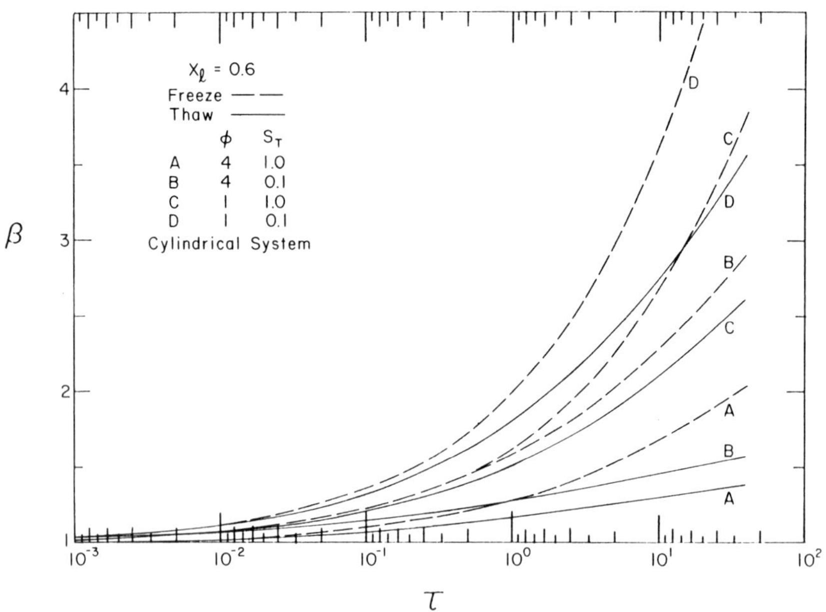

Fig. 10.32. Radius versus Time (k_{12} = .4426, α_{12} = .2742, thaw).

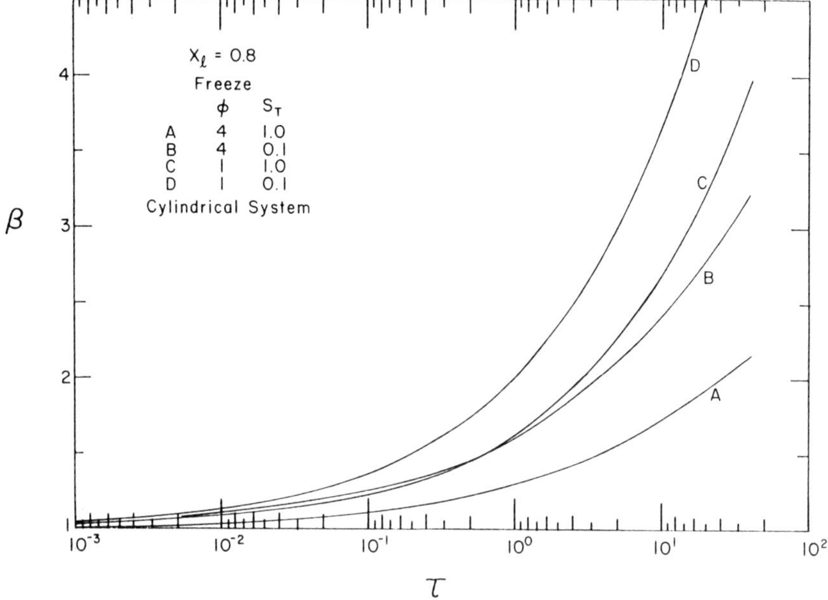

Fig. 10.33a. Radius versus Time, $k_{12} = 2.965$, $\alpha_{12} = 5.391$.

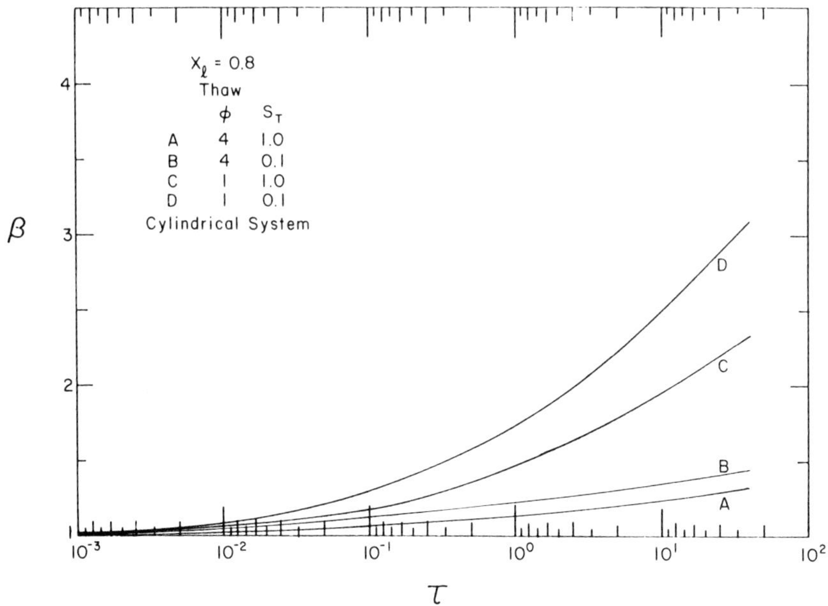

Fig. 10.33b. Radius versus Time, $k_{12} = .3373$, $\alpha_{12} = .1855$.

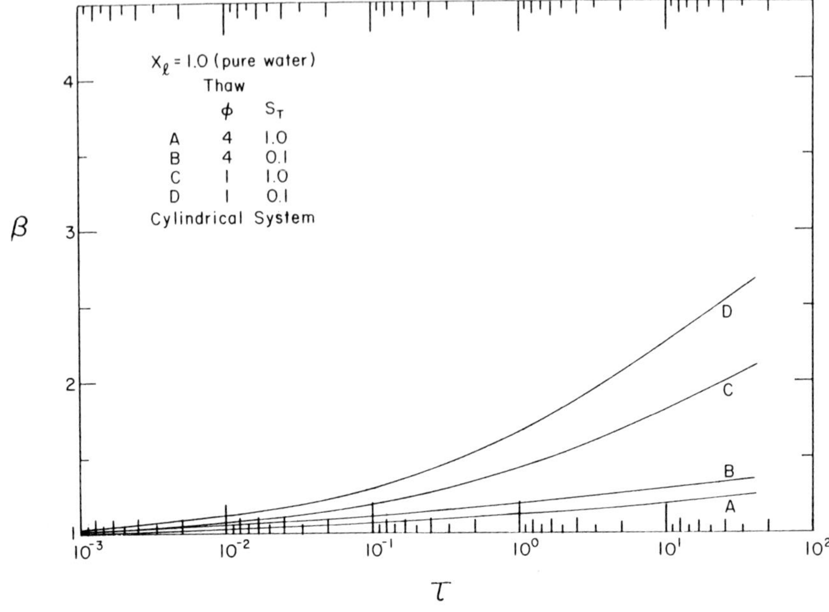

Fig. 10.34a. Radius versus Time, $k_{12} = .2571$, $\alpha_{12} = .1167$.

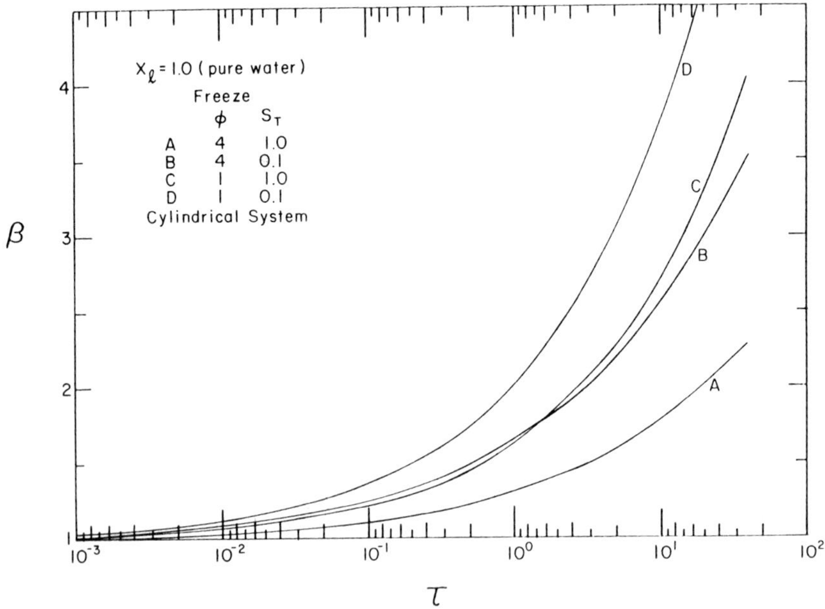

Fig. 10.34b. Radius versus Time, $k_{12} = 3.890$, $\alpha_{12} = 7.869$.

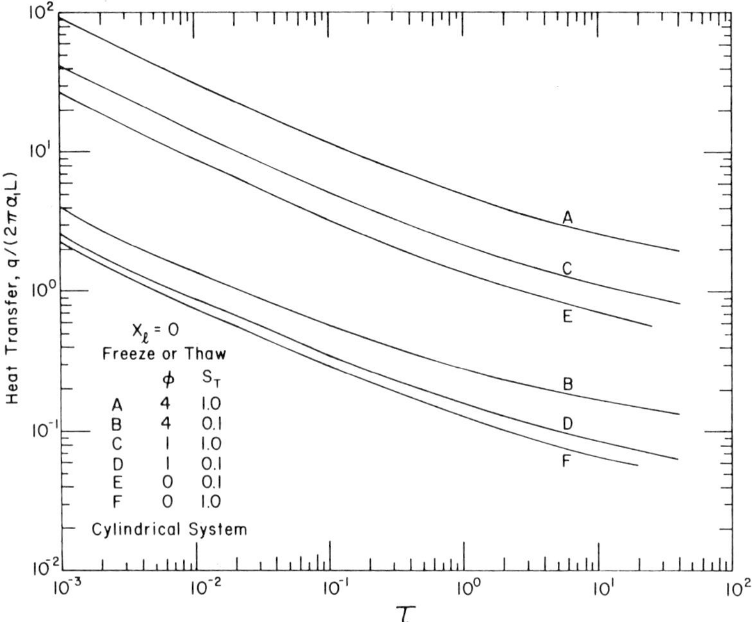

Fig. 10.35. Surface Heat Transfer Rate versus Time, $x_l = 0$.

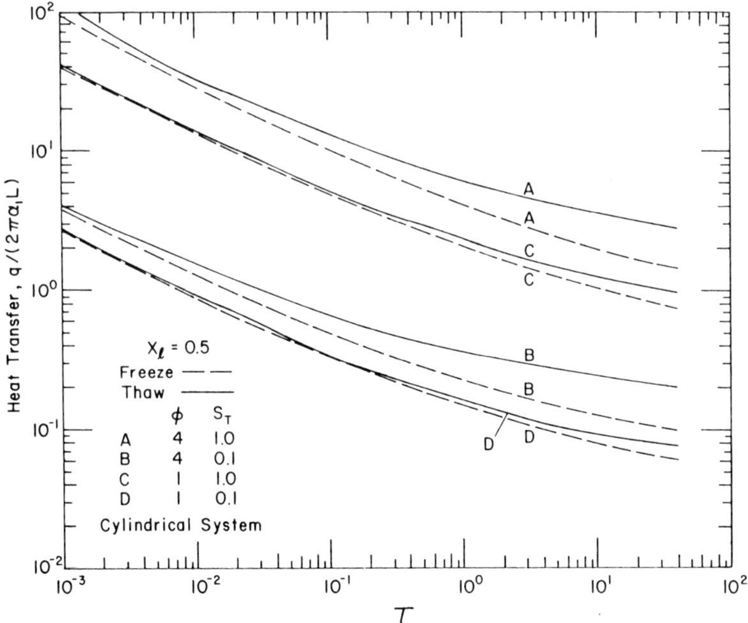

Fig. 10.36. Surface Heat Transfer Rate versus Time, $x_l = .5$.

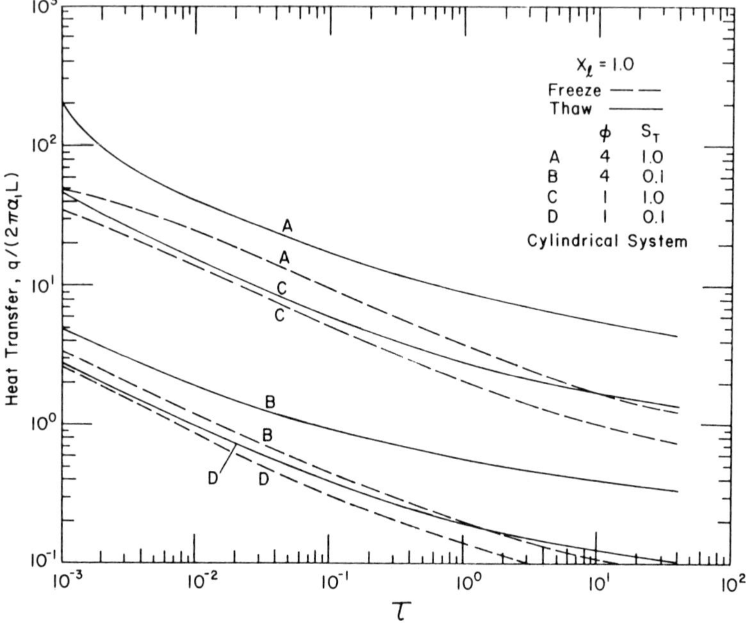

Fig. 10.37. Surface Heat Transfer Rate versus Time, x_l = 1.0, water.

The phase change location is

$$\beta = \left[1 - \left(\frac{\phi}{\phi + 1}\right)^{.05}\right] e + 1$$

$$e = (a + d)^{1/3} + (a - d)^{1/3} - 5.5$$

where

$$a = 6930 \frac{\alpha_{ef}}{\alpha} \frac{\tau}{S_T} - 166.375$$

$$d = [a^2 - 27680.6406]^{1/2}$$

$$\alpha_{ef} = \alpha_1 \left[\frac{\gamma}{erf^{-1}\left\{\frac{1}{\phi + 1}\right\}}\right]^2$$

The pipe surface heat transfer rate is

$$q^* = \frac{qc_l}{2\pi\, k_l l} = \frac{20\, S_T\,(1+\phi)}{e}$$

where

$$\alpha_{eq} = \alpha_l\,[(\phi + 1)\; erf\gamma]^2$$

The integrated heat transfer is

$$Q_t = \int_0^t q\, dt$$

This can be written in nondimensional form as

$$Q_t^* = \frac{Q_t}{2\pi r_0^2 \pi\, \rho_l l} = (1+\phi)\,\frac{S_T}{21}\left(\frac{e^2}{22} + e\right)$$

The parameter γ is given by Eq. (8.26) or Eq. (8.104).

10.6.2 Finite Depth of Burial

The solution for an infinite medium gives results that indicate too fast a growth rate of the thaw around a buried pipe when compared with the numerical results of Lachenbruch (1970) and others. Lachenbruch has presented a numerical solution for the transient thaw depth around a hot pipe buried in permafrost.

There are no solutions, even approximate, for the complete problem as described above, in the semi-infinite domain. Nevertheless, the quasi-static approximation has been used when the original ground temperature is a constant value of T_G, which is also the ground surface temperature. The problem is as follows (Fig. 10.38)

$$\frac{\partial^2 T_1}{\partial x^2} + \frac{\partial^2 T_1}{\partial y^2} = \frac{1}{\alpha_1}\frac{\partial T_1}{\partial t}$$

$$\frac{\partial^2 T_2}{\partial x^2} + \frac{\partial^2 T_2}{\partial y^2} = \frac{1}{\alpha_2}\frac{\partial T_2}{\partial t}$$

$$T(x,\, y,\, 0) = T(x,\, 0,\, t) = T_G$$

$$T(x,y,t) = T_p \qquad \text{(on the pipe surface)}$$

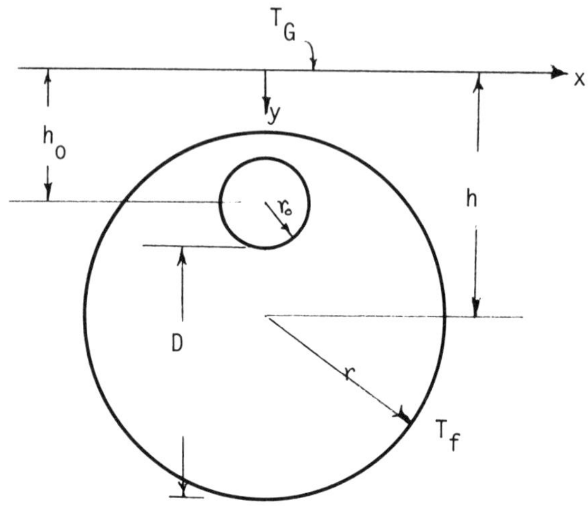

Fig. 10.38. Pipe Buried at Finite Depth.

Porkhayev (1963) assumed that the phase change energy balance at the centerline below the pipe was the controlling mechanism for the phase change. Equation (10.86) then is

$$k_1 \frac{\partial T_1}{\partial y} - k_2 \frac{\partial T_2}{\partial y} = L \frac{dY}{dt} \tag{10.89}$$

where the phase change is at $Y(t)$. Using the well-known steady state solution—see Chap. 7—it is possible to solve Eq. (10.89)

$$\int_{1+\mu}^{Y/r_0} \frac{f(\xi) - 1}{1 - \alpha \left[1 - \frac{1}{f(\xi)} \right]} \frac{d\xi}{f'(\xi)} = k_1 \frac{(T_p - T_f)\, t}{L} \tag{10.90}$$

where $\xi = y/r_0$.

$$\alpha = \frac{k_2 (T_G - T_f)}{k_1 (T_p - T_f)}$$

$$f(\xi) = \frac{\ln \dfrac{(\xi + \sqrt{\mu^2 - 1})}{(\xi - \sqrt{\mu^2 - 1})}}{\ln (\mu + \sqrt{\mu^2 - 1})}$$

$$\mu = h_o / r_o$$

Hwang (1977) has numerically evaluated Eq. (10.90), and Figs. 10.39 through 10.48 show the variation of the thaw beneath the pipe center.

Porkhayev's solution yields an underestimate of the thaw depth, and Hwang (1977) suggests that the mean of the Porkhayev solution and the infinite domain solution—Eq. (10.87)—will give a value adequate for engineering purposes but this is not usually valid.

Thornton (1976) also used the steady-state solution, but used the average value of the heat flux around the circular thaw interface in satisfying the energy condition of Eq. (10.86). This method led to the thawbowl decreasing as the pipe burial increases, which is contrary to expectations. Thornton's results should, thus, be used with some caution.

The infinite domain technique of Khakimov (1957) was used with a correction for the effect of finite burial depth by Lunardini (1977*b*). Looking at Fig. 10.49, it may be seen that when the thaw radius R reaches the ground

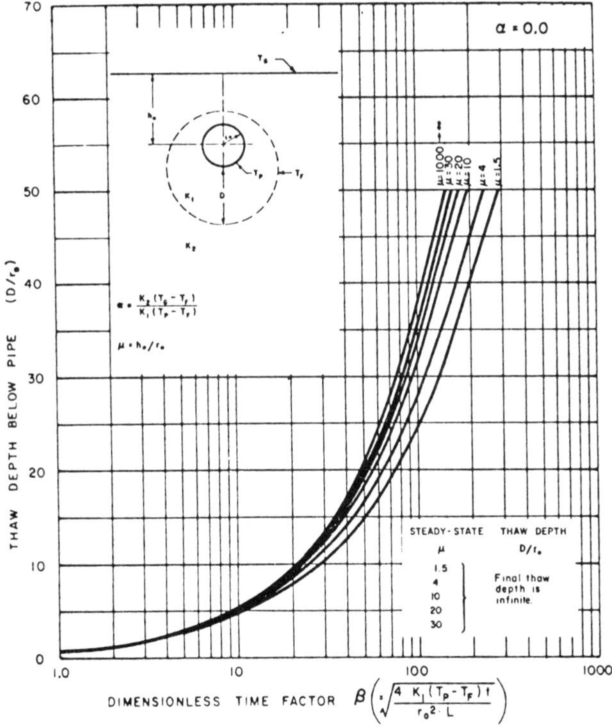

Fig. 10.39. Thaw Depth versus Time Factor for $\alpha = 0.0$. Reproduced by permission of the National Research Council of Canada from the *Canadian Geotechnical Journal* 14:180–192, 1977.

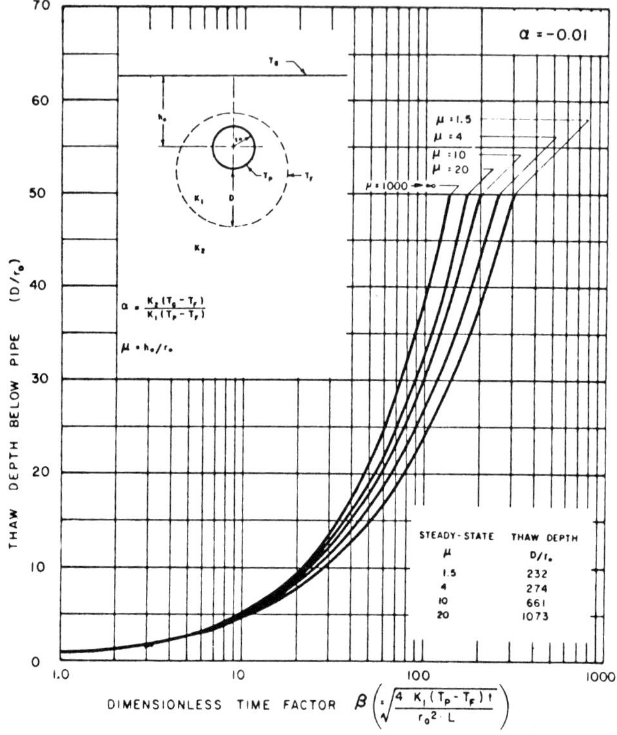

Fig. 10.40. Thaw Depth versus Time Factor for $\alpha = -0.01$. Reproduced by permission of the National Research Council of Canada from the *Canadian Geotechnical Journal* 14:180–192, 1977.

surface, i.e., $R > h_o$, there will be a conductive loss directly to the surface. If R is infinite, the total heat loss from a buried pipe is given by the well-known expression

$$q_\infty = \frac{2\pi\, k_1 (\theta_s - \theta_a)}{\ln\left(2\dfrac{h_o}{r_o}\right)} \tag{10.91}$$

where $\theta_a = T_G - T_o$ is the mean annual ground surface temperature. The conductive heat loss when $R > h_o$ is assumed to be proportional to the thawed surface area, the temperature gradient from the surface to the pipe, and the thermal conductivity of the thawed soil. The surface heat flux may be written as

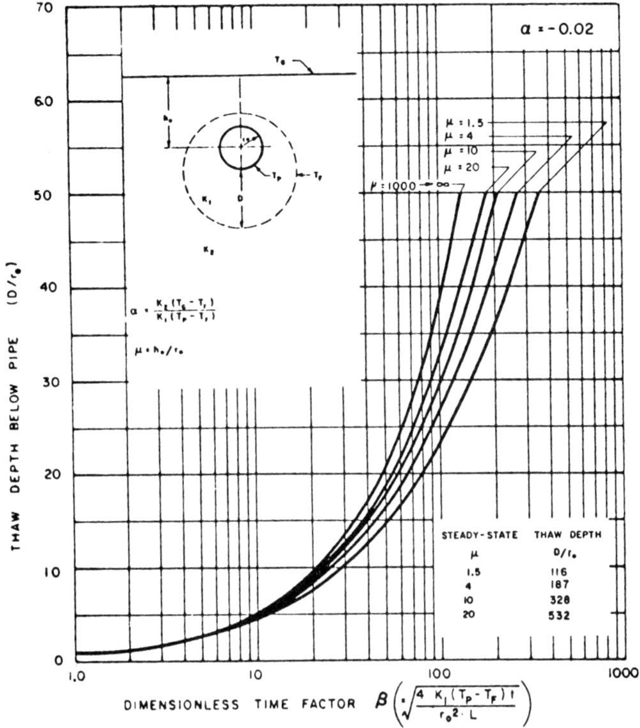

Fig. 10.41. Thaw Depth versus Time Factor for $\alpha = -0.02$. Reproduced by permission of the National Research Council of Canada from the *Canadian Geotechnical Journal* 14:180–192, 1977.

$$q_s = \frac{\sqrt{R^2 - h_o^2}\, k_l\,(\theta_s - \theta_a)}{2\left(\dfrac{h_o + R}{2}\right)}$$

where the heat flow area is taken as half the thawed surface area and the distance for the gradient is the mean of R and h_o (Fig. 10.50). This is a crude estimate, but the results are close to values obtained from an electrical analogy study of the problem. The surface heat flux may be written as

$$q_s = 2k_l(\theta_s - \theta_a)\frac{\sqrt{\beta - \mu}}{\sqrt{\beta + \mu}} \tag{10.92}$$

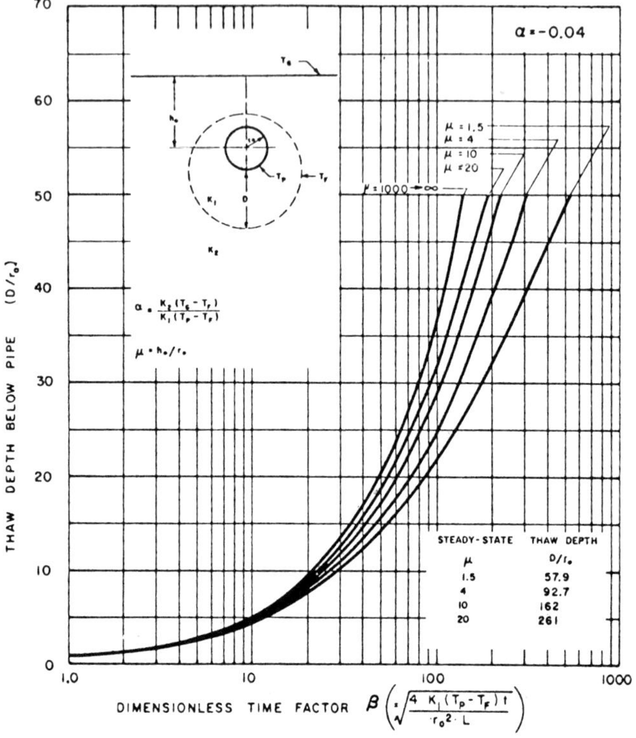

Fig. 10.42. Thaw Depth versus Time Factor for $\alpha = -0.04$. Reproduced by permission of the National Research Council of Canada from the *Canadian Geotechnical Journal* 14:180–192, 1977.

The change in the energy absorbed by the system will now equal the energy added by the pipe minus the conductive loss to the surface

$$\frac{dQ}{dt} = -k_1 A_o \left(\frac{d\theta_1}{dr}\right)_{r=r_o} - q_s \tag{10.93}$$

After substitution and rearrangement, the differential equation for the phase change interface is

$$u\left[1 - \frac{v \ln \beta}{\pi}\sqrt{\frac{\beta - \mu}{\beta + \mu}}\right] = \left[m\,\beta \ln \beta + \frac{\beta}{2} - \frac{\beta^2 - 1}{4\beta \ln \beta}\right]\frac{d\beta}{dt} \tag{10.94}$$

where

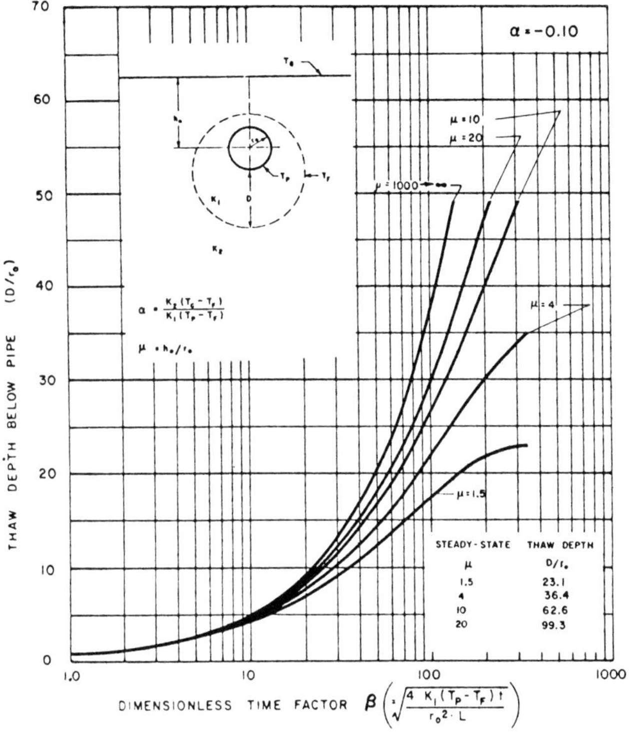

Fig. 10.43. Thaw Depth versus Time Factor for $\alpha = -0.10$. Reproduced by permission of the National Research Council of Canada from the *Canadian Geotechnical Journal* 14:180–192, 1977.

$$v = \frac{\theta_s - \theta_a}{\theta_s - \theta_f}$$

$$u = \frac{k_1}{C_1 \, r_o^2}$$

Equation (10.94) cannot be integrated in a closed form, but simple numerical quadrature leads to the following solutions

$$ut = \begin{cases} \tfrac{1}{4}\,[2m\,\beta^2 \ln \beta + (1-m)(\beta^2 - a) - \psi(\beta)] & 1 < \beta < \mu \\ m\, I_1(\beta) + I_2(\beta) + u\, t(\mu) & \beta > \mu \end{cases} \tag{10.95}$$

The functions $\psi(\beta)$, $I_1(\beta)$, and $I_2(\beta)$ are given by

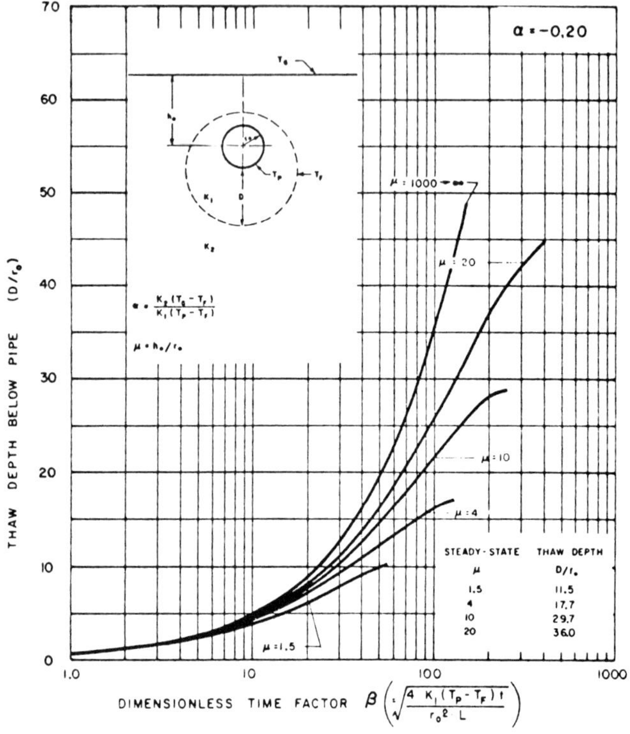

Fig. 10.44. Thaw Depth versus Time Factor for $\alpha = -0.20$. Reproduced by permission of the National Research Council of Canada from the *Canadian Geotechnical Journal* 14:180–192, 1977.

$$\psi = \int_1^\beta \frac{(x^2 - 1)}{x \ln x} dx$$

$$I_1(\beta) = \int_\mu^\beta \frac{x \ln x}{1 - \dfrac{v \ln x}{\pi} \sqrt{\dfrac{x - \mu}{x + \mu}}} dx$$

$$I_2(\beta) = \int_\mu^\beta \frac{\dfrac{x}{2} - \dfrac{(x^2 - 1)}{4x \ln x}}{1 - \dfrac{v \ln x}{\pi} \sqrt{\dfrac{x - \mu}{x + \mu}}} dx$$

These functions are plotted in Fig. 10.51 for some values of μ with a temperature ratio $v = 1.01$ used for the functions I_1 and I_2. With these

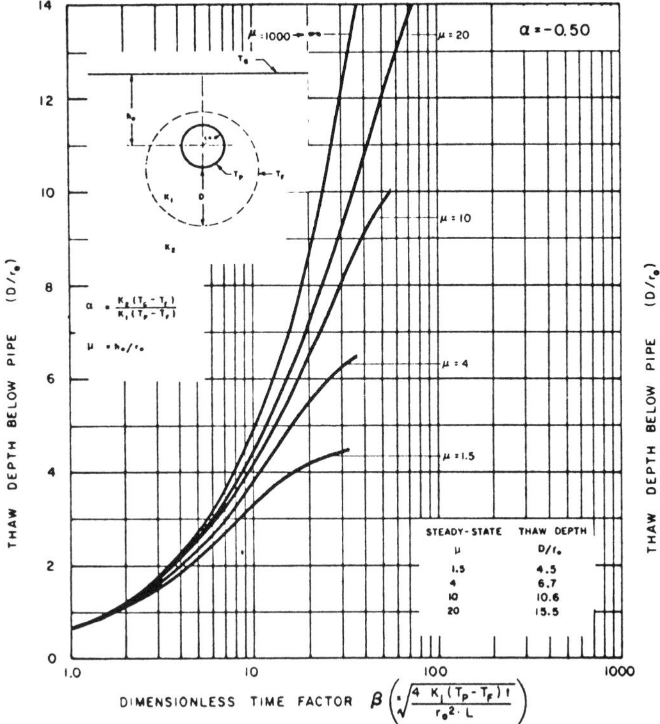

Fig. 10.45. Thaw Depth versus Time Factor for $\alpha = -0.50$. Reproduced by permission of the National Research Council of Canada from the *Canadian Geotechnical Journal* 14:180–192, 1977.

functions, the phase change effect can be evaluated quite simply for a variety of practical problems. Equation (10.95) compared favorably with numerical solutions, such as Lachenbruch's (1970).

The heat transfer rate per foot of pipe can be calculated easily from

$$q_p = \frac{2\pi\, k_l\, (\theta_s - \theta f)}{\ln \beta} \tag{10.96}$$

The values of β versus time are known, and thus the heat transfer rate at any time can be found. In the early stages of thaw, the heat transfer rate is considerably larger than for the case of a buried pipe with no phase change. Thus the heat loss from a hot pipe buried in permafrost will, initially, be considerably greater than calculations using the infinite, or no phase change,

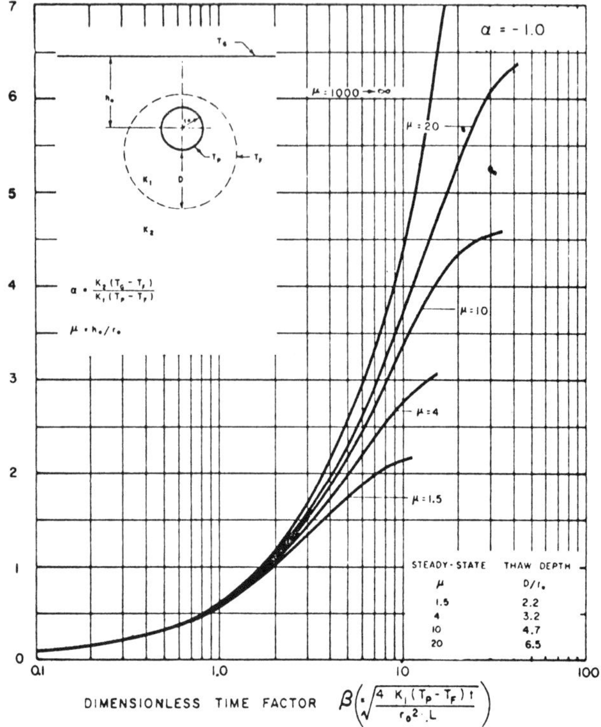

Fig. 10.46. Thaw Depth versus Time Factor for $\alpha = -1.0$. Reproduced by permission of the National Research Council of Canada from the *Canadian Geotechnical Journal* 14:180–192, 1977.

solution would indicate. The rate of heat loss and the thaw depth depend strongly on the soil conditions surrounding the pipe.

10.6.2.1 Buried Pipe Thaw Data

There is very little published information on the thawing of permafrost due to buried pipes. Rowley et al. (1973) and Watson et al. (1973) report on a 61-cm pipe circulating hot oil at Inuwik, NWT. The data showed that the vertical distance from the pipe bottom to the 0°C isotherm, for the first few months, was given by

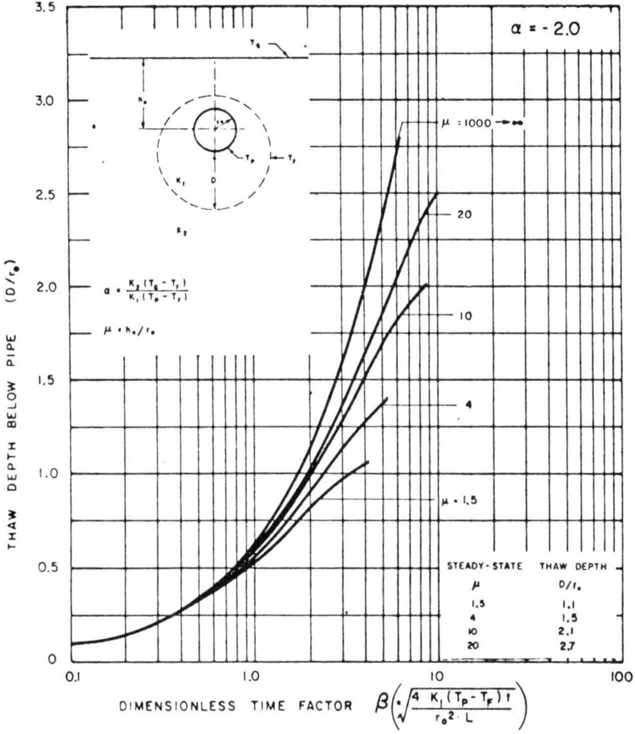

Fig. 10.47. Thaw Depth versus Time Factor for $\alpha = -2.0$. Reproduced by permission of the National Research Council of Canada from the *Canadian Geotechnical Journal* 14:180–192, 1977.

$$X = .046\sqrt{t}$$

where X = meters and t = hours.

It should be noted that the phase change about a cylinder will always be considerably slower than the equivalent plane problem (Neumann solution). Morgenstern and Nixon (1973) calculated a coefficient of 0.044 using the Stefan equation, but this prediction is probably fortuitous. It can be shown that the thaw beneath a cylinder will be considerably less than for the equivalent Stefan problem (plane case). The pipe in the experiment was surrounded by very distinct horizontal layers making the choice of thermal properties difficult and somewhat arbitrary. Finally the pipe centerline was moving downward rapidly, which will make the use of the standard solutions dubious.

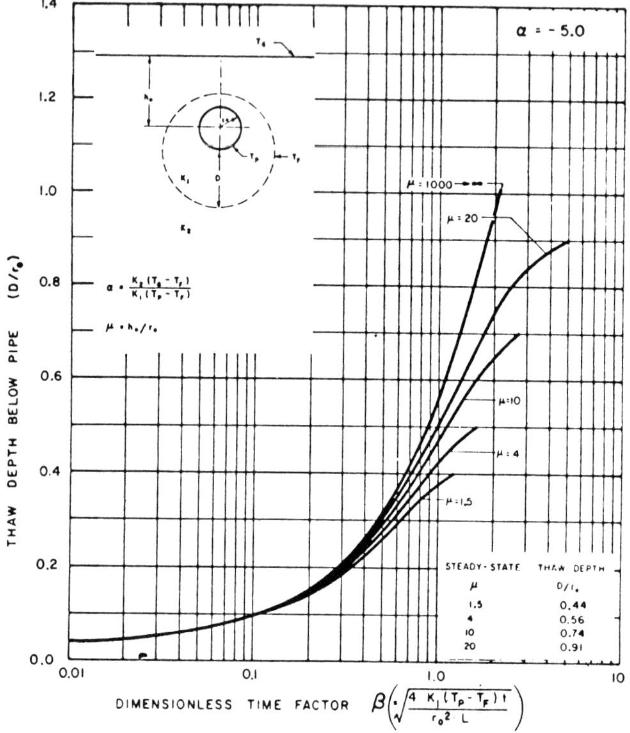

Fig. 10.48. Thaw Depth versus Time Factor for $\alpha = -5.0$. Reproduced by permission of the National Research Council of Canada from the *Canadian Geotechnical Journal* 14:180–192, 1977.

Example 6: Consider a 61-cm radius pipe buried very deep. The following properties apply. Calculate the thaw beneath the pipe after 129.2 days.

$$T_s = 71°C$$
$$T_G = T_F = 0°C$$
$$L = 52.54\ cal/cm^3$$
$$k_u = .0025\ cal/s\text{-}cm\text{-}°C$$
$$C_u = .74\ cal/cm^3\text{-}°C$$
$$\alpha_u = 12\ cm^2/hr$$

Thus, $S_T = 1.0$, $\phi = 0.0$ (no subcooling), $\tau = 10$. Using the various methods described leads to the following results:

Method	Thaw depth, cm
Quasi-static, no sensible heat, Eq. (10.87)	213.5
Quasi-static, sensible heat, Fig. 10.28, $S_T = 1.0$	195.2
Heat balance integral, Fig. 10.29	196.4
Quasi-static, no sensible heat, Eq. (10.90), Fig. 10.39	240.3
Quasi-static, sensible heat, Eq. (10.88)	189.1

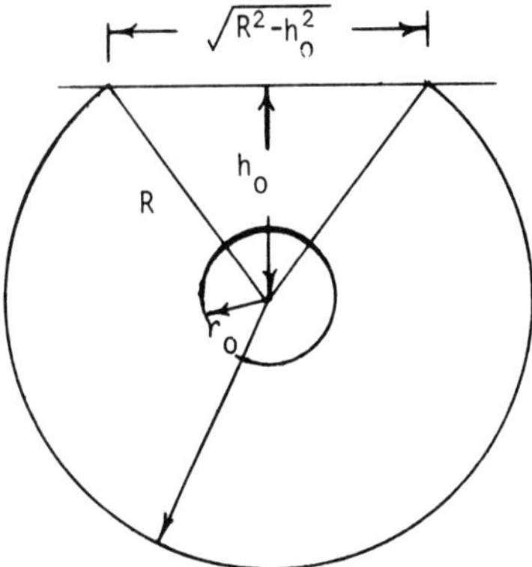

Fig. 10.49. Pipe Thaw Approximation.

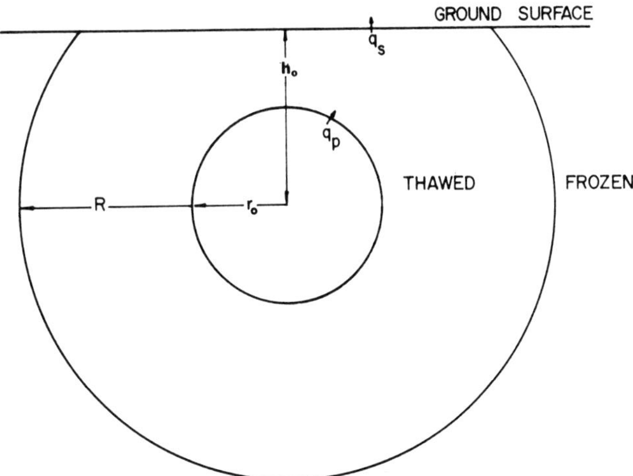

Fig. 10.50. Heat Flow Geometry.

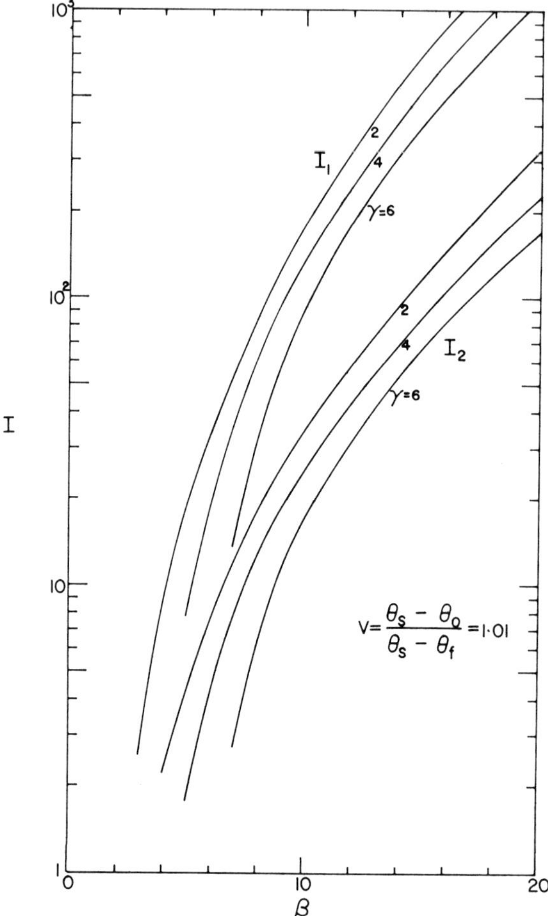

Fig. 10.51. Functions for Thaw beneath Buried Pipe.

Because the heat balance integral method is essentially exact, this example illustrates the relative accuracy of the methods. Contrary to Hwang's (1977) surmise, the quasi-static method using Eq. (10.90) overestimates the thaw depth, at least for the infinite burial problem.

10.7. ICE FORMATION IN PIPES

The question of the freezing of water in pipes is of importance in terms of water supply, sewage removal, etc. Figure 10.52 graphically describes the

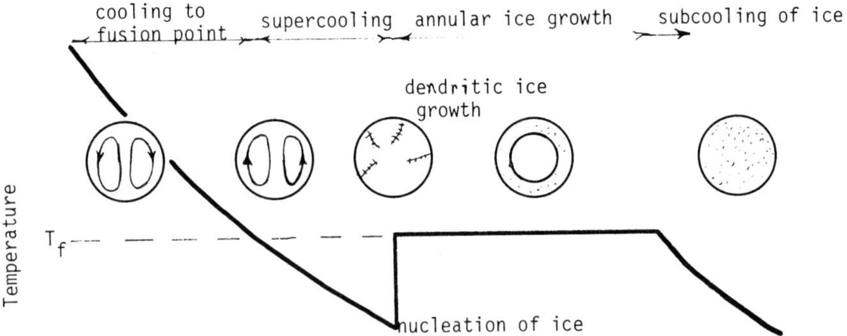

Fig. 10.52. Cooling of Water in Pipes.

growth of ice in a tube. The density of water is a maximum at about 39°F, and as the water near the pipe surface cools to 39°F, the cooler, denser water sinks as shown. As the temperature drops below 39°F, the pattern is reversed, and the cooler lighter water rises. A certain amount of subcooling is necessary before nucleation of ice will occur. Zarling (1977) and Gilpin (1977) report values of 8–9°F of subcooling, but this will vary from system to system. Nucleation will be followed by dendritic ice growth and a rapid return of the water to the fusion temperature, after which the usual annular growth will begin.

The effects of dendritic ice are usually ignored, but they may have significance for pipe blockage, as will be seen.

10.7.1 Cooling of Water to Fusion Temperature

The time required to cool water in a pipe from an arbitrary temperature to the fusion value can be easily estimated. Consider an analysis where the water in the tube does not vary spatially and the effects of the density maximum are neglected. For a unit length of insulated pipe, with the notation of Fig. 10.53, an energy balance of the water yields

$$\frac{2(T_w - T_a)}{\dfrac{1}{r_i h_i} + \dfrac{1}{k_p}\ln\dfrac{r_o}{r_i} + \dfrac{1}{k_k}\ln\dfrac{r_k}{r_o} + \dfrac{1}{r_k h_o}} = -r_i^2 C_w \frac{dT_w}{dt} \tag{10.97}$$

The solution to this equation is

$$T_w = (T_o - T_a)e^{-t/s_1} + T_a \tag{10.98}$$

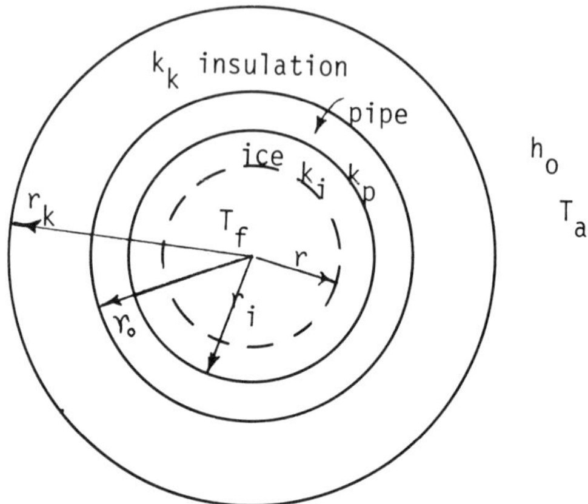

Fig. 10.53. Ice Growth Inside Pipe.

where

$$S_1 = \frac{r_i^2 \rho_w c_w}{2}\left[\frac{1}{k_p}\ln\frac{r_o}{r_i} + \frac{1}{k_k}\ln\frac{r_k}{r_o} + \frac{1}{r_k h_o} + \frac{1}{r_i h_i}\right]$$

T_o = initial temperature of water and
T_a = ambient air temperature.

The time for the water to cool to T_f is

$$t_c = S_1 \ln\left(\frac{T_o - T_a}{T_f - T_a}\right) \tag{10.99}$$

Example 7: Calculate the time required for a 2-in. copper pipe containing water to cool from 42°F to 32°F. The pipe has ½ in. of fiberglass insulation, and the ambient temperature is −10°F.

The following data apply:

$T_a = -10°F$ $\quad T_f = 32°F$
$r_o = 1.00$ in. $\quad r_i = .92$ in. $\quad r_k = 1.5$ in.
$k_i = 1.3$ Btu/hr-ft-°F $\quad k_p = 224$ $\quad k_k = .022$ $\quad h_o = 1.6$ Btu/hr-ft²-°F

Ignore the thermal resistance of the inside water layer. Then

$$S_l = 4.30 \text{ hr}$$
$$t_c = 0.92 \text{ hr}$$

10.7.2 Freezing of Water

Consider the growth of an annular ice layer as shown in Fig. 10.53. If the water temperature is T_f and the air temperature is T_a, then the heat liberated by the ice formation must be transferred outward through the layers of ice, pipe, and insulation. The basic equation is

$$\frac{(T_f - T_a)}{\dfrac{\ln\left(\dfrac{r_i}{r}\right)}{k_i} + \dfrac{\ln\dfrac{r_o}{r_i}}{k_p} + \dfrac{\ln\dfrac{r_k}{r_o}}{k_k} + \dfrac{1}{r_k h_o}} + L\,r\frac{dr}{dt} = 0 \qquad (10.100)$$

The solution to this equation for the time required to form an ice layer of thickness r^* is

$$t = \frac{L}{k_i\,(T_f - T_a)}\left\{\left[\frac{1}{2} + \Sigma\right]\frac{r_i^2 - r^2}{2} + r^{*2}\ln\frac{r^*}{r_i}\right\} \qquad (10.101)$$

where

$$S = k_i\left\{\frac{1}{k_p}\ln\frac{r_o}{r_i} + \frac{1}{k_k}\ln\frac{r_k}{r_o} + \frac{1}{r_k\,h_o}\right\}$$

The water will freeze completely when $r^* = r_i$ and $r = 0$

$$t_f = \frac{L}{k_i\,(T_f - T_a)}\frac{r_i^2}{2}\left[\frac{1}{2} + S\right] \qquad (10.102)$$

Lock (1970) presented the same solution, but neglected the insulating effect of the ice and the pipe.

Equation (10.102) can be written in nondimensional form as

$$\tau_f = \frac{1}{2}\left(\frac{r_i}{r_k}\right)^2\left[\frac{1}{2} + \frac{k_i}{k_p}\ln\frac{r_o}{r_i} + \frac{k_i}{k_k}\ln\frac{r_k}{r_o} + \frac{1}{B_i}\right] \qquad (10.103)$$

where

$$\tau_f = \frac{\alpha_i t_f}{r_k^2} S_T$$

$$S_T = \frac{C_i}{L}(T_f - T_a)$$

$$B_i = \frac{h_o r_k}{k_i}$$

Equation (10.103) is interesting since, for the special case of a thin-walled pipe with no insulation, it reduces to

$$\tau_f = \frac{1}{2}\left[\frac{1}{2} + \frac{1}{B_i}\right] \tag{10.104}$$

Note that this is exactly the theoretical result described by Eq. (8.288).

Example 8: Calculate the time required for the pipe of Example 6 to freeze solid; $L = 8326.8$ Btu/ft³.
Then

$$S = 1.3\left\{\frac{1}{224}\ln\frac{1}{.92} + \frac{1}{.022}\ln\frac{1.5}{1} + \frac{12}{1.5 \times 1.6}\right\} = 30.46$$

Thus, from Eq. (10.6)

$$t_f = 13.73 \ hr$$

The total time to go from 42°F to complete freeze is then 14.65 hours.

10.7.3 Dendritic Ice in Horizontal Pipes

Dendritic ice is that formed when supercooled water suddenly nucleates and rapidly changes phase. The latent heat evolved raises the surrounding water temperature back to the normal fusion temperature where the dendritic ice is in the form of thin platelike crystals that can block the flow of water in a pipe even if only a small part of the water volume has changed to ice. This is due to the interlocking of the plates, later further cemented in place by annular ice growth. The possibility of flow blockage depends upon the nucleation temperature, cooling rate of water at the center of the pipe, pipe density, and flow velocity.

Typical nucleation temperatures of water are listed below:

Water source	Nucleation temperature, °C	Reference
Cold tap	−4 to −5	Gilpin (1977*a*)
Hot tap	−5 to −6	" "
Fresh lake or river	−3 to −5	Gilpin (1977*b*)

The nucleation temperature is independent of container material and cooling rate for rates less than 1°C/min.

A well-insulated pipe will be most susceptible to dendritic blockage, as the water center temperature can cool down to −2°C before the colder surface water causes nucleation. If the center temperature is higher than −2°C, at nucleation, the dendritic ice formed will tend to disappear due to convection effects at the dendritic growth location.

Gilpin (1977*a*) found that, with dendritic growth, a definite pressure gradient was required to start the water in the pipe flowing. This relation is shown in Fig. 10.54. If the cooling of the water is less than a critical value, then dendritic blockage is possible.

The critical cooling rate is given by

$$H_c = C\left(\frac{-2 - T_n}{D}\right)^{5/4} \tag{10.105}$$

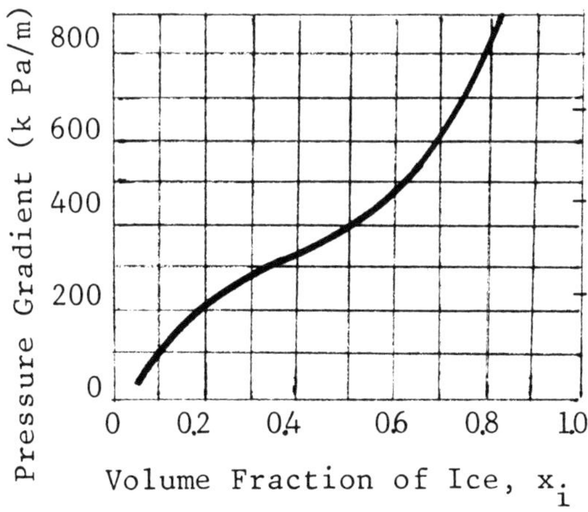

Fig. 10.54. Changes in Pressure Gradient to Start Flow during the Freezing History of a Pipe (adapted from Gilpin (1977*a*). Reprinted with permission from *International Journal of Heat and Mass Transfer* 20, R. R. Gilpin, "The Effects of Dendritic Ice Formation in Water Pipes," 1977, Pergamon Press Ltd.

where

$$C = \begin{cases} 21\text{–}29 \text{ copper tubes (uniform wall temperature)} & (10.106a) \\ 6.3\text{–}13.7 \text{ plastic pipes (uniform heat flux)} & (10.106b) \end{cases}$$

H = cooling rate of water temperature, °C/min
D = pipe inside diameter, mm and
T_n = nucleation temperature, °C.

For most cases relation (10.106*b*) will be most appropriate. If the cooling rate is less than the critical value, Fig. 10.54 can be used to find the ice volume fraction at blockage. Then the time for blockage to occur is estimated as

$$t_{df} = x_i \, t_f \tag{10.107}$$

where t_f is given by Eq. (10.102). These relations are only estimates, as the dendritic ice relations are not sufficiently known, but it appears that dendritic blockage may be significant.

The mean cooling rate for a pipe can be estimated from

$$H = \frac{(T_n - T_o)}{S_l \ln\left(\dfrac{T_o - T_a}{T_n - T_a}\right)} \tag{10.108}$$

where T_o = initial temperature of water.

Example 9: Consider the effect of dendritic growth in Example 7.

From Eq. (10.108), the cooling rate is H = 10.1°F/hr = .0933°C/min. The critical cooling rate is calculated from Eq. (10.105) with $T_n = -5$°C, C = 6.3, H_c = .18°C/min.

Thus, dendritic blockage should occur because $H < H_c$. The blockage time can be estimated as follows. Suppose the main pressure available is 400 kPa (60 psi) and the pipe length is 1.5 m (4.5 ft). Then the pressure gradient available is 267 kPa/m and, from Fig. 10.54, $x_i \sim 0.3$. Note that most of this ice is due to annular growth that strengthens the dendritic plates to block the flow. Then the blockage time is

$$t_{df} = .3(13.73) = 4.12 \text{ hr}$$

and the total freeze time is 5.04 hr. This is only about 35% of the time needed to freeze the pipe solid without considering dendritic effects.

Example 10: Suppose the pipe of Example 7 has no insulation. Calculate the total freeze time.

The cooling time to 32°F is

$$t_c = .20 \text{ hr}$$

and the freeze time is

$$t_f = 3.10 \text{ hr}$$

Thus, the total freeze time is 3.3 hours. The cooling rate is .47°C/min, and dendritic blockage should not occur. Note that the insulated pipe requires only 1.74 hours more to freeze than the uninsulated pipe, if dendritic blockage does exist in the insulated system.

The safeguarding of pipes from freezing with insulation only can be a risky business.

10.7.4 Freezing and Pipe Flow Rate

Water will not freeze in pipes if the flow is maintained. The minimum flow rate to keep the pipe from freezing can be estimated without too much difficulty (Fig. 10.55). The equation governing the water temperature is

$$\dot{m}\, c_w \frac{dTw}{dx} + \frac{Tw - Ta}{R_T} = 0 \tag{10.109}$$

The total resistance to heat flow is

$$R_t = R_1 + R_2 + R_3 + R_4 \tag{10.110}$$

where

$$R_1 = \frac{1}{2\pi r_i h_i}$$

is the inside convective resistance. This resistance is plotted on Fig. 10.56

$$R_2 = \frac{1}{2\pi k_p} \ln \frac{r_o}{r_i}$$

is the conduction resistance of the pipe, usually ignored

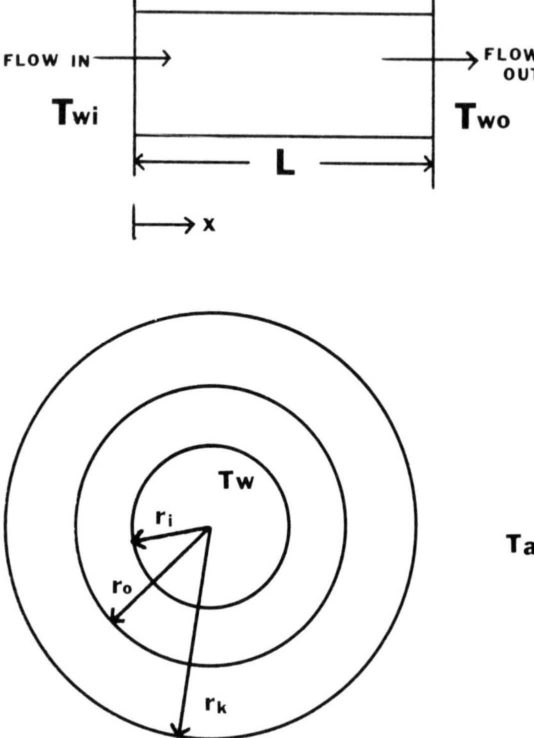

Fig. 10.55. Freezing in Horizontal Pipes.

$$R_3 = \frac{1}{2\pi k_k} \ln \frac{r_k}{r_o}$$

is the conduction resistance of the insulation, normally the most important resistance and

$$R_4 = \frac{1}{H_c + H_r}$$

is the outside convective and radiative resistance. H_c is plotted as Fig. 10.57. The radiative resistance is

$$H_r = 4\pi\sigma\epsilon\, d_k\, T_a^3 \tag{10.111}$$

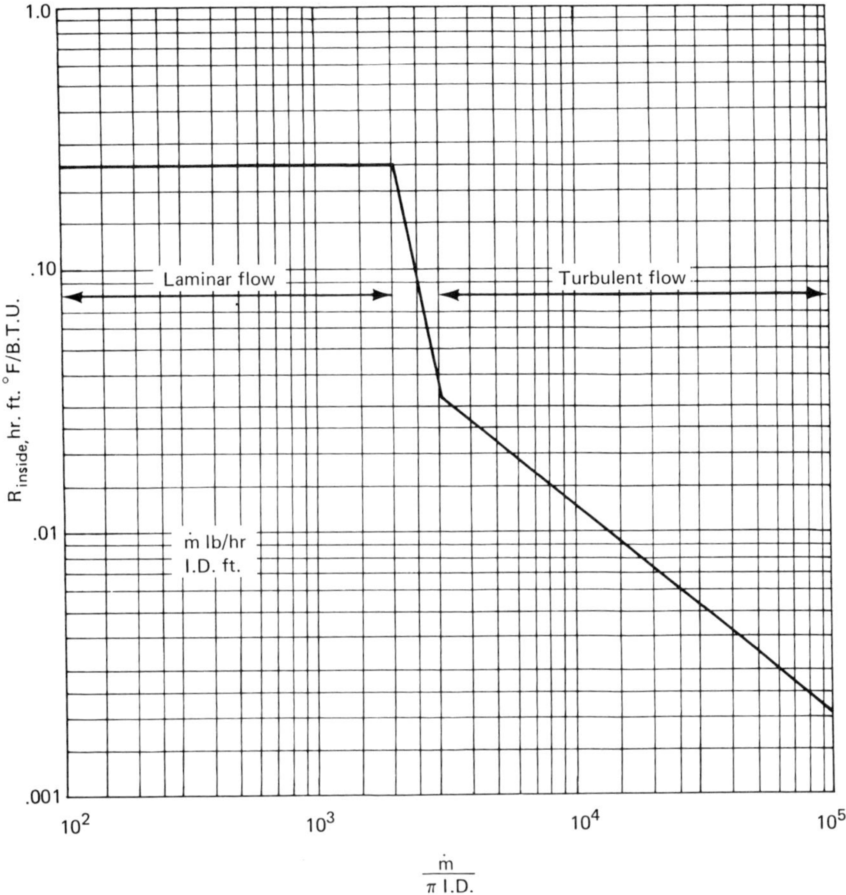

Fig. 10.56. Thermal Resistance between Pipe and Water Flowing through it.

This value can be approximated, for the low ambient temperatures important here, as

$$H_R = 2\, d_k \tag{10.112}$$

The solution to Eq. (10.109) is

$$T_w = (T_{wi} - T_a) \exp\left(-\frac{x}{\dot{m}\, c_w R_T}\right) + T_a \tag{10.113}$$

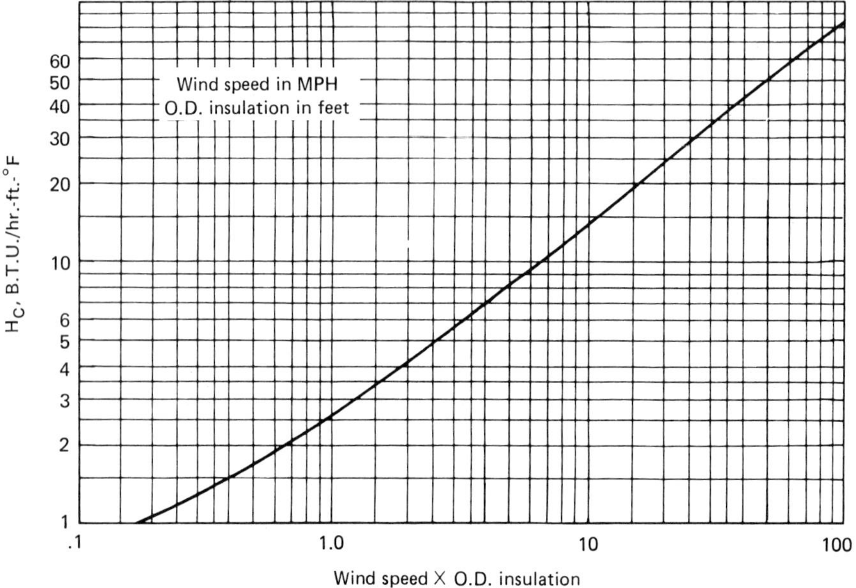

Fig. 10.57. Convective Heat Transfer from Cylinder to Air in Crossflow.

If the water is to be maintained above freezing, the pipe surface temperature at the outlet should be held at 32°F or higher. Then the minimum outlet water temperature is

$$T_{wom} = \frac{R_I}{R_T - R_I}(32 - T_a) + 32 \tag{10.114}$$

The flow rate needed for this minimum water temperature is

$$\dot{m}_m = \frac{L/(c_w R_T)}{\ln\left[\dfrac{R_T - R_a}{R_T}\dfrac{T_{wi} - T_a}{32 - T_a}\right]} \tag{10.115}$$

Example 11: The pipe of Example 7 is 100 ft long and the ambient wind speed is 15 mph. The water inlet temperature is 42°F. This is a trial-and-error solution.

Guess

$$R_I = 0.26 \text{ (laminar flow)}$$

Now

$$R_2 \cong 0 \qquad R_3 = \frac{1}{2\pi\,(.022)} \ln \frac{1.5}{1} = 2.933$$

$$H_c = 6.5 \text{ (Fig. 10.42)} \qquad H_R = 0.5$$

$$R_4 = .143$$

$$R_T = 3.336$$

$$\dot{m}_m = \frac{100/(1)(3.336)}{\ln\left(\dfrac{3.336 - .26}{3.336}\;\dfrac{42 + 10}{42}\right)} = 226.4 \text{ lbm/hr}$$

Now $\dot{m}/\pi d_i = 470$. Thus flow is laminar (see Fig. 10.56). If $\dot{m}/\pi d_i > 3000$, the turbulent range for R_1 would be used. Because we assumed laminar flow, the solution is correct.

The outlet water temperature is

$$T_{wom} = \frac{.26}{3.336}(32 + 10) + 32 = 35.3\,°\text{F}$$

REFERENCES

Aldrich, H. P. 1956. *Frost penetration below highway and airfield pavements.* Highway Research Board Bulletin 135.

———, and Paynter, H. M. 1953. *Frost investigations. Fiscal year 1953, first interim report. Analytical studies of freezing and thawing of soils.* ACFEL Tech. Rept. 42.

Barber, E. S. 1957. *Calculations of maximum temperatures from weather reports.* Highway Res. Board Bull. 168.

Berg, R. L. 1977. Unpublished data, CRREL.

———, and Quinn, W. F. 1976. *Use of a light-colored surface to reduce seasonal thaw penetration beneath embankments on permafrost.* Second Int. Symp. Cold Regions Engineering. Univ. Alaska.

———, and Aitken, G. W. 1973. *Some passive methods of controlling geocryological conditions in roadway construction.* Permafrost 2nd International Conference. Yakutsk, USSR, pp, 581–586, NAS, Washington, D.C.

Berggren, W. P. 1943. Prediction of temperature distribution in frozen soils. *Transactions, American Geophysical Union* 24 (3):71–77.

Beskow, G. 1935. *Soil freezing and frost heaving with special application to roads and railroads.* Swedish Geological Society, Series C, No. 375.

Bolsenga, S. J. 1964. *Daily sums of global radiation for cloudless skies.* CRREL R.R. 760.

Braun, J. S. 1957. A study of frost penetration into fine grained soils beneath bituminous pavements in southeastern Minnesota. M.S. Thesis, Univ. of Minnesota.

Brown, J. L., and Galate, J. 1971. *Literature review of N-factors used to determine depth of frost penetration beneath pavements.* CRREL unofficial memo.

Calkins, D. J. 1979. *Accelerated ice growth in rivers.* Rept. 79–14, U.S. Army Cold Regions Research and Engineering Lab., Hanover, New Hampshire.

Carlson, H., and Kersten, M. S. 1953. *Calculation of depth of freezing and thawing under pavements.* Highway Res. Board Bull. 71.

———. 1952. Calculation of depth of thaw in frozen ground. Research Board Spec. Rept. 2, *Frost action in soils,* pp. 192–223.

Carroll, C., et al. 1966. Digital simulation of heat flow in soils. *J. Soil Mech. Found.* Div. ASCE, ASTM 92 (SM4):31–49.

Carslaw, H. S. and Jaeger, J. C. 1959. *Conduction of heat in solids.* Oxford: Clarendon Press, p. 282.

Crabb, G. A., and Smith, J. L. 1953. *Soil temperature comparisons under varying covers.* Highway Research Board Bulletin 71, pp. 32–80.

Deacon, E. L. 1949. Vertical diffusion in the lowest layer of the atmosphere. *Q. J. R.* Meteorol. Soc. 75:81–103.

Dempsey, B. J., and Thompson, M. R. 1969. *A heat transfer model for evaluating frost action and temperature related effects on multilayered pavement systems.* Prof. Rept. IHR 401, Univ. of Illinois.

Esch, D. C. 1973. Control of permafrost degradation beneath a roadway by subgrade insulation. *Permafrost 2nd International Conference Yakutsk USSR.* NAS, Washington, D.C., pp. 608–622.

Gilpin, R. R. 1977*a*. The effects of dendritic ice formation in water pipes. *Int. J. Heat Mass Transfer* 20 (6):693–699.

———. 1977*b*. The effect of cooling rate on the formation of dendritic ice in a pipe with no main flow. *J. Heat Transfer* 99:419–424.

Gold, L. W., Johnston, G. N., Slusarchuk, W. A., and Goodrich, L. E. 1973. Thermal effects in permafrost. *Proc. Can. Northern Pipeline Res. Conf.,* pp. 25–45.

Gupta, J. P. 1973. An approximate method for calculating the freezing outside spheres and cylinders. *Chem. Eng. Sci.* 28:1629–1633.

Heiersted, S. R. S. 1975. Thermal climate regime on road and ground surface. *Frost I Jord* NR 10.

Hwang, C. T. 1977. On Quasi-static solution for buried pipes in permafrost. *Can. Geotech. J.* 14(2):180–192.

———, Murray, D. W., and Brooker, E. W. 1972. A thermal analysis for structures on permafrost. *Can. Geotech. J.* 9(2):33–46.

Janson, Lars-Eric. 1963. Frost penetration in sandy soil. Stockholm, Ph.D. Thesis.

Jumikis, A. R. 1955. *The frost penetration problem in highway engineering.* New Brunswick, New Jersey: Rutgers Univ. Press.

Kersten, M. S. 1949. *Thermal properties of soils.* Univ. of Minnesota Experiment Station, Bulletin No. 28.

———. 1959. *Frost penetration: Relationship to air temperature and other factors.* Highway Research Board Bulletin 225, pp. 45–80.

———, and Johnson, R. W. 1955. *Frost penetration under bituminous pavements.* Highway Res. Board Bull. 111.

Khakimov, K. R. 1957. *Voprosy teorii i praktiki iskusstvennogo zamorazhivaniya grantov.* Isdat el'stvo Akademii Nauk SSSR, Moskva, (TT66–51051 U.S. Dept. of Commerce).

Lachenbruch, A. H. 1959. *Periodic heat flow in a stratified medium with application to permafrost problems.* U.S. Geol. Survey Bull. 1083-A.

———. 1970. *Some estimates of the thermal effect of a heated pipeline in permafrost.* U.S. Geological Survey Circular No. 632, Washington, D.C.

Lobacz, E. F. 1977. Personal communication.

Lock, G. S. H., Freeborn, R. D. J., and Nyren, R. H. 1970. *Analysis of ice formation in a convectively cooled pipe.* 4th Int. Heat Transfer Conf., Paris, France.

Lunardini, V. J. 1971. *Presentation of some thermal properties of soil systems.* CSME paper 71-CSME-38, EIC General Meeting, Quebec City.

———. 1977*a*. *Geothermal aspects of N-factors and calculation methods.* Internal Report NRC, Contract 032–579, Ottawa, Canada.

———. 1977*b*. Thawing of permafrost beneath a buried pipe. *J. Can. Pet. Technol.,* 16 (4):34–37.

———. 1978*a*. Theory of N-factors and correlation of data. *Proceedings, Third International Conference on Permafrost.* Vol. 1, pp. 40–46, Edmonton, Alberta.

———. 1978*b*. A correlation of N-factor data. Cold Regions Engineering and Construction, ASCE, Anchorage, Alaska, May 17–19, *Proceedings.* Vol. 1, pp. 233–244.

McCormick, G. 1971. Frost penetration factors. *The Northern Engineer* 3 (3):14–16.

McRoberts, E. C. 1975. Field observations of thawing in soils. *Can. Geotech. J.* 12:126–130.

Michel, B. 1971. *Winter regime of rivers and lakes.* CRREL Monograph III-BIA, Hanover, New Hampshire.

Morgenstern, N. R., and Nixon, J. F. 1975. An analysis of the performance of a warm-oil pipeline in permafrost, Inuwik, N.W.T. *Can. Geotech. J.* 12:199–208.

Moulton, L. K. 1968. *A study of the relationship between air temperatures and depth of frost penetration as related to pavement performance of West Virginia highways.* Final Report. Res. Proj. No. 7, Eng. Exp. Stn., W. Virginia Univ.

Oosterbaan, M. D., and Leonards, G. A. 1965. Use of insulating layer to attenuate frost action in highway pavement. *Highw. Res. Rec.* 101:11–27.

Penner, E., et al. 1966. Performance of city pavement structures containing foamed plastic insulation. *Highw. Res. Rec.* 128:1–17.

Porkhayev, G. V. 1963. Temperature fields in foundation. *Proceedings First International Conference on Permafrost.* NRC No. 1287, pp. 285–291.

Pryer, R. W. 1959. Frost action and railroad maintenance in the Labrador Penninsula. *Highw. Res. Board Bull.* 218:34–48.

Quinn, W. F., and Lobacz, E. F. 1962. Frost penetration beneath concrete slabs maintained free of snow and ice with and without insulation. *Highw. Res. Board Bull.* 331:98–115.

Rhode, J. J., and Esch, D. C. 1976. *Kotzebue Airport runway insulation over permafrost.* 2nd Int. Cold Regions Engineering, Univ. Alaska.

Rhodes, E. M. 1974. Ice crossings. *The Northern Engineer* 5 (1):19–24.

Rowley, R. K., Watson, G. N., Wilson, T. M., and Auld, R. G. 1973. Performance of a 48-in warm oil pipeline supported on permafrost. *Can. Geotech. J.* 10:282–303.

Ruckli, R. 1950. Der Frost in Baugrund. Wien: Springer-Verlag.

Sanger, F. J. 1959. Discussion of frost penetration: Relationship to air temperatures and other factors, Kersten, M.S. *Highw. Res. Board Bull.* 225.

———. 1963. Degree days and heat conduction in soils. *Proceedings International Permafrost Conference.* NRC Pub. No. 1287.

Stefan, J. 1889. Uber Die Theorien Des Eisbildung in Polarmere. *Wien Sitzunsber, Acad. Wies., Ser. A,* vol. 42, pt. 2, pp. 269–286.

———. 1891. *Ann. Phys. Chem.* (Wiedemann) N.F. 42:269–286.

Thornton, D. E. 1976. Steady-state and quasi-static thermal results for bare and insulated pipes in permafrost. *Can. Geotech. J.* 13(2):161–170.

Thompson, M. R. 1973. *Environmental factors and pavement systems. State of the Art,* Construction Engineering Res. Lab. T. Rp. 5, pp. 61–186.

U.S. Army. 1966. *Arctic and subarctic construction-calculation method for determination of depths of freeze and thaw in soil.* U.S. Army TM5–852–6.

———. 1972. *Met. Data Report, Farmers Loop Road Test Site.* Fairbanks, Alaska.

U.S. Army Corps of Engineers. 1949. *Addendum No. 1, 1945–47, Report on Frost Investigation.* Frost Effects Laboratory, New England Div., Boston, Mass.

———. 1950*a*. *Comprehensive report investigation of military construction in arctic and subarctic regions 1945–1948.* St. Paul District.

———. 1950*b*. *Frost effects laboratory. Report of pavement surface temperature transfer study.* New England Division, Boston, Mass.

Watson, G. N., Rowley, R. K., and Slusarchuk, W. A. 1973. Performance of a warm-oil pipeline buried in permafrost. *Proceedings, Second International Conference on Permafrost, Yakutsk, Siberia,* pp. 759–766.

Yoder, E. J., and Lowrie, C. R. 1952. Some field measurements of soil temperatures in Indiana. *Highw. Res. Board, Spec. Rep.* 2:41–50.

Zarling, J. P. 1978. Growth rates of ice. *Proceedings of Conference on Applied Techniques for Cold Environments.* Vol. 1. ASCE pp. 100–111.

PROBLEMS

1. Find the thaw depth in Example 5 if the air strip was located at Churchill, Manitoba. Thaw season = 141 days; T_a = 18°F; thawing index = 2056°F-days.
2. Consider the thaw depth at Churchill and at Thule if seven layers are considered as below.

	Layer	γ	W	
.4	$\overline{AC}$			*a*
1	GW–GP	155	2.4	b
2	*GP*	157	1.8	c
3	*GW*	151	3.2	d
4	*GW*	151	3.2	d
5	*GP*	152	2.1	e
6	*SM*	136	6.5	f
7	*SM*	144	4.6	g
8	*SC*	143	4.6	g

3. A pipe 2 ft in radius is buried at a depth of 12 ft, at Inuvik, NWT. If the surrounding soil is a silt at γ_d = 1.3 g/cm³, T_p = 80°C, and W = 20%, calculate the thaw depth beneath the pipe after 10 years. Calculate the total heat loss from a foot of pipe during this time period. How does this compare with the heat loss from the pipe in a completely thawed soil?
4. At Inuvik, NWT, a 4-in. moss layer with a resistance of 2.0 hr-ft²-°F/Btu overlies a silt with γ_d = 136 lbm/ft³ and W = 6.5%. What is the depth of the permafrost table?
5. For the details of problem 4, assume a 4-in. asphalt airstrip is placed over a gravel pad (γ_d = 151, W = 2.5), which is laid directly over the moss. The resistance of the moss is reduced to ⅕ its original value. What depth of gravel will keep the permafrost table constant?

6. In problem 5, suppose that the original moss and active layer are stripped off. What depth of gravel is now needed?
7. A semi-infinite slab of the above soil is initially thawed at 0°C, and the surface temperature is dropped to −10°C and held there. Calculate the depth of freeze after 156 days. Clay: $\gamma_d = 1.3$ g/cm³, $W = 28\%$.
8. Assume a soil is initially at a uniform temperature T_0. The surface temperature is suddenly dropped to T_s and remains there. It can be assumed that the temperature distribution in the frozen layer is linear. The problem is to evaluate in the energy balance the energy removed in cooling the layer to be frozen from the initial temperature to the final temperature of the layer. Write an energy balance at the freezing interface X that will include the sensible heat effect of the cooling process.
9. Discuss the physical basis for the modified Berggren equation and explain how to apply it to multilayer systems.
10. The table below lists some of the properties of a soil that is covered by 1 ft of peat and 2 ft of snow. The location is at Churchill, Manitoba. Find the depth of freezing with and without the peat layer. Note that the snow does not change phase.

Layer	Thickness, ft	W, %	γ_d, lbm/ft³	Specific Heat, Btu/ft³-°F	Thermal Conductivity, Btu/hr-ft² °F/in.
Snow	2	–	–	9.0	1.75
Peat	1	175	10	16	1.00
Soil	50	6.5	136		

11. Predict the thickness of ice on a medium-sized lake with moderate snow cover where the annual freezing index is 2500°F-day.
12. Estimate the ice thickness in a body of water after 10 days of −10°F temperatures. The water temperature is at the freezing point, and the wind speed is 5 mph.
13. Calculate the time required for a 2-in. diameter water pipe initially at 42°F to cool to 32°F with an ambient temperature of −18°F. The pipe is insulated with 2 in. of fiberglass ($k = .025$ Btu/hr-ft-°F).
14. Estimate the frost penetration for a snow covered sand embankment overlying a silty soil when the air freezing index is 60,000°C-hr, the mean annual temperature is −5°C, and the duration of the freezing period is 225 days (5400 hr). The characteristics of the materials are:

Given	Snow	Sand	Silt
k_t, cal/h·m·°C	–	2.0	1.2
k_f, cal/h·m·°C	0.2	2.2	1.6
γ_d, kg/m³	300	2000	1600
W, %	–	5	20
d, m	0.1	1.0	indefinite

15. Calculate the expected maximum ice thickness for a medium-sized lake with moderate snow cover when the annual air freezing index is 2085°F-day.
16. Calculate the input temperature and the rate of heat loss (heat input) for a 9500-ft recirculating water system for various flow rates if the water temperature is to be maintained at a minimum of 40°F when the ground temperature at the pipe depth is 15°F. The pipe is plastic with an outside diameter of 6.5 in. and an inside diameter of 5.4 in., with 2 in. of polyurethane foam insulation.
17. Calculate the design freezeup time and the safety factor time and the complete freezing time for the pipe design used in problem 16 if the water ceases to flow.

11
Heat Transfer for Structures in Cold Regions

11.1 INTRODUCTION

We will limit ourselves to problems of structures as opposed to those of roads, airfields, etc. The heat transfer relations for roads, airfields, etc., in northern climates have been discussed in Chap. 10. The heat transfer problems in structures differ more in degree than in kind with regard to northern problems. In general, the design problems may be divided into two categories: foundations and superstructures. Without doubt the problems associated with foundations lead to the greatest divergence in design procedure when compared with southern construction (nonpermafrost zones). This is due to the thermal interaction of the structure and the underlying permafrost. The upper soil layers undergo cycles of freezing and thawing that result in a loss of strength of the bearing medium. Even in seasonally frozen ground regions, such as the northern United States, the frost may penetrate to more than 2 m. The design of cold regions foundations thus requires a more careful consideration of thermal effects than is normally the case.

11.2 DESIGN PHILOSOPHY IN PERMAFROST ZONES

Under natural conditions, the level of the permafrost is determined by the dynamic thermal equilibrium of the system. Any heat transferred into the region during the thaw season is removed during the freezing season, and a stable permafrost table will exist. A disturbance that tends to introduce more energy during the yearly cycle than was introduced under natural conditions, without a compensating heat removal, will cause the permafrost to reach a new equilibrium at a lower depth or to disappear entirely. Such a process may take many years to reach the new equilibrium. When a heated structure, or any effect that changes the thermal regime, is placed over permafrost, a continuous introduction of excess thermal energy into the permafrost will eventually cause melting and a drop in the permafrost level. It is also possible to alter the thermal regime so that the permafrost level increases, but this is not likely without careful thermal design. Simply insulating the lower part of a structure will not suffice to preserve the bearing strength of the permafrost. Insulation may slow down the thaw rate, but over a number of years the permafrost will slowly degrade.

In the north, during the winter, there is usually ample cooling capacity available, and this excess may be used to offset the added heat addition due to the thermal load of the structure. One must then design so that the cooling effect available during the winter is increased until it at least offsets the thawing that will occur during the summer. Even with a poor design, the heat load is usually not detrimental during the winter season, but when coupled with the natural effects of the thawing season, it can cause a slow but continuous destruction of the underlying permafrost.

For moderate heat loads, the design may merely require proper insulation of the floor with a minimum of ventilation to carry off some of the heat load from the structure. It is desirable, and sometimes necessary, to stop the ventilation effect during the summer, as this would then promote thaw rather than freeze. This may be done with shutters or covers or the use of sunshades. With heavy heat loads, it may be necessary to increase the winter ventilation effect by the use of ducts with stacks or even forced-draft systems. Because forced draft systems require additional maintenance, they should be avoided where possible. Each case will require special consideration as to the exact techniques used, but the overall philosophy will not vary.

As far as heat transfer is concerned, the foundation of a structure can be considered to include the bottom floor of the superstructure, that is to say, the lower region or floor that is directly involved with heat transfer to the ground.

One of the major design problems is the lack of reliable feedback from the operating structure as designed. Many problems could be avoided if the structures were instrumented to monitor performance. This lack of operating data is recognized, but economics seems to preclude large-scale implementation of monitoring equipment.

When building in permafrost regions, it is usual to design on the basis of one of the four following approaches.

11.2.1 Disregard the Thermal Regime of the Underlying Permafrost

This method assumes that thawing of the permafrost will have no detrimental effect upon the structure. The calculated depth of thaw would involve only soils or rock that do not change their load-bearing ability after thaw, such as well-drained, granular soils. This may be acceptable for very small or short-lived structures, but is a rather risky design approach in general. Disastrous failures have occurred when this method was used simply in ignorance of permafrost problems. If a very good site investigation has been carried out, large structures may be so built, but even structures built over rock have failed, where the rock was fractured and ice-filled. In effect, this method is simply the application of conventional techniques in permafrost regions.

11.2.2 Thawing of Permafrost Prior to Construction

If the underlying material (expected to thaw in a 10-year period of normal use) is relatively small, then it may be thawed and compacted or removed and replaced with more suitable fill. Early construction, especially in the Soviet Union, used this method, but it is very expensive and is not now likely to be used. It has been used extensively in Alaska (Sanger 1969) and Russia (Bakakin 1959) in the mining industry. High-velocity water jets have been used to both thaw and remove the permafrost overburden of large-scale gold-mining operations (U.S. Army Corps of Engineers 1958). Soil removal rates of 6 in. per 24 hours are noted, with 20–30 ft of overburden removed. Linell (1973) notes that removal of 6 in. of silts over thaw-stable sands has been used in Alaska and Canada, and also prethawing of up to 9 in. of sandy gravels has been done to ensure consolidation of looser upper strata prior to construction. Sanger (1969) points out that thawing would not be considered for a frozen island more than 40 ft across and 20 ft thick or if the frozen layer thickness were more than 60% of the expected 10-year thaw depth. This procedure would probably be limited to the discontinuous permafrost regions, with no case of complete thawing to remove permafrost reported in North America. Replacement of small volumes of soil containing excessive ice, with non-frost-susceptible fill is possible, where permafrost preservation is impractical.

Some methods that have been used for thawing are noted in U.S. Army (1958). The use of solar energy is cheap but least efficient from the point of view of time. One to four inches of thaw per day is normal. Modification of the surface albedo has been used effectively in Russia. Steam jets that introduce steam into the ground through 1-in. pipes that are forced into the ground as thaw proceeds have been used to thaw localized areas or small excavations. Water needles, which use hot water, are good for larger areas, but require extensive, although cheap, equipment. Thaws to the 4-ft level have been accomplished in several days.

11.2.3 Allowing for Thawing during Construction and Operation of the Structure

In the discontinuous, fringe, permafrost areas where consolidation and thaw must be expected, the permafrost may, under very careful conditions, be allowed to thaw during and after construction. The method should be used only where the foundation materials are thaw-stable or where expedient or short-term construction is involved.

The Soviet Building Code (1960) allows the use of this method if the calculated settlement (magnitude or differential amount) and the rate of settlement do not exceed recommended values. Generally, the design must allow for a decrease in differential thawing, and the structure must be designed to accommodate differential settlements. Flexible foundations that adjust to eliminate deformation as differential settlement occurs as needed. Settle joints, which allow part of the building to settle without disturbing adjacent parts of the structure, are also possible (AINA 1973). Although the Soviet code lists the depth and rate of thaw allowable for various types of structures, designing in this fashion is not wise if it is at all avoidable. The prediction of settlement rates and of differential settlement is so unreliable that the entire procedure must be looked upon as risky.

11.2.4 Maintaining the Permafrost in a Frozen State

It has long been known that if the thermal integrity of permafrost can be maintained, the permafrost will usually be an adequate foundation material from the viewpoint of mechanical strength. As the temperature rises to the fusion point, the strength decreases rapidly, and the soil may be unable to support the design load. Thus, one of the most common design techniques now used is to prevent thawing of the permafrost. This technique, which is vital in continuous permafrost, particularly for fine-grained soils with high ice content, is now widely used in Russia, Canada, and the United States. As the working load of the structure is carried by the permafrost according to the design, it is essential that the permafrost table not be lowered during the life of the structure. Thus, for best results in this method, the temperature of the permafrost should be monitored on a continuous basis for at least the first several years and preferably for the life of the structure. This can be done reliably and simply with thermocouple or thermistor strings. The best way to protect the thermal regime must be decided in each case. Seasonal movement may occur in the permafrost at least to 10 m, and placement of foundation supports to a depth of 20 m may be required (Linell 1973).

11.3 FOUNDATIONS

We shall now examine some preliminary details for achieving the goal of a proper permafrost foundation. At this point, it might be well to note some details of construction as carried out in North America and Russia.

Leggett (1959) notes that until the 1920s almost all building in the Canadian north was done by Hudson's Bay Co., religious missions, or the Royal Canadian Mounted Police. Sherwood (1974) relates some elements of early construction in the Canadian north dealing mainly with small log structures. It is only in the last 40 years that large buildings have been constructed, and even now they are not common. One of the first large buildings was built on piers, for the Royal Canadian Navy, at Churchill, Manitoba, in 1949–1950 (Dickens 1960); see Fig. 11.1. At the time of this construction, there was virtually no information on this type of foundation in Canada. Notice the 4-ft air space to isolate the structure's heat load from the foundation and the use of moss for insulation. A building code for the Canadian North (Assoc. Comm. Nat. Blg. Code Canada 1968*a*, 1968*b*), has been written and contains a number of recommendations. It contains general information, but does not deal with design specifics.

A great deal of information has been published in the Russian literature concerning foundations in the north. The Soviet building code (1960) outlines

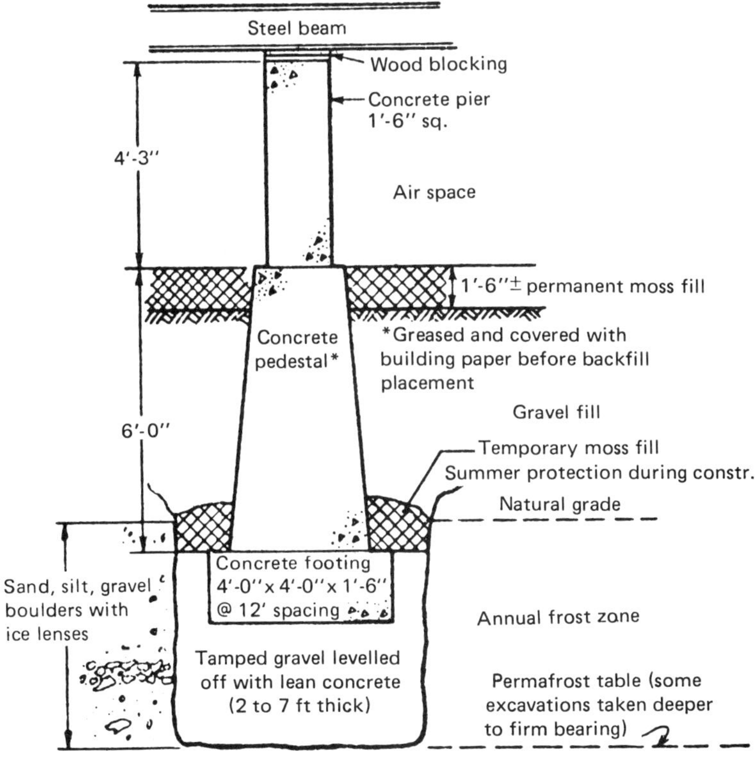

Fig. 11.1. Pedestal and Footing on Permafrost near Churchill, Manitoba (adapted from Sanger 1959).

the theoretical approaches and gives specific details, including tables, charts, and equations to be applied in foundation design. All the methods discussed in Section 2 are examined in detail. It is not possible to apply most of the information directly to North America, but the basic ideas are sound.

11.3.1 Surface Foundations

Surface foundations are laid directly upon the ground surface with little prior preparation.

These may be log mudsills, timber pads, etc., and are only for small or temporary buildings (buildings of one story and 500 ft² area). The following recommendations are taken from the Canadian code:

(i) The gravel pad shall consist of granular material of not more than 3 in. size.

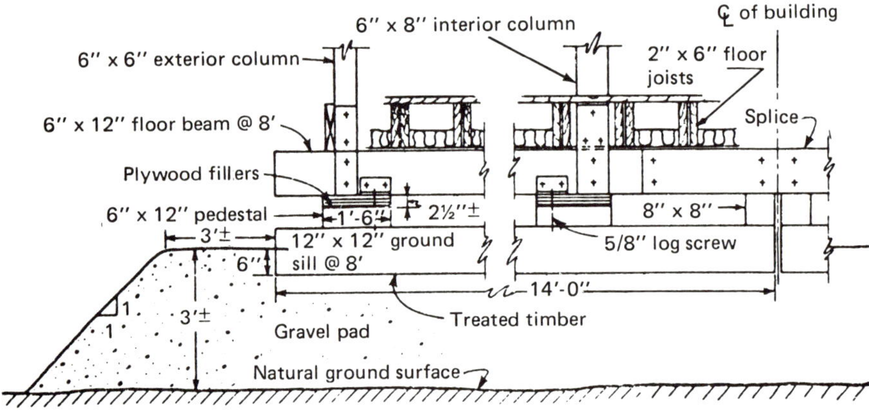

Note the well-insulated floor and the 3-ft gravel pad, both involved in the thermal design necessary to preserve the permafrost. No special care has been taken to ventilate the floor, probably a building with a moderate thermal load.

SITE CONDITIONS
Permafrost Silty, sandy, GRAVEL
(Temp 12°F to 32°F)
Mean Thawing Index 700°F-Day
Mean Freezing Index 8000°F-Day
Mean Annual Temp 12°F

Fig. 11.2. Foundation for Men's Barracks, Thule, Greenland (adapted from Sanger 1969).

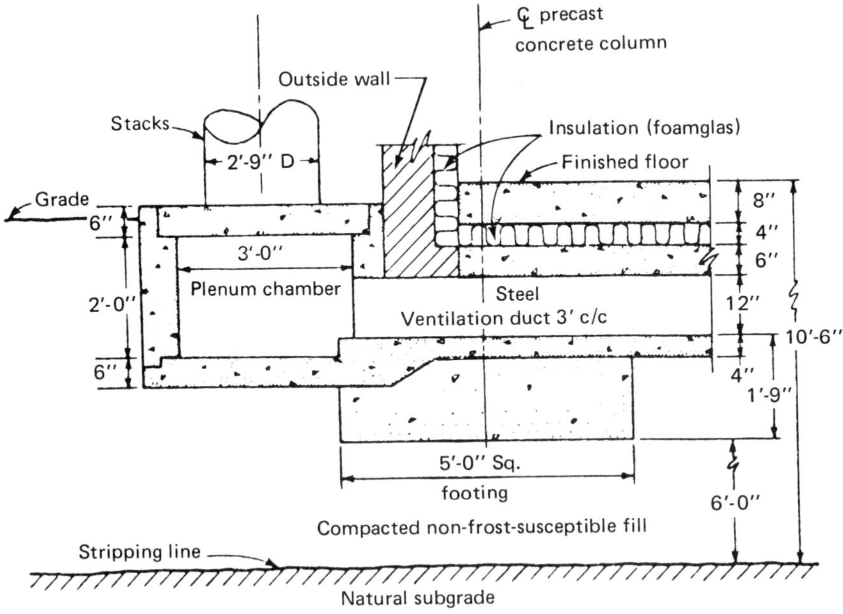

SITE CONDITIONS
Annual Frost Zone: SILT to silty SAND
(Approx. thickness: 8 ft)
Permafrost: GRAVEL through SAND to SILT
(Temp. 28°F to 32°F)
Mean Thawing Index: 1900°F-Days
Mean Freezing Index: 5100°F-Days
Mean Annual Temp: 24°F
Temp. Range: −49°F to +73°F

Internal ventilating ducts connecting to stacks (chimneys). These ducts may employ forced ventilation, but it is better to design for natural draft and avoid use of machinery (pumps, fans) that requires power and maintenance. A design such as this will accommodate heavy thermal and structural loads.

Fig. 11.3. Pan-Slab Foundation Showing Ventilation Duct, Greenland (adapted from Sanger 1969).

(ii) It shall be at least 24 in. above the existing grade. Buildings over 1000 ft^2 shall have at least 36 in. total gravel thickness.

(iii) The top surface of the pad shall extend a minimum of 3 ft beyond the edge of supporting footings or a minimum of 1 ft beyond exterior walls, whichever is greater.

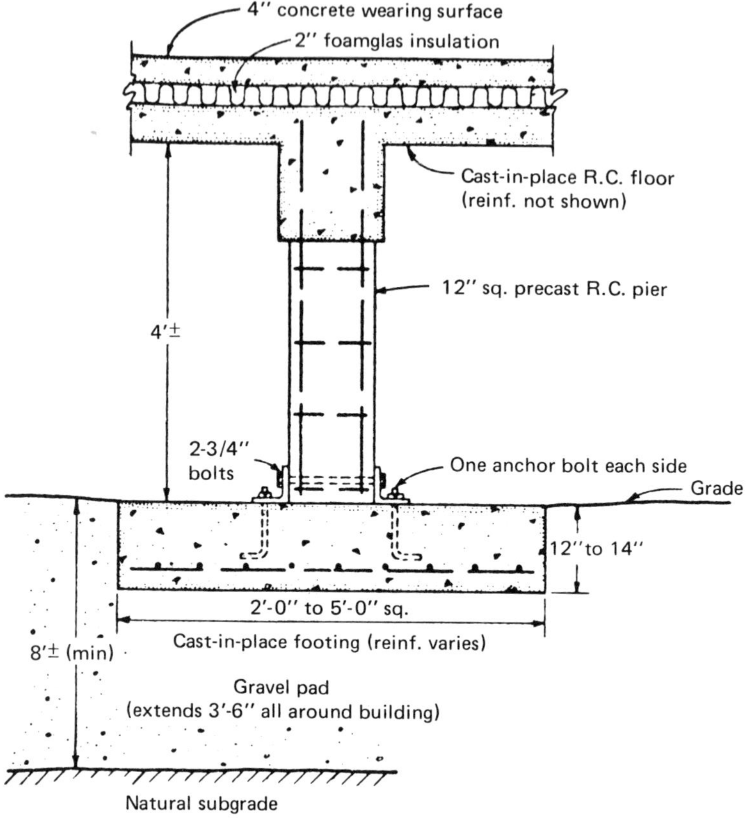

Floor insulated, with 4-ft ventilation space. The natural surface layer has been replaced by about 8 ft of gravel.

SITE CONDITIONS
Permafrost: Sandy SILT (SM-ML)
(Temp: 29°F to 32°F)
Mean Thawing Index: 3200°F days
Mean Freezing Index: 7200°F days
Mean Annual Temp: 20°F

Fig. 11.4. Post-on-Pad Foundation for Composite Building for Yukon, Alaska (adapted from Sanger 1969).

Johnston (AINA 1973) notes that for small structures that can sustain movements 1–2 ft of gravel fill is adequate, though larger structures may require up to 10 ft of appropriate fill. Wedge-leveling is common for houses, but

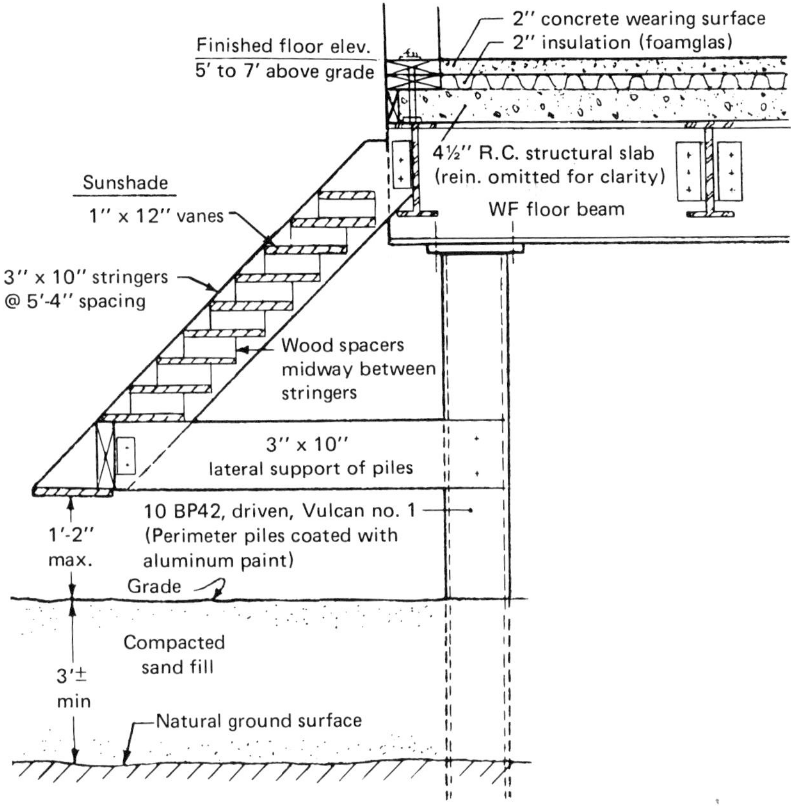

SITE CONDITIONS
Marginal Permafrost
Permafrost: Silty SAND
(Temp: 28°F to 32°F)
Mean Thawing Index: 2700°F days
Mean Freezing Index: 4000°F days
Mean Annual Temp: 30°F
NOTE:
Sunshade along south side of building and extended around S.W. corner. Painted white, to reduce solar load

Fig. 11.5. Steel-Pile Foundation for Utility Building, Bethel, Alaska (after Sanger 1969).

neoprene or vinyl-coated nylon air bags have been used to level structures that are often moved (Platts 1960).

Figures 11.2 through 11.4, taken from Sanger (1969), illustrate successful

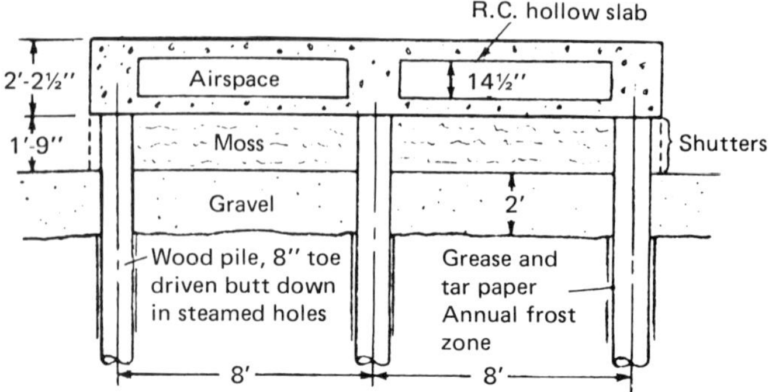

Wooden piles in steamed holes, moss insulation with air shutters to allow cold air to flow in during winter and keep out warm air in summer. Air space in concrete floor increases thermal resistance.

SITE CONDITION
Sporadic Permafrost
Permafrost: Silt CLAY with layers of fine SAND
Mean Thawing Index: 2910°F days
Mean Freezing Index: 5490°F days
Mean Annual Temp: 25°F

Fig. 11.6. Wood-Pile Foundation for Heating Building, Hay River, NWT, Canada (after Sanger 1969).

foundation designs that have been used in the northern regions of various countries, most of which employ some form of thermal control.

11.3.2 Buried Foundations

A considerable number of foundations are buried in a non-frost-susceptible fill, such as well-graded gravel. Some examples of buried foundations that have been used in permafrost regions are given below in Figs. 11.5 through 11.7. The Canadian Building Code notes some suggestions for buried foundations.

(i) If subject to frost action in the active layer the foundation shall be embedded in permafrost of sufficient depth to ensure a firm anchorage.
(ii) The foundation shall not be loaded until adfreezing has developed to provide anchorage against frost heaving.

Robinsky (AINA 1973) notes the use of insulating concretes that contain polystyrene beads mixed into cement and sand. These are not as good as sheet insulation, but give better application methods. Also, it is mentioned that insulated foundations may reduce the fill depth to half the usual amount.

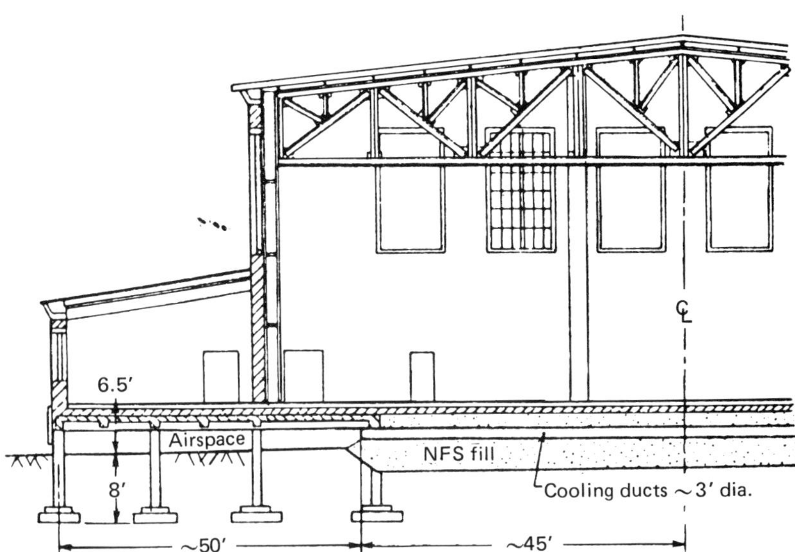

Note the use of a ventilation space and cooling ducts, needed to assure adequate ventilation under the central portions of a large building. It is possible to employ ducts without the use of stacks and chimneys.

Fig. 11.7. Combined Air Space and Large Cooling Ducts, USSR (after Sanger 1969).

11.3.3 Foundations on Piles

The most common foundation method on thaw-unstable permafrost in the north of Canada or Alaska is the ventilated air space on piles (Linell 1973). If wood is used, it must be pressure-creosoted even in dry areas, although in the Arctic Islands untreated wood may last indefinitely (Platts 1960). Steel pipes or H-section piles are also used, precast concrete is used much less, and cast-in-place concrete is only occasionally used.

Piles are placed by slurrying in drilled holes or driving. Steaming is rarely used now, but in Inuvik some 20,000 spruce piles were placed in steam-thawed holes and driven to refusal. The piles were placed 6 months in advance of use (Crawford 1971). Generally the piles are placed 15–30 ft deep. Piles should be embedded in the permafrost to a depth of at least twice the thickness of the active layer or a minimum of 10 ft. Each case should be analyzed separately, but the minimum depth of 10 ft must be used. Piles are well suited to heavy floor loads. The forces acting on a pile during winter loading can be noted in Fig. 11.8.

The design must be such that the adfreezing force plus the load is greater than the frost uplift force. The length of embedment in the permafrost is

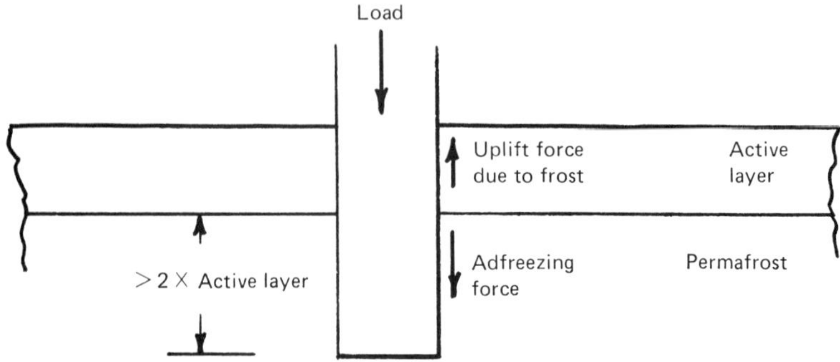

Fig. 11.8. Forces on Pile during Winter.

based on summer design conditions, as the adfreezing force is usually a minimum at that time. Complete loading design details for piles are given in Andersland and Anderson (1978).

If a ventilated space is allowed between the ground surface and the bottom of the building, it should have a clearance of at least 2 ft. This will allow adequate ventilation and access for cleaning of snow, debris, etc; this maintenance is very important.

Piles may be placed by steaming holes with 30–100 psia steam for depths of less than 20 ft and greater than 100 psia for deeper piles. The steaming rates are about 6–8 ft/hr (Sanger 1969) for most soils, although steaming is no longer commonly used. Crory (1963) notes that artificial refrigeration may be needed when thawed holes are used in warm permafrost.

The emplacement of piles into the permafrost must follow structural limitations, as noted earlier. In addition, the act of placing the piles may itself cause a change in the thermal regime of the permafrost if care is not taken. The common method of placing piles in North America is drilling or augering oversize holes and then backfilling with a slurry to freeze the piles into place.

11.3.3.1 Thermal Aspects of Pile Spacing

The most economical procedure is to allow the slurry to freeze naturally by interaction with the permafrost. Mechanical refrigeration is used in special problem situations or for remedial purposes. It is assumed that the permafrost temperature will not rise above the fusion temperature; the mean temperature of the permafrost will rise, but this rise can be compensated for during the winter following pile placement. There are two points to consider: spacing of the piles and the time required for freezeback.

A simple estimate of the spacing of the piles can be made as follows. Consider the system of a pile and an augered, backfilled hole, as shown in

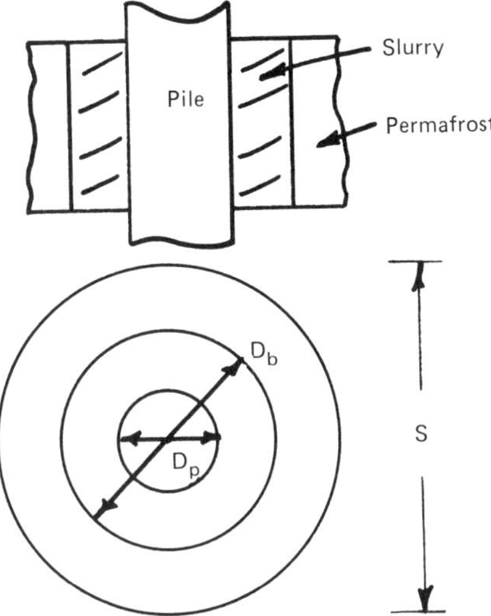

Fig. 11.9. Pile Geometry.

Fig. 11.9. One can use the simple, steady-state application of the first law of thermodynamics, from an initial (unfrozen) to a final (frozen) state, of the backfill and permafrost. It is desired to find the diameter of influence S that the freeze of the backfill will exert upon the permafrost. The first law of thermodynamics is

$$Q - W = \Delta U$$

Because the pressure is constant and there is no heat transfer from the entire system, this is a constant enthalpy process, $\Delta H = 0$. Writing the initial and final enthalpies of the backfill and permafrost, the first law is

$$m_{Bf}\, h_{Bf} + m_{pf}\, h_{pf} = m_{Bi}\, h_{Bi} + m_{pi}\, h_{pi} \tag{11.1}$$

where the subscripts f, i denote final and initial states of backfill and permafrost.

As the masses remain constant

$$m_B(h_{Bf} - h_{Bi}) = m_p(h_{pi} - h_{pf})$$

Now the backfill will freeze, and thus the enthalpy change must take into account the latent heat effect, as follows

$$\rho_B V_B[-h_{if} + c_B(T_{Bf} - T_{Bi})] = \rho_p V_p c_p(T_{pi} - T_{pf})$$

or

$$V_B[-L_B + C_B(T_{Bf} - T_{Bi})] = V_p C_p(T_{pi} - T_{pf}) \tag{11.2}$$

where

h_{if} = latent heat of fusion
L_B = volumetric latent heat of backfill and
C, c = volumetric and mass specific heats.

The volumes per foot of pile length are

$$V_B = \frac{\pi}{4}(D_B^2 - D_p^2)$$

$$V_p = \frac{\pi}{4}(S^2 - D_B^2)$$

The final equilibrium temperature of the system is $T_{Bf} = T_{pf}$. Solving for the influence diameter S, one has

$$S = \sqrt{(D_B^2 - D_p^2)\left\{\frac{L_B + C_B(T_{Bi} - T_{pf})}{C_p(T_{pf} - T_{pi})}\right\} + D_B^2} \tag{11.3}$$

or

$$S = \sqrt{(D_B^2 - D_p^2)\left\{\frac{1.8L_B + C_B(T_{Bi} - T_{pf})}{C_p(T_{pf} - T_{pi})}\right\} + D_B^2} \tag{11.3a}$$

where

D = ft
L = cal/cm^3
C_p = cal/cm^3 − °C
T = °F and
S = ft.

The temperature rise of the permafrost, for a given pile spacing, is

$$T_{pf} - T_{pi} = \frac{L_B + C_B(T_{Bi} - T_{pi})}{KC_p + C_B} \tag{11.4}$$

where

$$K = \frac{(S/D_B)^2 - 1}{1 - \left(\frac{D_p}{D_B}\right)^2}$$

Example 1: A 1-ft diameter, wooden pile is placed in a 1.5-ft augured hole in permafrost at an initial temperature of 27°F. The final permafrost temperature is to be 31°F. If the permafrost and backfill properties are as follows, calculate the thermal disturbance radius about the pile.

Permafrost			*Backfill*
$T_{pi} = 27°F$	$C_p = .453$ cal/cm³-°C	$T_{Bi} = 40°F$	$L_B = 42$ cal/cm³
$\gamma_d = 0.97$ g/cm³		$\gamma_d = 1.28$	$C_B = 0.72$ cal/cm³-°C
$W = 60\%$		$W = 40\%$	
$D_B = 1.5$ ft		$T_{pf} = 31°F$	
$D_p = 1$ ft			

Then the diameter of influence is

$$S = 7.67 \text{ ft}$$

Note that a considerable diameter of influence is exerted by the freezeback process. For best design, if the loading can be justified, a spacing of at least 5 ft should be allowed. Excessively wet slurries should be avoided, and a high quality backfill is desirable. If the structural needs require a smaller value of S (more piles), then placing the piles when the permafrost is at its minimum temperature—early spring—may help or artificial refrigeration could be used. If the initial permafrost temperature in Example 1 is 24°F, the spacing is 5.8 ft.

11.3.3.2 Time for Natural Freezeback

The time required to reach a new thermal equilibrium is important, as the foundation must not be loaded before freezeback is complete. Generally the process is considered complete when the temperature reaches 90% of the natural permafrost temperature. This may be very long or may never occur if the permafrost is near 32°F or if the holes have been steamed initially. It is often desirable to place the piles a year before construction, in the spring, when the permafrost temperature is at a minimum. In this case, care must be taken to avoid heaving of the unloaded piles. The process of freezeback is a complicated heat transfer problem. Consider the transient behavior of the slurry and permafrost materials (Fig. 11.10). The slurry, initially at T_o, drops in temperature, while the permafrost at T_p increases in temperature. When the temperature between the slurry and permafrost, T_s, drops to T_f, the slurry starts to freeze, and the process continues until $R = r_p$, when the slurry is completely frozen. The problem is one of three regions, permafrost, frozen slurry, and thawed slurry, complicated by a variable boundary temperature T_s. Unfortunately, even a simplified quasi-steady analysis cannot be completed, taking into account the freezing process.

An approximate solution to this problem has been completed by Lee (1969). The analysis does not consider the actual phase change, but assumes that the latent heat of the slurry acts as a constant temperature heat source for radial flow to the infinite permafrost sink. With this simplification the problem reduces to a standard transient problem (Carslaw and Jaeger 1959). This solution forms the basis of the freezeback design used in the United States (Crory 1963). Figure 11.11 presents the solution graphically.

The time for the slurry temperature to drop to within 10% of the original permafrost temperature is about double the freezeback time. Thus a factor of safety of 3 should allow the pile to be firmly frozen in place.

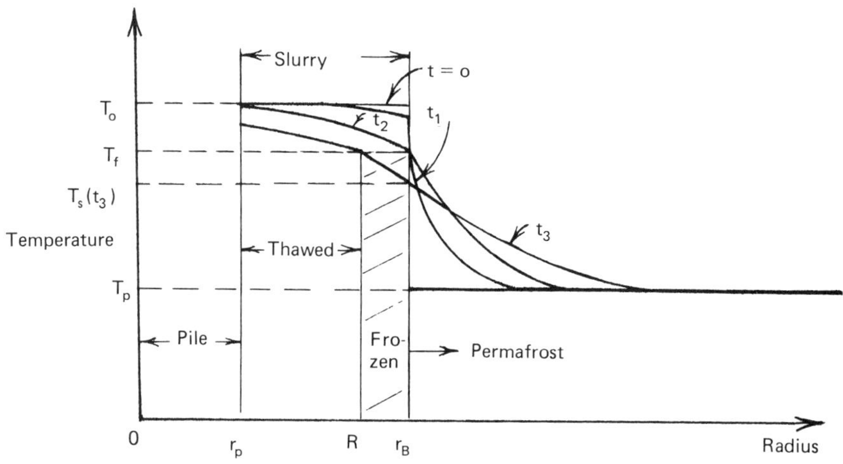

Fig. 11.10. Transient Temperatures During Freezeback.

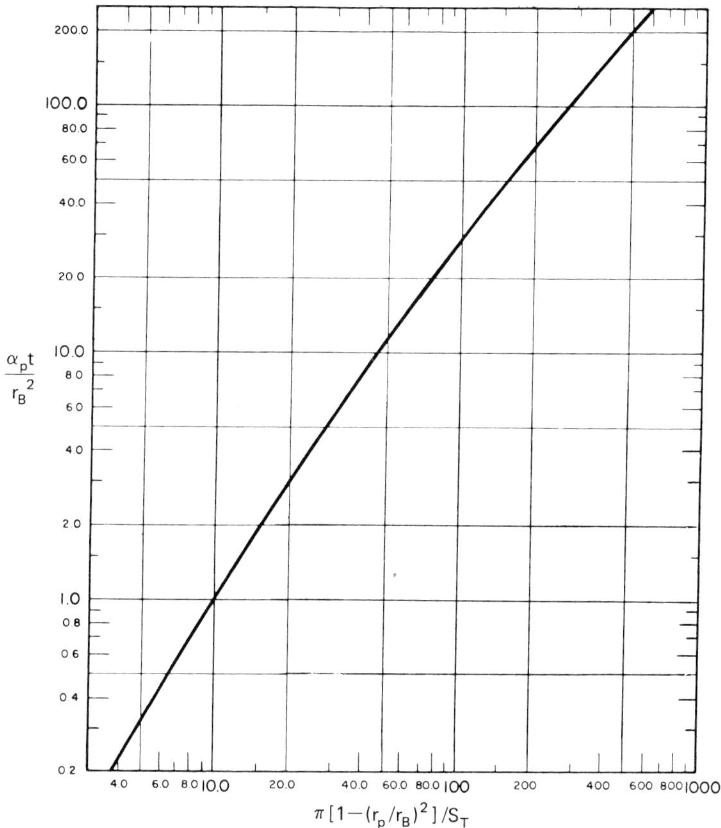

Fig. 11.11. General Solution of Slurry Freezeback (adapted from U.S. Army 1966).

Example 2: A steel pile 25 cm in diameter is set in a 46-cm augered hole. The slurry has $\gamma_d = 1.92$ g/cm³, $W = 15\%$, and the permafrost is a saline, ice-rich, sandy silt with $\gamma_d = 1.1$ g/cm³, W = 100%. The permafrost temperature at depth is as follows:

Depth, m	Permafrost initial temperature, °C
2.7	−4.0
3.9	−7.0
5.1	−9.0

Estimate the freezeback time at the various depths. The thermal properties of the permafrost are $k_t = 2.0$ kcal/hr-m-°C, $C_f = .73$ cal/cm³-°C, the latent heat of the slurry is 23 cal/cm³.

Depth, m	T_p, °C	$S_T = \frac{C_{pf}(T_f - T_p)}{L_B}$	$\pi\left[1-\left(\frac{r_p}{r_B}\right)^2\right]/S_T$	$\frac{\alpha_p t}{r_B^2}$ [a]	Freezeback time, t, hr	Measured time, hr [b]
2.7	−4	.127	17.4	2.4	46.3	38
3.9	−7	.222	10.0	1.0	19.3	22
5.1	−9	.286	7.7	.62	12.0	8

[a] From Fig. 11.11.
[b] Tobiasson and Johnson (1978).

With a factor of safety of 3, an allowed time of 6 days seems adequate.

11.3.3.3 Refrigeration

If natural freezeback is too long, then artificial cooling can be used for rapid freezeback of piles or when permafrost temperature is high and any increase would be critical. Refrigeration tubing around a pile is frequently used in the marginal permafrost areas, but is not necessary if the mean annual ground temperature is less than −4°C (Linell 1973).

Example 3: Consider the piles described in Section 11.3.3.1. If the length of a pile is 31.6 ft, then the slurry content is 31 ft³/pile. The slurry properties are: $C_s = 30$ Btu/ft³-°F; $L_s = 3000$ Btu/ft³; $T_{Bi} = 40$°F; $T_{BF} = 25$°F. Assume the refrigeration available = 18.5 tons = 222,000 Btu/hr and, there are 80 piles. The total heat to be extracted from the backfill is

$$Q_s = 80 \times 31\,[(40 - 25)(30) + 3000] = 8.56 \times 10^6\ \text{Btu}$$

Then, with the refrigeration capacity available, the time to freeze back is

$$\text{time} = \frac{8.56 \times 10^6}{.222 \times 10^6} = 38.6\ \text{hr (allow 2 days)}$$

The brine to be circulated must be specified. The ground temperature is 25°F, and the highest brine temperature should be 10°F less than this, and with a 5°F rise in the brine temperature this gives a minimum brine temperature of 10°F. A 21% NaCl brine will be suitable with specific gravity = 1.159, $c = .8$ Btu/lbm-°F, 1 gallon = 9.65 lbm. It is assumed that the air temperature will not drop below 0°F. If this is not true a different brine will be needed to avoid freezing of the brine.

The circulation rate of the brine is found from

$$Q = \dot{m}\, c\, \Delta T \tag{11.5}$$

and with a $\Delta T = 4°\text{F}$ for the brine (this is the design temperature rise of the brine)

$$\dot{m} = \frac{222{,}000}{(.8)(4)(60)} = 1156.3 \text{ lb}_\text{m}/\text{min} = 120 \text{ gal/min}$$

Four 30-ft coils of ¾-in. nominal tubes are used for each pile (various numbers can be used). The flow velocity is found from the relation $\dot{m} = \rho A V$, with the velocity calculated as 55 ft/min.

The pressure drop is found, as usual, from the friction factor relation (ASHRAE 1979). The pressure drop $\Delta p = 0.6$ psi per coil (allow 20% for bends, or calculate equivalent lengths of bends). Based on a flow velocity in the headers of 5.3 ft/sec, the header tube diameter is 3 in., and the length is 1000 ft, allowing for bends. Using the same procedure, the header pressure loss is

$$\Delta p = 22 \text{ psi}$$

Thus, the total pressure loss is

$$\Delta p = 23 \text{ psi} = 54 \text{ ft } H_2O$$

Allow 35 ft head loss in the brine cooler; then $\Delta p_t = 89$ ft H_2O = 38 psi. The power requirements are

$$Q - W = \Delta H \tag{11.6}$$

$$-W = \dot{m}\left(\frac{\Delta p_t}{\rho}\right)$$

$$W = \dot{m}\frac{\Delta p_t}{\rho}\frac{144}{550}$$

$$W = \frac{(19.3)(38)}{72}\frac{144}{550} = 2.7 \text{ hp, say 3 hp, required in the compressor}$$

11.3.3.4 Thermosyphons as Support Members

It is possible to incorporate refrigeration into the piles without using mechanical effects. This is done by using various forms of thermosyphons, or heat pipes, which operate by gravity flow. These have been used in installations requiring an absolute minimum of movement, such as radar installations (Long 1966). For this case, sketched in Fig. 11.12*a*, a steel pipe was used with propane as the working fluid, at about 4 kg/cm². During the winter $T_a < T_g$, and the heat flow into the tube, below grade, causes the propane to boil and carry the energy as latent heat in the vapor. The vapor condenses along the tube, above grade, where the temperature is less than the boiling point of the propane. The latent heat given up is delivered to the atmosphere through radiators designed to augment the convective effect. The condensate then returns, by gravity, to the boiler section. The ground temperature is thus lowered, during the winter, in excess of natural cooling. The thermosyphon acts as a one-way heat valve, as gravity will oppose the fluid motion during the summer when the temperature gradient is reversed. Figure 11.12b shows a simple thermosyphon used by the Russians (Tsytovich 1975). A closed circuit of tubes used kerosene as a working fluid. The density of the fluid increases in the exposed section causing the kerosene to flow down one leg and up the other. The kerosene velocity was 4 cm/sec with $(T_g - T_a) = 40°C$ and 1 cm/sec for a 10°C temperature difference.

The largest use of heat pipes for permafrost protection has been in the construction of the Trans-Alaska Pipeline (Wheeler 1978). The heat pipes, sketched in Fig. 11.12c, are similar in operation to those described above.

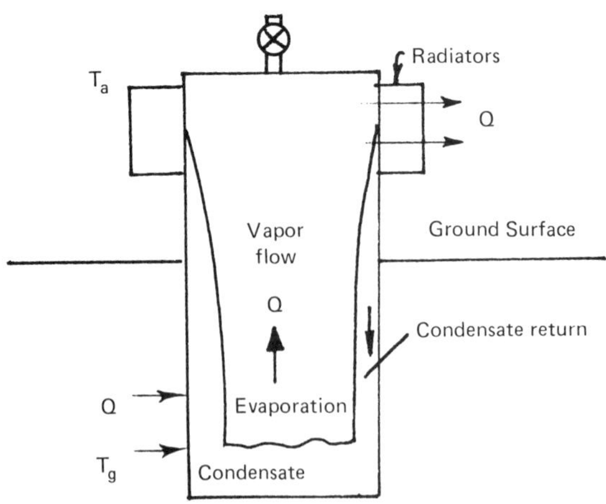

Fig. 11.12*a*. Heat Pipe Used as Thermopile.

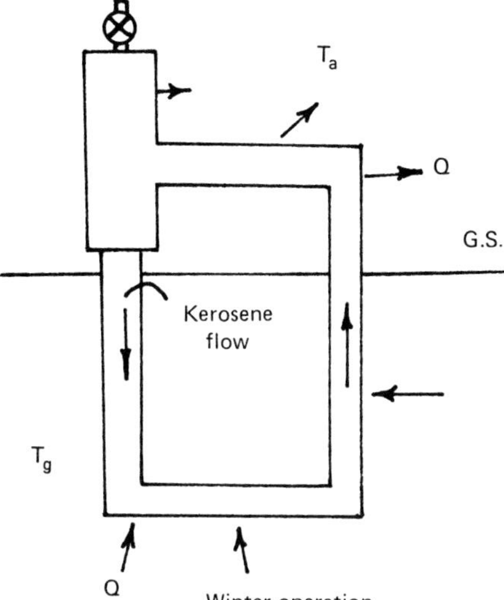

Fig. 11.12*b*. Thermosyphon Used to Cool Ground.

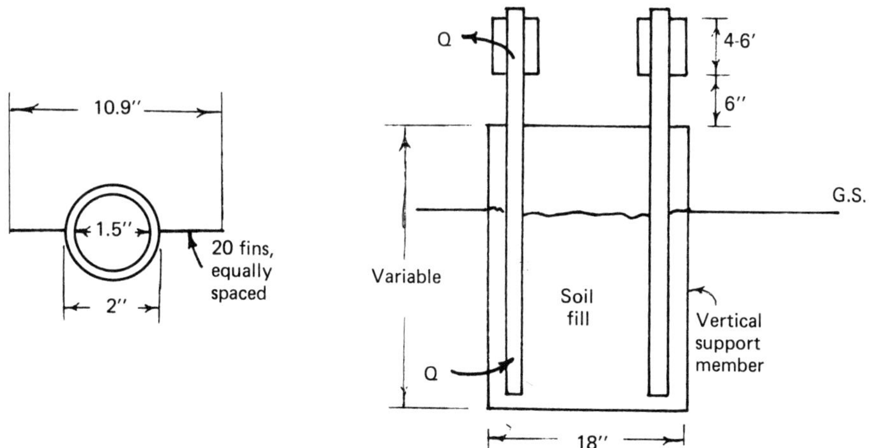

Fig. 11.12*c*. Heat Pipes in Vertical Support Members for Oil Pipe above Permafrost.

Ammonia was used as the working fluid. Two heat pipes were used in each of 61,000 vertical support members. Data indicate that these heat pipes have maintained the soil near the VSM at less than the freezing value for most of the year.

11.3.4 Cooling Ducts

When heavy heat loads are encountered, the heat exchange due to air gaps beneath structures, or simple ventilation, may be inadequate to prevent permafrost damage. In this case, when the foundation is buried, cooling ducts may be incorporated into the foundation itself. The cost of the duct system can easily become excessive if care is not taken in the design. Experience has shown that ducts placed below ground level tend to collect ice and sediment, and artificial thawing may cause ground temperature disturbances (Linell 1973). The ducts should be placed above ground or be self-draining during the summer to get rid of any winter accumulations. The design and function of ducts are best described with a specific example.

Example 4: Duct Design. Consider the heat transfer design of the structure shown in Fig. 11.13*a*. It is required to calculate the depth of gravel fill and the duct system, as well as the floor thermal resistance. No consideration will be given to the mechanical loads on the foundation.

The following information is given:

$L = 220$ ft, length of a duct (determined by size of building)
$T_F = 60°$F year-round temperature of the floor

Properties of gravel pad

$$\gamma_d = 2.0\ g/cm^3 \qquad W = 2.5\% \qquad x_l = .05$$

$$k_u = 1.10\frac{Btu}{hr\ ft°F} \qquad C_u = 23.4\frac{Btu}{ft^3 °F} \qquad L_s = 450\frac{Btu}{ft^3}$$

FI = 6400°F-days freezing index (The mean freezing index at the site is 8000, but the minimum value expected is about 80% of this.)

FS = 228 days, length of freezing season
TS = 137 days, length of thawing season
$k_i = .0333$ Btu/hr-ft-°F, conductivity of foam glass insulation
$k_c = .47 - .81$, conductivity of concrete and
$k_F = .11$, conductivity of floor finish layer (wood).

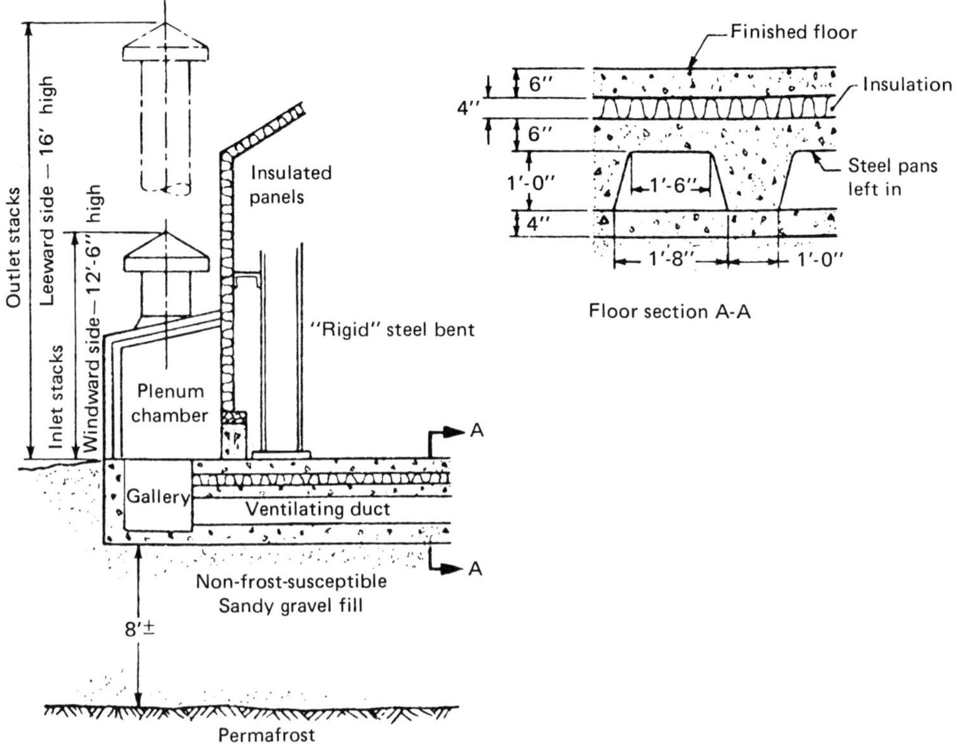

Fig. 11.13*a*. Ducted Foundation for Example 4 (adapted from Sanger 1969).

The operating procedure is as follows. During the thawing season, the gravel pad will thaw to a certain depth. The air ducts are closed during this time to prevent warmer air from outside augmenting the thaw effect. During the freezing season, the thawed layer of gravel must be refrozen so as to keep the permafrost table stationary; thus the ducts are open allowing free circulation of cold outside air by natural convection. The first task is to calculate the thaw depth that will occur. It is necessary to determine the thermal resistance of the floor, and we must guess a thickness of insulation to use. We may carry out this first guess by the use of the following empirical relation (from Sanger 1969)

$$H = \frac{1}{140}\left[\frac{L}{v_s(t+1)}\right]^2 (L + 250) \tag{11.7}$$

where H = stack height (ft) and t = thickness of foamglass insulation (or equivalent). It is important to note that this first guess need not be too accurate and will be checked later by a more precise method.

Now v_s = FI/FS = 6400/228 = 28.1°F. This is the average daily temperature below freezing for the freezing season. The average air temperature during the freezing season is 3.9°F. If the insulation thickness is 4 in., then the initial stack height is 8.3 ft. Now the pad thickness must be greater than the depth of thaw given by

$$X = k_u R \left\{ \sqrt{1 + \frac{48\lambda^2 I_f}{k_u L_s R^2}} - 1 \right\}$$

where

R = resistance of the structure above the pad
I_f = thawing index at the floor
L_s = latent heat of soil and
X = depth of thaw penetration (pad thickness needed).

The thawing index at the floor is calculated as follows

$$I_f = (T_F - 32)\,\text{TS}$$

For our case

$$I_f = (60 - 32)(137) = 3836 \text{ degree-days}$$

Different values of T_F, TS will, naturally, alter this accordingly. The mean temperature of the pad at the start of thawing will be about 26°F, and if $T_s = T_F = 60°$, then T_o = 26°F and

$$\alpha = \frac{T_f - T_o}{T_s - T_f} = \frac{32 - 26}{60 - 32} = 0.21$$

$$\mu = \frac{C_u}{2L_s}(T_s - T_f) = \frac{(23.4)(28)}{2(450)} = 0.73$$

Then from Fig. 10.4, λ = 0.72.

The resistance of the floor is calculated in the usual way for a series wall, assuming that the resistance of the air layer is the same as that for concrete (actually k air space ≃ 1.0 Btu/hr-ft-°F). The resistance is 13.5 < R < 16.1. We shall use R = 13.5 ft²-hr-°F/Btu. Then the pad thickness should be $X = 6.4$ ft. It may be noted that in this calculation there is some uncertainty about λ, due to the uncertainty in T_o. If λ is 1.0, then the pad thickness would be 10.8 ft. Thus, we will choose a value of 8.6 ft, which is nearly the average value.

Heat Added to Pad during Thawing. If we return to our basic equations (see Section 10.5.3) we can note that the temperature of the top of the gravel pad after complete thaw is

$$T_{d_l} = T_s + \frac{(T_f - T_F)d_l}{d_l + k_l X/k_u} \tag{11.8}$$

where k_1, d_1 refer to the conductivity and thickness of the floor: $d_1 = 2.67$ ft, $k_1 = (d_1/R)$.

Solving for our case, we find that the approximate mean temperature of the frozen pad is 26°F, and that of the unfrozen pad is 37.4°F. The temperature profiles at the beginning and end of thaw are as shown in Fig. 11.13*b*.

The heat added to the pad is made up of sensible and latent heat. Consider a column of the pad 1 ft² in area

$$Q_A = X(L_s + C\Delta T) = 8.6[450 + 23.4(10.3)] = 5935 \text{ Btu/ft}^2$$

Note that the sensible heat is about 54% of the latent heat. The percentage is high because the water content is quite low. During the freezing season, when the ducts are opened, this amount of heat must be withdrawn from

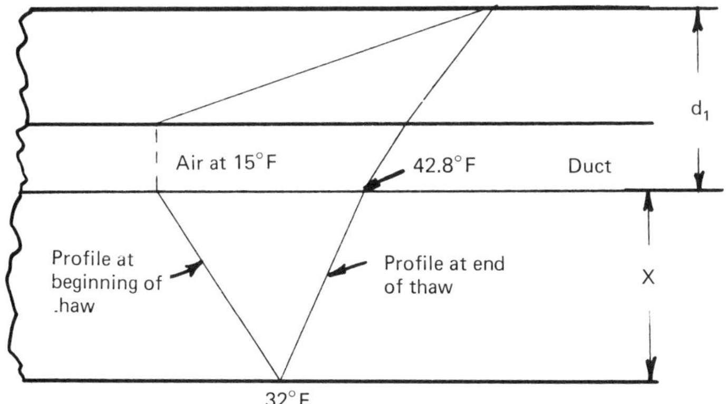

Fig. 11.13*b*. Temperature Profiles in Duct System.

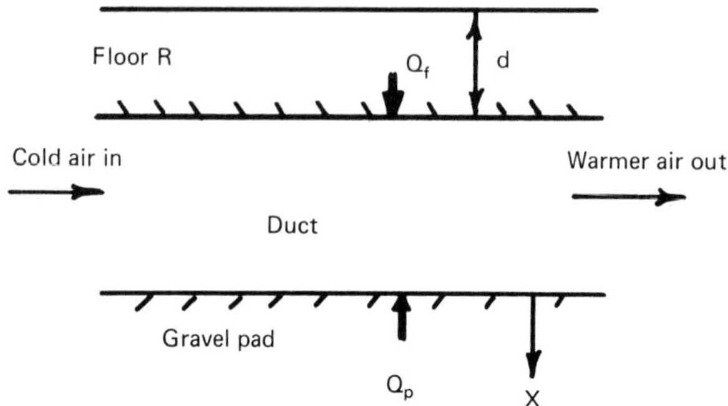

Fig. 11.13*c*. Heat Flows in Duct During Freeze Season. Q_f = heat flux rate into duct from the heated floor and Q_p = heat flux rate into duct from the pad.

the pad to refreeze 8.6 ft of gravel. Examine Fig. 11.13*c* showing the heat transfer flow during the cooling season. The freezing season is 228 days, and therefore the average rate of heat withdrawal must be

$$Q_p = \frac{5935}{228(24)} = 1.09 \frac{\text{Btu}}{\text{hr ft}^2}$$

The freezing index at the gravel pad surface can be calculated from

$$I_p = \frac{L_s X^2}{48\lambda^2 k_f} = 1602°\text{F-days} \tag{11.9}$$

where

$$k_f = 0.73 \qquad C_f = 21.8 \qquad T_o = 36.7 \qquad T_s = \frac{T_{st} + T_{sf}}{2} = \frac{40.5 + 3.9}{2} = 22°F$$

$$\alpha = 0.4 \qquad \mu = 0.3 \qquad \lambda = 0.77$$

Thus, the average surface temperature of the duct should be

$$32 - T_{\text{out}} = \frac{1602}{228} = 7.0$$

$T_{\text{out}} = 25°F$ (This is conservative because it gives a lower mean temperature. It assumes that outlet air is at the same temperature as the duct surface.)

The average inlet temperature of the air is 3.9°F, as was found earlier, and the mean temperature of the air is 14.4°F. Thus

$$Q_f = \left(\frac{60 - 14.4}{R}\right) = \left(\frac{60 - 14.4}{12}\right) = 3.80 \frac{\text{Btu}}{\text{hr ft}^2}$$

The total heat flow to be picked up by the air is $Q_T = Q_p + Q_f = 4.89$. Now for air at 14.4°F, $\rho = .084\ \text{lb}_\text{m}/\text{ft}^3$, $c_p = 0.24\ \text{Btu/lb}_\text{m}\text{-°F}$. Each duct covers a 32-in. strip of floor and must take up

$$Q_D = \left(\frac{32}{12}\right)(1)(220)\ 4.89 = 2867 \frac{\text{Btu}}{\text{hr}}$$

The cross-sectional area of a duct is 1.58 ft² and

$$Q_d = \dot{m}\ c_p\ \Delta T = \rho A V c_p \Delta T \tag{11.10}$$

Thus, the mean air velocity is

$$V = 71.1\ \text{ft/min} = 1.19\ \text{ft/sec}$$

Check of Design. The design must now be checked to see if the stack height will deliver the required pressure differential. Assume $\epsilon = .003$, duct roughness. The equivalent length of duct bends can be found from reference tables (ASHRAE 1977) as

$\dfrac{L_e'}{D}$	*Right Angle Bend*	*Outlet*	*Inlet*
	65	10	10

Therefore $\dfrac{L_e'}{D} = 150$ for each duct. This is the equivalent length of duct due to fittings and bends.

Let the stack efficiency be 80%. Thus, $L = 220 + 20 + 5 = 245$ ft, length of duct plus stacks; $D_e = 4A/p = 1.22$ ft, equivalent diameter of the duct; $L_e' = 183$ ft; and $L_e = 428$ ft.

Now the Reynolds number is

$$R_e = 12{,}020 \qquad \epsilon/D_e = \frac{.003}{1.22} = .0025$$

The friction factor can be found from ASHRAE (1977)

$$f = .0055\left[1 + \left(20{,}000\frac{\epsilon}{D_e} + \frac{10^6}{R_e}\right)^{1/3}\right] \tag{11.11}$$

Thus the friction factor is $f = .035$. Therefore, the pressure drop in the duct is

$$h_f = f\frac{L_e}{D_e}\frac{V^2}{2g_c}\frac{\rho_a}{\rho_w}$$

$$h_f = 4.33 \times 10^{-3}\ in\ H_2O$$

A pressure increase of this amount must be generated by the stack height. The pressure rise for a stack may be calculated as follows, for the starting condition and for the steady flow condition.

For starting conditions, the inside temperature is uniform, and the density is uniform such that $\rho_o > \rho_i$. The inside air is heated by the floor, etc. Now

$$p_1 = p_o + \rho_i g h_i - \rho_o g(h_o - h_i)$$
$$p_2 = p_o + \rho_i g h_o$$

Then the pressure rise is

$$p_1 - p_2 = \frac{g}{g_c}(\rho_o - \rho_i)(h_o - h_i) \tag{11.12}$$

Note that the starting pressure rise depends upon the inlet and outlet stacks having different heights (Fig. 11.14).

In the case of steady-flow conditions, the air is heated as it flows through the stack

$$p_1 - p_2 = \frac{g}{g_c} h_o(\rho_o - \rho_i) \tag{11.13}$$

In this case the pressure rise does not depend upon the difference in the inlet and outlet stack heights

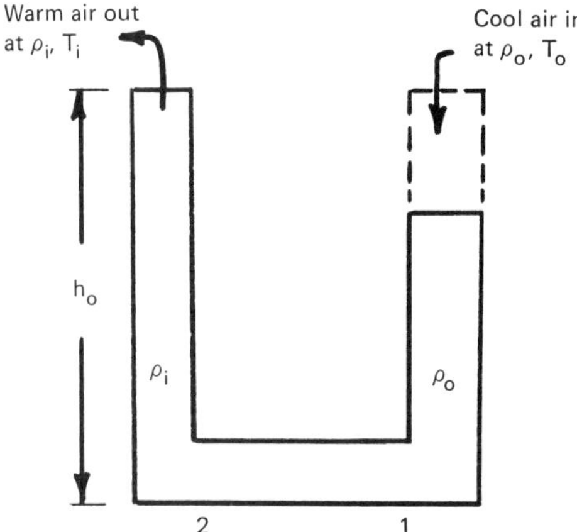

Fig. 11.14. Pressure Rise in Stacks.

$$h_D = H(\rho_o - \rho_i) = H\frac{P}{R}\left(\frac{1}{T_o} - \frac{1}{T_i}\right)$$

$$H = .131\ h_D \frac{T_o}{\left(1 - \frac{T_o}{T_i}\right)}$$

But

$$h_d = h_f$$
$$T_o = 463.9°R$$
$$T_i = 485°R$$
$$H = 6.05 \times 1.2 = 7.3 \text{ ft}$$

A stack with a height of 8 ft would probably be suitable. Notice that our original estimate of 8.2 ft was quite good, but is not really necessary for the calculation. If needed, additional insulation thickness could be used.

11.4 TEMPERATURES BENEATH HEATED STRUCTURES

We have already considered the temperature beneath structures, without any phase change, and the effect of phase change on the rate of thaw or freeze beneath an infinite medium. The rate of thaw beneath a building will be influenced by the thawing index at the bottom of the building and the freezing index outside the building.

Unless special precautions are taken, the permafrost beneath a heated structure will continue to thaw for many years. Figure 11.15 shows the progressive effect of thawing beneath a heated building in Russia. Figure 11.16 shows a similar effect for a building in Alaska. Both these buildings are for designs that did not attempt to reduce the thaw rate. Notice that in both cases the thaw continues as long as measurements are made; the Russian building has a decreasing yearly rate of thaw with time, and the Alaskan building has a more or less constant thaw rate.

No exact solutions are available for phase change beneath two- or three-dimensional structures, but there are some useful approximate relations.

11.4.1 Quasi-Steady Solutions

Consider the steady solution, given in Chap. 7, for the case when a surface is divided into two half-planes at different temperatures, as shown in Fig. 11.17. The temperature at any location is

$$T - T_o = (T_p - T_o)\frac{\tan^{-1}(z/x)}{\pi} = (T_p - T_o)f(x,z) \qquad (11.14)$$

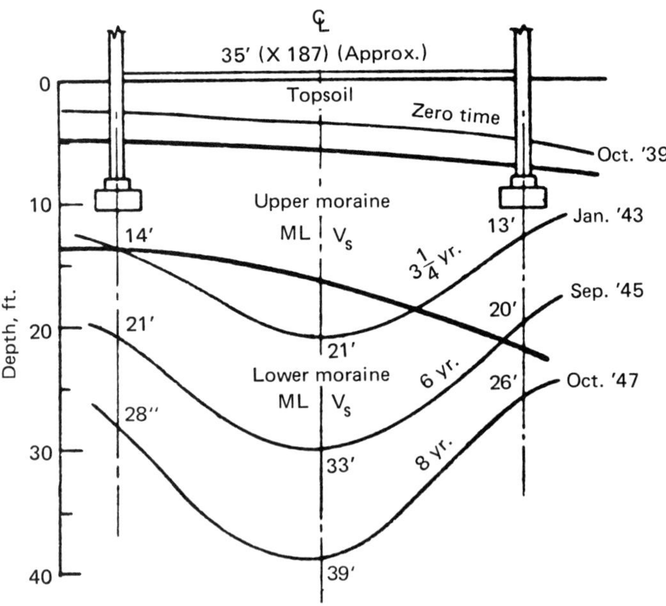

Fig. 11.15. Thaw Progression beneath a Heated Building, USSR (adapted from Sanger 1969).

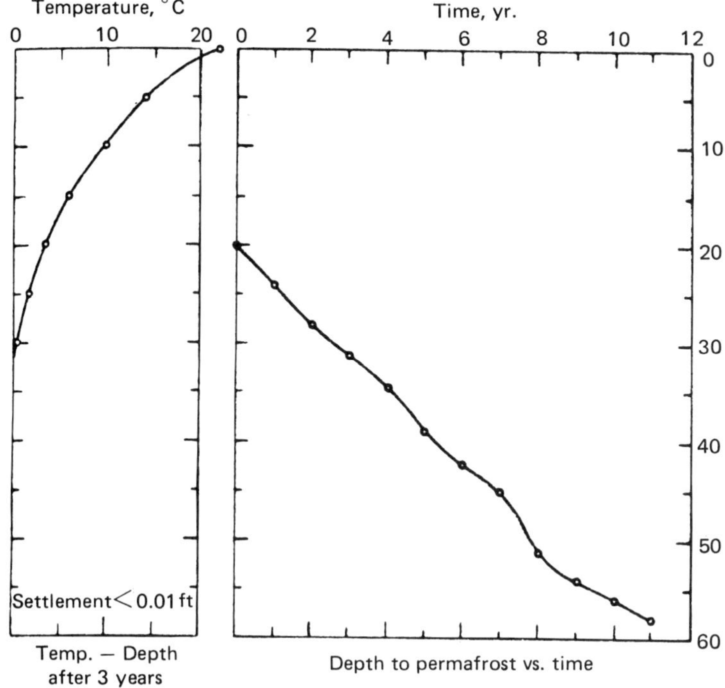

Fig. 11.16. Degradation of Permafrost under a Hospital near Fairbanks, Alaska (adapted from Sanger 1969).

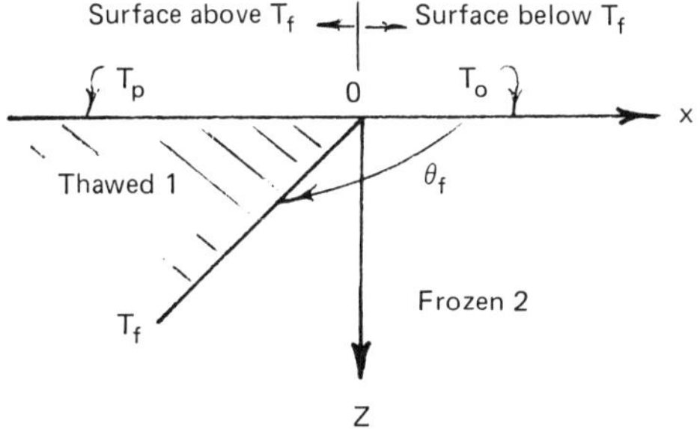

Fig. 11.17. Steady-State Temperatures.

If $T_p \geq T_f$, and $T_o < T_f$, then the freezing isotherm is the radius (or plane) located at θ_f where

$$\theta_f = \left(\frac{T_f - T_o}{T_p - T_o}\right)\pi \tag{11.15}$$

If we are interested in the transient problem where θ_f starts at π and gradually moves to the steady-state location given by Eq. (11.15), we can imagine T_p and T_o changing with time in such a way that this is accomplished.

Then a fictional freezing isotherm can be written as

$$\theta_{fB} = \left(\frac{T_f - T_{oB}}{T_{pB} - T_{oB}}\right)\pi \tag{11.16}$$

where T_{oB} and T_{pB} are fictitious boundary temperatures. Because θ_{fB} should approach θ_f as time approaches infinity, T_{oB} and T_{pB} should also approach the steady-state values.

Thus, the temperatures in the thawed and frozen zones can be written as

$$T_1 - T_{oB} = (T_p - T_{oB})f(x,z) \tag{11.17}$$

$$T_2 - T_o = (T_{pB} - T_o)f(x,z) \tag{11.18}$$

Let f_o be the value of the function f where the phase change temperature exists. The temperatures T_{oB}, T_{pB} can be found by evaluating Eqs. (11.17) and (11.18) for the phase change temperature T_f. Then

$$T_{oB} = \frac{T_p f_o - T_f}{f_o - 1} \tag{11.19}$$

$$T_{pB} = \frac{T_f - T_o}{f_o} + T_o \tag{11.20}$$

Equations (11.17) and (11.18) are then

$$T_1 = T_p f + \frac{(T_p f_o - T_f)(1 - f)}{f_o - 1} \tag{11.21}$$

$$T_2 = \frac{(T_f - T_o)f}{f_o} + T_o \tag{11.22}$$

The boundary condition along the phase change interface is given by the familiar expression

$$\int_S \left(k_1 \frac{\partial T_1}{\partial n} - k_2 \frac{\partial T_2}{\partial n}\right)_{fo} ds = \pm L \frac{dV}{dt} \tag{11.23}$$

where V = volume contained between the phase interface and the outer thawed boundary.
S = Surface of the phase change interface.

Equations (11.21) through (11.23) can now be used to evaluate the location of the phase change interface. The evaluation of Eq. (11.23) is so elaborate, however, that an approximate method can be used for many problems.

Assume a surface of symmetry exists, as it will for most practical problems, and the coordinate axes are located on the symmetry surfaces. The depth to the interface along the symmetry surface will be $z = \xi_o$. Equation (11.23) may then be written as

$$\int_S \left\{ k_1 \left[T_p f'(\xi) - (T_p f(\xi) - T_f) \frac{f'(\xi)}{f(\xi) - 1} \right] - k_2 [T_f - T_o] \frac{f'(\xi)}{f(\xi)} \right]\right\} ds = -L \frac{dV}{dt} \tag{11.24}$$

where

$$f(\xi) = f(0,0,\xi) \text{ and}$$

$$f'(\xi) = \frac{df(0,0,\xi)}{dz}.$$

Now if the integrand in Eq. (11.24) is assumed constant over the phase change interface, the following equation for the phase change depth along axes of symmetry results

$$k_1 \left[T_p - \frac{(T_p f - T_f)}{(f-1)} \right] - k_2 \left[\frac{T_f - T_o}{f} \right] = -\frac{L}{f'} \frac{d\xi}{dt} \tag{11.25}$$

Integration of Eq. (11.25) gives

$$\int_o^{\xi_o} \frac{f-1}{f'\left[1 + \beta\left(\frac{f-1}{f}\right)\right]} d\xi = k_1 \frac{(T_p - T_f)t}{L} \tag{11.26}$$

where

$$\beta = -k_{21}\frac{(T_o - T_f)}{(T_p - T_f)}$$

With the value of ξ_o, evaluated from Eq. (11.26), the phase change interface surface is given by letting $T_2 = T_f$ in Eq. (11.22). Then

$$f(x,y,z) = f(\xi_o) \tag{11.27}$$

The maximum growth will occur when $t \to \infty$ in Eq. (11.26). If a finite maximum exists, then the denominator in Eq. (11.26) must be zero, or

$$f(\xi_\infty) = q_l = \frac{(T_o - T_f)}{(T_o - T_f) - k_{12}(T_p - T_f)} \tag{11.28}$$

This relation is merely the expression for the phase change temperature isotherm, in the steady-state solution, but with the correction $k_{12} = k_t/k_f$ to account for the difference in the thawed and frozen conductivities. The steady-state solution did not include property differences.

These relations can be applied to some practical examples.

11.4.1.1 Semi-Infinite Strip

The steady-state temperature is given in Section 7.1.2

$$T - T_o = \frac{1}{\pi}(T_p - T_o)\tan^{-1}\left(\frac{2a\,z}{x^2 + z^2 - a^2}\right) \tag{11.29}$$

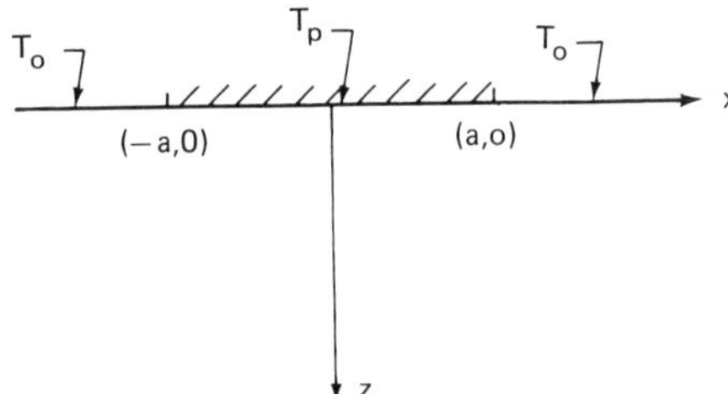

Thus

$$f(x,z) = \frac{1}{\pi}\tan^{-1}\left(\frac{2a\,z}{x^2 + z^2 - a^2}\right) \tag{11.30}$$

$$f(0,z) = \frac{1}{\pi}\tan^{-1}\left(\frac{2a\,z}{z^2 - a^2}\right) \tag{11.31}$$

The limiting thaw depth is obtained with Eq. (11.28)

$$\frac{1}{\pi}\tan^{-1}\left(\frac{2\,\xi_\infty/a}{\left(\frac{\xi_\infty}{a}\right)^2 - 1}\right) = q_1 \tag{11.32}$$

This leads to

$$\frac{\xi_\infty}{a} = \cot\left(\frac{\pi q_1}{2}\right) \tag{11.33}$$

The interface surface is given by Eq. (11.27)

$$\tan^{-1}\left(\frac{2a\,z}{x^2 + z^2 - a^2}\right) = \tan^{-1}\left(\frac{2\,\xi_\infty/a}{\left(\frac{\xi_\infty}{a}\right)^2 - 1}\right) \tag{11.34}$$

Thus the x coordinate of the steady-state interface is

$$x = \pm\sqrt{\frac{\xi_x}{\xi_\infty}\left(\xi_\infty^2 - a^2\right) + a^2 - \xi_x^2} \tag{11.35}$$

where ξ_x is the z coordinate of the interface.

The transient interface is found from Eq. (11.26)

$$\int_o^\gamma \frac{1 - \frac{2}{\pi}\tan^{-1}\left(\frac{1}{u}\right)}{1 + \beta\left[1 - \frac{\pi}{2\tan^{-1}\left(\frac{1}{u}\right)}\right]}\left(u^2 + 1\right)du = \frac{2k_1(T_p - T_f)t}{\pi\, a^2 L} \tag{11.36}$$

where $\gamma = \xi_o/a$.

11.4.1.2 Rectangular Area

The steady-state temperature is given by Eq. (7.19)

$$T - T_o = \frac{(T_p - T_o)}{2\pi}\left\{\tan^{-1}\frac{(x+a)(y+b)}{z\sqrt{z^2+(x+a)^2+(y+b)^2}} - \tan^{-1}\frac{(x-a)(y+b)}{z\sqrt{z^2+(x-a)^2+(y+b)^2}} - \tan^{-1}\frac{(x+a)(y-b)}{z\sqrt{z^2+(x+a)^2+(y-b)^2}} + \tan^{-1}\frac{(x-a)(y-b)}{z\sqrt{z^2+(x-a)^2+(y-b)^2}}\right\} \quad (11.37)$$

Thus

$$f(\xi) = \frac{2}{\pi}\tan^{-1}\left[\frac{n}{\frac{\xi}{a}\sqrt{\left(\frac{\xi}{a}\right)^2 + n^2 + 1}}\right] \quad (11.38)$$

where $n = b/a$.

The limiting interface position is

$$\frac{\xi_\infty}{a} = \left[\sqrt{n^2\cot^2\left(\frac{\pi q_l}{2}\right) + \frac{1}{4}(n^2+1)^2} - \frac{1}{2}(n^2+1)\right]^{1/2} \quad (11.39)$$

The transient location of ξ is given by

$$\int_0^\gamma \frac{1-p}{1+\beta\left(1-\frac{1}{p}\right)} \frac{[u^2(u^2+n^2+1)+n^2]\sqrt{u^2+n^2+1}}{n(2u^2+n^2+1)}\,du = \frac{2k_l(T_p - T_f)t}{\pi a^2 L} \quad (11.40)$$

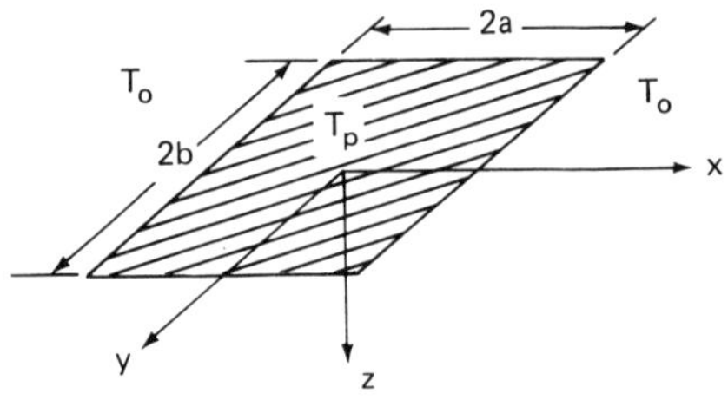

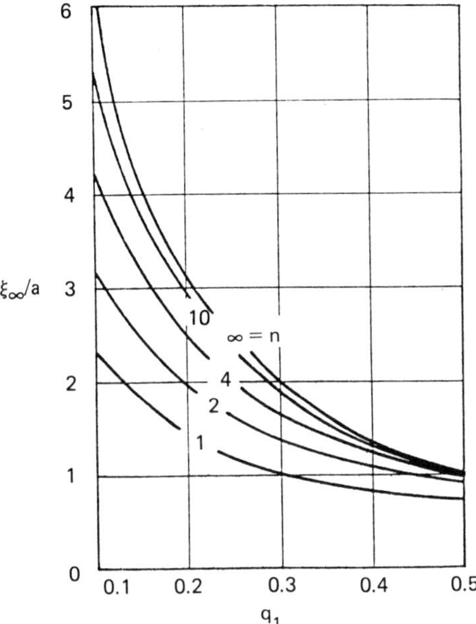

Fig. 11.18. Foundation Thawing Limit under Center of Structure in Two- and Three-Dimensional Problems (from Porkhayev 1963).

where

$$p = \frac{2}{\pi} \tan^{-1}\left(\frac{n}{u\sqrt{u^2 + n^2 + 1}}\right)$$

These relations are graphed in Figs. 11.18 and 11.19.

The temperatures beneath the structure can be calculated at any time by finding ξ_o, from Eq. (11.36) or Figure 11.19, evaluating $f(\xi_o)$ with Eq. (11.38), and then using Eqs. (11.37), (11.21), (11.22).

The geothermal gradient, G, can also be included, but then a trial-and-error solution would be necessary for the maximum thaw depth. The relation, analogous to Eq. (11.39) is

$$\tan^{-1}\left(\frac{n}{\frac{\xi_\infty}{a}\sqrt{\left(\frac{\xi a}{a}\right)^2 + n^2 + 1}}\right) + \frac{\pi}{2}\frac{Ga}{(T_p - T_o)}\frac{\xi_\infty}{a} - \frac{\pi}{2}q_1 = 0 \quad (11.41)$$

The inclusion of the geothermal gradient will increase the thaw depth at any time (Demchenko 1975).

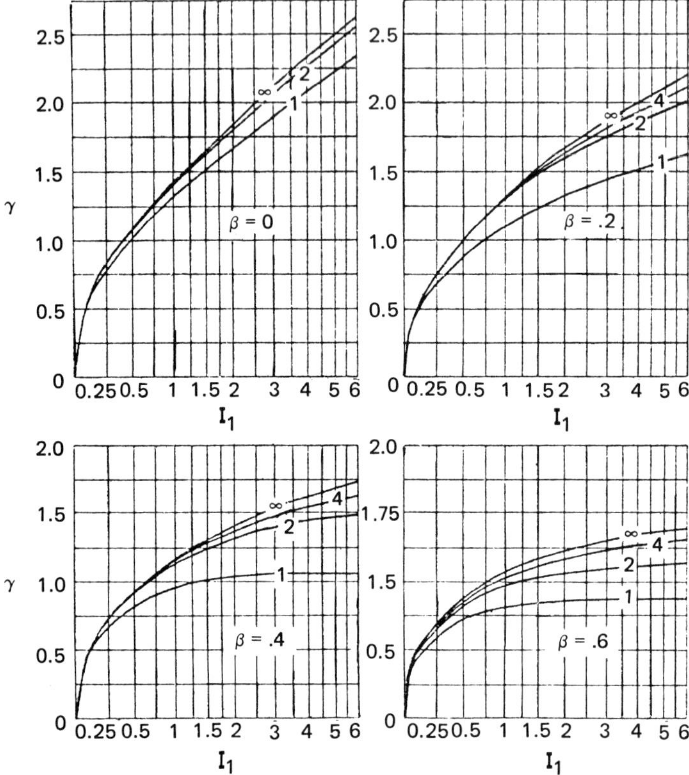

Fig. 11.19. Thawing Depth with Time under the Center of a Structure in Two- and Three-Dimensional Problems (from Porkhayev 1963). Figures 11.18 and 11.19 reproduced from *Permafrost Second International Conference,* 1963, with the permission of the National Academy of Sciences, Washington, D.C.

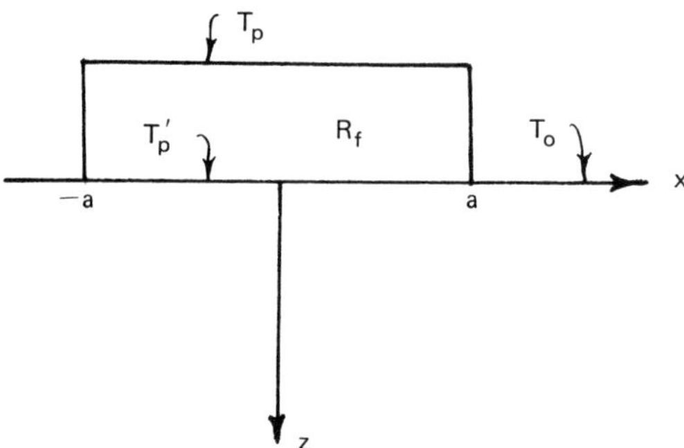

Fig. 11.20. Insulation under Structure.

11.4.1.3 Effect of Insulation

For most cases, the structure will have a layer of insulation between the ground and the bottom of the structure. Then the ground will see an effective temperature T_p', in place of the temperature at the floor surface T_p, as shown in Fig. 11.20.

Using Eq. (11.29), with T_p replaced by T_p', the heat flow through the insulation can be equated to the heat flow from the heated strip (evaluated for $-.95a \le x \le .95a$) to the ground. This gives the following expression for the equivalent surface temperature T_p'

$$\frac{T_p' - T_o}{T_p - T_o} = \frac{1}{1 + \dfrac{7\alpha}{3}} \tag{11.42}$$

where $\alpha = k_t R_f / 2a$. The thickness of a layer of earth equivalent to the insulation is $k_t\ R_f$.

Equation (11.33) for the maximum thaw, beneath the center of the building (assume n = b/a is large), is then

$$\frac{\xi_\infty}{a} = \cot\left[\frac{\pi}{2}\,\frac{1 + \dfrac{7\alpha}{3}}{1 + \dfrac{1}{\beta} + \dfrac{7}{3}\alpha(1 - k_{12})}\right] \tag{11.43}$$

Porkhayev (1970) gives graphical solutions to the equations for the thaw beneath a building. These are given in Tables 11.1 and 11.2, and Figs. 11.21 through 11.23. The centerline thaw is

Table 11.1. Value of Coefficient k_1 (adapted from Porkhayev (1970), with permission of USSR Copyright Agency).

	Coefficient β									
I	0	0.4	0.8	1.2	2.0	0	0.4	0.8	1.2	2.0
		$n = 1.0$						$n = 2$		
0.10	1.00	0.93	0.87	0.83	0.80	1.00	1.00	0.99	0.97	0.96
0.25	0.95	0.85	0.78	0.74	0.70	1.00	0.97	0.92	0.89	0.88
0.50	0.94	0.78	0.68	0.66	0.70	0.99	0.95	0.88	0.86	0.88
1.00	0.92	0.70	0.63	0.66	0.70	0.97	0.90	0.84	0.86	0.88
1.50	0.90	0.64	0.63	0.66	0.70	0.96	0.87	0.84	0.86	0.88

Table 11.2. Value of Coefficient k_2 (adapted from Porkhayev (1970), with permission of USSR Copyright Agency).

	Value of β				
n	0.2	0.4	0.8	1.2	2.0
1	0.45	0.56	0.63	0.66	0.70
2	0.62	0.74	0.84	0.86	0.88
3	0.72	0.84	0.91	0.93	0.96
4	1.00	1.00	1.00	1.00	1.00

$$\frac{\xi_{oc}}{2a} = k_1(\xi_c - k_c) \tag{11.44a}$$

$$\frac{\xi_{\infty c}}{2a} = k_2\xi_{cu} \tag{11.44b}$$

The thaw at the edge is

$$\frac{\xi_e}{2a} = \begin{cases} k_1\xi_m & \alpha = 0 \quad (11.45a) \\ k_1(\xi_m - k_m - .1\beta\sqrt{I}) & \alpha \neq 0 \quad (11.45b) \end{cases}$$

$$\frac{\xi_{e\infty}}{2a} = k_2\,\xi_{mu} \tag{11.46}$$

where

$$I = \frac{k_t(T_p - T_f)t}{4a^2L}$$

11.4.2 Empirical Correction Factor

The thaw beneath a heated structure can also be estimated by a simple method using empirical correction factors to the infinite thaw case.

First calculate the thaw depth, for the length of time of interest, beneath the structure as if it were of infinite extent. This can be done by using one of the formulae derived earlier. Then, using Fig. 11.24, correct for the finite dimensions of the structure. It should be pointed out that the rationale for Fig. 11.24 is based on Russian field data as reported in the Russian Building Code (Tsytovich 1975).

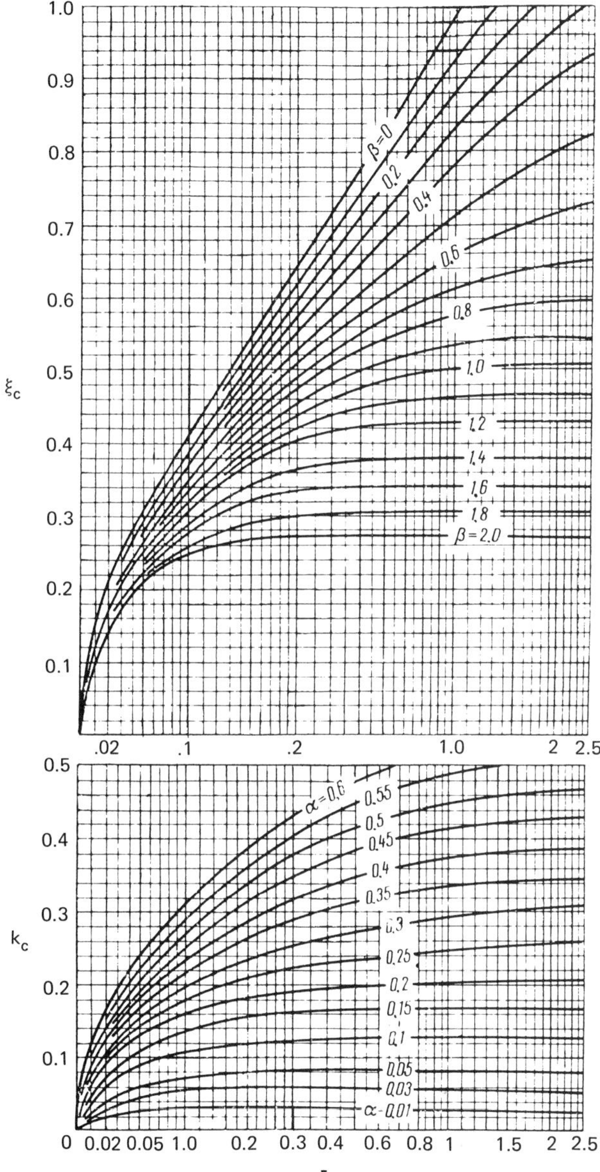

Fig. 11.21. Ground Thaw Depth Coefficients for Center of Building (adapted from Porkhayev (1970), with permission of USSR Copyright Agency).

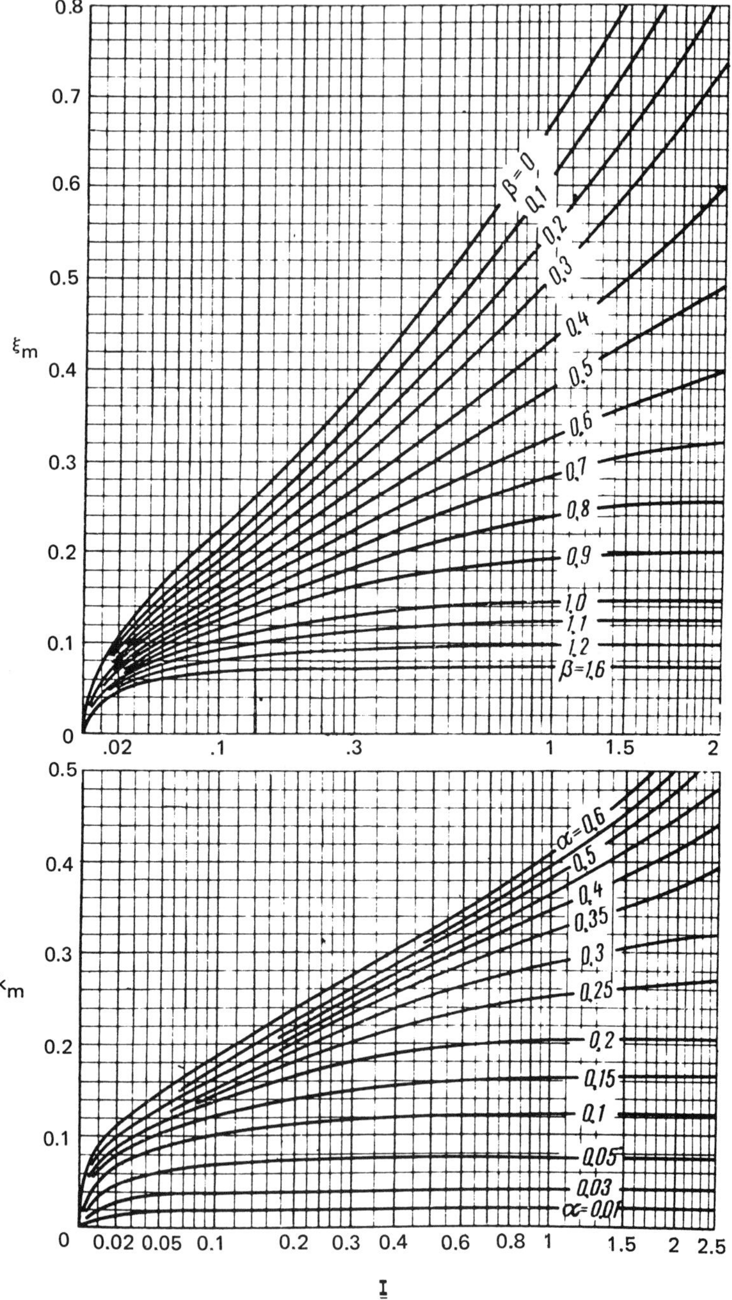

Fig. 11.22. Ground Thaw Depth Coefficients under Edge of Building (adapted from Porkhayev (1970), with permission of USSR Copyright Agency).

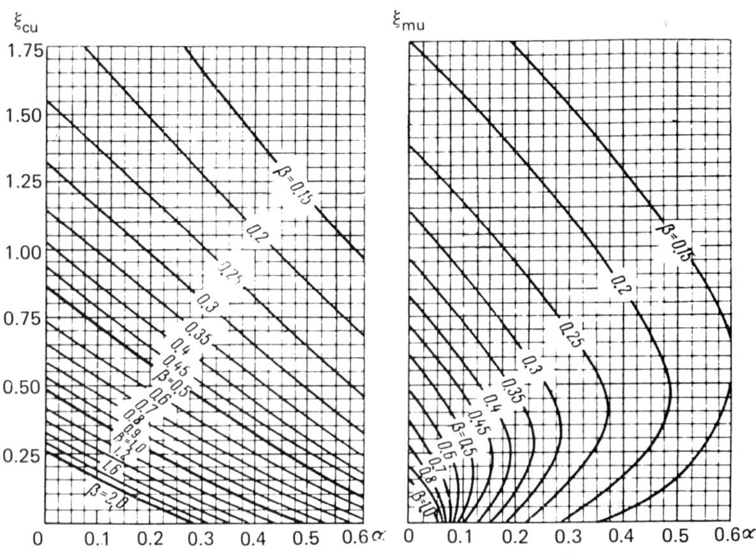

Fig. 11.23. Maximum Thaw Coefficients under Buildings (adapted from Porkhayev (1970), with permission of USSR Copyright Agency).

Example 5: Consider a 70 × 210-ft building at Chesterfield, NWT. The following data are given: $\gamma_d = 1.85$ g/cm³, $W = 15\%$, $T_a = 11°F$, $T_F = 60°F$ (year-round), FI = 8750°F-day. Calculate the centerline thaw after 8.7 years.

$k_1 = 2.23$ Btu/hr-ft-°F $\quad k_2 = 2.63 \quad k_{21} = 1.18$
$C_1 = 30$ Btu/ft³-°F
$L = 2480$ Btu/ft³

Method 1: Use the correction factor

$v_o = 32 - 11 = 21 \quad v_s = 28 \quad TI = (60 - 32)(365)(8.7) = 89{,}000$
$\alpha = 0.75 \quad \mu = 0.34 \quad \lambda = 0.75$
$X = 46.5$ ft of thaw for an infinite system

From Fig. 11.24

$K_c = 0.79 \quad K_E = 0.6$
$X_c = 36.7$ ft
$X_E = 22.0$ ft

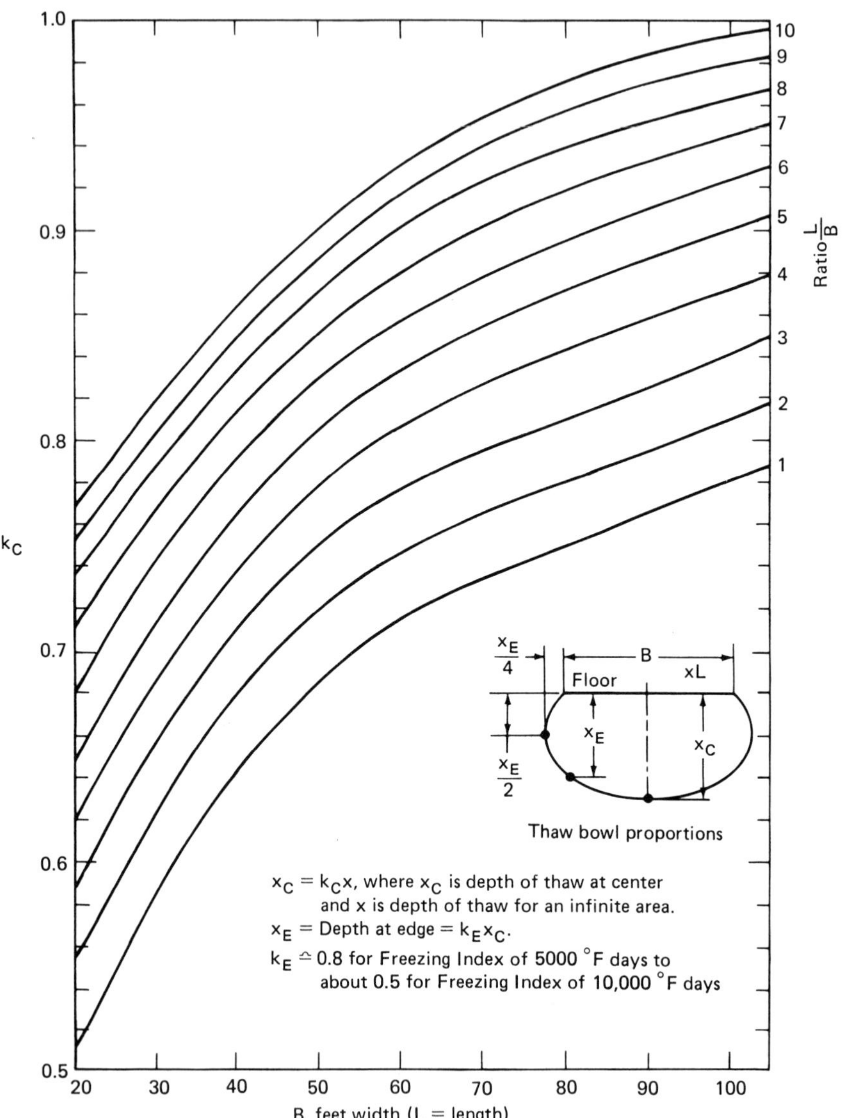

Fig. 11.24. Thaw Bowl beneath a Heated Building on Permafrost (adapted from Sanger 1969).

Method 2: Use quasi-steady procedure

$$n = \frac{b}{a} = 3$$

$$I = k_1 \frac{(T_p - T_f)t}{4a^2L} = .392$$

$$\beta = -k_{21} \frac{T_o - T_f}{T_p - T_f} = .89$$

From Fig. 11.21

$$\xi_c = .483$$

$$\xi_{oc} = X_c = .483(70) = 33.8 \text{ ft}$$

The two methods give very close results, but the quasi-steady method is preferred, especially as the permafrost temperature decreases.

A more realistic calculation of the thaw depth beneath the building would include the thermal resistance of the floor structure. Assume that the floor resistance R_i = 13.5 ft²-hr-°F/Btu. Then using

$$X = k\,R_i\left(\sqrt{1 + \frac{48\,\lambda^2 I}{k\,L\,R_i^2}} - 1\right)$$

one obtains

$$X = 25.3 \text{ ft}$$

and

$$X_c = 20.0 \text{ ft}$$
$$X_E = 12.0 \text{ ft}$$

For this case, I = .392, β = .89, α = 0.43, and k_1 = 1.0. From Fig. 11.21, ξ_c = .483, k_c = .357

$$\xi_{oc} = (.483 - .357)(70) = 8.8 \text{ ft}$$

From Fig. 11.22, k_m = .270, ξ_m = .175

$$\xi_e = [.175 - .270 - (.1)(.89)\sqrt{.392}]\,70 < 0, \textit{ implies no thawing.}$$

These calculations no longer agree well with the simple method. The simple method assumes that the entire upper surface is insulated which drastically retards the cooling effect of the ambient surface. The quasi-steady method is preferred.

If the building width is more than 100 ft, one may extrapolate K_c linearly, up to 1.0

Example 6: Find the correction factor for the following area

$$B(width) = 200 \qquad L(length) = 200 \qquad L/B = 1.0$$

$$K_c = 0.78 + \frac{\Delta K_c}{\Delta B}\,\Delta B = 0.78 + \left(\frac{.78 - .65}{80}\right) 100 = .94$$

Clearly the larger the building, the closer the centerline thaw will be to that of the infinite case.

Example 7: For the problem of Example 5, calculate the temperature at $y/a = 0$, $x/a = 0.5$, and $z/a = .5$ and 1.5 for $t = 8.7$ years and also after an infinite time.

At $t = 8.7$ years

$$\frac{\xi_o}{a} = .966$$

Then from Eq. (11.38)

$$f_o = f(\xi_o) = .481$$

From Eq. (11.37)

$$f(.5, 0, .5) = .63$$
$$f(.5, 0, 1.5) = .32$$

Then, from Eqs. (11.21) and (11.22)

$$T(.5, 0, .5) = (60)(.63) + \frac{(60)(.481) - 32}{.481 - 1}(1 - .63) = 40.1°F\,\text{(thawed)}$$

$$T(.5, 0, 1.5) = (32 - 11)\frac{.32}{.481} + 11 = 25.0°F\,\text{(frozen)}$$

After steady state is reached, $q_1 = .375$, and $\xi_\infty = 45.9$ ft. Then $f(\xi_o) = .375$

$$T(.5, 0, .5) = 53.0°\mathrm{F}$$
$$T(.5, 0, 1.5) = 28.9°\mathrm{F}$$

11.5 SUPERSTRUCTURES

The heat transfer problems associated with superstructures are not significantly different in the north or south. The main difference is one of degree, due to the longer duration of the severe ambient conditions of the north. The climate in the north of Canada, for example, is not much different from that of the prairie provinces. Lack of skilled labor, short construction season, and cost dictate off-site prefabrication for much of the northern structures. The traditional Eskimo structure, the igloo, is admirably suited to its task. It is actually a fine example of natural engineering design and can maintain temperature differences between inside and outside of 70°F with a very small heat input. It has the minimum surface area to volume and uses easily available materials. It is seasonal, however, and tends to become cold and damp with continued occupation. This is not a disadvantage because it was originally used only for periods of a few months. The caribou skin tent is also damp, seasonal, and small, having only 100 ft^2 for three to five occupants. Neither of these structures is suitable for permanent housing in fixed communities.

The present native housing in the Canadian north uses partially prefabricated, modular construction structures of about 750 ft^2. These are well engineered, but the following user complaints were noted (AINA 1973).

(a) Window icing (condensation) has not been solved.
(b) Emergency doors needed; outside door should seal over the jamb; locks tend to freeze up and are too complicated. Ice tends to build up on hinge side of open door, which then causes hinges to pull out as door is forced shut.
c) Porches are needed; these are the northern equivalent of cellars and can be used for cold storage of food, equipment, tools, and as an air lock.
(d) Doors tend to warp, bend, frost over, break, and rip off in the wind. Hinges on one side only would cut conduction problems.
(e) Entrances below floor level would create a trapped air bubble of warm air and prevent loss of heat. Double doors should be standard.
(f) Cold air drafts tend to blow up through floor from ventilation space (pile construction).

(g) Lack of furnace backup system could lead to severe problems if a power shortage occurred.

Clearly not all the problems have been solved, despite the considerable advances. This is largely a question of lack of feedback from the user to the designer. Rounthewaite (AINA 1973) mentions the following general design questions:

(a) Insulation and heating design.
(b) Minimize surface area to floor area.
(c) Snow, wind, rain—details of roofs, entrances, joints, and the external envelope.
(d) Orientation and elevation of buildings.
(e) Lateral structural strength.
(f) Cladding material.
(g) Windows and blinds.
(h) Building finishes—effect of intense, continuous, ultraviolet radiation.
(i) Low winter humidity—must humidify makeup air, a problem in all cold climates.

In particular, care must be used in dealing with water vapor and water transmission, which can lead to condensation; excessive heat loss due to inadequate insulation; and problems of glazing.

Mobile homes are normally neither designed nor suited for arctic or subarctic use. Modular constructed buildings set on permanent foundations can meet all building codes and are designed for the arctic. Log houses can use local material, but require skilled labor and may have heating bills 30% greater than 4-in. bulk insulated structures. Prefabs, woodframe and stressed skin, account for the majority of new northern housing in Canada (Platts 1960). This includes Canadian Armed Forces structures, airport facilities, radio, weather, and administrative buildings, DEW line, schools, hospitals, and even entire mining towns in northern Quebec.

11.5.1 Vapor Transmission

The flow of water vapor through the walls of buildings is a process that can lead to problems in the north as the temperature within the wall will often drop below the dew point level, and condensation can occur. This leads to a buildup of ice that may structurally damage the wall, or the moisture may degrade the thermal performance of the unit. Nevertheless, the flow of water carried with air flows, through cracks and leaks, is probably a greater problem than vapor diffusion as far as severe condensation is concerned

(Platts 1962). Unfortunately, it is far easier to calculate the diffusion of water vapor through walls, than it is to predict the flow of moist air through cracks. The prefabricated walls are commonly made up of two sheets of plywood sealed over a 2 $\times$ 4 frame. No special vapor barrier on the inside seems necessary. Each panel has a permeance of about 1 perm (see following relations). Windows, joints, and connectors cause condensation problems, but with the best systems no problems occur if the inside relative humidity is less than 30%. Relative humidities this low may lead to discomfort for individuals, however, and have an adverse effect upon furniture and other glued systems. Thus, simply dropping the inside humidity is not an absolute solution to the problem of moisture migration. Breaks in the walls for services can allow the wall to fill with frost, but closed cell cores (polystyrene, polyurethane) have worked well in the north (Platts 1962).

Joints should prevent vapor entrapment on the inside and penetration of rain and snow from the outside. They should resist wind racking and ground movements. The T and batten joint was developed for the Canadian Armed Forces GPH and is now commonly used (Platts 1960). The thermal performance of open rain screen joints, like the above, is as good as a wooden stud (Platts 1966).

The structure must have a flow of fresh air, of course, for aesthetic purposes and to control humidity. Kent (1961) notes that 4.5 ft^3/min/person was acceptable in Canadian trials, but 10 ft^3/min/person is recommended (Platts 1960). With the present stress on energy conservation, the upper ventilation standards may be relaxed downward.

The flow of water vapor in a medium is given by Fick's law

$$\omega = -\mu \frac{\partial p}{\partial x} \tag{11.47}$$

where

ω = mass of water vapor transmitted per unit area per unit time
p = vapor pressure
x = coordinate in direction of vapor flow and
μ = permeability of medium for water vapor.

The water vapor pressure is a function of temperature, and thus the diffusive flow of vapor will occur in the same direction as the flow of energy, from the warm side to the colder side. For this reason, vapor barriers must be placed on the warm side of the wall to prevent moisture from entering the wall. Placing the vapor barrier on the cold side is worse than having no barrier at all, since it will actually trap the moisture within the wall.

Equation (11.47) is of the same form as Fourier's law for conduction, and the governing equations are similar. For average conditions, Eq. (11.47) can be written

$$W = \bar{\mu}\, A \frac{\Delta p}{l} \tag{11.48}$$

where

$\bar{\mu}$ = average permeability, perm-in.
W = mass transfer rate, grains/hr, (7000 grain = 1 lb_m.)
A = cross-sectional area
l = thickness of medium and
Δp = pressure drop, in. of Hg

The units for $\bar{\mu}$ are the perm-in., or some compatible units, where

$$1\ perm = \frac{1\ grain}{hr\ ft^2\ in.\ Hg\ vapor\ pressure\ difference}$$

The permeance coefficient M is often used with $M = \bar{\mu}/l$, and the units of M being the perm.

Then Equation (11.48) is

$$W = M\, A\, \Delta p \tag{11.49}$$

If one considers a series wall made up of slabs with a given thickness and permeability (Fig. 11.25), then the mass flow across each slab is the same

$$\frac{W}{A} = M_1(p_1 - p_2) = M_2(p_2 - p_3) = M_3(p_3 - p_4) \tag{11.50}$$

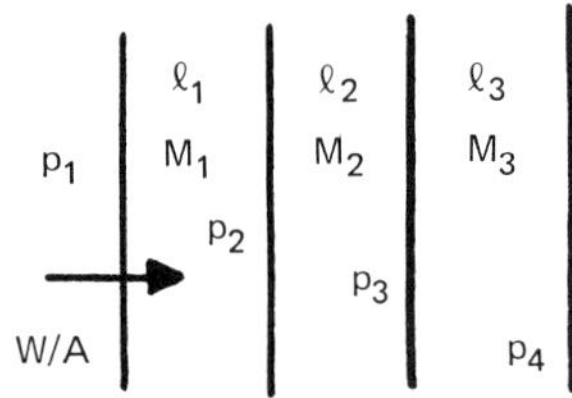

Fig. 11.25. Vapor Diffusion in Series Flow.

Also, an overall permeance coefficient can be defined as

$$\frac{W}{A} = M_t \, \Delta p_t \tag{11.51}$$

Equating Equations (11.50) and (11.51) gives, for a series wall

$$M_t = \frac{1}{\frac{1}{M_1} + \frac{1}{M_2} + \frac{1}{M_3}} = \frac{1}{\sum_i^N \frac{1}{M_i}} \tag{11.52}$$

Similar reasoning gives, for a parallel wall

$$A_T M_T = M_1 A_1 + M_2 A_2 + M_3 A_3 = \sum_i^N M_i A_i \tag{11.53}$$

At a given total pressure, air can hold some maximum amount of water vapor. This is the saturation condition, and the relative humity is 1.0 (100%). The temperature of the mixture at this condition is called the "dew point temperature." If the temperature of an air-water vapor mixture drops, the ability of the air to hold water as vapor drops until saturation occurs. At this temperature (dew point) the water will start to condense out as a liquid. The condensation may occur on exterior surfaces, such as walls or windows, or in the interior of walls. The problem with vapor flow arises when the temperature in the wall drops below the equilibrium value for the vapor pressure: the dew point temperature. At this point condensation will occur, and the resultant water may freeze and accumulate in the wall. The problem becomes more severe as the climate becomes colder due to the longer heating season. The wall may have little opportunity to dry out, and its thermal performance may be severely reduced. This will be particularly true if the dew point temperature is low enough so that the condensed vapor freezes into ice.

Consider the wall shown in Fig. 11.26. The temperature profile in the wall can be calculated as described earlier. Associated with this temperature profile is a saturated vapor pressure curve, as sketched in Fig. 11.27. The actual vapor pressure variation can be found using Eqs. (11.49) through (11.51); this flow equilibrium curve is also shown.

Note that condensation will occur when the actual vapor pressure is less than the saturation value. This occurs here, somewhere in the insulation layer, which will then be wet and possibly destroyed. This can be alleviated by a vapor barrier on the warm side of the wall. Due to the severe consequences

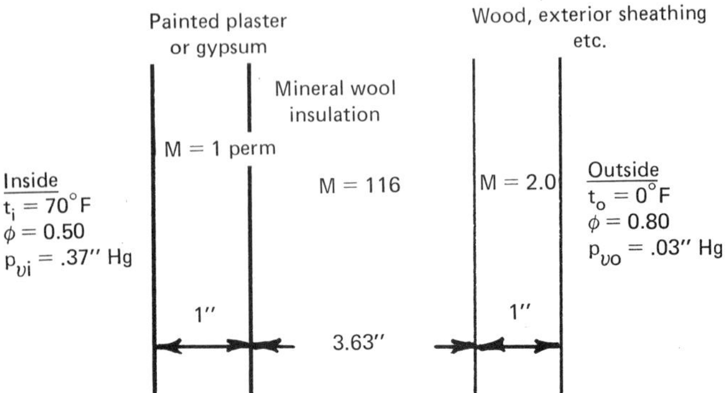

Fig. 11.26. Steady-State Vapor Diffusion.

of condensation, it is important to use a continuous vapor barrier. Single sheet rolls of wide polyethylene are commonly used to cover insulation and studs with the joints overlapped. Once the dew point temperature is reached, both the temperature and the vapor pressure curves will change from the ideals sketched in Fig. 11.27.

Whether the condensed moisture is due to diffusive flow or to the leakage of moist air, the problems are essentially the same. As most of the moisture is carried by air leakage, a continuous vapor barrier is essential. The condensed

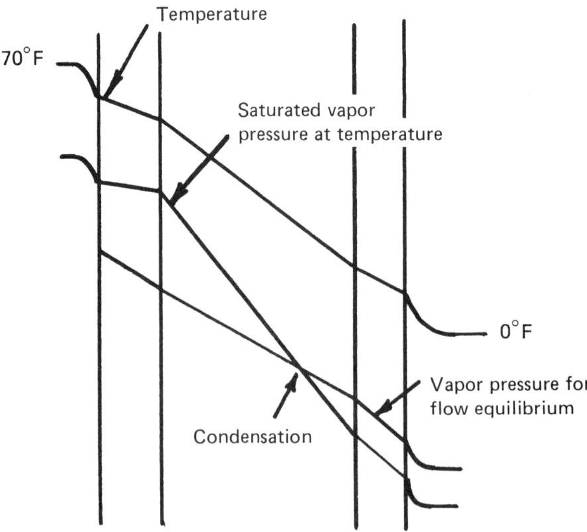

Fig. 11.27. Temperature and Pressure Profiles.

moisture, having frozen, will remain immobile in a severe climate until it thaws out, perhaps during a mild spell. The water can now flow out at various locations simulating an outside leak. More seriously, it may simply remain within the structure causing mildew, delamination of plywood, rot, etc. The basic cure is to prevent the ingress of warm, moist air and allow the egress of cool, moist air.

11.5.2 Insulation

The heating load for a structure in the north is typically two or three times that of more southerly cold climate regions. Thus, heating costs of more than $1000 for a hut are possible. For this reason, insulation deserves careful attention in northern structures. Until recently, however, the advantages of heavy insulation have been neglected.

Construction in the prairie provinces of Canada showed that reflective type insulation led to trouble with sealing, integrity, and condensation. Thus, bulk type insulation is most appropriate and is now most used.

Caribou moss (*cladonia* sp.), a fairly common ground cover in northern regions, can provide insulation about equal to mineral wool if it is dry and compressed to one-half volume (Dickens 1960). This insulation has been used, but it is doubtful if sufficient supplies exist for large-scale use. Another local material that has been used for insulation and superstructures is peat sod. The use of these materials is, of course, limited and possibly short-term.

The commonly used insulations are mineral wool (fiberglass, etc.), polystyrene, and polyurethane. The latter two are valuable for their ability to foam in place and the fact that their volume expands up to 50 times. In addition, when used as cores, they prevent moisture or vapor flow at wall openings. They are more expensive than mineral wool, and there is some question as to their durability. Polyurethane, which uses a heavy inert gas, will reduce to the insulative value of polystyrene under low temperature conditions and is probably not worth the additional cost. Polystyrene is unsurpassed for use in contact with the ground, and foamed glass is excellent for chemical resistance.

Fiberglass is cheap (relatively) and has the advantage that toxic fumes are not produced when burned (AINA 1973). Double-faced plywood walls, with mineral wool, will give about 1 hour of fire resistance as opposed to 6–12 minutes without the insulation (Platts 1960). Because fire is an extreme and constant hazard in the north, this may allow cheap fire walls to be incorporated into the building. Foil-backed bulk insulations are probably not cost effective in severe climates. The radiative reflection is lost if the cavity is filled (as it should be) and a better vapor barrier must be added, in any case.

The amount of insulation to use is a problem that is changing continuously due to increasing fuel costs. The cost of all fuels, and in particular oil, is changing so dramatically that almost yearly updates are needed for "optimum" insulation thicknesses. It would appear that the maximum insulation thickness that can be reasonably incorporated into the design will be economic. Economic analyses have been carried out at various times, and it now appears that in the Canadian north, 12–15 in. of insulation should be used, though in Alaska, at 1973 prices, 6–8 in. of insulation in the walls and 12–15 in. in the roof are economic (AINA 1973). Rarely is more than 4 in. (actually about 3¼ in.) used in northern construction. The reason, as it is in the south, is mainly a failure to adapt construction techniques to a wall thicker than the basic 2 × 4 stud. In addition, prefabricated panels are usually 4 in. thick (rarely 6 in.), but it would seem possible to use two layers of such panels. Fortunately, this philosophy is rapidly being eliminated by the explosive costs of energy. These costs exemplify the futility of quoting optimum thicknesses. The use of adequate insulation will increase comfort and drastically reduce fuel consumption whether it is economic or not.

Platts (1959) found that insulation up to 8 in. thick was certainly justified. The following table, from Platts, is of interest in detailing the effect of insulation on fuel cost and consumption.

Insul. Thick.	U Btu/hr-ft^2-°F	10^6 Btu/ yr	Gal. oil	($) Cost	Total Operating Cost ($)
0	.30	266	2300	1450	1635
2	.108	96	830	525	710
4	.080	71	610	385	570
6	.052	46	400	250	435
8	.043	38	330	210	395

This example was for a 16 × 35 ft hut in Pangnirtung, Quebec, on piles, with 1950 ft^2 of insulated surface and 30 ft^2 of glazing (triple). Interest rates were 5%, and fuel costs were 63¢/gal. It was found that calculations for existing huts showed actual operating costs about double those calculated. But in controlled huts the calculations were correct. Thus, proper maintenance and operation are critical.

The optimum economic thickness of insulation depends upon the cost of materials, labor, transportation, and fuel. The cost may be stated as

Total yearly cost = operating cost (fuel) + amortized cost of insulation including extra materials and labor

The operating cost decreases while the "capital" costs increase with increasing insulation thickness. Thus, an economic optimum thickness will exist for any design, assuming costs can be estimated well into the future.

It would seem, with the cost of fuel rising continuously, that insulation thicknesses should increase throughout the cold climates. The analysis in the north is complicated by the widely different transportation and fuel costs in different parts of the arctic. It is possible, but not likely, that the fossil fuel costs in the north will stabilize due to the recent exploration and development work on fossil fuels.

11.5.3 Glazing

Among the most difficult areas to protect, in terms of temperature and condensation, are glazed areas, or windows. The advantages of windows must be very great to outweigh their disadvantages. Windows admit light and solar energy and provide a view or connection to the exterior, which seems to be a powerful human psychological need. This need is intensified where the climate is associated with a long heating season and short or nonexistent days. Windows also provide emergency exits, which can be critical in northern communities. Unfortunately windows are extremely poor thermally; even a triple-glazed window will lose energy at 8–10 times the rate of an equal area of 6-in. insulated wall. Windows are cool surfaces upon which condensation can occur, with the effect that the visual advantages of the window are negated. Finally, the visual effects of windows occur only during daylight, which itself may be lacking for several weeks in cold climates.

Recommendations commonly call for triple-glazing, using standard glass and ¼-in. air gaps, but most structures use double-glazing (AINA 1973). Triple-glazing should utilize a sealed inner pane, with the outer two panes allowed to breathe to the outside. This will avoid condensation between the panes, which can freeze and is difficult to remove. The inner surfaces of the window frame should have minimum protrusions to allow a free flow of inside air. This will minimize the condensation effects. Beyond this, condensation on the inside must be prevented by humidity control. Condensation is likely if the relative humidity exceeds 30%, while a value of 40–50% is more healthful and comfortable.

The number and size of windows should be minimized, with windows facing south to take psychological advantage of the winter sun, weak though it may be.

11.5.3.1 Window Heat Loss

The loss of energy through windows, exclusive of air leakage, follows simple heat transfer relations. The heat loss is

$$q = \frac{A\Delta T}{R_T} = UA\Delta T \tag{11.54}$$

$$R_T = \frac{1}{h_i} + \frac{\Delta x}{k_g} + R_c + \frac{1}{h_o} = \sum_{i=1}^{4} R_i \tag{11.55}$$

where

$\Delta T = T_i - T_o$ total temperature difference across the window
A = window area
R_T = total thermal resistance of window
U = overall coefficient of heat transfer
R_i = thermal resistance of each "layer" of the window
h_i = 1.46 Btu/hr-ft^2-°F, inside surface conductance of air
h_o = 6.0 Btu/hr-ft^2-°F, outside surface conductance of air; this is a function of wind speed, the value here is for 15 mph
Δx = total thickness of all glass
k_g = thermal conductivity of glass, ~ .44 Btu/hr-ft°F and
R_c = total conductance of all air gaps trapped between glass.

The conductance of the air gap is a complicated function of mean air gap temperature, temperature difference, and thickness of the air gap. Table 11.3 gives reasonable estimates of R_c for each air gap.

Table 11.3. Resistance of Vertical Air Layers to Horizontal Heat Flow, adapted from ASHRAE (1977). Reprinted with permission from the 1977 Fundamentals Volume, ASHRAE Handbook and Product Directory.

	Air Gap Resistance			
	$\bar{T}$ = 0°F	0	−50	−50
Thickness of Gap (in.)	Δ = 20°F	10	20	10
.5	1.13	1.15	1.39	1.46
.75	1.18	1.26	1.39	1.56
1.50	1.12	1.23	1.33	1.48
3.50	1.14	1.23	1.37	1.50

$\bar{T}$ = mean temperature of the gap; Δ = temperature change across gap.

The temperature drop across any part of a series system can be evaluated from

$$\frac{R_i}{R_T} = \frac{\Delta T_i}{\Delta T} \tag{11.56}$$

where ΔT_i = temperature drop across a region and R_i = thermal resistance of the region.

Using Equations (11.54) and (11.55) and Table 11.3, the resistance and overall coefficient can be evaluated for any number of glass layers. For a mean temperature of 0°F, a temperature difference of 20°F, and a ¾-in. air gap, the following values are obtained for ⅛-in. glass.

No. of Glass Layers	R (hr-ft²-°F)/Btu	U Btu/hr-ft²-°F
1	.69	1.45
2	1.89	.528
3	3.10	.323
4	4.30	.233
5	5.50	.182
6	6.71	.149
7	7.91	.126
8	9.12	.110

A 6-in. insulated wall will have a resistance of about 25.9, and even an eight-layer window will lose heat at 2.8 times the wall rate. Thus, windows are certain to be thermally expensive unless insulation is added either inside or outside. If insulated covers are to be used, the proper location is on the exterior to avoid condensation when the cover is removed.

The economic number of glazing sheets to use is determined in the same way as for the choice of insulation thickness. Triple-glazing should now be the minimum for the northerly cold climate regions.

Example 8: A single sheet of glazing exists for a house. Calculate the possibility of condensate occurring inside the window. T_i = 70°F, T_o = −20°F, ϕ_i = 30%, T_{dp} = 37.2°F, and the glass is ⅛-in. thick.

$$\frac{\Delta T_i}{\Delta T} = \frac{R_i}{R_T}$$

$$\frac{T_i - T_S}{T_i - T_o} = \frac{R_i}{R_T}$$

$$R_i = \frac{1}{h_i} = .685$$

$$R_T = \frac{1}{h_i} + \frac{\Delta x}{k_g} + \frac{1}{h_o} = .685 + \frac{.125}{(12)(.44)} + \frac{1}{6.0} = .875$$

Thus

$$\frac{70 - T_s}{90} = \frac{.685}{.875} = .783$$

$$T_S = -.43° F$$

This is below the dew point; thus condensate (ice) will form.

Example 9: What inside relative humidity will prevent fogging of the window in Example 8? $T_i = 70°F$, $T_{dp} = -.43°F$; thus $\phi_i = 5\%$ (far too low for comfort).

Example 10: How many layers of glazing will prevent fogging at $\phi_i = 30\%$; 40%?

$$T_S \geq 37.2° F$$

$$\left(\frac{70 - 37.2}{90}\right) = \frac{R_i}{R_T}$$

$$R_i = .685$$

Then

$$R_T = 1.879$$

Thus, from Table 11.3, two layers will be satisfactory if $R_T = 1.89$. If $\phi_i = 40\%$, $T_{dp} = 44.6$, $R_T = 2.43$. Thus, three layers ($R_T = 3.10$).

The effect of the outside wind is to increase the inside surface temperature if the wind decreases below 15 mph and to decrease T_S if the wind goes above 15 mph.

REFERENCES

American Society of Heating, Refrigeration, and Air-Conditioning Engineers. 1977. Fundamentals. *ASHRAE handbook.* New York.

Andersland, O. B., and Anderson, D. M. 1978. *Geotechnical engineering for cold regions.* New York: McGraw-Hill.

Arctic Institute of North America. 1973. *Man in the north.* Technical paper, Conference on building in northern communities.

Associate Committee on the National Building Code of Canada. 1968*a*. *Canada building code for the north.* Ottawa, NRC 9945.

———. 1968*b*. *Supplement to the building code for the north.* Ottawa, NRC 10368.

Bakakin, V. P. 1959. *Principles of geocryology.* Part 2, Chapter 5, G. V. Porkhayev. Academy of Sciences USSR, NRC TT-1250.

Crawford, C. B., and Johnston, G. H., 1971. Construction on permafrost. *Can. Geotech. J.* 8 (2):236–251.

Crory, F. E. 1963. Pile foundations in permafrost. *Proceedings first international conference on permafrost.* NAS-NRC Pub. No. 1287, pp. 467–476.

Demchenko, R. Ya. 1975. Approximate calculation of the boundaries of thawing zones in permafrost under heat sources. *Heat Transfer Sov. Res.* 7 (6):109–118.

Dickens, H. B., and Gray, D. M. 1960. Experience with a pier-supported building over permafrost. *J. Soil Mech. Found.* 86 (5):1–14.

———, and Platts, R. E. 1960. Housing in northern Canada; Some recent developments. *Polar Rec.* 10 (66):223–230.

Jumikis, A. R. 1978. Graphs for disturbance-temperature distribution in permafrost under heated rectangular structures. *Proceedings third international conference on permafrost.* Vol. 1, pp. 590–596, Ottawa, Canada.

Kent, A. D., and Wilson, A. G. 1961. *Arctic hut heating and ventilating trials.* DBR Paper 109.

Lee, T. M. 1962. *Note on freezeback time of slurry around piles in permafrost.* Tech Note 174a, U.S. Army, CRREL.

Leggett, R. F., and Dickens, H. B. 1959. *Building in northern Canada.* DBR-NRC No. 62.

Linell, K. A., and Johnston, G. H. 1973. Engineering design and construction in permafrost regions: A review. *Second international conference on permafrost.* Yakutsk, Siberia, pp. 553–575.

Long, E. L. 1963. The long thermopile. *Permafrost: Proceedings of international conference.* NAS No. 1287 Washington, D.C., pp. 487–491.

Platts, R. E. 1959. *Insulation in northern building.* Tech. Paper No. 75, DBR, NRC Ottawa.

———. 1960. *Prefabrication in northern housing.* DBR, TP 10, Ottawa.

———. 1962. Condensation control in stressed skin and sandwich panels. *For. Prod. J.* 12 (9):429–30.

———. 1966. The Angirraq: Low cost prefabrication in arctic houses. *Arctic* 19 (2):192–200.

Porkhayev, G. V. 1970. *Thermal interaction between buildings, structures, and perennially frozen ground.* NAUKA, Publisher, Moscow.

Sanger, F. J. 1969. *Foundations of structures in cold regions.* Cold Regions Research and Engineering Laboratory, Hanover, New Hampshire, Monograph III-C4.

Sherwood, A. 1974. Building in the north. *Bull. Assoc. Preservation Technology* 6 (3):1–25.

Soviet Building Code. 1960. *Technical consideration in designing foundations in permafrost.* State Committee of the Council of Ministers (USSR) for Building Problems. (SN91–60), NRC Translation TT-1033.

Tobiasson, W., and Johnson, P. 1978. The details behind a typical alaskan pile foundation. *Proceedings third international conference on permafrost.* NRC Ottawa, Canada, pp. 892–897.

Tsytovich, N. A. 1975. *The mechanics of frozen ground,* ed. G. K. Swinzow. New York: McGraw-Hill.

U.S. Army. 1958. *Review of frozen ground excavation methods.* USA-SIPRE TR51.

———. 1966. *Calculation methods for determination of depths of freeze and thaw in soils.* TM5–852–6, Washington, D.C.

Wheeler, J. A. 1978. Permafrost thermal design for the trans-Alaska pipeline. *Moving boundary problems,* eds. D. G. Wilson, A. D. Solomon, P. T. Boggs, pp. 267–284. New York: Academic Press.

PROBLEMS

1. A warehouse 60 ft × 240 ft in Frobisher Bay, N.W.T., has a floor, as shown below, directly over a gravel pad.
   (a) Calculate the depth of thaw after 2.5 years at the center of the building.
   (b) If this building is to use air ducts and 8″ of insulation $\left(k = .035 \dfrac{\text{Btu}}{\text{hr-ft-°F}}\right)$, what depth of gravel pad is required?

T 55°F

2" wood $k = 0.11$ $\dfrac{\text{Btu}}{\text{hr - ft - °F}}$

6" concrete $k = 0.6$

Gravel pad

$\gamma_d = 1.5$ g/cc

$W = 6\%$

2. Air ducts are to be used in the floor of a building to keep the thaw confined to a specific layer. Explain how the air ducts accomplish this. In particular discuss the heat transfer, in detail, during the thawing season and during the freezing season; write down and explain the important equations used.
3. If the gravel pad of problem 1 has $W = 12\%$ find the depth of thaw in one year and in 10 years at the center of the building.
4. A building is to be located at Churchill, Man. The original site has moss overlying a silt. The properties are

	W(%)	γ_d(g/cc)	depth (cm)
Moss	75	0.8	4
Underlying Silt	25	1.3	—
Fill	3	1.6	—

   Find the active layer depth.
5. A concrete footing in problem 4 is placed with its top 12 inches below the natural grade. Fill with 1.5 ft of dried moss on top of it is built up over the footing. What depth of fill will move the permafrost table to the level of the top of the footing?
6. In problem 5 suppose that the moss absorbs water until $W = 50\%$. Where is the permafrost table now located?

7. For the foundation shown, find:
   (a) time for thaw penetration to reach natural ground surface beneath center of building.
   (b) time to reach the permafrost table,
   (c) estimate annual thaw rate.

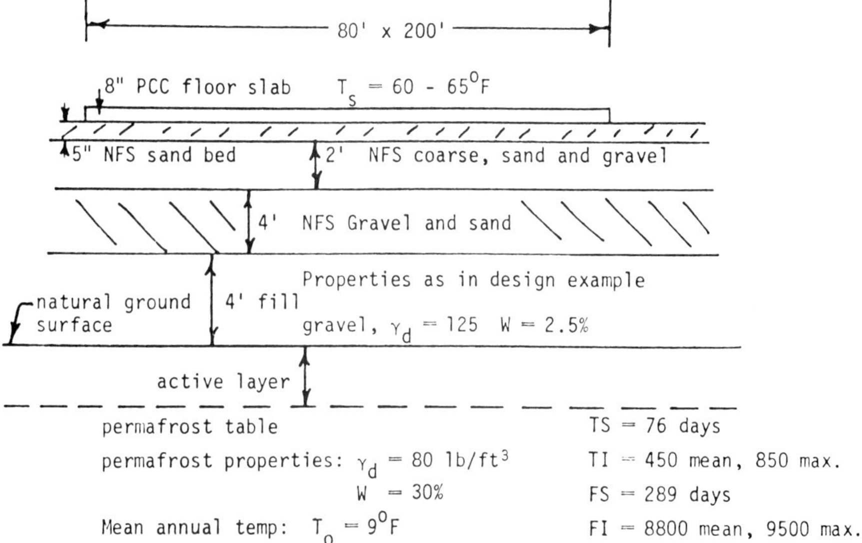

8. Repeat the duct design example problem (Example 4) but find the stack height needed for an insulation thickness of 6 inches.
9. Determine the economical thickness of insulation for an above ground 6.625-in. OD steel water pipe that is maintained at 41°F for the temperature conditions given below. The cost of fuel oil, which has a heat content of 140,000 Btu/US gallon, is 75¢/US gallon, and the efficiency of the heating plant is 85%. The installed costs of polyurethane foam insulation ($k = 0.014$ Btu/ft·hr·°F) for various available thicknesses are given below. The economic life is 20 years, and the cost of capital (discount rate) is 8% net of inflation.
   The ambient temperatures (typical of Barrow, Alaska) are:

Mean Monthly Temperatures, °C

	Mean Annual	J	F	M	A	M	J	J	A	S	O	N	D
Air	−12.8	−28.4	−24.9	−24.0	−16.8	−11.7	−1.2	4.5	2.6	−0.2	−8.8	−15.2	−21.4

Insulation Thickness, in.	Installed Cost, $/ft
1	3.80
2	4.80
3	5.70
4	6.50

12
Man in Cold Climates

12.0 INTRODUCTION

From a physiological viewpoint, man is classified with the homeotherms—animals that maintain a fairly constant deep body, or core, temperature regardless of ambient conditions. The core temperature of a normal person is 36.4–37.5°C and oscillates daily over this range about a mean value of 37.0°C (98.6°F). An unexplained and interesting fact is that the entire temperature range of homeotherms, which includes animals as varied as mice, elephants, and birds, is 36°C (elephant) to 41°C (birds) (Pembrey 1898). This excludes the hibernating state of true hibernants that can maintain core temperatures far lower than 36°C. Man regulates his temperature in accordance with the usual laws of heat transfer; however, the system for doing this must be thought of as a very complicated, variable-geometry, variable-flow, feedback-controlled heat exchanger with energy generation. The physiological details and mechanisms are generally known, but these details are beyond the scope of this discussion. Interested readers are referred to Newburgh (1949) as an excellent starting point for further details. One should not expect to find, at this time, all the complicated details and questions answered on the control of body temperature.

12.1 HEAT TRANSFER RELATIONS

The heat flow from man, or animals in general, is schematically presented in Fig. 12.1. In this chapter, we shall be concerned with the heat loss in cold climates, rather than the heat gain in hot regions. The body produces energy by combustion of food or body tissue in a process called "metabolism." This energy is used to maintain the body in thermal equilibrium with the surroundings by adjusting to the heat losses from the body. Part of the heat flow from the body is sensible heat from the core (deep layers or regions of the body) by conduction and convection, which traverse three thermal resistance layers: the body tissue itself, the resistance of clothing, and the resistive layer of air on the surface of the body or clothing. A second part of the energy flow is by evaporation, which in turn is lost as moisture during breathing (respiratory tract) or as an insensible perspiration flow up to the skin and evaporation at the skin surface.

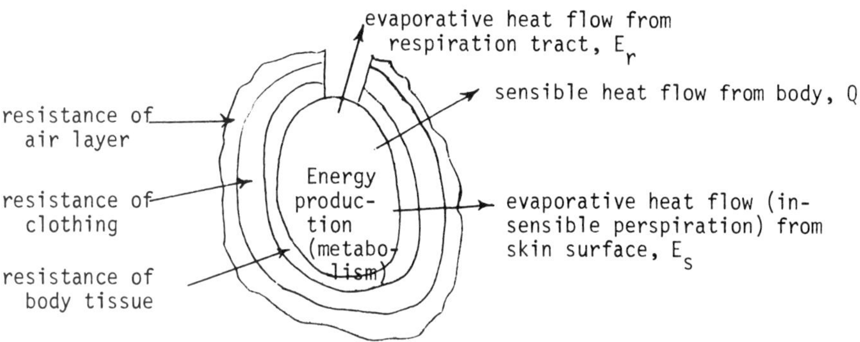

Fig. 12.1. Schematic of Bodily Heat Flow.

The general energy balance for this system can be written as

$$Q + E_s + E_r + M - W = S \tag{12.1}$$

where

Q = sensible heat flow rate through resistive layers
E_s = evaporative heat flow rate from skin surface
E_r = evaporative heat flow rate during respiration
$E = E_r + E_s$, total evaporative rate
M = metabolic rate
S = rate of change in stored energy of the body and
W = work done; normally this term can be neglected.

If S is a positive quantity, the core temperature of the body will be increasing; a negative value would be manifested by a decreasing core temperature. The evaporative heat flow is usually negative, the metabolism is always positive, and the sensible heat flow can be positive or negative, i.e., a gain or loss of energy by the body.

12.1.1 Stored Energy

The energy contained in the body at any instant of time is normally called the "stored energy." The change of this energy is due to the net effect of the instantaneous transfer of heat, work, and mass by body. A change of stored energy in the body shows up as a change of the total body temperature (a weighted average of the surface layer temperature and the core temperature), as there can be no phase change of the living body tissue itself. Ultimately, the core temperature itself will change if the stored energy continues to change for a long enough time. The terms on the left-hand side of Eq. (12.1) are in terms of the body surface area, and thus the change in stored energy can be written

$$S = \frac{m\, c_b}{A_s} \frac{dT}{dt} \tag{12.2}$$

where

m = mass of the body
c_b = specific heat of the body
A_s = surface area of the body
T = average body temperature (sometimes erroneously equated with the core temperature) and
t = time.

Values to use in Eq. (12.2) are given for a hypothetical man by Newburgh (1949) as body mass equals 63 kg, and surface area equals 1.8 m. The skin area is given by Du Bois (1916) as

$$A_s = .202\, m^{.425}\, h^{.725} \tag{12.3}$$

where

A_s = m^2
m = kg and
h = average height (1.73 m).

With m = 70 kg, this gives A_s = 1.83 m^2.

The average temperature of the body can be expressed as

$$T = (1 - x)\, T_b + x\, T_s \tag{12.4}$$

where T_b = core temperature and T_s = average surface temperature.

This relation merely attempts to weight the average body temperature between the core and surface values. This is not simply on a mass basis, however, as about 54% of the body mass lies within 1 in. of the surface, and the core temperature is reached only at a depth of 1 in. or more. Burton (1935) gives a value of x = 0.36, Hardy and DuBois (1938) give a value of 0.2, and Burton and Edholm (1955) give a "best value" of x = 0.3. The latter value can be used as there is very little difference in the stored energy change when x is between 0.2 and 0.4. Livingston (1968) showed that x is a function of time and ambient temperature, with values of x between 0.21 and 0.38. Therefore

$$T = 0.7\,T_b + 0.3\,T_s \tag{12.5}$$

Gagge et al. (1969) noted that the skin temperature is independent of the metabolic rate up to about five times the rest value; for unclothed subjects

$$T_s = 24.85 + .332\,T_o - .00165\,T_o^2 \tag{12.6}$$

and for clothed subjects

$$T_s = 25.8 + .217\,T_o \tag{12.7}$$

The operative temperature, T_o, is the mean of the ambient air temperature and mean radiative temperatures, °C. The averaging is done on the basis of convective and radiative heat transfer. For cold climates, the operative temperature is about 3–5°C above the ambient for a winter day and about 3°C below the ambient during a winter night. These values refer to ambient temperatures below 0°F. For further information see Stoll (1967).

Because the body is composed largely of water, one might expect the specific heat to be near that of water. Burton and Edholm (1955) give a value of $c_b = 0.83$ kcal/kg-°C based on data from Pembrey (1898). Greenfield et al. (1927) found a value of the specific heat of fingers as 0.75, which is in agreement with the value of 0.83, as the finger very likely has a higher proportion of bone and muscle than the body as a whole. The following values can then be used as reasonable estimates:

body mass—63 to 70 kg
body surface area—1.82 m^2
body specific heat—0.83 $\frac{kcal}{kg\text{-}^\circ C}$
average tissue temperature—Eq. (12.5)

12.1.2 Sensible Heat Flow

The flow of sensible heat from the core of the body to the surrounding air is an example of series flow of heat through composite structures. Figure 12.2 shows this flow of heat for the human body. Here

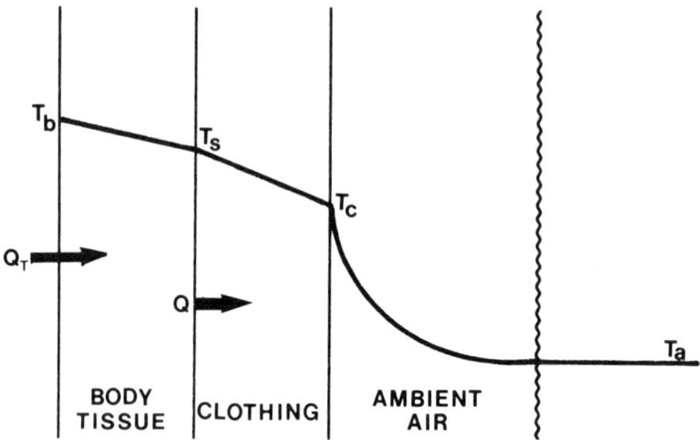

Fig. 12.2. Body Heat Flow through Insulative Layers.

T_b = core temperature of body
T_s = surface or skin temperature
T_c = temperature of outer surface of clothing and
T_a = air temperature.

The flow of energy through the body tissues is made up of the sensible heat flow Q plus the evaporative heat flow E_s. The latter energy flow need not traverse the clothing because there may be condensation between the skin and clothing or in the clothing layer itself, with a subsequent release of latent heat. One can write the series heat flow equations as

$$Q_T = \frac{T_b - T_s}{R_b} \tag{12.8}$$

$$Q = \frac{T_s - T_c}{R_{cl}} \tag{12.9}$$

$$Q = \frac{T_c - T_a}{R_a} \tag{12.10}$$

where R is the usual thermal resistance of a layer. In Eq. (12.10), the heat flow is due to the convection and radiation. If, however, there is direct solar radiation, this must be added as an additional heat flow term at the surface. Now

$$Q_t = Q + E_s \tag{12.11}$$

and

$$\frac{Q_T}{Q} = 1 + \frac{E_s}{Q} \tag{12.12}$$

DuBois (1927) has found that the evaporative heat loss is from 23 to 27% of the total heat loss $Q_t = Q + E$. This proportion is fairly constant at various levels of activity and may be taken as 25%. This energy is partitioned at about one-third for the respiratory tract E_r and the remainder for the skin evaporation term E_s. Note that $Q_t = Q_T + E_r$.
Then

$$E_s = \left(\frac{2}{3}\right)(0.25)\, Q_t = \frac{2}{3}\frac{(.25)Q}{.75} = \frac{2}{9}Q \tag{12.13}$$

as Q is the difference between the total heat Q_t and the evaporative heat flow E.
Thus, from Eq. (12.13)

$$\frac{E_s}{Q} = 0.22 \tag{12.14}$$

Using Eqs. (12.14) and (12.12), one has

$$Q_T = 1.22\, Q \tag{12.15}$$

and Eq. (12.8) may be rewritten

$$Q = \frac{T_b - T_s}{1.22\, R_b} \tag{12.16}$$

Eliminating the internal temperatures T_s and T_c, as is usual for series heat flow, using Eqs. (12.9), (12.10), and (12.16)

$$Q = \frac{T_b - T_a}{(1.22\, R_b + R_{cl} + R_a)} \tag{12.17}$$

The total thermal resistance of the body plus clothing plus air is

$$R_T = 1.22\, R_b + R_{cl} + R_a \tag{12.18}$$

The relations of this section are valid if there is no sensible perspiration at the skin surface. If the energy removal demands are high (large metabolic rates), then profuse sweating may be present, with E making up a high percentage of the total heat flow. See Section 12.1.4.

12.1.3 Insulation and Units

The thermal resistance is a commonly accepted and used concept in engineering. This resistance is often given the name "thermal insulation" by investigators with a background in physiology or clothing science. As the term is so common in the literature of clothing science and animal heat loss, we shall adopt the terminology and use I to denote the thermal insulation, keeping in mind that this is really the thermal resistance. There are two units, which sometimes cause confusion, that must be defined: the clo and the met.

Consider a resting, sitting man wearing indoor clothing, located in a room with an air temperature of 21°C and an air velocity of 10 cm/sec. Under these conditions, what value of clothing insulation will keep the subject comfortable without metabolic increase? As explained earlier, under these conditions, the heat flow through the clothing will equal 76% of the resting metabolic rate M of 50 kcal/m^2-hr or

$$Q = 0.76M = \frac{T_s - T_a}{I_{cl} + I_a} \tag{12.19}$$

where I_{cl} = insulation of clothing and I_a = insulation of air.
Now a value of the skin temperature under comfort conditions is 33°C (Burton 1955), and the insulation required is then

$$I_{cl} + I_a = \frac{(33 - 21)}{(.76)\ 50} = 0.32 \frac{°\mathrm{C} - \mathrm{hr} - \mathrm{m}^2}{\mathrm{kcal}}$$

Under these same conditions, the thermal resistance or insulation of air is about 0.14, and thus the clothing insulation is

$$I_{cl} = 0.32 - 0.14 = 0.18 \frac{°\mathrm{C} - \mathrm{hr} - \mathrm{m}^2}{\mathrm{kcal}} \tag{12.20}$$

Gagge et al. (1941) reasoned that an individual with normal indoor clothing[1]

[1] The normal indoor clothing of Europeans typically has a greater insulation value than that of North Americans.

is comfortable in this situation, and the insulation of indoor clothes should be defined as 1 clo. Therefore

$$1\ \text{clo} \equiv 0.18 \frac{^{\circ}\text{C} - \text{hr} - \text{m}^2}{\text{kcal}}$$

The same authors also defined a heat transfer unit as the heat flow equal to the normal resting metabolism

$$1\ \text{met} \equiv 50 \frac{\text{kcal}}{\text{hr} - \text{m}^2}$$

It is clear that these are merely definitions of heat transfer and resistance units and in no way depend upon experiment and will not vary with future experiments. Of course, this does not mean that all indoor clothes or even any certain combination will have an insulation value of 1 clo. A change in the rather subjective comfort indoor conditions could lead to other definitions of the value of the clo, but this will not change the accepted definition.

With these definitions and concepts, one can examine the various insulation values to use in Eqs. (12.17) and (12.18).

12.1.3.1 Body Tissue Insulation in Man

It can be seen from Eq. (12.1) that the rate of change of the energy stored in the body can be varied in two ways: a variation of the heat transfer or a variation of the metabolism. Under cold stress, man can alter his heat transfer rate to a certain extent by varying the total thermal insulation of the body. For a given type of clothing and ambient conditions, the insulation of the clothing and the ambient air cannot be varied, and thus the sole term in Eq. (12.18) that can be altered is the insulation value of the body tissue, I_b. Variation in this quantity is called "vasomotor regulation" and is affected by either decreasing the blood flow to the surface tissues (vasoconstriction), which will increase the tissue insulation, or by increasing the flow of surface blood (vasodilation), which will decrease the tissue insulation or increase its conductance. The insulation value of the body tissues is due to the combined effect of the conductivity through tissues and the convection associated with the blood flow. Unfortunately, man cannot vary the value of I_b very much. Burton and Edholm (1955) used the experimental effective thermal conductivity values reported by Burton and Bazett (1936) to express the tissue insulation range from 0.15 clo for full vasodilation to 0.90 clo for full vasoconstriction. Although this range of values will somewhat slow down the rate of body heat loss, it cannot protect man during the cold stress normally encountered in cold climates. Not surprisingly, animals adapted to cold climates show a far greater vasomotor control of body insulation and temperature.

12.1.3.2 Insulation in Animals

It has been found that most homeotherms can maintain their body temperature without an increase in resting metabolism if the ambient temperature is above a critical value (Scholander et al. 1950). This can be understood by referring to Fig. 12.3. As the ambient temperature decreases, the metabolic rate, which in the steady state must equal the total body heat loss, remains constant until the outside temperature drops to a critical value T_{crit}. This is accomplished by a decrease in the surface conductance or an increase in the insulation as the ambient temperature drops (vasomotor control). When the maximum insulation possible is reached, the animal must increase its metabolic rate or suffer a decrease in body temperature. Unlike man, arctic animals can control the thickness of their fur layer by vasomotor control and thus have a much wider range of total insulation values.

It was shown that for equilibrium the metabolic rate must equal the surface heat loss plus the heat loss due to evaporation. Thus

$$M \sim \frac{(T_b - T_a)}{I} = 1.333 \frac{\Delta T}{I} = \frac{1.333(T_b - T_a)}{1.22\, R_b + R_{cl} + R_a} \tag{12.21}$$

Equation (12.21) assumes that the total heat loss is proportional to $\Delta T/I$, I being the total insulation with the evaporative loss at 25%. Now if the

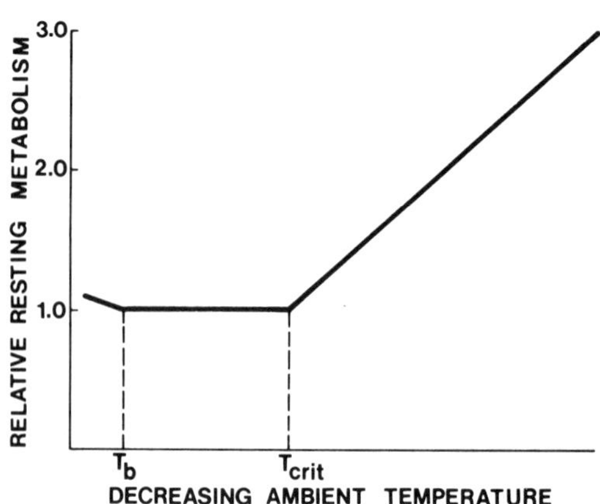

Fig. 12.3. Effect of Ambient Temperature on Metabolism.

outside temperature is at the critical value, then the metabolism is related to the insulation by

$$M \cong \frac{\Delta T_c}{I_c} \tag{12.22}$$

where

M = rest metabolism
ΔT_c = temperature difference between body and the critical value and
I_c = insulation at critical temperature (maximum value to prevent a metabolic increase).

Each species has a set of values satisfying Eq. (12.22). Figure 12.4 demonstrates two different animals, one adapted to a cold climate, and the other to a mild climate. Animal 2, adapted to cold, has a critical temperature of −20°C, and the first has a value of 20°C. These curves explain why arctic animals can survive severe ambient temperatures without an inordinate energy intake. Of course, the metabolism of the two need not be the same. The cold climate animal, in this hypothetical case, can withstand temperatures down to −20°C without any extra energy input (food) by altering its insulation. Note that insulation refers not only to the effective thermal conductivity of the surface layer, but also to the effect of geometry changes on the insulative

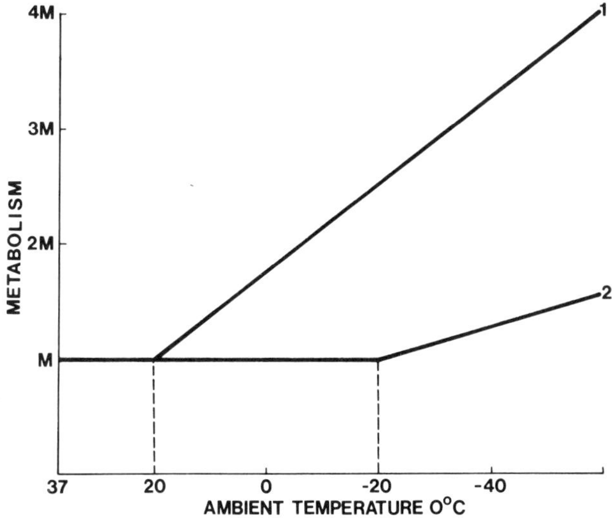

Fig. 12.4. Example of Heat Loss from Two Animals.

value. It can then withstand a further drop of 20°C, down to −40°C, by increasing its metabolic rate by 50%. The mild-climate animal must start to increase its metabolism at 20°C, and to survive at 20°C lower than this, i.e., 0°C, it must multiply its metabolic rate by about 3 times. Clearly it may not be able to survive for any length of time at −40°C. Man would fall into this category and survives in cold regions by artificially increasing his insulation (with clothing) so that the critical temperature is lowered.

Example 1: Suppose the following information is available for an average man:

Body temperature 37°C
Lower critical temperature 28°C
Surface area 1.8 m^2
Weight 63 kg
Resting metabolism $\dfrac{50 \text{ kcal}}{\text{hr} - \text{m}^2}$

If the outside air temperature drops below 28°C, man must increase his insulation or his metabolic rate. Calculate the insulation needed in clo if a man is to exist in the arctic at −20°C without an increase in metabolic rate.

Solution: At the normal lower temperature

$$I_{c_0} = \frac{1.333 \Delta T_c}{M} = \frac{1.333(37-28)}{50(.18)} = 1.33 \text{ clo}$$

At the new, lower critical temperature

$$I_{c_1} = \frac{\Delta T_{c_1}}{M} = \frac{T_B - T_{c_1}}{T_B - T_{c_0}} I_o = \frac{37-(-20)}{37-28} I_o = 8.44 \text{ clo}$$

Thus, one would require 8.44/4.0 = 2.1 in. of arctic clothing, as arctic clothing has an insulation value of about 4 clo/in. This is somewhat above the limit of clothing thickness that can be comfortably worn without restrictions. A better solution to this problem will be examined later.

12.1.3.3 Insulation of air

The air layer surrounding a body has an insulation value, or resistance, that is a function of the geometry of the body, air movement, and the temperature of the body and the air. This resistance is due to either free or forced convection, or a combination of the two, and radiation. At the usual low environmental temperatures (<40°C), the radiation may be combined with the convection as

$$Q = \frac{T_c - T_a}{I_a} \tag{12.23}$$

It is well known (McAdams 1954; Kreith 1966) that the resistance for forced convection varies inversely with a power of the wind speed, where the power can be from 0.33 to 0.9. For relatively low velocities the insulation could be given by

$$I_a \propto \frac{1}{\sqrt{u}} \tag{12.24}$$

where u = ambient wind speed.
The free convection case is more complicated, but a typical value might be

$$I_a \propto \sqrt{u} \tag{12.25}$$

where, in this case, the value of u is not usually known a priori. Burton (1955) corrected the data of Winslow et al. (1939) for heat loss from the human body for temperature and obtained the following relation

$$I_a = \frac{1}{0.61\left(\frac{T}{298}\right)^3 + .19\sqrt{u}\left(\frac{298}{T}\right)} \tag{12.26}$$

where

I_a = insulation of air in clo
u = wind speed in cm/sec and
T = ambient temperature in °K

This relation is adequate for our case, although the values at very low velocities will not reflect the combined action of free and forced convection. The value of I_a will be almost independent of temperature, from −40°C to 25°C, due to the opposing effects of radiation and convection. Thus, I_a will vary from 0.1 clo for wind speed of 2280 cm/sec to 0.85 clo at 15 cm/sec (essentially still air).

It has long been known that the insulating effect of insulators is mainly due to the air trapped in voids in the insulator and only secondarily due to the conductivity of the material itself, which is also true of clothing and animal fur. If the air gaps are large, however, convective currents will increase the heat flow and thus decrease the insulation. The maximum practical insulation of an air gap is about 4.7 clo/in., which indicates the maximum values one could expect with good clothing.

12.1.4 Evaporation

As has been noted previously, a significant percentage of the total heat loss is due to evaporation. This amounts to about 25% if there is no noticeable sweating, that is, insensible evaporation. If there is obvious sweating, this percentage will increase rapidly. Under cold conditions, heavy sweating is usually not present, unless the body is active in heavy clothing, but the evaporation effect is still important. Calculation of the evaporative heat transfer is not too reliable, but some details may be noted.

12.1.4.1 Respiratory Tract

Consider a steady-flow process of drawing in air at ambient conditions and expiring saturated air at body temperature. The energy balance for this process is

$$Q_r = \dot{m}_a[(c_{pa} + .45\ W_1)(T_b - T_a) + (W_2 - W_1)\ h_{fg}] \qquad (12.27)$$

where

$\dot{m}_a$ = mass flow rate of respiratory air, $\frac{\text{g}}{\text{hr} - \text{m}^2}$

c_{pa} = specific heat of air (.2403 $\frac{\text{cal}}{\text{g} - {}^\circ\text{C}}$ for $-20^\circ\text{C} < \text{T} < 100^\circ\text{C}$)

W_1 = specific humidity of air at T_a

W_2 = specific humidity of saturated air at body temperature, T_b

h_{fg} = latent heat of vaporization at the body temperature T_b, $\frac{\text{cal}}{\text{g}}$ and

Q_r = respiratory heat flux $\frac{\text{cal}}{\text{hr} - \text{m}^2}$

The first term on the right-hand side of Eq. (12.27) is the sensible heat involved in respiration. This term is often neglected unless the ambient air temperature is very low. Nevertheless, the sensible heat will be about 15% of the evaporative at $T_a = 20°C$ and 75% at $T_a = -40°C$. If the sensible heat is neglected, then the evaporative heat flow from the respiratory tract can be written

$$E_r = \dot{m}_a (W_2 - W_1) h_{fg} \tag{12.28}$$

It is convenient to express E_r in terms of the partial pressure of the water vapor

$$E_r = \frac{R_a}{R_v} \left(\frac{P_{vb}}{P_{ab}} - \frac{P_{va}}{P_{aa}} \right) h_{fg} \dot{m}_a \tag{12.29}$$

where R_a, R_v = gas constants for air and water vapor, and P_a, P_v = partial pressures of air and water vapor. Since the partial pressure of the air is very nearly equal to the barometric pressure, a good approximation is

$$E_r = \frac{R_a h_{fg}}{R_v P_a} (P_{vb} - P_{va}) \dot{m}_a \tag{12.30}$$

where P_{vb} = partial vapor pressure of saturated air at T_b, and P_{va} = partial pressure of water vapor in the ambient air.

12.1.4.2 Skin Evaporation

The process of skin evaporation appears quite different from the respiratory tract details, but the equations can be put into a similar form. There is a flow of water (perspiration) to the skin surface where evaporation will take place in accordance with the ambient conditions. The flow of water to the skin surface is probably regulated by the internal body conditions and not directly by the ambient properties, but this removal of water will involve only a modest heat transfer due to the enthalpy flow if no evaporation at the surface is possible. Thus, the efficient removal of heat by perspiration requires an evaporative process at the surface. The mass transfer can be described in a way analogous to convective heat transfer as

$$\dot{m}_s = \frac{h_D}{R_v T_m} A (P_{vb} - P_{va}) \tag{12.31}$$

where

$(P_{vb} - P_{va})$ = the same as for Eq. (12.30)
h_D = mass transfer coefficient
A = surface area across which evaporation actually occurs (wetted surface) and
T_m = average temperature of air-water vapor mixture near surface.

The mass transfer coefficient is not a transport property but a function of geometry, ambient conditions, and fluid properties. The evaporative heat transfer is

$$E_s = \frac{h_D A_w}{R_v T_m} h_v (P_{vb} - P_{va}) \tag{12.32}$$

where A_w = the fraction of the body area that is completely wet, and h_v is the enthalpy change of the water from a saturated liquid at the skin temperature to a superheated vapor at the ambient air temperature and pressure. This may be calculated from

$$h_v = h_{fg} + R\,T_a \ln\left(\frac{P_{vs}}{P_{va}}\right) \tag{12.33}$$

where R = universal gas constant of steam, and P_{vs} = partial pressure of vapor when saturated at T_a. Equations (12.30) and (12.32) may be combined as

$$E_s = B(P_{vb} - P_{va})\,A_s \tag{12.34}$$

where

$$B = \dot{m}_a \frac{R_a h_{fg}}{R_v P_a} + \frac{h_D A_w h_v}{R_v\,T_m}$$

The coefficient B is, of course, not a constant, and it is best obtained from measurements on the body. Measurements by Gagge (1957) show that B varies from 28.5 (kcal/hr-m^2-cmHg) for 100% wetted skin to a constant value of about 4.3 for insensible perspiration at temperatures of 16–30°C.

12.2 METABOLISM AND ACCLIMATIZATION TO COLD

As discussed in Section 12.1.3.1, when the limit of the vasomotor regulation of insulation is reached, the body must increase its metabolic rate if thermal equilibrium is to be maintained. This is usually manifested first by involuntary shivering or contraction of the muscles, liberating energy. Voluntary increased activity of the body, which is really a transfer of work into heat by certain muscle combinations, will also greatly increase the rate of energy production of the body. Ultimately these activities must result in increased food intake if the activity is maintained for long periods of time without loss of body mass. There has been considerable controversy over whether man has made long-term adjustments, in terms of metabolism or other physiological responses, to living in cold climates. Hart (1950) suggested the following definitions:

acclimatization: changes in the responses of the organism produced by continued alterations in the environment.
acclimation: alterations related to changes in a lifetime.
adaptation: change occurring during a period of several generations.

There is no convincing evidence that man has become acclimatized to cold to any significant physiological degree. Table 12.1 lists metabolic rates for various activities. These values are averages and will not be valid for specific individuals. The length of time that these metabolic levels can be maintained decreases as the metabolic rate increases. It is not likely that man could indefinitely maintain a metabolic rate much over 2–3 met. Various measurements of the basal metabolism of Eskimos (Krogh 1913; Heinbecker 1928; Crile 1939) show rates 15–35% above the value of 0.8 met, the standard value for Europeans in temperate zones, but this can be explained due to a high fat and protein diet, rather than a physiological adaptation. Most elevated basal metabolisms in cold climates are consistent with the extra effort related to wearing heavy clothing. At any rate, an increase of 30% in the basal metabolism would not be sufficient to alleviate severe cold stress. Likewise, the increased caloric intake of military personnel in cold regions can be attributed to the extra effort required in wearing heavy arctic clothing. These studies show that man's survival in cold climates is not so much a question of physiologic adaptation as it is cultural or behavioral adaptation, i.e., the development of suitable shelter and clothing for cold climates. Man reacts physically in cold climates, even after thousands of years, in essentially the same manner as is the case for men existing in warmer climates, although there may be a limited increase in resistance to cold.

Table 12.1. Metabolic Levels for Different Activities (adapted from Newburgh 1949).

Activity	Metabolism kcal/m²/hr	Met	Percentage Basal Metabolism
Sleeping, postdigestive	36	.72	90
Sleeping and digesting	40	.80	100
Lying quietly, postdigestive	40	.80	100
Lying quietly and digesting	45	.90	112
Sitting	50	1.00	125
Standing	60	1.20	150
Walking, 1½ mph	90	1.80	225
Level walking 3 mph (light work)	130	2.60	325
Level walking, 4 mph (moderate work)	180	3.60	450
Walking, 5% grade, 3½ mph	220	4.4	550
Walking, 10% grade, 3½ mph	340	6.8	850
Level running, 10 mph (exhausting work)	500	10.0	1250
Uphill running, 7 mph, 10% grade	700	14.0	1750
Sprinting (estimated values)	2000	40.0	5000

12.3 CLOTHING IN COLD CLIMATES

It is clear from the previous discussions that man must artificially increase his insulation if he is to survive in cold climates. The clothing should have adequate insulation so that the body temperature can be maintained without excessive heat loss. There are problems here, however, as adequate clothing can lead to overheating and sweating, when the body is active, even when the ambient temperature is quite low. This effect is well known to people in cold climates, and care is taken by the experienced so that exertion does not lead to sweating. This means that clothing must, in general, be discarded or rearranged in a coordinated way when physical exertion is to be undertaken. Irving (1972) notes: "It is certainly the impression from experience and observation that hard work in arctic cold increases the feeling of warmth to an uncomfortable degree unless clothing can be discarded or ventilated." Arctic clothing should be capable of a rapid and easy reduction of insulation due to the great variation of heat production for different physical activities.

12.3.1 Insulation Requirements

The value of insulation required for different activities and conditions can be easily obtained from the basic equations. From Eq. (12.19) the insulation of clothing and air required to dissipate 76% of the metabolic energy can be written as

$$I = \frac{0.1461(33 - T_a)}{M} \tag{12.35}$$

where the comfort value of the skin temperature is taken as 33°C and

I = total insulation of clothing and air, clo
T_a = ambient air temperature, °C and
M = metabolic rate, met.

Figure 12.5 is a plot of Eq. (12.35) for various levels of activity. As was mentioned earlier, the maximum insulation of still air is about 4.7 clo/in. (using the thermal conductivity of air gives an absolute limit of 7.3 clo/in.) and the maximum practical value for arctic clothing is about 4.0 clo/in.

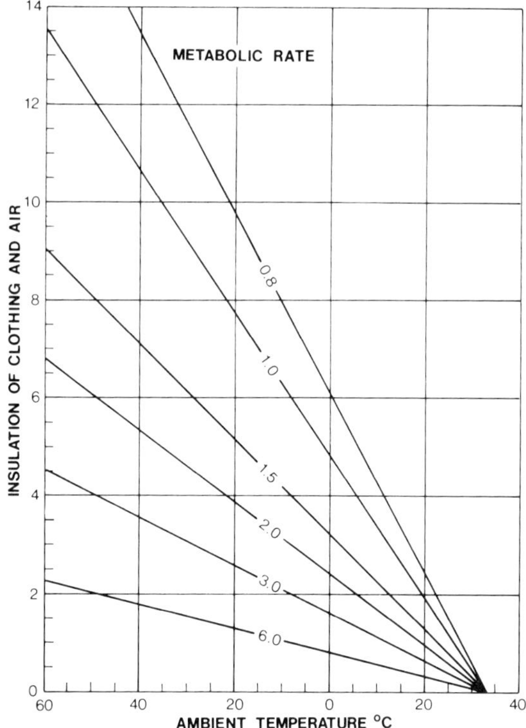

Fig. 12.5. Total Insulation Needed for Different Metabolic Rates.

Burton (1955) notes that 1½ in. is the limit of thickness for arctic clothing without seriously reducing the mobility of the wearer. This gives an insulation value of 6.0 clo, which actually exceeds most arctic uniforms. At −32°C (−25°F) an arctic uniform of 6.0 clo will allow a man to remain comfortable if his metabolic rate is maintained at 1.6 met, that is, somewhat more than very light work. This level of activity could be continued for a long period of time with an adequate food intake. For the same clothing, if the metabolic rate is the resting value of 1.0 met, the subject could withstand only an ambient temperature of −8°C, without a change of body temperature. A temperature of −35°C is about the lower limit for useful outdoor activity, even with the best insulative clothing.

One can estimate the time that can elapse before the body temperature falls to an uncomfortable level. Studies show that a heat debt of 25 kcal/m² leads to discomfort and that a loss of 150 kcal/m² is acutely uncomfortable. These correspond to body temperature drops of 2.6°C, for an average man. A sleeping individual awakes after a loss of 75 kcal/m² and a drop of skin temperature of 3°C. A deep body temperature less than 35°C threatens loss of control of the body temperature regulating mechanism, and a value of 28°C is considered critical for survival. Tests showed that a skin temperature of 33.3°C was comfortable, 31.0°C was uncomfortably cold, 30°C was shivering cold, and 29°C was extremely cold. A value of 25°C is the lower limit without numbness. A hand-skin temperature of 20°C was uncomfortably cold, 15°C was extremely cold, and 5°C was painful.

From Eq. (12.1), the following equation can be derived, based upon 76% sensible heat flow.

$$I = \frac{0.1461(T_s - T_a)}{(M + Q_e/\Delta t)} \qquad (12.35)$$

where

Q_e = heat loss, met-hr
Δt = time of exposure, hr
I = clo
T = °C and
M = met-hr.

Again with a uniform of 6.0 clo, an ambient temperature of −32°C, and a metabolism of 1 met, a man could remain out-of-doors for about 3 hours before reaching the discomfort temperature level (a loss of 80 kcal/m²). In actual fact, a value of 6 clo is too high for present arctic clothing. Table 12.2 lists some insulation values realized for an actual cold climate uniform.

Table 12.2. Insulation Values of an Actual Cold Climate Uniform (after Newburgh 1949).

Activity	Metabolism, met	Cold climate uniform insulation, clo
Sitting	1.0	2.9
Standing	1.6	2.6
Walking, 2.3 mph	2.9	1.6
Walking, 3.5 mph	4.0	1.4
Walking, 4.5 mph	5.2	1.3

12.3.1.1 Effect of Solar Radiation on Insulation

Shortwave or solar radiation will increase the amount of energy that must be dissipated. The sensible heat flow Q through the layer of clothing is the same as previously, that is, 76% of the metabolic rate. Solar radiation of amount q_r falls on the surface of the clothing and $Q_r = q_r \alpha$ is absorbed, as shown in Fig. 12.6. The energy loss to the air is then $Q + Q_r$. The energy flows may be written as

$$Q = \frac{T_s - T_c}{I_{cl}} \tag{12.36}$$

$$Q + Q_r = \frac{T_c - T_a}{I_a} \tag{12.37}$$

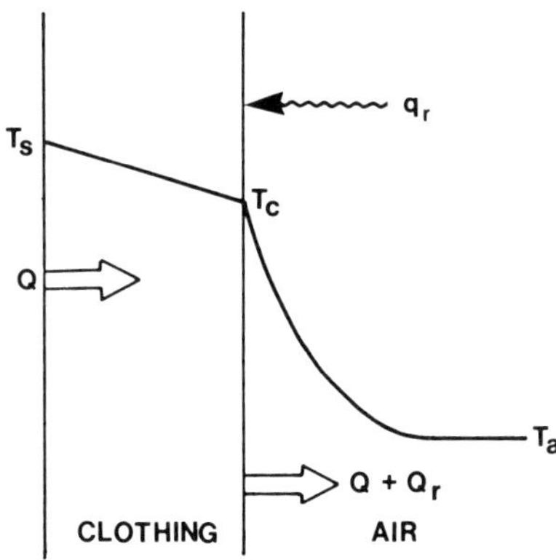

Fig. 12.6. Radiative Flux on Clothing.

These equations can be solved for the insulation requirements as follows

$$I_{cl} + I_a = \frac{.1416(T_s - T_a)}{M} - 1.316\frac{Q_r}{M} I_a \tag{12.38}$$

$$I_{cl} = \frac{.1416(T_s - T_a)}{M} - \frac{1.316}{M}(Q_r + .76\,M)\, I_a \tag{12.39}$$

where

$I =$ clo
$Q_r =$ met
$M =$ met and
$T =$ °C.

Obviously the solar radiation will decrease the insulation requirements for a given ambient temperature, wind speed, and level of activity.

Equation (12.35) for the allowable heat loss can be modified for radiation

$$I = \frac{.1461\,(T_s - T_a)}{\left(M + Q_r + \dfrac{Q_e}{\Delta t}\right)} \tag{12.40}$$

The time to reach the discomfort level is now

$$\Delta t = \frac{Q_e}{\dfrac{.1461(T_s - T_a)}{I} - (M + Q_r)} \tag{12.41}$$

The units are as above, with the time in hours. The radiation load must be calculated for each instant or for some average time period. There will be no solar radiation during the night or during that part of the year without daylight (above the Arctic Circle). Calculations of this type are given in Chap. 6. Auliciems et al. (1973) give comprehensive maps of clothing requirements in Canada during the winter season.

12.3.2 Moisture

There must be a continuous flow of water from the skin to the exterior to account for the body perspiration. Thus, as is well known in buildings, there can be condensation problems within the clothing layers. Unfortunately, it

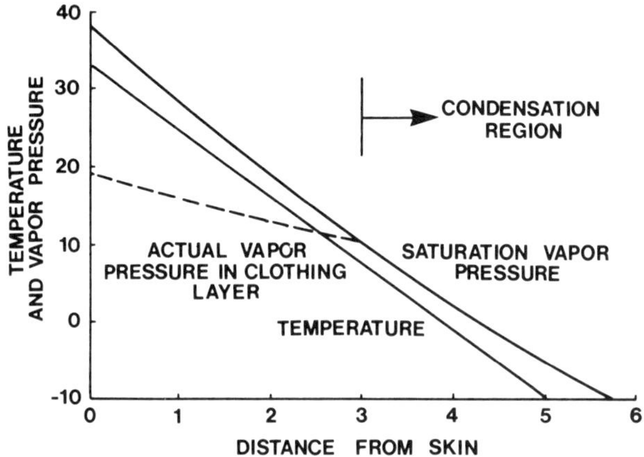

Fig. 12.7. Moisture Flow in Clothing.

is not possible to utilize vapor barriers with clothing, as this would create severe problems of moisture accumulation at the skin surface. Figure 12.7 explains how and where condensation can occur within the clothing layers. With a skin temperature of 33°C and an ambient temperature of −10°C, the figure shows the curve for the saturated vapor pressure, which is simply a function of the temperature. Due to the continuity of the flow of moisture from the skin to the exterior, there is also an actual vapor pressure curve as shown. When the actual vapor pressure equals the saturation value, the relative humidity is 100%, and condensation can occur. Thus, in this example, there would be a condensation layer at about 3 cm. In this example, it would be water because the temperature at this point is above freezing, but the moisture could well be condensed out as frost if the temperature is below 0°C. If the clothing is not dried out, there will be a loss of insulation and a buildup of weight that could render the clothing useless. Some of the heat of condensation may be registered as a heat gain by the body, thus reducing the efficiency of the evaporative process as a heat transfer mechanism. The clothing should be permeable to water but impermeable to air for optimal results.

12.3.3 Wind

The wind plays a major role in the insulation value of clothing. If the clothing is permeable to air the insulation can fall significantly due to the wind. Breckenridge (1950) notes that a standard U.S. Army uniform of 3.5 clo in still air had an insulation of 2.0 clo in a 24-mph wind. The internal movement of air due to walking or other activity is also significant. Belding (1949) estimates that the insulation value while at rest may drop to one half that value if the subject is walking. This decrease of insulation is actually an advantage as it allows the increased heat production of the activity to be more easily dissipated. The automatic decrease of insulation with activity is an almost necessary characteristic of good arctic clothing and may explain the superior qualities of Eskimo clothing.

12.3.4 Eskimo Clothing

Almost from the time of the first northern explorers, it has been recognized that the Eskimos had developed clothing that was admirably suited to the harsh environment. Although this may seem an obvious necessity, there is evidence (Stefansson 1921) that some Indian tribes of northern Canada did not have good clothing and had to remain continuously active when outdoors in the winter. Most of the following details are taken from Stefansson. The success of Eskimo clothing can be attributed to three things: materials, design, and use. The typical winter costume consists of an outer garment of fur—usually caribou, which has an insulation of about 4.3 clo/in. (Hamel 1955), and which is superior in terms of wind resistance—turned hair side out and an inner unit with the fur turned in. Boots are of caribou uppers, fur out, and soles of caribou or seal fur turned in. If waterproof boots are needed, then sealskin is favored, with the hair removed. Pads of dry grass or moss were often used in the boots and mittens for added insulation. The proper use of the clothing is very important as far as comfort and utility are concerned. The clothing is loose-fitting but individually tailored. It is gathered at the waist by a belt for warmth. To cool off, during activity, one first loosens the belt, allowing air to circulate, then removes mittens, then throws back the hood, and finally, if necessary, removes the garment entirely. This sequence of events is based on sound physiological principles to avoid sensible perspiration. Moisture lessens the insulation, increases the weight, and will cause the deterioration of fur if it persists for 2 or 3 days. The Eskimo is careful to beat out any frost in the fur layers each day to avoid accumulation. The arms can be withdrawn inside the clothing against the body leaving the sleeves empty. All these details make the Eskimo clothing extremely adaptable in terms of insulation and maneuverability. In terms of durability, however, there may be some question. Also each garment is individually fitted to the wearer's contours, something that is not possible with mass produced garments. The fur garments of Eskimos are at least four times as warm per pound as European costumes. An Eskimo costume will weigh 12–18 lb and may have a clo value as high as 8–10, as opposed to a good 3 to 4 clo European uniform, which will weigh 20–30 lb. Nevins et al. (1974) have presented some data on the insulative values of typical, modern European garments (Table 12.3).

12.4 COLD INJURY

The effect of prolonged exposure to cold can be lethal, and lesser effects might be temporary or permanent loss of function of various parts of the body, usually the extremities. The heart cannot survive at temperatures lower than 10–18°C, and this must be considered an extreme lower limit for the core temperature. Nevertheless, the skin, muscles, and nerves can take short

Table 12.3*a*. Fabric Data and Insulation Values in Clo, Women's Garments (adapted from Nevins et al. 1974).

Description	Fabric construction	Total weight, oz	Fabric weight, oz/yd²	Clo value
Cool dress	Knit		3.7 Dress	0.17
	Plain weave	5.3	2.2 lining	
Warm dress	Plain weave	41.7	17.8	0.63
Warm long sleeve blouse	Plain weave	5.9	2.8	0.29
Warm skirt	Twill weave	14.9	15.5 Skirt	0.22
	Plain weave		2.8 lining	
Cool long sleeve blouse	Plain weave	3.0	1.9	0.20
Cool slacks	Knit	5.7	3.7	0.26
Warm slacks	Twill	21.9	13.5	0.44
Warm jacket or blazer	Weave	25.0	–	0.43
Cool jacket or blazer	Weave	18.0	–	0.31
Cool sleeveless sweater	Knit	6.1	–	0.17
Warm long sleeve sweater	Knit	10.6	8.2	0.37
Cool vest	Weave	6.8	5.7	0.20
Warm vest	Weave	8.0	8.5	0.30
Warm shawl (over shoulders)	Knit	12.0	–	0.40
Cool shawl (over shoulders)	Knit	8.0	–	0.30
Cool short sleeve blouse	Plain weave	3.1	–	0.17
Pantyhose	–	–	–	0.01
Warm "tights"	Knit	5.0	–	0.25
Warm knee-high socks	Knit	6.0	–	0.08
Cool knee-high socks	Knit	4.0	–	0.06
Bra & panties	–	–	–	0.04
Girdle	–	–	–	0.04
Full slip	–	–	–	0.19
Half slip	–	–	–	0.13
Long underwear—tops	Knit	5.0	–	0.25
Long underwear—bottom	Knit	7.0	–	0.25
Long thermal underwear—tops	Waffle knit	6.0	–	0.35
Long thermal underwear—bottoms	Waffle knit	8.0	–	0.35
Nylon-cotton coverall	50/50	16–18	–	0.55
Shoes	–	10.0	–	0.03
Knee-high fashion boots	–	16.0	–	0.25
Knee-high boots, leather lined	–	32.0	–	0.30

Table 12.3*b*. Fabric Data and Insulation Values in Clo, Men's Garments (adapted from Nevins et al. 1974). Reprinted from *ASHRAE Journal* © April 1974. Appearance of this article in *ASHRAE Journal* does not necessarily signify or imply endorsement by the American Society of Heating, Refrigerating and Air-Conditioning Engineers, Inc.

Description	Fabric construction	Total weight, oz	Fabric weight, oz/yd^2	Clo value
Briefs	Knit	2.0	–	0.05
Sleeveless undershirt	Knit	3.0	–	0.08
Short sleeve undershirt	Knit	3.5	–	0.09
Long underwear—tops	Knit	5.0	–	0.25
Long underwear—bottom	Knit	7.0	–	0.25
Long thermal underwear—tops	Waffle knit	6.0	–	0.35
Long thermal underwear—bottoms	Waffle knit	8.0	–	0.35
Shoes, low	–	16.2	–	0.04
High-shoes and side-zips	–	20.0	–	0.15
Knee-high boots, leather lined	–	32.0	–	0.30
Cool socks	–	2.0	–	0.03
Warm socks	–	4.0	–	0.04
Cool knee-high socks	–	4.0	–	0.06
Warm knee-high socks	–	6.0	–	0.08
Short sleeve woven shirt	Plain weave	7.1	4.0	0.19
Long sleeve woven shirt	Plain weave	5.9	2.8	0.29
Cool trousers	Plain weave	11.7	5.4	0.26
Warm trousers	Twill weave	18.1	8.4	0.32
Short sleeve cool knit shirt	Knit	7.1	5.6	0.22
Short sleeve warm knit shirt	Knit	6.8	6.8	0.25
Long sleeve cool knit shirt	Knit	6.9	6.1	0.14
Long sleeve warm knit shirt	Knit	10.6	9.2	0.37
Warm sports jacket or suit coat	Twill weave	29.9	–	0.49
Cool sports jacket or suit coat	Twill weave	20.0	–	0.35
Nylon-cotton coverall	50/50	16–18	–	0.55
Vest, cool	Weave	7.1	5.7	0.20
Vest, warm	Weave	8.2	8.5	0.30

exposures to temperatures of 0–5°C without severe injury. The effects of cold can be discussed in terms of localized effects and general body cooling.

12.4.1 Localized Cooling

The effect of cold on localized areas of the body is usually not fatal, but may lead to loss of members or loss of function of the affected part. Trench foot or immersion foot is injury due to exposures at temperatures well above freezing, but with damp or wet conditions. Exposures of several days at

8°C in seawater has resulted in damage to feet and legs, with pain, loss of sensation, swelling, and, in severe cases, gangrene and ultimate tissue destruction. There need not be great amounts of water, as damp boots worn for several days will present the same conditions. A more severe form of local injury is frostbite, which involves the actual freezing of tissues. Lewis (1941) suggests that the freezing temperature of skin normally is −1° to −2°C. There is considerable question as to whether the freezing of the tissues is itself a cause of tissue damage, but there is no doubt that the ultimate effects can be severe. Usually there will be a loss of blood flow to the extremities, and often gangrene and loss of tissue will occur. Whether the gangrene is due to the lack of blood flow or the freezing itself is not clear. Rapid rewarming (immersion in water at 42°C) of the frozen limb seems to reduce the tissue loss to a minimum (Crimson 1951). Surprisingly, there is a very low incidence of frostbite in northern countries (Irving 1957), and this is probably due to the fact that inhabitants of cold climates know of the danger and take the necessary precautions.

12.4.2 Hypothermia

It is well known that exposure of the body, with clothing or without, to low-temperature water will lead to a fairly rapid drop in the core temperature. In general, a drop in the core temperature of the body, with serious or fatal effects, is called "hypothermia." Many survivors of exposure to low temperatures are known to expire shortly after removal from the water. Thus it is thought that many deaths attributed to drowning in northern waters are in fact due directly to the adverse effects of low temperature. When the core temperature drops to a certain level (10–18°C), the heart fails to maintain its proper electrical rhythm, and death results. As expected, a human exposed to severe, prolonged cold attempts to maintain the core temperature by vasoconstriction, that is, the surface blood flow is decreased and the surface tissue temperature falls to a value close to the ambient. Continued exposure will, however, cause a fall in the core temperature, with cardiac arrest being a possibility. If the victim is rescued from the cold and allowed to slowly warm up, the result can be, and often is, a further rapid drop in core temperature with fatal results. The reason for this apparent contradiction is that in response to the warmer ambient conditions, the surface blood flow is increased. This blood will be cooled by contact with the very cold surface tissues, and this drop in temperature cannot be offset by contact with room temperature air, for example. Consequently, the cold blood flowing back to the core from the surface layers causes a further drop in core or heart muscle temperature, which may be sufficient to cause heart failure. Many, perhaps most, deaths due to immersion in cold water are due to hypothermia rather than drowning.

It would appear that the best procedure for hypothermia is rapid rewarming of the victim in a hot water bath. Obviously every precaution must be taken to avoid a severe drop in body temperature by the availability and use of proper clothing and shelter in cold climates.

12.5 WIND-CHILL

The comfort level or sensation experienced by a person under cold stress is rather subjective and not easily expressed in simple form. The state of feeling cold is related to the rate of heat loss from the body, and this rate of heat loss is discussed in Section 12.1. Siple (1945) defined the term "wind-chill" to describe the rate of cooling of a heated body exposed to the wind. Siple's empirical relation, based upon experiments carried out under difficult-to-control natural conditions on the freezing of cylinders of water, is

$$K_o = (10\sqrt{u} + 10.45 - u)(33 - T_a) \tag{12.42}$$

where K_o = wind-chill, kcal/m²/hr, and u = wind speed, m/sec.

Although this equation uses the standard comfort level value for the skin temperature, 33°C, it does not really represent the heat loss from a human. At a wind speed of 9 m/sec and an air temperature of 20°C, Siple's relation gives K_o = 408 kcal/m²/hr. If one uses Eqs. (12.23) and (12.26) for the human body with no clothing, the result is 174 kcal/m²/hr. Equation (12.42) is considerably higher than this, but as the relation is generally used for comparison purposes it can be useful. Siple (1949) has also devised a relation between the wind-chill and the comfort sensation, as noted in Table 12.4. This table is often used for estimating the degree of cold stress expected for a given region, although the quantitative values are no doubt incorrect.

Table 12.4. Wind-Chill as a Discomfort Index.

Wind-chill, kcal/m²/hr	Sensation
50	Hot
100	Warm
200	Pleasant
400	Cool
600	Very cool
800	Cold
1000	Very cold
1200	Bitterly cold
1400	Exposed flesh freezes
2500	Intolerable

The danger value of 1400 kcal/m^2/hr is listed as the value at which exposed flesh freezes, but this clearly must be related to some time span because values less than this must also lead to freezing given sufficient time. Also it should be noted that Table 12.4 really applies to naked subjects, and this clearly does not describe clothed individuals. It is still useful because the clothed individual's tolerance to cold is often limited by the exposed face and the poorly protected extremities, such as fingers and hands. Thus, in this regard, the subject is effectively unclothed. Thomas and Boyd (1957) have used Siple's relation and the wind-temperature data for Canada to arrive at wind-chill charts. The probability of any wind-chill level can be found from Fig. 12.8. Simply obtain the average wind-chill from Eq. (12.42), and note the percentage of the time this value would be exceeded from Fig. 12.8. This chart is based on the relation between the mean values for a month and the standard deviations.

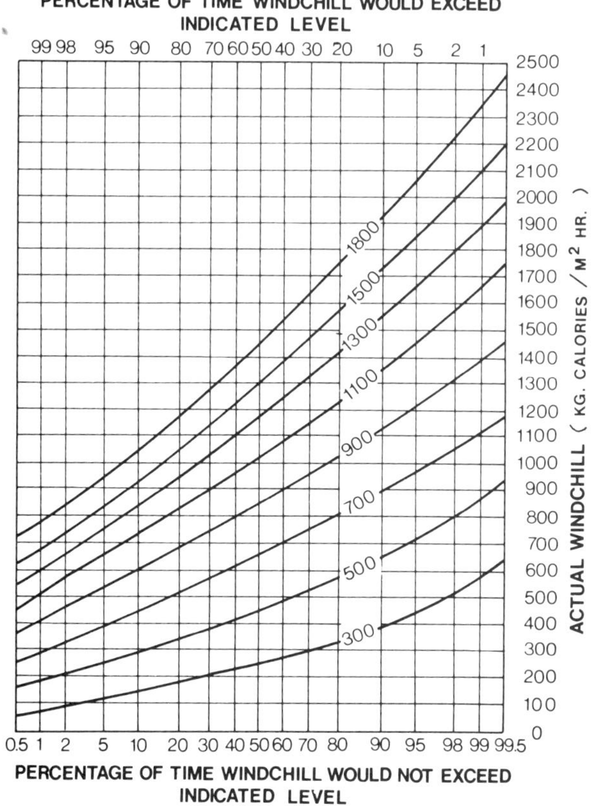

Fig. 12.8. Wind-Chill Probability (adapted from *QMREC Annual Report* 1958).

12.6 WIND-CHILL TEMPERATURE

In order to change the wind-chill, which is a heat transfer rate, into an index that is easier to grasp physically, various investigators have suggested a temperature index. The wind-chill temperature, also called the "effective temperature" or "equivalent temperature," is defined as that value of ambient temperature that in still air would give the same wind-chill as produced by the actual ambient temperature and wind speed. Thus, the wind-chill temperature is a defined quantity and cannot be determined experimentally. It is another way of expressing the wind-chill, and, like the wind-chill, it is a useful but subjective index. It is also based upon a constant value of the exposed skin temperature that does not accord with reality. Certainly under frostbite conditions, the skin temperature will be far lower than 33°C, and consequently the wind-chill will be far less than that denoted by Eq. (12.42). From the definition, at a wind speed of zero, the wind-chill temperature should be exactly the same as the actual air temperature; however, Eq. (12.42) is not correct at zero velocity, and thus a value of $u = 2.23$ m/sec is often used. Then

$$h(33 - T_{eq}) = (K_o)_{2.23}(33 - T_a) \tag{12.43}$$

where h = heat transfer coefficient at reference wind velocity, and T_{eq} = wind-chill temperature.

It is clear that, based on Eq. (12.42), the value of $h = (K_o)_{u=2.23}$ is 23.16. The wind-chill temperature is then

$$T_{eq} = 33 - [.432\sqrt{u} + .451 - .0432\, u]\,(33 - T_a) \tag{12.44}$$

Equation (12.44) can be used to compute the temperature that will give the same wind-chill as the actual combination of ambient temperature and wind speed.

12.7 HUMAN ACTIVITY IN COLD CLIMATES

A question of some significance for engineering activities in cold climates relates to the lowest ambient temperatures at which man and machines can function properly. Our previous discussion shows that temperature alone is not the entire problem for men, but it is the usual climatic reference used. The U.S. Army Cold Regions Research and Engineering Laboratory, Hanover, New Hampshire, surveyed a number of engineering firms operating in Alaska to provide some answers to the above question. Table 12.5 is an abstract of this survey. The values listed in Table 12.5 are subjective in

Table 12.5. Temperature Extremes for Man and Machines in Cold Climates.

Activity	Absolute low temperature, °F Low	High	Average	Low temperature for efficient operation, °F Low	High	Average
Machine excavation	−70	−20	−37	−30	+32	− 1
Hand excavation	−70	+35	−22	−30	+33	+18
Earth moving & grading	−45	−20	−35	−20	+32	+13
Paving	+15	+45	+32	+25	+60	+41
Concrete formwork	−50	−10	−25	−10	+33	+10
Concrete placement	−50	−25	−31	+20	+40	+32
Steel erection	−50	−20	−32	−30	+30	− 1
Block & stone masonry	−50	0	−25	0	+40	+31
Roofing	−30	+15	−13	0	+60	+24
Finish carpentry	−50	−10	−27	+15	+50	+36
Painting	−50	+40	+13	+32	+60	+42
Electrical	−70	−10	−31	+20	+40	+28
Piping & mechanical	−50	−10	−27	+20	+40	+29
Pipe welding	−50	0	−21	−30	+33	+ 8
Prebuilt components	−50	−10	−28	−30	+33	+ 9
Layout & surveying	−50	0	−33	−30	+20	+ 2
Loading & unloading	−60	−10	−39	−35	+20	− 1
Subsurface exploration	–	–	−40	–	–	–

that no actual measurements were made, but the opinions of experienced, cold-climate contractors should aid in evaluating the problems associated with these effects. Economics will dictate the extreme temperatures that can be withstood because it is definitely possible to design and build systems that will operate under any forseeable climatic conditions, if cost is no object.

REFERENCES

Auliciems, A., De Freitas, C. R., and Hare, F. K. 1973. *Winter clothing requirements for Canada.* Climatological Studies No. 22. Environment Canada, Toronto.

Belding, H. S. 1949. *Protection against dry cold.* Office of the Quartermaster General Environmental Protection Service, Report No. 155.

Breckenridge, J. R., and Woodcock, A. H. 1950. *Effects of wind on insulation of arctic clothing.* Environmental Protection Service, Quartermaster Climatic Research Laboratory, Report No. 164.

Burton, A. C. 1935. The average temperature of the tissues of the body. *J. Nutr.* 9:261.

———, and Bazett, H. C. 1936. A study of the average temperature of the tissues, of the exchange of heat and vasomotor responses of man in a bath calorimeter. *Am. J. Physiol.* 117:36.

———, and Edholm, O. G. 1955. Man in a cold environment. London: Edward Arnold.

Crile, G. W., and Quiring, D. P. 1939. Indian and Eskimo metabolism. *J. Nutr.* 18:361.

Crimson, J. M. 1951. Pathology and physiology of frostbite. *Bull. Vascular Surgery* 12:110.
DuBois, E. F. 1927. Basal metabolism in health and disease. Philadelphia: Lea and Febiger.
DuBois, D., and DuBois, E. F. 1916. A formula to estimate approximate surface area if height and weight are known. *Arch. Intern. Med.* 17:863.
Gagge, A. P. 1937. A new physiological variable associated with sensible and insensible perspiration. *Am. J. Physiol.* 120:277.
______, Burton, A. C., and Bazett, H. C. 1941. A practical system of units for the description of the heat exchange of man with his environment. *Science* 94:428.
______, Stolwijk, J. A. J., and Nishi, Y. 1969. The prediction of thermal comfort when thermal equilibrium is maintained by sweating. *ASHRAE Trans.* 75 (II):108.
Hamel, H. T. 1955. Thermal properties of fur. *Am. J. Physiol.* 182:369–376.
Hardy, J. D., and DuBois, E. F. 1938. Basal metabolism, radiation, convection, and evaporation at temperatures of 22 to 35°C. *J. Nutr.* 15:477.
Hart, P. 1950. Defence Research Board Conference on Cold, Kingston, Ontario. Reported in Burton (1955).
Heinbecker, P. 1928. Studies on the metabolism of Eskimos. *J. Biol. Chem.* 80:461.
Irving, L. 1957. Animal adaptation to cold. In *Cold injury,* Ferrer, M. I., ed. Transaction of 5th conference, Arctic Aeromedical Pub., Alaska.
______. 1972. *Arctic life of birds and mammals.* New York: Springer-Verlag, p. 101.
Kreith, F. 1966. *Principles of heat transfer.* 2nd ed. Scranton: International Textbook.
Krogh, A., and Krogh, M. 1913. *A study of diet and metabolism of Eskimos.* Copenhagen.
Lewis, T. 1941. Observation on some normal and injurious effects of cold upon the skin and underlying tissues. Br. Med. J. 795:837.
Livingston, S. D. 1968. Calculation of mean body temperature. *Can. J. Physiol. Pharmacol.* 46:15–17.
McAdams, W. H. 1954. *Heat transmission.* 3rd ed. New York: McGraw-Hill.
Nevins, R. G., Minall, P. E., and Stolwijk, J. 1974. How to be comfortable at 65 to 68 degrees. *ASHRAE J.* 16 (4):41–43.
Newburgh, L. H., ed. 1949. *Physiology of heat regulation and the science of clothing.* Philadelphia: Saunders.
Pembrey, M. S. 1898. Animal heat. In *Textbook of physiology,* E. A. Schaefer. Vol. 1. London: Hodder and Stoughton.
Scholander, P. F., Hock, R., Walters, V., Johnson, F., and Irving, L. 1950. Heat regulation in arctic and tropical mammals and birds. *Biol. Bull.* 99:237–258.
Siple, P. 1945. Measurements of dry atmospheric cooling in subfreezing temperature. *Proc. Am. Philos. Soc.* 89:177–199.
______. 1949. Clothing and climate. In *Physiology of heat regulation.* ed. L. H. Newburgh. Philadelphia: Saunders.
Stefansson, V. 1921. *The friendly arctic.* New York: Macmillan.
______. 1944. *Arctic manual.* New York: Macmillan.
Stoll, A. M. 1967. Heat transfer in biotechnology. In *Advances in heat transfer,* eds. J. P. Hartnett and T. F. Irvine. Vol. 4. New York: Academic Press.
Thomas, M. K., and Boyd, D. W. 1957. Wind-chill in northern Canada. *Can. Geogr.* 10:29–39.
Winslow, C.-E. A., Gagge, A. P., and Herrington, L. P. 1939. The influence of air movements upon heat losses from the clothed human body. *Am. J. Physiol.* 127:505.

PROBLEMS

1. When the wind-chill is such that exposed flesh freezes (according to Siple), evaluate the wind velocity for air temperatures of 0°C, −10, −20, and −30. What are the equivalent temperatures at these conditions?

2. Compare the heat transfer calculated directly for an unclothed human with the values in Problem 1.
3. Derive Eq. (12.35).
4. The outside temperature is −25°F. Find the clo value needed to maintain body temperature for a man walking at 1.5 mph. Evaluate the time needed to reach the discomfort temperature level.
5. Consider an arctic uniform as given in Table 12.2. Evaluate the time a man could walk at 2.3 mph before reaching the discomfort temperature level. Outdoor temperature is −25°F. The clo value is for the clothing only.

Index